RECENT HURRICANE RESEARCH - CLIMATE, DYNAMICS, AND SOCIETAL IMPACTS

Edited by **Anthony R. Lupo**

Recent Hurricane Research - Climate, Dynamics, and Societal Impacts
Edited by Anthony R. Lupo

Published by InTech

Janeza Trdine 9, 51000 Rijeka, Croatia

Publishing Process Manager Ivana Lorkovic

Technical Editor Teodora Smiljanic

Cover Designer InTech Design Team

Additional hard copies can be obtained from orders@intechopen.com

Recent Hurricane Research - Climate, Dynamics, and Societal Impacts,
Edited by Anthony R. Lupo
p. cm.
ISBN 978-953-307-238-8

For EU product safety concerns:
IN TECH d.o.o., Prolaz Marije Krucifikse Kozulić 3, 51000 Rijeka, Croatia,
info@intechopen.com or visit our website at intechopen.com.

Meet the editor

Dr. Anthony R. Lupo is a Professor and Chair of Atmospheric Sciences in the Soil, Environmental, and Atmospheric Sciences Department at the University of Missouri – Columbia. He earned his BS in Meteorology from the State University of New York at Oswego in 1988, and his M.S. and Ph.D. degrees from Purdue University in 1991 and 1995, respectively. His research is in atmospheric dynamics, and climate dynamics and change, including tropical meteorology. He is a member of the American Meteorological Society, and National Weather Association. He was a Fulbright Scholar, studying climate change at the Russian Academy of Sciences in Moscow. Additionally, he has served as an expert reviewer and contributing author to the Intergovernmental Panel on Climate Change (IPCC) and the Nongovernmental International Panel on Climate Change (NIPCC) Assessment Reports. The IPCC shared the Nobel Peace Prize with Vice President Albert Gore, Jr. He won the University of Missouri Kemper Foundation Award for Excellence in Teaching, the University of Missouri professor of the year, and the Most Distinguished Scientific Achievement Award by the Missouri Academy of Science.

Contents

Preface

Tropical cyclones are one of nature's most destructive phenomena. In the past, these storms would often strike without warning and bring physical- as well as economic disruption, to societies in tropical and subtropical regions, and even into the mid-latitudes. Historically, these storms have caused as many problems for developed nations as for less developed nations and this has invariably prompted further research to establish the characteristics of tropical cyclones such as frequency of occurrence, common genesis regions, intensities, and the physical factors that maintain them, which has been proven to be invaluable in helping meteorologists understand the evolution and movement of them. This, of course, led to greater forecast skill.

The development of RADAR and satellite technology combined with advances in numerical models, has resulted in a greater understanding of the structure and dynamics of tropical cyclones. As the understanding of these topics grew, and the monitoring of tropical cyclones increased, forecast techniques for track and intensity have been developed and improved. The understanding of the climatological occurrence of tropical cyclones has also been refined, along with the establishment of the interannual and interdecadal variations in their occurrences. Today we understand that these powerful storms represent the cooperative interaction between atmospheric and oceanic conditions, and these storms will occur only when conditions in both are favorable. Climate change may also influence their frequency of occurrence and intensity.

These storms will continue to disrupt the regular functioning of modern societies and be responsible for the loss of life and property. Thus, newer research, examining the impact of tropical cyclones on societal functions such as evacuation procedures, public and ecological health, and the assessment of damage is now being done by interdisciplinary teams. These researchers are taking advantage of observation tools such as light detection and ranging (LIDAR) and satellite imagery and using them in innovative ways in order to assess and quantify impact. They are using also geographic information systems (GIS) to build platforms that decision makers can use in the future to minimize the loss of life and/or property.

This book represents a compilation of cutting edge research on tropical cyclones and their impact from researchers at many institutions around the world. A wide range of topics is also covered. The book itself has been divided into six sections, and each section organized by topic.

The first part, climate and climatological factors, contain contributions that explore the global occurrence of tropical cyclones over the past 30-70 years, and establishes the interannual and interdecadal variability in their occurrence and intensity. Other chapters explore terrestrial factors that may contribute to the variability and long-term trends in tropical cyclone occurrence. New work also examines tropical cyclone occurrences and intensity as related to solar activity, while other chapters feature research which discusses the impact climate change may have on tropical cyclone frequency and intensity.

The second part of the book looks at the dynamics and structure of these powerful storms, examining the background conditions in the atmosphere and ocean which may contribute to the maintenance and the structure of tropical cyclones. Some researchers examine the structure of storms using traditional diagnostics, while others took a more modern approach to examining the thermo-dynamics associated with these storms. Several use numerical models in order to look at the structure of these storms in finer detail. A model for wind-wave interaction was also developed.

In the third part, the focus is on Forecasting of Tropical Cyclones with research demonstrating that advancements in short range predictions of tropical cyclone tracks and intensities can be made with improvements to the physics and incorporation of satellite information.

The fourth and fifth part examine the societal impacts of tropical cyclones. In part four, Risk and Damage Assessment, authors investigate of the use of GIS technology for anticipating, and later evaluating, the amount of damage that may be done by land-falling storms. The economic vulnerability of coastal regions is also studied, and a storm surge impact scale is developed. Section five examines Evacuation Risks and Hazards, examines through case studies, the potential for chemical release hazards and subsequent evacuation procedures and hazards. Other researchers take a close look at the procedures and potential problems associated with the evacuation of medical facilities. This followed by a comprehensive case study looking at demographic and societal factors in evacuation decisions, and another examining road hazards that are the result of the increased road congestion associated with hurricane evacuations provide a potential basis for developing improved evacuation techniques.

Finally, part six, Ecological Impacts and Modeling, begins with research that quantifies the potential for wind to damage forest ecosystems and the time it may take for these systems to recover from tropical cyclone damage. Other chapters study the assessment of, and damage to, landscapes and watersheds using LIDAR and image analysis techniques. Another chapter considers the growth of phytoplankton blooms that may result associated with the passage of a tropical cyclone. They demonstrate the utility of satellite technology for monitoring these as well as the use of a numerical model to simulate their growth.

Anthony R. Lupo
Department of Soil, Environmental, and Atmospheric Sciences 302 E Anheuser Busch Natural Resources Building University of Missouri – Columbia MO, USA

Part 1

Climatology and Climatological Factors

1

The Interannual and Interdecadal Variability in Hurricane Activity

Anthony R. Lupo
Department of Soil, Environmental, and Atmospheric Sciences
302 E Anheuser Busch Natural Resources Building
University of Missouri – Columbia
MO, USA

1. Introduction

While the interannual variability of hurricane activity has been studied extensively over the last few decades, there has been a recent increase in the interest of the *interdecadal* variability of hurricane activity (e.g., Henderson-Sellers et al. 1998; Emanuel, 2005; Webster et al. 2005; Schultz, 2007; Lupo et al. 2008; Ng and Chan, 2010) worldwide. This interest has been prompted by several active hurricane years in the Atlantic Ocean basin, the occasional occurrence of tropical cyclones in the South Atlantic in the last decade (e.g. Reagan et al. 2005; McTaggart-Cowan et al. 2006), as well as concern about changes in hurricane frequency that may be a result of changes in Earth's climate. These active years in the Atlantic which garnered much media attention included the 2004 season during which several storms struck the Southeast United States including Florida, and the 2005 season which was the most active hurricane season since 1933, and included the deadly hurricane Katrina.

Some studies, e.g., Lupo and Johnston (2000) (*hereafter LJ00*), and then Lupo et al. (2008), suggested that there is significant interdecadal variability in the occurrence and intensity of Atlantic region hurricane activity, and that this variability may be linked to the Pacific Decadal Oscillation (PDO). More specifically, these studies found that there was a change in the behavior of interannual variability of tropical cyclone (TC) occurrences in the North Atlantic as related to the El Niño and Southern Oscillation (ENSO). During the negative phase of the PDO (1947-1976), they found little ENSO related variability in the number and intensity of hurricanes basin wide. However, during the positive phase of the PDO (1977 – 1999) they found that there was strong interannual variability in TC occurrences between El Niño and La Niña years. During La Niña years, there were more and stronger hurricanes overall even though there were fewer months with significant hurricane activity. Also, there were more Caribbean and Gulf region hurricanes during La Niña years. Additionally, studies have shown that since 1999 (see Lupo et al. 2007, 2008, and Birk et al. 2010), the Pacific Ocean basin region has reverted back to the negative phase of the PDO.

Recently, Vimont and Kossin (2007) (and Kossin and Vimont, 2007) found a strong correlation in the North Atlantic between accumulated cyclone energy (which is positively correlated with TC activity) and the Atlantic Meridional Mode (AMM). In their work, they also find that the Atlantic Multi-decadal Oscillation (AMO) may excite the AMM on a decadal time scale. Thus, these indexes, which quantify the climatological background state, correlate with TC activity on the decadal time scale (AMO) or both the interannual and decadal (AMM) time-scales. Additionally, interdecadal variability as related to the PDO in the tropical and mid-latitude background climatological state have also been demonstrated in other studies (e.g., Deser et al. 2004, 2006, and Deser and Phillips 2006), and additional studies have discussed the synergistic relationship between oceanic and atmospheric conditions needed for an active tropical season (e.g., Zuki and Lupo 2008).

Studies of TC activity in the Eastern Pacific have shown that this area is generally more active than the Atlantic (e.g. Schultz, 2007; Collins, 2007; Lupo et al. 2008; Collins, 2010). The major findings for East Pacific TC activity have demonstrated that the long term trend in this region is decreasing, but the decrease is not statistically significant. Lupo et al. (2008) show that a slightly higher percentage of intense hurricanes occurred in the East Pacific region as compared to the Atlantic region. Collins (2007) shows that the eastern part of the East Pacific ocean basin was the most active in producing TCs, and that the western part was more variable with respect to ENSO due to thermodynamic considerations (atmospheric relative and absolute humidity, sea surface temperatures, vertical motions). Lupo et al. (2008) demonstrate that the seasonal activity (season beginning, peak, and end) was about 15 to 30 days earlier than that in the Atlantic Ocean basin.

These studies referenced above also demonstrated that there were more East Pacific TCs during El Niño years, and that this was mainly accounted for by more TCs becoming intense hurricanes than during La Niña years. The TC season was one or two months longer in El Niño years (Lupo et al. 2008), while more storms formed in the southeast and southwest part of the East Pacific Ocean basin (Collins, 2007). This is likely due to the fact that ENSO years bring warmer waters to the East Pacific region. When breaking down the ENSO years by phase of the PDO, the ENSO-related differences in occurrence and intensity and geographic formation region are accentuated in PDO1 years, but were blurred in PDO2 years. This ENSO and PDO related variability is similar to that occurring in the Atlantic (LJ00).

While the most extensive studies have been performed in the Atlantic Region, there have been studies demonstrating ENSO variability in West Pacific and Northern Indian Ocean region TC occurrence as well (e.g., Chan, 1985; Wu and Lau, 1992; Ng and Chan, 2010). These studies showed similar results for the Pacific Ocean basin as a whole. However, Ramage and Hori (1981), Lander (1994), and Chan and Liu (2004) found no ENSO variability in the overall TC numbers throughout the North Pacific as a whole, but significant variability in the primary TC genesis regions. Since then, studies have shown that the focus of tropical storm activity moves southeastward (northwestward) during El Niño (La Niña) years (e.g., Chan, 1985; Wu and Lau, 1992; Lander, 1994; Wang and Chan, 2002; and Chan 2007, 2009).

Recently, there has been concern that TC frequencies and intensities may be increasing because of climatic (whether attributed to natural or anthropogenic forcing) warming, and several studies have investigated the issue (Henderson-Sellers et al., 1998; Pielke et al., 2005; Emanuel, 2005; Webster et al., 2005; Chan 2009). These studies, however, have shown inconsistent results, and the study of Chan (2009) suggests that the relationship between the atmospheric variables and sea surface temperatures (SSTs) are too complex to determine that warmer SSTs alone can drive increased TC activity.

There has also been some disagreement in the literature regarding the relationship between the QBO and TC activity over the western north Pacific, the North Atlantic or the East

Pacific. Chan (1995) found that the westerly phase of the QBO was positively correlated with increased tropical cyclone activity over the western north Pacific, which is similar to the results of Gray (1984a) for the Atlantic Ocean basin. However, Lander and Guard (1998) showed that the global TC activity was not significantly correlated to the QBO.

Additionally, Gray (1968) and Gray (1979) discuss the important atmospheric and oceanic factors needed in the development of tropical cyclones. Since then, Zuki and Lupo (2008) and Chan (2009) demonstrate that the interannual variability in TC frequencies is not determined solely by SSTs and their variations. Rather, it appears that SST variations may provide a necessary but not sufficient condition for variations in TC occurrences and/or intensities. These studies demonstrate that the atmospheric conditions must also be favorable whether one examines dynamic variables (e.g., vorticity or upper level divergence), or thermodynamic variables (latent heating), and that it is possible to have warm SSTs but little or no TC activity (Zuki and Lupo, 2008). Chan (2009) demonstrates that dynamic and thermodynamic considerations may provide a sufficient condition, but that these variables may also be impacted by ENSO. Thus, the effect of ENSO and the PDO on TC activity is to alter the oceanic and atmospheric variables in a way that may be synergistically more or less favorable to TC activity.

The goal of this work is two-fold. The first is to examine TC activity and trends in all the global ocean basins where it occurs, including the South Atlantic. Where appropriate this work will be placed into context with respect to the earlier work of LJ00 and Lupo et al. (2008) and others. Second, the methodologies of LJ00 and Lupo et al. (2008) will be applied to study interannual and interdecadal TC activity globally, as a similar comprehensive study of TC internannual and interdecadal activity, including a partition of such activity into sub-basins, is not available elsewhere. Additionally, such a study for Southern Hemisphere activity is not yet available.

2. Data and methods

The data and methods are similar to those used by LJ00 and Lupo et al. (2008), and more detail regarding some of these subjects can be found there and in references therein. This study has been expanded to include all the globe's ocean basins and includes tropical storm occurrences as well. This study breaks down the global ocean basins as follows; the North Atlantic, East Pacific, West Pacific, North Indian, Southern Hemisphere (includes South Indian and the South Pacific), and the South Atlantic. A study of the background atmospheric and oceanic variables contributing to TC formation is not performed here as it is beyond the scope of this work.

2.1 Data

The TC occurrence and intensity data for all basins were downloaded via the UNISYS website (http://weather.unisys.com), and these data can also be found in other archives (e.g., LJ00). A complete description of these data sets and their reliability (e.g., Landsea 1993) can be found in Jarvinen et al. (1984) for the Atlantic, and Davis et al. (1984) for the East Pacific. In the West Pacific, these data were provided to UNISYS through the Joint Typhoon Warning Center (JTWC) best track data set. As stated by Lander (1994), TCs in the West Pacific Ocean have been monitored by the United States Navy since 1945 and by the United States Air Force since 1947 using aircraft reconnaissance. In 1959, these entities were combined under one command, the JTWC, to track TCs over the western North Pacific.

Weather satellite observations became routine in 1965, and these observations contributed to all the data sets as well. Lastly, the JTWC also monitors TC activity over the Northern Indian Ocean and the Southern Hemisphere as well. The data are also available via the International Best Track archive for Climate Stewardship Project (http://www.ncdc.noaa.gov/oa/ibtracs/) (e.g., Ng and Chan, 2010).

In this study, we use the term TC to include both hurricanes and tropical storms (TS) following Lupo et al. (2008), and here tropical storms refer only to those entities that obtained maximum wind speeds between 35 – 64 kt. TC data in the Atlantic basin before 1938 were not used in the study as explained in LJ00, and following their lead, this study only uses Saffir-Simpson (Simpson 1974) hurricane intensity data. In the East Pacific, data before 1970 were not included since all hurricanes before that date were classified as a category one (e.g., see Schultz 2007). In the Indian Ocean and Southern Hemisphere, tropical cyclones were not classified with an intensity value early in the record (1949 to the 1970s). TCs were not regularly categorized until 1977 in the North Indian Ocean and 1980 in the Southern Hemisphere. Thus, the results here will not include TCs before these dates in those regions. In order to eliminate problems with some of the data as discussed in LJ00, we combined hurricane intensity categories (category 1 and 2 - weak; category 3, 4, and 5 – intense) following Landsea (1993). Additionally, in some parts of the discussion tropical storms will be referred to as category – 0.

2.2 Methodology

The complete archive for the years discussed above for all the ocean basins were examined in this study. The Atlantic Ocean Basin (Fig. 1a) and the East Pacific Ocean Basin (Fig. 1b) are shown here. Arithmetic means and correlations were analyzed, and means were tested for statistical significance using a two-tailed z-score test, assuming the null hypothesis (e.g., Neter et al. 1988). Intensity distributions were also tested using a χ^2 statistical test. These distributions were tested using the total sample climatology as the expected frequency and a sub-period as the observed frequency. It has been hypothesized that using the climatological frequency as the "expected" frequency is more appropriate than using an approximated distribution since such analytical distributions (e.g., Poisson distribution) may not adequately represent real-world distributions (e.g., Lupo et al. 1997; LJ00). It should be cautioned that while statistical significance reveals strong relationships between two variables, it does *not* imply cause and effect. Conversely, relationships that are found to be strong, but not statistically significant may still have underlying causes due to some atmospheric forcing process or mechanism (e.g., Lupo et al. 1997; LJ00).

Finally, the data were stratified by calendar year in the Northern Hemisphere (NH). In the Southern Hemisphere (SH), the tropical cyclone year is defined as the period beginning on 1 July and ending on June 30. We then analyzed the annual and monthly distributions of TC occurrence in order to find both long and short-term trends in season length within both the total sample and each intensity category. The sample was also stratified by sub-basins in the Atlantic Ocean basin following LJ00. The East Pacific Ocean basin was stratified into quadrants divided by 125° W and 20° N following Collins (2007) and Lupo et al. (2008). The West Pacific Ocean was divided in quadrants by the lines 140° E and 20° N. The North and South Indian Ocean was dived into an east and west basin along 75° N. In the South Pacific, the ocean basin was divided into west and east along 180° longitude. TCs were assigned to the basin in which they first reached tropical storm status.

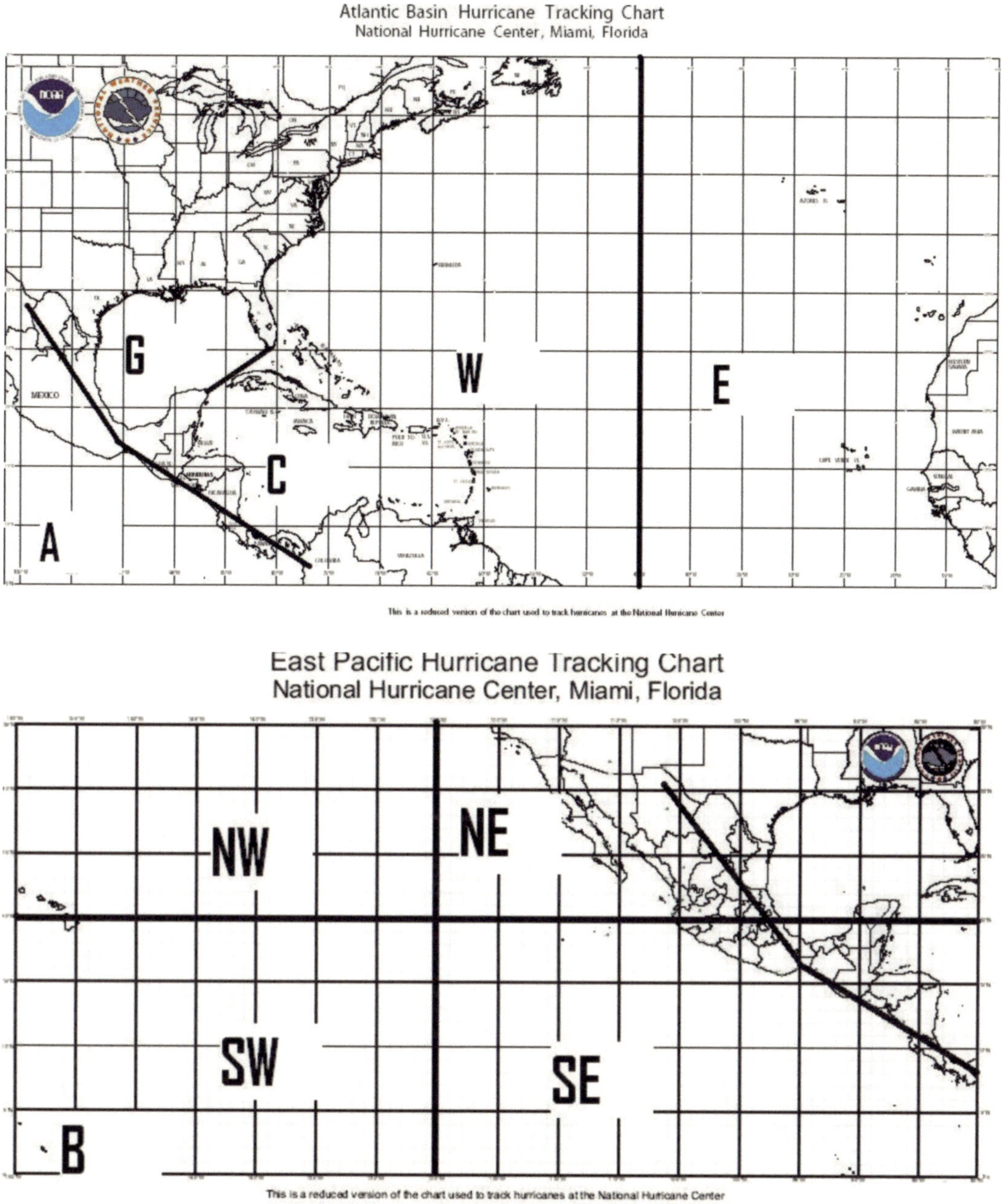

Fig. 1. The maps of a) the Atlantic Ocean Basin (G = Gulf, C = Caribbean, W (E) = West (East) Atlantic), and the b) Pacific Ocean Basin (NW = Northwest, NE = Northeast, SW = Southwest, and SE= Southeast) used in this study and provided by the National Hurricane Center. The heavy lines represent the boundaries between sub-basins used in this study. Their boundaries are outlined in section 2b and LJ00. The diagonal line across Mexico represents a dividing line between the Atlantic and Pacific Basins.

2.3 El Niño and Pacific decadal oscillation definitions

As in Lupo et al. (2008) and references therein, the Japan Meteorological Agency (JMA) ENSO Index was used in this study. A list of El Niño (EN), La Niña (LN), and Neutral (NEU) years used here are displayed in Table 1. In addition to consulting LJ00 or Lupo et al. (2008), a more detailed description of the JMA ENSO Index can be found on the COAPS website (http://coaps.fsu.edu/jma.shtml). In summary, the index uses long-term running mean sea surface temperature (SST) anomalies from the Niño 3 and 3.4 regions in the central and eastern tropical Pacific (e.g., Pielke and Landsea 1999). The SST anomaly thresholds used to define EN years are those greater than +0.5° C, less than -0.5° C for LN years, and NEU otherwise. The JMA ENSO criterion defined the El Niño year as beginning on 1 October of the previous year. This definition, however, was modified for the Northern Hemisphere following LJ00 so that the El Niño year coincided with the Atlantic and East Pacific TC seasons. No modification was needed for the SH.

The PDO (also known as the North Pacific Oscillation [NPO] in some studies) is a 50 to 70 year oscillation recently described (e.g., Minobe 1997; Mantua et al. 1997) within the eastern Pacific Ocean basin. As defined by Gershanov and Barnett (1998), the positive (warm) phase of the PDO (Fig. 2) is characterized by an anomalously deep Aleutian Low. Cold western and central north Pacific waters, warm eastern Pacific coastal waters, and warm tropical Pacific waters also characterize this phase of the PDO. Following Lupo et al. (2008) this phase is referred to as PDO1. The reverse conditions characterize the negative (cool) phase of PDO (Fig. 2) and these conditions are referred to as PDO2 (e.g., LJ00; Lupo et al. 2008). Each PDO period, or era, is defined using calendar years (see Table 2) and this information can also be found in Gershanov and Barnett (1998) and Lupo et al. (2007, 2008).

La Niña (LN)	Neutral (NEU)	El Niño (EN)
1938	1939	1940
1942	1941	1951
1944	1943	1957
1949	1945-1948	1963
1954 - 1956	1950	1965
1964	1952 – 1953	1969
1967	1958 – 1962	1972
1970 - 1971	1966	1976
1973 - 1975	1968	1982
1988	1977 – 1981	1986 – 1987
1998 - 1999	1983 – 1985	1991
2007	1989 – 1990	1997
	1992 – 1996	2002
	2000 – 2001	2006
	2003-2005	2009
	2008	

Table 1. A list of years examined in this study separated by ENSO phase. The current year is not officially classified yet.

PDO Phase	Period of Record
Phase 1	1933 - 1946
Phase 2	1947 - 1976
Phase 1	1997 – 1998
Phase 2	1999 - present

Table 2. Phases of the Pacific Decadal Oscillation (PDO) since 1933.

Landsea (1993), Goldenberg and Shapiro (1996), Gray et al. (1997), and Landsea et al. (1999) have demonstrated that hurricane activity is tied to changes in the long-term pressure patterns in the Atlantic Ocean basin (e.g., the North Atlantic Oscillation, or NAO). Deser et al. (2004) also demonstrate that sea level pressure and other background variables (e.g., wind shear) have displayed interdecadal variations, and Vimont and Kossin (2007) correlate these to Atlantic tropical cyclone activity. LJ00 and Lupo et al. (2008) found that the influence of the PDO was manifested by changes in the ENSO-related variability, specifically, that there was little or no ENSO related variability during PDO2, and significantly fewer and less (more and more) intense hurricanes during EN (LN) years during PDO1. As stated by LJ00, the NAO-related variations in TC activity can make interpretation of PDO-related hurricane variability more difficult given the overlap in the time scales between the PDO and long-term variations in the NAO. Many studies have speculated that these decade-to-decade variations in Atlantic hurricanes are forced by a combination of both. In the East Pacific Ocean, there were more TCs during EN years, and these in general were more intense (Lupo et al. 2008). As was found in the Atlantic, the PDO variability in the Pacific Ocean was primarily found in PDO1, while there was little or no variability in PDO2. Thus, ENSO is the main driver of interannual TC activity as demonstrated by many studies, and since PDO can be shown to modulate ENSO behavior, the focus here will be on these teleconnections.

3. Global tropical cyclone activity and trends up to the present

As stated in Lupo et al. (2008), there have been some changes to the information related to some storms due to reanalysis in the original data sets used by LJ00, and it is possible this would impact more than just the Atlantic region. Additionally, the latest TC seasons have been added for the Atlantic and East Pacific since the original publication of LJ00 or Lupo et al. (2008), and tropical storm information is included here. The analysis of TC activity in the other global ocean basins is new, including that of the South Atlantic.

3.1 Atlantic tropical cyclone activity: 1938 – 2009

The data set used here for the Atlantic Ocean has the longest period of record at 72 years. Since 1938, there have been a total of 752 TCs, including 326 tropical storms, and 165, 78, 86, 70, and 27 category 1 - 5 hurricanes (Table 3a, 4). During the 2000-2009 tropical cyclone seasons, there were 72 Atlantic hurricanes (Table 3a), which means that the years since the LJ00 study were the most active during the 1938 - 2009 record. However, the Atlantic TC seasons since 1995 have been relatively active. During the six year period from 1995 - 2000, there were 49 hurricanes (81 TCs), while during the years from 2001 - 2006, there were been 47 hurricanes (94 TCs), in spite of the fact that 15 hurricanes occurred during the 2005 season. Two other six-year periods were compared to these two recent eras and both of these earlier periods occurred before the satellite era. In particular, 43 (37) hurricanes and 62 (74) TCs occurred from 1884-1889 (1932-1937). Thus, it would be reasonable to consider the activity of these pre-satellite era periods as comparable to the modern eras, since it is very likely that some TCs were missed during the pre-satellite era.

Additionally, during the 2000 – 2009, there were 35 intense hurricane (category 3 – 5), which represented 49% of all storms (Table 3a). This compares with the LJ00 study in which 42% of all storms were intense during the previous 62 years. This indicates an increase in the ratio of intense storms overall. The increased activity of the latest period, however, is similar to that of the number of intense storms for the 30-year period of 1940-69 (48% - Table 3). Thus, as shown in Tables 3a and 4 for the entire 72-year record, the inclusion of the last ten years has not made any noteworthy change in the ratio of hurricane intensities for the overall Atlantic Ocean basin activity.

Pacific Decadal Oscillation

Positive (warm) phase (PDO1) Negative (cool) phase (PDO2)

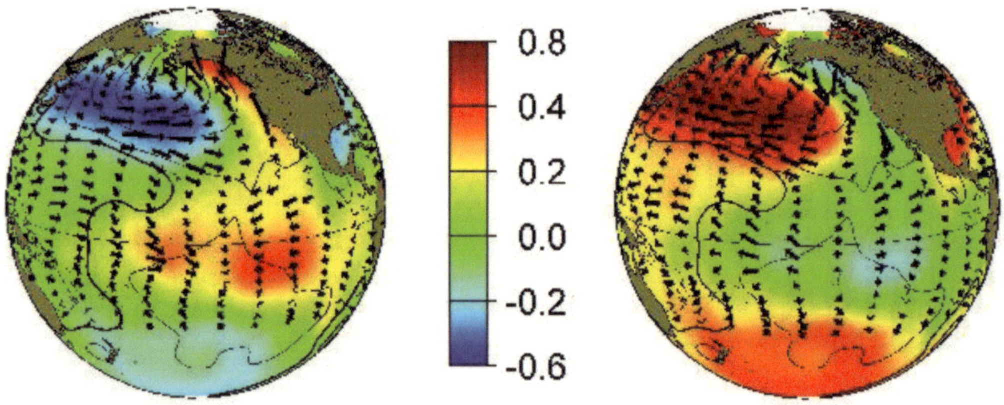

Fig. 2. The phases of the PDO as described in the text. SST anomalies are shown in color in accord with the scale shown. The arrows show anomalies in wind speed and direction. Reproduced with permission from the Joint Institute for the Study of Atmosphere and Ocean (JISAO – http://jisao.washington.edu) at the University of Washington.

LJ00 reported that there were no statistically significant trends in the Atlantic Ocean basin hurricane activity. Adding the 2000-2009 hurricane and tropical storm occurrences does not result in any change in the trends for the overall occurrence of category 1 – 5 storms (Fig. 3a). Separating these by category demonstrates that the trends for category 2-5 storms were not statistically significant (not shown). The trend in tropical storms (Fig 3b) and category 1 hurricanes (not shown) show an increase stronger than that in Fig. 3a, which shows the overall trend for the total number of hurricanes (significant at the 99% confidence level). Whether this upward trend in category one storms or tropical storms is real is clouded by two issues. The trend could be the result of 1) the problem of the "fortuitous" placement starting and ending points in a long time series (the recent activity could be the peak in a long-term oscillation), or 2) that the earlier part of the 72-year period was in the pre-satellite era and storms may have been undercounted. It is likely that both of these factors are contributing to the overall increase. Nonetheless, if the 1970-2009 period only was used, there was an increase in the occurrence of Atlantic basin tropical storms and category one hurricanes and the trends were also statistically significant (not shown) at the 95% confidence level. However, an examination of Fig. 3 suggests significant interannual variability as found by LJ00 and Lupo et al. (2008).

3.2 East Pacific tropical cyclone activity: 1970 – 2009

Including tropical storms, there were 656 East Pacific TC events included in the 40-year sample. This resulted in an average of 16.1 TCs per season (Table 3b, 5). This includes 298 tropical storms (7.3 per year) and 358 hurricanes (8.9 per year). As expected, the number of category 1 storms was the largest (125), and there were 166 intense storms (category 3-5), which represented 46.4% of the hurricane activity. This compares to 42.8% overall in the Atlantic Ocean basin overall.

a. Atlantic

Category	1940-69	1970-89	1990-99	2000-09	All
TS	3.7	4.3	4.6	7.7	4.5
Cat. 1,2	3.1	3.5	3.9	3.7	3.3
Cat. 3-5	2.9	1.7	2.5	3.5	2.5
Cat. 4,5	1.4	0.9	1.4	2.1	1.3
Tot hur.	6.0	5.1	6.4	7.2	5.8
TS+Hur.	9.7	9.4	11.0	14.9	10.3

b. East Pacific

Category	1970-89	1990-99	2000-09	All
TS	7.6	5.6	9.1	7.3
Cat. 1,2	5.2	4.5	4.4	4.6
Cat. 3-5	4.2	5.5	2.8	4.1
Cat. 4,5	2.1	3.9	1.8	2.4
Tot hur.	9.3	10.0	7.2	8.8
TS+Hur.	16.9	15.6	16.3	16.1

c. West Pacific

Category	1940-69	1970-89	1990-99	2000-09	All
TS	7.2	10.0	10.3	10.4	8.9
Cat. 1,2	7.5	8.2	9.1	6.5	7.7
Cat. 3-5	10.4	7.8	9.0	9.5	9.1
Cat. 4,5	6.6	5.1	7.3	8.0	6.4
Tot hur.	17.8	16.0	18.1	16.0	16.8
TS+Hur.	25.1	26.0	28.4	26.4	25.7

d. North Indian

Category	1940-69	1970-76	1977-89	1990-99	2000-09	All
Unknwn	14.6	16.0	N/A	N/A	N/A	14.9
TS	N/A	N/A	3.7	2.9	2.9	3.4
Cat. 1,2	N/A	N/A	0.6	1.3	0.7	0.9
Cat. 3-5	N/A	N/A	0.3	1.0	0.3	0.5
Cat. 4,5	N/A	N/A	0.1	0.7	0.3	0.4
Tot hur.	N/A	N/A	0.9	2.3	1.0	1.4
TS+Hur.	N/A	N/A	4.6	5.2	4.9	4.8

e. Southern Hemisphere

Category	1940-69	1970-79	1980-89	1990-99	2000-09	All
Unknwn	27.8	30.2	N/A	N/A	N/A	28.5
TS	N/A	N/A	14.6	12.8	13.0	13.5
Cat. 1,2	N/A	N/A	8.0	7.8	5.6	7.0
Cat. 3-5	N/A	N/A	5.0	7.3	7.2	6.6
Cat. 4,5	N/A	N/A	1.6	4.8	4.5	3.8
Tot hur.	N/A	N/A	13.0	15.1	12.9	13.5
TS+Hur.	N/A	N/A	27.6	27.9	25.8	27.0

Table 3. The mean annual number of hurricane and tropical storm (TS) events within the a) Atlantic, b) East Pacific, c) West Pacific, d) North Indian, and e) Southern Hemisphere Basins over decadal time periods corresponding with similar studies. Hurricanes are stratified by weak hurricanes (Cat. 1 and 2), intense hurricanes (Cat. 3, 4, and 5), and Cat. 4 + Cat. 5. Categories are based on the Saffir-Simpson scale.

	All	TS	Cat 1	Cat 2	Cat 3	Cat 4	Cat 5	T_{hur}
LN	10.9	5.0	1.9	1.4	1.1	1.1	0.4	5.9
NEU	10.8	4.4	2.6	1.0	1.3	1.1	0.4	6.4
EN	8.8	4.1	2.1	0.9	1.1	0.6	0.1	4.7
Total	10.3	4.6	2.3	1.1	1.2	1.0	0.4	5.8

Table 4. The average annual occurrence of Atlantic Ocean Basin tropical storms and hurricanes versus El Niño / La Niña phase for each Saffir-Simpson category from 1938 - 2009.

Phase	All	TS	Cat 1	Cat 2	Cat 3	Cat 4	Cat 5	T_{hur}
LN	15.1	7.3	3.0	1.7	1.6	1.4	0.1	7.8
NEU	16.4	7.2	3.3	1.7	1.5	2.6	0.1	9.1
EN	17.6	8.0	2.9	1.7	2.1	2.2	0.7	9.6
Tot	16.1	7.3	3.0	1.7	1.7	2.2	0.3	8.8

Table 5. As in Table 4, except for the East Pacific from 1970 – 2009.

The overall trend demonstrates that there was a decrease in the number of East Pacific Ocean hurricanes (Fig. 4a), but an upward trend in the number of tropical storms (Fig. 4b). However, neither trend was statistically significant. The overall trend in hurricane numbers was opposite to the Atlantic basin where there was a statistically significant increase in activity over the same period. If the two basins are combined for the period, there was no statistically significant trend upward or downward. It is noted here that trend lines shown in Fig. 3a and Fig. 4a are simple linear best fit (regression) models. The two time series (from 1970 for the Atlantic region) correlated to each other negatively (correlation coefficient of -0.385, significant at the 90% confidence level), and this will likely be explained by the interannual variability in hurricane activity for both basins being opposite one another as discussed below. The correlation between the two increases to -0.405 when all TCs are included in Figs. 3 and 4. An examination of the trends in each East Pacific hurricane category would demonstrate that only the category 3 (5) storms showed a significant downward (upward) trend, while the rest of the categories showed no significant trend.

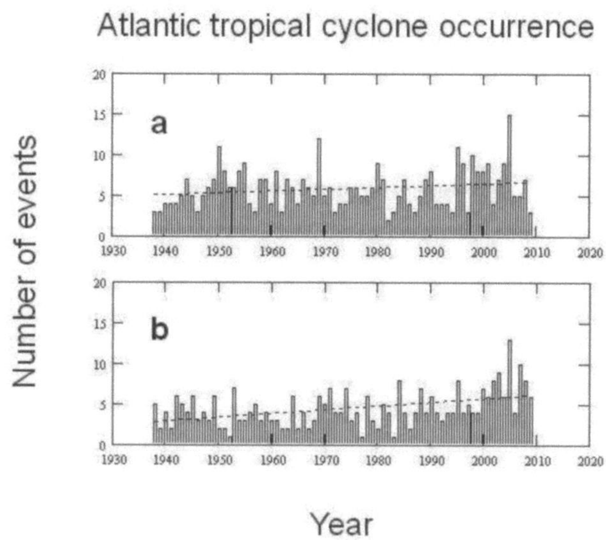

Fig. 3. The annual frequency of hurricane occurrence in the Atlantic Ocean basin for a) all categories, and b) tropical storms for the period 1938 - 2009. The dashed line is the linear trend line.

3.3 West Pacific tropical cyclone activity: 1945 – 2009

The West Pacific Ocean data set used here was the second longest period of record of the global ocean basins where TCs form at 65 years. This basin is arguably the most active of the global ocean basins. Since 1945, there have been a total of 1694 TCs, including 588 tropical storms, and 312, 195, 179, 257, and 163 category 1 - 5 hurricanes (Table 3c, 6). This represents a mean of 16.9 hurricanes and 25.7 total TCs per year in the West Pacific. In this region, the proportion of intense hurricanes (599) is greatest at 54% of the total number of hurricanes (1106). Of these intense hurricanes, Category 4 storms represented the second largest total (257) of hurricanes in the entire sample, which represented 23.2% of all TCs. It should not be surprising that this basin should contain the greatest fraction of intense hurricanes given that some of the warmest ocean temperatures in the world are found within the basin (e.g., Hurrell et al. 1995).

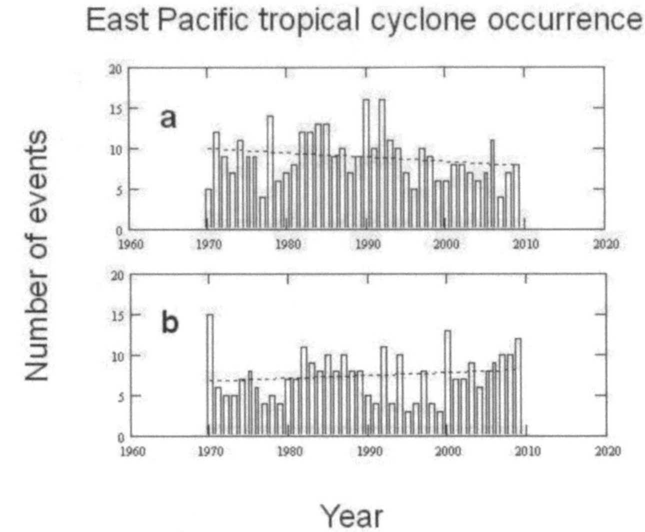

Fig. 4. As in Fig. 3, except for the East Pacific Ocean Basin and the years 1970 - 2009.

Phase	All	TS	Cat 1	Cat 2	Cat 3	Cat 4	Cat 5	T_{hur}
LN	25.5	9.6	5.0	3.0	2.5	3.3	2.2	15.9
NEU	26.4	9.2	5.1	3.1	2.7	3.9	2.4	17.2
EN	25.9	8.2	3.9	2.7	3.2	4.8	3.1	17.7
Tot	25.7	8.9	4.7	3.0	2.7	3.9	2.5	16.8

Table 6. As in Table 4, except for the West Pacific from 1945 – 2009.

As in the East Pacific Ocean Basin, the number of hurricanes since 1945 were showing a slight downward trend (Fig. 5a), but this trend is not statistically significant. Table 3c demonstrates that within this trend, the number of category 1 and 2 storms were higher during the 1970 - 1990s, while the number of intense hurricanes were higher during the 1940's - 1969 and during the most recent decade. This suggested there may be variability within this basin on the interdecadal time scale and this will be investigated in section 4. The number of tropical storms increased throughout the period (Fig. 5b), and this trend was shown to be statistically significant (at the 99% confidence level) as well. This result is similar to the result found for the Atlantic Ocean basin, but this is also likely to be the result of better monitoring for these storms. Recall the satellite era began in the mid-1960's and Table 3c suggests that there has been no significant trend in West Pacific tropical storm activity since 1970. Lastly, West Pacific Ocean basin hurricane activity correlates positively to East Pacific hurricane activity (0.375) and this correlation is significant at the 90% confidence level. Like the East Pacific, this basin also correlates negatively to Atlantic Ocean hurricane activity, but the correlation is not statistically significant.

3.4 North Indian tropical cyclone activity: 1977 - 2009

In this region, tropical storm activity has been recorded since 1945 but storms were not assigned an intensity rating regularly until 1977. This is a similar situation to that of the East Pacific where TCs were classified as tropical storms or category 1 hurricanes from 1949 to 1969. During the early period there was an average of 14.9 TCs per year in the North Indian Ocean Basin. From 1977 - 2009 there were only 161 North Indian TCs found in this basin or an average of 4.8 events per year (Tables 3d, 7). In the next section, it will be demonstrated that most of this change took place in the Bay of Bengal region (Eastern half of the study region). This is likely to be the result of a change in the monitoring of these storms, in that during the earlier period, it is possible that tropical depressions and tropical waves were also counted. The North Indian Ocean Basin TCs include 116 tropical storms (3.4 per year) and 45 hurricanes (1.4 per year). As expected, the number of category 1 storms was the largest (23), and there were only 17 intense storms (category 3-5), which represented only 37.8% of the hurricane activity, or an average of 1 intense hurricane every two years. This is the lowest proportion of intense hurricanes of the NH Ocean basins.

The overall trend since 1977 demonstrates that there was little or no trend in either hurricanes or the number of tropical storms within the North Indian Ocean Basin (Fig. 6). There was little trend in the pre-1977 overall numbers as well (not shown). Additionally, there is little or no correlation between TC occurrence in the Northern Indian Ocean and the other NH Basins. However, the lack of a trend does not mean there are no interannual or interdecadal variations to describe and these will be discussed below.

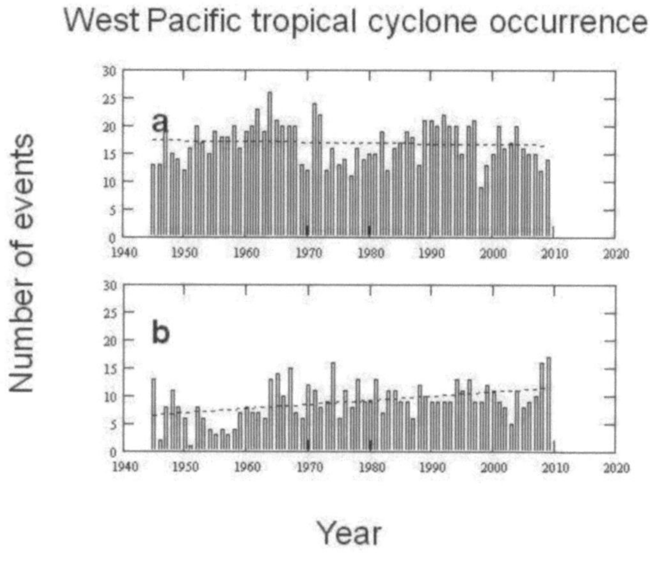

Fig. 5. As in Fig. 3, except for the West Pacific Ocean Basin and the years 1945 - 2009.

3.5 Southern Hemisphere tropical cyclone activity: 1980 - 2010

In this region, tropical storm activity has also been recorded since 1945 but storms were not assigned an intensity rating regularly until 1980. In the SH, the tropical season peaks during the NH winter and spring. Since the tropical season does not follow the calendar here, the tropical year 2010 actually began 1 July, 2009 and ended June 30, 2010. During the period before 1980, there was an average of 28.5 TCs per year in the SH. From 1980 – 2010 there were 837 SH TCs found for an average of 27.0 events per year (Tables 3e, 8). Thus, there was not a large change between the means of the pre- and post- 1980 era. Additionally, this is only one more TC than that of the West Pacific Ocean basin only on an annual basis.

Phase	All	TS	Cat 1	Cat 2	Cat 3	Cat 4	Cat 5	T_{hur}
LN	5.5 / 14.8	3.0	1.0	0.3	0.8	0.3	0.3	2.5
NEU	4.8 / 15.4	3.4	0.8	0.1	0.1	0.2	0.0	1.3
EN	4.9 / 14.0	4.0	0.3	0.1	0.0	0.4	0.1	0.9
Tot	4.8 / 14.9	3.4	0.7	0.2	0.2	0.3	0.1	1.4

Table 7. As in Table 4, except for the North Indian Ocean from 1977 – 2009 (all – second number is pre-1977 era).

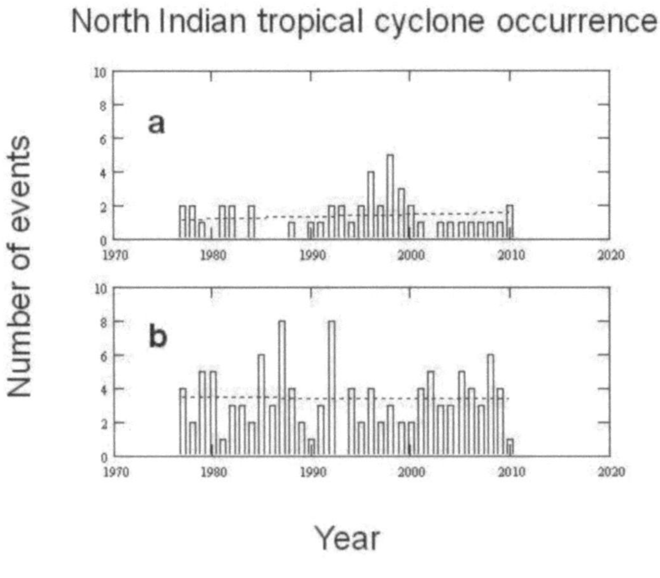

Fig. 6. As in Fig. 3, except for the Northern Indian Ocean Basin and the years 1977 - 2010.

Phase	All	TS	Cat 1	Cat 2	Cat 3	Cat 4	Cat 5	T_{hur}
LN	29.3 / 30.3	13.3	5.5	2.5	3.3	4.3	0.5	16.0
NEU	26.9 / 25.8	13.9	4.8	1.8	3.9	2.8	0.5	12.9
EN	26.1 / 31.1	12.3	5.1	2.1	2.3	3.3	1.0	13.8
Tot	27.0 / 28.5	13.5	5.0	2.0	2.8	3.1	0.6	13.5

Table 8. The mean number of SH tropical storms and hurricanes separated by ENSO phase and intensity (Saffir – Simpson scale) from 1980 – 2010 (except all – the second number is the pre-1980 era).

In this region, it is likely that better records were kept following 1980, and if tropical depressions are added in post-1980, the means are nearly identical (not shown). In the SH, there were 417 tropical storms (13.5 per year) and 420 hurricanes (13.5 per year). Like the West Pacific Ocean basin, category 4 storms constituted the second largest number of hurricanes (cyclones) after category 1 storms with 97 total events, and this represented 23.1% of all hurricanes. The percentage of intense storms (category 3-5, 204 cyclones) was 48.6%, which was similar to that of the Atlantic region.

The overall trend since 1980 demonstrates that there was little or no trend in either hurricanes (cyclones) or the number of tropical storms within the SH (Fig. 7), which is similar to the North Indian Ocean Basin. There was little trend in the pre-1980 overall numbers as well (not shown). Additionally, there is little or no correlation between TC occurrences in the SH basins with those in the NH basins, except for those in the Northern Indian Ocean basin. However, this correlation did not rise to the level of statistical significance. As will be shown later, there are interannual or interdecadal variations and these will be discussed below.

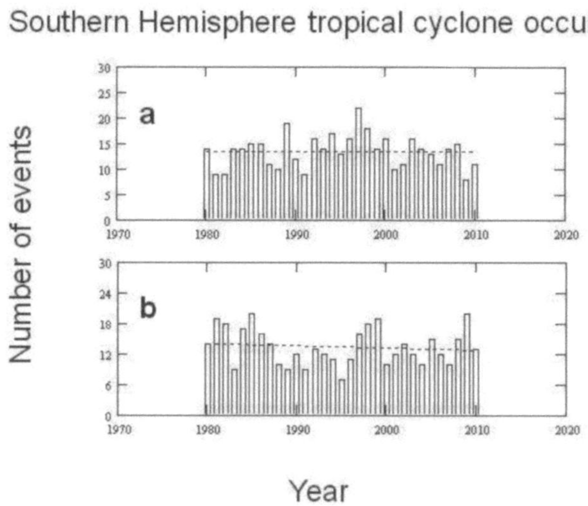

Fig. 7. As in Fig. 3, except for the Southern Hemisphere and the years 1980 - 2010.

3.6 South Atlantic tropical cyclone activity: 1960 - 2010

While there are very few instances of TC occurring in the South Atlantic because of cold sea surface temperatures and stronger tropospheric wind shears, there have nonetheless been occurrences worthy of note. The most famous event was Hurricane Catarina which occurred in March 2004, and is recognized as the only storm to become of hurricane strength in this basin (e.g. Reagan et al. 2005; McTaggart-Cowan et al. 2006). The most recent occurrence was a tropical storm (designated 90Q, or Tropical Storm Anita) documented by the National Aeronautics and Space Administration (NASA) during March 10-12, 2010 (http://www.nasa.gov/mission_pages/hurricanes/archives/2010/h2010_90Q.html). There have been five other possible events, all tropical depressions or tropical storms, in this ocean basin occurring during March 1974, April 1991 (McAdie and Rappaport, 1991), January 2004, February, 2006, and January, 2009. There may be some question whether or not some of these reached tropical storm strength, however, there is no doubt that with better observational technology, there have been more possible tropical events observed during the last decade in this region than at any other time. The tropical season in this region would be primarily January to March, which would correspond to July through September in the North Atlantic. Also, given the small number of events, the interannual and interdecadal variability cannot be assessed.

4. Interannual and interdecadal variability in tropical cyclones.

Then, in order to identify the significant periodicities within the times series (e.g., Figs. 3 - 7), power spectra were constructed and analyzed. Fig. 8 displays the power spectrum analysis of the Atlantic Basin annual TC occurrence time series constructed using Fourier coefficients. Also shown in the Fig. 8 is the 95% confidence level against a continuum background of white or red noise (Wilks 2006). Not surprisingly, periods of significant variability emerge in the analysis. In Fig 8a (b), significant interannual and interdecadal variability is evident, and the time periods are at roughly 5, 9, and 24 (3, 5, 12, and 18) years for Atlantic hurricanes (TC). The higher frequency variability is consistent with the time scale of ENSO, while the lowest frequency variations are suggested here to be representative of interdecadal PDO-related variability (Minobe, 2000). In addition, significant interdecadal variability at time periods of approximately 9 and 12-15 years are similar to that of Enfield and Mestas-Núñez (1999), Mestas-Núñez and Enfield (1999), or Birk et al. (2010), and attributed to the PDO or PDO - ENSO related interactions. Similar variability was found in the other ocean basins (not shown), especially the NH Pacific Ocean basins which have longer time series available.

4.1 Atlantic Ocean Basin variability

In Table 4, there were more TCs in LN and NEU years in the 72-year dataset as well as in the number of hurricanes and intense hurricanes than in EN years as in LJ00 and Lupo et al. (2008). None of these results rise to the level of statistical significance when testing the means only. Additionally, if these results are tested as a probability distribution (Fig. 9), and using the chi-square test (e.g., Neter et al. 1988) also reveals no statistically differences between the distribution in ENSO years. However, the distributions were not the same at the 95% confidence level either.

In examining differences within the Atlantic Ocean Basin by sub-region (Fig. 1 and Table 9) reveals that there were some differences in which regions were more active. While the West Atlantic showed little difference between individual years in the number of TCs, the Caribbean was more (less) active in LN (EN) years. But, these differences in TC activity do not rise to the level of statistical significance. Also, if the annual distribution is examined (Fig. 10), there were some differences in the duration of the TC season as shown in LJ00 and Lupo et al. (2008). The nominal TC season in the Atlantic is recognized as 1 June to 30 November. This can be demonstrated by a certain level of observed TC activity occurring during a particular month. Here, 0.6 events per year was chosen as this represents a frequency greater than once every two years for a particular month. In Fig. 10, the TC seasons for the total sample and neutral years were 1 June to 30 November. However, in LN years, the season typically begins later (July), exhibits a stronger peak, and is more active later into the season (November). However, EN years tended to be more active earlier (June), and less active later, and these results were found by LJ00 and Lupo et al. (2008).

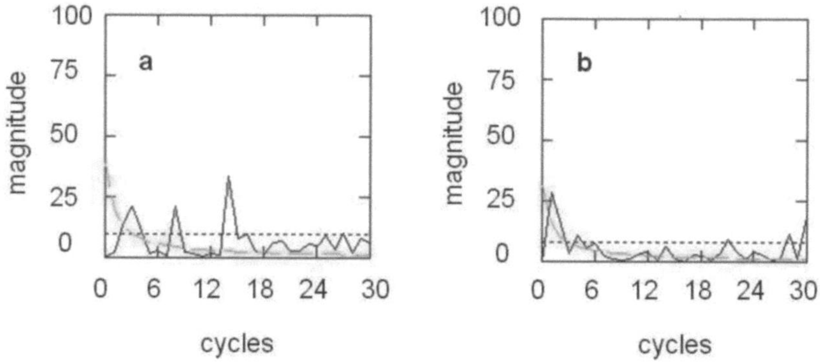

Fig. 8. The fourier power spectra for Atlantic Basin a) hurricane, and b) total tropical cyclone frequencies. The abscissa is the number of cycles during the 72 year period, and the ordinate is the magnitude of the fourier coefficients. The dotted (dashed) line is the 95% confidence level assuming a white (red) power spectrum (e.g., Wilks, 2006).

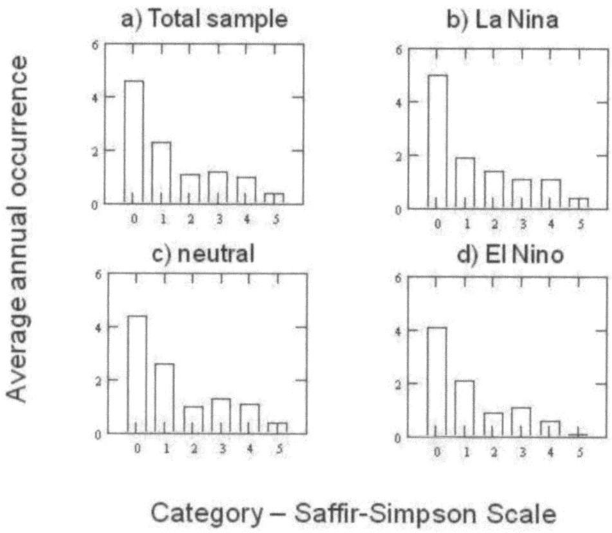

Fig. 9. Histograms of the average annual hurricane frequency in the Atlantic separated by category (0 is tropical storm) for a) the total sample (1938-2009), b) La Niña years, c) neutral years, and d) El Niño years.

	All	Crbn	Gulf	W Atl	E Atl
LN	10.9	2.1	2.3	5.0	1.4
NEU	10.8	1.7	2.2	5.1	1.8
EN	8.8	0.8	1.8	5.2	1.2
Total	10.3	1.6	2.1	5.1	1.5

Table 9. The average annual occurrence of Atlantic hurricanes and tropical storms by sub-ocean basin as stratified by ENSO phase. The regions are the Caribbean (Crbn), Gulf of Mexico (Gulf), West Atlantic (W Atl), and East Atlantic (E Atl) sub-basins.

While many of the differences above did not rise to the level of statistical significance, these results are examined then by separating into PDO phase (Table 2) following LJ00 and Lupo et al. (2008) then some interesting results emerge. The TC activity during PDO2 (Table 10a) displayed little ENSO-related variability. Also, the variability as measured by the standard deviations during these years was also larger (not shown). Thus, predictability during these years would be more difficult. During PDO1 (Table 10b), there was stronger ENSO-related variability than found in the dataset overall. LJ00 and Lupo et al. (2008) found that the number of hurricanes was significantly greater in LN years than during EN years and this was significant at the 90% confidence interval, and this is true here as well. The greatest disparity was found in intense hurricanes, and during PDO1 LN years there were three times as many intense hurricanes.

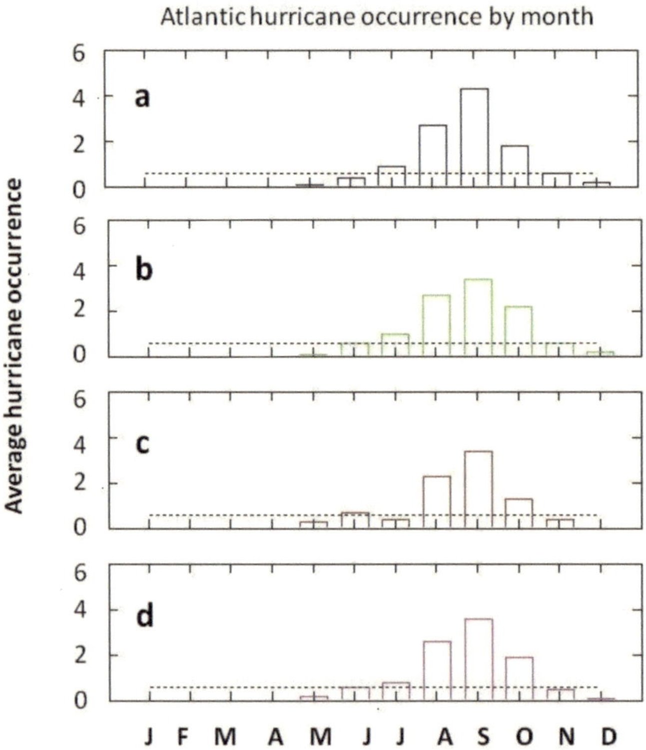

Fig. 10. The annual mean tropical cyclone occurrence by month for a) La Niña, b) neutral, c) El Niño, and d) total sample for the Atlantic Ocean Basin. The dotted line represents 0.6 TC events per month.

If these results were separated by sub-basin (not shown), the West Atlantic showed little ENSO related variability in either PDO phase. However, during PDO1, the other sub-basins were more active in LN years and in the Caribbean the difference was significant at the 90% confidence level. The seasonal differences are also magnified during PDO1 years (not shown). During LN (EN) years, the season is July through November (July through October). Finally, there is some differences in the length of season among each of the sub-basins. The west Atlantic has the longest season, while in the East Atlantic is basically active during August and September and this shows little variability by ENSO or PDO year.

a. PDO2

	All	TS	Cat 1	Cat 2	Cat 3	Cat 4	Cat 5	T_{hur}
LN	11.0	4.9	1.9	1.3	1.2	1.2	0.4	6.0
NEU	12.1	4.9	2.6	0.8	1.8	1.3	0.7	7.2
EN	9.8	4.3	2.1	1.1	1.4	0.7	0.2	5.5
Total	11.1	4.8	2.3	1.0	1.5	1.1	0.5	6.4

b. PDO1

	All	TS	Cat 1	Cat 2	Cat 3	Cat 4	Cat 5	T_{hur}
LN	11.0	5.2	1.8	1.8	0.8	0.8	0.6	5.8
NEU	9.7	4.1	2.6	1.1	0.9	0.9	0.3	5.7
EN	7.2	3.8	2.0	0.5	0.5	0.3	0.0	3.3
Total	9.4	4.2	2.3	1.1	0.8	0.7	0.3	5.2

Table 10. The average annual occurrence of Atlantic tropical cyclones stratified by ENSO phase and Category (Saffir-Simpson scale) during a) PDO2 (1947 - 1976, 1999-present), and b) PDO1 (1938-46, 1977-98).

4.2 East Pacific Ocean variability

A breakdown by ENSO phase (Table 5) indicates that there is some interannual variability in East Pacific region TC activity. Traditionally, May is the first month that significant TC activity begins in the East Pacific. In this study, August was the most active month with 165 total TC's occurring over the 40 year climatology (4 tropical cyclones per year), and by November, this region is relatively inactive. Thus, the East Pacific season was roughly one calendar month ahead of the Atlantic season, a result similar to Davis et al. (1984) and Schultz (2007). By geographic region (Table 11), the southeast part of the East Pacific was the most active, and 84% of the TC occurrences happened within this region. Only five TC, including two hurricanes occurred in the northwest region.

Phase	NE	SE	SW	NW	Tot
LN	1.0	12.3	1.4	0.3	15.1
NEU	1.0	12.8	2.4	0.1	16.4
EN	0.5	13.7	3.4	0.0	17.6
Tot	0.9	12.7	2.4	0.1	16.4

Table 11. As in Table 9, except for the East Pacific basin geographic region (along 20° N and 125° W).

During EN years, there were more TC in the East Pacific (Table 5), and this included more storms reaching hurricane strength, especially intense hurricanes, in this region than in LN years. This might be expected as the waters in the East Pacific are warmer during EN years. While the greater overall frequency of EN year TC is not statistically significant, it is opposite of the ENSO variability in the Atlantic Ocean basin. When separating by PDO phase (Table 12), it can be shown that there was little overall ENSO variability in PDO2 years, while the ENSO variability was accentuated in PDO1 years (two more hurricanes and four more TC overall during EN years) as it was in the Atlantic region (LJ00), except here EN years were more active (a result significant at the 90% confidence level). Additionally, there was more TC activity in PDO1 years (e.g. Lupo et al., 2008). These were years characterized by stronger warm ENSO SST anomalies located closer to the Americas (see Lupo et al. 2007), or over the East Pacific region, especially the southeast portion of our basin of study. This provided the rationale for the breakdown of the East Pacific Region into sub-basins.

a. PDO2

	All	TS	Cat 1	Cat 2	Cat 3	Cat 4	Cat 5	T_{hur}
LN	15.4	7.7	3.1	1.9	1.4	1.1	0.1	7.7
NEU	15.7	8.8	3.0	2.0	0.7	1.2	0.0	6.8
EN	16.8	7.8	2.0	1.8	2.2	2.2	0.8	9.0
Total	15.9	8.1	2.8	1.9	1.4	1.4	0.3	7.8

b. PDO1

	All	TS	Cat 1	Cat 2	Cat 3	Cat 4	Cat 5	T_{hur}
LN	14.0	6.0	2.5	1.0	2.0	2.5	0.0	8.0
NEU	16.7	6.6	3.4	1.5	1.9	3.1	0.1	10.1
EN	18.4	8.2	3.8	1.6	2.0	2.2	0.6	10.2
Total	16.8	6.9	3.4	1.5	1.9	2.9	0.2	9.9

Table 12. As in Table 10, except for the East Pacific.

In order to determine if there were variations in the intensity of storms, LJ00 and Lupo et al. (2008) used histograms and compared their distributions, testing these for statistical significance. For the overall East Pacific sample here, there were no significant differences in the distributions (Fig. 11), even though the differences in the hurricane frequencies (Table 12) demonstrate that there were differences in the overall averages for the occurrence of intense storms. During PDO2 (there was little ENSO-related interannual variability), and PDO1 (EN years significantly more active), the results were different even if there were no statistically significant differences in the distributions (which can be inferred from Table 12). There were clearly more intense storms during EN years (5.2 versus 2.6 for LN) for PDO2 years. However, the real substantive differences between hurricane occurrences in EN and LN years during PDO1 were the result of a greater number of weaker hurricanes and tropical storms. The PDO1 result is consistent with the results of Schultz (2007), who used a different methodology but found that during more active years, there were a greater number of weaker storms. This is, however, different from the Atlantic were LJ00 (and Lupo et al. 2008) found that there were more intense storms during LN years, especially in the PDO1 phase.

Comparing the length of the East Pacific season by examining the monthly occurrence (Fig. 12), shows that during EN years, TC activity is greatest from June through November, while in LN years, the season is one month shorter (primarily June - October). The seasonal variations in tropical storms versus hurricanes were similar overall. However, the seasonal peak in each ENSO phase occurred in different months. During EN years, the peak in tropical cyclone activity was clearly August, while in neutral years the peak was later (August / September), and earlier (July / August) in LN years. There was little difference in the ENSO-related seasonal cycles between the phases of the PDO here (not shown).

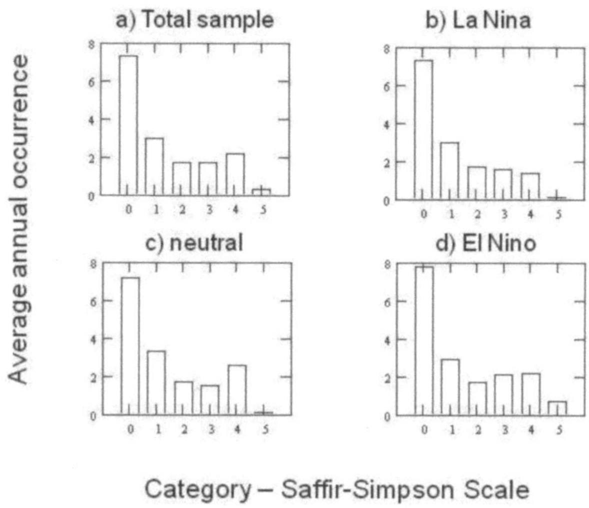

Fig. 11. As in Fig. 9 Except for the East Pacific

A breakdown of TC activity by geographic region (see Lupo et al., 2008) demonstrated that the southeast region was the most active, and varied similarly to that of the total sample discussed above. The northeast region, however, was relatively more active in LN and NEU years, while the southwest region was more active in EN and NEU years (Table 11). During the EN years, this reflects the fact that warmer SSTs were located over the East Pacific formation regions (see Lupo et al. 2007), especially in the southern sectors. While nearly all of the intense hurricanes were in the southwest and southeast regions, neither region had a significantly higher proportion of category 4 and 5 storms. Only one hurricane which formed north of 20° N achieved category 3 status. A separation of these data by phase of the PDO (not shown) would reveal that the ENSO variability described above is independent of the phase of the PDO in the East Pacific.

4.3 West Pacific Ocean variability

As shown in section 3, West Pacific Ocean TC activity correlated positively with that in the East Pacific. Unlike the results of section 4.1 and 4.2, Table 6 shows that indicates that there is little ENSO related variability in West Pacific region TC activity overall. This would agree with the results of Chan (1985), Wu and Lau (1992), Lander (1994), Wang and Chan, (2002), and Chan (2007, 2009). Traditionally, in this basin it has been recognized that TC activity is possible throughout the year, however, the NH spring (fall) are times where this basin is minimally (maximally) active. In this basin, August and September were the most active months with 663 total TC's occurring over the 65 year climatology (about 5 TCs in each month per year). January through March in this region is relatively inactive using 0.6 storms per year as the benchmark for defining an active month. By geographic region (Table 13), the southwest part of the West Pacific (dividing along 20° N latitude and 140° E Longitude) was the most active, but this region accounted for only 46% of the West Pacific TC occurrences. The southeast part accounted for 32%, while the northeast and northwest regions accounted for 11% each of the remaining TC occurrences.

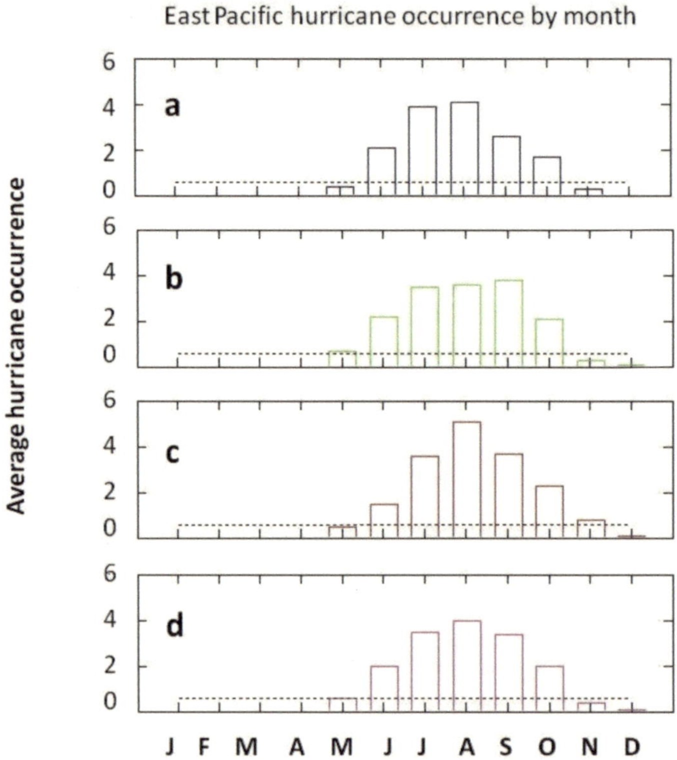

Fig. 12. As in Fig. 10 except for the East Pacific.

Phase	NE	SE	SW	NW	Tot
LN	3.3	5.1	13.1	3.5	25.5
NEU	3.0	8.2	12.0	3.0	26.4
EN	1.7	11.6	10.4	2.1	25.9
Tot	2.7	8.2	11.7	2.9	25.7

Table 13. As in Table 9, except for the West Pacific by geographic region (along 20° N and 140° E).

While there was little difference in the overall TC numbers in this basin, a more detailed look would find some significant variability within the West Pacific Basin. Similar to the results of the East Pacific, there were more storms reaching hurricane strength, and especially more intense TCs (11.1 versus 8.0) during EN years (Table 6), while there were more weaker storms during LN years. The intense hurricane difference was significant at the 90% confidence level. As in the East Pacific, the expanded pool of warmer waters in the tropical Pacific overall may be part of the reason for this result. When separating by PDO phase (Table 14), the results are similar to the East Pacific and Atlantic. There was little overall ENSO variability in PDO2 years, while there was strong ENSO variability in PDO1 years. In PDO2 years, the ENSO variability mimicked that described above, however, during PDO1 years, there was very strong ENSO related variability, which was significant at nearly the 95% confidence level. There were roughly six more TCs during PDO1 EN years. This difference was found primarily among intense storms where there were 12 (6) intense storms during EN (LN) years, and this result is significant at the 98% confidence level.

a. PDO2

	All	TS	Cat 1	Cat 2	Cat 3	Cat 4	Cat 5	T_{hur}
LN	26.2	9.5	5.2	3.2	2.7	3.2	2.4	16.7
NEU	26.1	8.3	4.7	2.9	3.2	4.2	2.8	17.8
EN	25.1	8.3	3.6	1.9	3.1	5.5	2.7	16.8
Total	25.9	8.7	4.6	2.8	3.0	4.2	2.7	17.2

b. PDO1

	All	TS	Cat 1	Cat 2	Cat 3	Cat 4	Cat 5	T_{hur}
LN	21.5	10.5	3.5	1.5	1.0	4.0	1.0	11.0
NEU	26.6	10.1	5.5	3.4	2.2	3.5	1.9	16.5
EN	27.4	8.0	4.4	4.2	3.4	3.4	4.0	19.4
Total	26.4	9.7	5.1	3.4	2.3	3.5	2.3	16.7

Table 14. As in Table 10, except for West Pacific Ocean.

In order to confirm further the variations in the intensity of TCs, here we compare histograms and their distributions, testing these for statistical significance (Fig. 13). In the overall results, the distribution for LN years was different from that of the distribution in EN years, but this result was not significant at the 90% confidence level. It does rise to the level of statistical significance (90%) if only PDO1 years were used (not shown).

Comparing the length of the West Pacific season by examining the monthly occurrence (Fig. 14), shows that during EN years, the TC season is longest extending from March - January, while in LN and NEU years, the season is two months shorter (primarily April – December). The seasonal peak in each ENSO phase occurred in August and September just as in the total sample. During EN years, the peak in TC activity was clearly August, while in neutral years the peak was later (August / September), and earlier (July / August) in LN years. There was little difference in the ENSO-related seasonal cycles for PDO2 years, but interestingly, in PDO1 EN seasons were one month shorter. However, PDO1 LN years were considerably shorter than that of other years (not shown) extending from only June - December, and with a pronounced September peak not found in any other subsample.

A breakdown of TC cyclone activity by geographic region demonstrated that the southwest region was the most active (Table 13), and but the interannual and interdecadal variations were opposite that of the total sample. In particular, there were more TCs in LN years in the overall sample and during PDO2, and little TC variability during PDO1. As pointed out by others, the region of maximum activity changes as ENSO phase changes. During LN years, the southeast region is much less active, while the other three basins are more active. This may be partly due to the fact that there are cooler waters in the tropical eastern Pacific (see Lupo et al. 2007). Thus, the southeast region of the West Pacific basin and the southwest region of the East Pacific are relatively inactive confining Pacific Ocean activity overall closer to the continental areas during LN years. During EN years, the tropics across the entire Pacific are active, while the northern basins are less active. These results are also consistent with Zuki and Lupo (2008) whose study was performed within the southwest basin of this study.

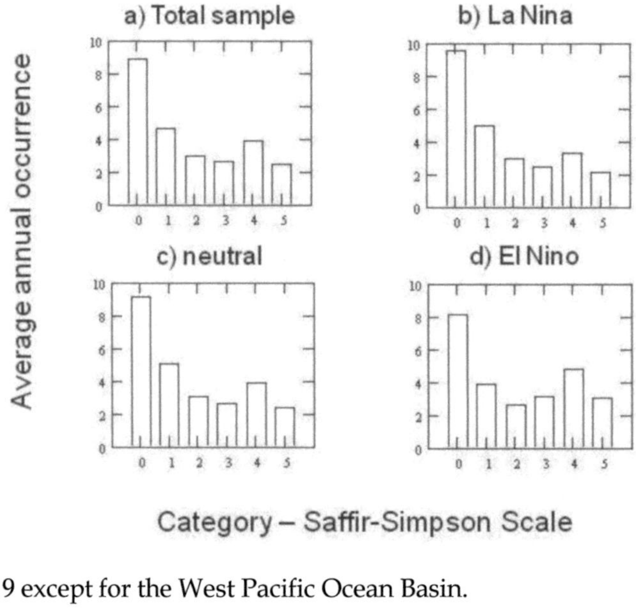

Fig. 13. As in Fig. 9 except for the West Pacific Ocean Basin.

As in the East Pacific, nearly all of the intense hurricanes were in the southwest and southeast regions, and neither region had a significantly higher proportion of category 4 and 5 storms. Of the 599 TCs which were intense storms only 59 TCs (10%) formed north of 20° N. As the sample is limited to category 4 TCs, the number of storms in the northern two quadrants was only 5%, and only four (0.6%) TCs which were rated category 5 formed north of 20° N. This is consistent with the results above that demonstrate that northern areas are less active in EN years, and TCs are less intense in LN years.

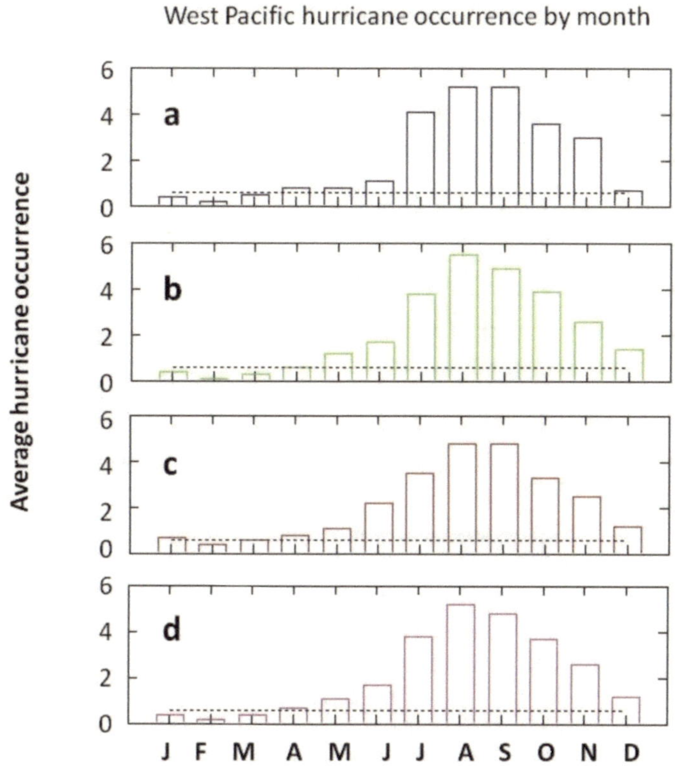

Fig. 14. As in Fig. 10 except for the West Pacific Ocean Basin.

4.4 Northern Indian Ocean variability

In section 3.4, it was shown that the data set in this region is only reliable after 1977. While this is fortuitous and coincides with an abrupt shift to PDO1 (Table 2), the result is that there is only a small sample of PDO2 TCs available, and, thus, any conclusions regarding interdecadal variability may be suspect. However, ENSO variability in the TC occurrences was significant at the 95% confidence level in the power spectra analysis (not shown) and there was significant variability at the 16 year period in the hurricane (cyclone locally) occurrences.

As shown in section 3, Northern Indian Ocean TC activity did not correlate with that of any other NH basin. Table 7 shows that the ENSO variability results in maybe one more TC per year in LN years, however, this basin has less activity than the other NH basins. In this basin, it has been recognized that TC activity maximizes at two particular times, late spring and fall with a summer minimum in TC activity (e.g., Ng and Chan, 2010). October and November are the most active months with 76 total TC's occurring over the 33 year climatology (about 1 TC in each month per year). February, March, July, and August are the two inactive periods in this basin. This is likely related to the semi-annual passage of the Inter-tropical Convergence Zone (ITCZ) through this region (Hurrell et al., 1995) similar to that of the TC season in Malaysia (Zuki and Lupo, 2008). The passage of the ITCZ through the region would provide favorable atmospheric conditions for the generation of TCs.

Separating this region into west and east regions (Table 15), the east part of the North Indian Ocean (dividing along 75° E Longitude) was the more active, accounting for about 75% of the North Indian TC occurrences. In section 3.4, it was speculated that the counting of tropical depressions and tropical waves likely provided for the severe over-count for TCs in this basin and this is especially true in the eastern half. There was almost no ENSO related variability in the east half, but in the west half, the variability was stronger with double the number of TCs in a LN years during the current era.

Phase	West Indian	East Indian	Total
LN	2.3 / 2.5	3.8 / 12.2	5.5 / 14.8
NEU	1.4 / 1.8	3.4 / 13.6	4.8 / 15.4
EN	1.0 / 2.7	3.9 / 11.3	4.9 / 14.0
Tot	1.4 / 2.3	3.5 / 12.6	4.8 / 14.9

Table 15. As in Table 9, except for the North Indian Ocean by geographic region (along 75° E) (second number is pre-1977 era).

An examination of the ENSO related variability in the North Indian region would reveal differences in the intensity of storms as well. While the results here were not similar to those of the West Pacific overall, there were similarities to the activity in the South China Sea (see Zuki and Lupo, 2008). More storms reached hurricane strength (2.5 to 1.0), and more intense TCs (1.4 versus 0.5) during LN years (Table 7), than during EN years. The intense hurricane difference was significant at the 95% confidence level. When separating by PDO phase (Table 16), the results are similar to the total sample for PDO1 years, while there was little variability in the small sample of PDO2 years. In order to further examine the variations in the intensity of storms, here we compare histograms and their distributions, testing these for statistical significance (Fig. 15). In the overall results, the distribution for La Niña years was different from that of the distribution in EN years, but this result did not rise to significance at the 90% confidence level.

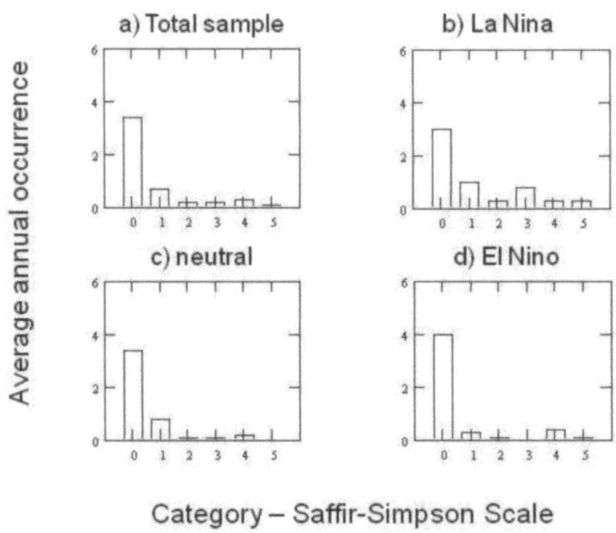

Fig. 15. As in Fig. 9, except for the Northern Indian Ocean.

Comparing the length of the North Indian season by examining the monthly occurrence (Fig. 16), shows that during LN years, the TC season can be defined as May and June, and then again in October and November. In NEU years, the season is defined as May, and then October to December, while in EN years the TC season is basically May and November. The yearly maximum in each ENSO phase occurred in the late fall. During EN years, the peak in TC activity was clearly November, while in NEU and LN years the peak was October and November. There was little difference in the ENSO-related seasonal cycles for PDO1 years. However, for PDO2 years, the sample is similar to that of the total sample but the sample size is probably too small as of yet to comment on differences.

a. PDO2

	All	TS	Cat 1	Cat 2	Cat 3	Cat 4	Cat 5	T_{hur}
LN	4.5	2.5	0.5	0.0	0.5	0.5	0.5	2.0
NEU	5.0	3.8	0.8	0.0	0.0	0.3	0.0	1.2
EN	5.0	4.3	0.3	0.0	0.0	0.3	0.0	0.7
Total	4.9	3.7	0.6	0.0	0.1	4.2	0.1	17.2

b. PDO1

	All	TS	Cat 1	Cat 2	Cat 3	Cat 4	Cat 5	T_{hur}
LN	6.5	3.5	1.5	0.5	1.0	0.0	0.0	3.0
NEU	4.7	3.3	0.8	0.2	0.2	0.2	0.0	1.4
EN	4.8	3.8	0.2	0.2	0.0	0.4	0.2	1.0
Total	4.9	3.4	0.7	0.2	0.2	0.2	0.1	1.4

Table 16. As in Table 10, except for the North Indian Ocean.

A breakdown of TC activity by geographic region demonstrated that the eastern region was the most active (Table 15), but there was little interannual and interdecadal variation. In the western half, there were more TCs in LN years in the overall sample and during PDO1, but little TC variability during PDO2. This difference between PDO1 and PDO2 activity is similar to that of all the other NH basins.

In the Northern Indian Ocean, there were only 17 intense hurricanes (cyclones), and many of these were in the eastern half (13) of the basin which represents roughly 75% of the total. Thus, the proportion of intense TCs in each basin is comparable to that of the total sample. Thus, the threat of an intense storm developing is similar in each half of the Northern Indian Ocean.

4.5 Southern Hemisphere variability

As shown above, there was a slight, but not statistically significant downward trend in TC activity in this hemisphere. It was shown also that the data set in this region is only reliable after 1980. The variability in this hemisphere was impacted strongly by ENSO, especially in the Southwest Pacific where the South Pacific Convergence Zone (e.g., Vincent, 1994; Hurrell et al. 1995), in particular the tropical portion, can provide the seed disturbances for

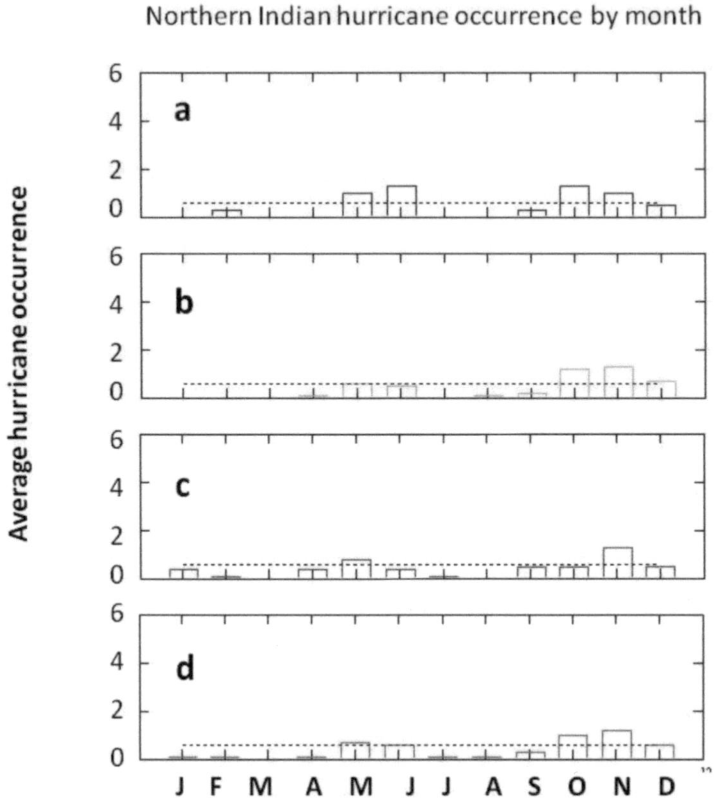

Fig. 16. As in Fig 10, except for the Northern Indian Ocean Basin.

TC formation. As was the case for the Northern Indian Ocean, any statements regarding interdecadal variability must be viewed with some caution in light of the fact that there was only 31 years worth of reliable records. Nonetheless, the power spectrum analysis (not shown) revealed statistically significant variability in the three, six and ten year periods in the SH. The SH was subdivided into four basins, the West and East Indian Ocean (along 75° E as in the NH), and the Southwest and Southeast Pacific (divided along the dateline).

TC activity in the SH as a whole only positively correlated to activity in the North Indian Ocean, and this did not rise to the level of statistical significance. Table 8 shows that during LN years there was three (two) more TCs (hurricanes) overall during the annual cycle. This is not statistically significant difference. In the period before 1980, there was little ENSO related variability found in the SH Basin. The annual cycle for this basin was described in section 3, but the period of peak activity is clearly during the summer and early fall months of the SH (December – March). The winter months (June – August) constitute the inactive period in this basin.

Separating the Indian Ocean part of the SH basin into the west and east Indian regions (Table 17), demonstrates the west part of the South Indian Ocean (dividing along 75° E Longitude) was less active than the east part, but this difference is much smaller than the difference between the activity in the west and east North Indian Ocean Basin. In spite of the fact that the activity was three times as great in the Southern Indian as in the Northern, the ratio between the west and east was roughly two to three.

Phase	West Indian	East Indian	SW Pacific	SE Pacific	Total
LN	7.3 / 8.2	9.8 / 9.2	7.0 / 10.5	5.0 / 2.0	29.3 / 29.5
NEU	6.4 / 8.6	11.2 / 7.8	6.3 / 7.2	2.9 / 1.8	26.9 / 25.8
EN	5.0 / 8.6	12.6 / 10.7	6.8 / 12.3	2.8 / 1.6	26.1 / 33.7
Tot	6.1 / 8.5	8.8 / 11.3	6.4 / 9.3	3.3 / 1.8	27.0 / 28.5

Table 17. As in Table 9, except for the Southern Hemisphere (along 75° E, Indian, and along dateline in Pacific, second number represents pre-1980 era).

In the west South Indian Ocean, there was strong ENSO variability in that there was more TC events in LN years in the west Indian Ocean, a result significant at the 90% confidence level. There was also less ENSO variability in the Eastern half of the South Indian Ocean basin but with a slight tendency toward more events in EN years. The ENSO related variability in each half compared favorably to the ENSO related variability in the Northern Indian Ocean, and this likely accounted for the positive correlation between the SH as a whole and the North Indian Ocean basin (compare Table 15 and Table 17). In the Southwest Pacific, there was little ENSO related variability in TC occurrence, but in the Southeast Pacific, there were roughly two more TCs in LN years than in EN years, however, this result was not statistically significant.

An examination of the ENSO related variability in the SH region would reveal little difference in the intensity of TCs (Table 8, Fig. 17). This result is not similar to those in other regions, where there was a tendency for more intense storms in one phase or another of the ENSO cycle. The differences in the distributions in Fig. 17 did not show a difference in intensity either when tested using the Chi-square test. Even the number of intense hurricanes across each phase of ENSO was similar (52% LN years versus 49% EN years). In the SH as a whole, up to the time of this study, there was little difference in the overall

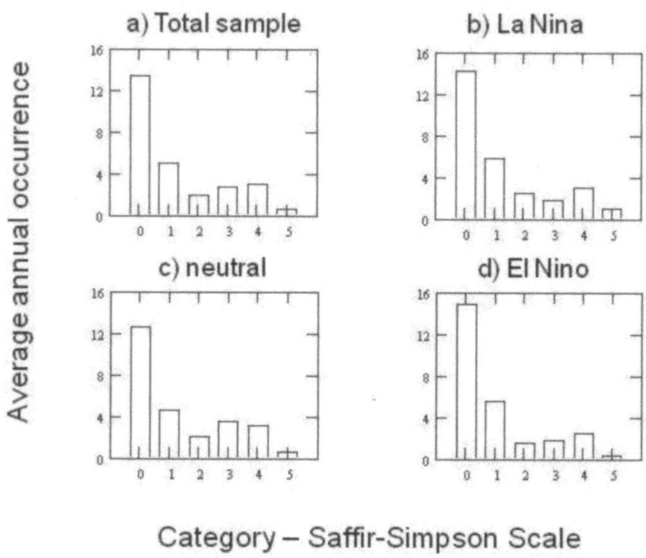

Fig. 17. As in Fig. 9, except for the Southern Hemisphere

a. PDO2

	All	TS	Cat 1	Cat 2	Cat 3	Cat 4	Cat 5	T_{hur}
LN	28.0	12.5	6.0	2.0	3.0	4.5	0.0	15.5
NEU	25.0	13.8	3.6	0.8	2.3	3.2	1.2	11.2
EN	25.3	11.6	4.0	1.0	2.7	4.7	1.3	13.6
Total	25.6	13.0	4.2	1.1	2.5	3.8	1.0	12.6

b. PDO1

	All	TS	Cat 1	Cat 2	Cat 3	Cat 4	Cat 5	T_{hur}
LN	30.5	14.0	5.0	3.0	3.5	4.0	1.0	16.5
NEU	27.7	14.0	5.3	2.3	3.2	2.7	0.2	13.8
EN	26.6	12.8	5.8	2.8	2.0	2.4	0.8	13.8
Total	27.8	13.7	5.4	2.5	3.0	2.8	0.5	14.1

Table 18. As in Table 10, except for the Southern Hemisphere.

ENSO variability across each phase of the PDO (Table 18). This differed from all the NH basins where ENSO variability was enhanced in PDO1. However, a closer examination of Table 18 reveals that the number of intense hurricanes was much greater in the decade of the 2000's during ENSO years (64% of all hurricanes, 34% of TCs) versus ENSO years during the period 1980-1999 (37% of all hurricanes, 20% of all TCs). While this may be a function of the small sample as of yet, given the interdecadal TC variability in the NH, it would not be unexpected that there may be some interdecadal variations.

Comparing the length of the SH season by examining the monthly TC occurrence (Fig. 18), shows that the season can be defined as October to May overall (using 0.6 as the monthly guide), with the peak activity occurring during January and February. During EN and NEU years, the TC season could be defined similarly (but with a peak in February and January respectively), while in LN years, the season was shorter (November to April). However, in LN years, the most active months which constitute the broadest peak for any of the global ocean basins (January through March) are consistently more active (19) than the corresponding TC activity in EN and NEU years (16 and 17, respectively).

5. Forecasting based on this work

5.1 Hindcasting the 2000-2005 seasons in the Atlantic and East Pacific

Using the results of LJ00 and Lupo et al. 2008, forecasts have been made using this information. When forecasting (and hindcasting) TC activity in each basin for the upcoming season (Table 19) from 2000-2010 in a manner similar to that demonstrated by Lupo et al. (2008) (and LJ00 for 1999), it is shown that this work can be used to produce reliable forecasts. Since a forecast of the geographical distribution of storm genesis regions was given using their method, we evaluated the forecast using the same calculation used to determine χ^2 for the statistical test (Neter et al. 1988). Thus, a lower score represents a better forecast, or shows an annual distribution closer to that of the climatological contingency forecast used by LJ00. The years 2000 - 2005 presumed that the forecast of the ENSO phase was correct as LJ00 did for 1999. In the Atlantic, the poorest hindcast was that for the 2005 season, in which every basin was quite active and, thus, large errors in TC occurrence were recorded in each category. However, an active year did not guarantee a high score as the 2003 and 2004 seasons were similarly active, but the hindcast for the 2004 season was much better. In the East Pacific, the TC hindcasts were better overall and this is likely due to the fact that the southeast region TCs dominate the activity in this basin. Thus, the threshold for what constitutes a good forecast should probably be lower in this region, and several more years would be needed in order to establish a recommendation.

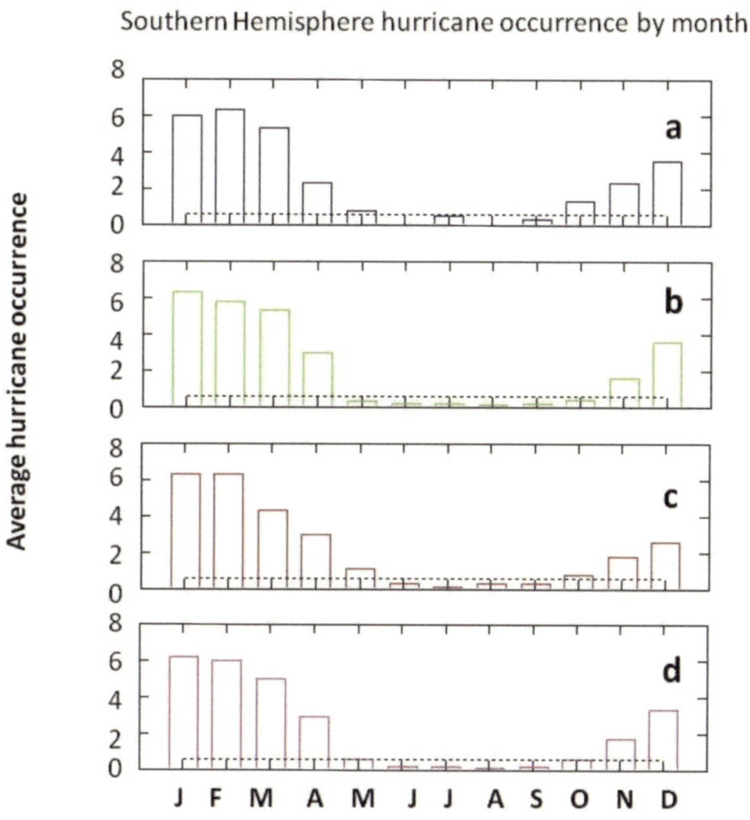

Fig. 18. As in Fig. 10, except for the Southern Hemisphere.

For each of these years, the ratio of tropical storms, weak hurricanes, and strong hurricanes was similar to what should be expected (see Tables 4 and 5). This would result in a forecast of a decreasing frequency in each category of stronger winds, except for PDO2 EN years in the East Pacific region where more strong TCs than weak storms are expected. In the Atlantic, only 2004 did not conform reasonably to the expected TC intensity distribution (we observed three weak hurricanes and six strong). In the Pacific and for each of these years, TC intensities were reasonably hindcast (not shown).

Atlantic Region

Year		Carribean	Gulf	West. Atl	East Atl.		Tot Score
2000		1 (1)	3 (1)	8 (3)	3 (1)		6 (3.30)
2001		4 (2)	2 (0)	8 (3)	1 (1)		6 (4.50)
2002		2 (0)	3 (1)	6 (1)	1 (1)		3 (2.20)
2003		1 (1)	6 (4)	4 (1)	5 (3)		9 (13.20)
2004		1 (1)	2 (0)	6 (1)	5 (3)		5 (5.20)
2005		7 (5)	6 (4)	11 (6)	4 (2)		17 (31.70)
2006		1 (1)	1 (1)	7 (2)	1 (1)		5 (2.30)
2007		2 (0)	4 (2)	7 (2)	2 (0)		4 (2.80)
2008		4 (2)	1 (1)	3 (2)	3 (1)		6 (3.80)
2009		1 (0)	1 (1)	4 (1)	3 (2)		4 (4.70)
2010		5 (2)	1 (1)	7 (2)	6 (4)		9 (9.97)

East Pacific Region

Year		northeast	southeast	southwest	Northwest		Tot Score
2000		0 (1)	15 (3)	3 (1)	1 (1)		6 (3.25)
2001		1 (0)	9 (3)	5 (3)	0 (0)		6 (5.25)
2002		0 (1)	11 (2)	4 (1)	0 (0)		4 (1.64)
2003		3 (2)	12 (0)	1 (1)	0 (0)		3 (4.50)
2004		0 (1)	10 (2)	2 (0)	0 (0)		3 (1.33)
2005		1 (0)	13 (1)	1 (1)	0 (0)		2 (0.58)
2006		1 (0)	16 (3)	2 (1)	0 (1)		5 (2.69)
2007		1 (0)	10 (3)	3 (2)	0 (1)		6 (5.69)
2008		0 (1)	11 (2)	1 (0)	0 (1)		4 (2.31)
2009		0 (1)	14 (1)	6 (3)	0 (0)		5 (3.08)
2010		1 (0)	6 (6)	0 (2)	0 (0)		8 (5.00)

Table 19. The annual hindcasts for tropical cyclone frequencies based on the work of LJ00 (for the 2000-2005 seasons), and forecasts (for 2006 - 2010) based on the results shown here. A simple point scoring system (| predicted - observed |) is used here. The table displays the observed value (| difference from observed |). The value in parenthesis beside the total score is a χ^2 score for that season (a lower score is a better forecast, or is a better "fit").

5.2 Forecasting the 2006 – 2010 seasons

For the 2006 through 2010 seasons, a true forecast was made in the January and February time frame based on forecasts made for the upcoming ENSO phase. For example in 2010, the forecast was for the weak EN to die out and the onset of a LN-type season based on the SSTs (see Lupo et al. 2007, their table 1 and Fig. 19 here). A weak EN SST pattern set in during the late summer-to-early autumn during the 2009 hurricane season. In the Atlantic, the 2006 through 2009 forecasts were fairly good compared to the previous years, and this is likely due to the fact that during PDO2 years, the distribution of tropical cyclones was fairly similar across each phase of ENSO. Thus, an incorrect ENSO phase would not be as potentially damaging to the forecast as during PDO1 years where there were distinct differences across each ENSO phase. In 2010, the season was particularly active as a strong LN set in during the fall. The hurricane season peaked strongly in September, which is typical of a LN year, but the overall forecast was not as good due to the active East Atlantic.

In the East Pacific, the 2006 and 2008 forecasts were relatively strong compared to the previous hindcasts, but 2007 and 2009 were weaker. The 2007 forecast was not very good even though the overall number of storms was close to what would be expected. There were more (fewer) TCs in the southwest (southeast) part of the East Pacific basin than expected. The 2006 (seven TS, five weak, three strong) and 2008 (nine TS, five weak, and two strong) TC intensity forecasts for this region was somewhat good, except that there were more strong storms observed in 2006 (eight TS, five weak, six strong) than expected. In 2008, there was one more tropical storm than expected. In 2007 and 2009, the intensity forecasts were more reliable; even if the overall numbers were lower (2007: 10 TS, three weak, one strong, and 2009: 12 TS, three weak, and five strong). In the East Pacific, the forecast in 2010 was a little worse as well. This was due to the fact that the southeast region and the East Pacific Basin as a whole were less active this year than at any other time since 1970.

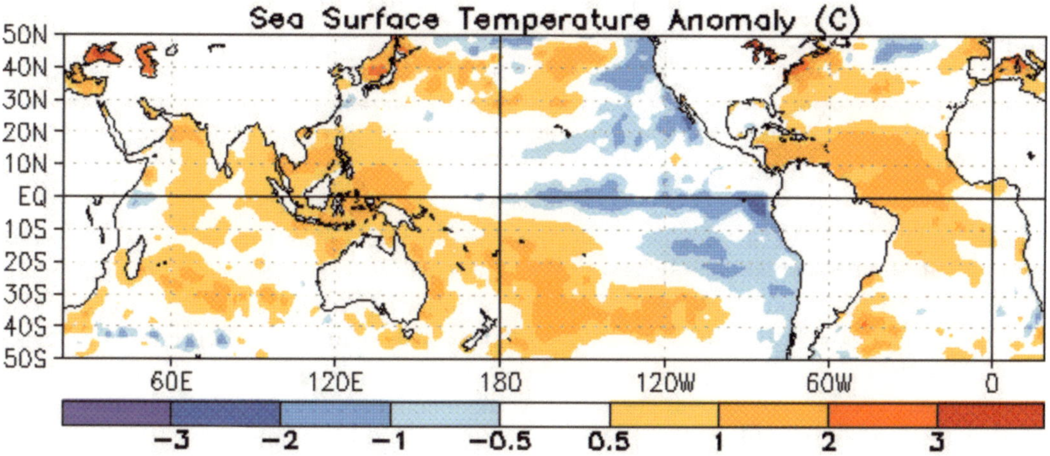

Fig. 19. A sample sea surface temperature anomaly (°C) map from the month of July 2010 (source: Climate Prediction Center: Climate Diagnostics Bulletin). The anomalies are relative to the 1971 - 2000 base period.

Similar TC forecasts for the upcoming seasons for the other basins based on this work could be formulated and would look like the following; West Pacific (northeast four, southeast five, southwest 13, northwest four, or ten TS, and eight each for the weak and intense hurricanes), North Indian (two west, three east, or three TS, and one each for the weak and intense hurricanes), and Southern Hemisphere (west Indian seven, east Indian ten, southwest Pacific seven, southeast Pacific five, and South Atlantic none, or 13 TS and eight each for the weak and intense hurricanes).

6. Discussion, summary and conclusions

The climatological behavior of TC activity for the entire globe was examined using the methodologies of LJ00 and Lupo et al. 2008. In the Atlantic and East Pacific, only an update to include the 2008 and 2009 seasons, as well as any re-categorization of hurricanes (e.g., Andrew - 1992) or ENSO years (e.g., 1974 becomes a LN year) was needed. In the West Pacific, a thorough breakdown of the hurricane and tropical storm activity from 1945 - 2009 was examined, as well as the activity in the North Indian Ocean (1977 - 2009), and the

Southern Hemisphere (1980 – 2010). We also briefly examined the activity in the South Atlantic (1960 – 2010).

The major finding from this work that was common across the entire globe was that it is not enough to look at the interannual variations of TC occurrence and intensity, especially if one is to forecast the occurrence of TCs in a given region. In each of the ocean basins, there was some ENSO related variability in the number and or intensity of TCs found. In many cases, this variability did not rise to the level of statistical significance until the data were stratified by phases of the PDO. In general, the ENSO related variability was smaller during the cool phase of the PDO. This type of behavior was noted in studies of other phenomena as well (e.g., Berger et al. 2003; Lupo et al. 2005). In contrast, during PDO1 years, ENSO variability was enhanced. For, example, in the Atlantic and Indian Ocean basins, there were more TC, and more intense TC during LN years than during EN years, and these results were statistically significant. In the East Pacific and West Pacific, EN years produced more TCs, and this was especially noteworthy in PDO1 years. In the East Pacific, there were more TCs in EN years among the weaker hurricanes and TS, while in the West Pacific, more of the TC were intense hurricanes. In the SH, the interdecadal variability was more subtle and there was more TC during PDO1 years. Additionally, there was an increase in the proportion of intense ENSO year TCs during the latest two decades.

Also, it was found that there were no statistically significant trends in global TC activity, neither in any of the basins nor over the globe as a whole. Several studies (e.g. Zuki and Lupo, 2008, and Chan, 2009) demonstrate that TC formation and maintenance are not solely a function of favorable oceanic or atmospheric conditions only. Thus, it is not clear how climate changes may influence TC numbers. In most basins, interannual variability resulted in a shift in the primary TC genesis regions rather than large changes overall. In the Atlantic, the increased LN TC activity was the result of more active sub-regions such as the Gulf of Mexico, the Caribbean, and the East Atlantic. In the combined Pacific regions, LN (EN) years tended to have more TC activity along the continental coastal margins than the Central Pacific (across the entire tropical Pacific). In the North and South Indian Ocean regions, as well as the Southeast Pacific, there were more TCs in LN years, while there was little TC variability in the East Indian Ocean or the southwest Pacific.

An examination of the length of TC seasons and when the peak would occur reveal interannual differences. In the Atlantic Ocean basin, the TC season extends from June to October (July to November) in EN (LN) years, with the weaker (stronger) peak in September. In the East Pacific basin, the EN TC season is longer than LN TC season by roughly one month as both begin in June and EN TC seasons extend into November. In this basin, the peak is stronger during EN years and was clearly in August, while the LN peak was in July and August. In the West Pacific, the seasonality was similar to the East Pacific. The EN (LN) TC year was two months longer (shorter) and peaked in August (July and August). While this ocean basin is considered to be active all year, there was clearly a minimum in TC activity during February and March, and in LN PDO1 years, the season was considerably shorter than any other period in the West Pacific (June – December, with a peak in September).

The Indian Ocean Basin was different from all other global basins in that there were two periods of strong TC activity in May and June, and again in October to December. In LN (EN) years, the maximum activity periods were stronger and longer (weaker and shorter), and encompassed May/June (May) and October/November (November). Finally, in the SH, the seasonal activity was two months longer (shorter) for EN (LN) years, extending from October to May (November to April). However, the LN year had a much stronger and longer period of peak activity.

Finally, it is shown that this work may have some forecasting utility by using the same forecasting / hindcasting methods of LJ00 and Lupo et al. (2008), except here we assigned each season from 2000 - 2005 a forecast score based on a statistical distribution calculation. This approach included assuming that the ENSO forecast was correct. While the TC hindcasts for most of these years were reasonable, more seasons will be needed in order to evaluate what would constitute a good score in each basin. TC forecasts have been made since the 2006 season; and in the Atlantic, the forecasts were relatively good. In the Pacific, the TC season forecasts were relatively poor for the 2007 and 2009 seasons, while the forecasts for 2006 and 2008 have been relatively good. The poor forecasts may have been due to a poor ENSO forecasts. Additionally, the year 2010 forecasts were made and will be evaluated later. This work, however, is preliminary and data from additional seasons will be used to verify our approach.

7. Acknowledgements

The authors would like to thank the reviewers for their helpful comments in making this work stronger. We would like to also thank Dr. Nathan Mantua at the Joint Institute for the study of Atmosphere and Ocean (JISAO - http://jisao.washington.edu) for providing us with Figure 2.

8. References

Berger, C.L., Lupo, A.R., Browning, P., Bodner, M., Rayburn, C.C., Chambers, M.D. (2003). A climatology of Northwest Missouri snowfall events: Long term trends and interannual variability. *Phys. Geog.*, Vol. 14, 427 - 448.

Birk, K., Lupo, A.R., Guinan, P.E., and Barbieri, C.E. (2010). The interannual variability of Midwestern temperatures and precipitation as related to the ENSO and PDO. *Atmofera,* Vol. 23, 95 - 128

Chan, J. C. L. (1985). Tropical cyclone activity in the northwest Pacific in relation to the El Niño/Southern Oscillation phenomenon, *Mon. Wea. Rev.*, Vol. 113, 599- 606.

Chan, J.C.L. (1995). Tropical cyclone activity in the western north Pacific in relation to the stratospheric Quasi- Biennial Oscillation, *Mon. Wea. Rev.*, Vol. 123, 2567- 2571.

Chan, J.C.L. (2007). Interannual variations of intense typhoon acitivity. *Tellus,* Vol. 59A, 455-460.

Chan, J.C.L. (2009). Thermodynamic control on the climate of intense tropical cyclones. *Proc. R. Soc., A,* Vol. 495, 3011- 3021.

Chan, J.C.L. and Liu, K.S. (2004). Global warming and western North Pacific typhoon activity from an observational perspective. *J. Climate,* Vol. 17, 4590 - 4602.

Collins, J.M. (2007). The relationship of ENSO and relative humidity to interannual variations of hurricane frequency in the North-East Pacific Ocean. *Papers of the Appl. Geog. Conf.*, Vol. 30, 324-333.

Collins, J.M. (2010). Contrast High North-East Pacific Tropical Cyclone Activity with Low North Atlantic Activity. *Southeastern Geog.*, Vol. 50, No. 1, 83-98.

Davis, M.A.S., Brown, G.M., and Leftwich, P. (1984). A tropical cyclone data tape for the Eastern and Central North Pacific Basins, 1949-1983: Contents, Limitations, and Uses. NOAA Technical Memorandum NWS NHC 25, *Coral Gables, Florida, 20 pp.*

Deser, C., and Phillips, A.S. (2006). Simulation of the 1976/1977 climate transition over the North Pacific: Sensitivity to Tropical Forcing. *J. Clim*, Vol. 19, 6170 – 6180.

Deser, C., Capotondi, A., Saravanan, A.R., and Phillips, A.S. (2006). Tropical Pacific and Atlantic climate variability in CCSM3. *J. Clim*, Vol. 19, 2451 – 2481.

Deser, C., Phillips, A.S., and Hurrell, J.W. (2004). Pacific interdecadal climate variability: Linkages between the tropics and North Pacific during the boreal winter since 1900. *J. Clim*, Vol. 17, 3109 – 3124.

Elsner, J.B., Kara, A.B., and Owens, M.A. (1999). Fluctuations in North Atlantic hurricane frequency. *J. Clim*, Vol. 12, 427 – 437.

Emanuel, K. (2005). Increasing destructiveness of tropical cyclones over the past 30 years. *Nature*, Vol. 436, 686.

Enfield D. B. and Mestas-Nuñez, A.M. (1999). Multiscale variabilities in global sea surface temperatures and their relationships with tropospheric climate patterns. *J. Clim.* Vol. 12, 2719-2733.

Gershanov, A., and Barnett, T.P. (1998). Inderdecadal modulation of ENSO teleconnections. *Bull. Amer. Meteor. Soc.*, Vol. 79, 2715 - 2725.

Goldenberg, S.B., and Shapiro, L.J. (1996). Physical mechanisms for the association of El Niño and West African rainfall with Atlantic major hurricane activity. *J. Climate*, Vol. 9, 1169 - 1187.

Gray, W. M. (1968). Global view of the origin of tropical disturbances and storms. *Mon. Wea. Rev.*, Vol. 96, 669– 700

Gray, W.M. (1984a). Atlantic season hurricane frequency. Part 1: El Niño and 30 mb Quasi Biennial Oscillation influences. *Mon. Wea. Rev.*, Vol. 112, 1649 - 1668.

Gray, W.M. (1984b). Atlantic season hurricane frequency. Part 2: Forecasting its variability. *Mon. Wea. Rev.*, Vol. 112, 1669 - 1683.

Gray, W.M. (1979). Hurricanes: Their formation, structure, and likely role in the tropical circulation. *Meteorology over the Tropical Oceans*, pp. 155 – 218, D.B. Shaw, ED., Royal Meteorologocal Society.

Gray, W.M., Sheaffer, J.D. and Landsea, C.W. (1997) Climate trends associated with multidecadal variability of Atlantic hurricane activity. *Hurricanes, Climate, and Socioeconomic Impacts*, 15 - 53. Springer, Berlin, H.F. Diaz and R.S. Pulwarty, Eds.

Henderson-Sellers, A., Zhang, H., Berz, G., Emanuel, K., Gray, W.M., Landsea, C.W., Holland, G., Lighthill, J., Shieh, S.L., Webster, P., and McGuffie, K. (1998) Tropical cyclones and global climate change: A post – IPCC assessment. *Bull. Amer. Meteor. Soc.*, Vol. 79, 19 – 38.

Hurrell J. W., van Loon, H., and Shea, D.J. (1995). The mean state of the troposphere. NCAR Tech Memo # NCAR/CAS/95-08, D. Karoly and D. Vincent eds., 83 pp.

Jarvinen, B.R., Neumann, C.J., and Davis, M.A.S. (1984). A tropical cyclone data tape for the North Atlantic Basin, 1886 - 1983: Contents, limitations, and uses. *NOAA Tech. Memo. NWS NHC 22, Coral Gables, Florida, 21 pp.*

Kossin, J.P., and Vimont, D.J. (2007). A more general framework for understanding Atlantic hurricane variability and trends. *Bull. Amer. Meteor. Soc.*, Vol. 88, 1767-1781.

Lander, M. A. (1994). An exploratory analysis of the relationship between tropical storm formation in the western North Pacific and ENSO, *Mon. Wea. Rev.,* Vol. 122, 636-651.

Lander, M. A., and C. P. Guard (1998). A look at global tropical cyclone activity during 1995: Contrasting high Atlantic activity with low activity in other basins. Mon. Weather Rev., Vol. 126, 1163- 1173.

Landsea, C.W. (1993). A climatology of intense (or Major) Atlantic hurricanes. *Mon. Wea. Rev.*, Vol. 121, 1703 - 1713.

Landsea, C.W., Pielke Jr., R.A., Mestas-Nuñez, A.M., and Knaff, J. (1999) Atlantic Basin hurricanes: Indicies of climate changes. *Climatic Change,* Vol 42, 89 - 129.

Lupo, A.R., Oglesby, R.J., and Mokhov, I.I. (1997) Climatological features of blocking anticyclones: A study of Northern Hemisphere CCM1 model blocking events in present-day and double CO_2 atmospheres. *Clim. Dyn.*, Vol. 13, 181-195.

Lupo, A.R., and Johnston, G.J. (2000). The variability in Atlantic Ocean Basin hurricane occurrence and intensity as related to ENSO and the North Pacific Oscillation. *Nat. Wea. Dig.*, Vol. 24, No. 1,2, 3 – 13.

Lupo, A.R., Albert, D., Hearst, R., Market, P.S., Akyuz, F.A., and Allmeyer, C.L. (2005). Interannual variability of snowfall events and snowfall-to-liquid water equivalents in Southwest Missouri. *Nat. Wea. Dig.,* Vol. 29, 13 – 24.

Lupo, A.R., Kelsey, E.P., Weitlich, D.K., Mokhov, I.I., Akyuz, F.A., Guinan, P.E., Woolard, J.E. (2007). Interannual and interdecadal variability in the predominant Pacific Region SST anomaly patterns and their impact on a local climate. *Atmosfera,* Vol. 20, 171- 196.

Lupo, A.R., Latham, T.K., Magill, T., Clark, J.V., Melick, C.J., and Market, P.S. (2008). The interannual variability of hurricane activity in the Atlantic and East Pacific Regions. *Nat. Wea. Dig.,* Vol. 32, No. 2, 119-135.

Mantua, N.J., Hare, S.R., Zhang, Y., Wallace, J.M., and Francis, R.C. (1997). A Pacific Interdecadal Climate Oscillation with Impacts on Salmon Production. *Bull. Amer. Meteor. Soc.,* Vol. 78, 1069–1079.

McAdie, C.J. and Rappaport, E.N. (1991). *Diagnostic Report of the NHC,* Vol. 4, No. 1, NOAA, NHC, Coral Gables, FL, 45 pp

McTaggart-Cowan, R., Bosart, L. F., Davis, C. A., Atallah, E. H., Gyakum, J. R., and Emanuel, K. A. (2006). Analysis of Hurricane Catarina (2004). *Mon. Wea. Rev., Vol.* 134, 3029–3053.

Mestas-Nuñez, A. M., and Enfield, D.B. (1999). Rotated global modes of non-ENSO sea surface temperature variability. *J. Clim.* Vol. 12, 2734-2746.

Minobe, S. (1997). A 50 - 70 year climatic oscillation over the North Pacific and North America. *Geophys. Res. Lett.*, Vol. 24, 683 - 686.

Minobe S. (2000). Spatio-temporal structure of pentadecadal variability over the North Pacific. *Prog. Oceanogr.* Vol. 47, 381-408.

Neter, J., Wasserman, W., and Whitmore, G.A. (1988). *Applied Statistics, 3rd edition*. Boston: Allyn and Bacon, 1006 pp.

Ng, K.W.E., and Chan, J.C.L. (2010). Interannual variations of tropical cyclone activity over the Northern Indian Ocean. *Int. J. Clim., under review.*

Pielke, R.A., and Landsea, C.W. (1999). La Niña, El Niño, and Atlantic hurricane damages in the United States. *Bull. Amer. Meteor. Soc.,* Vol. 80, 2027 - 2033.

Pielke, R. A., Jr., Landsea, C.W., Mayfield, M., Laver, J., and Pasch, R. (2005). Hurricanes and global warming. *Bull. Amer. Meteor. Soc.*, Vol. 86, 1571- 1575.

Ramage, C. S., and Hori, A.M. (1981). Meteorological aspects of El Niño. *Mon. Wea. Rev.*, Vol. 109, 1827- 1835.

Reagan, M.R., Franklin, A.D., Market, P.S., and Lupo, A.R. (2005). The South Atlantic hurricane ("Catarina") of March 2004. *The 30th Annual Meeting of the National Weather Association, 15 – 20 October, 2005, St. Louis, MO.*

Schultz, L.W. (2007). Some climatological aspects of tropical cyclones in the eastern north Pacific. *Nat. Wea. Dig.*, Vol. 32, No. 1, 45-54.

Simpson, R.H. (1974). The hurricane disaster potential scale. *Weatherwise*, Vol. 27, 169 and 186.

Vimont, D.J., and Kossin, J.P. (2007). The Atlantic meridional mode and hurricane activity. *Geophys. Res. Lett.*, Vol. 34, *L07709, doi:10.1029/2007GL029683.*

Vincent, D.G. (1994). The South Pacific Convergence Zone (SPCZ): A Review. *Mon. Wea. Rev.*, Vol. 122, 1949-1970

Wang, B., and Chan, J.C.L. (2002). How strong ENSO events affect tropical storm activity over the western north Pacific. *J. Clim.*, Vol. 15, 1643- 1658.

Webster, P.J., Holland, G.J., Curry, J.A., and Chang, H.R. (2005). Changes in tropical cyclone number, duration, and intensity in a warming environment. *Science,* Vol. 309, 1844 - 1846.

Wilks D. S. (2006). *Statistical Methods in the Atmospheric Sciences,* 2nd ed. Int. Geophys Series, Vol. 91, Academic Press, 627 pp.

Wu, G., and Lau, N.C. (1992). A GCM simulation of the relationship between tropical storm formation and ENSO, *Mon. Wea. Rev.*, Vol. 120, 958- 977.

Zuki, Md. Z., and Lupo, A.R. (2008). The interannual variability of tropical cyclone activity in the southern South China Sea. *J. Geophys. Res.,* Vol. 113, D06106, doi:10.1029/2007JD009218 - 14.

2

Influence of Cosmophysical Phenomena and African Dust on Hurricanes Genesis

Jorge Pérez-Peraza[1], Víctor Manuel Velasco Herrera[1] and Igor Libin[2]
[1]*Instituto de Geofísica, Universidad Nacional Autónoma de México, Ciudad Universitaria, Coyoacán, 04510, México D.F.,*
[2]*International Academy of Appraisal and Consulting, Moscow,*
[1] *México,*
[2]*Russia*

1. Introduction

The links between the Space Weather and Meteorological Weather have been often discussed not only for the last century ((70) ; (72)), but also for several centuries ago ((99)), and even before some thousands of years ago ((74)). Correlational works between solar and climatic parameters give often interesting results ((4)). A great deal of efforts have been done to clarify the mechanism of all complicated interconnections between the cosmophysical and climatic phenomena at earth, some of them recently summarized by (1), (41), (32) and (23). In the last years more and more investigations show that the solar activity have noticeable impact on the meteorological parameters ((75), (29); (50), (51); (60); (113), (115); (114) and cosmic rays ((67); (52); (24); (13); (71), (95), (94), (130)). Besides some indications appeared that several purely meteorological processes in the terrestrial atmosphere are connected with the changes in the Cosmic Ray (CR) intensity, and influenced by solar activity, and magnetosphere variations ((53), (67), (50)).
One of the main goals of Space climate research is to know how and when the periodicities of space phenomena do modulate terrestrial Climatic changes. Some insights have been obtained: for one side, the solar Hale cycle (20 − 25 years): changes in solar activity for the last 300 years have been studied ((98)), with the aim of revealing a possible contribution of solar activity to climatic variability. On the other hand, quasiperiodic climatic oscillations with periods of 20 − 25 years have been revealed in the analysis of parameters such as ground surface temperatures, drought rhythm, variations in sea surface temperature, precipitation periodicity, etc. ((78); (8); (92); (93); (128); (100); (97), (48)). To understand the involved physical mechanisms it is required, as a first step, of confident observational or experimental facts. Nevertheless, to clarify the mechanism of all the complicated interconnections between the cosmophysical phenomena and climatic phenomena at earth, a great deal of efforts have been done; for instance ((23), (32), (1) and (41)).
One of the principal difficulties in quantifying the role of the space phenomena on climate changes has been the absence of long-term measurements of both, the climatic and space phenomena. Consequently, people often recur to the use of proxies. In the last years more and more investigations show that the solar activity ((115), (50)) and Cosmic Rays, have noticeable impact on the meteorological parameters. However, the influence of

cosmophysical phenomena on climatic phenomena is currently debated (e.g. (31), (104)). Previous work by means of a correlational analysis ((17); (44), (45); (80)), seems to indicate that certain extraterrestrial phenomena could have some kind of relation with the occurrence of Hurricanes. It is even speculated that such kind of correlations could seat the basis of deeper studies to use the results as indicators of hurricanes precursors. To give to those results a higher meaning, it is convenient to carry out spectral studies of the different involved times series to delimitate with more preciseness the existence of those potential relationships. That is, to find incident cosmophysical periodicities that may modulate terrestrial phenomena. Though the Atlantic Multidecadal Oscillation (AMO) has been linked with the frequency of Atlantic hurricanes, however, in the present context, little attention has been given to such a large scale climatic phenomena: as the question about the role of the Sun in modulating these phenomena has not been clarified, it requires further assessment.

So, special mention must to be done regarding the Links between Geo-external forcing and Hurricanes: correlational works between cosmophysical and climatic parameters may be done, giving often interesting results (e.g. (4)). Within this context, in a series of works ((17); (44),(45); (80),(81)) several efforts were addressed to find possible interconnections between the appearance and development of Atlantic hurricanes and changes in solar activity, geomagnetic disturbances and Cosmic Ray intensity. Changes in geomagnetic activity and sharp Cosmic Rays (CR) decreases, namely Forbush Events (FE), are well known to be related to Solar Activity changes (SS index). These facts provoked the interest of the previously mentioned authors for a detailed study of such collateral phenomena and their statistical comparison with hurricane phenomena. Within that context they were looking for signatures of coadjutant hurricane forecasting to conventional meteorological models, well before the period at which those models usually produce their predictions: it had been analyzed the behavior of the (CR) intensity, Sunspots (SS) and geomagnetic indexes (AP) and (KP) in long intervals preceding the development of the North Atlantic Cyclones, that means, before the first observations of the just born cyclonical system. The authors tried to examine the eventual connection of CR, SS, AP and KP changes with the processes developing in the atmosphere, far before the formation of North Atlantic hurricanes.

Their main hypothesis is that any specific changes of the collateral parameters during the days, preceding the cyclone appearance could be used as an indicative precursor for an approaching hurricane event. In order to reveal any possibility for immediate (not delayed) relationship between them, it was used simultaneous data to find statistical dependencies between the specific sharp changes in the geomagnetic field and cosmic rays intensity and the corresponding values of the hurricane intensification. Their efforts were especially concentrated not only in North Atlantic hurricanes, but in particular those which struck the East Coast of Mexico. All such hurricanes, recorded in the period 1950 – 2007 were analyzed. Trying to generalize the big amount of information obtained from this kind of analysis, it can be said that CR, SS, AP and KP showed much more intensive disturbances in the periods preceding and following the hurricane appearance. For SS this disturbance gradually increase with the hurricane strength. A characteristic peak in the CR intensity appears before the hurricane start. But its place varies between 5 and 20 days before that start. Specific changes were observed in the SS. For major hurricanes they begin sometimes more than 20 days in advance. The AP and the KP show series of bursts, spread over the whole period of 35 preceding days. The chosen long preceding period of days permits to reveal the behavior of these parameters long before the cyclone appearance. Specific precursors exist persistently before the cyclone start.

During the time of major cyclone development specific changes are also noticeable: a considerable change in the solar activity and the depending on it, CR intensity and Geomagnetic field activities, precede the appearance of intense cyclones, though the preceding time fluctuates considerably from event to event. For instance, investigating the daily intensification of KATRINA it has been noticed that a strong geomagnetic change (through changes in KP and AP indexes), was recorded 5 days before its maximum value. From the other side, some repeatedly observed coincidences between the Hurricane appearances and preceding Forbush Effects (FE) (as for example Hurricane ABBY 1960 and CELIA 1970 suggested a possibility for closer relationship between the FE and hurricane intensification. The obtained interconnections show that these parameters should be taken in consideration, when complicated processes in the upper atmosphere are used to determine the hurricane formation, that potentially will contribute to hurricane development forecasts.
Obviously the parameters SS, CR, AP, KP are not the basic driving factor for hurricane appearance and development. At present, it cannot be claimed an accurate forecast of cyclone activity only on the basis of preceding KP and CR, Forbush Effects data drastic changes, However, the results confirmed the preliminary hypothesis suggestion that there is some kind of interconnection between these parameters and the appearance of tropical cyclones - especially with the most powerful of them: Though, it is not pretended by the moment to forecast the creation of a dangerous vortex on the basis of peculiar behavior preceding of those data changes, however, looking at the presented results it could be strongly alerted if a package of large SS and a sharp rise in KP index is recorded together with Forbush decreases appearing during the summer (i.e. trend of the deeper minimum in CR intensity and the higher maximum in SS, for higher hurricane categories). Then investigating all the parallel atmospheric data we could be closer to a most probable prediction.
Finally, it should be emphasized that the results described in those works are limited to a correlational analysis. Though, very interesting results were obtained, however, it should be reminded that that kind of analysis, is only the necessary first step to be done when it is to be determined whether or not there is connection between two time series characterizing to different physical; in fact, correlational analysis only provides global information about the degree of linear dependence between two time series but does not gives information when the correlation dependence is of non-linear nature. Even if the global correlation coefficient is low, that does not means that there is no a physical relation. In fact, there is the possibility that such a relation could be of non-linear nature, or that there is a strong phase shift between the cosmophysical phenomenon and the plausible associated terrestrial effect, or, there is a time delay between one series time (input) and the system reaction (output).
Nevertheless, the obtained results are enough exciting for motivating us to jump to a next step, that is, to reinforce such analysis with collateral methods, by means of a more precise statistical analysis technique, namely the Wavelet analysis (Section 7) which furnishes not only global but also local information in time and frequency band ((35); (116), (117)), that is, it provides the coherence between two series, by means of the evolution of the relative phase between both series, determining whether their correlation is linear or not in different band widths. It may occur that the global correlation coefficient is low, but in some periods of the studied time interval the coherence be, however relatively high, indicating a non-linear correlation (a complex one) in those periods The important feature of wavelet spectral analysis is that it gives the evolution of the synchronization in time-frequency space. In contrast the so called Pearson correlation coefficient does not provides the evolution of the common synchronized periodicities, nor the evolution of the relative phase between two time series.

2. Atlantic hurricanes

Hurricanes are considered one of the most astonishing meteorological phenomena in our planet. Strong winds, clouds of great size and intense storms unite to advance for the ocean and to reach mainland, razing with everything to their passage, fallen trees, damages in buildings, changes in the natural landscape and, fatal victims, are alone some of the consequences that these unpredictable events can generate. Due to the great intensity that they reach, with winds that can overcome the 350 km/hr, they are classified as true natural disasters: whole towns disappear under the force of the impetuous winds. No device achieves, nowadays, to counteract its force; the man has become a simple spectator. Hurricanes have always been associated with the damage that causes mainly translated in human and material losses, but it is necessary to stand out that they also brings such benefits as the increase in precipitation in regions where the agricultural development depends on the precipitation, as well as the recharge of the dams and the bodies of water, vital for the development of the populations. The word hurricane has its origin in indigenous religions of the old civilizations. The Mayan named to their god of the storm as Hunraken. Taino people, a culture of the Caribbean, called Hurricane to a God that they considered malicious. Nowadays, hurricanes are not considered wicked forces but due to their great force, and great potential for loss of lives and material damages, they are considered one of the most powerful phenomena in the nature.

The tropical hurricanes are the only natural disasters to which are assigned a name. These names are known well before they happen and as well as their possible effects, contrary to other natural phenomena as the earthquakes, tornados and floods. Gilberto, Katrina, Mich and Isidoro, to name some of the most recent, are examples that remember a very particular image, for the severe damages that caused. These phenomena present common characteristics, although each one shows particular features. The destruction caused by the hurricanes in the Caribbean and Center America is a force that has modified the history and that it will follow it making in the future in these regions. The danger is born from a combination of factors that characterize to the tropical cyclonal storms: elevation of the level of the sea, violent winds, and strong precipitation. For their location, an example on this, is the Peninsula of Yucatan which is affected in a direct or indirect way by most of the hurricanes that are formed in the Western Caribbean. The hurricanes can have a diameter as long as the peninsula itself, so that practically any hurricane that is formed affects in more or smaller measure the oriental coast of the peninsula ((129)).

To give an idea of hurricanes power, the energy concentrated in the vortex system is estimated to be $> 10^{16} J$ if we consider that the air over a surface with a diameter of $\sim 800km$ has a mass of $\sim 2x10^{12}$ tons, turning with average velocity of $\sim$ 15 to 20 ms^{-1}, we could easily calculate an energy of $\sim 1011KWh$). This corresponds to the energy released during the explosion of more than 2000 Atomic bombs of the Hiroshima type. That explains the devastating effect of the hurricanes when it touches a populated area. North Atlantic hurricanes frequently strike the Caribbean islands, Mexico, and the United States. Only one hurricane hitting over their coasts could take hundred human lives and can cause damages for billions dollars and practically every year one or two such hurricanes devastate these regions. They rank at the top of all natural hazards ((16)). The hurricane KATRINA, (Fig. 1), destroying not only the city of New Orleans but vast areas from the states of Louisiana and Alabama is an example of that. The tropical hurricanes are sometimes driven by weak and erratic winds, that makes even more difficult to predict them. Published warnings have substantially improved entailing a decrease of the deaths.

Fig. 1. Hurricane KATRINA, taken from a NASA satellite, it was the deadliest category-5 hurricane to strike the North Atlantic coast

So, today a lot of efforts are devoted to understand better the hurricane formation and intensification to unveil connections with other physical processes, as could be for instance the fact that there has been a low Atlantic hurricane activity in the 1970*s* and 1980*s* compared to the past 270 years, but increasing destructiveness over the past 30 years. Also, even if, since the beginning of the 1990 there is a general trend to increase the frequency of tropical cyclones, there are however some years with do not follow such a tendency, as it was the case of 2006. It is hope that the conglomeration of different researches with very different focus contribute, in overall, to the task of improving the prediction of its complicated trajectories for a better foresee of hurricane appearances, prediction of their probable devastation , and then, to warn with enough time the threatened population.

Hurricanes are perturbations that take place in tropical regions where the waters of the ocean are relatively warm (temperatures around the $26-27^{o}C$). They are characterized by a great center of low pressure, around which the air at great speed rotates embracing an extension of several hundred of kilometers. Hurricanes have a certain anatomy and their classification depends on the intensity of the winds, on the atmospheric pressure and on the potential damages that they may cause. Powered by the intensive solar heating, producing fast evaporation in the second semester of each year, large upward hot high velocity circular wind streams are born over the hot equatorial waters of the oceans, with a velocity higher than $60Km/hr$, and reaching rotational velocities beyond $350Km/hr$. Tropical Cyclone is the scientific term for a closed meteorological circulation of enormous mass of atmospheric air rotating intensely, that is developed on tropical waters. These systems to great scale, non frontals and of low pressure happen on areas of the world that are known as tropical basins of hurricanes.

Therefore, the Tropical Cyclone is a low-pressure system that is located over hot waters of tropical oceans (between the tropics of Cancer and Capricorn and at least $4^{o}-5^{o}$ away from equator). The intensive heating, low pressure and resulting powerful evaporation increase fast the rotational wind speed. This huge system moves generally from East to West and slightly to the North, but deviations to the East are not exceptions. These exceptions are dangerous especially for the West coast of Mexico and USA. Generally these cyclones are known under the name of hurricanes. If they are formed over the Atlantic and North-eastern Pacific Oceans.

We use hereafter indifferently the terms cyclone and hurricane. If they are born over the western Pacific Ocean, they are called Typhoons. Because of the earth rotation, they rotate counter clockwise in the North Hemisphere and clockwise in South Hemisphere.

In the first moments of the formation of Tropical Cyclones, when the circulation of the closed isobar reaches a speed of 18 ms^{-1} (i.e. < 34 kt or 61 km/h), the system is denominated as Tropical Depression (TD). This is considered as a tropical hurricane in formative phase. If the sustained speed of the wind ranges from 18 to 32 ms^{-1} (34 till 63 kt, i.e. $62 - 115$ km/h) it is called a Tropical Storm (TS) and a certain name is given. Likewise, when the speed of the wind exceeds the 119 km/h or > 33 ms^{-1} (> 64 kt) the system takes the name of Hurricane (or Typhoons). That is the speed accepted to define the beginning of a hurricane over the Atlantic and a typhoon over the Pacific. They have a defined nucleus of pressure in very low surface that can be inferior to 930 hpa. Every year develops an average of 10 tropical storms in the Ocean Atlantic, the Caribbean or the Gulf of Mexico, and about 6 of those which end up becoming hurricanes. In a three year-old period, the North Atlantic coasts receive an average of five hurricanes, two of those which are considered bigger hurricanes. In general, the tropical depressions and tropical storms are less dangerous than the hurricanes; however, they can still be mortal. The winds of the depressions and tropical storms are not the most dangerous thing.

The intense rains, floods and the severe natural phenomena, as the tornados, are the biggest threat. Hurricanes can then be described as turbulence phenomena caused by a current of hot air that is formed in the summer in the tropic and that it goes to the North Pole compensating the difference in temperature between the Ecuador and the Pole. One counter current of the north to the south, compensates the difference in pressure. This circulation of winds north-south and south-north at level of the north hemisphere, together with the daily circulation of the earth that causes the trade winds, are the main factors from the point of view of the winds to create situations that can form hurricanes. Another condition for the formation of a hurricane is the temperature of the surface of the ocean, as energy source to give form to the phenomenon, which should be $>$ to 26^oC. Under these conditions, it is the column of hot and humid air originated in the ocean the one that becomes the nucleus around which rotate the winds and form later the so called "eye" of the hurricane.

The adjacent air is gradually involved in the rotation and the diameter of the whole vortex spreads to $500 - 1000$ km. With the further increase of the circular velocity, reaching sometimes $150 - 160$ kt (80 ms^{-1}), the whole vortex spread out to a gigantic ring with a diameter of several hundred kilometers. As we said before, in its center there is a relatively calm region called the "Eye" of the hurricane. Around it, the rotational velocity is the greatest and decreases out of the center. With the increase of the circular velocity, the whole vortex spread out to a gigantic ring with a diameter of several hundred kilometers (Table 1). In his East-West motion the whole system sweeps a lane about 1000 km wide. It gradually intensifies its rotational wind velocity, simply cooling the hot oceanic surface (e.g. (47)). Lingering over the ocean sometimes $20 - 30$ days, these systems describe complicated trajectories. The lost of energy of the phenomenon usually happens when the hurricane moves inside coastal areas and it goes into to the continent.

The energy that requires a hurricane to maintain its activity comes from the liberation of heat that takes place in the process of condensation of the vapor of water that it evaporates from the surface of the ocean, forming nebulosity and intense precipitation. When a hurricane enters in the continent it loses intensity quickly when stopping the process of strong evaporation from the surface. The hurricane works like a vapor machine, with hot and humid air providing its fuel. When the sunrays heat the waters of the ocean, the humid air warms; it expands and begins to rise as they make it the globes of hot air. More humid air replaces that air and that same process begins again. The rotation of the earth eventually gives it a circular movement to this system, the one that begins to rotate and to move as a gigantic spinning top. As in all hurricane, this turn is carried out in having felt clockwise in the south hemisphere and counter clockwise in the north hemisphere.

Parameter	Range	Average	Unit
Diameter (D)	$200-1300$	500	km
Eye	$6-80$	50	km
Rot. velocity (V)		0	ms^-1
Duration	130	8	$days$
Kinetic Energy	4-8	6	Twh
Surface winds:		> 33	ms^-1
Energy Source		Latent Heat Release	
Equivalent Energy		2000 Bombs Hiroshima Type	
Lives (North Atlantic)		200,000 from year 1700	
Damages (North Atlantic)		1180 billion dollars from year 1900	

Table 1. Basic Hurricane Parameters

All the tropical depressions that grow deriving in hurricanes originate practically under the same conditions, and they conserve the same meteorological characteristics to the long of the life. The physical differences that can be presented from an event to other reside in the speeds that each event can reach and the time that these they can stay.

Recent studies on the formation of hurricanes point out like cause, the violent circulation of air in them and the transformation of the liberated caloric energy when it condenses the vapor of water contained in the air that ascends from the surface in a very extensive area. Such a condition implies having an appropriate provision of latent heat, and of some mechanism that trigger and maintain the upward vertical movement required to produce the condensation of the vapor, and with it the liberation of that latent heat. These requirements are satisfied when the temperature of the seawater in a specific area is as we said before, equal same or higher to 26^oC, when the distance of the same one to any coast or island is superior to 400 km, and when inside that same region, convergence associated to any perturbation exists, be tropical wave, polar water-course, inter-tropical convergence line or area. The conditional instability is an atmospheric state that favors the formation of a hurricane in a potential region; it has been a clear relationship between the presence of the instability and the favorable months for the formation of the tropical hurricanes.

The temperature in the ocean and the high relative humidity in the stocking and low troposphere are also requirements for the development of the hurricane. In Fig. 2 figure shows a map of the superficial temperature of the sea for the summer in the north hemisphere . The yellow, the orange, and the red colors demonstrate the temperatures of the quite hot water to sustain hurricanes. Another necessary condition for the organization of the circulation inside the region in which ascents of air and the liberation of latent heat of vaporization take place, is that they happen in a superior latitude at 5°, since in an inferior latitude the organizing effect of coriolis (rotate of the earth) it has very low values. It is for this reason that the hurricanes are formed and are intensified when they are located on tropical or subtropical oceans in both hemispheres where the force of rotation of the earth is sufficiently strong, so that the rotation movement begins around the center of low pressure and whose temperatures of water at surface level are around 26.5^o or warmer. The main development region for tropical cyclones

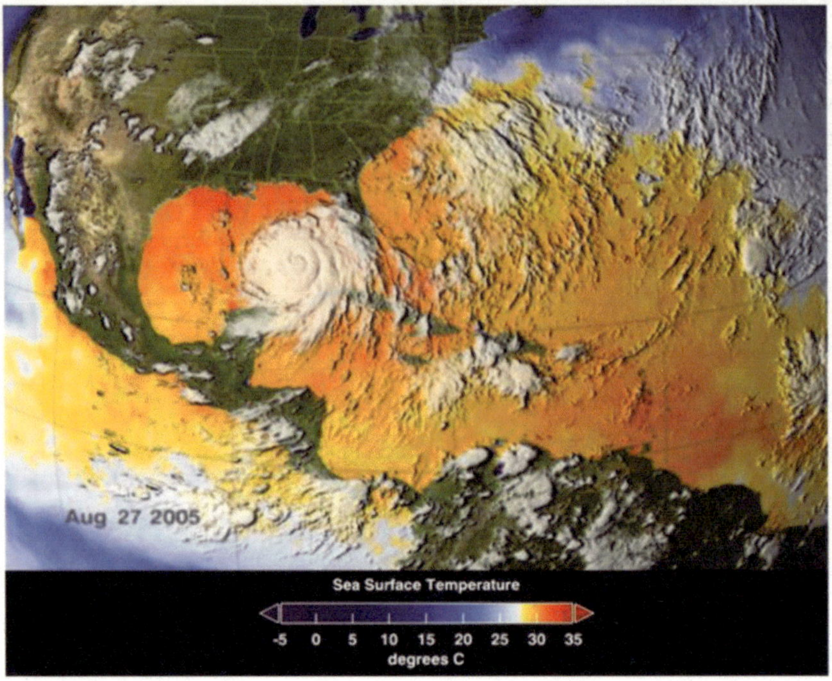

Fig. 2. Hurricane Katrina and the Sea Superficial Temperature

is the basin bounded by 25 and 60 degrees W longitude and by 8 and 23 degrees N latitude where effects of Ocean heat content on hurricane genesis are almost constant (Fig. 3).
Depending on the rotational velocity, which extreme exceeds 165 *knots* ($> 300\ km/h$), the hurricanes themselves are classified in several ways, generally based upon the vortex wind velocity and their destructive power. In 1969, the Organization of United Nations requested the evaluation of the damages generated by the passage of the hurricanes in a certain type of housings. Starting from it, the North American engineer Herber Saffir and the then director of the National Center of Hurricanes of United States, Robert Simpson, developed a mensuration scale to qualify the potential damages that it can cause a hurricane, considering the minimum pressure, the winds and the tide after its passage. This is now know as the Saffir-Simpson scale and consists of seven categories: Tropical Depression, Tropical Storm and five categories of hurricanes going from hurricanes type-1 up to type-5 (Table 2). Independently of hurricane category, the damages they potentially may cause are more intense when their translation speed is small or almost zero, provided they stay longer time over one location.
According to this scale, hurricanes evolution is as follows:
- Birth (tropical depression): first it is formed a peculiar atmospheric depression because the wind begins to increase in surface with a maximum speed of 62 km/h or less; the clouds begin to be organized and the pressure descends until near the 1000 hectopascales (hpa).
- Development (tropical storm): the tropical depression grows and it acquires the characteristic of tropical storm, what means that the wind continues increasing to a maximum speed of 63 to 117 km/h; the clouds are distributed, in hairspring form and it begins to be formed a small eye, almost always in circulate form, and the pressure decreases to less than 1000 *hpa*. It is in this phase when it receives a name corresponding to a list formulated by the World Meteorological Organization (Committee of Hurricanes). Formerly, each hurricane

Fig. 3. Control Geographic Basin, where the water surface temperature is practically constant.

Range Storm Knots	Range km/h	Range m/s
TD $30-34$	$56-62$	$15-17$
TS $35-64$	$63-118$	$18-32$
H1 $65-82$	$119-153$	$33-42$
H2 $83-95$	$154-177$	$43-49$
H3 $96-113$	$178-209$	$50-58$
H4 $114-135$	$210-249$	$59-69$
H5 > 135	> 249	> 69

Table 2. Saffir-Simpson scale. MAX Rot. Wind Velocity

was denominated with the name of the saint of the day in that it had been formed or it had been observed. It fits to clarify that if a hurricane causes an important social and economic impact to a country, the name of that hurricane doesn't appear in the list again.
- Maturity (hurricane): the tropical storm is intensified and it acquires the characteristic of Hurricane, that is to say, the wind reaches the maximum of the speed, being able to reach even 370 km/h, and the cloudy area expands obtaining its maximum extension between e 500 and 900 km of diameter, producing intense precipitations. The eye of the hurricane, whose diameter varies from 24 to 40 km, is a calm area free of clouds. The intensity of the hurricane in this stage of maturity graduates by means of the scale of $1-5$ of the Saffir-Simpson scale.
- Dissipation (Final Phase): The pressure in the center of the system begins to increase and the winds fall gradually accompanied by a weakening of the system. In this stage the hurricanes that penetrate to land become extra-tropical hurricanes. A central factor in the end of a hurricane is the lack of energy sustenance provided by the warm waters. Another is that when arriving to earth, the friction with the irregular surface of the land provokes cloudy expansion of the meteor and it causes its detention and dissipation in strong rains. An additional factor is that the hurricane meets with a cold current.

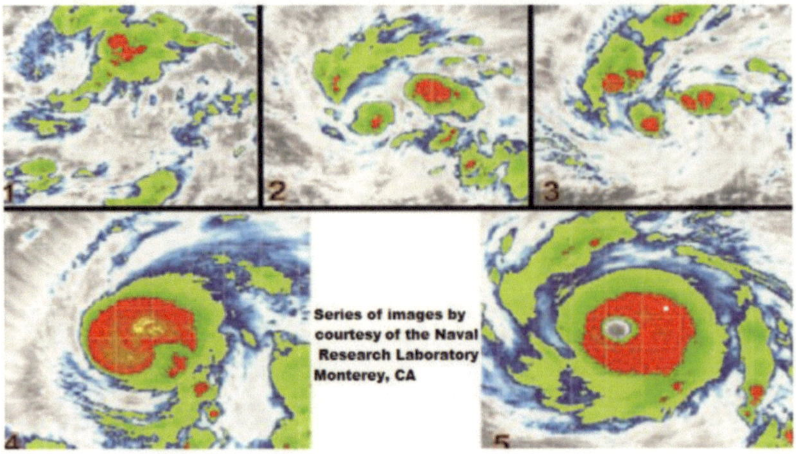

Fig. 4. Series of images of Hurricane Floyd, 1999 (Naval Research Laboratory)

Figure 4 presents some images taken in a period of 6 days, during the stages of development of a hurricane: in photo 1 - we see a good tropical disturbance that favors a tropical depression 12 hours later, in photo 2, we see a tropical depression that continuous with their escalation. In photo 3, the hurricane Floyd has been intensified in a tropical storm. In photo 4 one can already observe to Floyd like a hurricane of category-1 and in photo 5 it is already observed as category-4.

Tropical cyclone intensification depends on many factors ((9)) including oceanic heat content and proximity to land, etc. The hurricanes are formed and are intensified when they are located on tropical or subtropical oceans in both hemispheres where the force of rotation of the earth (Coriolis) it is sufficiently strong so that the rotation movement begins around the center of low pressure and whose temperatures of water at level of the surface are quite warm. The main regions are not stable as for their location, since this obeys the position of the centers of maximum marine heating, those that in turn are influenced by the cold currents of California and the equatorial warm counter current in the Ocean Pacific, as well as for the drift of the ramifications of the warm current of the "Gulf Stream". Also, they do not stay for themselves on land, independently of the superficial temperature.
An analysis of the trajectories of tropical hurricanes shows that there is not coastal area of Mexico that is free of the threat of the tropical depressions that arrive in many cases to the hurricane intensity. In the Gulf of Mexico and in the Pacific the coast of the country is vulnerable to the effects of the tropical storms, although their behavior in both coasts is something different. The depressions that are generated in the southeast of Mexico, specifically in the Bank of Campeche, they generally go toward the north, while those of the Caribbean travel toward the west until touching the costs of Central America, or those of the Peninsula of Yucatan. When they cross it, they vanish, but not enough to be annulled, due to the narrowness of the peninsula, so when arriving to the Gulf of Mexico they find the warm water again that reefed them, recovering their fury and continuing their devastating work.
In a study on the activity of the depressions in the North Atlantic during the first half of last century, some investigators found that more than 78% of those happened in the Gulf of Mexico took place starting from 1932, and only 36% has reached the hurricane force; the duration of these depressions has been of 4.4 days and that of the hurricanes of 2.2 days. The closed form

of the Gulf conditions their short duration and low frequency, since the storms reach rapidly the land and then vanish. The Peninsula of Yucatan is the most affected by the depressions and, of the total previously mentioned a 46% affected the peninsula. In the last two decades it has been increased the frequency and intensity of the hurricanes in this region; they should stand out Gilberto in September, 1988 and Mitch in October, 1998.
The season of hurricanes gives principle when the climatic equator moves in direction of the poles carrying with it high temperatures that heat the air and the seawater, giving place this way to the emergence of an area of low pressure. This generally happens between the months of May and November. Summarizing, for a hurricane may be formed there need to be present certain elements:
- Pressure: witnesses or preexistence of a convergence zone in the low levels and low superficial pressure, of synoptic scale.
- Temperature $> 80^oF$: At this temperature, the water of the ocean is evaporating at the required quick level so that the system is formed. It is that evaporation process and the eventual condensation of the vapor of water in form of clouds the one that liberates the energy, that gives the force to the system to generate strong winds and rain. And since in the tropical areas the temperature it is usually high, they constantly originate the following necessary element.
- Humidity: As the hurricane needs the evaporation energy like fuel, there must to be quite a high humidity, which happens with more easiness on the sea, so their advance and increment in energy happen there more easily, weakening when arriving to mainland.
- Wind: The presence of warm wind near the surface of the sea allows there to be a lot of evaporation and that it begins to ascend without big setbacks, originating a negative pressure that crawls to the air in hairspring form toward inside and up, allowing that the evaporation process continues. In the high levels of the atmosphere the winds should be weak so that the structure stays intact and the cycle is not interrupted - Gyre (or Spin): The rotation of the earth gives a circular movement to this system, which begins to rotate and to move as a gigantic spinning top. This turn is carried out in sense contrary to that of the pointers of the clock in the north hemisphere, and in favorable sense in the south hemisphere.
Finally it is worth mentioning that Meteorologists have records of North Atlantic hurricanes that date back into the $19th$ century. Over the last half-century, these records are based on a wide range of measurements including ship and land reports, upper-air balloon soundings, and aircraft reconnaissance. Lately, it was also included radar imaging and satellite photographs. The geographical position of the Eye center and the rotational velocity is measured and published every 12 and lately every $6h$.

3. African dust

Great quantities of dense dust supply from the great North African and Asian deserts are often carried over huge areas of the Caribbean, the tropical North Atlantic, the temperate North Pacific and Indian oceans during much of the year, with different effects in those regions (Fig. 5). The environmental conditions of Earth, including the climate, are determined by physical, chemical, biological, and human interactions that transform and transport materials and energy; this "Earth system" is a highly complex entity characterized by multiple nonlinear responses. One important part of this system is soil dust which is transported from land through the atmosphere to the oceans, affecting ocean biogeochemistry and hence having feedback effects on climate and dust production ((40)).
Dust production arises from saltation and salt blasting, when winds above a threshold velocity transport soil grains horizontally, producing smaller particles, a small proportion of which get

Fig. 5. Dust Plumes off Western Africa,September 15, 2010 (Earth Observatory, NASA)

carried up into the atmosphere for long-range transport processes. These processes depend on rainfall, wind, surface roughness, temperature, topography, and vegetation cover, which are interdependent factors linked to aridity and climate in a highly nonlinear way. Such a production depends on the supply of wind-erodible material, which ironically usually requires fluvial erosion, often from adjacent highlands, followed by subsequent drying out and the loss or absence of vegetative protection. Desert dust aerosol is dominated by particles of diameter 0.1 to $10\mu m$, with the mean size being around $2\mu m$. Such aerosols have a lifetime of hours to weeks, allowing long range transport over scales of thousands of kilometers ((14), (25)) but producing strong gradients of dust deposition and concentrations that vary substantially on time scales of $\sim$ 1day.

Dust winds show large interannual changes that are highly anticorrelated with rainfall in the Soudan-Sahel east African droughty regions. The annual emissions budgets over Sahara (North of 21.25^{o}N) and Sahel (South of 21.25^{o}N) in North Africa indicates that from Sahara are twice the corresponding emissions from Sahel, and represents 64% of North Africa and 42% of global emissions. For both regions, the year 1996 has the lowest annual emission. At global scale it is know that dust concentrations were sharply lower during much of the 20th century before 1970, when rainfall was more normal. The interannual variability of African dust transport over the north tropical Atlantic is monitored using in situ surface concentrations measurements performed at Barbados since 1966, along with the Total Ozone Mapping Spectrometer (TOMS) and Meteosat dust optical thickness (DOT) records covering the last two decades.

Much of the transport of dust occurs at altitudes of several kilometers, with subsequent removal by wet deposition. Dust deposition estimates are in the order of $1.7x10^(15)$ gr year^{-1} varying substantially from year to year, with almost two-thirds from North Africa and 26% of the dust reaching the oceans. Dust production, transport, and deposition to the oceans depend on climatic factors, particularly atmospheric structure, which regulates uplift, and wind speed and precipitation, which influence removal. Over large areas of the Earth, the atmospheric aerosol composition is dominated by mineral dust. Dust storms and dust plumes are the most prominent, persistent, and widespread aerosol features. The great variability of African dust transport has broader implications : Iron associated with dust is an important micronutrient for phytoplankton ((21)). Thus, variations in dust transport to the oceans could modulate ocean primary productivity and, consequently, the ocean carbon cycle and atmospheric CO_2. Dust could play a positive role in reducing global warming by greenhouse gas CO_2. Carbon fixation by phytoplankton in the oceans acts as a sink for CO_2. Aeolian dust deposition is the primary source of bio-available iron in the iron-limited open oceans and effectively controls hytoplankton blooming ((68)). Another important effect of dust particles is their role in the photochemical production of tropospheric ozone by reducing by as much as 50% the photolysis rates (e.g. (12); (63); (69)) and by providing reaction sites for ozone and nitrogen molecules (e.g. (84); (11)). Additionally, dust particles affect air quality ((87)) and are potential vectors for long range transport of bacteria.

The great variability in dust transport demonstrates the sensitivity of dust mobilization to changes in regional climate and highlights the need to understand how dust, in turn, might affect climate processes on larger scales: Mineral dust, emitted by wind erosion of arid and semiarid areas of the Earth, is thought to play an important role in climate forcing. However, it has been difficult to quantify because of the relatively complex and highly uncertain effect of dust on radiative forcing ((38); (106)). Because of the great sensitivity of dust emissions to climate, future changes in climate could result in large changes in dust emissions from African and other arid regions that, in turn, could lead to impacts on climate over large areas.

In order to understand the forcing involved in past climate trends and to improve estimates of future dust-related forcing, it is necessary to characterize the variability of dust emissions in response to climate-change scenarios and to distinguish between natural processes and human impacts. Aerosols, including mineral dust, can affect climate directly by scattering and absorbing solar radiation and indirectly by modifying cloud physical and radiative properties and precipitation processes ((42)). Theoretical and experimental studies have shown that the effect of the mineral dust component in the atmospheric radiation budget is comparable to the greenhouse effect gases but opposite in sign. In fact, dust can have either a net positive or negative radiative effect depending on the surface albedo and the aerosol single scattering albedo [Liao and Seinfeld, 1998]. The Saharan Dust Experiment (SHADE) experiment, held off the coast of West Africa during September 2000, shows that the net radiative impact of African dust, if extrapolated to all sources of the entire Earth, would be approximately -0.4 Wm^{-2}. However, because of the complexities of the competing solar and terrestrial radiative forcings, even the sign of the net effect is unknown ((38)), suggesting a dust radiative forcing in the range $+0.4$ to -0.6 Wm^{-2}.

Mineral dust may also exert an indirect radiative effect by modifying cloud properties and precipitation process ((101); (102)). The quantification of the dust impact on climate change is however particularly uncertain because of the lack of knowledge about the natural variability of dust emissions and the temporal and spatial variability of transported dust. Another major uncertainty is due to the lack of reliable estimates of the anthropogenic fraction of mineral dust in the atmosphere ((33)). Recent estimate of this anthropogenic fraction using climate models is of about 10% of the global dust load ((112)).

Several studies have shown that dust particles, by absorbing and scattering solar radiation, modify the atmospheric radiative budget (e.g. (Tegen & Lacis 1996); (105); (124)).

Nonetheless, dense dust clouds over the oceans reduce insolation at the ocean surface, thereby reducing the heating of ocean surface waters ((10)) and sea-surface temperatures, which in turn affects the ocean-atmosphere transfer of water vapor and latent heat, which are important factors in climate ((61)). Reduced heating over the tropical Atlantic could contribute to the interhemispheric, tropical Atlantic, sea-surface temperature anomaly patterns that have been associated with Soudan-Sahel drought ((54); (127)). Thus, increased dust could conceivably lead to more intense or more prolonged drought. Dust could also affect climate through cloud microphysical processes, possibly suppressing rainfall and conceivably leading to the perpetuation and propagation of drought ((101)). Over south Florida, clouds are observed to glaciate at relatively warm temperatures in the presence of African dust ((102)), an effect that could alter cloud radiative processes, precipitation, and cloud lifetimes. Besides, the frequency and intensity of Atlantic hurricanes have been linked to East African rainfall ((55)), showing decreased activity during dry phases. Although there is no evidence that exposure to dust across this region presents a health problem, it does demonstrate how climate processes can bring about changes in our environment that could have a wide range of consequences on intercontinental scales. It is thus important to understand the long-term variability of dust distribution, in order to determine which processes are controlling such variability.

(36) has shown that the circulation and precipitation over Europe and the North Atlantic is modulated by the North Atlantic Oscillation (NAO) with a period of about 8 years.(26) found that in winter a large fraction of the North Atlantic and Africa dust loading is correlated with the North Atlantic Oscillation (NAO) index. They show that a controlling factor of such correlation can be attributed to dust emission from the Sahel. The Bodele depression is the major dust source in winter and its interannual variability is highly correlated with the NAO. Studies based on Meteosat/visible light spectrometer (VIS) and Total Ozone Mapping Spectrometer (TOMS) observations (22 years from 1979 to 2000) have established the link between Sahel drought, dust emissions in Sahel and summer dust export over the Atlantic ((73)), and have shown the role of the North Atlantic Oscillation (NAO) on winter dust transport ((5)).

Dust export to the Atlantic ((73)) and to Barbados ((89)) are most highly correlated with Sahel rainfall of the previous year (i.e., the rainy season preceding the dust occurrence). The significant correlations obtained between interannual variability of summer surface dust concentrations at Barbados, and that of TOMS/DOT over both Sahel and northeastern tropical Atlantic suggest that Sahel sources significantly contribute to the dust transport over the western Atlantic. Thus the Sahelian region, if not of first importance in terms of intensity of dust emissions compared to the Saharan sources ((88)), is probably critical in controlling the year-to-year variability of dust export, which should allow to progress in the understanding of the mechanisms of influence of these climatic parameters, and to provide more accurate estimates of the anthropogenic fraction of mineral dust. In spite of coverage limitations of TOMS and DOT, their agreement among them and with different kinds of ground-based measurements (aerosol optical thickness) (73) and mineral dust concentrations ((6)) gives a picture of dust dynamics with a reasonable confidence level. Finally, it should be emphasized that the frequency and intensity of Atlantic hurricanes have been linked to east African rainfall ((55)), showing decreased activity during dry phases.

4. Relationship between North Atlantic cyclone activity and African dust outbreaks

The recent increase since 1995 in Atlantic tropical cyclones (including both hurricanes and tropical storms) affecting North America has raised the awareness of their impact on society and the economy. Currently, there is a debate surrounding the cause of this observed

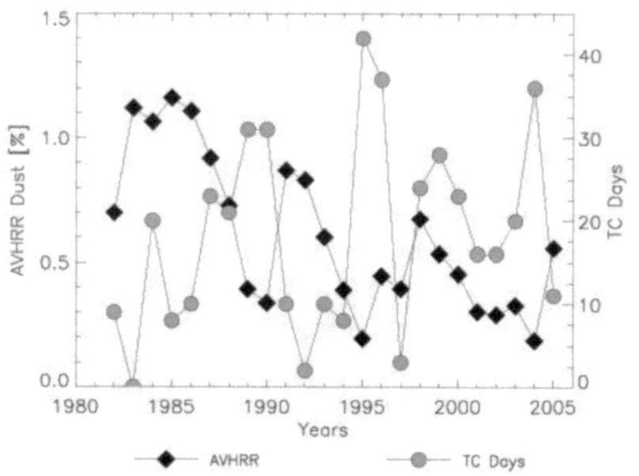

Fig. 6. Time series of North Atlantic tropical cyclone days and Saharan-induced dust cover for 19822005. The black line and diamonds represents the detected dust cover and the gray line and circles represents the tropical cyclone days, over the region of 030*N* and 1560*W* for the time period of August 20 through September 30 ((20)).

increase in cyclone activity. Several recent studies have explored the relationship between long-term trends in tropical cyclone activity (either in terms of their number or intensity) and environmental factors that may or may not be influenced by global warming ((18), (19); (56); (118); (125)). Other studies, however, have concluded that different environmental factors -not necessarily related to global warming- control trends in cyclone activity ((27);(49)).

The role of atmospheric dust as possible contributor to changing North Atlantic tropical cyclone activity was suggested by (15), who showed that tropical cyclone activity may be influenced by the presence of the Saharan Air Layer, which forms, as previously mentioned, when a warm, well-mixed, dry and dusty layer over West Africa is over imposed to the low-level moist air of the tropical North Atlantic ((3), (85)). In fact, historical data indicates that Saharan dust may have a stronger influence than El Niño on hurricane statistics in the subtropical western Atlantic/Caribbean region, while El Niño influence may be stronger in the tropical eastern Atlantic.

It was also mentioned in the previous sections that Atlantic tropical cyclone activity varies strongly over time, and that summertime dust transport over the North Atlantic also varies from year to year, but any connection between tropical cyclone activity and atmospheric dust has been only recently examined. (20) reported a strong relationship between interannual variations in North Atlantic tropical cyclone activity and atmospheric dust cover as measured by satellite, for the years 1982 – 2005 (Fig. 6).

Due that dust observations are a good proxy for the Saharan Air Layer, ((20)) showed that the contrast between the presence of dust and the lack of tropical cyclone activity for the 1983 and 1985, in particular when dust activity during the 1980*s* was more intense than during any other period in the record; the appearance of cyclone days where dust is lacking in the years 1995 and 2004, suggests an inverse correlation between dust and tropical cyclone activity, what is consistent with the hypothesis of (15). (20) suggest the possibility that the intense activity of the Saharan Air Layer indicates the presence of an environment less conducive to deep convection and tropical cyclogenesis, whereas the lack of Saharan Air Layer activity demonstrates the opposite situation. It can be speculated that the anomalous low Hurricane activity during the decades 1970*s* and 1980 was due to a general high dust activity. In fact, that is true during the 1980*s*.

A correlation coefficient of $\sim$ 0.51 is observed between tropical cyclone and dust activities time series, significant at $\sim$ 99% during the last decade, Goldenberg et al, (2001). Sea surface temperature is important in shaping the interannual variability of North Atlantic tropical cyclones ((27); (57)). However, over at least the last 26 years regional tropical cyclone activity and sea surface temperature exhibit an upward trend, while dust activity shows a downward one. The partial correlation coefficients of the de-trended time series are both -0.50, significant at 98.5%. Implying that Saharan dust activity can account for variance in the tropical cyclone record that cannot be attributed to ocean temperature. The Accumulated Cyclone Energy (ACE) time series for the topical Atlantic is also well correlated with dust cover series, with a correlation coefficient of 0.59, significant at 99.5%, possibly reflecting the effects of the Saharan Air Layer on cyclone intensity as well as genesis, as suggested by (15). ACE index is defined as the sum of the squares of the maximum sustained surface wind speed (knots) measured every six hours for the Hurricane Best Track Files ((39)).

Although the mean dust coverage and tropical cyclone activity are strongly (inversely) correlated over the tropical North Atlantic, this does not provide conclusive evidence that the dust itself is directly controlling tropical cyclone activity. It has been mentioned that a link exists between Sahel precipitation and North Atlantic hurricanes: increases in Sahel precipitation are thought to cause increases in North Atlantic hurricane activity through enhancement of African easterly waves, and reductions in Sahel precipitation and North Atlantic hurricane activity have been tied together through the associated changes in wind shear across the Atlantic basin ((28); (55)).

Therefore, it is possible that if precipitation changes in the Sahel alter West African dust outbreaks, then this variability in rainfall may be the cause of our observed correlations. However, it has been shown that, at least for the summertime months, interannual changes in dustiness over the North Atlantic are related to changes in Sahel precipitation from the previous year and are not strongly correlated with same-year Sahel precipitation events ((73)). (20) suggest that because dust is a good tracer for the Saharan Air Layer, these observed correlations may result from the effect of the Saharan Air Layer acting as a control on cyclone activity in the Tropical Atlantic, consistent with the hypotheses of (15). It is worth noting that the variability in the dust time series may not only reflect variations of the presence of the Saharan Air Layer, but it may also reflect changes in dust loadings within the Saharan Air Layer itself, which could also have important meteorological implications. It is interesting to note that a 5 to 8 year oscillating behavior is also seen in the dust record, superimposed over a downward trend in dustiness.

In contrast to the dynamic effects of the Saharan air layer (SAL) in suppressing cyclogenesis, (58) suggest an extensive cooling over the subtropical North Atlantic may be related to the shielding of' solar radiation (the so-called solar dimming effect) by dust. They exemplify with the particular behavior of the hurricane season in 2006, when it was expected a continuation of the trend of nine preceding years of above- normal hurricane seasons, however, the 2006 hurricane season was near normal with four tropical storms and five hurricanes, but decidedly much fewer as compared with the record numbers of 12 tropical storms and 15 hurricanes in 2005, including Katrina. Given the recent warming tendency in the Atlantic and the prevailing favorable preseason conditions: sea surface temperature (SST) was above normal, vertical wind shear was low, and sea level pressure was reduced over the tropical Atlantic, at the beginning of the season, all signs indicated that it would be more active than the 2005 season. Though in overall, the dust loading in 2006 was higher than in 2005, by the beginning of June there was a major increase in dust loading, and two 2 weeks later occurs a major episode of SST cooling, concomitant with a long-term trend to warming . Even if the 2006 SST in the

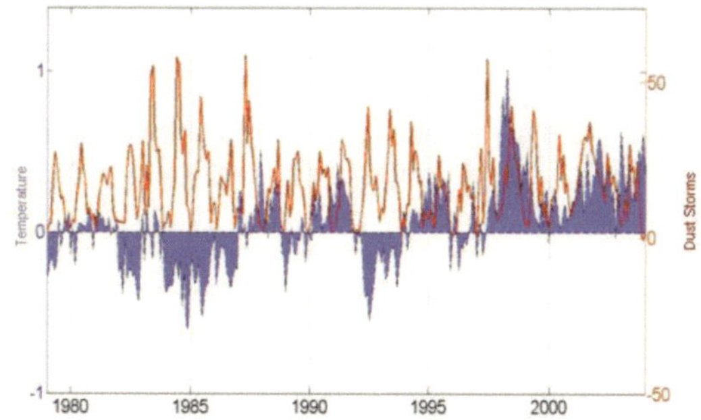

Fig. 7. Superficial Sea Temperature vs Dust Storms (1979 – 2004)

Atlantic Ocean was above normal compared with the long-term climatology, there was an abrupt cooling of the Atlantic from 2005 to 2006 independently of the long-term variation. It can be seen in Fig. 7 that AD storms could be seen as a cooling factor of the Atlantic ocean superficial waters except during the periods 1990 – 1992, 1994 – 1996, 1998 – 2004, when both signals AD and SST do increase. The most pronounced SST cooling began in mid-June 2006, reaching a maximum in late June and mid-July, until the end of September after which the SST returned to the 2005 level. The cooling was widespread, covering most of the subtropical and equatorial North Atlantic, with the strongest signal ($\sim 0.6^o - 0.8^oC$) over the WAC (western Atlantic and Caribbean). The SST cooling of the North Atlantic appeared to be closely related to the variation of Sahara dust over the region: a substantial increase in atmospheric dust loading covering nearly all the northern tropical and subtropical Atlantic and western Africa with the oceanic maximum over the western Atlantic and Caribbean, where the negative SST anomaly was most pronounced. The major dust episode lasted for about a month, until the end of June. In 2006, no hurricanes were found over the WAC and the Gulf region.

(20) has objected the role of Sahara dust in triggering a series of rapid feedback processes in the ocean-atmosphere system resulting in unfavorable conditions for hurricane formation in the Atlantic in 2006. The main point is, how much initial dust radiative forcing is enough to trigger the feedback process?. By an analysis of the aerosol optical thickness of the region in 2005 and 2006, (59) estimated an increase of 28 – 30% in 2006 relative to 2005, what is translated in a reduction of surface solar shortwave radiation flux of 4.3 – 8 $watts\ m^{-2}$. Such values could explain the required forcing. Whether there is a direct or indirect link remains elusive, since no direct causality has been yet established. Some authors suggests the variability in dust (and variability in the presence of the Saharan Air Layer), and others claim a solar dimming effect by Saharan dust as linking mechanisms for the changes in North Atlantic tropical cyclone activity. However, if up to day there is not a conclusively direct causal relationship, there is conclusive evidence of robust link between tropical cyclone activity and dust transport over the Tropical Atlantic.

5. Signal theory as a tool to find common connections among different phenomena

We have mentioned in the previous sections some work that, by means of a correlational analysis, seems to indicate that certain extraterrestrial phenomena could have some kind of relation with the terrestrial phenomena, and particularly with the occurrence of Hurricanes ((17), (43), (44),(45) and (46)). It is even claimed that such kind of connections could seat the basis of deeper studies to use the results as indicators of hurricanes precursors. To give to those results a higher meaning, it is convenient to carry out spectral studies of the different involved times series to delimitate with more preciseness the existence of those potential relationships. With this goal in mind we use here Signal Theory tools, searching to determine the most prominent signals between North Atlantic hurricanes of all categories, the Sea-Surface Temperature (SST), the Atlantic Multidecadal Oscillation (AMO), African Dust (AD) for one side, and Solar Activity (SS) and Galactic Cosmic Rays (GCR) for other side. Such analysis leads to establish the evolution in frequency and time, as well as the phase between two time series of those phenomena, allowing inferring the nature of any connection among them. That is, to find incident cosmophysical periodicities that may modulate terrestrial phenomena. As the question about the role of the Sun in modulating these phenomena has not been clarified, it requires further assessment. In the next sections we describe the investigations done on the behavior of the main common periodicities among the AMO, SST, AD, solar activity phenomena (SS), Galactic Cosmic Rays (GCR) and hurricanes. Here, as before, in the correlational analysis, it is assumed that, if there is a good interconnection between the studied terrestrial phenomena and hurricanes, and on the other hand, there is a good interconnection between these terrestrial phenomena and cosmophysical phenomena, therefore it should be also a good interconnection between hurricanes and cosmophysical phenomena.

6. Methods of analysis

The simplest and widely known technique to investigate common periodicities between two series of data is the Fourier Transform, the Fast Fourier transform and Regression Analysis. However, while useful for stationary time series, these methods are not the best for time series that are not of stationary nature ((35); (116), (117)), as those analyzed in this work. A nice description of the limitations and drawbacks of Fourier Analysis, (including the short-time-Garbor or the windowed Fourier transform, and the kindness of the wavelet transform was given by (83). In contrast to those methods, one of the most powerful tools to work with non-stationary series in Signal Theory is the so called Wavelet Spectral Analysis: within this context in order to find the time evolution of the main frequencies within a simple non stationary series at multiple periodicities the Morlet-Wavelet technique is a useful technique for analyzing localized variations of power spectrum ((116); (30)). A way to analyze two non-stationary time series, to discern whether there is a lineal or non-linear relation is by means of the Coherence-Wavelet method which furnish valuable information about when and which periodicity do coincide in time, and then about its nature, lineal or non-linear relation between the given series (for instance, solar and terrestrial phenomena), provided there is not a noticeable diphase among them. The Wavelet coherence is especially useful in highlighting the time and frequency intervals where two phenomena have a strong interaction.

6.1 The Wavelet Tranformer

In order to analyze local variations of power within a single non-stationary time series at multiple periodicities, such as the dust or hurricanes series, we apply the Wavelet (WT) using the Morlet wavelet ((116)). The Morlet wavelet consists of a complex exponential modulated by a Gaussian $e^{i\omega_o t/s}\, e^{-t^2/2s^2}$ where t is the time, with $s = 1/frecuency$ is the wavelet scale and ω_o is a non-dimensional frequency. Here it is used $\omega_o = 6$ in order to satisfy the admissibility condition ((22)). (116) defined the wavelet power $|W_n^x|^2$, where W_n^x is the wavelet transform of a time series X and n is the time index. The power spectra for each one of the parameters described in the study of next sections was computed using a Morlet wavelet as a mother wave. For the Morlet Wavelet spectrum, the significance level is estimated for each scale, using only values inside the cone of influence (COI). The COI is the region of the wavelet spectrum where edge effects become important: it is defined as the e-folding time for the autocorrelation at each scale of the wavelet power. This e-folding time is chosen such that, the wavelet power for a discontinuity at the edge drops by a factor e-2, and ensures that the edge effects are negligible beyond that point ((116)). Wavelet Power Spectral Density (WPSD) is calculated for each parameter; the contour of this cone marks the interval of 95% confidence, that is, within the COI. To determine significance levels of the global wavelet power spectrum, it is necessary to choose an appropriate background spectrum.

6.2 The Coherence and the cross wavelet

For analysis of the covariance of two time series x and y, such as the dust and hurricanes series we used the cross wavelet $W_k^{xy}(\psi)$ (XWT), which is a measure of the common power between the two series. The cross wavelet analysis was introduced by Hudgins et al (1993). Torrence and Compo, (1998) defined the cross wavelet spectrum of two time series X and Y, with wavelet transforms (W_n^x) and (W_n^y), as $W_n^{xy} = W_n^x W_n^{y*}$, where $(*)$ denotes complex conjugation. The cross wavelet energy is defined as $\left|W_n^{xy}\right|$. The complex argument is the local relative phase between X and Y in time-frequency space. Torrence and Webster, (1999) defined the cross-wavelet power as $\left|W_n^{xy}\right|^2$. The Coherence $(R^2(\psi))$ is a number between 0 and 1, and gives a measurement of the cross-correlation between two time-series and a frequency function. Statistical significance level of the wavelet coherence is estimated using Monte Carlo methods with red noise to determine the 5% significance level ((117)). The coherence significance level scale appears at the bottom of the figures of next sections.

6.3 The wavelet transformer coherence

The wavelet-squared transform coherency (WTC) is especially useful in highlighting the time and frequency intervals, when the two phenomena have a strong interaction ((116), (117)). The wavelet square coherency $(R^2(\psi))$ is defined as the absolute value squared of the smoothed cross-wavelet spectrum XWT , normalized by the smoothed wavelet power spectra. Unlike the cross wavelet power, which is a measure of the common power, the wavelet square coherency is a measure of the intensity of the covariance of the two series in time-frequency space ((116)).

The WTC measures the degree of similarity between the input (X) and the system output (Y), as well as the consistency of the output signal (X) due to the input (Y) for each frequency component. When $R^2(\psi) = 1$ or ~ 1, this indicates that all frequency components of the output signal (Y) correspond to the input (X) and means that there is synchronization between output signal (X) and input signal (Y). The synchronization can be in phase, frequency and/or amplitude. If $R^2(\psi) \ll 1$, then output Y is not related to input X because of the presence of noise, nonlinearities and time delays in the system.

6.4 The signal/noise wavelet coherence transformer

The coherence of the system can be calculated through the relation signal/noise, defined as ($WTC_{s/n}$) in (122). The $WTC_{s/n}$ is just what we are using in this work, because it allows us to find linear and nonlinear relationships, while verifying that the periodicities of cross-wavelet are not spurious, in order to minimize the effects of noise. If the XWT and the $WTC_{s/n}$ of two series are enough high, the arrows in the XWT and $WTC_{s/n}$ spectra show the phase between the phenomena:

a) Arrows at 0° (pointing to the right) indicate that both time series are correlated (in phase) and arrows at 180° (pointing to the left) indicate that they are anticorrelated (in anti-phase). It is important to point out that these two cases imply a linear relation between the considered phenomena.
b) Non horizontal arrows indicate an out of phase situation, meaning that the two studied phenomena do not have a linear relation but a more complex relationship ((122)).

6.5 The wavelet global spectra

On the right blockls of figures of next sections, it is shown the global spectra, which is an average of the power of each periodicity in both the wavelet and the coherence spectra (GXWT and $GWTC_{s/n}$). It is usually used to notice, at a glance, the global periodicities of either the time series, or, of the coherence analysis. The significance level of the global wavelet spectra is indicated by the dashed curves, they refer to the power of the red noise level at the 95% confidence that increases with decreasing frequency ((30)). It is a way to show the power contribution of each periodicity inside the COI, delimiting the periodicities that are on, or, above the red noise level. The uncertainties of the periodicities of both global wavelet and coherence spectra are obtained at the half maximum of the full width peak. As in the Morlett wavelet, the 95% confidence level of the coherence is inside the black contour (the COI).

7. Coherence analysis of terrestrial and cosmophysical forcings

Data.- To assess the long-term relations between space phenomena and indicators of the global climate it is often necessary to use reconstructions of Galactic Cosmic Rays (CR), solar activity (SS) and climate phenomena. Direct measurements of solar activity based on sunspot numbers exist since 1749, but trustable CR data is only available since the 1950*s* decade when the neutron Monitor Stations began to operate. Records of climatic phenomena exist from the end of the 19th century.
For the AMO the annual time series between 1851 − 1985, data from the World Data Center for Paleoclimatology is used here: (*http : //www.ncdc.gov/paleo/*). For Solar Activity we use the daily number of Sun spots number (SS): http://www.ngdc.noaa.gov/stp/SOLAR/ftpsunspotnumber.html#american). Concerning the ^{10}Be, there is a polemic about whether it can be considered as a Proxy of Galactic Cosmic Rays (see for instance. (107), (108)). Nevertheless, a good number of researchers still support the use of ^{10}Be as a Proxy of CR (e.g. (123); (64); (64); (119), (120)) etc. Under this last context we have considered the ^{10}Be concentration in the Dye 3 ice core ($65.2N, 43.8W, 2477m$ altitude) from (2), which data for the period 1851-1985 were offered to us by the author. Regarding data of Hurricanes the WEB page *http : //weather.unisys.com/hurricane/* has been considered.
The analysis results are displayed through the next panels. The upper block of each figure shows the time series of the data involved. The power level color code used throughout this paper is indicated at the bottom of each panel. Areas inside black contours correspond to the 95% significance level. As we are working with two time series, the wavelet coherence and phase difference are obtained. As the blocks of the wavelet coherence indicate the time and frequency intervals where two phenomena have a strong interaction. The global spectra (on the right blocks of each figure) allows us to notice at a glance the global periodicities of either the time series or the coherence analysis. The significance level of the global spectra is indicated by the dashed curves. It refers to the power of the red noise level: peaks below the line implies a global periodicity with a confidence lower than 95% at the corresponding frequency, whereas peak for above indicates a confidence level higher than 95% at the given frequency. The Spectral power (abscissa axis) is given in arbitrary units.

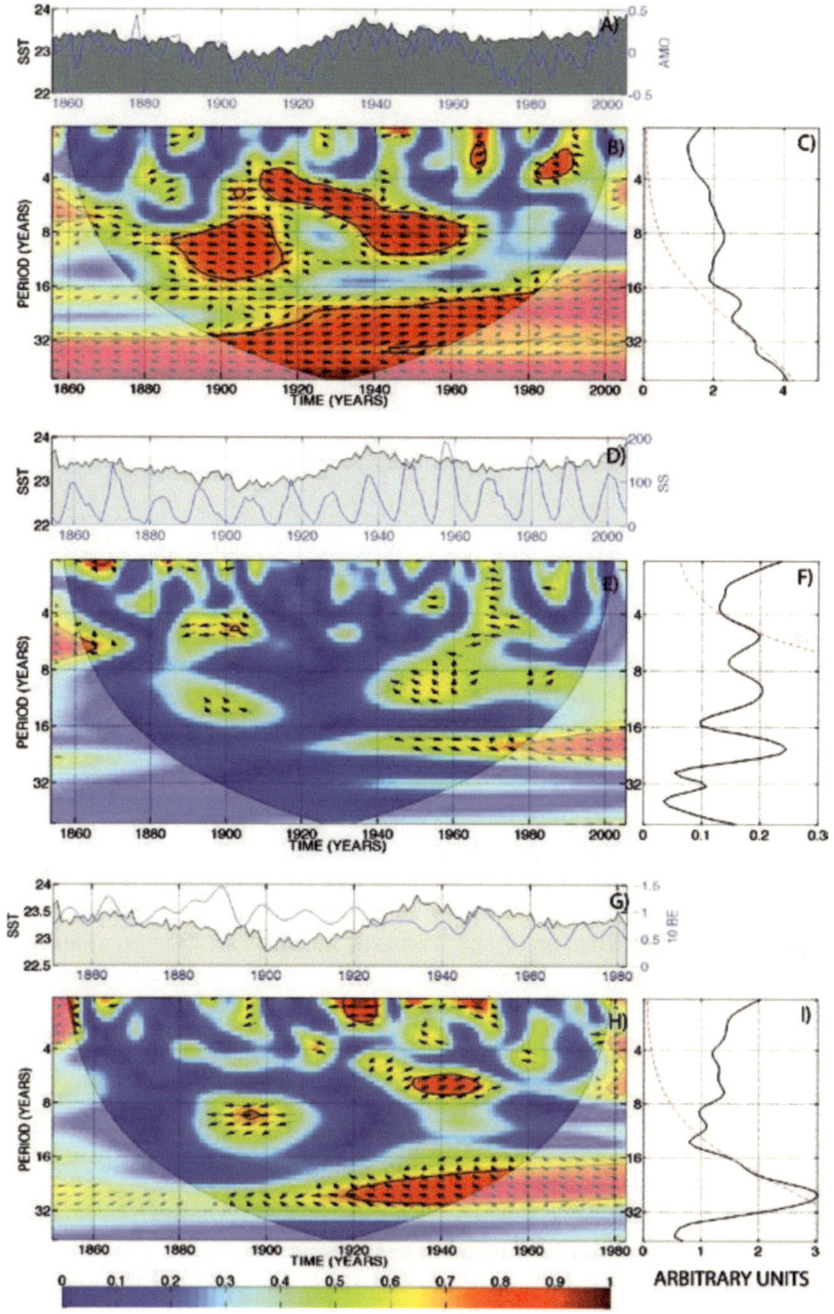

Fig. 8. (A) Time series of AMO (blue line) and SST (gray area). (B) Coherence between SST and AMO. (C) Significance level of the global spectra of SST and AMO. (D) Time series (blue line)of SS and SST. (E) Coherence between SST and SS. (F) Significance level of the global spectra of SST and SS. (G) Time series of SST and CR (^{10}Be) (blue line). (H) Coherence between SST and CR (^{10}Be). (I) Significance level of the global spectra of SST and CR (^{10}Be)

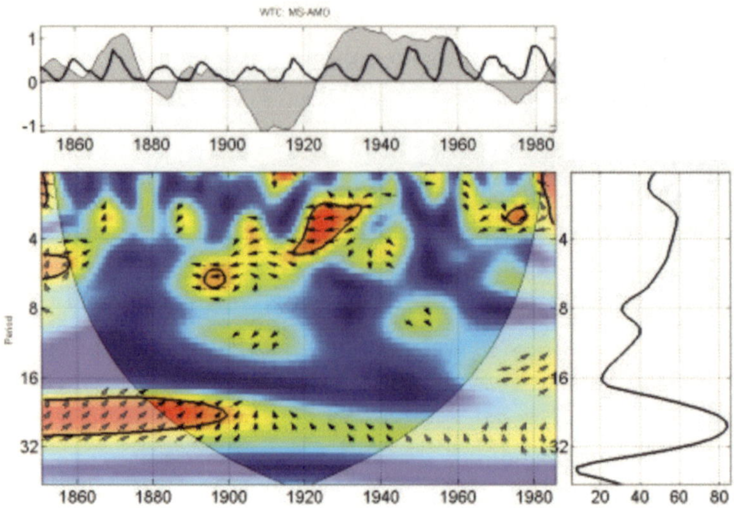

Fig. 9. Upper block: Time series of AMO (gray area) and SS (black line) . Lower block: Coherence between SS and AMO . Right Block: Global Spectrum

The main results that can be drawn from (Fig. 8) can be summarized as follows:
1) There is a coherence of 0.95 inside the COI between the AMO and SST anomalies through the band of 15 – 32 years, in the time interval 1900 – 1980 (Figs. VI.2.B-C). The oscillation in the 30 years frequency is completely in phase, indicating a lineal relation among both phenomena, which is not surprising because is something very well known by climate specialists.
2) There is a coherence of 0.6 between the SST anomalies and SS, also limited to short intervals, 1895 – 1910, 1945 – 1960 at the frequency of 11 years with tendency to be in anti-phase, and 1940 – 1980 at the 22 years frequency, with tendency to be in phase (Figs. VI.2E-F). It can be seen from these figures that no frequency at the 30 years periodicity was found for Solar Activity (at least through the use of SS), by means of the wavelet spectral analysis.
3) There is a coherence of 0.90 inside the COI between SST anomalies and ^{10}Be (the CR proxy) at the 30 years frequency, in the time interval 1920 – 1950 (Figs. VI.2H-I) for the case of SST anomalies. The oscillations have a tendency to be quasi-perpendicular, indicating a complex relation among both terrestrial phenomena and CR. It can be mentioned that the same frequency is found among AMO and ^{10}Be in the period 1870 – 1950, but with a lower coherence of $\sim$ 0.75 (Fig. 6a in (121)).
From Fig. 9 it can be seen that there is a non-linear coherence of 0.90 inside the COI between the AMO and SS near the 30 years frequency, in the time interval 1875 – 1895. This is gradually attenuated during the minimum of the modern secular solar cycle (1890 – 1940), with a coherence of 0.55 and a complex phase between the oscillations. Also, there is a quasi linear anti-phase linear coherence < 0.80 at the frequency band 3 – 7 years in the period 1890 – 1915 reaching a coherence > 0.9 from 1915 – 1930.

8. Coherence analysis of terrestrial and cosmophysical forcings vs cyclones

Results from panel (Fig. 10) can be summarized as follows:
1) There is a coherence of 0.9 inside the COI at the 7 years frequency, between the total number of hurricanes (i.e., including all magnitudes from TS to all together) and the SST anomalies

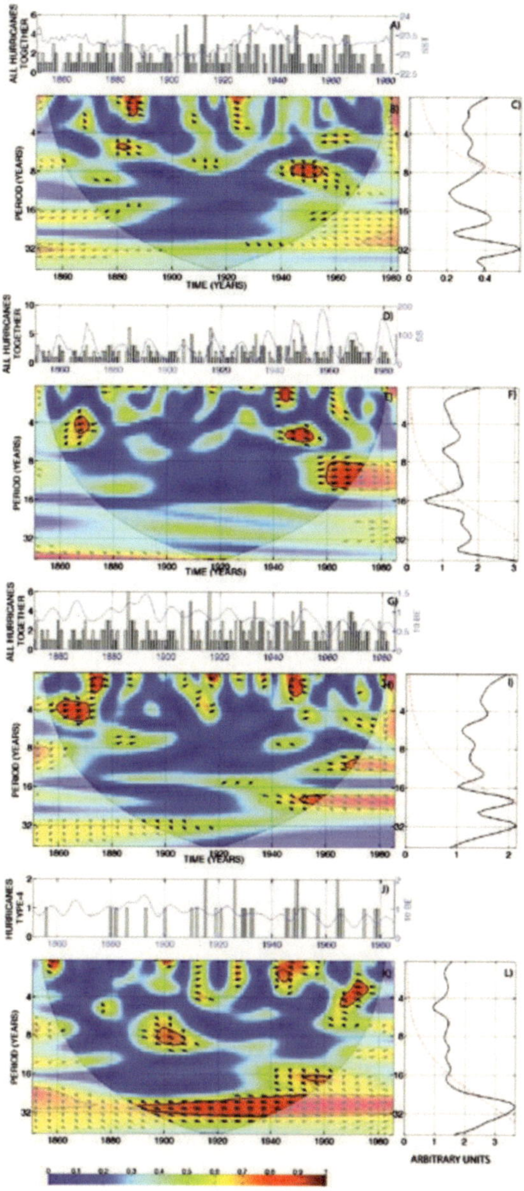

Fig. 10. (A) Time series of SST and all Hurricanes together. (B) Coherence between SST and all Hurricanes together. (C) Significance level of the global spectra of SST and all Hurricanes together. (D) Time series of SS and all Hurricanes together. (E) Coherence between SS and all Hurricanes together. (F) Significance level of the global spectra of SS and all Hurricanes together. (G) Time series of CR (^{10}Be) and all Hurricanes together. (H) Coherence between CR (^{10}Be) and all Hurricanes together. (I) Significance level of the global spectra of CR (^{10}Be) and all Hurricanes together. (J) Time series of CR (^{10}Be) and Hurricanes of magnitude-4. (K) Coherence between CR (^{10}Be) and Hurricanes of magnitude-4. (L) Significance level of the global spectra of CR (^{10}Be) and Hurricanes of magnitude-4.

However this is only limited to the time interval 1945 – 1955. This is illustrated in Figs. VII.1.B-C, for the case of SST anomalies. The oscillations have a tendency to be in anti-phase, indicating a lineal relation of both phenomena with hurricanes. 2) The coherence between SS and Hurricanes of all magnitudes together is about 0.9, limited at the frequency of 11 years during the period 1955 – 1965, (E-F in Fig. 10). However, the analysis of SS vs hurricanes of individual magnitudes gives relatively low values of coherence inside the COI.
3) The coherence at the frequency of 30 years, between CR (through the proxy (^{10}Be) with hurricanes of all magnitudes together is about 0.6 in the period 1890 – 1940 (H-I in Fig. 10). In contrast those of 5, 11, 22 reach a coherence of 0.9 in the intervals 1860 – 1870, 1960 – 1970 and 1950 – 1960 respectively.
4) For some hurricanes, as for instance those of magnitude-4, the coherence with CR (^{10}Be) is > 0.9 at the 30 years frequency, during a relatively long period, 1890 – 1950 (K-L in Fig. 10). In these cases there is a tendency of the oscillations to be in-phase, indicating a linear relationship among both phenomena.

9. Coherence of African Dust vs Cyclones

In order to rise the relevance of extraterrestrial influence on hurricane phenomena, it is needed a frame of reference. We consider here, as such a frame a terrestrial phenomenon which is well established to be related with cyclone development, as is the case of the Dust Cover originated in African Dust Outbreaks (e.g., (20), (58), (7)). Hence, the results shown on the next figures correspond to the spectral analysis of coherence, by the wavelet method mentioned before, for Atlantic tropical cyclones of all categories, versus African Dust Outbreaks ((122)).
Though there has existed satellite data on AD since the 1980, here we use only multidecadal continuous in-situ monthly data available from Barbados from 1966 up to date ((87), (90)). Monthly data of hurricanes was taken from the national Weather Service and transformed into a dates series of pulses ((122)) as: n = number of hurricanes, where 0 = no hurricane. Both data series are shown in blocks (a) of Fig. 11 and Fig. 12. On the next figures the global spectra are shown on the left side blocks.
The GXWT [panel (b) of Fig. 11] between dust and Category 5 hurricanes shows a very prominent annual periodicity with confidence higher than 95%. As can be seen in the XWT [panel(c)], this periodicity presents a high variability; it is not continuous and doesn't have the same intensity throughout the 1966-2004 period, becoming more intense during the 1970s and at the beginning of the 1980s, just when Category 5 hurricanes were, average, the most intense. This annual periodicity is presumably related to the dust cycle in North Africa and to seasonal changes in atmospheric circulation ((37)). Additionally, there is a decadal periodicity (11-13 years), which is present throughout the entire time interval with an anti-correlation tendency [panel(c)]. Its temporal tendency can be observed in panel (a) of Fig. 10 with a dotted line obtained by means of a Daubechies type modified wavelet filter (122)). This decadal periodicity is presumably related to the Atlantic trade wind variations and the dominant meridional mode of SST variability in the tropical Atlantic ((103)), as well as with with solar activity ((34) and cosmic rays ((81)) . The interaction of solar activity and cosmic rays with hurricanes is probably accomplished through the modulation of the Atlantic multidecadal oscillation (Perez08a).
This decadal variation shows that Category 5 hurricanes occur around the decadal minimum [panel (a)], because the local vertical wind shear (Vz > 8m/s) is unfavorable for the genesis of tropical cyclones ((9)). This would explain why from 2008 up to the present, there have not been any Category 5 hurricanes, since it was precisely during this time that African dust in the atmosphere has been increasing. This would imply that if such a tendency continues, the

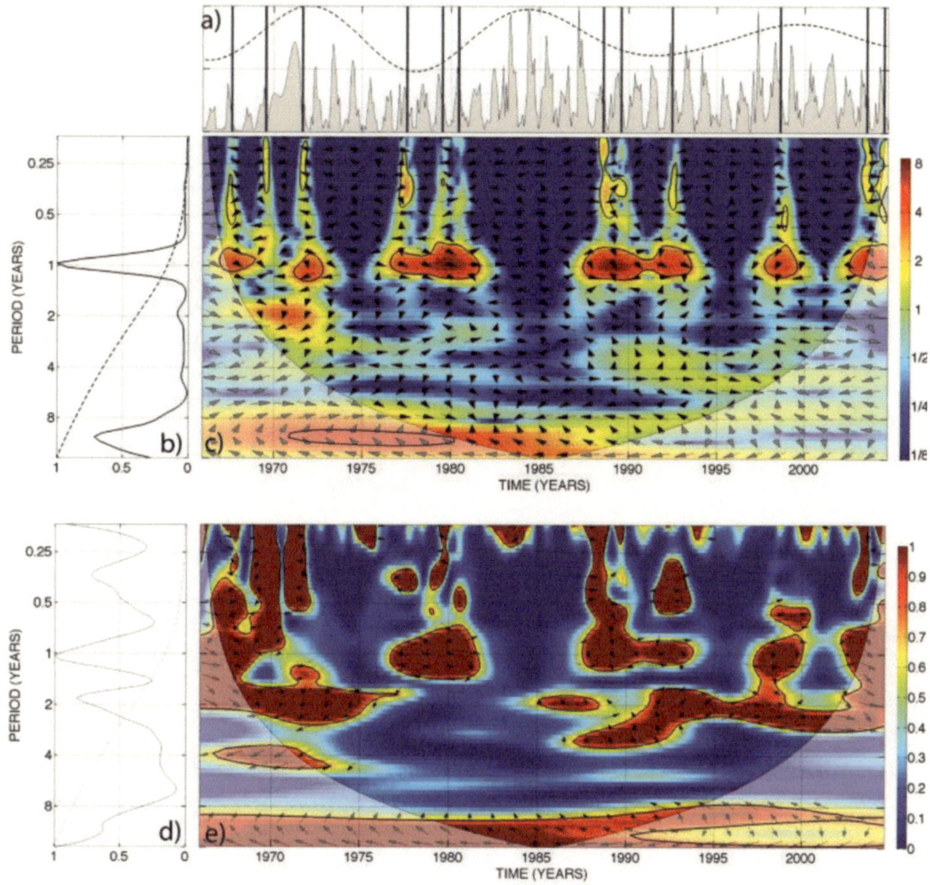

Fig. 11. (a) Time series of African Dust (shaded area). - Category 5 hurricanes (black bars) - decadal tendency of dust (dotted line). (b) global cross-wavelet coherence, GXWT. (c) cross-wavelet coherence XWT, (d) global wavelet-squared transform coherency (signal/noise), GWTCs/n. (e) wavelet-squared transform coherency (signal/noise), WTCs/n. ((122))

next group of tropical cyclones will not evolve to category 5 until the next decadal minimum of African dust occurs. To confirm if the annual and decadal periodicities obtained with the cross-wavelet are intrinsically related to the modulation of African dust on Category 5 hurricanes, we also obtained the modified wavelet coherence (WTCs/n) and found, in addition to these two periodicities, two others, of 125 days and 1.8 years with a confidence level higher than 95% [panels (d) and (e) of Fig. 11].

Fig. 12 shows the GXWT (panel b) between dust and tropical storms where it can be seen that the most prominent periodicity is that of 1 year. These annual variations do not have the same intensity throughout the period studied, as can be observed in the XWT (panel c) when the periodicity was low and parsimonious from the second half of the 1960s up to the first half of the 1970s, due to an increase in precipitation in North Africa. At this point, the coherence becomes very intense due to the very severe droughts in West Africa, known as the Sahel

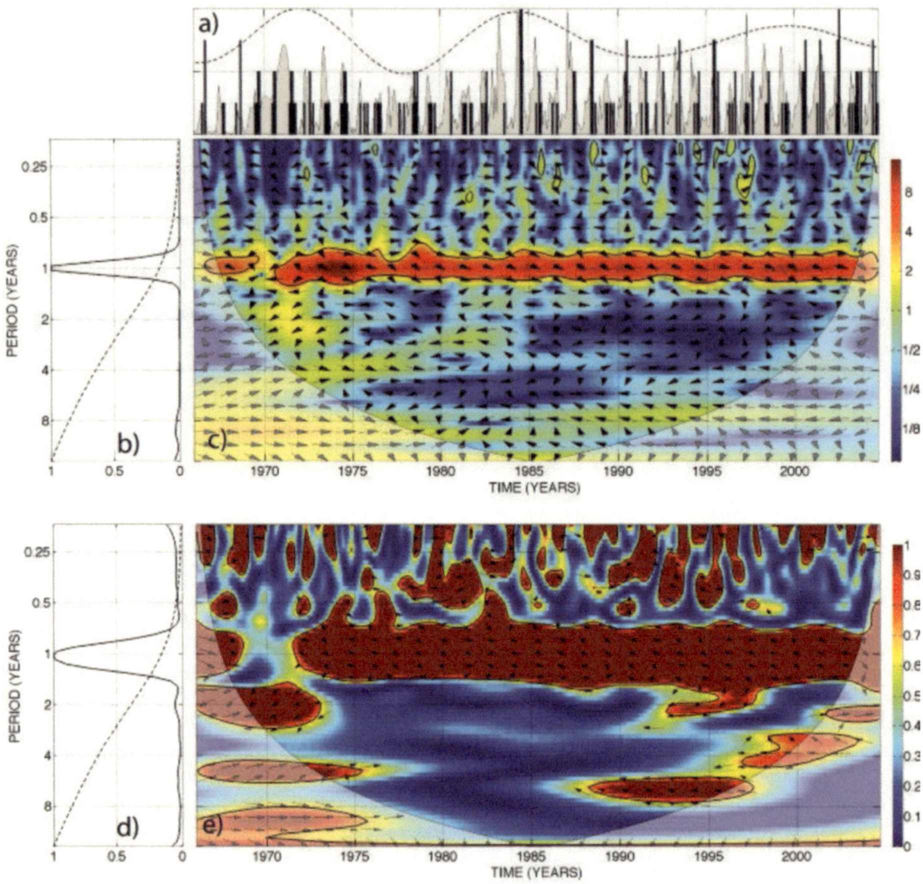

Fig. 12. (a) Time series of African Dust (shaded area). - Tropical Storms (black bars) - decadal tendency of dust (dotted line). (b) global cross-wavelet coherence, GXWT. (c) cross-wavelet coherence XWT, (d) global wavelet-squared transform coherency (signal/noise), GWTCs/n. (e) wavelet-squared transform coherency (signal/noise), WTCs/n. ((122))

drought, that began in the middle of the 1970s and lasted for several decades. The in-phase behavior of this annual periodicity with a linear tendency seems to indicate that variability in dust has a quasi-immediate effect on the genesis and evolution of tropical storms. There are other periodicities lesser and greater than 1 year but with inconsistent patterns. This may be interpreted as that dust concentration and the evolution of tropical storms are the result of many external and internal factors occurring on different time scales. The XWT and WTCs/n show that the multiannual periodicities, including the decadal periodicity, are not essential factors in the formation of tropical storms.

The global GXWT and WTCs/n (panels b and d), as well as the cross-wavelet and coherence (panels c and e) between dust and Category 1 hurricanes (Fig. 4 in (122)) show a great deal of similarity with tropical storms, though with different intensities. These differences may be due to the increase of precipitation in North Africa, as indicated by the Sahel rainfall index. Decadal periodicity is absent for Category 1.

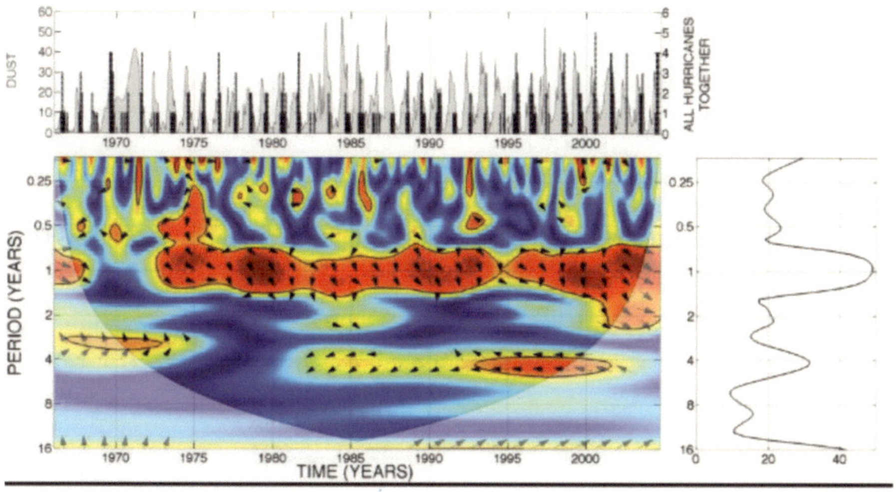

Fig. 13. Time series of African Dust (shaded area) - all hurricanes categories together (black bars). The lower panel is wavelet-squared transform coherency (WTC), and the right panel is the Global spectrum GXWT.

It can be observed in Fig. 5 in (122) that the GXWT and GWTCs/n (panels b and d), as well as the XWT and WTCs/n (panels c and e) between dust and hurricanes for category 2 show an annual periodicity that has a less continuous time interval as compared to Category 1 hurricanes and tropical storms (Fig. 12), becoming more intense from 1972-1982 and 1988-2004. Additionally, there exist less prominent periodicities of 3.5-5.5 years that are associated with the El Niño Southern Oscillation (ENSO) . Decadal periodicity is absent for Category 2 hurricanes.
The GXWT and GWTCs/n (panels b and d), as well as the XWT and WTCs/n (panels c and e) in Fig. 6 and Fig. 7 (in (122)) between dust and Category 3 and 4 hurricanes, respectively, show that the annual periodicity is also not continuous over the period studied. It can also be observed in panel (e) that the periodicity of 3.5 years for Category 3 hurricanes becomes more intense during the interval 1980-1992. For Category 4 hurricanes the periodicities of 3.5-5.5 years (panel e) are anti-correlated during 1988-2002. Decadal periodicities for Category 3 and 4 hurricanes are practically absent.
It should be emphasized that the observed correlations show not only a direct effect of African dust on hurricane activity but also reflect an indirect relationship between the wind, SST, AMO, the Modoki cycle, El Niño, la Niña, precipitation, solar activity and cosmic rays that to a greater or lesser degree modulate the evolution of Atlantic hurricanes.

10. Coherence between cosmic rays and cyclones

For comparison of the influence of CR on Cyclones with the influence of AD on Cyclones (for which the period time is relatively short 1966 – 2005), it is not necessary to use a Proxy for CR, since data of NM stations is quite confident since the later 50*s*. Therefore we use data from the worldwide NM station network. in units of counts/min that we transform to monthly and annual data. Here below we present results corresponding to annual data.
The results obtained in section 11 as compared with those of section 10 can be summarized as follows:

1) It can be seen on Figs. 12 and 14 for Tropical Storms, that the coherence is higher and more continuous in time with AD than with CR.
2) It can be appreciated from Figs. 4 (in (122)) and 14 for hurricanes of magnitude-1, that the coherence with Dust is very high, of the order of 1, almost during all the studied period, but a shot , in the period 1996 – 1999 is only observed with CR, with a coherence higher than 0.9 at the 2 years frequency.
3) For Hurricanes of magnitude-2 the coherence with CR is 0.7 years, in short periods, with a coherence near 0.8, again stronger than with Dust (Figs. 5 in (122)) . Fig. 14 (G) - (H)), show that the coherence with AD is > 0.9 at the periodicity of 1 year in anti-phase all the studied period.
4) For Hurricanes of Magnitude-3, it can be seen from Fig. 14 (J)-(K) that the dominant periodicity after 1973 is in 0.6 years, for short time periods, whereas for AD it can be seen from Fig. 6 (in (122)) that the coherence is > 0.9 at the 1 year periodicity, in anti-phase all the time .
5) For Hurricanes of Magnitude-4, (Fig. 14 (L)- (O)) shows that the coherence with CR is of the order of 0.95 and is in anti-phase at the periodicity of 1.7 years in limited time intervals . Coherence with AD at the frequency of 1 year, with a coherence > 0.9 after 1973 (Fig. 7 in (122)).
6) For the more dangerous hurricanes, those of magnitude-5 we can see from Fig. 14 (P)-(R) that the coherence is of complex nature, in the 0.7 and 1.7 years periodicities, in short time intervals, with a coherence around 0.9. From Fig. 11 we can see that coherence is of the order of 1 at the 1 and 10 – 11 years periodicity always in anti-phase. The 1 year periodicity occurs mainly in limited periods, occurring every 10 years.
7) It can be seen from Figs. 14 (T) and 14 (S)-(U) that the coherence between all kind of Atlantic hurricanes together (from Tropical Storms to magnitude-5) and cosmic rays is at the 1.3 and 1.7 yrs frequencies. Variations seem to be in phase from 1987 to 1991 with coherence of 0.6, and suddenly they switch in anti-phase during 1996 – 2002, with coherence higher than 0.9.
8) In contrast to previuos statement (7) on Fig. 13 it is shown that the coherence between all kind of Atlantic hurricanes together and the African Dust Outbreaks is more continuous than with CR , with coherence higher than 0.9, and it is concentrated around the 1.3 and 4 years periodicities, for long time periods, with complex non-linear phases. Moreover, it should be mentioned that the red noise of the confidence level of the global spectrum in Fig. 13 is not shown, because it is far above the frequency picks.

11. On the genesis of the category-5 Atlantic Hurricanes: "The geography of the marine bottom"

Fig. 15 shows a map of the barometric distribution of the eyes of Category 5 hurricanes illustrated for six of these kind of hurricanes, between 2005 and 2007: Wilma (2005), Rita (2005), Katrina (2005), Emily (2005), Dean (2007) and Felix (2007). The black circles at the right side column indicate the evolution of each hurricane, from Category 1 to 5, during their trajectory. ((122)).
By tracing the trajectories of all hurricanes, we have delimited the existence of four areas of deep water in the Atlantic Ocean where the eye of the hurricane has the lowest pressure (<920mb): I) the east coast of the United States, II) the Gulf of Mexico, III) the Caribbean Sea and IV) the Central American coast. This must be pointed out because it implies that, in addition to the required climatological conditions for the genesis of this kind of Category 5 hurricanes to take place, the geography of the marine bottom also plays an important role, and that these hurricanes do not originate in hazardous places (Fig. 14). ((122)).

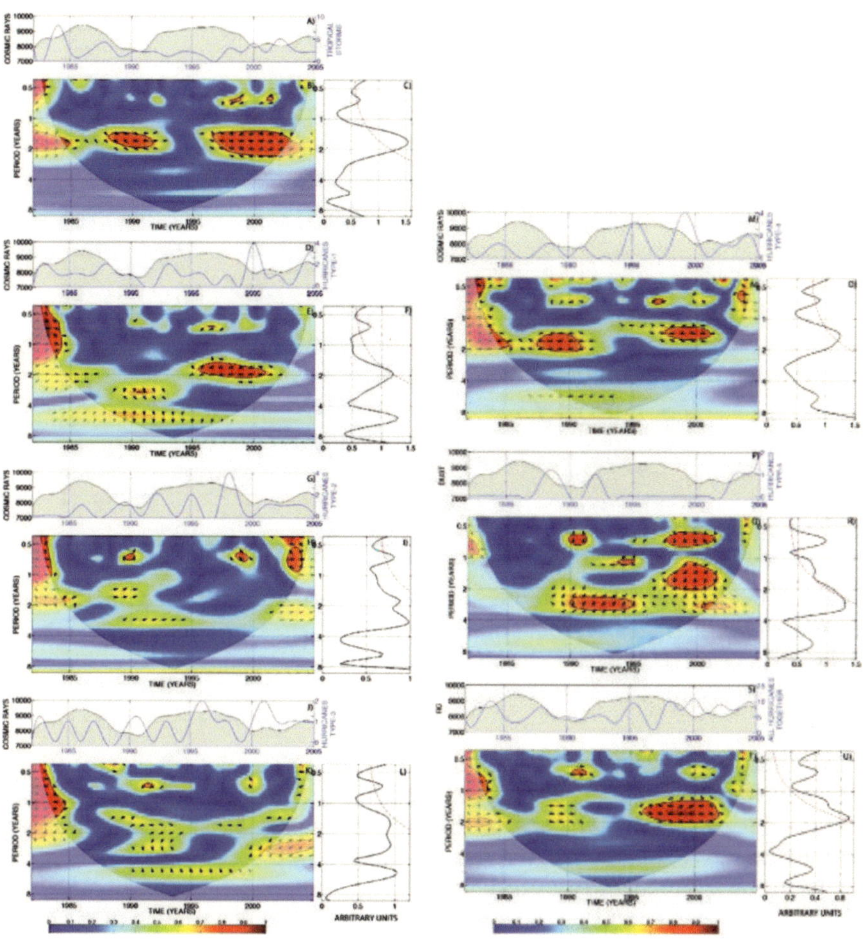

Fig. 14. A) Time series of CR and Tropical Storms. (B) Coherence between CR and Tropical storms. (C) Significance level of the global spectra of CR and Tropical Storms. (D) Time series of CR and Hurricanes of magnitude-1. (E) Coherence between CR and Hurricanes of magnitude-1. (F) Significance level of the global spectra of CR and Hurricanes of magnitude-1. (G) Time series of CR) and Hurricanes of magnitude-2. (H) Coherence between CR and Hurricanes of magnitude-2. (I) Significance level of the global spectra of CR) and Hurricanes of magnitude-2. (J) Time series of CR and Hurricanes of magnitude-3. (K) Coherence between CR and Hurricanes of magnitude-3.(L) Significance level of the global spectra of CR and Hurricanes of magnitude-3. (M) Time series of CR and Hurricanes of magnitude-4. (N) Coherence between CR and Hurricanes of magnitude-4. (O) Significance level of the global spectra of CR and Hurricanes of magnitude-4. (P) Time series of CR and Hurricanes of magnitude-5. (Q) Coherence between CR) and Hurricanes of magnitude-5. (R) Significance level of the global spectra of CR and Hurricanes of magnitude-5. (S) Time series of CR and Hurricanes of all magnitudes together. (T) Coherence between CR and Hurricanes of all magnitudes together. (U) Significance level of the global spectra of CR and Hurricanes of all magnitudes together

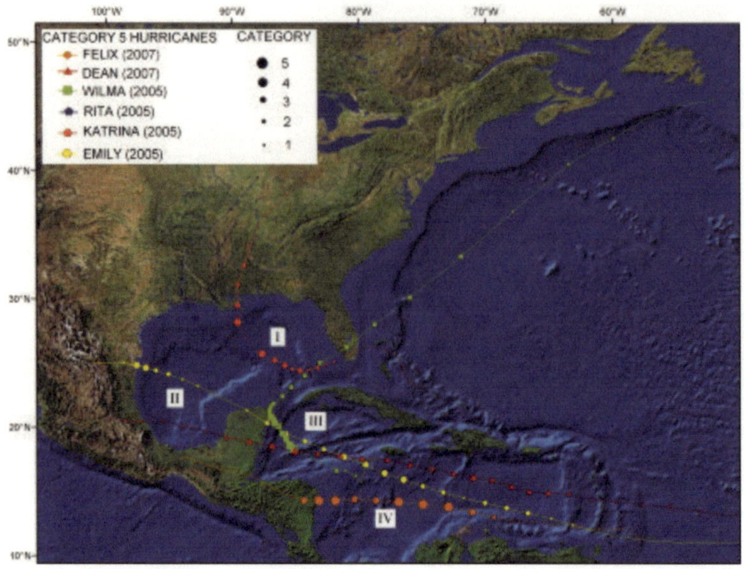

Fig. 15. Map of the Atlantic Ocean centers of cyclogenesis of Category 5 hurricanes: I) the east coast of the United States, II) the Gulf of Mexico, III) the Caribbean Sea and IV) the Central American coast ((122)).

12. Discussion

Keeping in mind that the quasi-anticorrelation between SST and Dust Storms (Fig 8) is not a perfect one and that a linear correltaion between SST and North Atlantic Cyclones (Fig. 10 (B)-(C)) is also not perfect, but the anti-phase situation is systematically found between AD and hurricanes, then we assume here, in first instance, that the more abundant the AD storms the fewer the number of North Atlantic Cyclones, and conversely, a linear correlation would mean that the more abundant the AD storms the higher the occurrence of cyclones, at least they were completely independent phenomena, which seems not to be the case.

In this context, the spectral analysis carried out here for the study of common periodicities among Cosmic Rays and phenomena which are presumably associated with Hurricanes (AMO and SST-anomalies) is in agreement with previous results ((80), (81)). We have found that there are common periodicities between some extraterrestrial and terrestrial phenomena, namely 3.5, 5.5, 7, 11 years and the more prominent one, that of 30 yrs. is often present (as can be seen in most of Figs. 9, 10) with exception of SS. It should be mentioned that no figure with Dust shows the 30 year periodicity, because data 13 Dust only cover 40 years. Such a periodicity is also present in CR vs Hurricanes of Type-4 as was shown in (81), where the applied Morlet-wavelet technique allowed to put in evidence, for the first time from a Coherence wavelet study, the periodicity of 30 years in cosmic ray fluctuations. The most dangerous Hurricanes do not show that periodicity, since confident data on category-5 exist only since 1980. It should be mentioned that this frequency is also found in other indexes of hurricane activity, as has been previously mentioned. Preliminarily, it can be speculate that this 30 years cycle may be associated to a semi-phase (either of the maximum or the minimum) of the Secular Cycle of 120 years of Solar Activity, that is, half of the so called Yoshimura-Gleissberg cycles ((131), (121)).

If the coherences found in this work among the studied phenomena, may be interpreted as a modulator factor, then, from the analysis of the previous results it could be speculated that

the modulator agent of terrestrial phenomena is the open solar magnetic field , translated in GCR (via the ^{10}Be). This modulation seems to be more important in the period 1880 – 1960. That does not mean there is no modulation after and before, but according to the coherence wavelet technique the significance level is lower than 95%.
It seems then that GCR are modulating in some way both the SST and AMO (Fig. 9) and Fig. 6a in (121)), and these in turn modulate in some way hurricanes as it can be seen from the Coherence wavelet analysis (Fig. 10 (B)-(C)) which confirms the conventional statement of hurricanes to be linked to warmer oceans. We then assume here that the action of GCR on the clouds is an additional warming factor of the SST and AMO, as it does the greenhouse gases on the earthÕs surface temperature (95), which in some way is to be translated in the development of hurricanes. In contrast, the indicator of closed solar magnetic field (via SS) presents, within the COI, a lower and attenuated coherence with the terrestrial phenomena.

13. Conclusion

The spectral analysis carried out here for the study of common periodicities among Cosmic Rays and phenomena which are presumably associated with Hurricanes (AMO and SST-anomalies) is in agreement with previous results ((80), (81)). We find again that the frequency of 30 years is often present with exception of the SS graphs. Though, it is well known that cosmic rays and solar activity phenomena are inversely related in time, such a relation is not directly translated on their influence on hurricanes development. The temporal scale of their influence is certainly different: cosmic rays influence is a relatively prompt effect, whereas solar activity seems to act as a result of a slower buildup effect ((79)); it should be appreciated the good coherence between CR and hurricanes of all magnitude, particularly with those of magnitude-4. In contrast, the indicator of closed solar magnetic field (via SS) does not present the 30 year periodicity, but it presents (within the COI) a very low and attenuated coherence with terrestrial phenomena, at the frequencies of 3.5, 5, 7, 11 and 22 years (Figs. 9, 10 (E)-(F)).
Nevertheless, the analysis must be extended with data of other solar indexes, (as for instance, radio in 10.3 cm and coronal holes) versus specific parameters of the hurricanes (vorticity, linear velocity, duration, energy, power destruction index, PDI, accumulated cyclone energy, ACE, and storm intensity). There is also the possibility that the periodicity of 30 years could be associated in a non-linear way to the solar Hale Cycle ((98)) with a certain phase shift.
Also, we would like to state that though we cannot say in a conclusive way, that CR modulates the AMO and SST, we must keep in mind that the AMO has intrinsic periodicities (at least since 1572) at 30, 60, 100 years ((121)) and the AMO is in turn a modulator of the SST ((109)). Because, the only other phenomena that we know that present such periodicities are SS and CR, we infer that such a modulation of AMO and SST may be related to one or both cosmophysical phenomena.
In order to estimate the relevance of cosmophysical influence on hurricane activity we have compared it with the African Dust outbreaks (sections 8-11) and found that the coherence, at similar frequencies, is in many cases higher with AD than with CR, though the influence of CR is not at all negligible. However, Cosmophysical influences cannot be disregarded, and may eventually become of the same order of importance of some terrestrial effects, thought it seems not to be the case for the dust cover originated in African dust outbreaks. Furthermore, the fact that two data series (cosmophysical and climatic) have similar periodicities does not necessarily imply that one is the cause and the other the effect: a physical mechanism must be behind, able to give an explanation of such coincident signals. Since such interconnections

are still in the stage for establishment of trustable evidences, it is perhaps still no time to look for physical mechanisms.
Finally, special mention must be done to the results of section 10, from where we can conclude that African dust influence on the genesis and evolution of Atlantic hurricanes varies in two main ways: (a) annually for tropical storms and hurricanes of all categories and (b) decadally for Category 5 hurricanes ((122)).
Category-5 hurricanes develop during the minimums of decadal cycles of African dust, when winds are lower than $8m/s$, and Ocean temperatures are around $26^{o}C$.
Recent studies ((122)) seems to indicate that category-5 Atlantic hurricanes are not random events, neither in time nor in space, and that in addition to peculiar climatological conditions, the geography of the Ocean bottom is an important factor in the development and evolution of Category 5 tropical cyclones, since at least in fout Atlantic dee-water regions, hurricane eyes have the lowest pressure. Every category 4 hurricane crossing in some segment of its trajectory throughout one or more of these four regions will evolve to a category 5 hurricane. Finally, it is worth mentioning that if climatologically tendencies continue as they have done in recent decades, according to our analysis [see panel (a) of 9] future hurricanes will not be able to develop into category 5 until the next decadal minimum, that will begin around 20152 years .

14. Acknowledgments

The authors would like to thank Prof. Prospero for the time series of African Dust data.
The authors wish to thank the Universidad Nacional Autónoma de México (DGAPA-UNAM) for its support under grants PAPIIT-IN119209 and IXTLI -IX100810

15. References

[1] Benestad, R.E. *Solar Activity and Earth's Climate*, Springer, Praxis Publish. Ltd., UK, 2006.
[2] Beer, J., Blimov, A., Bonani, G, Hofman, H.J. and Finkel, *Nature 347, 164, 1990.*
[3] Carlson, Toby N. and Prospero, Joseph M. *The Large-Scale Movement of Sahara Air Outbreaks over the Northern Equatorial Atlantic, J. Appl. Meteor., 11, 283-297. 1972.*
[4] Chernavskaya, M. M., Kononova, N. K., Val'Chuk, T. E., *Advances in Space Research, 37(8), 1640- 1645, 2006.*
[5] Chiapello, I., and C. Moulin, *Geophys. Res. Lett., 29(8), 1176, doi:10.1029/2001GL013767, 2002.*
[6] Chiapello, I., J. M. Prospero, J. Herman, and C. Hsu, *J. Geophys. Res., 104, 9277-929, 1999.*
[7] Chronis, T., Williams, E. and Petersen, W. *EOS, Transactions, AGU, 88(40), 397, 2007.*
[8] Cook, E. R., D. M. Meko and C. W. Stockton, *J. Climate 10, 1343, 1997.*
[9] DeMaria, M., Balk, J. J. and Kaplan, J. J. *Atmos. Sci. 50, 1133-1147 (1993).*
[10] Diaz, J. P. et al., J. Ceophys. *Res. 106, 18403 (2001).*
[11] Dentener, F.J., Carmichael, G.R., Zhang, Y., Lelieveld, J., Crutzen, P.J. *J. Geophys. Res. 101, 22. 1996.*
[12] Dickerson, R.R., Kondragunat, S., Stenchikov, G., Civerolo, K.L.,Doddridge, B.G., Holben, B.N. *Science 278, 827-830. 1997.*
[13] Dorman L.I. *Advances in Space Research 37 (8), 1621-1628, 2006.*
[14] Duce, R. A. *in Aerosol Forcing of Climate, Eds. R. J.Charlson, J. Heintzenberger, (Wiley, Chichester,UK,), pp. 43-72, 1995*
[15] Dunion, J.P. and Velden, C.S. *Bull. Amer. Meteor. Soc 85, 353-365 (2004).*

[16] Elsner, J. B., Kara, A. B., *Hurricanes of the North Atlantic Climate and Society, New York, Oxford University, 1999.*

[17] Elsner J. B. and S. P. Kavlakov, *Atmosph. Science Letters, 2. 86-93. 2001.*

[18] Emanuel, K, *Nature, 438(7071), E13, 2005a.*

[19] Emanuel, K., *Nature 436(7051), 686-688., 2005b.*

[20] Evan, A.T., Dunion, J., Foley, J.A., Heidinger, Velden, C.S. A.K. *Geophys.Res.Lett., 33, L19813, doi.10.1029/2006GL026408, 2006.*

[21] Falkowski, P. G, R. T. Barber, V. Smetacek, *Science 281, 200 (1998).*

[22] Farge, Marie, *Wavelets and turbulence in 'Geophysical Fluid Dynamics' , ed. Bernard Legras, publication CNRS-INSU, 25-32, 1992.*

[23] Fastrup, E. et al., *Los Alamos National Laboratory arXiv:physics/0104048 v1, 1-111, 2001.*

[24] Gierenes K. and Ponater M. *JASTP, 61(11), 795-797, 1999.*

[25] Ginoux, P. et al., *J. Geophys. Res. Atmos. 106, 20255, 2001.*

[26] Ginoux, P., J. M. Prospero, O. Torres, and M. Chin, *Long-term simulation of global dust distribution with the GOCART model: Correlation with North Atlantic Oscillation, Environ. Model. Software, 19(21),113-128, (2004).*

[27] Goldenberg, S.B. ,C.W. Landsea, A.M. Mestas-Nuez, and W.M. Gray, L.J. *Science 293, 474-479 (2001).*

[28] Gray, W.M , *Science 249, 1251-1256 (1990).*

[29] Gray,L.J., Haigh,J.D., Harrison,R.G., *The influence of solar changes on the Earth's climate, Hadley Centre technical note 62, Publisher: MET Office, Pages: 1- 81, 2005.*

[30] Grinsted, A., Moore J. and Jevrejera S., *Nonlinear Process, Geophys. 11, 561-566 (2004).*

[31] Haigh, J. *Science, 294:2109-2111, 2001.*

[32] Haigh, J. D., M. Lockwood, M. S. Giampapa, *The Sun, Solar Analogs and the Climate, Saas-Fee. Advanced Course 34, Swiss Society for Astrophysics and Astronomy, Springer-Verlag, Berlin, 2005.*

[33] Haywood, J., and Boucher, O. *Estimates of the direct and indirect radiative forcing due to tropospheric aerosols: A review, Rev. Geophys.,38, 513Ð543, 2000.Society for Astrophysics and Astronomy, Springer-Verlag, Berlin 2005.*

[34] Hodges, R.E. and Elsner,. J. B., Evidence linking solar variability with US hurricanes, International journal of Climatology, DOI: 10.1002/joc.2196, Article first published online: 14 JUL 2010.

[35] Hudgins, L. ,. Friehe, C and Mayer, M. E. *Phys. Rev. Lett. 71, 3279-3282 (1993).*

[36] Hurrell, J. W., *Science, 269, 676Ð 679, (1995).*

[37] Husar, R.B., Prospero, J.M. and Stowe, L. L. *J. Geophys.Res. 102, 16889 (1997).*

[38] IPCC, Intergovernmental Panel on Climate Change Climate Change 2001: *The Scientific Basis: Contribution of Working Group I to the Third Assessment Report of the Intergovernmental Panel on Climate,edited by J. T. Houghton et al., Cambridge Univ. Press, New York (2001).*

[39] Jarvinen, B. R., Neumann, C. J. and Davis, M. A. *A tropical cyclone data tape for the North Atlantic Basin, 1886-1983: Contents, limitations, and uses, NOAA Tech. Memo. NWS NHC 22, 21 pp., Coral Gables, Fla. (1984).*

[40] Jickells, T.D., et al, (Review paper) *Science, 30, 69-71, 2005.*

[41] Kanipe, J. *Nature 443,141-143, 2006.*

[42] Kaufman, Y. K. , D. Tanré, O. Boucher, *Nature 419, 215 (2002).*

[43] Kavlakov S., Intern., *J. Modern Physics., 20(29), 6699-6701, 2005.*

[44] Kavlakov, S., Elsner, J.B., Pérez-Peraza, *J. Proc. 30th ICRC Mérida, . Vol. 1 (SH), p. 693-696, 2008a.*

[45] Kavlakov, S., Elsner, J.B., *Proc. 30th ICRC Mérida, Vol. 1 (SH), p. 697-700, 2008b.*

[46] Kavlakov, S., Elsner, J.B. and Pérez-Peraza, *J., Geofisica Internacional 47, 207-213 2008c.*

[47] Kerry, E., *Physics Today 59-8, 74- 75, 2006*

[48] Khorozov, S. V., V. I. Budovy, I. M. Martin, V. A. Medvedev and V. S. Belogolov, *COSPAR, Beijing , TCI- 0232; C4.2-0056-06, 2006.*

[49] Knutson, T. R., and Tuleya, R. E.), *J. Clim.,17(18), 3477-3495, 2004.*

[50] Kristjansson J.E. et al. *GRL 29 .(23), 2017, doi:10.1029/2002GL015646, 2002.*

[51] Kristjansson, J. E. et al. *GRL. 29, 23, p. 2007.*

[52] Kudela K, M. Storini , M.Y.Hofer and A. Belov, *Space Sci. Rev. 93, 153-174, 2000.*

[53] Kudela, K. and M. Storini, *JASTP, 67, 907-912, 2005.*

[54] Lamb, P. J. and Peppier, R. A. *J. Clim. 5. 476 (1992).*

[55] Landsea, C.W. and Gray, W.M. *J. Climate 5, 1528-1534, (1992)*

[56] Landsea, C. W., Meteorology-Hurricanes and global warming, *Nature, 438(7071), E11-E13, 2005.*

[57] Landsea, C. W., R. A. Pielke, A. Mestas-Nunez, and J. A. Knaff, *Clim. Change, 42(1), 89-129, 1999.*

[58] Lau, W.K.M. and Kim, K-M, *EOS Trans. AGU, 88(9), 105-107,2007a.*

[59] Lau, W.K.M. and Kim, K-M, *EOS Trans. AGU, 88 (26), 271, 2007b.*

[60] Laut Peter, *JASTP, 65. 801-812, 2003.*

[61] Lelieveld, J. et al., *JScience 298, 794 (2002).*

[62] Liao, H. and Seinfeld, J. *J. Geophys. Res., 103, 31,637Ð 31,646. (1998).*

[63] Liao, H., Yung, Y.L., Seinfeld, J.H. *J. Geophys. Res. 104, 23. 1999.*

[64] McCracken, K.G., *JProc. of 27th ICRC, Hamburg, Germany, 4129-4132, 2001.*

[65] McCracken, K. G., and McDonald, F. B *27th ICRC 2001, Hamburg, Germany, 3753-3756 . 2001b*

[66] McCracken, K. G., Beer, J. and. McDonald, F. B *J Advances in Space Research 34(2), 397-406, 2004.*

[67] Marsh, Nigel D. and H. Svensmark, *Phys.Rev.Lett. 85, 23. 5004-5007, , 2000.*

[68] Martin, J.H., Gordon, R.M,. *Deep Sea Res. 35,177-196. 1988.*

[69] Martin, R.V., Jacob, D.J., Logan, J.A., Bey, I., Yantosca, R.M., Staudt, A.C., Li, Q., Fiore, A.M., Duncan, B.N., Liu, H., Ginoux, P., Thouret, V., *J. Geophys. Res. 107 (4351), doi:10.1029/2001JD001480, 2002.*

[70] Mason S. J and . Tyson, P. D, *J. Geophys. Res. Atmos. D5, 5847 (1992).*

[71] Mavromichalaki, H. et al. *Advances in Space Research, 37 (6), 1141-1147, 2006.*

[72] Mazzarella A., F. Palumbo, *Theoret. And Applied Climatol. 45 (3), 201-20, 1992.*

[73] Moulin, C., and I. Chiapello, *Geophys. Res. Lett., 31(2), L02107,doi:10.1029/2003GL018931,2004.*

[74] Neff, U. S., Burns, J., Mangini, A., Mudelsee, M., Fleitmann, D., *Matter A411 (6835): 290-293, 2001.*

[75] Ney, E P, *Nature 183 451, 1959a*

[76] Ney, E P, J. R. Winckler, and P. S. Freier, *Phys. Rev. Lett. 3, 183-185,1959b*

[77] Ney, E P, *CERN COURIER, February 24, 2010.*

[78] Ol, A. I., *Reports of Arctic and Antarctic Research Institute (Trudy Arkticheskogo i Antarkticheskog), 1969.*

[79] Pérez-Peraza J. *Space Conference of the Americas: Perspectives of Cooperation for the Development, PNUD, San José de Costa Rica, 1, 96-113, 1990.*

[80] Pérez-Peraza, J., S. Kavlakov, V. Velasco, A. Gallegos-Cruz, E. Azpra-Romero, O. Delgado-Delgado, and F. Villicana-Cruz Solar, *Advances in Space Research 42, 1601-1613, 2008a.*
[81] Pérez-Peraza, J ., V. Velasco, and S. Kavlakov, *Geofisica Internacional 47, 231- 244 2008b.*
[82] Pérez-Peraza, J. et al, *30th ICRC, Merida, Mexico, Vol. 1 (SH), p. 785-788 , 2008c*
[83] Polygiannakis, J., Preka-Papadema, P., andMoussas, X., *Mon. Not. R. Astron.Soc., 343, 725-734, 2003.*
[84] Prospero, J.M., Schmitt, R., Cuevas, E., Savoie, D.L., Graustein, W.C.,Turekian, K.K., Volz-Thomas, A., Diaz, A., Oltmans, S.J., Levy, II, H. *Geophys.Res. Lett. 22, 2925-2928, 1995.*
[85] Prospero J.M. and T.N. Carlson, *Pure Appl. Geophys 119,677- 691 (1981).*
[86] Prospero, J.M and R.T. Nees, *Nature 320, 735-738 (1986).*
[87] Prospero, J.M., *Journal of Geophysical Research 104(D13):15,917-15,927, 1999*
[88] Prospero, J.M. ,P. Ginoux, O. Torres, S. Nicholson, and T. Gill, Rev. *Geophys. 40, 1002 (2002).*
[89] Prospero J.M , and P. J. Lamb, *Science 302, 1024-1027 (2003).*
[90] Prospero, J. M., *Oceanography Vol. 19, No. 2, June 2006.*
[91] Prospero, J.M., Schmitt, R., Cuevas, E., Savoie, D.L., Graustein, W.C., Turekian, K.K., Volz-Thomas, A., Diaz, A., Oltmans, S.J., Levy II, H. *Geophys.Res. Lett. 22, 2925Ð2928. 1995.*
[92] Pudovkin, M. I. and A. A. Lyubchich, *Geomagnetism and Aeronomy 29(3), 359, 1989.*
[93] Pudovkin, M. I. and O. M. Raspopov, *Geomagnetism and Aeronomy 32(5), 593,1992.*
[94] Raisbeck, G. M. and F. Yiou. *Phys Rev Lett. 92, 199001, 2004.*
[95] Ram and Stolz (2009). *EOS, Trans AGU, RAN, Vol 90, 44, 3 Nov. 2009.)*
[96] Ran and Zhang, *Chinese Journal of Eco-Agriculture, DOI CNKI:SUN:ZGTN.0.2007-02-037, 2007.*
[97] Raspopov, O. M., O. I. Shumilov, E. A. Kasatkina, E. Turunen, M. Lindhol m and T. Kolstrom. *Geomagnetism and Aeronomy 41(3), 407, 2001.*
[98] Raspopov, O.M., Dergachev, V.A. and Kolstrom, T. *Solar Physics 224, 455, doi10.1007/s11207-005-5251-8, 2005.*
[99] Rodrigo F. S., Esteban-Parra , M. J., Pozo-Vazquez, D.,Castro-Diez, Y. *Int. Journal of Climatology, 20 (7), 721-732, 2000.*
[100] Roig, F. A., C. Le-Quesne, J. A. Boninsegna, K. R. Briffa, A. Lara, H. Grudd, Ph. Jones and C. Villagran, 2001. *Nature 410, 567. 2001.*
[101] Rosenfeld, D., Y. Rudich, and R. Lahav, *Desert dust suppressing precipitation: A possible desertification feedback loop, Proc. Natl. Acad. Sci. U.S.A., 98, 5975- 5980, (2001).*
[102] Sassen, K. ,. DeMott, P. J ,. Prospero. J. M ,. Poellot, M. R. *Geophys. Res. Lett. 30, 1633,10.1029/2003GL017371, (2003).*
[103] Shanahan, T.M. , J.T. Overpeck, K.J. Anchukaitis, J.W. Beck, J.E. Cole, D.L. Dettman, J.A. Peck, C.A. Scholz, and J.W. King *Science 324, 377-380 (2009).*
[104] Shindell, D.T., G.A. Schmidt, M.E. Mann, D.Rind, and A. Waple, *Science, 294, 2149-2152, 2001.*
[105] Sokolik, I. N., and O. B. Toon, *Nature, 381, 681-683, (1996).*
[106] Sokolik, l, N. et al., *J. Ceophys. Res. 106, 18015 (2001).*
[107] Stozhkov Y. I., V. P. Okhlopkov and N. S. Svirzhevsky. *Solar Phys. 224(1), 323, 2004.*
[108] Stozhkov, Y. I. *Bulletin of the Lebedev Physical Institute 34(5), 135, 2007.*
[109] Sutton, R.T. and Hudson, D.R.L. *Science 290, 2133, 2005.*

[110] Tanre, D., J. Haywood, J. Pelon, J. F. Leon, B. Chatenet, P. Formenti,P. Francis, P. Goloub, E. J. Highwood, and G. Myhre, *J. Geophys. Res., 108(D18), 8574,doi:10.1029/2002JD003273, (2003).*

[111] [Tegen & Lacis 1996] Tegen, I., Lacis, A .A. 1996 and fung, I. *Nature 380, 419-422, 1996.*

[112] Tegen, I., M. Werner, S. P. Harrison, and K. E. Kohfeld, *Geophys. Res. Lett., 31, L05105, doi:10.1029/2003GL019216, (2004).*

[113] Tinsley B.A. 1996. *J. Geophys. Res, 101, 29701-29714, 1996.*

[114] Tinsley B.A. and K.V. Beard. *Bull. Amer. Meteor. Soc., 78, 685-687, 1997.*

[115] Tinsley B.A. *Space Sci. Rev. Vol. 94, 231, 2000.*

[116] Torrence C. and Compo G., *Bull Am. Meteorol. Soc. 79, 61-78, 1998.*

[117] Torrence C. and Webster P., *J. Clim. 12, 2679-2690. 1999.*

[118] Trenberth, K., *Science, 308(5729), 1753-1754. (2005).*

[119] Usokin, Ilya G., Schussler, M., Solanki, S.and Mursula K. *J. Geophys. Res. 110, A101102, doi: 10.1029/2004JAA010946, 2005.*

[120] Usokin, I. G and Kovaltsov, G. *Review Paper, Comptes Rendus Geoscience, doi: 10.1016/ j.crte.2007.11.001, 2007.*

[121] Velasco Herrera, V.M. and Mendoza, B. *Advances in Space Research 42, 866878 (2008).*

[122] Velasco Herrera, V. M. Pérez-Peraza, J., Velasco, G. and Luna G. *African Dust Influence on Atlantic Hurricane Activity and the Peculiar Behavior of Category-5 Hurricanes. Los alamos Laboratoty, http://arxiv.org/abs/1003.4769.).*

[123] Wagner, G., J. Beer, C. Laj, C. Kissel, J. Masarik, R. Muscheler and H.A. Synal, *Chlorine-36 evidence for the Mono Lake event in the Summit GRIP ice core, Earth Planet. Sci. Lett., 181,1-6, 2000.*

[124] Weaver, C. J., Ginoux, P., Hsu, N. C., Chou, M. D., and Joiner, J., *J. Atmos. Sci., 59, 736-747,2002.*

[125] Webster, P. J., Holland, G. J, Curry, J. A. Chang, H.-R. *Science 309 (5742),1844-1846, 2005.*

[126] Webster, P.J., Cury, J.A., Liu, J., Holland, J. *Science 311, 1713c, 2006.*

[127] Ward, M. N. *J. Clim. 11, 3167 (1998).*

[128] White, W. B., M. D. Dettinger and D. R. Cayan. *The Solar Cycle and Terrestrial Climate, Proceedings of 1st Solar and Space Weather Euroconference, Santa Cruz de Tenerife, Spain, ESA-SP-463, 125, 2000.*

[129] Wilson N. Mulokwa and Mankin Mak, *Monthly Weather Review, American Meteorological Society108(10), 1533-1537, 1980.*

[130] Yiou, F., G. M. Raisbeck, S. Baumgartner, J. Beer, C. Hammer, S. Johnson, et al. *J. Geophys. Res. 102, 26783-26794, 1997.*

[131] Yoshimura, H. *Astrophys. J. 227, 1047, 1979.*

The Impact of Hurricanes on the Weather of Western Europe

Dr. Kieran Hickey
Department of Geography
National University of Ireland, Galway
Galway city
Rep. of Ireland

1. Introduction

Hurricanes form in the tropical zone of the Atlantic Ocean but their impact is not confined to this zone. Many hurricanes stray well away from the tropics and even a small number have an impact on the weather of Western Europe, mostly in the form of high wind and rainfall events. It must be noted that at this stage they are no longer true hurricanes as they do not have the high wind speeds and low barometric pressures associated with true hurricanes. Their effects on the weather of Western Europe has yet to be fully explored, as they form a very small component of the overall weather patterns and only occur very episodically with some years having several events and other years having none.
This chapter seeks to identify and analyse the impact of the tail-end of hurricanes on the weather of Western Europe since 1960. The chapter will explore the characteristics and pathways of the hurricanes that have affected Western Europe and will also examine the weather conditions they have produced and give some assessment of their impact. In this context 23 events have been identified of which 21 originated as hurricanes and two as tropical storms (NOAA, 2010).

Year	End Date	Name
1961	September 17	Hurricane Debbie
1966	September 6	Hurricane Faith
1978	September 17	Hurricane Flossie
1986	August 30	Hurricane Charley
1987	August 23	Hurricane Arlene
1983	September 14	Hurricane Floyd
1995	September 6	Hurricane Iris
1996	October 29	Hurricane Lili
1998	September 9	Hurricane Danielle
1998	September 27	Hurricane Ivan
1998	September 28	Hurricane Karl
1998	October 1	Hurricane Jeanne
1998	November 5	Hurricane Mitch

2003	April 24	Tropical Storm Ana
2004	August 9	Hurricane Alex
2004	October 4	Hurricane Lisa
2005	September 14	Hurricane Maria
2005	September 14	Hurricane Nate
2005	October 11	Hurricane Vince
2006	September 20	Hurricane Gordon
2006	September 24	Hurricane Helene
2009	August 26	Hurricane Bill
2009	October 6	Tropical Storm Grace

Table 1. List of Hurricanes and Tropical Storms which have affected NW Europe from 1960 onwards considered in this chapter.

2. Historical background

It has been long established that the tail-end of hurricanes and tropical storms made their way across the Atlantic and had an impact on Western Europe, but it is only from the 1860's onwards with the advent of modern standardised meteorological instruments and the development of national meteorological networks that proper tracking of events could take place. This initial plotting was pretty crude and was reliant on ships' observations from the major trans-Atlantic shipping lanes for the information at sea. As a result one of the first events to be plotted was that of Hurricane II of the 1848 season which formed in the mid Atlantic, skirted the eastern side of the Caribbean and then headed across the Atlantic towards Western Europe tracking just northwest of Ireland and Scotland (Chenoweth 2006). In addition one of the events in this survey Hurricane Vince has a remarkable almost identical event dating from 1842 when event IX which was a Category 2 hurricane eventually struck Spain and Portugal and travelled inland as far as Madrid (Vaquero et al. 2008). Numerous events have since being recorded as affecting Western Europe right up to the present day.

3. Seasonality

Hurricanes in the North Atlantic generally occur between late summer and late autumn especially the months of August, September and October with peak activity in September. However hurricanes have been recorded in the North Atlantic in virtually every single month of the year. In the case of the 20 hurricanes and 3 tropical storms identified as having affected W Europe 12 occurred in September not unsurprisingly, 5 in October although Hurricane Jeanne dissipates on October 1st so almost completely occurs in September, 4 in August and 1 each in November and April. In the latter case Tropical Storm Ana was the first tropical storm of 2003 and made it all the way across the Atlantic to NW Europe.

4. Tracks

The tracks taken by the hurricanes and tropical storms are also very variable but some commonalities can be picked out in terms of source area, track and where they reach W. Europe. In terms of source areas nine of the events originated off West Africa, eight developed in the mid-Atlantic and off the east coast of North America and the remaining six developed in the Caribbean from off the north coast of South America to west of Florida. However, these loose groupings do not work when their subsequent paths are examined and where they impact W. Europe.

Two main paths were taken by the nine events which originated off the West Africa, four travelled west/northwest to the mid-Atlantic before turning and travelling east/northeast under the influence of the Coriolis Force including Hurricane Debbie in 1961. Four travelled all the way across the Atlantic to North America before eventually turning eastwards and heading for Western Europe including Hurricane Faith in 1966 which was the largest hurricane by diameter ever recorded in the Atlantic at 973km in diameter. One event originating off West Africa headed straight east to Portugal and Spain but Hurricane Vince was one of the more unusual hurricanes in this survey and only lasted 4 days in duration.

The eight events that formed in an area stretching from the mid-Atlantic to just off the East coast of North America show much less clustering of tracks and contain nearly all of the most erratic tracks of all those surveyed. Four of the events have full loops in their tracks including Hurricane Arleen in 1987, Tropical Storm Ana in 2003, Hurricane Gordon in 2006 which does a most unusual large loop of the west and south coasts of Ireland and Tropical Storm Grace. Others initially track west, north or south before eventually turning and travelling in a generally east/northeast direction towards Western Europe. Only Hurricane Karl in 1998 tracks straight east without any other detours.

The remaining six events which originate in the Caribbean also generally trend northeast towards Western Europe, however four of the six most notably Hurricane Mitch in 1998 tracked west or northwest before eventually turning and heading eastwards across the Atlantic to Western Europe.

In terms of where they have impacts on the weather of Western Europe this is also very variable and stretches from Iceland, the Faroe Islands to Norway, Denmark and Russia, to Scotland, Ireland, Wales, England and to Belgium, France, Spain, Portugal (excluding the Azores), Italy and Cyprus with the effects on the weather rapidly diminishing inland on the continent of Europe.

Overall and not unsurprisingly given its position in the North Atlantic Ireland was the country most recorded as being affected by the hurricanes and tropical storms in this survey and its weather was identifiably affected by 15 of the 23 events, England was next with 12 followed by Scotland with 9 (Table 2). In all 15 countries of Europe were recorded as having their weather identifiably affected by these events.

Using the same tripartite system concerning the origins of the hurricanes and tropical storms the most affected receipt areas can be examined. Hurricanes and tropical storms which originate just west of Africa have the largest impact zone and are recorded as having affected the weather of 12 countries along Western Europe from Iceland in the north to Portugal and Spain in the south and even as far inland as Russia via the Baltic, however most weather effects are recorded in Ireland with 6 events, followed by England and Scotland with 4 each.

Hurricanes and tropical storms which originated in an area stretching from the mid-Atlantic to just off the East coast of North America have a slightly different profile affecting 11 countries from Iceland in the north to Portugal in the south but also penetrating into the European mainland particularly along the Mediterranean as far as Cyprus and including Italy. Ireland and England are still most affected with four events each followed by Scotland with three events.

Country	Incidence	Percentage
Ireland	15	65%
England	12	52%
Scotland	9	39%
Wales	7	30%
Norway	5	22%
Portugal	5	22%
Spain	4	17%
Denmark	2	9%
France	2	9%
Iceland	2	9%
Russia	2	9%
Belgium	1	4%
Cyprus	1	4%
Faroe Islands	1	4%
Italy	1	4%

Table 2. List of Countries Affected by the Hurricanes and Tropical Storms from this chapter.

The final grouping which is hurricanes and tropical storms which originate in the Caribbean affect only 6 countries and are confined to a much tighter geographical area stretching from Norway to France, however, again Ireland is the most affected country with 5 events followed by England with 4 and 2 each for Wales and Scotland.

5. Duration

The duration of the hurricanes and tropical storms in this survey is very variable ranging in length from 3 to 17 days, four events were 5 days or less including the shortest which was Tropical Storm Grace in 2009, six events lasted between 6 and 10 days, most events, 11 in total lasted between 11 and 15 days and this was the most common duration category. Only two events were 16 days or more in duration and these were Hurricane Lili in 1996 which lasted 16 days and Hurricane Faith in 1966 which lasted 17 days and was the longest of this survey.

6. Maximum strength

All categories of hurricane strength are recorded as having an effect on W. Europe but most are at the lower values of category 3 hurricanes or less. Only one event originated in a Category 5 hurricane and this was Hurricane Mitch in 2005, only one event was category 4 strength and this was Hurricane Bill in 2009, seven events were category 3, five events category 2, seven were category 1 hurricanes although Hurricane Lisa only achieved this status for a brief time and there was 2 tropical storms in this survey which failed to reach hurricane strength.

7. Barometric pressure

The lowest Western European barometric pressure reading for any of the events in the survey is 950 hPa which was recorded between Ireland and Scotland and came from Hurricane Debbie in 1961. Other low values include Hurricane Floyd in 1993 with 960 hPa and Hurricane Maria in 2005 with 962 hPa. Most of the events in this survey produced barometric readings in the 970 hPa plus category, similar to normal mid-Atlantic depressions, even the values in the 950's and 960's occasionally occur associated with these mid-latitude depressions (Rohan, 1986).

8. Wind speed

Clearly once of the main effects of the arrival of the tail-end of hurricanes and tropical storms is severe winds, both in terms of gusts and sustained winds and this is one of the principal agents of damage associated with these events. Typical maximum recorded wind speeds on land from Western Europe for the events in this survey vary from 70km/h gusts all the way up to nearly 200km/h gusts with values in excess of 100km/h not being uncommon. The highest recorded value of any of the events in this survey was from Hurricane Debbie in 1961. At Malin Head on the extreme NW tip of Ireland a gust of 182km/h was recorded. The next highest was at Fair Isle off Northern Scotland between the Orkneys and the Shetlands where Hurricane Flossie in 1978 produced a gust of 167km/h. Hurricane Lili in 1996 produced the next highest gust value of 148km/h and this was recorded at North Hessary Tor in Devon, England. More recently Hurricane Gordon in 2006 produced a gust of 130km/h at Truro in Cornwall, England and this was the next highest recorded value of the events in this survey. Five other events had maximum gust values in excess of 100km/h.

In terms of sustained winds the values are obviously lower than that of the gusts but equally important in terms of generating all sorts of damage from the coast moving inland. Unsurprisingly Hurricane Debbie in 1961 also generated the highest sustained values and from the work of MacClenahan et al. (2001) the detailed hourly wind values for Malin Head, NW Ireland can be outlined. Figure 1 shows the 10 minute mean values of sustained wind

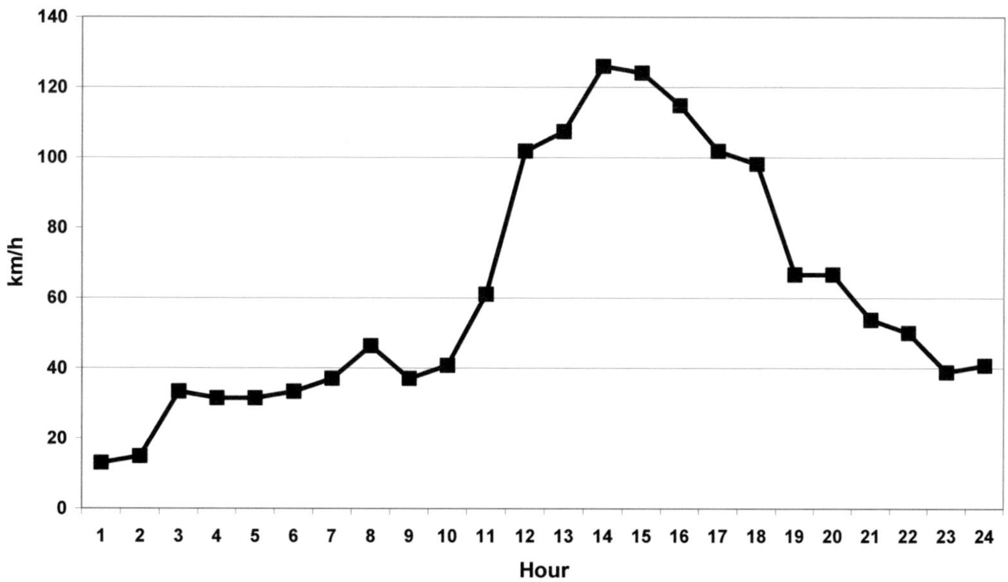

Fig. 1. Sustained Winds Per Hour from Hurricane Debbie on the 16th of September 1961 at Malin Head, NW Ireland (after MacClenahan et al., 2001).

speed as the hurricane travels offshore along the west coast of Ireland. Initially values at Malin Head were quite low and did not exceed 40km/h until 8am. At 9am at Valentia Observatory in the extreme SW of Ireland the hurricanes passing produced a maximum sustained wind speed of 107km/h but this was exceeded as the hurricane passed much closer to Malin Head later that day (Hickey, in press). From 10am onwards wind speeds rapidly rose at Malin Head and reached their maximum value at 2pm with a sustained value of 126km/h, an hour late the wind speed had barely dropped to 124km/jh, thereafter as the hurricane moved away the windspeeds started to decline but even at 4pm and 5pm the wind speed was above 100km per hour and at 6pmk the wind had just dropped below 100km/h, thereafter the wind speed dropped more rapidly and by midnight was hovering around 40km/h.

9. Precipitation

The tail ends of hurricanes and tropical storms bring significant pulses of precipitation when they approach or cross parts of Western Europe. Typical values in the events surveyed in this chapter were of the order of 25-75mm which would also be typical of large mid-latitude storms. These events also produce flooding on occasion which can vary from being localised to much more widespread.

However, much higher values of precipitation are also recorded, occasionally in excess of 100m and even in excess of 200mm and these can values can cause significant flooding.

Hurricane Charley in 1986 produced the highest precipitation values of this survey. In the mountains south of Dublin, Ireland in excess of 200mm was recorded over a 24 hour period, which set and new one day record for Irish rainfall and unsurprisingly caused significant flooding in the city. In places water heights reached 2.4m and over 450 buildings were affected by the flooding. When the remnants of Hurricane Charley moved on to Wales it also produced excessive rainfall but not quite so high as Dublin. At Aber in Gwynedd in Wales 135mm was recorded and this is the second highest vale of events in this survey. One other event of note produced precipitation in excess of 100mm and this was Hurricane Maria in 2005 where on the Isle of Skye in Western Scotland 131mm of rainfall was recorded.

10. Other meteorological effects

Given the nature of hurricanes and tropical storms and the inherent instability of the air masses other meteorological effects can also be generated including substantial thunderstorm activity, tornadoes and rises in temperature in the latter case due to the tropical origin of the air masses. Hurricanes Danielle in 1998 and Tropical Storm Lisa in 2004 both produced tornadoes in England and Hurricane Danielle also produced significant thunderstorm activity during one of them an aircraft was hit by lightning. Given that most events affect Western Europe in September and October they can be very important in ending the summer drought associated with countries such as Portugal, Spain and Italy and bring much needed rainfall and this would include Hurricane Arlene in 1987.

11. Impacts

All sorts of damage are recorded as a result of the impact of the weather associated with the hurricanes and tropical storms in this survey. Given the high winds and flooding fatalities and injuries are not uncommon but not on the scale of original hurricanes and tropical storms when they make landfall. Hurricane Mitch in 1998 which struck Honduras and Guatemala in particular killed over 11,000 people and caused damage of $6.2billion (at 1998 prices) caused no deaths or injuries when it hit Western Europe despite it being the second mostly deadly hurricane on record. The damage that hurricanes and tropical storms generate include losses of ships and other vessels, damage and destruction of coastal defences, coastal erosion, coastal and estuarine flooding caused by high waves and storm surges (Cooper and Orford 1998), river and other flooding caused by high rainfall, damage and destruction of infrastructure, electricity cables, buildings, crops and trees and others caused by high winds and flooding. Virtually any sort of damage can be caused by these events in much the same way that large mid-latitude storms can have the same effects.

12. Fatalities

Remarkably only 27 fatalities were recorded as directly resulting from the weather generated from the hurricanes and tropical storms which affected Western Europe. The 27 fatalities are associated with just five of the 23 events in this survey. The first of these five events is Hurricane Debbie which in September 1961 caused 11 fatalities in Ireland and injured atleast 50 people. This was partly due to the fact that Hurricane Debbie was still a hurricane when it made landfall on Achill Island, Co. Mayo in the West of Ireland and was the only known hurricane to do so. It then tracked over Co. Mayo and onto the rest of the mainland of Ireland where the fatalities occurred.

The weather emanating from the remnants of Hurricane Faith in September 1966 caused one offshore drowning in Denmark. The weather remnants of Hurricane Charley in August 1986 also caused 11 fatalities, six in Ireland and five in England, most of these fatalities were as a result of drowning in flooded rivers, the event being described by the UK Meteorological Office as being of a 1 in 600 year return interval. In October 1996 the weather associated with Hurricane Lili caused 2 fatalities in England. Sadly an elderly couple were killed when a large tree fell on their car. The final event that caused fatalities was the weather remnants of Hurricane Maria and Hurricane Nate in 2005 which combined together and caused the two fatalities to occur in Norway. The two people were killed as a result of being buried by mudslides caused by the torrential rain caused by these two hurricanes.

13. Damage

In terms of the monetary cost of the damage caused by the events in this survey there is very little information to go on. This is for a number of reasons, some of the events caused little or no actual damage or the damage was so minimal that it was not worthwhile compiling the costs if these were even known. Other events which caused damage have had no research carried out on their economic impact and as such the cost is unknown even though in some events it could be quite substantial. Only three events in this survey have some sort of economic costs compiled.

The first event is Hurricane Charley from 1986 and the only figure available is for Ireland but this figure is by no means complete for Ireland because the figure available of $8.65 million (1986 prices) was what the Irish Government allocated just for road and bridge repairs. The true figure would be much higher especially if other affected Western European countries were included.

A better assessment of the potential economic damage that these events can cause can be seen from the work done on Hurricane Lili where it is estimated that $300 million (1996 prices) in damages occurred all across the United Kingdom. The final figure that is available is from Norway and represents the payment of $3.1 million (2005 prices) in compensation as a result of the impact of Hurricane Maria and Hurricane Nate which had combined together at this point.

14. Conclusions

This chapter has shown that the tail-end of hurricanes and tropical storms affect Western Europe far more commonly than was previously known. Along with high winds, high precipitation events and flooding they can also cause significant numbers of fatalities and injuries and can cause extensive damage running to hundred of millions of dollars in the most extreme events, similar to the biggest and severest of the mid-latitude storms.

In the context of global warming and in particular the rise in sea surface temperatures in the tropics the likelihood of more of these events arriving in Western Europe and being more tropical when they arrive will increase as the century progresses (El Niño not withstanding). As a result more attention needs to be paid to these events and the frequency with which they affect Western Europe.

15. References

Bedford R. (2008) Yesterday's Weather, West Yorkshire, Self-Published.

Chenoweth M. (2006) A Reassessment of the Historical Atlantic Tropical Cyclone Activity: 1700-1855, Journal of Climatic Change, 76 (1-2), 169-240.

Cooper J.A.G. and Orford J.D. (1998) Hurricanes as Agents of Mesoscale Coastal Changes In Western Britain and Ireland, Journal of Coastal Research Special Publications 26, 71-77.

Eden P. (2008) Great British Weather Disasters, London, Continuum.

Hickey K.R. (in press) The Hourly Gale Record from Valentia Observatory, SW Ireland 1874-2008 and some Observations on Extreme Wave Heights in the NE Atlantic, Journal of Climatic Change.

MacClenahan., McKenna J., Cooper J.AG. and O'Kane B. (2001). Identification of highest magnitude coastal storm events over Western Ireland on the basis of wind speed and duration thresholds, International Journal of Climatology 21, 829-842.

National Oceanic and Atmospheric Administration (NOAA), USA (2010) Atlantic Hurricane Database, Retrieved 6th September 2010.

Rohan P.K. (1986) The Climate of Ireland, 2nd Edition, Dublin, The Stationery Office.

Vaquero J.M., Garcia-Herrera R., Wheeler D., Chenoweth M. and Mock C.J. (2008) A Historical Analog of 2005 Hurricane Vince, Bulletin of the American Meteorological Society 89 (2), 191-201.

4

The Effects of Space Weather on Hurricane Activity

Blanca Mendoza
Instituto de Geofísica
Universidad Nacional Autónoma de México
México

1. Introduction

The Sun is a variable star which emits electromagnetic radiation, mass in the form of transient and continuous phenomena (the solar wind) and energetic particles. The solar wind is plasma that drags the solar magnetic field throughout the interplanetary space beyond the Pluto orbit (which is at ~40 AU, being 1 AU the Sun-Earth mean distance), forming a region that is known as the heliosphere. Observations of the Voyager 1 spacecraft indicate that the heliospheric termination shock is at ~94 AU (e.g. Webber et al., 2009).
The bodies embedded in the heliosphere react to the impact of solar activity according to their characteristics, i.e. whether or not they have intrinsic magnetic fields, ionosphere or neutral atmosphere. In particular the Earth responds to solar variability through geomagnetic activity, variations of the high atmosphere, and possibly changes of weather, climate and biota.

1.1 The impact of solar activity on planetary bodies

The changes in solar activity modulate all the space phenomena that impinge on the planetary bodies, including a flux of charged high energy particles coming from outside the heliosphere, the galactic cosmic rays (CR). The interplanetary solar magnetic field (IMF) constitutes a barrier for the CR, which consist mostly of high energy protons (energies up to ~10^{11} GeV) produced by various stellar processes in our galaxy, mainly supernova remnants; although CR of energies ~ 10^{10}-10^{11} GeV are thought to be extragalactic. At the time scales of the 11-years solar cycle the CR show an approximate anticorrelation with the sunspot numbers (SN) (e.g. Caprioli et al., 2010).
The main large-scale solar phenomena that most affect the planetary bodies are the solar flares, the coronal mass ejections (CMEs) and the fast solar wind coming out of coronal holes. They impact our planet at different times scales: the short-time scales of a solar storm (hours to few days), or the solar cycles in the scale of years like the 11 or 22 years.
During a solar storm (see Figure 1), solar flares send electromagnetic radiation (X-rays, EUV, visible, radio bursts) that reach eight minutes after being produced the terrestrial atmosphere and affect it during 1 to 2 days. The combined effect of flares and CMEs produce high energy solar particles (~ 100 Mev) (also called solar cosmic rays or proton events) that arrive between 15 minutes and few hours after being produced and affect the

atmosphere for some hours to few days; some proton events have very high energies (few GeV) similar to those of the CR, they are the so-called ground level enhancements. The CMEs and solar wind produce low and medium energy particles (up to 100 keV) arriving after 2 to 4 days and affecting the atmosphere for some days. In Figure 1 we can appreciate the effect that these emissions cause on man-made technology.
Also the Forbush decreases (decreases of the CR fluxes along few hours) are produced by CMEs and high speed wind streams.

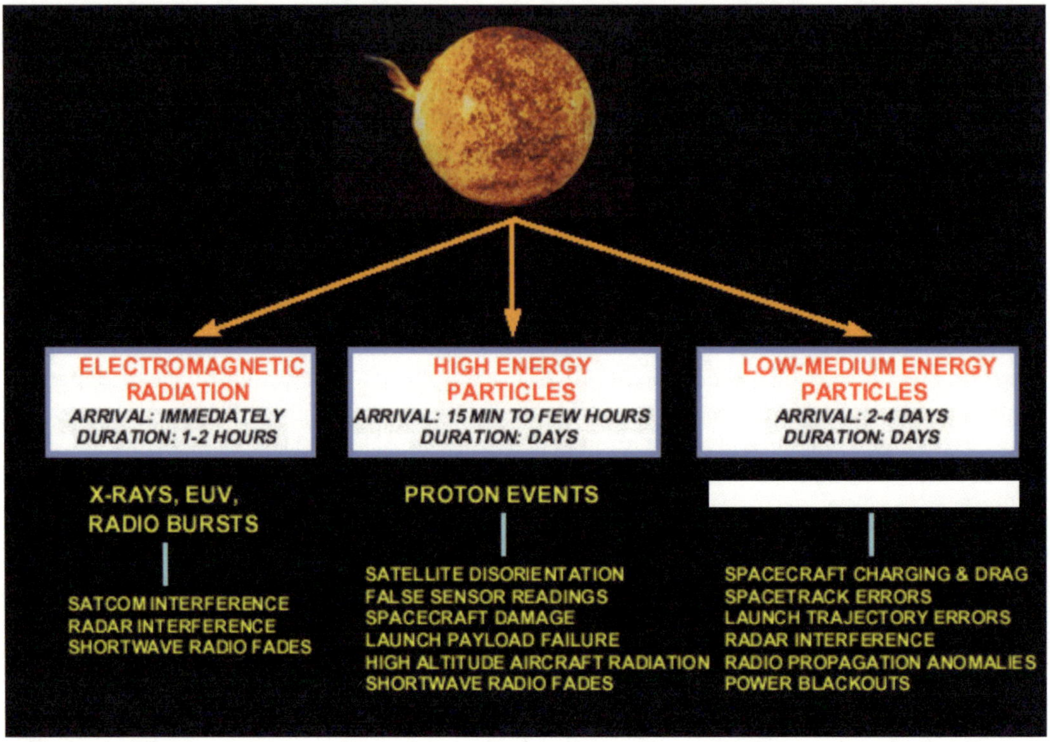

Fig. 1. A scheme relating the solar storm with their impact on the Earth. Radiation and particles (Schwenn, 2006).

1.2 Geomagnetic storms

The emergence of the magnetic field at the photosphere (the lowest part of the solar atmosphere) and its interaction with the solar atmosphere produce both the CME and the flare (see Figure 2). The high speed CMEs develop into interplanetary shocks. When they arrive at Earth, and if they have an IMF negative component ($-B_z$), produce a geomagnetic storm. Also high speed winds produce geomagnetic storm but they are in general less intense that those produced by the CMEs.
The term *geomagnetic storm* must not be confounded with the term *solar storm*. A geomagnetic storm is a prolonged perturbation of the geomagnetic field produced by the impact of the solar activity. They are observed in registers of the time variations of the magnetic field components, these registers are named magnetograms. Usually a geomagnetic storm is accompanied by auroral activity.

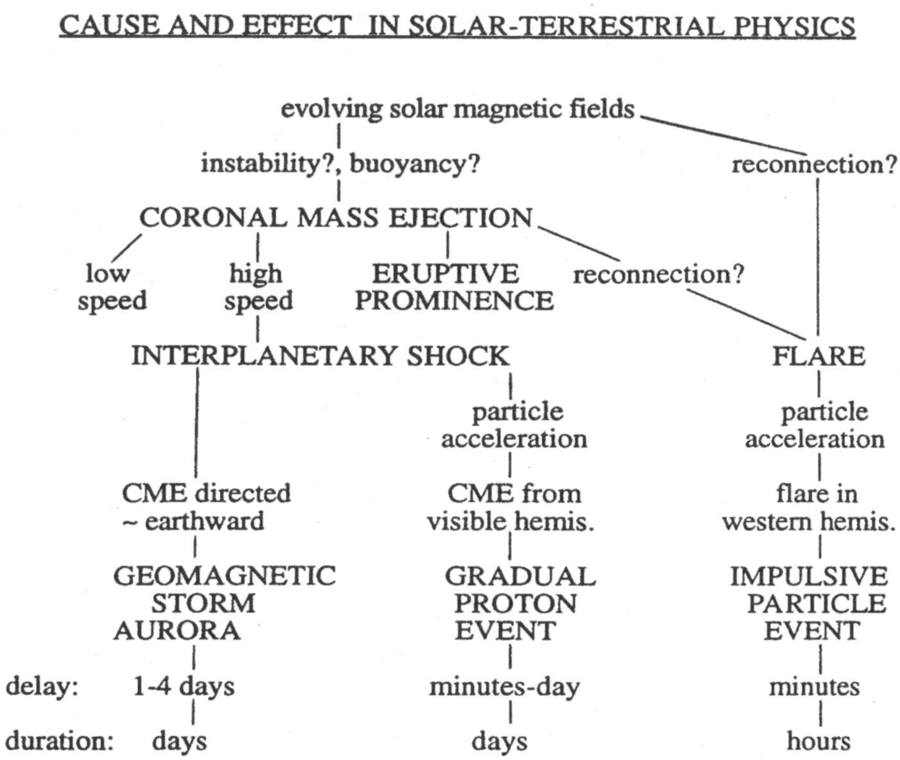

Fig. 2. A scheme relating the solar storm with their impact on the Earth. CMEs and flare (Schwenn, 2006).

A classical geomagnetic storm presents three parts (Tsurutani and Gonzalez, 1997) according to Figure 3.

The first part is a sudden commencement due to the arrival of the shock associated to a high speed CME or to a high speed wind stream. There is a compression of the geomagnetic field. It is observed at low latitude observatories as an increment in the horizontal (H) component of the magnetic field lasting for few hours and presenting variations of several nanoteslas. It is observed in all the planet. Some times there is only the sudden commencement but the rest of the variations in Figure 3 are not present, in this case we only have a sudden impulse. In other occasions, there is no sudden commencement but the other two phases are present.

The second part is the main phase. The entry of particles feeds an equatorial ring current that generates a field opposite to that of the planet. It requires the presence of a $-B_z$ component of the IMF. In this way there is a reconnection between the interplanetary and the geomagnetic fields that allows the entry of particles inside the magnetosphere. At low latitudes there is a decreasing of the field that reaches around -100 nanoteslas or less. Lasts several hours up to a day.

The third and final part is the recovery phase. Due to the loss of particles in the ring current, the field increases along few days until it reaches the value it had before the storm.

At the time scales of the 11-years solar cycle, the geomagnetic perturbations are produce by the interplanetary CMEs mainly during solar maximum and the corotating high-speed streams mainly during solar minimum (Tsurutani and Gonzalez, 1997).

Teoloyucan Observatory, México. Magnetic coordinates: co-latitude 70.254 , longitude 260.8E. Geographic coordinates: 99° 11' 36'' longitude W and 19° 44' 45'' latitude N

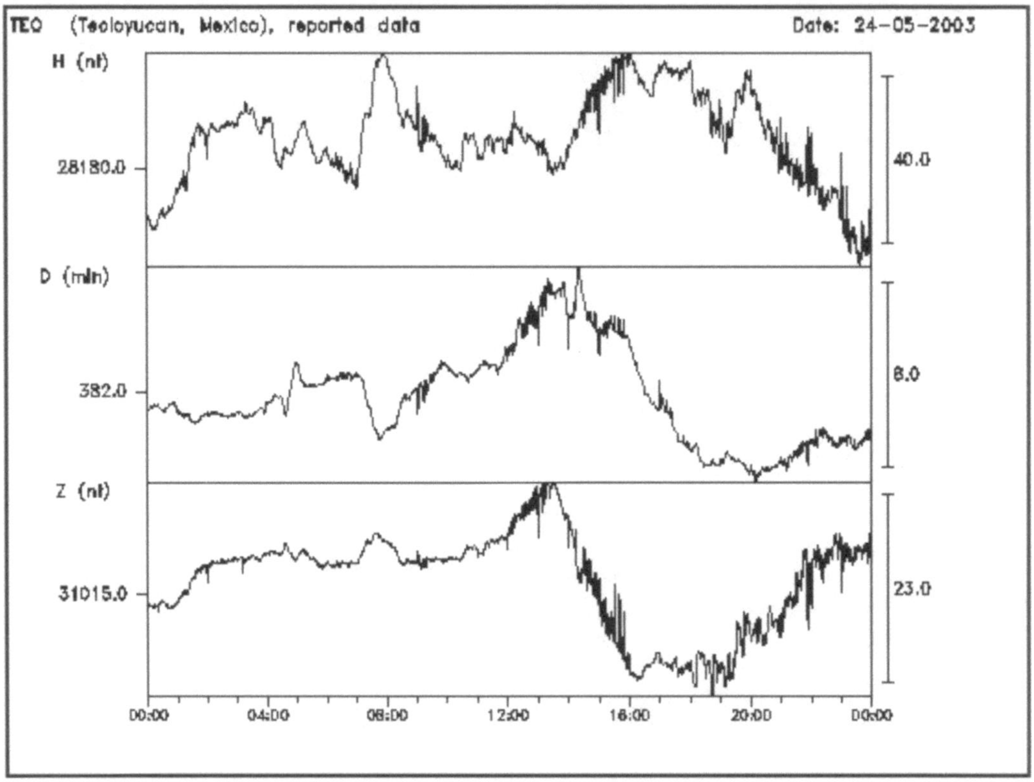

Fig. 3. Magnetogram showing the geomagnetic storm of the 24 of May, 2003 recorded in the Mexican station of Teoloyucan (http://geomaglinux.geofisica.unam.mx).

1.3 Geomagnetic indices

They measure the intensity of the geomagnetic activity. Three indices commonly used are the Kp, AE and Dst. The K index is obtained from the H component of the magnetic field, that usually it is the most disturbed; it is defined for intervals of 3hrs. The K_p or planetary index is probably the most used. It is constructed with the average K indices of 12 observatories located between the magnetic latitudes of 48° to 63°. The Dst index or ring current index is the most used to study the magnetic perturbations at low latitudes and measures the intensity of the ring current. The high latitude AE index measures the activity of the auroral electrojet (Campbell, 1997). In Figure 4 we present the distribution of some of the observatories contributing to the construction of the geomagnetic indices.

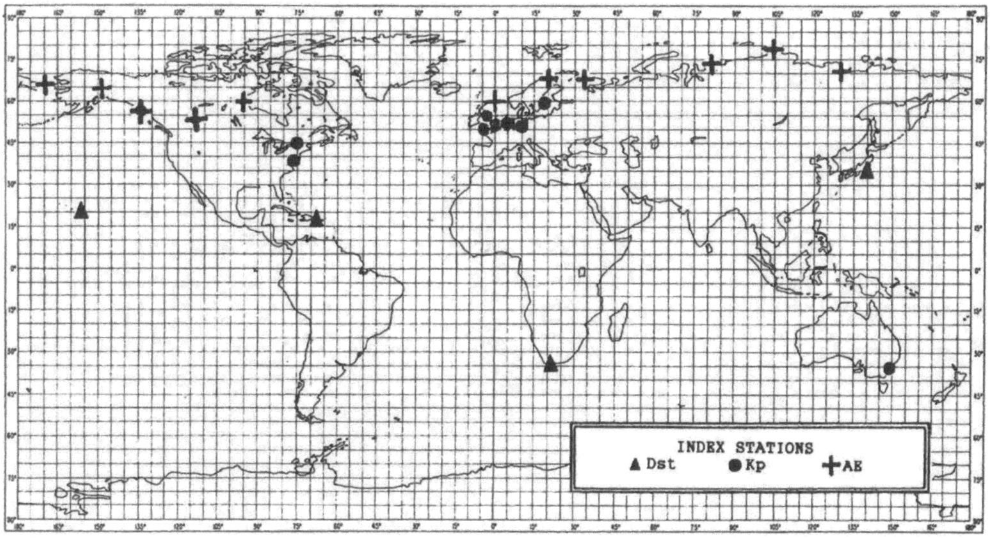

Fig. 4. Global map showing the location of some of the observatories contributing to the formation of geomagnetic indices (Campbell, 1997).

1.4 The concept of space weather

The effect of solar activity on the Earth´s environment not only affects the natural phenomena on Earth but also the man-made technology, impacting profoundly the globalized society that exists today. Now it is necessary to forecast the solar activity to quantify the geophysical reaction of the magnetosphere, the atmosphere and the incidence on technology. This is the science of Space Weather.

2. The impact of the space weather on climate

The relation between meteorological phenomena and space weather has been investigated in many reports for over two hundred years. Moreover, an average global warming of 0.6° ± 0.2°C has been measured along the 20th century (e. g. IPCC, 2007). It has been attributed preponderantly if not solely to the anthropogenic influence on climate. The current interest of the role of the Sun in climate change stems from the possibility that solar activity is also contributing to the observed temperature increase.

Two mechanism have been proposed for the space weather-climate relations: variations in the total and ultraviolet (UV) solar radiation, and the solar wind modulation of energetic particles.

2.1 Mechanisms relating solar radiation and climate

Changes in the total solar irradiance produce changes in the energy input into the Earth's atmosphere where weather and climate occur and the biosphere exists. The 1367 W/m^2 of solar irradiance arriving to the Earth's orbit is distributed over the planet, then the average solar radiation at the top of the atmosphere is 1/4 of this: ~ 342 W/m^2. As the Earth has a planetary albedo of 0.3, the incoming radiation is further reduced to 239 W/m^2. Also, upon

entering the atmosphere wavelengths shorter than 300 nm are absorbed in the stratosphere and above. It is at shorter wavelengths that the total solar irradiance varies the most. For the last two solar cycles, the portion of the total irradiance that actually arrives at the troposphere presents a change with the solar cycle of ~0.1% (see review by Lean, 2000) that seems too small to have an appreciable effect on surface climate, unless amplifying mechanisms exist.
The solar UV spectral emission range shows a strong change between solar maximum and minimum (between 1 and 100%) (see review by Lean, 2000). It has been proposed as forcing because the UV is absorbed by the stratospheric ozone raising the temperature there. The warming of the lower stratosphere produces stronger winds, and penetration of these winds into the troposphere alters the Hadley circulation modifying the equator-to-pole energy transport and the lower atmosphere temperature (Haigh, 1999; Shindell, et al., 1999). The models show the observed 11-year variation in the stratosphere but the amplitude of the simulated changes is still too small compared to the observations (e.g. Larkin et al., 2000). Even more, Foukal (2002) compared his reconstructed UV solar irradiance with global temperature along 1915-1999, finding a poor correlation of r = 0. 46. This result suggests that the interaction of UV irradiance and climate could be indirect and/or that an amplifying mechanism is at work.

2.2 Mechanisms relating the solar wind modulation of energetic particles and climate

2.2.1 The CR modulate directly the production of clouds

After finding a good correlation between cloud cover changes and CR along 1983-1994 and adopting the mechanism proposed by Dickinson (1975), Svensmark and Friis-Christensen (1997) suggested a direct CR modulation of clouds on time scales of decades and longer. Air ions produced by CR may act as sites for the nucleation of condensation nuclei, in turn these nuclei may grow into cloud condensation nuclei, this step is enhanced by the particles being charged.

2.2.2 Varying electric fields affect ice cloud processes

At temperatures between -40°C and 0°C liquid water cannot freeze spontaneously but requires a surface on which to start freezing. Only a small portion of atmospheric aerosols are suitable sites. In laboratory experiments it has been observed that large electric fields produce freezing of supercooled water droplets. The precipitation of CR, relativistic electrons and the ionospheric potential distribution in the polar cap, each of these affects the ionosphere-Earth current density and therefore the electric fields. Other work indicates that charged aerosol particles are more effective than neutral aerosol particles as ice nuclei. These processes are collectively known as electrofreezing. The latent heat released during freezing is available for modifying the weather systems and therefore ice clouds (Tinsley, 2000).
At the present time the question of whether CR modulate climate directly through cloud changes is not yet settled. The spectrum of opinions goes from the view that the CR are the main contributor to radiative forcing through clouds (e.g. Svensmark, 2007) or that CR can partially affect cloud formation (e.g. Voiculescu et al., 2006), to consider that they have a negligible effect on climate (e.g. Kristjánsson et al., 2008; Erlykin et al., 2009). However, it is also possible that the correlation between CR and clouds is due to the fact that CR fluxes are a proxy of another phenomenon which may influences climate, the total solar irradiance, that anticorrelates with CR (Lockwood, 2002). Pallé-Bagó and Butler (2000) found that the low cloud cover annual means present a slightly higher anticorrelation with SN compared

to the correlation with CR (although they did not comment on that). Kristjánsson et al. (2002) showed that the anticorrelation of the low cloud cover with the total solar irradiance and SN is higher and more consistent than the correlation with cosmic rays. Yet, due to the well known anticorrelation between CR and SN (e.g. Heber et al., 2006) it is difficult to tell which of the two mechanisms is at work, or what combination of the two.
Even more, it is still uncertain whether these mechanisms take place in the atmosphere and if their effect on clouds are measurable.

3. Examples of recent work relating space weather and hurricanes

In this section we review the related papers appearing in the literature in the last 5 years. This review is no intended to be exhaustive; its aim is just to give an idea of the type of work that is currently being developed. But first we present some basic information concerning hurricanes and several of the first attempts to relate space weather and hurricanes.

3.1 Hurricane main facts and data characteristics

Among the meteorological events, tropical cyclones have attracted much attention, due to the devastating effects that produce when they touch populated areas. Tropical cyclones are large-scale circular flows occurring within the tropics and subtropics (Anthes, 1982). Depending upon the magnitude of the maximum sustained winds, cyclones in the Atlantic and Northeastern Pacific oceans are called Tropical Storms (17-32 m/s) and Hurricanes (exceeding 32 m/s). For hurricanes only, there is the Saffir/Simpson scale classification of speed categories: 1 (33-42 m/s), 2 (43-49 m/s), 3 (50-58 m/s), 4 (59-69 m/s) and 5 (exceeding 69 m/s). Categories 1 and 2 are weak hurricanes and categories 3 to 5 are considered major or intense hurricanes.
Atlantic tropical cyclones occur in the regions that include the North Atlantic Ocean, the Caribbean Sea and the Gulf of Mexico. Hurricanes in the Atlantic Ocean are mainly due to easterly waves forming in West Africa, influenced by local convection and mesoscale systems. Another possible source of Atlantic tropical cyclogenesis is subtropical cyclones, which typically form at the equatorward extreme of a mid-latitude frontal zone (Guishard et al., 2007). Tropical cyclogenesis in the eastern Pacific is associated to moist easterly waves intruding from the Atlantic Ocean and the Inter Tropical Convergence Zone, providing low-level cyclonic vorticity maximum in weak vertical wind shear and deep convection, necessary elements for the development of tropical cyclones (Dickinson and Molinari, 2002; Zhender, 1991).
The dominant factors for hurricane formation are the magnitude of the local vertical wind shears (V_z) and the sea surface temperature (SST). The favorable conditions for the development of hurricanes are $V_z < \sim 8$ ms^{-1} and SST~ 27°C or more (e.g. Goldenberg et al., 2001).
Atlantic hurricane information is available since 1851, while Northeast Pacific hurricane data is available since 1949. The hurricane data can be obtained from the National Weather Service via the NOAAPORT satellite data service (http://weather.unisys.com/hurricane/). Data before 1944 is not very reliable. Without aircraft reconnaissance and satellite imagery beginning in 1944, some tropical cyclones were either not counted or were miss-assigned as to their correct intensity. It is after the mid 1960's that continuous satellite coverage started. However, reliable data coverage for hurricanes that have affected the United States is available since 1899 (e.g. Landsea, 1993).

3.2 First attempts to relate space weather and hurricanes

Since the 19 century, there have been numerous attempts to link hurricane numbers and tracks to space weather. Some of the most representative ones have been the following: In 1872 Meldrum found a correlation between Indian cyclones and the SN. In 1873 Poey found the same relation for Caribbean cyclones. However after 1910, as Figure 5 indicates, the correlation became an anticorrelation (see review by Hoyt and Schatten, 1997).

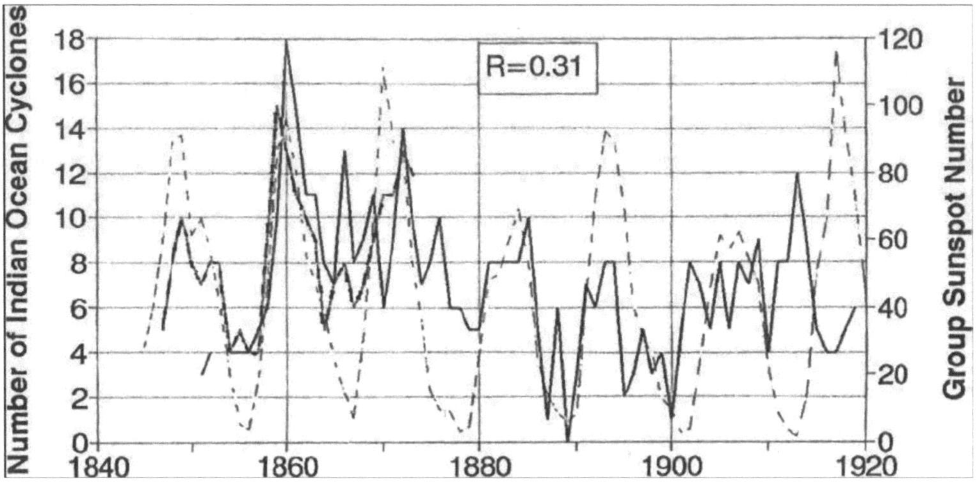

Fig. 5. The number of Indian Ocean cyclones obtained by Meldrum (continuous line with dashes) and Visher (updating Meldrum´s work in 1924, continuous line), and the Group Sunspot Number (dashed line) (Hoyt and Schatten, 1997).

Moreover, Wilcox et al., (1973) found that crossings of the IMF sector were strongly associated to decreases of the vorticity area index as shown in Figure 6. Cohen and Sweetser (1975) found confident periodicities of ~11, 22 and 52 years in the number of Atlantic hurricanes, that coincide with those of the solar cycles. Brown and John (1979) examined storm tracks over the North Atlantic and Europe and found an anticorrelation with the SN cycle.

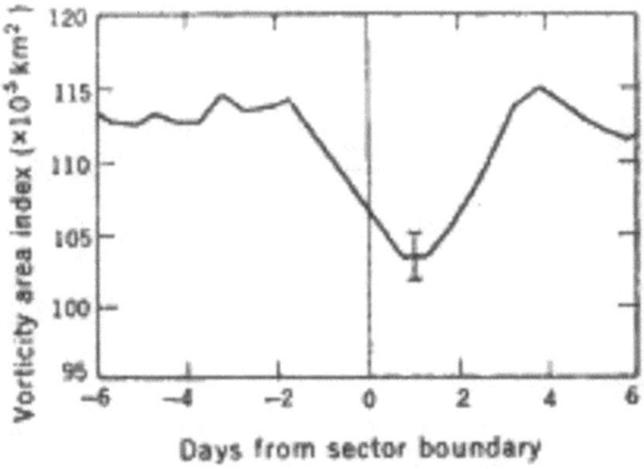

Fig. 6. The decrease of the vorticity area index after the crossing of the IMF sector (day 0) (Wilcox et al., 1973).

3.3 Examples of current work relating space weather and hurricanes

The SN, being the longest direct series of the solar activity has been used profusely in this type of associations. Recently, an inverse relationship between hurricane activity over the Caribbean and the Gulf of Mexico and the SN using a time series between 1866 and 2006, has been reported after considering also the North Atlantic Oscillation (NAO), the Southern Oscillation Index (SOI) and the SST (Elsner and Jagger, 2008). Extending this finding to changes in UV radiation, that presents a rough correlation with SN, a statistically significant anticorrelation was also identified, after accounting for annual SST variation and the El Nino cycle (Elsner et al, 2010). A previous work also suggested that peaks and trends of major north Atlantic hurricane activity concur with lower total solar irradiance, i.e. lower solar activity, and vice versa, in several period between 1730 and 2005 (Nyberg et al., 2007).

Some papers include besides the SN other solar activity-associated phenomena. For instance, correlations between hurricane occurrence, using the available Northeastern Pacific and Atlantic complete hurricane series, and the total solar irradiance, CR and the Dst index of geomagnetic activity have been obtained (Mendoza and Pazos, 2009). The results indicate that the highest significant correlations occur between the Atlantic and Pacific hurricanes and the Dst index (see Figure 7). Most importantly, both oceans present the highest hurricane-Dst relations during the ascending part of odd solar cycles and the descending phase of even solar cycles (Type 2). Conversely, for the ascending parts of even cycles and descending parts of odd cycles (Type 1) the relations are low. This shows the existence of a 22yrs cycle. The cause of such behaviour can be explained as follows: the aa index is ~ 20% higher during the descending phase of even cycles and ascending phase of odd cycles. These epochs present enhanced transients such as CMEs during the ascending phase of odd cycles, and enhanced high-speed corotating steams, i.e. stronger poloidal fields, during the descending part of even cycles (Cliver et al., 1996). Furthermore, Mendoza and Pazos (2009) found that the Atlantic hurricanes anticorrelate with the Dst index while Pacific hurricanes correlate for Type 2 epochs and they correlate for Type 1 epochs in both oceans. The authors suggest that this behaviour could be due to differences in cyclogenesis and to the presence of large scale climatic phenomena such as the NAO, SOI, etc.

Another work used data for all hurricanes born over the Atlantic and Pacific waters in the last 55 years that hit the Mexican borders (Pérez-Peraza et al., 2008a). The author considered as basic hurricane parameters the maximum rotational velocity and the estimated total energy. The behaviour of the CR intensity, the SN, and the geomagnetic indices Ap and Kp 35 days prior and 20 days after the cyclone start were investigated. The CR, SN, Ap and Kp showed much more intensive disturbances in the periods preceding and following the hurricane appearance.

A statistically positive correlation was found between geomagnetic activity and tropical cyclone intensification over the tropical Atlantic where major hurricanes form. The result is consistent with an earlier study showing a connection between geomagnetic activity and tropical cyclone intensity. It expands on this earlier work by focusing on intensification rather than intensity and by examining hourly data. No significant relationship was found with Forbush decreases (Kavlakov et al., 2008).

The correlation between variations in geomagnetic activity and tropical cyclogenesis during the complete solar activity cycle 23 was studied. The correlations were calculated for four cyclogenesis regions: the Atlantic, northeastern and central Pacific, northwestern Pacific, and water areas of oceans and seas in the Southern Hemisphere. Such coefficients changed

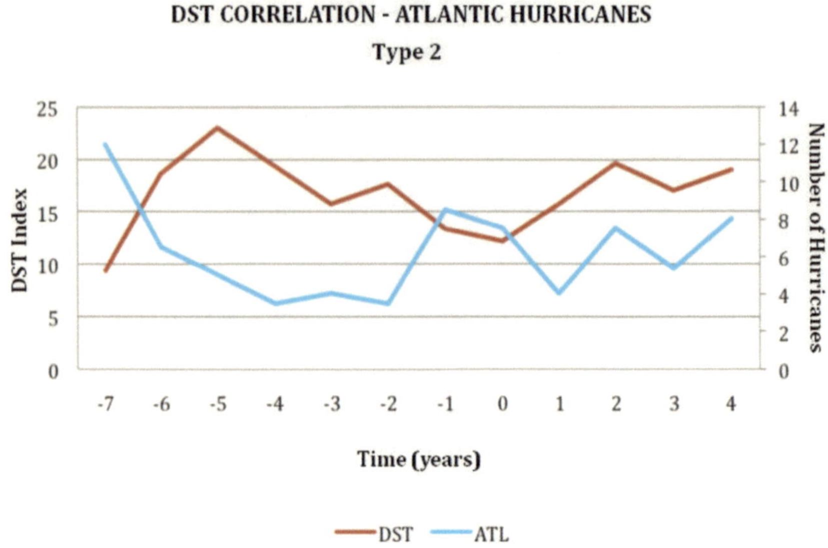

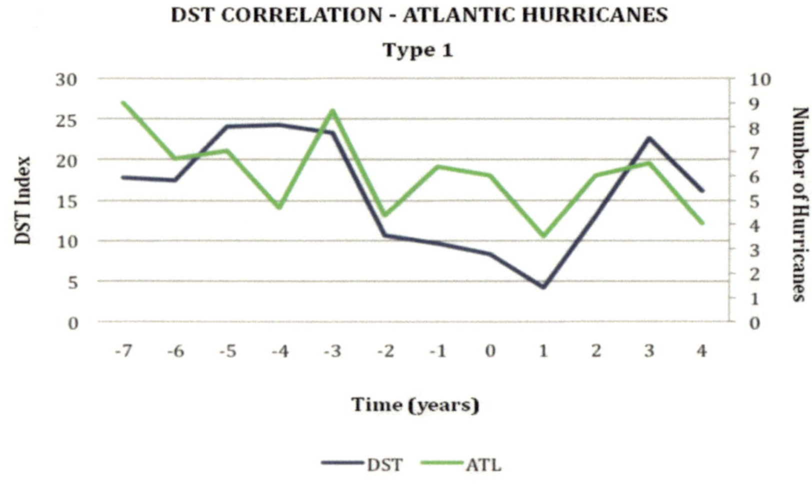

Fig. 7. Superpose epoch analysis for the Dst geomagnetic index and mayor Atlantic hurricanes (1851-2007). The figure shows the difference during Type 1 and Type 2 epochs of the solar activity cycle (Mendoza and Pazos, 2009).

in different regions from positive to negative values and some of them were significant: 0.55, 0, −0.50, and −0.50, respectively (Ivanov, 2007).

IMF crossing are again found to be associated with hurricanes: The generation of the Katrine hurricane was associated to the geomagnetic extrastorm of August 24, 2005, at a repeated crossings of the strongly disturbed IMF sector boundary (see Figure 8). Also few hours after the IMF crossing and before the generation of the hurricane, the flux of 1MeV protons increased. High values of Ap were also observed during the hurricane generation (Ivanov, 2006).

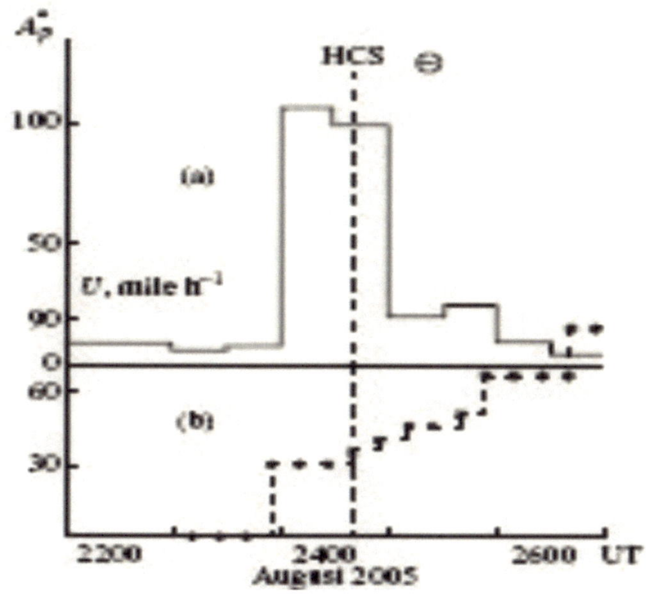

Fig. 8. (a) Variations in the average diurnal $A^*{}_p$ values of the a_p planetary index. (b) wind velocity variations during the generation of the hurricane Katrine: depression (dots), storm (crosses) and hurricane (circles). The dashed line (heliospheric current sheet, HCS) marks the instant of the last repeated crossing of the sector boundary and entry into the negative IMF sector (Ivanov, 2006).

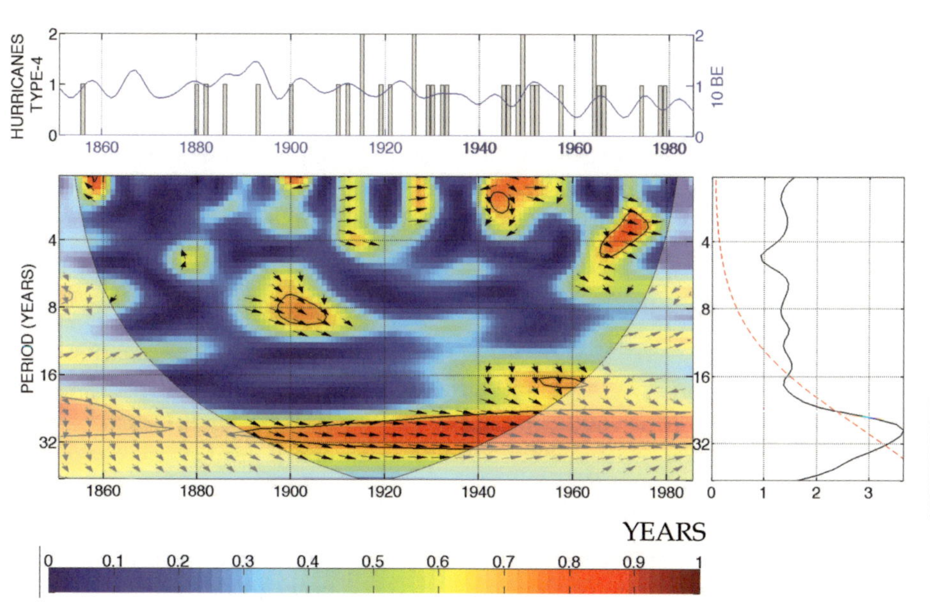

Fig. 9. Wavelet coherence between hurricanes magnitude 4 and CR. Upper panel: time series. Lower left panel: wavelet coherence spectra. Lower right panel: global coherence spectra, the dashed line is the red noise level; a 30 years peak is clearly above the noise. The color bar at the bottom indicates the significance level (Pérez-Peraza et al., 2008b).

Concerning periodicities, a common frequency of 30 years (see Figure 9) was found in the number of tropical storms landing in Mexico, in their average rotational wind velocity and in their total cyclone energy (Pérez-Peraza et al., 2008a). Also from a coherence wavelet study, a common frequency of 30 years was identified between phenomena presumably associated to hurricanes, such as the Atlantic Multidecadal Oscillation (AMO) and the SST versus CR, and on the other hand CR versus Atlantic hurricanes (Pérez-Peraza et al., 2008b). Also applying spectral analysis to the available Atlantic and Pacific hurricane time series, periodicities that coincide with the main SN and magnetic solar cycles were identified: ~11 and 22 years (Mendoza and Pazos, 2009). Another work indicates that Atlantic hurricane activity shows coincidences with several solar cycles (Kane, 2006).

4. Discussion and conclusions

Here we present the comparison of some results, the discussion concerning some proposed mechanism and a general conclusion about the review.

4.1 Comparing results

For the long-time series of the Atlantic hurricanes, SN and UV radiation anticorrelate. Also, other solar activity-related phenomena such as CR, SN, and various geomagnetic indices show large perturbations before and after Atlantic and Pacific hurricanes (Pérez-Peraza et al, 2008a). However, work by Mendoza and Pazos (2009) and Ivanov (2007) have pointed out that the correlations between Atlantic and Pacific hurricanes and solar activity-related phenomena are different. Moreover, Mendoza and Pazos (2009) have also shown that these relations change according to the phase of the solar cycle when divided in the ascending part of odd solar cycles and the descending phase of even solar cycle (Type 2), and the ascending parts of even cycles and descending parts of odd cycles (Type 1).

Mendoza and Pazos (2009) found that the Atlantic hurricanes anticorrelate with the Dst index while Pacific hurricanes correlate. While Ivanov (2007) found that Atlantic hurricanes correlate and northeastern and central Pacific hurricanes have no correlation. We point out that Ivanov (2007) worked only with the odd solar cycle 23, then according to the results of Mendoza and Pazos (2009) during the ascending phase of this cycle a good anticorrelation is expected and oppositely for the descending phase, then the results for the whole cycle cannot be compared with those of Mendoza and Pazos (2009). Moreover, Ivanov (2007) found that the hurricane Katrina, that occurred during the descending phase of odd cycle 23 (in 2005) was intensified after multiple crossings of the IMF sector and in the presence of high geomagnetic activity, as it should have been a low link between Katrina and the Ap index, it is likely that the IMF sector crossing was more important concerning the intensification of the hurricane.

Hurricanes show periodicities that coincide with the 11, 22 and 30 years solar cycles. The 11 years can be clearly associated to the evolution of the SN and other solar phenomena; the 30 years cycle could be a harmonic of the SN Gleissberg cycle (80-90 years). However the association to the 22 years is more complicated, as several possibilities exist. The 22yrs cycle has been identified in solar and geomagnetic activity. The polarity changes of the general solar magnetic field shows this cycle. In SN this periodicity appears in spectral analysis, it is attributed to the low-high alternation of even-odd sunspot maxima. This alternation is

ultimately linked to the solar magnetic field and its changing polarity via the interaction with a relic constant magnetic field in the convection zone (e.g. Mursula et al., 2001; Prestes et al., 2006; Demetrescu and Dobrica, 2007). In CR the 22yrs cycle is associated to the inversion of the Sun's general magnetic field around the maximum of activity (Kota and Jokipii, 1983). The aa geomagnetic index seems to present this periodicity (e.g. Cliver et al., 1996; Demetrescu and Dobrica, 2007), associated to a polarity reversal/geometrical coupling mechanism (Rosenberg and Coleman, 1969; Russell and McPherron, 1973) plus an intrinsic variation of the polar magnetic field of the Sun (Cliver et al., 1996). Climatic phenomena also present a 22 yrs cycle, for instance droughts (e.g. Cook et al., 1997; Mendoza et al., 2005; Mendoza et al., 2006), Russian tree ring width (Raspopov et al., 2004), temperature records (e.g. Dobrica et al., 2008), tree ring width from Brazil and Chile (Rigozo et al., 2007), or total annual rainfall series in southern Brazil (Souza-Echer et al., 2008). Moreover, large-scale climatic phenomena such as the NAO, the AMO, the Pacific Decadal Oscillation or the Southern Oscillation, show quasi-bidecadal periodicities (e.g. Velasco and Mendoza, 2008).
Comparing cyclonic activity at middle and subpolar latitudes, the North Atlantic cyclones also show a relation with solar activity: Long-period variations in the cyclonic activity along 1874–1995 indicate oscillations of the surface pressure with periods close to the main periods of solar activity (~80 and ~11 years) (Veretenenko et al., 2007).

4.2 Mechanisms

At present there are two main solar-related proposed mechanisms to account for the relation space weather-climate: The solar and UV radiation, and the solar-modulated energetic particles. For the specific case of the space weather-hurricanes these two proposals also apply:

4.2.1 The total and UV solar radiation

A hurricane´s maximum potential energy is inversely related to the temperature at the top of the convective clouds in the central core (Emanuel, 1991; Holland, 1997). An active Sun warms the lower stratosphere and upper troposphere through ozone absorption of additional UV radiation. A warming response in the upper troposphere to increased solar UV forcing decreases the atmosphere convective available potential energy leading to a weaker cyclone (Elsner and Jagger, 2008). General circulation models suggest a mechanism that impacts the baric atmospheric properties; the models show that changes in the stratosphere, induced by interactions between UV and ozone may penetrate down to the troposphere affecting winds and sea level pressure (Shindell et al., 2001); during solar maximum the increased solar radiation increases sea level pressure, which would result in weaker easterly winds and therefore weaker vertical wind shear promoting more hurricanes (Nyberg et al., 2007). In this context, the good relations found between hurricanes and SN would be due to the close relation between SN and solar radiation.

4.2.2 The solar wind-modulated energetic particles

a) There is a possible triggering mechanism for condensation and freezing within the convective clouds of the cyclone (see review by Tinsley, 2000): the ionization of the upper extent of the storm vortex leads to additional latent heat release and subsequent warming of

the core region of the cyclone. Central core-warming is associated with lowering of the surface pressure and thus with intensification of the cyclone (Kavlakov et al., 2008). This ionization can be due to the precipitation of particles coming from either solar and cosmic rays, or to the precipitation of energetic electrons or protons (of MeV energies) from the radiation belts towards the atmosphere; the latter precipitation is associated to the IMF sector crossings (i.e. Wilcox, 1979; Tinsley, 2000; Ivanov, 2006). b) Through changes in the atmospheric transparency impacting on the baric atmospheric properties, caused by variations of the stratospheric chemical composition produced by energetic protons (solar or galactic). These particles may ionize and dissociate nitrogen molecules and produce nitrogen compounds (NO) which in turn determines the concentration of NO_2. The NO_2 intensely absorbs solar radiation in the visible, while the NO exerts a destructive influence on the stratospheric ozone concentration, decreasing the UV absorption in that layer. Then during solar maximum times, more visible and less UV radiation reach the low atmosphere, the inverse happens at solar minimum. The changing amounts of radiation entering the low atmosphere should lead to temperature changes and therefore atmospheric pressure changes that promote cyclonic activity (Pudovkin and Raspopov, 1993).
As mentioned in Section 1, after a solar storm is produced, the energetic particles will arrive to the atmosphere in few minutes-hours. Few days afterwards, the arrival of the CME or the high speed solar wind will cause a Forbush decrease. Then decreases of pressure and then increases will be observed. These baric changes will again modulate the appearance and/or intensification of cyclones.

4.3 General conclusion

As a general conclusion we may say that hurricane activity studied trough the number of hurricanes, category, intensification, maximum rotational velocity, total energy, etc., presents a relation with space weather-associated phenomena.
The relations depend on the ocean basin and on the phase of the solar activity cycle. It is likely that various of the proposed mechanisms are at work.

5. Acknowledgements

This work was partially supported by DGAPA-UNAM-IN103209-3 and CONACYT-F282795 grants.

6. References

Anthes, R. A. Tropical cyclones: their evolution, structure and effects, Boston: American Meteorological Society, 1982.

Brown, G.M. and John, J.I. Solar cycle influences in tropospheric circulation. J. Atm. and Solar-Terr. Phys. 41, 43-52, 1979.

Campbell, W. H. Introduction to the geomagnetic fields. Cambridge Univ. Press, 286 pp, 1997.

Caprioli D., Amato E. and Blasi P. The contribution of supernova remnants to the galactic cosmic ray spectrum. Astroparticle Phys. 33, 160-168, 2010.

Cliver, E.W., Boriakoff,V. and Bounar, K.H. The 22-year cycle of geomagnetic and solar wind activity, J. of Geophys. Res., 101, 27091-27109, 1996.

Cohen, T.J. and Sweetser E.I. The "spectra" of the solar cycle and data for Atlantic tropical cyclones. Nature 256, 295-296, 1975.

Cook, E.R., Meko, D.M. and Stockton, C.W. A new assessment of possible solar and lunar forcing of bidecadal drought rhythm in the western United States, J. Climate, 10, 1343-1356, 1997.

Demetrescu, C., and Dobrica, U. Signature of the hale and Gleissberg solar cycles in the geomagnetic activity, J. Geophys. Res., 113, doi: 10.1029/2007JA012570, 2007.

Dickinson,R.E. Solar variability and the lower atmosphere, Bull. Am. Meteorol. Soc. 56, 1240-1248, 1975.

Dickinson, M. and Molinari, J. Mixed rossby-gravity waves and western Pacific tropical cyclogenesis Part I: Synoptic Evolution, J. Atmos. Sci.,59, 2183-2196, 2002.

Dobrica, U., Demetrescu, C., Boroneant, C. and Maris, G. Solar and geomagnetic activity effects on climate at regional and global scales: case study- Romania, J. Atm. and Solar-Terr. Phys., doi: 10.1016/j.jastp.2008.03.022, 2008.

Elsner, J.B. and Jagger, T.H. United States and Caribbean tropical cyclone activity related to the solar cyle. G. Res. Lett. 35, L18075, 2008.

Elsner, J.B., Jagger, T.H. and Hodges R.E. Daily tropical cyclone intensity response to solar ultraviolet radiation. G. Res. Lett. 37, L09701, 2010.

Emanuel, K.A. The theory of hurricanes, Ann. Rev. Fluid Mech. 23, 179-196, 1991.

Erlykin, A.D., Gyalai,G., Kudela,K., Sloan, T. and Wolfendale, A.W. On the correlation between cosmic ray intensity and cloud cover. J. Atm. Solar-Terr. Phys., doi.org/10.1016/j.jastp.2009.06.012, 2009.

Foukal, P. A. Comparison of variable solar total and ultraviolet irradaince outputs in the 20th century. Geophys. Res. Lett. 29, 2089-2092, 2002.

Goldenberg, S.B., Landsea, C.W., Mestas-Nuñez, A.M. and Gray, W.M. The recent increase in Atlantic hurricane activity: causes and implications, Science, 293, 474-479, 2001.

Guishard, M. P., Nelson, E.A., Evans,J.L., Hart, R.E. and O'Connell, D.G. Bermuda subtropical storms, Meteorology and Atmospheric Physics, 97, 239-253, 2007.

Haigh, J.D. A GCM study of climate change in response to the 11-year solar cycle. Q.J.R. Meteorol. Soc. 125, 871-892, 1999.

Heber B., Fichtner H., and Scherer, K. Solar and heliospheric modulation of galactic cosmic rays. Space Sci. Rev 125:81-93, 2006.

Holland, G.J. The maximum potential intensity of tropical cyclones. J. Atm. Sci. 54, 2519 2541, 1997.

Hoyt, D.V., Schatten, K.H. The role of the sun in climate change, Oxford Univ. Press, 1997.

IPCC (2007). Contribution of Working Group 1 to the Fourth Assessment Report. The Physical Science Basis.

Ivanov, K.G. Generation of the Katrine hurricane during the geomagnetic extrastorm at crossing of the heliospheric current sheet: Is it an accidental coincidence or physical essence? Geom. and Aeron. 46, 609-615, 2006.

Ivanov, K.G. Correlation between tropical cyclones and magnetic storms during cycle 23 of solar activity. J. Geomag. and Aeron. 47, 394–398, 2007.

Kane, R.P. Spectral characteristics of Atlantic seasonal storm frequency, J. of India Meteor. Dept. (MAUSAM), 57, 597-608, 2006.

Kavlakov, S., Pérez-Peraza, J. and Elsner, J.B. A statistical link between tropical cyclone intensification and major geomagnetic disturbances, Geofisica Internacional, 47, 207-213, 2008.

Kota, J. and Jokipii, J.R. Effects of drifts on the transport of cosmic rays, VI. A three dimensional model including diffusion, Astrophys. J. 265, 573-581, 1983.

Kristjánsson, J.E., Staple, A. and Kristiansen, J. A new look at possible connection between solar activity, clouds and climate, Geophys. Res. Lett. 29, 2107- 2110, 2002.

Kristjánsson, J.E., Stjern, C.W., Stordal,F., Fjaeraa, A.M. and Jónasson, K. Cosmic rays, cloud condensation nuclei and clouds-a reassessment using MODID data, Atm. Chemistry and Physics 8, 7373-7387, 2008.

Landsea, C.W. A climatology of intense (or major) Atlantic hurricanes, Month. Weath. Rev. 121 1703-1713, 1993.

Larkin, A., Haigh, J.D., Djavidnia, S. The effect of solar UV irradiance variations on the Earth´s atmosphere. Space Sci. Rev. 94, 199-214, 2000.

Lean, J. Short-term direct indices of solar variability. Space Sci. Rev. 94, 39-51, 2000.

Lockwood, M. An evaluation of the correlation between open solar flux and total solar irradiance, Astron. Astrophys. 382, 678-687, 2002.

Mendoza, B., Jáuregui, E., Díaz-Sandoval, R., García-Acosta,V., Velasco,V. and Cordero, G. Historical droughts in central Mexico and their relation with El Niño, J. Appl. Met., 44, 709-716, 2005.

Mendoza, B., Velasco, V. and Jáuregui, E. A study of historical droughts in south eastern Mexico, J. Climate, 19, 2916-2934, 2006.

Mendoza, B. and Pazos, M. A 22-yrs Hurricane Cycle and its Relation to Geomagnetic Activity. J. Atm. Solar-Terr. Phys., 2009.

Mursula, K., Usoskin, I.G. and Kovaltsov, G.A. Persistent 22-year cycle in sunspot activity: Evidence for a relic solar magnetic field, Solar Phys., 198, 51-56, 2001.

Nyberg, J., Malmgren, B.A., Winter, A., Jury, M.R., Halimeda Kilbourne, K. and Quinn, T.M. Low Atlantic hurricane activity in the 1970s and 1980s compared to the past 270 years, Nature, 447, 698-702, 2007.

Pallé-Bagó, E., Butler, C.J. The influence of cosmic rays on terrestrial clouds an global warming. Astron. Geophys. 41, 418-422, 2000.

Pérez-Peraza, J., Kavlakov, S., Velasco, V., Gallegos-Cruz, A., Azpra-Romero, E., Delgado-Delgado, O. and Villicaña-Cruz, F. Solar, geomagnetic and cosmic ray intensity changes, preceding the cyclone appearences around Mexico, Adv. in Space Res., 42, 1601-1613, 2008a.

Pérez-Peraza, J., Velasco, V. and S. Kavlakov. Wavelet coherence analysis of atlantic hurricanes and cosmic rays. Geofísica Internacional, 47, 231-244, 2008b.

Prestes, A., Rigozo, N.R., Echer, E., Vieira, L.E.A. Spectral analysis of sunspot number and geomagnetic indices (1868-2001). J. Atm. Solar-Terr. Phys., 68, 182-190, 2006.

Pudovkin, M.I and Raspopov, O.M. The mechanism of action of solar activity on the state of the lower atmosphere and meteorological parameters (A review). J. Geomag. and Aeron. 32, 593-608, 1993.

Raspopov, O.M., Dergachev, D. A. and Kolström, T. Hale cyclicity of solar activity and its relation to climate variability Sol. Phys. 224, 455-463, 2004.

Rigozo, N. R., Nordemann, D.J., Echer, E., da Silva, H.E., de Souza-Echer, M.P. and Prestes, A. Solar and climate imprint differences in tree ring width from Brazil and Chile, J. Atm. Solar-Terr. Phys., 69, 449-458, 2007.

Rosenberg, R.L. and Coleman, P.J. Heliographic latitude dependence of the dominant polarity of the interplanetary magnetic field, J. Geophys. Res., 74, 5611-5622, 1969.

Russell, C.T. and McPherron, R.L. Semiannual variation of geomagnetic activity, J. Geophys. Res., 78, 92-108, 1973.

Schwenn, R. Space weather: The solar perspective. Living Reviews in Solar Physics, 2006. http://www.livingreviews.org/lrsp-2006-2.

Shindell, D., Rind, D., Balachandran, N., Lean, J., Lonergan, P. Solar cycle variability, ozone and climate. Science 284, 305-308, 1999.

Shindell, D.T., Schmidt, G.A., Mann, M.E., Rind, D., Waple, A. Solar forcing of regional climate change during the Maunder minimum. Science 294, 2149-2152, 2001.

Souza-Echer, M.P., Echer, E., Nordemann, D.J., Rigozo, N.R. and Prestes, A. Wavelet analysis of a centennial (1895-1994) southern Brazil rainfall series (Pelotas, 31°46'19"S 52°20' 33"W), Clim. Change 87, 489-497, 2008.

Svensmark, H. Cosmoclimatology: a new theory emerges, Astron. and Geophys. 48, 18-24, 2007.

Svensmark, H., Friis-Christensen, E. Variations of cosmic ray flux and global cloud coverage-a missing link in solar-climate relationships. J. Atm. Solar-Terr. Phys. 59, 1225-1232, 1997.

Tinsley, B.A. Influence of solar wind on the global electric circuit, and inferred effects on cloud microphysics, temperature and dynamics in the troposphere. Space Sci. Rev. 94, 231-258, 2000.

Tsurutani, B.T. and Gonzalez, W.D. The interplanetary causes of magnetic storms: A review, in Magnetic Storms, Eds. B.T. Tsurutani, W.D. Gonzalez, Y.

Kamide, and J.K. Arballo, American Geophysical Union, Geophysical Monograph, 98, 77-89, 1997.

Velasco, V. and Mendoza, B. Assessing the relationship between solar activity and some large scale climatic phenomena, Adv. in Space Res., 42, 866-878, 2008.

Veretenenko, S.V., Dergachev, V.A. and Dmitriyev, P.V. Solar activity and cosmic ray variations as a factor of intensity of cyclonic processes at midlatitudes. J. Geomag. and Aeron. 47, 399-406, 2007.

Voiculescu, M., Usoskin,I.G. and Mursula, K. Different response of clouds to solar input, Geophys. Res. Lett. 33, L21802, 2006.

Webber, W.R., Cummings, A.C., McDonald, F.B., Stone, E.C., Heikkila. B. and Lal, N. Transient intensity changes of cosmic rays beyond the heliospheric termination shock as observed at Voyager 1. J. Geophys. Res. A: Space Physics 114, art. no. A07108, 2009.

Wilcox, J.M., Scherrer, P.H., Svalgaard, L, Roberts, W. O. and Olson R.H. Solar magnetic sector structure: relation to circulation of the earth´s atmosphere. Science, 180, 185-186, 1973.

Wilcox, J.M. Tropospheric circulation and interplanetary magnetic sector boundaries followed by MeV proton steams. Nature 278, 840-841, 1979.

Zehnder, J. A. The interaction of planetary-scale tropical easterly waves with topography: A mechanism for the initiation of tropical cyclones, J. Atmos. Sci., 48, 1217-1230, 1991.

5

Hurricane Intensity, Sea Surface Temperature, and Stochastic Variation

Kenneth T. Bogen[1], Larry E. Fischer[2], and Edwin D. Jones[2]
[1]*Exponent, Inc., Oakland, CA 94607*
[2]*Lawrence Livermore National Laboratory, Livermore, CA 94550*
USA

1. Introduction

"Hurricanes" are North Atlantic Ocean (NAO) cyclones that attain a maximum sustained surface-wind speed, v_{max}, of ≥119 km/h (Bell et al., 2000), which corresponds to Saffir-Simpson storm-intensity categories ≥1 (NOAA, 2006). A key climate-change concern involves the question of whether an increasing trend in sea surface temperature (SST)—considered a signal of global warming by being induced or enhanced by increased atmospheric greenhouse-gas concentrations—portends increased tropical cyclone intensities. Historical data, physical theories, and modeling all appear to link increasing SST to increased cyclone activity, intensity, or potential destructiveness in the NAO and other ocean basins (Saunders & Harris, 1997; Emanuel 2000, 2004; Goldenberg et al., 2001; Knutson & Tuleya, 2004; Emanuel, 2005; Trenberth, 2005; Webster et al., 2005; Hoyos et al., 2006; Mann & Emanuel, 2006; Santer et al., 2006; Trenberth & Shea, 2006). Correlations between cyclonic storm intensity, in particular, and increased SST beginning around 1950 to 1970 (Emanuel, 2000, 2004; Goldenberg et al., 2001; Knutson & Tuleya 2004; Emanuel, 2005; Trenberth, 2005; Webster et al., 2005; Hoyos et al., 2006; Mann & Emanuel, 2006; Santer et al., 2006; Trenberth & Shea, 2006) are claimed to support the "SST hypothesis" that local SST directly affects cyclonic intensity, just as predicted by models of storm-related heat-transfer dynamics (Emanuel, 2000, 2004; Knutson & Tuleya, 2004). For example, tropical cyclone frequency, duration, and intensity over the past 35 years showed an increased number and proportion of hurricanes reaching Saffir-Simpson categories 4 or 5 (most notably in the North Pacific, Indian, and Southwest Pacific Oceans; less so in the NAO), but a decreased number of cyclones and cyclone days in all ocean basins except the North Atlantic during the past decade (Webster et al., 2005). This conclusion is consistent with those from studies that focused instead on recent trends toward increased hurricane destructive potential, largely reflecting hurricanes of category 4 or 5 (Emanuel, 2005; Trenberth, 2005).

Other studies have questioned a direct, causal link between SST and hurricane intensity, and support an alternative, "extreme-value hypothesis" that recent strong hurricanes like Katrina are simply extreme samples from an essentially stationary intensity distribution that has not increased appreciably over at least the last half century (Landsea et al., 1996; Michaels et al., 2005; Landsea et al., 2006; Bogen et al., 2007). Among factors that are thought to influence hurricane activity—such as multidecadal oscillations in oceanic thermohaline circulation, upper tropospheric high-pressure regions, dips in tropospheric vertical wind shear that inhibit vortex formation, atmospheric stability, and equatorial shear winds that favor hurricane development—only some are clearly correlated with SST (Elsner et al., 2000; Goldenberg et al., 2001; Elsner, 2003; Chu, 2004; Trenberth, 2005; Gray, 2006).

Studies supporting the SST hypothesis (e.g., Goldenberg et al., 2001; Emanuel, 2005; Trenberth, 2005; Webster et al., 2005; Hoyos et al., 2006; Mann & Emanuel, 2006; Santer et al., 2006; Trenberth & Shea, 2006) share a focus on annual or multi-decadal mean SST estimates averaged over large ocean areas, such as the NAO Main Development Region (MDR), also referred to as the "Atlantic cyclogenic region," in which most hurricanes develop (here assumed to denote the region within 6° to 18° N latitude by 20° to 60° W longitude). Relatively elevated MDR SST during 1995 was hypothesized to increase the frequency of hurricane formation by transferring a relatively greater amount of heat to the usual number of easterly waves of wind (typically ~60) that enter the MDR each year between May and November after leaving the west coast of Africa, each acting to stimulate shallow cumulus convection that may evolve into cyclonic behavior (Saunders & Harris, 1997). The SST hypothesis has extended this frequency-specific observation, based on NAO data gathered in a single year, to pertain generally to cyclone intensity. However, the physical basis for this hypothesized extension involving local SST effects is dubious, in view of the great distances typically traversed by these storms.

Over the last century, local NAO SST has varied in a large and predictable way along individual hurricane tracks, which typically extend many thousands of kilometers (km) from their points of origin (Fig. 1). The magnitude of this variation is sufficient to allow historical hurricane-activity records to serve as "natural experiments" that can be used to perform a direct test of the SST hypothesis, by examining the historical correlation between estimated hurricane intensities, and corresponding estimated historical changes in local SST (Smith & Reynolds, 2003, 2004), along individual hurricane tracks. To perform this test, we examined paired differences ($\Delta v_{\max}$) in estimated local hurricane intensity $v_{\max}$, and corresponding differences (ΔSST) in local (i.e., year-, month-, latitude-, and longitude-specific) SST, over various non-overlapping intervals $\Delta t = t_{i+1} - t_i$ over time t for adjacent positions (indexed by i) along each hurricane track. Over all such intervals that occurred prior to attainment of each peak hurricane intensity, Max($v_{\max}$), the causal SST hypothesis requires that any detectable correlation between $\Delta v_{\max}$ and ΔSST be positive. Below we describe methods and historical data on individual hurricane tracks that we used to implement this direct test of the SST hypothesis, and results we obtained that indicate that variation in SST is not causally linked to variation in hurricane intensities.

If SST variations do not explain the historically observed pattern of hurricane intensities, what other factor(s) might explain this pattern? Statistical and trend analysis for hurricane activity traditionally has focused on annual frequency, landfall likelihood, storm duration, storm intensity (maximum sustained storm-specific wind speed $v_{\max}$), or Saffir-Simpson-scale wind-intensity category (Fisher, 1958; Solow, 1989; Wilson, 1999; Emanuel, 2000; Elsner et al., 2000; Elsner & Bossak, 2004; Webster et al., 2005). Normalized values of $v_{\max}$, in particular, were shown to have a strikingly uniform statistical distribution in the North Atlantic and North Pacific (Emanuel, 2000). It has been argued that this pattern demonstrates that hurricanes opportunistically extract potential energy created by the difference between SST and upper-atmosphere temperature under relatively rare but continually (and randomly) recurring initial conditions, and so supports the hypothesis that increased SST is associated with increase hurricane intensities (Emanuel, 2004; Knutson et

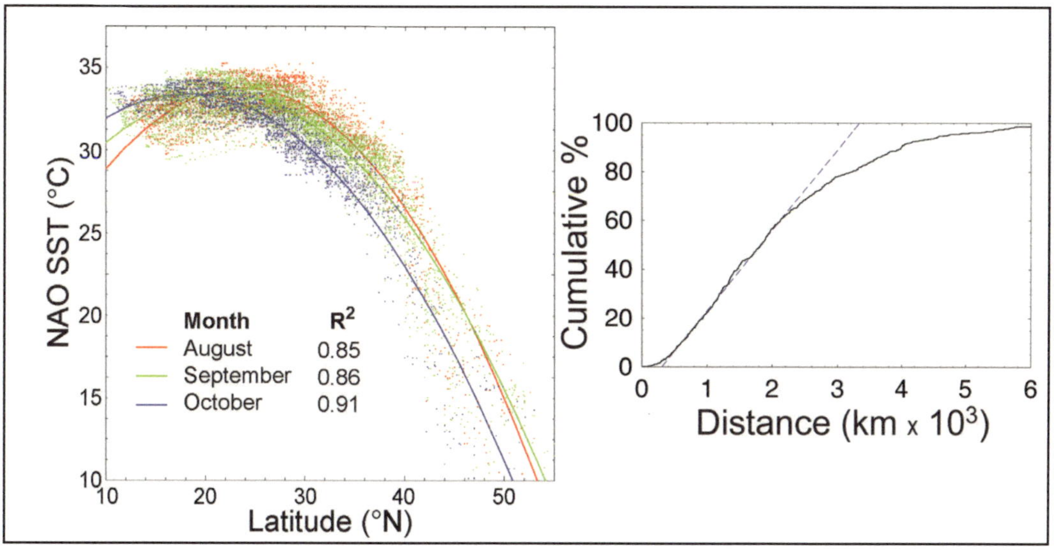

Fig. 1. **(left)** Historical NAO SST vs. corresponding latitude, at a total of n 6-hourly positions along tracks of all 593 recorded hurricanes in Aug (n = 2,598), Sept (n = 5,269) or Oct (n = 1,971) of 1880–2002, with corresponding least-squares parabolic fits and coefficients (R^2) indicating regression-explained fractions of month-specific SST variance. **(right)** Cumulative frequency distribution of distance travelled along each estimated track of 629 recorded hurricanes with increasing intensities during 1880–2002.

specific hurricane intensities and corresponding local (as opposed to regional) SST. Despite fundamental uncertainty about forces driving long-term trends in hurricane activity, which limit the ability to forecast trends in hurricane intensity, some of the fairly well-understood physical processes that underlie hurricane development exhibit substantial underlying regularity. Extreme-value theory is often used to characterize the statistical behavior of extreme events that arise from homogeneous physical processes. Therefore, we also explored the extent to which specific hurricane intensities and corresponding local (as opposed to regional) SST have been correlated historically, the stochastic pattern of historical hurricane intensities, and how the observed pattern can be used to forecast future extremes in hurricane intensity over the next century.

Methods we used to undertake this second aspect of our study, and the results we obtained, are presented below. Following the presentation of results obtained from our comparison of Δv_{max} and ΔSST, and from our stochastic characterization of historical SST variation, we discuss these results and summarize the overall conclusions implied by this study.

2. Methods

2.1 Hurricane data

Historical "best track" data on 6-hourly maximum-estimated velocity v and latitude/longitude position of NAO hurricanes during 1880 through 2002 were obtained from the National Hurricane Center HURDAT database, including updated "best track" data through 1910 (Jarvinen et al., 1984; Landsea et al., 2004). Enhanced data reliability begins in this record since 1944, when aerial reconnaissance of Atlantic tropical cyclones began, and yet higher reliability began in the mid-1960s with the onset of operational satellite detection (Neumann et al., 1999; Martin & Gray, 1993). HURDAT estimates of v (in m/s) between 1945 and 1970 were adjusted to new corresponding values, $v(1 - 2\times10^{-5}\,v^2)$, to account for systematic overestimation of v by aerial reconnaissance in the NAO during that period (Emanuel, 2000, 2005). The peak storm intensity Max(v) recorded for each storm within its corresponding HURDAT best-track data is hereafter denoted v_{max}.

2.2 Sea surface temperature (SST) data and local SST-specific maximum wind speed

The Extended Reconstruction SST (ERSST.v2) database (Smith & Reynolds, 2003, 2004) was used to obtain monthly SST estimates (in °C) at 2-degree NAO grid points during 1880 through 2005. At other positions, SST was estimated by linear interpolation from surrounding ERSST.v2 data points. The historical trend of annual v_{max}-weighted average values of SST was examined in relation to corresponding values of latitude and longitude estimated (for each storm) at their joint position (LL_{max}) at which v_{max} occurred. Storm intensities v_{max} were compared to storm-specific September SST values, with each contributing SST estimated either: (1) at the LL_{max} position, (2) at the position where the storm first attained tropical-storm status, or (3) averaged over all positions between positions (1) and (2). Storm-specific SST, and 1950–2005 NAO-basin September SST, were also compared to corresponding values of latitude and longitude. The NAO basin analysis involved 56 annual sets of September SST values at those 582 joint values of latitude (y) from 6 to 40 °N by 2° and longitude (x) from 2 to 90 °W by 2° that neither violate $x \leq (y+70°)$ nor fall on land, where the ith such set S_i of SST values S_{ij} (i = 1,…,56; j = 1,…,582) was normalized via scaling by the ratio (SST_o/SST_i) in which SST_i = Mean($S_{ij}\,|\,i$) over all j, and SST_o = Mean(S_{ij}) over all i and j = 26.7 °C. September SST was used because this month accounts for ~90% of all v_{max}-weighted average values of the month of hurricane occurrence, >50% of total dissipated hurricane power, and >40% of hurricane occurrence frequency (Bogen et al., 2007).

2.3 Statistical, extreme-value and trend analyses

Temporal and geographic patterns of individual historical hurricane tracks were characterized statistically. F-tests from analysis of covariance for linear regression (ANOCOVAR) (Selvin, 1995) were used to analyze change (Δv_{max}) in estimated v_{max}, vs. change (ΔSST) in corresponding estimated values of SST. SST values were estimated at track-specific pairs of non-overlapping positions separated by duration Δt hours along each of the recorded tracks of all hurricanes lasting ≥36 h that occurred in 1880–2002. Data sets were grouped by subsets of Saffir-Simpson wind-intensity categories 1–5 identified by ANOCOVAR to have least-squares linear fits with common slopes. Common regression slopes and intercepts within category subsets were taken in each case to be category-specific values determined not to differ significantly ($p \leq 0.01$) by a corresponding ANOCOVAR F-test, generating a corresponding $F_{df1,df2}$ statistic with df1 and df2 degrees of freedom. Reported values of R^2 denote coefficients of determination, i.e., fractions of total variance explained by regression.

Extreme values X generated by a homogeneous independent stochastic process can be modeled by fitting generalized extreme value (GEV) distributions (Coles, 2001; Coles & Pericchi, 2003). However, conditional on an estimated cumulative probability distribution function (cdf) $F_X(x) = \mathrm{Prob}(X \leq x)$ modeling the value of X at each occurrence, the cdf of the greatest X-value attained after m occurrences can be calculated as $[F_X(x_m)]^m$ (Ang & Tang,

1984). The latter approach was used in the present study to forecast extreme values of future hurricane intensity. To reduce inter-annual variability effects, annual time series were smoothed by double application of the transformation: $x_i' = 0.25(x_i-1 + x_i+1) + 0.5x_i$, where x_i is the ith-year value and x_i' is the corresponding smoothed value. The same 5-year symmetric moving-average filter was used by Emanuel (2005). The significance of linear trends, and of the difference between means or between nested linear-regression slopes or intercepts, were assessed respectively by 2-tail t-tests for non-zero regression slope, and by F-tests from either analysis of variance (ANOVA) (Selvin, 1995) or analysis of covariance (ANOCOVAR) mentioned above. The significance of the difference between a sample cdf and an estimated corresponding Gaussian cdf was assessed using approximate Kolmogorov, Cramer-von-Mises, and Watson statistics (Stephens 1970). Significance of the difference between a sample cdf and an estimated non-Gaussian cdf was assessed heuristically using the exact Kolmogorov 1-sample test (Friedrich & Schellhaas, 1998).
All p-values $<10^{-10}$ are reported as ~0. All data analysis was done using *Mathematica*® 5.2 (Wolfram Research, Champaign, IL; Wolfram Research, 2010) and related *RiskQ* computer software (Bogen, 2002).

3. Results

3.1 Historical relationship between ΔSST and Δv_{max} along individual hurricane tracks, 1880–2002

During 1880–2002, tracking records indicate that only 79 (7.1% of all) hurricanes regained hurricane status after having lost it. In even more rare cases in which ≥1 peak intensity attained the same peak (i.e., Max(v_{max})) value, the peak-intensity position was taken to be the first occurrence of this peak hurricane intensity value. Nearly all hurricanes during 1880–2002 persisted for ≤12 days; durations ≤6 days were approximately uniformly distributed, and the median duration was ~4 days. Most of these hurricanes therefore persisted sufficiently long to travel far from the MDR, with most (>90%) attaining peak intensity outside the MDR. Only six HURDAT records (for hurricanes during 1880–1898) show intensities that failed to increase along each recorded track. The distribution of cumulative track (i.e., hurricane-migration) lengths of each of the remaining 629 hurricanes was calculated from each corresponding track origin to the location at which peak hurricane intensity (i.e., Max(v_{max})) occurred (Fig. 1, right). This distribution has a mean (±1 SD) of 2,100 ± 1,400 km. Approximately 66% of these migration distances were nearly uniformly distributed between 300 and 2,300 km, approximately 90% of them were ≥600 km, and approximately 10% of them were ≥4,000 km (Fig. 1, right). The track lengths were also approximately linearly proportional to hurricane duration ($R^2 = 0.69$), covering an average of about 660 km/day, with 81% (±31%) of each distance occurring outside the MDR. The historical data thus clearly indicate that only a small portion of the total duration of growth of each MDR-origin hurricane is spent within the MDR.
Consequently, there is no clear physical basis through which historical variation in MDR-specific SST could have causally influenced the magnitude of hurricane-specific values of Max(v_{max}). Results described in the next paragraph address the question of whether elevations in local SST encountered over the course of specific hurricane tracks (as indicated above, mostly outside the MDR) may tend to act to increase hurricane intensity.
Table 1 presents ANOCOVAR results pertaining to track-specific {Δv_{max}, ΔSST} data for hurricanes lasting ≥36 h during 1880–2002, conditional on a specified non-overlapping time

interval Δt applied along each hurricane track. Data within subsets of Saffir-Simpson intensity category were grouped by common values of least-squares linear-regression slope, which were identified by ANOCOVAR as summarized in Table 1. This table lists only category subsets with common slope values, where common values of regression slope and intercept within category subsets were taken in each case to be category-specific values determined not to differ significantly ($p \leq 0.01$) by corresponding ANOCOVAR F-test. For each listed duration Δt, the "combined model" pertaining to the rightmost two columns of the table refers to all subset-specific linear fits obtained that have a common slope, together with corresponding common intercept values, unless the ANOCOVAR p-value listed for intercept homogeneity was ≤0.01, in which case the combined model included the corresponding category-specific intercept estimates (not listed). Results of this analysis corresponding to Δt = 6 h and 24 h are summarized in Fig. 2 (similar results not shown were obtained using Δt = 12 h). Common regression slopes and intercepts within category subsets were taken in each case to be category-specific values determined not to differ significantly ($p \leq 0.01$) by a corresponding ANOCOVAR F-test, as summarized in Table 1.

All of the estimated slopes listed in Table 1 and plotted in Fig. 2 are negative, indicating a consistent pattern of negative correlations between Δv_{max} and ΔSST that is just the opposite of the pattern of positive associations that is posited by the SST hypothesis. The binomial likelihood (i.e., p-value) of by chance alone estimating nine negative slopes under the null hypothesis that the slopes are in fact all zero or positive is <0.2%. For each of three values of Δt considered, highly significantly predictive fits (p = ~0) were obtained to corresponding paired-difference data grouped by hurricane categories, each exhibiting a common slope identified by ANOCOVAR. At least one estimated slope was found to be statistically significantly negative ($p \leq 0.05$) within each Δt-specific set of results listed in Table 1. A total of five of the nine slope estimates listed in Table 1 are significantly negative ($p \leq 0.05$). For Δt = 24 h in particular, the longest time period considered, both of the two slopes that together correspond to all five hurricane categories are significantly negative ($p \leq 0.05$) (Table 1). Interestingly, there is a general tendency among the estimated negative slopes to

Δt (h)	Number of hurricanes	{ΔSST, Δv_{max}} data		Common intercept (km/h)	Common slope (km/h-°C)	*p*-value for = slopes	*p*-value for = intercepts	2-tail *p*-value for slope = 0	Combined model	
		Category (*k*)	Number (*n*)						R^2	*p*-value
6	481	1	1,944	3.80	-0.304	–	–	0.43	0.091	~0
		2–3	2,094	7.99	-1.76	0.50	0.00021	0.028		
		4–5	217	12.5	-5.26	0.049	0.081	0.11		
12	449	1	755	5.49	-1.15	–	–	0.059	0.18	~0
		2	728	14.9	-1.79	–	–	0.011		
		3	394	15.5	-6.82	–	–	0.000018		
		4–5	135	25.3	-3.38	0.27	0.013	0.31		
24	372	1–3	788	19.8	-1.62	0.53	~0	0.0070	0.36	~0
		4–5	97	49.0	-7.07	0.51	0.0057	0.036		

Table 1. Results of covariance analysis of track-specific data for hurricanes persisting ≥36 h during 1880–2002.

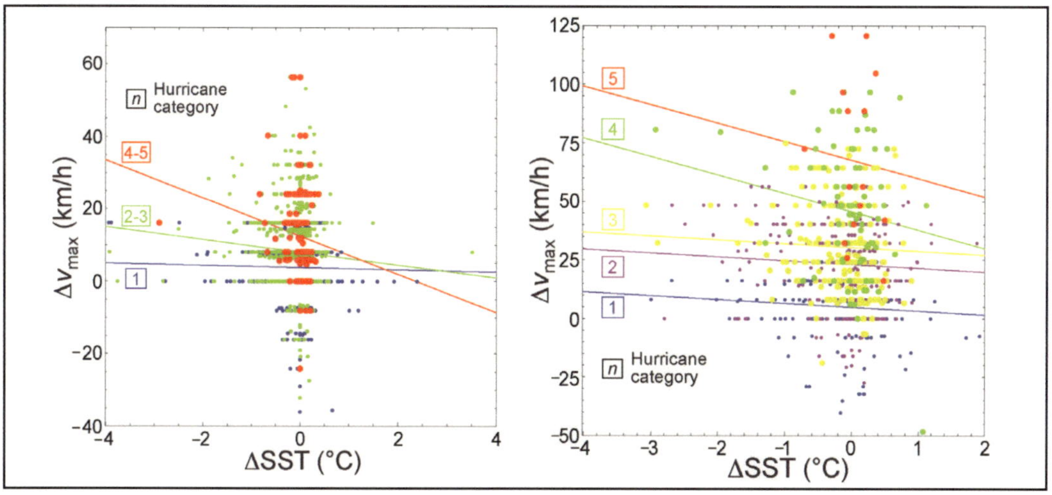

Fig. 2. Change in $v_{\max}$ for track-specific pairs of **(left)** adjacent 6-hourly positions ($\Delta t = 6$ h), and **(right)** daily positions ($\Delta t = 24$ h), along recorded tracks of 481 hurricanes during 1880–2002 that lasted ≥36 h, vs. corresponding change in estimated SST. Least-squares linear fits are shown to data grouped by Saffir-Simpson intensity category, using common, significantly negative ($p < 0.01$) slopes in categories 1, 2–3, and 4–5 identified by ANOCOVAR. The consistent pattern of negative slopes exhibited by the plotted historical data are not consistent with the SST hypothesis, which posits a causal link between increased SST and increased hurricane intensity.

increase in absolute magnitude in proportion to Saffir-Simpson hurricane category. This pattern is again contrary to the SST hypothesis and is statistically significant (by unweighted least-squares linear regression, $R^2 = 0.69$, $p = 0.020$) for those seven slopes listed in Table 1 that differ (at a significance level of $p \leq 0.15$) most significantly from zero.

3.2 Historical variation and covariance of hurricane-specific SST and intensity

Fig. 3 (left) shows the decadal-average percentage of all (1,089) NAO tropical storms, and of all (838) storms that attained a Saffir-Simpson category ≥1, since 1880, that subsequently attained a category ≥4. Over this period, the average fraction of all tropical or greater-intensity storms that never attained a category ≥1 remained about 40%, without any significant overall temporal trend. The decadal data are consistent with the hypothesis that the period 1880–1909 shown reflects censoring due to relatively fewer observations during that period prior to a subsequent surge in steam-powered transatlantic shipping, which occurred after the demise of major passenger cartels that operated prior to World War I (Deltas et al., 1999; Keeling, 1999). Excluding the three earliest points, the plotted decadal percentages show no significant temporal trend, but do show evidence of oscillation with a period of approximately 40 y. The annual frequency of all storms since 1910 for which $v_{\max} \geq 145$ km/h also shows no temporal trend (Fig. 3, right). The distribution of these annual frequencies has significantly extra-Poisson variance ($\chi^2 = 126.4$, df = 92, $p = 0.01$), and is consistent ($\chi^2 = 8.85$, df = 9, $p = 0.26$) with a negative binomial distribution with a mean of 3.4 events/year and an over-dispersion parameter of 0.106. Likewise, there is no temporal trend in $v_{\max}$ during this period ($p = 0.71$), this period prior to 1970 ($p = 0.89$), or since 1970 (p = 0.47). While there is a slight upward trend in annual Max(v_{max}) (R^2 = 0.074, p = 0.0083), no such trend is evident (p > 0.089) when examined during either 1909–1969 or 1970–2002.

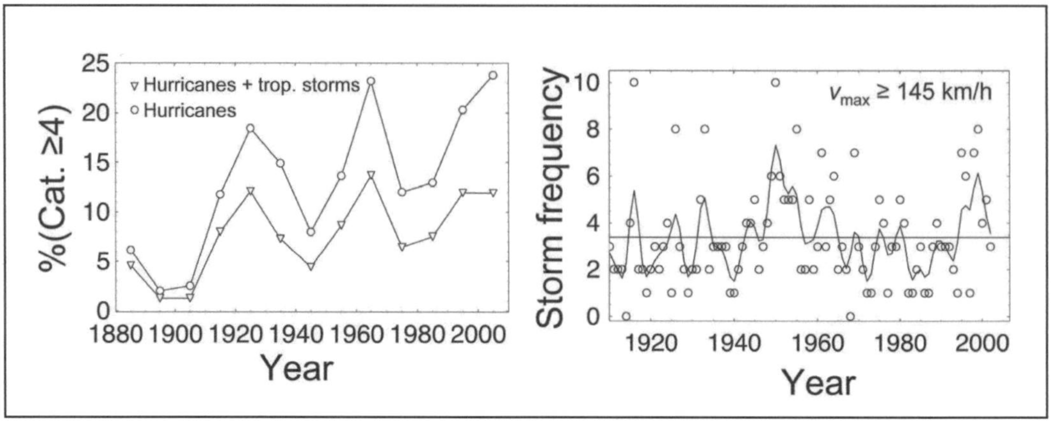

Fig. 3. **(left)** Decadal-average percent of all NAO tropical storms, and of all storms that attained a Saffir-Simpson category ≥1, that subsequently attained a category ≥4. Neither temporal trend is significant for the corresponding post-1909 data (R^2 = 0.088 with p = 0.41, and R2 = 0.15 with p = 0.26, respectively). **(right)** Annual frequency of all storms since 1910 in which v_{max} attained 145 km/h, together with smoothed trend (curve) and mean frequency of 3.4 (horizontal line).

Figure 4 contrasts the historical pattern of annual v_{max}-weighted mean values of storm-specific local SST (shown on the left), with that of storm-specific adjusted latitude (shown on the right), since 1880, using HURDAT and corresponding interpolated SST data for tropical storms and hurricanes (n = 1,079). The additive latitude adjustment (by -2.45°) was selected to make the median value of each data set coincide, which helps to reveal the significant negative correlation between these two variables (R^2 = 0.55, p = ~0), together with their approximate common ~80-year period of oscillation. This correlation is indicated more directly in Fig. 4 (right), which shows that the storm-specific data are generally consistent with the pattern exhibited by 32,592 normalized September SST grid-points covering the entire NAO basin during the period 1950–2005. The negative correlation shown was found by ANOCOVAR to be weaker at latitudes <30 °N for both storm data ($F_{1,\ 119}$ = 5.04, p = 0.027) and NAO basin data ($F_{1,\ 32588}$ = 1,407, p = ~0). A positive but somewhat weaker correlation was found between annual v_{max}-weighted means of storm-specific SST and corresponding longitude (R^2 = 0.37, p = ~0). Multiple linear regression yielded even better models predicting storm-specific SST from corresponding latitude (y), longitude (x), and year (t):

$$\mathrm{SST} = 28.8\ ^\circ\mathrm{C} + 0.0833x - 0.288y \quad (R^2 = 0.69), \text{ and} \tag{1}$$

$$\mathrm{SST} = 28.6\ ^\circ\mathrm{C} + 0.0709x - 0.249y + 0.547\ \mathrm{Sin}[2\pi\,(t - 1915)/78.9] \quad (R^2 = 0.75), \tag{2}$$

where x and y have a weak negative correlation (R^2 = 0.11, p = 0.00017). The latter regression was conditioned on a separately optimized displacement year (1915) and period (78.9 years), and all three of its linear coefficients are significantly positive by F-to-remove criteria ($F_{1,119} \leq$ 26.7, p < 10^{-6}).

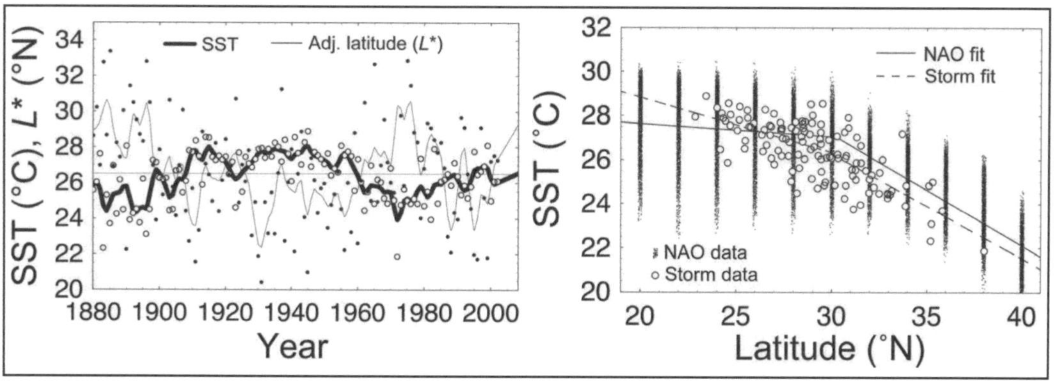

Fig. 4. **(left)** Annual v_{max}-weighted mean values of storm-specific local SST (open points, bold curve), and of corresponding values of adjusted-latitude (solid points, thin curve), are shown with corresponding smoothed trends ($R^2 = 0.55$ with p = ~0 for $n = 122$ annual data values; horizontal line shows joint median value). **(right)** Values of annual v_{max}-weighted mean SST values for storms (open points, with ordinates equal to those of the corresponding open points plotted in the left panel), and annually normalized September SST values at NAO-basin grid points (small solid points at 2° latitude intervals, which were randomly twittered to enhance data-cloud visibility), are compared to corresponding latitudes, together with corresponding bi-linear fits assuming a common breakpoint at 30 °N.

Data	SST_{peak} constraint	Value of 100% × R^2 between v_{max} and SST_x at SST-position x, and corresponding value of n[b]					
		x = peak	n	x = mean	n	x = start	n
All storms[a]	≥ 25°C	1.9[c]	822	0.4[e]	996	0.03	1025
	≥ 20°C	1.2[c]	993	2.0[c]	1065	0.3	1066
	None	0.1[d]	1074	2.5[c]	1074	0.7[e]	1074
All hurricanes	≥ 25°C	3.9[c]	493	4.5[c]	602	0.1	613
	≥ 20°C	8.8[c]	590	8.2[c]	631	1.0[e]	630
	None	11.[c]	633	8.3[c]	633	1.7[e]	633
Category ≥ 3 Hurricanes	≥ 25°C	9.3[c]	218	1.7[e]	232	0.2	230
	≥ 20°C	9.0[c]	231	1.7[e]	232	0.3	232
	None	8.9[c]	232	1.7[e]	232	0.3	232

[a] v_{max} = maximum wind speed. SST was estimated at each year- and month-specific position recorded in the HURDAT 6-hourly storm track database, for years 1880–2002. SST_{peak} = SST at v_{max}; SST_{start} = SST at the position at which the system first attained "tropical storm" (TS) status (wind speeds ≥ 62 km/h); SST_{mean} = arithmetic mean SST at all 6-hourly positions at or after the system first became a TS. Only storms that attained or exceeded TS status were included in this analysis.

[b] R^2 = squared Pearson product-moment correlation coefficient, or fraction of total variance explained by the corresponding linear regression.

[c] $p \leq 10^{-5}$.

[d] $p \leq 0.001$.

[e] $p \leq 0.05$.

Table 2. Correlation between storm-specific v_{max} and sea surface temperature (SST).[a]

Thus, substantial predictability was observed for SST throughout the NAO basin, and even more so for annual v_{max}-weighted mean SST estimated at v_{max}-occurrence points within individual storm tracks. In contrast, storm-specific v_{max} was predicted poorly or not at all by corresponding local SST, regardless of whether SST was estimated at the position (LL_{max}) of each v_{max} occurrence, estimated at the position where each storm first attained tropical-storm status, or estimated by averaging SST between these two positions over each storm-growth track (Table 2, on the previous page). The greatest of these correlations explains just 11% of v_{max} (Table 2). Small correlations detected between v_{max} and SST were a bit larger with SST estimated at LL_{max} than with SST estimated at the beginning of, or averaged over, each storm-growth track. Values of v_{max} were predicted by storm-specific latitude or longitude at least as poorly as they were predicted by corresponding local SST (data not shown).

The empirical cumulative distribution function (cdf) of v_{max} for all NAO storms during 1910–2002 was found to reflect a mixture of values containing ~62% distributed approximately uniformly below v_o = 145 km/h, and the remainder approximately gamma-distributed (Fig. 5, left). The relatively high v_{max} values ($\geq v_o$) have a sample mean and variance of μ = 193 km/h and σ^2 = 905 km^2/h^2, respectively. Parameters α and β of the gamma distribution used to model adjusted conditional intensities $(v-v_o)\,|\,(\mathrm{Max}(v_{max}) \geq v_o)$ were derived by the method of moments as the corresponding estimates μ'^2/σ^2 and σ^2/μ', respectively, with $\mu' = \mu - v_o$. The corresponding v_o-shifted gamma distribution shown in Fig. 5 (left) shall be denoted $F(v)$.

The distribution, $\mathrm{Prob}(\mathrm{Max}(v_{max}) \leq v)$ for all $v \geq 0$, estimated from historical hurricane data (Fig. 5, left) can be used to estimate the cdf of the most extreme $\mathrm{Max}(v_{max})$ after m seasons past a reference year such as the year 2000. Specifically, the cdf of extreme values v_m = $\mathrm{Max}(v_{max}))\,|\,m$, with $m = t-s$ for calendar year t and reference year s = 2000, is just $\mathrm{Prob}(v_m \leq v) = F(v)^m$ (see Methods), conditional on all storms randomly sampled from the high v_{max} category considered. Because the estimated distribution for relatively low v_{max} is approximately uniform with upper bound v_o (Fig. 5, left), the cdf of v_m can be estimated directly after scaling m by the fraction p of all high-v_{max} storms, estimated above as p = 100% - 62% = 38%. The temporal v_m model as a function of year t is therefore given by

$$\mathrm{Prob}(v_m \leq v) = \prod_{t=s}^{p(t-s)} F(v) = \left[F(v)\right]^{p(t-s)} . \tag{3}$$

Figure 5 (right) shows how this model predicts v_m over the 21st century. These predictions indicate that over the next 50 years, v_{max} is as likely as not to exceed about 310 km/h, and is unlikely to exceed about 380 km/h. By the end of this century, these median and upper-bound extreme-value estimates increase to about 330 and 400 km/h, respectively.

4. Discussion

Results involving combined SST and hurricane characteristics estimated for the NAO basin since 1880, or even since 1910, must be interpreted cautiously in view of uncertainty about the accuracy of reconstructed historical SST maps and best-track hurricane records used for this analysis. Greater confidence might be acquired in results of future similar analyses by incorporating existing estimates of SST uncertainty (Smith & Reynolds, 2004), and planned improvements to the HURDAT database for post-1910 hurricanes (Landsea et al., 2004) into analytic methods applied. Awaiting clearly warranted further research along these lines, the following results from the present study appear to be sufficiently coherent and robust as to be unlikely due to chance alone or to substantial bias in estimated historical data relied upon.

4.1 Lack of evidence for a causal relationship between SST and hurricane intensity

During 1880–2002, tracking records indicate that hurricanes persisted sufficiently long to travel far from the MDR, with most (>90% of) attaining peak intensity outside the MDR. Any direct physical link between MDR conditions (such as SST) at the time of cyclogenesis, and $v_{\max}$ behavior days later and hundreds to thousands of km away from the MDR, is not plausible. The SST hypothesis is clearly inconsistent with historical data on hurricanes lasting ≥36 h for which track statistics were estimated between 1880 and 2002 (Fig. 2). As hurricanes intensified, those that migrated into *colder* waters tended to experience significantly *greater* intensification regardless of absolute position within the NAO basin, and this trend was more pronounced for hurricanes of greater intensity. Although effects of global warming on hurricanes may be plausible theoretically, the most directly pertinent historical data, which compare changes in SST and $v_{\max}$ along hundreds of individual hurricane tracks, are clearly inconsistent with any direct link between increased SST *per se* and increased hurricane intensity.

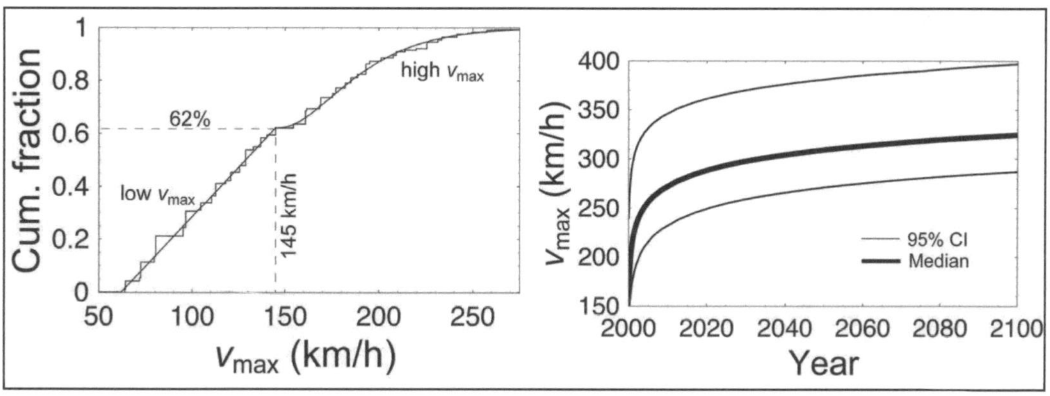

Fig. 5. **(left)** Cumulative empirical distribution of storm-specific $v_{\max}$, shown with a corresponding composite uniform-gamma distribution (solid line for low $v_{\max}$ region, curve for high $v_{\max}$ region). **(right)** Estimated median trend (bold curve) in Max($v_{\max}$) extremes, v_m = Max($v_{\max}$) | m with $m = t–s$ for calendar year t and reference year s = 2000, and corresponding 2-tail 95% confidence limits (light curves) predicted by the extreme-value model given by Eq. 3.

4.2 Stochastic characterization of historical hurricane intensities, and related forecasts

Despite clear evidence of a significant recent positive trend in NAO SST since about 1970 (Goldenberg et al., 2001; Knutson & Tuleya, 2004; Knutson et al., 2004; Emanuel, 2005; Trenberth, 2005; Bogen et al., 2007), a similarly significant trend was not found in HURDAT data pertaining to $v_{\max}$. While storm-specific SST was discovered to be largely predictable from corresponding data on storm latitude, longitude, and year of occurrence, this was not evident in the case of v_{max}. Rather, HURDAT estimates of historical v_{max} since 1910 appear to be random samples from two distinct distributions: one approximately uniform for low-v_{max} values <145 km/h, and the other more skewed and approximately gamma-distributed for high-v_{max} values ≥145 km/h (Fig. 5, left). Among all low-v_{max} storms (~62% of all storms attaining tropical or greater status), the magnitudes of v_{max} were found to be nearly uniformly distributed, consistent with similar observations made in previous studies (Emanuel, 2000, 2004).

The second part of our analysis demonstrated how extreme-value theory, combined with the average value and historical range of the fraction of NAO storms falling into each of the two v_{max} categories identified (Fig. 5, left), can be used to make predictions concerning Max(v_{max}) likely to be delivered by a single hurricane over a specified future period (Fig. 5, right). For example, the model we describe (Eq. 3) predicts that, during the present century, a NAO storm with v_{max} as great as 400 km/h may occur with non-negligible likelihood, and a storm with $v_{max} \geq 330$ km/h should be expected.

Results obtained using this model are similar to those we reported previously (Bogen et al., 2007) for storms classified by Power Dissipation Index (PDI, or integrated 3rd power of wind speed), rather than by the Max(v_{max}) metric used in the present study. Our earlier analysis examined intra-annual variation in storm-specific PDI, and inter-annual variation in annual accumulated PDI (APDI), as another way to investigate another version of the SST hypothesis, positing that the clearly positive trend in NAO sea surface temperature (SST) since 1970 has caused and will continue to cause increased hurricane destructive potential as measured by PDI and APDI. In that study, we applied a combination of statistical and probabilistic methods to the same sets of National Hurricane Center HURDAT best-track data on NAO hurricanes during 1880–2002, and corresponding NOAA Extended Reconstruction SST estimates, as those used for the present study, to compare hurricane behavior and corresponding hurricane-specific (i.e., spatiotemporally linked) SST. In contrast, previous similar comparisons considered only SST averaged over large NAO regions. Contrary to the SST hypothesis, we found that SST varied in a monthly pattern inconsistent with that of corresponding PDI or APDI, and was at best weakly associated with PDI or APDI, despite a strong correlation with corresponding mean latitude ($R^2 = 0.55$), and with combined mean location and a ~90-year periodic tend ($R^2 = 0.70$) (Bogen et al., 2007; cf. the ~80-year oscillatory period shown in Fig. 4-left above). Over the last century, the lower 75% of APDI values achieved appear to have been sampled randomly from a nearly uniform distribution, and the upper 25% of APDI values from a nearly lognormal distribution (Bogen et al., 2007). From the latter distribution, we derived a baseline (SST-independent) stochastic model predicting that over the next half century, APDI will not likely exceed its maximum value over the last half century by more than a factor of 1.5. This factor increased to 2 using a baseline model modified to assume SST-dependence conditioned on an upper bound of the increasing NAO SST trend observed since 1970 (Bogen et al., 2007). An additional model was developed that predicts PDI statistics conditional on APDI (Bogen et al., 2007). These PDI and APDI models can be used to estimate upper bounds on indices of hurricane power likely to be realized over the next century, under divergent assumptions regarding SST influence, analogous to the model (Eq. 3) predicting v_{max} extremes developed in the present study.

Applications of stochastic modeling approaches described here must consider the fact that the relation between v_{max} and seasonal destructive potential is weakened by stochastic variation of landfall likelihood, conditional on hurricane activity (Elsner & Bossak, 2004), as well as by other critical factors such as population density and state of economic development of affected coastal areas (Pielke & Landsea, 1998; Pielke & Sarewitz, 2005). Such estimates can be refined to focus on realized damage by expanding predictive models to incorporate landfall likelihoods and associated potential loss in relation to trends in coastal economic development.

5. Conclusions

Historical data, physical theories, and modeling have been used to claim that increased sea-surface temperature (SST) causes increased cyclone activity, intensity, or potential destructiveness in the North Atlantic Ocean (NAO) and other ocean basins (the "SST hypothesis"). Other studies, including this one, support an alternative, "extreme-value hypothesis," that recent strong hurricanes like Katrina are simply extreme samples from an essentially stationary intensity distribution that has not increased appreciably over at least the last half century. The SST hypothesis was tested using historical data on changes (Δv_{max}) in NAO hurricane intensity and corresponding changes (ΔSST) in estimates of local SST along tracks of individual hurricanes lasting ≥36 h, over equal time intervals set at a value between 6 and 24 h. The SST hypothesis requires a positive correlation between Δv_{max} and ΔSST. In contrast, tracking records during 1880–2002 indicate that, as hurricanes intensified, those that migrated into *colder* waters tended to experience significantly *greater* intensification regardless of absolute position within the NAO basin, and this trend was more pronounced for hurricanes of greater intensity. Consequently, although effects of global warming on hurricanes may be plausible theoretically, directly relevant historical data exhibit no evidence that increases in SST tend to induce increased hurricane intensities. To bound possible future trends in maximum intensity (v_{max}, km/h) of hurricane seasons, we performed stochastic and extreme-value analysis of National Hurricane Center HURDAT best-track data on hurricanes during 1880–2002, combined with NOAA Extended Reconstruction SST estimates. In contrast with a recent positive trend in NAO SST since about 1970, we detected no historical trend for v_{max}. Storm-specific SST was predicted fairly well from corresponding data on storm latitude, longitude, and year of occurrence, whereas v_{max} has had little or no direct association with SST or other local variables. Estimated v_{max} since 1910 appears to be randomly sampled from two distinct distributions: one (for v_{max} < 145 km/h) approximately uniformly distributed, and the other (v_{max} ≥ 145 km/h) approximately gamma-distributed. A corresponding extreme-value model developed in this study predicts Max(v_{max}) for any single hurricane over time. In particular, this model predicts that, during this century, a storm with v_{max} up to 400 km/h may occur with non-negligible likelihood, and a storm with v_{max} ≥ 330 km/h should be expected. Cautious interpretation of these modeling results is warranted, in view of uncertainty about the accuracy of reconstructed historical SST maps and best-track hurricane records used.

6. Acknowledgements

This work was prepared in part by Lawrence Livermore National Laboratory under Contract DE-AC52-07NA27344. We are grateful for reviewers' comments and suggestions that improved this manuscript.

7. References

Ang, A.H.-S. & Tang, W.H. (1984). *Probability Concepts in Engineering Planning and Design, Vol. 2 – Decision, Risk, and Reliability*. John Wiley & Sons, ISBN 0471-032-018, New York, NY, pp. 186–273

Bell, G.D.; Halpert, M.S.; Schnell, R.C.; Higgins, R.W.; Lawrimore, J.; Kousky, V.E.; Tinker, R.; Thiaw, W.; Chelliah, M. & Artusa, A.(2000). Climate Assessment for 1999. *Bull. Am. Meteorol. Soc.* 81:S1–S50, ISSN 1520-0477

Bogen, K.T. (2002). *RiskQ* 4.2: *An Interactive Approach to Probability, Uncertainty and Statistics for use with Mathematica®*. Lawrence Livermore National Laboratory, UCRL-MA-110232 Rev. 3, Livermore, CA

Bogen, K.T.; Jones, E.D.; Fischer, L.E. (2007). Hurricane destructive power predictions based on historical storm and sea surface temperature data. *Risk Anal* 27(6):1497–1517, ISSN: 0272-4332

Chu, P.-S. (2004). ENSO and tropical cyclone activity. In: Murnane, R.J. & K-b Liu, eds. *Hurricanes and Typhoons: Past, Present and Future*. Columbia University Press, ISBN 0-231-12388-4, New York, NY, pp. 297–332

Coles, S. 2001. *An Introduction to Statistical Modeling of Extreme Values*. Springer-Verlag, ISBN 1-85233-459-2, London, UK.

Coles, S. & Pericchi, L. (2003). Anticipating catastrophes through extreme value modelling. *J. Roy. Statist. Soc. Series C (Appl. Statist.)* 52:405–416, ISSN 0035-9254

Deltas, G.; Serfes, K. & Sicotte, R. (1999). American shipping cartels in the Pre-World War I era. In: Field, A.J.; Clark, G. & Sundstrom, W., eds., *Research in Economic History, Vol. 19*. Elsevier, ISBN: 0-7623-0575-4, New York, NY, pp. 1-38

Elsner, J.B.; Jagger, T. & Niu, X.-F. (2000). Changes in the rates of North Atlantic major hurricane activity during the 20th century. *Geophys. Res. Lett.* 27:1743–1746, ISSN 0094-8276

Elsner, J.B. (2003). Tracking hurricanes. *Bull. Am. Meteorol. Soc.* 84:352–356, ISSN 1520-0477

Elsner, J.B. & Bossak, B.H. (2004). Hurricane landfall probability and climate. In: Murnane, R.J. & K-b Liu, eds. *Hurricanes and Typhoons: Past, Present and Future*. Columbia University Press, ISBN 0-231-12388-4, New York, NY, p. 333–353.

Emanuel, K. (2000). A statistical analysis of tropical cyclone intensity. *Monthly Weather Rev.* 128:1139–1152, ISSN 0027-0644

Emanuel, K. (2004). Response of tropical cyclone activity to climate change. In: Murnane, R.J. & K-b Liu, eds. *Hurricanes and Typhoons: Past, Present and Future*. Columbia University Press, ISBN 0-231-12388-4, New York, NY, p. 395–407.

Emanuel, K. (2005). Increasing destructiveness of tropical cyclones over the past 30 years. *Nature* 436:686-688, ISSN 0028-0836

Fisher, E.L. (1958). The exchange of energy between the sea and the atmosphere in relation to hurricane behavior. *J. Meteorol.* 15:164-171, ISSN 0095-9634

Friedrich, T. & Schellhaas H. (1998). Computation of the percentage points and the power for the two-sided Kolmogorov-Smirnov one sample test. *Statist. Papers* 39:361–375, ISSN: 0932-5026

Goldenberg, S.B.; Landsea, C.W.; Mestas-Nuñez, A.M. & Gray, W.M. (2001). The recent increase in Atlantic hurricane activity: causes and implications. *Science* 293:474–479, ISSN 0036-8075

Gray, W.M. (2006). Comments on: "Increasing destructiveness of tropical cyclones over the past 30 years" by Kerry Emanuel, *Nature*, 31 July 2005, Vol. 436, pp. 686-68. CERN preprint physics/0601050, available at http://cdsweb.cern.ch/record/921137 (accessed Aug 10, 2007)

Jarvinen, B.R.; Neumann, C.J. & Davis, M.A.S. (1984). A Tropical Cyclone Data Tape for the North Atlantic Basin, 1886–1983: Contents, Limitations, and Uses. NOAA Technical Memorandum NWS NHC 22. National Hurricane Center, Miami, FL

Hoyos, C.D.; Agudelo, P.A.; Webster, P.J. & Curry, J.A. (2006). Deconvolution of the factors contributing to the increase in global hurricane intensity. *Science* 312:94-97, ISSN 0036-8075

Keeling, D. (1999). The transportation revolution and transatlantic migration, 1850-1914. In: Field, A.J.; Clark, G. & Sundstrom, W.; eds. *Research in Economic History, Vol. 19.* Elsevier, ISBN: 0-7623-0575-4, New York, NY, pp. 39-74

Knutson, T.R. & Tuleya, R.E. (2004). Impact of CO_2-induced warming on simulated hurricane intensity and precipitation: sensitivity to the choice of climate model and convective parameterization. *J. Climate* 17:3477–3495, ISSN 0894-8755

Knutson, T.R.; Tuleya, R.E.; Shen, W. & Ginis, I. (2004). Impact of climate change on hurricane intensities as simulated using regional nested high-resolution models. In: Murnane, R.J. & K-b Liu, eds. *Hurricanes and Typhoons: Past, Present and Future.* Columbia University Press, ISBN 0-231-12388-4, New York, NY, pp. 408–439

Landsea, C.W.; Nicholls, N.; Gray, W.M. & Avila, L.A. (1996). Downward trends in the frequency of intense Atlantic hurricanes during the past five decades. *Geophysical Res. Lett.* 23:1697-1700, ISSN 0094-8276

Landsea, C.W.; Anderson, C.; Charles, N.; Clark, G.; Dunion, J.; Fernandez-Partagas, J.; Hungerford, P.; Neumann, C. & Zimmer, M. (2004). The Atlantic Hurricane Database Re-analysis Project Documentation for 1851–1910 Alterations and Addition to the HURDAT Database. In: Murnane, R.J. & K-b Liu, eds. *Hurricanes and Typhoons: Past, Present and Future.* Columbia University Press, ISBN 0-231-12388-4, New York, NY. HURDAT data updated through 2005 were obtained from http://www.aoml.noaa.gov/hrd/hurdat/Documentation.html (accessed Aug 10, 2007)

Landsea, C.W.; Harper, B.A.; Hoarau, K. & Knaff, J.A. (2006). Can we detect trends in extreme tropical cyclones? *Science* 313:452-454, ISSN 0036-8075

Mann, M.E. & Emanuel, K.A. (2006). Atlantic hurricane trends linked to climate change. *Eos Trans. AGU* 87(24), 233-244, ISSN 0096-3941

Martin, J.D. & Gray, W.M. (1993). Tropical cyclone observation and forecasting with and without aircraft reconnaissance. *Weather and Forecasting* 8:519–532, ISSN 0882-8156

Michaels, P.J.; Knappenberger, P.C. & Landsea, C. (2005). Comments on "Impacts of CO_2-Induced Warming on Simulated Hurricane Intensity and Precipitation: Sensitivity to the Choice of Climate Model and Convective Scheme". *J. Climate* 18: 5179–5182, ISSN 0894-8755

National Oceanic and Atmospheric Administration (NOAA). 2006. The Saffir-Simpson hurricane scale. NOAA, National Hurricane Center, U.S. National Weather Service, http://www.nhc.noaa.gov/aboutsshs.shtml (last modified 22 June 2006; accessed Aug 10, 2007).

Neumann J.C.; Jarvinen, B.R.; McAdie, C.J. & Elms, J.D. (1999). *Tropical Cyclones of the North Atlantic Ocean, 1871-1998*. 5th Revision. National Climatic Center, Ashville, NC, pp. 11–15. Available at: http://www.preventionweb.net/files/5590_TropicalCyclones18711998.pdf (accessed Aug 10, 2007)

Pielke Jr., R.A. & Landsea, C.W. (1998). Normalized hurricane damages in the United States: 1925–1995. *Weather and Forecasting* 13: 351-361, ISSN 0882-8156

Pielke Jr., R.A. & Sarewitz, D. (2005). Bringing society back into the climate debate. *Population Environ*. 26:255-268, ISSN 0199-0039

Santer, B.D.; Wigley, M.L.; Gleckler, P.J.; Bonfils, C.; Wehner, M.F.; AchutaRoa, K.; Barnett, T.P.; Boyle, J.S.; Brüggenmann, W.; Fiorino, M.; Gillett, N.; Hansen, J.E.; Jones, P.D.; Klein, S.A.; Meehl, G.A.; Raper, S.C.B.; Reynolds, R.W.; Taylor, K.E. & Washington W.M. (2006). Forced and unforced ocean temperature changes in Atlantic and Pacific tropical cyclogenesis regions. *PNAS* 103:13905-13910, ISSN 0027-8424

Saunders, M.A. &Harris, A.R. (1997). Statistical evidence links exceptional 1995 Atlantic hurricane season to record sea warming. *Geophys. Res. Lett.* 24:1255-1258, ISSN 0094-8276

Selvin, S. (1995). Practical Biostatistical Methods. Duxbury Press / Wadsworth Publishing Co., Belmont, CA, pp. 139-168.

Smith, T.M. & Reynolds, R.W. (2003). Extended reconstruction of global sea surface temperature based on COADS data (1854-1997). *J. Climate* 16:1495–1510, ISSN 0894-8755

Smith, T.M. & Reynolds, R.W. (2004). Improved extended reconstruction of SST (1854-1997). *J. Climate* 17:2466-2477, ISSN 0894-8755. Corresponding SST data updated through 2002 were obtained from http://www.ncdc.noaa.gov/oa/climate/research/sst/sst.html (accessed Aug 10, 2007)

Solow, A.R. (1989). Statistical modeling of storm counts. *J. Climate* 2:131–136, ISSN 0894-8755

Stephens, M.A. (1970). Use of the Kolmogorov-Smirnov, Cramer-Von Mises and related statistics without extensive tables. *J. Royal Stat. Soc. Ser. B* 32:155–122, ISSN 0035-9246

Trenberth, K. (2005). Uncertainty in hurricanes and global warming. *Science* 308:1753–1754, ISSN 0036-8075

Trenberth, K.E. & Shea, D.J. (2006). Atlantic hurricanes and natural variability in 2005. *Geophys. Res. Lett.* 33:L12704(doi 10.1029/2006GL026894), ISSN 0094-8276

Webster, P.J.; Holland, G.J.; Curry, J.A. & Chang, H.-R. (2005). Changes in tropical cyclone number, duration, and intensity in a warming environment. *Science* 309:1844–1846, ISSN 0036-8075

Wilson, R.M. (1999). Statistical aspects of major (intense) hurricanes in the Atlantic basin during the past 49 hurricane seasons (1950–1998): Implications for the current season. *Geophys. Res. Lett.* 26:2957–2960, ISSN 0094-8276

Wolfram Research. (2010). Wolfram Mathematica® 7 Documentation Center. Wolfram Research, Inc., Champaign, IL (www.wolfram.com), http://reference.wolfram.com/mathematica/guide/Mathematica.html (accessed 26 January 2010).

6

Statistical Techniques for Exploring Possibly Increasing Trend of Hurricane Activity

Jesper Rydén
Uppsala University
Sweden

1. Introduction

Discussions on climate change are intense; the following quotation, by G.H. Liljequist (1949), is valid also today:

> The recent climatic fluctuation, which seems to embrace the whole earth, has been the object of many discussions and investigations. This is quite natural: its consequences have been far-reaching and have come to influence the everyday life of man in many ways.

Broadly speaking, one aspect is whether certain events, for instance storms or hurricanes, have been more common recently; another if such events (possibly in addition) have become more severe and violent. A difficult problem related to the latter issue (not discussed in this chapter) is the relation to climate change: are changes to occur statistically speaking, as a part of the inherent variability of nature, or are these an effect from human activity.
In this chapter, we present some statistical techniques for investigation of possible trends in time series, as applied to hurricane data: the yearly number of hurricanes as well as the strengths of hurricanes.

2. Background

Focussing on hurricanes in the North Atlantic, the season 2005 resulted in many broken records: 27 named storms, 15 hurricanes, 7 major hurricanes, 4 hurricanes of Category 5. Moreover, the hurricane Katrina implied very large and severe losses. Naturally, questions arise on possible trends of hurricane activity and this had been discussed even earlier. A comprehensive statistical analysis of hurricanes in the North Atlantic was given by Elsner and Kara (1999), and these authors state that "...caution is advised in interpreting the trends." Some researchers agree that a change in intensity has occurred in recent years, see for instance Webster et al. (2005), but careful definitions have to be made. When investigating a possible trend over time for the yearly number of hurricanes, the statistical analyses are often based on regression models where the number of hurricanes is supposed to be a random variable dependent on covariates like North Atlantic oscillation index (NAO) (see e.g. Elsner, Jagger and Niu 2000). Landsea et al. (1999) found that only weak linear trends can be ascribed to hurricane activity and that multi-decadal variability is more characteristic of the region.

The statistical tools in the literature when analysing possible trends of the total yearly number of hurricanes typically are based on some form of generalized linear model, out of which Poisson regression is a special case (clearly, such models are natural, since one wants to model counts of a quantity). As an example, see Solow (1989). Also, more recent statistical techniques have been employed, e.g. Bayesian methodology in a Markov Chain Monte Carlo approach was used by Elsner and Jagger (2004).

The study of damage caused by hurricanes constitute another aspect of hurricane research, where statistical methodology has been used to a large extent. For instance, Katz (2002) introduced a compound Poisson process with two components: one governing the occurrences of events, the other specifying damages associated with the individual events. Jagger et al. (2008) gave a review and presented a strategy for simulating annual insured losses based on pre-season values of NAO and Atlantic sea-surface temperature. The extreme losses were modelled with a generalized Pareto distribution. Further, hurricanes may cause severe power outages. A risk analysis was performed by Han et al. (2009), making use of a generalized additive model in the statistical modelling.

In this paper, we take a pragmatic approach and focus on the question of increasing trend of the yearly number of hurricanes by modelling only this quantity of interest, hence no covariates involved. The purpose of the paper (see also Rydén, 2010) is to point out what can be deduced by an exploratory graphical method with origin in scale-space theory and compare with more traditional statistical methods, and apply it in a case study of data sets related to hurricane activity. This method is known as the SiZer (Significance of Zero crossings of the derivative) map, which can be used to find significant features in data (Chaudhuri and Marron 1999). In environmental data, such features typically depend on the level of detail for which the time series is considered. As a benchmark, we compare with what a conventional method like Poisson regression implies; in that case, a statistical hypothesis test only tells whether to reject or not the hypothesis of a constant trend. The advantage of the SiZer approach is the possibility to investigate several time scales at the same time, presented in one single image. An application to climatology published recently was given by Weckström et al. (2006) and a review of the SiZer and related techniques was given by Godtliebsen, Olsen and Winther (2003).

The paper is organised as follows: Next, in Section 3, the data set of hurricanes is briefly presented. Further, a data series of an index measuring the strength of hurricanes (ACE index) is introduced. Thereafter, in Section 4, the SiZer methodology is reviewed and applied to these two data sets. In Section 5, a comparison is made with other methods, for instance a standard approach based on Poisson regression. Finally, some conclusions are given.

3. Data sets

The region of the Atlantic basin has been studied in this paper. Data are courtesy of Tropical Prediction Center, NOAA, and are conveniently presented at a homepage of Unisys Weather, http://weather.unisys.com/hurricane/atlantic/. In the sequel hurricanes from 1950–2009 are studied. Seasons officially run from 1 June to 30 November. Hurricane activity has been tabulated back to 1871, but normally the portion dating from about 1944 is considered reliable. Use of aircraft monitoring was then introduced. Before the advent of satellites and aircraft monitoring it was difficult to detect storms that did not affect land or ships. This should be kept in mind when analysing older data: the activity in some seasons before the middle of the 20th century might be under-estimated.

3.1 Hurricanes and categories

Hurricanes can be divided into 5 categories according to the Saffir–Simpson scale which is a 1-5 rating based on the present intensity of the hurricane. The scale is used to give an estimate of the potential property damage and flooding expected along the coast from a hurricane landfall. Wind speed is the determining factor in the scale. Major hurricanes are usually defined as those in Category 3 or higher. For reference, the definitions of categories are given in Table 1.

Category	1	2	3	4	5
Wind speed (km/h)	119-153	154-177	178-209	210-249	> 249

Table 1. Saffir–Simpson scale defining categories of hurricanes.

In this paper, the total yearly numbers of hurricanes in categories 1–5 are investigated. In Fig. (1), top panel, a times series is shown.

3.2 ACE index

The National Oceanic and Atmospheric Administration (NOAA) uses an Accumulated Cyclone Energy (ACE) index as a measure of total seasonal activity. This refers to the collective intensity and duration of named storms and hurricanes in the North Atlantic during a given season. The ACE index is defined as the sum of the squares of the maximum sustained surface wind speed measured every six hours for all named systems while they are at least tropical storm strength. The ACE for the 60 seasons 1950–2009 was investigated.
In Fig. (1), bottom panel, a times series of ACE is displayed. Moreover, in Fig. (2), a scatter plot is shown. Obviously a strong correlation is to be expected between total yearly number of hurricanes and ACE, and is also seen from the figure. However, the plot is interesting as also other factors influence the ACE; for instance, the number of named storms.

4. Analysis by SiZer

4.1 Statistical background

The SiZer technique can be considered a smoothing method. A central question to be posed when facing data is which features are significant, as opposed to sampling artifacts. We here give a brief review; see the original papers by Chaudhuri and Marron (1999), (2000) for full details.
In the simplest SiZer methodology, a random sample $(x_1, y_1), \ldots, (x_n, y_n)$ of independent observations is assumed (this assumption will be reasonable for our datasets under consideration, as will be discussed in the sequel). In a non-parametric regression framework, the data are smoothed to give a curve that can be regarded an estimated conditional mean $f(x) = \mathsf{E}[Y|X = x]$. For the data, the conditional expectation is estimated by

$$\hat{f}_h(x) = \underset{a}{\operatorname{argmin}} \sum_{i=1}^{n} [y_i - (a + b(x_i - x))]^2 K_h(x - x_i)$$

where h is a smoothing parameter called bandwidth and K_h results from a kernel function K, $K_h(\cdot) = (1/h)K(\cdot/h)$. A common choice of K is a Gaussian kernel. The estimate $\hat{f}_h(x)$ is wiggly when h is small and very smoothed when h is large.
The visual assessment of significance of peaks and valleys in a family of smooths is based on confidence limits for the derivative $\hat{f}_h'(x)$ in scale space. In the SiZer colour map, blue portions

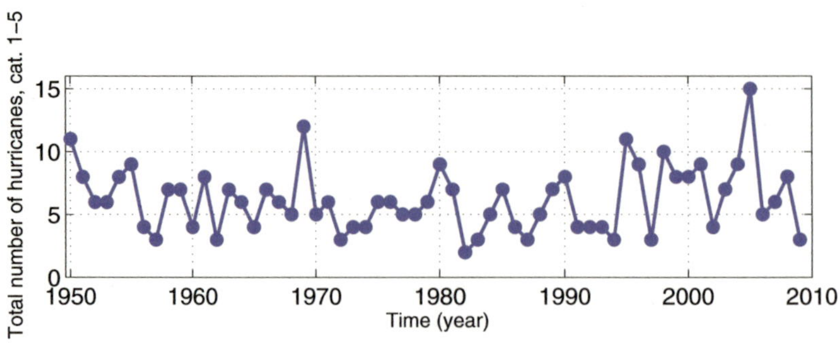

(a) Yearly hurricanes, 1950-2009.

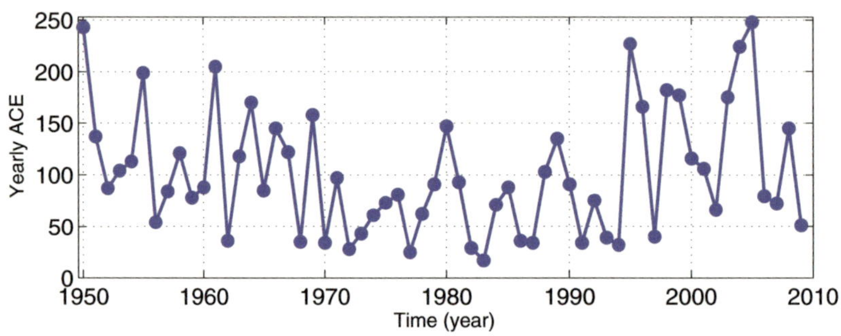

(b) ACE for hurricanes, 1950-2009.

Fig. 1. Time series: yearly hurricanes and ACE.

indicate locations of x and h where $\hat{f}_h'(x)$ is significantly positive, red portions locations where $\hat{f}_h'(x)$ is significantly negative. Furthermore, confidence limits for $f_h'(x)$ are computed (for details, see Chaudhuri and Marron 1999).

4.2 Check of assumptions

SiZer methodology handles independent observations but in time series, dependence is often present. We here study empirically if the assumption seems reasonable. Autocorrelation functions for the two data sets are shown in Figure 3.

The 95% confidence limits, marked as dotted lines, follow from approximative normality and are given by $-1/(n-1) \pm 1.96/\sqrt{n}$. None of the serial correlations fall outside the limits so as a reasonable approximation, we assume independence and proceed using the SiZer. Methodology for handling of dependent observations have been developed, see Rondonotti et al. (2007) or for a method denoted SiNos, Godtliebsen et al. (2003).

4.3 Analysis of data

In Figure 4 the data set of yearly number of hurricanes is analysed. In the top panel a family of smoothed curves are shown; the original observations are shown as dots and the solid lines show smooths resulting from various choices of bandwidth. In the bottom panel, the derivatives are visualized with respect to both location and scale (bandwidth). The purple colour in the upper part of that plot indicates that the derivative is not found to be significantly different from zero, while the darker grey colour at the lower part is used at locations with too

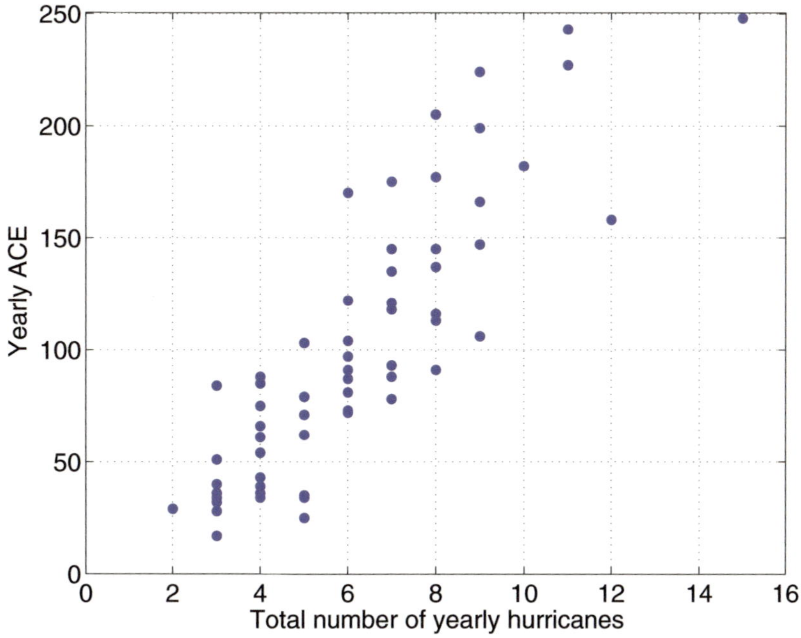

Fig. 2. Scatter plot: Total yearly number of hurricanes, yearly ACE.

few data available to do inference. Note that there are neither blue portions (corresponding to significant increases) nor red portions (corresponding to significant decreases). The solid horizontal bar in the SiZer plot shows a good data-driven bandwidth as suggested by Ruppert et al. (1995), here corresponding to approximately $h = 10^{0.66} = 4.6$ years.
The corresponding visualization for the ACE data is shown in Figure 5. The top panel shows original data and a family of smooths. Now, in the bottom panel, there are portions found with significant features: a decrease in the period 1950-1970 and an increase from about 1990 and onwards. These are valid for intermediate scales, values of $\log_{10}(h)$ in the interval $[0.9, 1.3]$. In summary, we are not able to conclude a significant change over longer time scales, which is of interest when discussing climate changes.

5. Other approaches

In this section we analyse the data sets by some other approaches. The purpose is to compare with the outcome of the SiZer technique; in statistical modelling of geological and environmental phenomena, it is often of interest to compare results from several methods and assumptions.

5.1 Poisson regression

Assume first that we study the total yearly number of hurricanes. Clearly, we model a discrete response variable, Y say, assumed to have a Poisson distribution. Suppose that the mean of Y is μ. In its most common form, the Poisson regression model is a special case of a generalized linear model which relates the mean to a linear function of covariates. In our case, we formulate the model

$$\mu_i = t_i \exp(\beta_0 + \beta_1 x_i), \quad i = 1, \dots, n \tag{1}$$

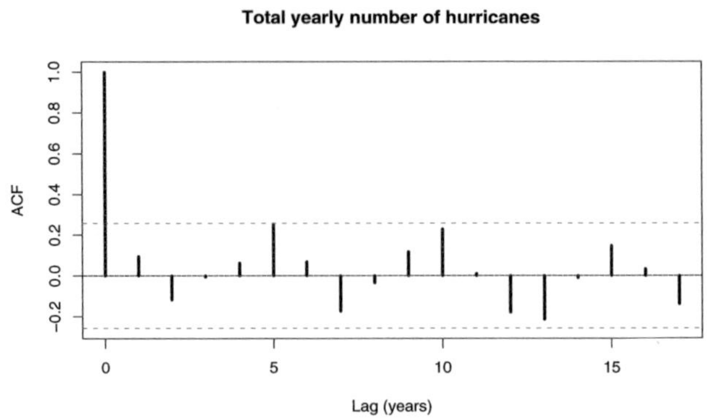

(a) Autocorrelation function for total yearly number of hurricanes.

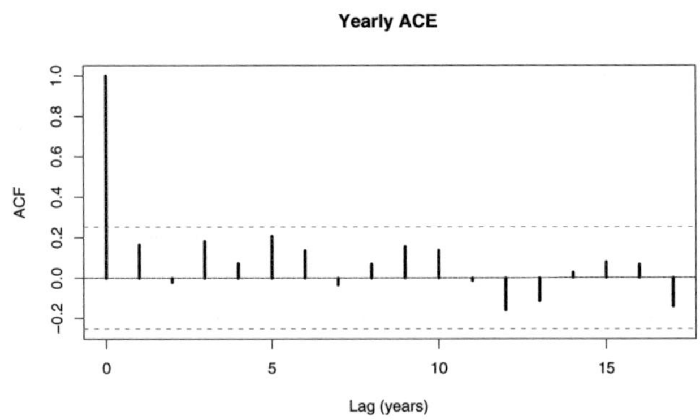

(b) Autocorrelation function for ACE.

Fig. 3. Autocorrelation functions for time series of total number of yearly hurricanes (top) and yearly ACE index (bottom)

where n is the number of years considered, t_i is the exposure time for each observation (one year, hence $t_i = 1$, all i), and x_i is an explanatory variable which is year. This explanatory variable is introduced since the purpose is to test for a possible trend over time: is the number of hurricanes increasing over the time period studied?
The statistical test for testing if $\beta_1 = 0$ is made by using the *deviance*. Parameter estimates are returned as ML estimates and l being the log-likelihood function, the deviance is given by

$$D = 2(l(\hat{\beta}_0, \hat{\beta}_1) - l(\beta_0, \beta_1))$$

This is used as follows: If $D > \chi_1^2(\alpha)$, reject the hypothesis. Comparison is made with table values $\chi_1^2(0.05) = 3.84$, $\chi_1^2(0.01) = 6.63$. The original data set results in $D = 0.35$, hence the hypothesis of zero slope cannot be rejected. However, when studying the regression diagnostics of the resulting model based on the original data, it is found that at least one outlier is present. Cook's distance is frequently used as a measure of leverage and values are presented in Figure 6. When again fitting a model, omitting years 1950 and 2005, the deviance $D = 0.103$ is found. Hence, again, the hypothesis is not rejected.

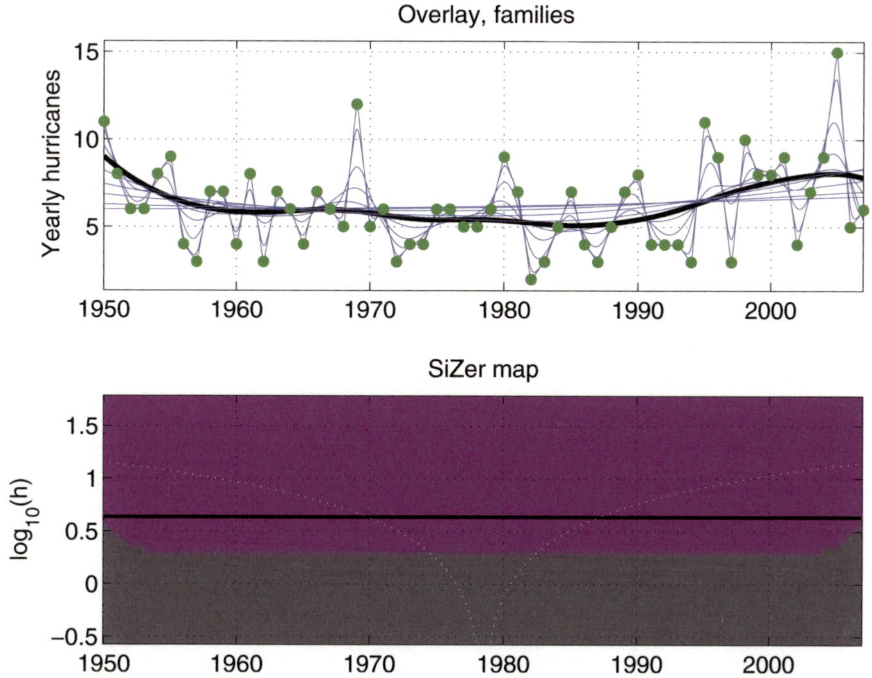

Fig. 4. SiZer analysis of total number of yearly hurricanes. Top: Overlay of family of smooths. Bottom: SiZer map.

To justify the use of Poisson regression, the possible presence of over-dispersion should be checked. A simple confidence interval for the ratio $\mathsf{V}[Y]/\mathsf{E}[Y]$ can be constructed following Rychlik and Rydén (2006), Section 7.3 (see also Brown and Zhao 2002). We find the 0.95 confidence interval $[0.61, 1.28]$ and hence the hypothesis of Poisson distribution is not rejected since 1 is within the interval.

5.2 Further models for count data

Several extensions of the model for Poisson regression have been presented in the literature, and although over-dispersion seems not to be a problem in our case, we use some of these. An expose is given by Zeileis, Kleiber and Jackman (2008), where hurdle and zero-inflated regression models are discussed along with the implementation in the R system for statistical computing. A model based on quasi-likelihood and a negative-binomial model were fitted to our data, but differences were small compared to the standard Poisson regression model.
In the literature, one also encounters zero-inflated Poisson regression (see e.g. Lambert, 1992). Focussing at only hurricanes of category 5, we face without doubt a data set with excess zeros. Using the package pscl in R, a zero-inflated model was fitted to the data of hurricanes of category 5, and the model was compared to the corresponding model obtained from conventional Poisson regression by a Vuong test. However, the zero-inflated model was not a significant improvement over the standard model.

5.3 Mann Kendall test

In environmental applications, a frequently used non-parametric test for trend detection is the Mann Kendall test. For our data sets, this results in the two-sided p-values 0.99 for the data

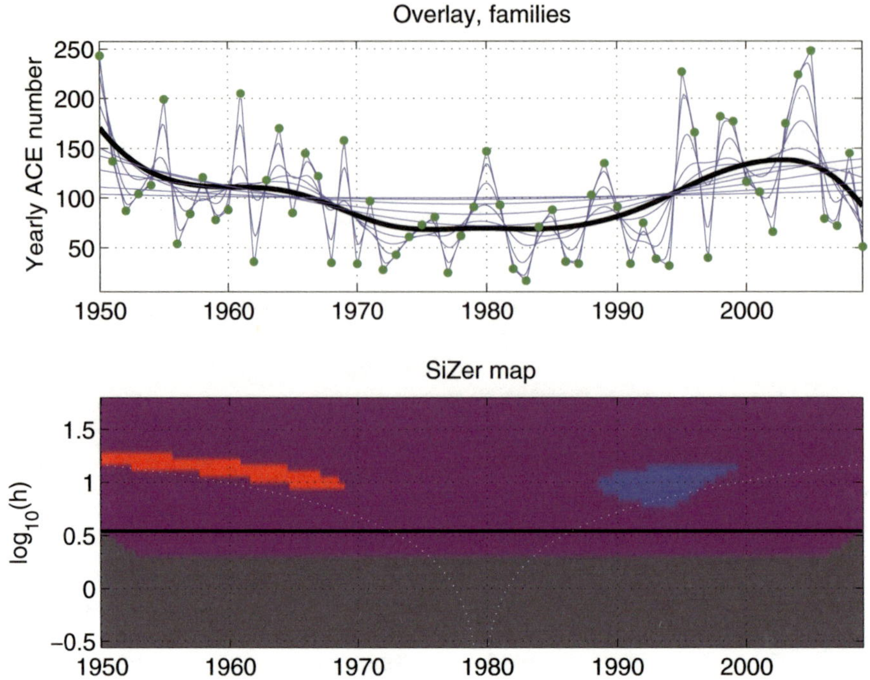

Fig. 5. SiZer analysis of ACE. Top: Overlay of family of smooths. Bottom: SiZer map.

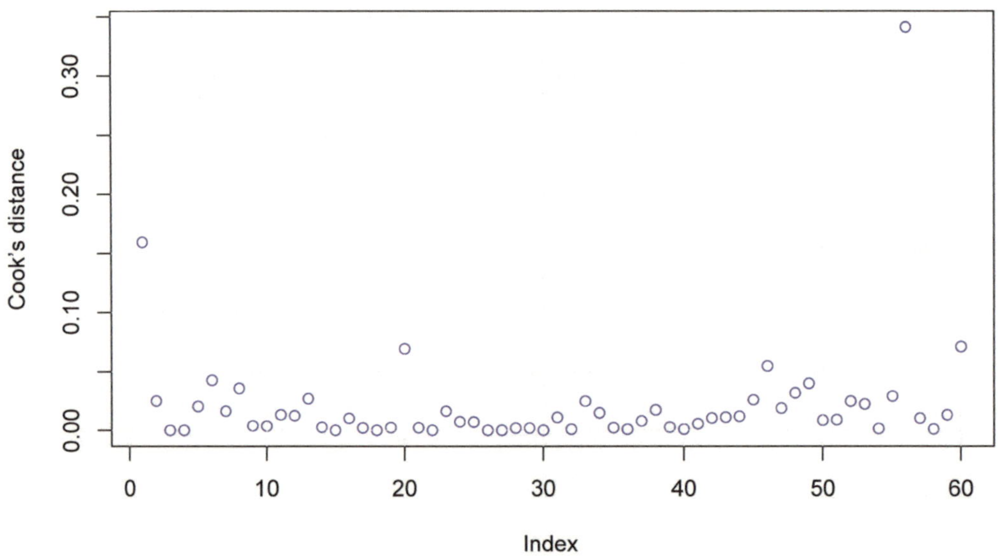

Fig. 6. Cook's distances after Poisson regression

of total yearly hurricanes, and 0.00014 for the ACE data. Again, for the total number of data, one cannot reject the hypothesis of no trend. For the ACE data, the test certainly indicates significant trend; however, keeping the result regarding time scales from the SiZer analysis in mind and recalling that the data set is quite small, there is not enough evidence to point at changes at longer time horizons.

6. Conclusions

Detection of trends are important and common problems in statistical analysis of environmental data. An exploratory analysis by the SiZer method provides a useful complement to other statistical techniques and deserves to be more well-known to statisticans and researchers in the field. The debate on increasing trend of hurricane activity has now and then been intense in the research community. The result provided by this tool, originating from kernel-estimation methodology and scale-space theory, renders a comprehensible visualization of data and in addition a statistical test for significance of slope. For applications of the type presented in this paper, by first applying the SiZer analysis when exploring data, a hint of interesting time periods can be obtained and further statistical tests in those regions can then be performed by other methods, for example techniques based on Poisson regression.

7. Acknowledgements

I am grateful for the publicly available software for SiZer analysis in the form of Matlab routines written by Prof. James S. Marron, Department of Statistics, University of North Carolina, Chapel Hill. These were used to a large extent for the analysis carried out in this paper.

8. References

Brown, L.D., Zhao, L.H. (2002). A test of the Poisson distribution. *Sankhya*, Vol. 64, 611-625.

Chaudhuri, P., Marron, J.S. (1999). SiZer for exploration of structures in curves. *Journal of the American Statistical Association*, Vol. 94, 807-823.

Chaudhuri, P., Marron, J.S. (2000). Scale space view of curve estimation. *Annals of Statistics*, Vol. 28, 408-428.

Elsner, J.B. & Kara, A.B. (1999). *Hurricanes of the North Atlantic: Climate and Society.* Oxford University Press.

Elsner, J.B., Jagger, T., Niu, X.-F. (2000). Changes in the rates of North Atlantic major hurricane activity during the 20th century. *Geophysical Research Letters* Vol. 27, 1743-1746.

Elsner, J.B. & Jagger, T.H. (2004). A hierarchical Bayesian approach to seasonal hurricane modeling. *Jouurnal of Climate*, Vol. 17, 2813-2827.

Godtliebsen, F., Olsen, L.R., Winther, J.-G. (2003). Recent developments in time series analysis: Examples of use in climate research. *Geophysical Research Letters*, Vol. 30, 1654-1657.

Han, S.-R., Guikema, S.D. & Quiring, S.M. (2009). Improving the predictive accuracy of hurricane power outage forecasts using generalized additive models. *Risk Analysis*, Vol. 29, 1443-1453.

Katz, R.W. (2002). Stochastic modeling of hurricane damage. *Journal of Applied Meteorology*, Vol. 41, 754-762.

Jagger, T.H., Elsner, J.B. & Saunders, M.A. (2008). Forecasting U.S. insured hurricane losses. In: *Climate Extremes and Society*, Diaz, H.F. & Murnane, P.J., 189-208, Cambridge University Press.

Lambert, D. (1992). Zero-inflated Poisson regression, with an application to defects in manufacturing. *Technometrics*, Vol. 34, 1-14.

Landsea, C.W., Pielke Jr, R.A., Mestas-Nunez, A.M., Knaff, J.A. (1999). Atlantic basin hurricanes: indices of climatic changes. *Climatic Change*, Vol. 42, 89-129.

Liljequist, G.H. (1949). On fluctuations of the summer mean temperature in Sweden. *Geografiska Annaler*, Vol. 31, 159-178.

Rondonotti, V., Marron, J.S., Park, C. (2007). SiZer for time series: A new approach to the analysis of trends. *Electronic Journal of Statistics*, Vol. 1, 268-289.

Ruppert, D., Sheather, M.J., Wand, M.P. (1995). An effective bandwidth selector for least squares regression. *Journal of the American Statistical Association*, Vol. 90, 1257-1270.

Rychlik, I. & Rydén, J. (2006). *Probability and Risk Analysis. An Introduction for Engineers.* Springer–Verlag.

Rydén, J. (2010). Exploring possibly increasing trend of hurricane activity by a SiZer approach. *Environmental and Ecological Statistics*, Vol. 17, 125-132.

Solow, A.R. (1989). Statistical modeling of storm counts. *Journal of Climate*, Vol. 2, 131-136.

Webster, P.J., Holland, G.J., Curry, J.A., Chang, H.-R. (2005). Changes in tropical cyclone number, duration, and intensity in a warming environment. *Science*, Vol. 309, 1844-1846.

Weckström, J., Korhola, A., Erästö, P., Holmström, L. (2006). Temperature patterns over the past eight centuries in Northern Ferroscandia inferred from sedimentary diatoms. *Quaternary Research*, Vol. 66, 78-86.

Zeileis, A., Kleiber, C. & Jackman, S. (2008). Regression models for count data in R. *Journal of statistical software*, Vol. 27, No. 8, 1-25.

7

Regional Increases in Landfall Frequency and Intensity of Atlantic Hurricanes in a Stochastic Model Forecast

Jessica K Turner
Risk Management Solutions
London,
UK

1. Introduction

Good assessments of hurricane landfall risk that are highly localized to particular geographic locations are needed for a range of applications, such as for insurance pricing, disaster planning and mitigation. Risk assessments are traditionally performed using the full historical record, even though hurricane activity is well-known to exhibit non-stationarity. The scientific literature addressing periodicity and secular changes in hurricane activity is usually focused on broad metrics, such as basin-wide frequency, which have only limited utility to local risk assessment. Stochastic hurricane models are often used for this purpose but are usually created to mimic the long-term average of historical data. This study will present results from a novel stochastic model capable of making predictions of near future shifts in hurricane activity, with a focus on geographical shifts in landfall frequency and intensity.

The existence of multidecadal variability in hurricane activity in the Atlantic, alternating between phases of high and low frequency, is well-known (Delworth and Mann 2000; Goldenberg et al. 2001; Elsner et al. 2004; Jewson and Penzer 2006; Holland and Webster 2007; Klotzbach and Gray 2008). The current high-activity phase began in 1995. Multidecadal phases of hurricane frequency are almost certainly linked to multidecadal phases of increased or decreased sea surface temperature (SST) anomalies in the main development region (MDR). Fluctuations in MDR SST have alternatively been classified as the Atlantic Multidecadal Oscillation (AMO) (Delworth and Mann 2000; Goldenberg et al. 2001; Gray et al. 2004; Knight et al. 2006; Enfield and Cid-Serrano 2009) or the Atlantic Meridional Mode (Kossin and Vimont 2007; Vimont and Kossin 2007). In addition to the multidecadal variability, some researchers have found an increasing trend in frequency (Emanuel 2005; Mann and Emanuel 2006; Emanuel 2008) and intensity (Webster et al. 2005; Elsner 2006; Elsner et al. 2008) linked with anthropogenic climate change. Recent papers have also emphasized the importance of the tropical mean SST, which is dominated by the tropical Indo-Pacific (IP), rather than SST strictly local to the developing storm (Vecchi and Soden 2007; Swanson 2008).

Dynamical models with coupled atmosphere-ocean physics have been used to examine future possible changes in hurricane activity in a warmer world (Bengtsson et al. 2007; Camargo et al. 2007; Caron and Jones 2008; Knutson et al. 2008; Zhao et al. 2009; Bender et al. 2010) and could in theory be used for near future predictions. However, since none of the global circulation models or downscaled regional models can simulate modes of unforced internal variability, they do not correctly capture the current multidecadal variability in the Atlantic. Numerical models also struggle to simulate realistic hurricane intensities (in general producing storms that are too weak) and they are too computationally expensive to generate more than few hurricane seasons. Typically these models are used for experiments simulating climate at the end-of-century or for an environment with doubled atmospheric $C0_2$ whereas this paper is focused on the near-term future.

Stochastic hurricane models have the ability to simulate very large numbers of synthetic hurricanes having similar statistical characteristics to the historical data. This increases the sample size and geographic coverage of events, allowing for risk assessment at a much finer spatial resolution than is possible using only historical events. It also enables the simulation of storms that are physically plausible but are not reproductions of the historical dataset. In general, stochastic models sample from distributions representing various aspects of hurricane activity, genesis density or translational velocities for example, with the parameters of the distributions fit to observations. These types of models are often proprietary and consequently only limited descriptions exist in the published literature. However, some examples can be found (Darling 1991; Chu and Wang 1998; Casson and Coles 2000; Vickery et al. 2000; James and Mason 2005; Rumpf et al. 2007). We will use a stochastic model to examine spatial shifts in hurricane activity characteristics given a prescribed MDR and IP SST which is higher than the historical mean but similar to the mean since 1995.

This chapter is presented in the following way. Section two describes the historical data used, a brief description of the model and the methodology of the experiment. Results are presented in section four along with some discussion of underlying physical mechanisms of the results. Section five gives the conclusions.

2. Data and methodology

a. *Data*

A database of historical, six-hourly observations, called HURDAT, is maintained by the United States' National Hurricane Center (Jarvinen et al. 1984; Landsea et al. 2004). This database was used in the creation and calibration of the stochastic model. Landfall information in the United States for hurricane categories 1-5 is considered accurate after 1900, when the coastal areas were relatively densely populated. However, because data quality away from the North American landmasses is known to be poor in the early years of the dataset, only use data after 1950 was used for hurricane characteristics over water, after which time aircraft reconnaissance became common.

Historical, global, gridded SSTs over the period 1950-2008 from the United Kingdom's Hadley Centre (Rayner et al. 2003) and SST predictions from the A1B scenario of the CMIP3 ensemble of global climate models were also used. The CMIP3 models were incorporated into the Intergovernmental Panel on Climate Change (IPCC) fourth assessment report (IPCC 2007). Finally, the NCEP/NCAR reanalysis was used for upper-air analysis (Kalnay et al. 1996).

b. *Model Description*

The stochastic model used for this paper's near future forecast is based on the track model described in Hall and Jewson (2007). It was developed with the HURDAT historical data, Hadley Centre SSTs and NCEP/NCAR reanalysis, as described in section 2a. The model is fast to run and capable of simulating millions of hurricane tracks from genesis through lysis, with zonal and meridional translational speed and central pressure simulated at six-hourly time steps. At genesis and thereafter at each time step until lysis, the characteristics of the hurricane are simulated from a distribution, the shape of which is dependent on predictors. If a hurricane characteristic is represented by y, then the mean of the distribution to be simulated from can be represented by

$$\bar{y} = f(\beta_0 + \beta_1 X_1 + ... + \beta_n X_n)$$

where $\bar{y}$ is the mean of the variable being simulated, X is a predictor variable and β is a coefficient determining the relationship between the predictor and $\bar{y}$. The β coefficients are scalar constants that vary spatially. They were calculated on a 1°x1° latitude-longitude grid using smoothed historical data to represent $\bar{y}$. Two predictor variables in the model are the area-averaged, July-September MDR (15-70 West, 10-20 North) and IP (IP, 40 East -80 West, 0-15 North) SSTs.

Simulation of an individual track begins with the latitude and longitude of genesis using a spatial Poisson distribution. After genesis location has been determined, the track path and pressure is simulated from

$$x(t+\Delta t) = x(t) + u(t)\Delta t$$

$$y(t+\Delta t) = y(t) + v(t)\Delta t$$

$$p_c(t+\Delta t) = p_c(t) + \frac{\Delta p_c}{\Delta t}\Delta t$$

where Δt is the 6-hourly time step, u and v are the zonal and meridional components of the translational speed and p_c is the central pressure. Axisymmetric wind fields are generated using a modified version of Willoughby et al. (2006). The moment of lysis is simulated using a regression method based on the historical difference between ambient atmospheric pressure and the storm's central pressure; the closer these two values are the more likely the storm will undergo lysis.

c. *Methodology*

Previously, the stochastic model used in this paper was described. Important components of the model are its predictors, two of which are the MDR and IP SST. The model was developed using historical values of these quantities; however, assuming that the β coefficients remain valid, explorations of hurricane activity can then be made using different SSTs. Predictions of the regionalization of near future hurricane activity using this model require only a prediction of future SSTs. We have chosen the five year period 2010-2014 as a prediction period. It is long enough to average over high frequency, unpredictable variability (El Niño variability for example) but short enough for stationarity to be assumed.

To create a prediction of average annual hurricane activity from 2010-2014, the mean and standard deviation of the MDR and IP SST over this period must be predicted.

The SST predictions are taken from a bias corrected ensemble mean of the CMIP3 climate models, concatenated with the twentieth-century radiative forcing runs without volcanoes. The twentieth-century runs do a good job of representing the observed upwards trend in SSTs, but the absolute values must be bias corrected. An optimal window over which to bias correct was found using out-of-sample hindcasts. Since the window length will depend on the magnitude of the trend and variability around the trend, the window length is different in the MDR and the IP. Greater detail on the bias correction can be found in Laepple et al. (2008). Out-of-sample, five-year mean SST hindcasts using the bias-corrected ensemble mean shows skill compared with persistence. For the prediction over 2010-2014, an assumption is made that the probability of a shift from the warm phase to cold in the MDR is very small. The MDR has been in the warm phase since 1995 whereas the length of both the AMO and AMM types of variability is several decades.

For the remainder of this chapter, the output of the stochastic model originally fit and run using the historical values of SST will be called the historical scenario and the output using predicted SSTs will be called the warm scenario. After the mean MDR and IP SSTs are calculated from the method described in the previous paragraph, the standard deviation must be calculated in order to characterize the distribution of SSTs for use in the stochastic model. A normal distribution is assumed, and the standard deviation is taken to be the root-mean-squared-error of the out-of-sample hindcasts of bias-corrected CMIP3 ensemble mean twentieth-century runs. This standard deviation is smaller than the historical standard deviation of the SST in the MDR and IP.

Table 1 shows the mean and standard deviation for the historical and warm scenario predictions. It can be seen that the MDR is historically cooler and has greater variability than the IP. In the warm scenario, the SST in the MDR is increased less than 1°C, but almost 2 standard deviations of its natural variability. The IP SST is also increased, but by a smaller fraction of its natural variability. Therefore the warm scenario has MDR SSTs high relative to the IP SSTs, as is true during active phases of the Atlantic.

		MEAN	STANDARD DEVIATION
MDR	Historical	27.34	0.31
	Warm	27.93	0.26
IP	Historical	28.05	0.27
	Warm	28.40	0.22

Table 1. The mean and standard deviation of the MDR and IP SST distributions for the historical and warm model scenarios.

Given an MDR and IP SST distribution, the stochastic model can be run using these predictors. Comparisons of the two scenarios, historical and warm, will be made, with an emphasis on the regionalization of landfall and intensity shifts. Although the model output can be examined at very high resolution, this paper will present results for the regions shown in Figure 1.

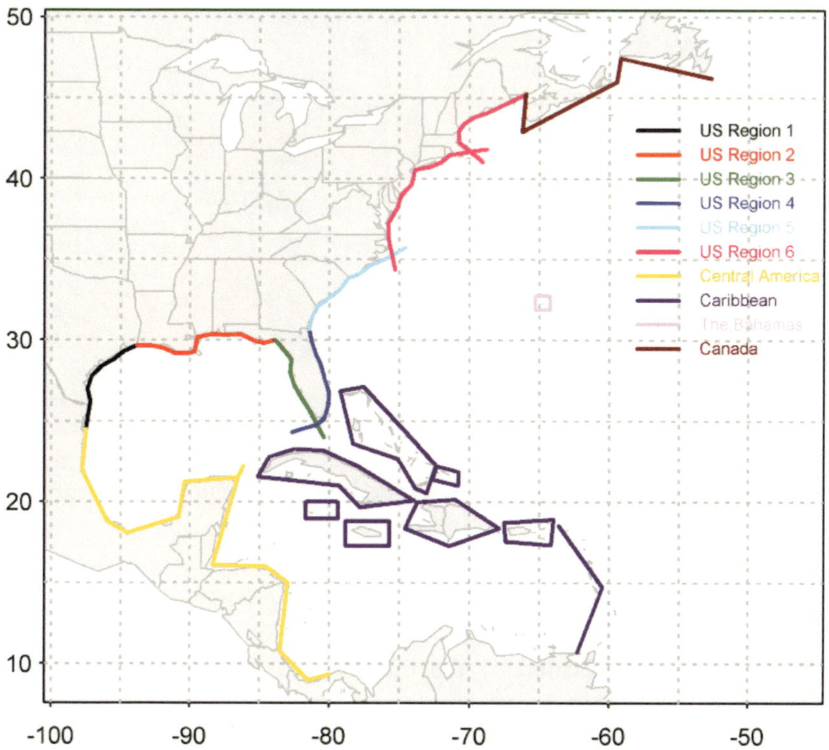

Fig. 1. Landfall comparison regions. There are extensions into the Atlantic to include by-passing storms in some areas.

3. Results

a. *Landfall rate regionalization*

One of the metrics of regionalization is the spatial distribution of landfall rates. In the warm scenario of the stochastic model genesis rates increase significantly in the main development region. As a consequence, landfall rate density increases everywhere, although with geographic regionalization of the increases. Figure 2 shows the percentage change in landfall rate by region (Fig. 1) for the warm scenario compared to historical, for categories 1-2 and 3-5. Landfall rate increases are larger for categories 3-5 than for categories 1-2, implying that intense storms are more strongly affected by increased SSTs. The Caribbean, Central America and the southeast United States from Atlantic Florida to North Carolina have the largest increases geographically.

Having discussed the regionalization shifts of landfall rate in the warm scenario compared to the historical, we will now examine the causes of this regionalization. That is, why the warm scenario has smaller increases in landfall rate along the Texas and northeastern coasts and larger increases from Louisiana to North Carolina. To address that question, Figure 3 shows the percentage change in genesis density in the warmed scenario compared to the historical, for hurricanes going on to make landfall in each of the six U.S. regions (Fig. 1), in 1°x1° latitude-longitude bins. Included in the figure are the mean, 10th and 90th percentile of tracks making landfall in each region, for both warm and historical scenarios. The track

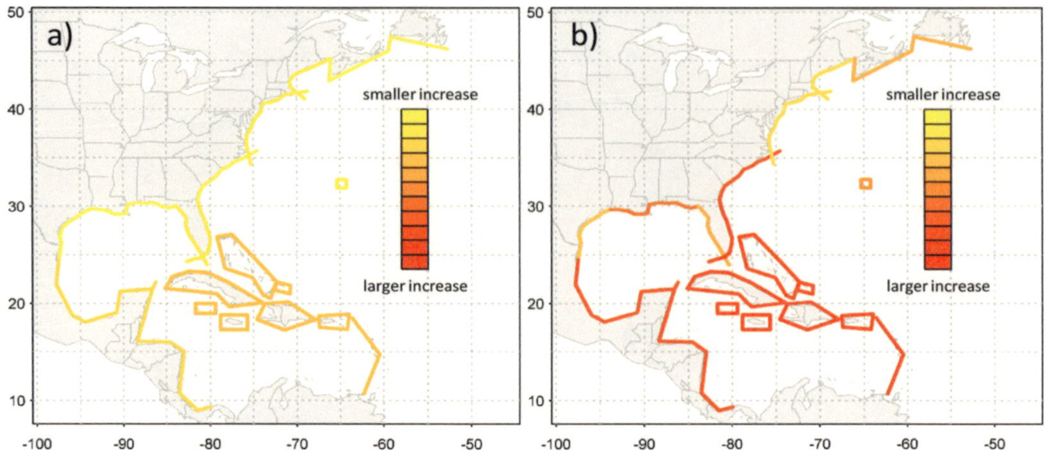

Fig. 2. Percentage increase in the warm scenario compared to historical for hurricane landfall at a) categories 1-2 and b) categories 3-5.

lengths represent the mean, 10th and 90th percentile of hurricane lifetime from genesis to landfall in the region. Average annual rate of genesis density in five boxes is given with the percentage change in inset text.

Mean hurricane lifetimes (indicated by the track length) increase in the warm scenario, another consequence of a shift in genesis density towards the eastern MDR and further from land. Other than increasing in length, the mean tracks do not change direction in the warm scenario. This indicates that the landfall rate regionalization patterns (Fig. 2) can be explained by changes in patterns of genesis rather than changes in track geometry.

Overall, genesis density in the eastern MDR (box 5) increases most in the warm scenario, even in regions where it is not the largest contributor to the total landfall rate (i.e. regions 1-3). In regions 4-6, storms originating in the eastern MDR become the largest contributors to landfall in the warm scenario while the western MDR (box 4) was the largest contributor to landfall in the historical scenario. Landfalls from hurricanes originating in the western MDR also increase in each region, although with smaller magnitude than those from the eastern MDR.

Unlike the increases in MDR genesis density, there are decreases in the eastern Gulf of Mexico and in the north Caribbean Sea. For example, region 1 (the Texas coast in Figure 3a) has one of the smaller magnitude increases in landfall rate (Fig. 2) in the United States. Texas receives a large proportion of its landfalls from storms with a genesis in the Gulf of Mexico (0.1 per year from box 2, compared with 0.07 from box 4 and 0.04 from box 5 in the historical scenario). However, genesis density in box 2 increases only a small amount compared to the MDR boxes because the eastern Gulf of Mexico has a significant decrease in genesis density.

Region 2 (Fig. 3b) receives a proportion of its landfalls from the Gulf of Mexico but there are also large increases in MDR hurricanes making landfall in region 2 in the warm scenario. Region 2 many landfalls also originate from box 3, the western Caribbean Sea around the Gulf of Honduras. Box 3 has an increase in genesis density in the warm scenario. The last Gulf of Mexico region is region 3, the west coast of Florida (Fig. 3c). Region 3 also has a large contribution to its landfalls from box 3. Storms with genesis in this box track westward on

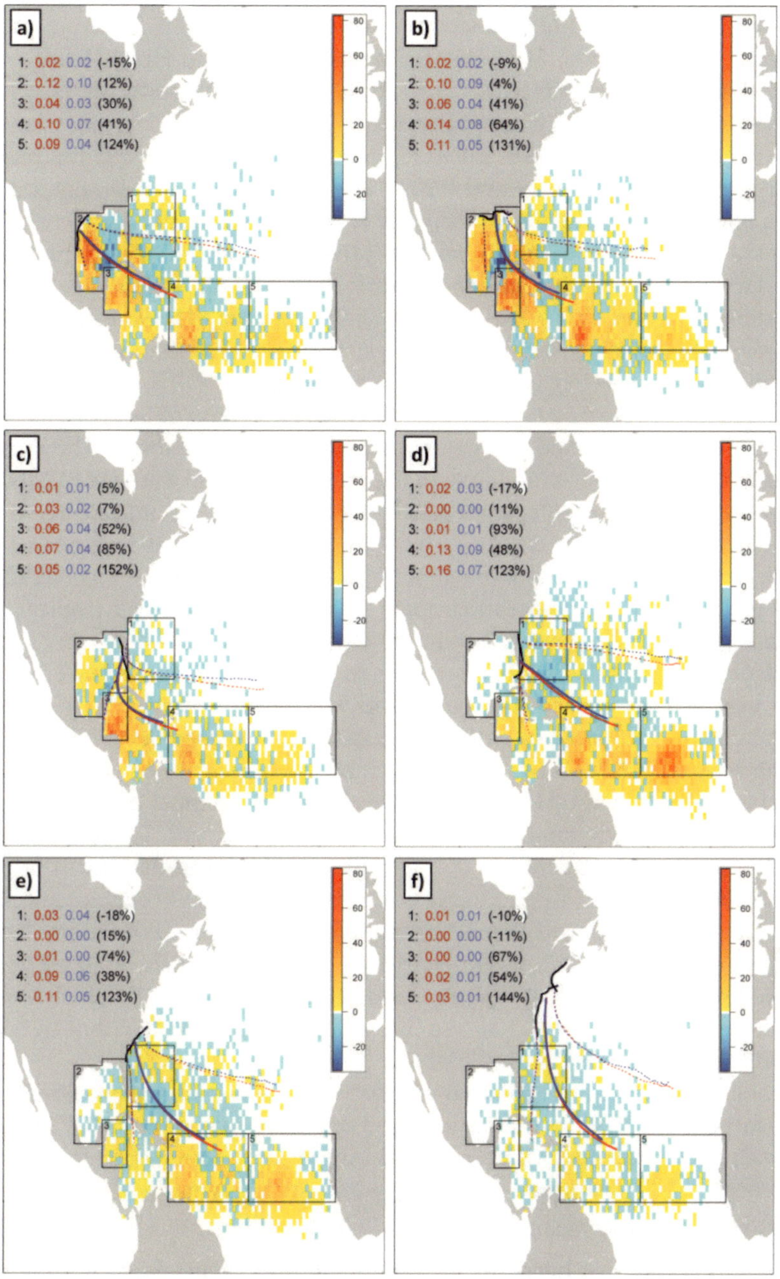

Fig. 3. Percentage change in hurricane genesis density of storms making landfall in a-f) regions 1-6 in the warm scenario compared to the historical. Warm (cool) colors indicate increased genesis density in the warm (historical) scenario. Mean, 10th and 90th percentile tracks are also shown for the warm (red) and historical scenario (blue). Average annual rate of genesis density in five boxes are given in inset text for the warm scenario (red) and historical scenario (blue) with the percentage change (black).

average, crossing into the Bay of Campeche and making landfall in Mexico. However, many of them also track northward to make landfall in western Florida.

Regions 4 (Fig. 3d) and 5 (Fig. 3e) have the largest increase in landfall rate in the United States (Fig. 2), especially for categories 3-5. These regions are very commonly hit by both western and eastern MDR-originating hurricanes. The combined annual landfall rate of storms from boxes 4 and 5 is 0.16 and 0.11 in the historical scenario for region 4 and 5 respectively. This increases to 0.29 and 0.20 in the warm scenario - almost a doubling. For region 5, this increase is somewhat reduced by the decreases in genesis density in the north Caribbean in the warm scenario. Finally, region 6, the northeast (Fig. 6f) experiences only small increases in landfall rate in the warm scenario (Fig. 2). This region does not experience frequent landfall in either the warm or historical scenarios. The landfalls it does receive originate largely in box 1, a box with decreased genesis density in the warm scenario.

b. *Physical mechanisms affecting genesis density regionalization*

It has been shown in the previous section that the regionalization of landfall rates in the warm scenario (Fig. 2) can be explained by regionalization of genesis density. The most significant change in the warm scenario was an increase in genesis density in the eastern MDR, the region of Cape Verde-type developments. If the MDR is very warm, African easterly waves leaving the coast of Africa can develop quickly. The shift towards the east in genesis with warmer MDR SSTs has been found in at least one dynamical modelling study. Wu et al. (2010) ran the ECHAM 4.8 global, atmosphere-ocean coupled model with a warmed tropical Atlantic and found an increase in hurricane formation in the eastern MDR. This shift can also be found in the HURDAT observational record. For example, over the period 1950-2009, 2.2 hurricanes had their genesis east of 25W on average, compared with 3 on average in the post-1995 active regime. The observed shift could be spurious because of increased observations in the eastern Atlantic in later years, although the hurricane count is believed to be complete after satellite observations begin in the 1960s (Holland and Webster 2007; Emanuel 2008).

The overall increase in MDR genesis and the shift towards the east in the warm scenario can be explained in part with reference to simple physical mechanisms known to be related to SST in the MDR and IP. If all other atmospheric variables are held constant and the local SST is increased, the potential intensity for a hurricane is also increased (Emanuel 1987). Figure 4 shows the correlation coefficient of July-September local SST in the Hadley Centre dataset and the July- September MDR predictor used in the fitting of the stochastic model's β coefficients. It can be seen that SSTs across the MDR and southern Caribbean Sea are highly correlated with the model's MDR predictor.

Local SST changes alone cannot explain the patterns of genesis density regionalization in the warm scenario because SSTs in the eastern and western half of the MDR are both highly correlated with the predictor MDR. Another important environmental factor affecting hurricane genesis and development is the wind shear, usually defined as the vector difference between monthly mean 850 hPa and 200 hPa winds ($V_s = |u_{850} - u_{200}|$). Local wind shear is known to act against hurricane development and is correlated with both MDR and IP SSTs. Aiyyer and Thorncroft (2006) regressed an MDR index (with a slightly different but similar areal definition of MDR to that used in this paper) and the Niño 3.4 index (a measure of IP warmth) on July-October vertical wind shear in European Centre for Medium-Range Weather Forecasts (ECMWF) 40-year reanalysis dataset (ERA-40). They showed that, whereas the MDR SSTs are associated with decreased wind shear in the MDR, the Niño 3.4 region SSTs are associated with increased wind shear, especially in the western MDR and

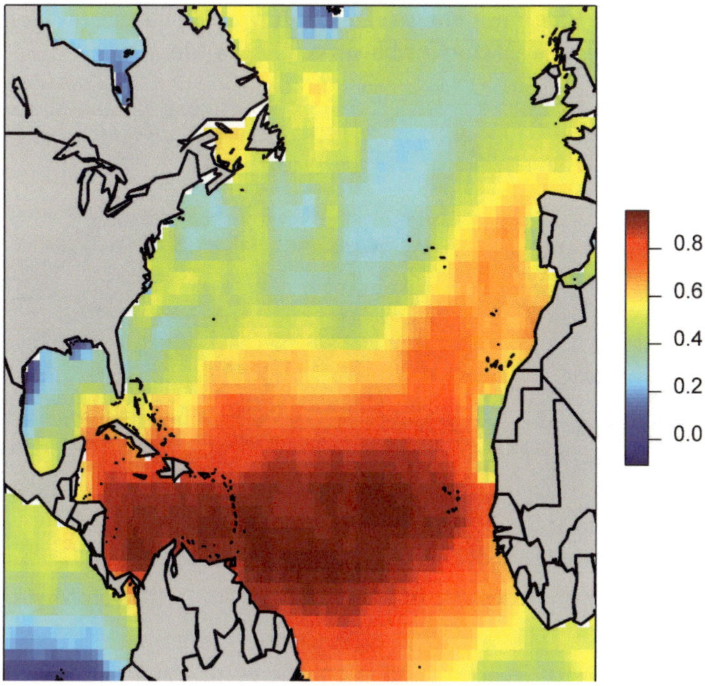

Fig. 4. Correlation coefficient of July-September local SST from the Hadley Centre dataset and the model's MDR SST predictor over the period 1880-2009.

southern Caribbean regions. We speculate that the genesis density regionalization found in the warm scenario is the result of a combination of the interaction between increased wind shear in the western MDR and south Caribbean owing to increased IP SST in the warm scenario, and an overall increase in genesis from increased local SST across the MDR.

Tropical transition-type storms with a mid-latitude synoptic precursor are most likely to form in the Gulf of Mexico, near the east coast of the United States and north Caribbean Sea, areas much less correlated with the MDR predictor (Fig. 4). Geneses of tropical transition storms are more dependent on atmospheric dynamics than local SST (Bracken and Bosart 2000; Hulme and Martin 2009). They often make landfall on the Gulf of Mexico or east coasts of the United States, especially from South Carolina to Maine. These regions had a smaller increase in landfall rate in the warm scenario (Fig. 2) and small increases or decreases in hurricane genesis density in the warm scenario (Fig. 3).

c. *Intensity Changes*

We have shown how landfall rates are regionalized in the warm scenario owing to regionalization changes in genesis density (Section 3). In the warm scenario the intensity distribution of hurricanes also shifts towards more intense storms. Figure 5 shows the 95th percentile of hurricane central pressure in storms passing through 1°x1° longitude-latitude bins in the warm and historical scenarios. It can be seen the central pressure decreases in most areas of the Atlantic in the warm scenario, especially in the central Caribbean Sea, Gulf of Mexico and around Atlantic Florida. Hurricane intensity, besides being intensified by local SST, is also determined by the length of time the storm has spent over warm water. In the warm scenario, the average lifetime of hurricanes is longer because their genesis shifts towards the east, further from land masses. This shift provides a greater length of time over which to intensify before reaching the Caribbean or Gulf of Mexico.

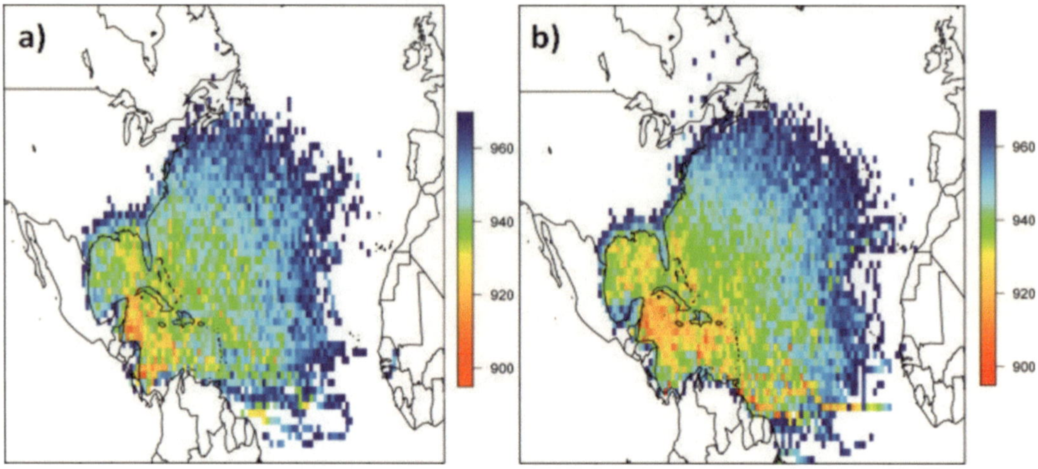

Fig. 5. 95th percentile of central pressure in 1^{0}x1^{0} latitude-longitude bins for the a) historical scenario and b) warm scenario.

Increased hurricane intensity in the Atlantic translates to increased intensity at landfall. Figure 6 shows the distribution of central pressure at landfall for all regions in the historical and warm scenarios. It can be seen the distribution has shifted towards higher intensities.

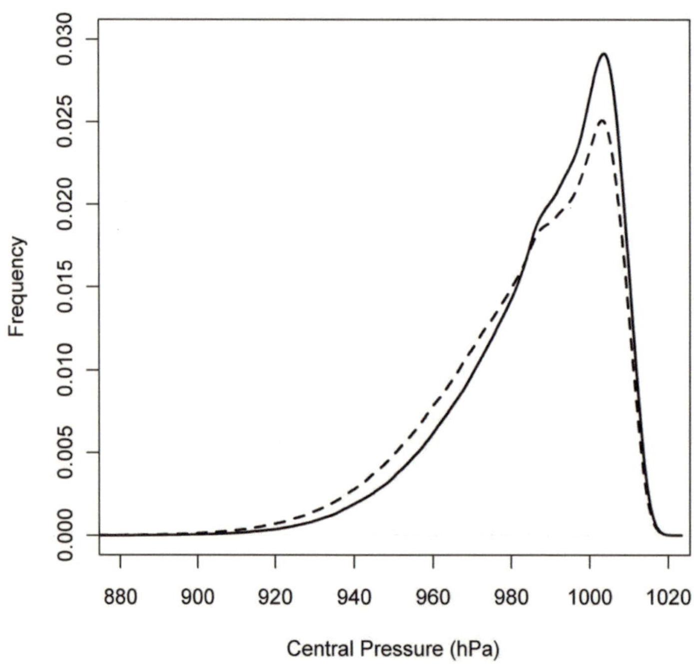

Fig. 6. Distribution at landfall of central pressure for all regions in the historical scenario (solid line) and warm scenario (dashed line).

Overall, central pressure at landfall shifts towards lower pressure in the warm scenario. However, it was seen in Figure 6 that there was basin-wide regionalization to the intensity increases; therefore, regionalization in landfall intensity can be expected. Table 2 gives the mean, 5th and 95th percentile of maximum 1 minute sustained, 10 meter height wind speed at landfall (maximum wind speed is strongly correlated with damage and storm surge height) for the regions in Figure 1. Regional patterns in percentage increases in intensity are similar to the patterns of frequency increases (Fig 2). There is no change in the 95th percentile of wind speed in Region 1, Texas, and only a 3% increase in the northeast of the United States. Atlantic Florida, Central America and the Caribbean experience the largest changes in the 95th percentile in the warm scenario, ranging from 5% to 8% increases. The pairing of frequency and intensity increases in the warm scenario is expected, since it was shown (Fig. 3) that the regionalization changes are driven by a shift in genesis density to the MDR, especially the eastern MDR. These Cape Verde storms hit Atlantic Florida, Central America and the Caribbean the most frequently, spend the longest time over warm, tropical water and have the highest wind speeds.

	Historical ($m\ s^{-1}$)			Warm ($m\ s^{-1}$)			Percentage Change ($m\ s^{-1}$)		
	5th Percentile	Mean	95th Percentile	5th Percentile	Mean	95th Percentile	5th Percentile	Mean	95th Percentile
Region 1	15	33	57	16	32	57	7%	-3%	
Region 2	15	31	55	15	33	58		6%	5%
Region 3	13	31	56	14	33	59	8%	6%	5%
Region 4	13	32	58	14	35	61	8%	9%	5%
Region 5	14	29	51	14	31	53		7%	4%
Region 6	11	24	40	12	24	41	9%		3%
Central America	14	32	61	14	34	65		6%	7%
Caribbean	13	32	60	13	35	65		9%	8%
Bermuda	14	30	52	14	32	54		7%	4%
Canada	12	22	33	12	23	34		5%	3%

Table 2. Mean, 5th and 95th percentile of peak wind speed at landfall by region for the historical and warm scenarios. The percentage change in the warm scenario compared to the historical scenario is given.

4. Summary and conclusions

An experiment was performed using a stochastic Atlantic hurricane model with main development region (MDR) and tropical Indo-Pacific (IP) sea surface temperatures (SSTs) as predictors. The model has genesis, track displacement and intensity components fit to these predictors using historical data. The stochastic model was run once using historical values of the SST predictors and then again using a forecast of SSTs for the period 2010-2014.

Regionalization changes in landfall rate in the warm scenario compared to the historical scenario were shown (Fig. 2). The largest increases in landfall rate were for the major

hurricanes (categories 3-5). Increased rates were found everywhere in the warm scenario but they were particularly high in the Caribbean, Central America and Atlantic Florida. This pattern of rate increases was shown to be owing to an increase in MDR originating storms, particularly the easternmost Cape-Verde type storms, compared to other types (Fig. 3). For some regions there was a smaller increase in landfall rate, including the United States' Texas and northeastern coasts. These regions receive a smaller percentage of landfalls from MDR and Cape Verde originating storms in the warmed and historical runs but receive more of their landfalls from storms having genesis in the Gulf of Mexico in the case of Texas and from the north Caribbean Sea in the case of the northeast. Both the Gulf of Mexico and north Caribbean Sea were shown to be decreasing in genesis density in the warm scenario (Fig. 3).
Changes in the intensity distribution of hurricanes in the warm scenario were also examined. Intensities were found to be increasing in the warm scenario compared to the historical scenario, especially in the Gulf of Mexico and Caribbean Sea (Fig. 5). Local SSTs act to increase hurricane intensity and in addition, a shift in genesis density towards the eastern MDR gives storms a longer period over water, increasing the time over which they have to intensify. This was found to have a consequence on both the distribution of central pressure (Fig. 6) and on wind speeds at landfall (Table 2). The geographic pattern of intensification follows that of the landfall rate pattern. The largest intensifications in mean wind speed at landfall are in Atlantic Florida, the southeastern United States and the Caribbean. These areas are particularly vulnerable to intense MDR and Cape Verde storms and are shown by our stochastic model to be at the highest risk of hurricane damage.

The author would like to acknowledge Enrica Bellone for her assistance running the stochastic model and Katie Coughlin.

5. References

Bender, M. A., T. R. Knutson, R. E. Tuleya, J. J. Sirutis, G. A. Vecchi, S. T. Garner and I. M. Held, (2010): Modeled Impact of Anthropogenic Warming on the Frequency of Intense Atlantic Hurricanes. *Science*, 327, 454-458.

Bengtsson, L., K. I. Hodges, M. Esch, N. Keenlyside, L. Kornblueh, J.-J. Luo and T. Yamagata, (2007): How may tropical cyclones change in a warmer climate? *Tellus A*, 59, 539-561.

Bracken, W. E. and L. F. Bosart, (2000): The Role of Synoptic-Scale Flow during Tropical Cyclogenesis over the North Atlantic Ocean. *Mon. Wea. Rev.*, 128, 353-376.

Camargo, S. J., A. H. Sobel, A. G. Barnston and K. A. Emanuel, (2007): Tropical cyclone genesis potential index in climate models. *Tellus A*, 59, 428-443.

Caron, L.-P. and C. G. Jones, (2008): Analysing present, past and future tropical cyclone activity as inferred from an ensemble of Coupled Global Climate Models. *Tellus A*, 60, 80-96.

Casson, E. and S. Coles, (2000): Simulation and extremal analysis of hurricane events. *Journal of the Royal Statistical Society: Series C (Applied Statistics)*, 49, 227-245.

Chu, P.-S. and J. Wang, (1998): Modeling Return Periods of Tropical Cyclone Intensities in the Vicinity of Hawaii. *J. Appl. Meteor.*, 37, 951-960.

Darling, R. W. R., (1991): Estimating Probabilities of Hurricane Wind Speeds Using a Large-Scale Empirical Model. *J Climate*, 4, 1035-1046.

Delworth, T. L. and M. E. Mann, (2000): Observed and simulated multidecadal variability in the Northern Hemisphere. *Climate Dyn.*, 16, 661-676.

Elsner, J. B., (2006): Evidence in support of the climate change-Atlantic hurricane hypothesis. *Geophys. Res. Lett.*, 33.

Elsner, J. B., J. P. Kossin and T. H. Jagger, (2008): The increasing intensity of the strongest tropical cyclones. *Nature*, 455, 92-95.

Elsner, J. B., X. Niu and T. H. Jagger, (2004): Detecting Shifts in Hurricane Rates Using a Markov Chain Monte Carlo Approach. *J Climate*, 17, 2652-2666.

Emanuel, K., (2005): Increasing destructiveness of tropical cyclones over the past 30-years. *Nature*, 436, 686-688.

Emanuel, K., (2008): The Hurricane-Climate Connection. *Bull. Amer. Meteor. Soc.*, 89, ES10-ES20.

Emanuel, K. A., (1987): The dependence of hurricane intensity on climate. *Nature*, 326, 483-485.

Enfield, D. B. and L. Cid-Serrano, (2009): Secular and multidecadal warmings in the North Atlantic and their relationships with major hurricane activity. *International Journal of Climatology*.

Goldenberg, S. B., C. W. Landsea, A. M. Mestas-Nunez and W. M. Gray, (2001): The Recent Increase in Atlantic Hurricane Activity: Causes and Implications. *Science*, 293, 474-479.

Gray, S. T., L. J. Graumlich, J. L. Betancourt and G. T. Pederson, (2004): A tree-ring based reconstruction of the Atlantic Multidecadal Oscillation since 1567 A.D. *Geophys. Res. Lett.*, 31.

Holland, G. J. and P. J. Webster, (2007): Heightened tropical cyclone activity in the North Atlantic: natural variability or climate trend? *Philosophical Transactions of the Royal Society A: Mathematical, Physical and Engineering Sciences*, 365, 2695-2716.

Hulme, A. L. and J. E. Martin, (2009): Synoptic- and Frontal-Scale Influences on Tropical Transition Events in the Atlantic Basin. Part I: A Six-Case Survey. *Mon. Wea. Rev.*, 137, 3605-3625.

James, M. K. and L. B. Mason, (2005): Synthetic tropical cyclone database. *Journal of Waterway, Port, Coastal and Ocean Engineering*, 131, 181-192.

Kalnay, E., M. Kanamitsu, R. Kistler, W. Collins, D. Deaven, L. Gandin, M. Iredell, S. Saha, G. White, J. Woollen, Y. Zhu, A. Leetmaa, R. Reynolds, M. Chelliah, W. Ebisuzaki, W. Higgins, J. Janowiak, K. C. Mo, C. Ropelewski, J. Wang, R. Jenne and D. Joseph, (1996): The NCEP/NCAR 40-Year Reanalysis Project. *Bull. Amer. Meteor. Soc.*, 77, 437-471.

Klotzbach, P. J. and W. M. Gray, (2008): Multidecadal Variability in North Atlantic Tropical Cyclone Activity. *J Climate*, 21, 3929-3935.

Knight, J. R., C. K. Folland and A. A. Scaife, (2006): Climate impacts of the Atlantic Multidecadal Oscillation. *Geophys. Res. Lett.*, 33.

Knutson, T. R., J. J. Sirutis, S. T. Garner, G. A. Vecchi and I. M. Held, (2008): Simulated reduction in Atlantic hurricane frequency under twenty-first-century warming conditions. *Nature Geosci*, 1, 359-364.

Kossin, J. P. and D. J. Vimont, (2007): A More General Framework for Understanding Atlantic Hurricane Variability and Trends. *Bull. Amer. Meteor. Soc.*, 88, 1767-1781.

Landsea, C. W., C. Anderson, N. Charles, G. Clark, J.Dunion, J. Fernandez-Partagas, P. Hungerford, C.Neumann and M. Zimmer, (2004): The Atlantic hurricane database re-analysis project: Documentation for the 1851–1910 alterations and additions to the HURDAT database. *Hurricanes and Typhoons: Past, Present and Future.* R. J. Murname and K.-B. Liu. Columbia Univ. Press, 177–221.

Mann, M. E. and K. A. Emanuel, (2006): Atlantic Hurricane Trends Linked to Climate Change. *Eos, Trans. Amer. Geophys. Union*, 87.

Rayner, N. A., D. E. Parker, E. B. Horton, C. K. Folland, L. V. Alexander, D. P. Rowell, E. C. Kent and A. Kaplan, (2003): Global analyses of sea surface temperature, sea ice, and night marine air temperature since the late nineteenth century. *Geophys. Res. Lett.*, 108, 4407.

Rumpf, J., H. Weindl, P. Höppe, E. Rauch and V. Schmidt, (2007): Stochastic modelling of tropical cyclone tracks. *Mathematical Methods of Operations Research*, 66, 475-490.

Swanson, K. L., (2008): Nonlocality of Atlantic tropical cyclone intensities. *Geochem. Geophys. Geosyst.*, 9.

Vecchi, G. A. and B. J. Soden, (2007): Effect of remote sea surface temperature change on tropical cyclone potential intensity. *Nature*, 450, 1066-1070.

Vickery, P. J., P. F. Skerlj and L. A. Twisdale, (2000): Simulation of Hurricane Risk in the U.S. Using Empirical Track Model. *Journal of Structural Engineering*, 126, 1222-1237.

Vimont, D. J. and J. P. Kossin, (2007): The Atlantic Meridional Mode and hurricane activity. *Geophys. Res. Lett.*, 34.

Webster, P. J., G. J. Holland, J. A. Curry and H.-R. Chang, (2005): Changes in Tropical Cyclone Number, Duration, and Intensity in a Warming Environment. *Science*, 309, 1844-1846.

Wu, L., L. Tao and Q. Ding, (2010): Influence of Sea Surface Warming on Environmental Factors Affecting Long-Term Changes of Atlantic Tropical Cyclone Formation. *J Climate*, 23, 5978–5989.

Zhao, M., I. M. Held, S.-J. Lin and G. A. Vecchi, (2009): Simulations of Global Hurricane Climatology, Interannual Variability, and Response to Global Warming Using a 50-km Resolution GCM. *J Climate*, 22, 6653-6678.

IPCC (2007). Climate Change 2007: Synthesis Report. Contribution of Working Groups I, II and III to the Fourth Assessment Report of the Intergovernmental Panel on Climate Change. R. K. Pachauri and A. Reisinger. Geneva, Switzerland: 104.

Jarvinen, B. R., C. J. Neumann and M. A. S. Davis (1984). A Tropical Cyclone Data Tape for the North Atlantic Basin 1886-1983: Contents Limitations and Uses, U.S. Department of Commerce.

Jewson, S. and J. Penzer (2006). An objective change-point analysis of historical Atlantic hurricane numbers.

Part 2

Tropical Cyclone Dynamics and Structure

8

Barotropic Aspects of Hurricane Structural and Intensity Variability

Eric A. Hendricks[1], Hung-Chi Kuo[2], Melinda S. Peng[3] and Daniel Hodyss[4]
[1,3,4]*Marine Meteorology Division, Naval Research Laboratory, Monterey, CA*
[2]*National Taiwan University, Taipei*
[1,3,4]*USA*
[2]*Taiwan*

1. Introduction

Hurricanes are intense atmospheric vortices characterized by extreme winds, torrential rain, and destructive storm surges. When a major hurricane makes landfall one or more of these processes can cause immense property damage and loss of life. Considerable progress has been made in recent decades unlocking the physical and dynamical mechanisms by which hurricanes form, and by which they change their structure and intensity. In spite of this progress, intensity forecast skill has shown little to no improvement in the past twenty years. During this same time period, global numerical weather prediction models have steadily improved in their ability to predict the large-scale environment. However, better knowledge of the environment has not led to significant intensity forecast skill improvements. The reason for this is that hurricane intensity and structure change is controlled to a large degree by internal dynamical processes, which are less well understood and predictable. Although these internal processes interact with the large-scale environment in complicated ways, much insight can first be obtained by examining these processes in the absence of environmental and ocean surface forcing, and in a two-dimensional framework.

We define internal dynamical processes here as those processes that occur within the region defined by twice the radius of maximum wind, or typically within approximately 150 km of the center of the hurricane, and whose dynamics is constrained to the hurricane vortex itself and not the environment. The internal processes of principle importance for the evolution of the hurricane vortex are the formation and evolution of spiral rainbands, mixing between the eye and eyewall, and concentric eyewall cycles. While in the real atmosphere, these processes are inherently three-dimensional and influenced by moist processes and friction, much insight into the dynamics can first be obtained in a simpler unforced barotropic framework. Strictly speaking, *barotropic* means that atmopheric pressure depends only on density, while *baroclinic* means that pressure depends both on density and temperature. Since tropical cyclones are surface based vortices whose winds decay with increasing height, the vertical shear of the vortex tangential winds are balanced by radial temperature gradients, making it baroclinic and giving it a warm-core signature at upper levels. However, if one is considering the behavior of a tropical cyclonce (hereafter TC) at lower levels where the vortex tangential winds are not changing significantly with height, a barotropic model is a good starting point.

The outline of this chapter is as follows. In section 2, we present the dynamical model used for this study, as well the fundamental principles associated with it. In section 3, the dynamical model and observations are used to examine barotropic aspects of inner and outer spiral rainbands, instabilities and mixing processes between the eye and eyewall, and concentric eyewall cycles. A summary of the results is given in section 4.

2. Dynamical model and fundamentals

2.1 Divergent barotropic model

Our dynamical model here is the divergent barotropic (or shallow water) equations on an f-plane in polar coordinates (r,ϕ,t), where r is the radius from the origin, ϕ is the azimuthal angle, and t is the time. The equations governing radial momentum u and tangential momentum v are

$$\frac{Du}{Dt} - fv - \frac{v^2}{r} + g\frac{\partial h}{\partial r} = 0, \tag{1}$$

$$\frac{Dv}{Dt} + fu + \frac{uv}{r} + g\frac{\partial h}{r\partial \phi} = 0, \tag{2}$$

where $D/Dt = \partial/\partial t + u\partial/\partial r + (v/r)\partial/\partial\phi$ is the material derivative. Equations (1) and (2) indicate that in a Lagrangian reference frame, an air parcel may be accelerated by the Coriolis, centrifugal, and pressure gradient forces, respectively. By taking the curl and divergence of equations (1) and (2), the equations governing the relative vorticity $\zeta = \partial(rv)/r\partial r - \partial u/r\partial\phi$ and divergence $\delta = \partial(ru)/r\partial r + \partial v/r\partial\phi$ are

$$\frac{\partial \zeta}{\partial t} + \frac{\partial[r(f+\zeta)u]}{r\partial r} + \frac{\partial[(f+\zeta)v]}{r\partial\phi} = 0, \tag{3}$$

$$\frac{\partial \delta}{\partial t} - \frac{\partial[r(f+\zeta)v]}{r\partial r} + \frac{\partial[(f+\zeta)u]}{r\partial\phi} + \nabla^2\left[gh + \tfrac{1}{2}\left(u^2+v^2\right)\right] = 0. \tag{4}$$

The conservation of mass is

$$\frac{\partial h}{\partial t} + \frac{\partial(rhu)}{r\partial r} + \frac{\partial(hv)}{r\partial\phi} = 0, \tag{5}$$

where h is the fluid depth. It is convenient to express u, v, ζ, δ in terms of the velocity potential χ and the streamfunction ψ as

$$u = \frac{\partial\chi}{\partial r} - \frac{\partial\psi}{r\partial\phi}, \qquad v = \frac{\partial\chi}{r\partial\phi} + \frac{\partial\psi}{\partial r}, \tag{6}$$

$$\zeta = \nabla^2\psi, \qquad \delta = \nabla^2\chi, \tag{7}$$

where $\nabla^2 = (\partial/r\partial r)(r\partial/\partial r) + \partial^2/r^2\partial^2\phi$. By combining (3) and (5) while eliminating the divergence, the potential vorticity principle is obtained:

$$\frac{DP}{Dt} = 0, \tag{8}$$

where $P = (f+\zeta)/h$ is the potential vorticity. Equation (8) indicates that for shallow water motions in the absence of friction, individual fluid parcels conserve their PV as they are advected in the flow.

2.2 Diagnostic balance equation

By setting $\partial\delta/\partial t = 0$ in (4), and neglecting terms involving the velocity potential χ, the relationship

$$f\nabla^2\psi + 2\left[\frac{\partial^2\psi}{\partial r^2}\left(\frac{1}{r^2}\frac{\partial^2\psi}{\partial\phi^2} + \frac{1}{r}\frac{\partial\psi}{\partial r}\right) - \left(\frac{1}{r}\frac{\partial^2\psi}{\partial r\partial\phi} - \frac{1}{r^2}\frac{\partial\psi}{\partial\phi}\right)^2\right] = g\nabla^2 h \tag{9}$$

between ψ and h is obtained. This is the *nonlinear balance equation* in polar coordinates. Equation (9) indicates that given ψ or h with appropriate boundary conditions, the other field which is in quasi-balance with that field may be obtained. Since $P = (\nabla^2\psi + f)/h$, equations (8) and (9) are sufficient to describe the quasi-balanced evolution of a shallow water vortex. By substituting $h = (\nabla^2\psi + f)/P$ into the right hand side of (9), an invertibility relationship is obtained, where ψ, h, ζ, u and v may be obtained diagnostically, given P. This invertibility principle is the reason P is such a useful dynamical quantity. The shallow water dynamics can be succintly speficied by one prognostic PV equation and a diagnostic relationship for the other quasi-balanced fields. For the special case of axisymmetric flow in which $v = \partial\psi/\partial r$ and $u = -\partial\psi/r\partial\phi = 0$, equation (9) reduces to

$$\frac{\partial}{r\partial r}\left[r\left(fv + \frac{v^2}{r} - g\frac{\partial h}{\partial r}\right)\right] = 0, \tag{10}$$

implying that, when enforcing the boundary conditions,

$$fv + \frac{v^2}{r} - g\frac{\partial h}{\partial r} = 0, \tag{11}$$

which is the gradient wind balance equation. In this context, (9) is a more general formulation of the gradient wind balance equation when the flow is both asymmetric and balanced.

2.3 Linearization

Equations (3)-(5) cannot be solved analytically due to the nonlinear terms. However, under the assumption that perturbations are small in comparison to their means, a modified set of equations can be obtained that can be solved analytically. This process is referred to as *linearization*. In order to linearize (3)-(5), a basic state chosen here is that of an arbitrary axisymmetric vortex $\bar{\omega}(r) = \bar{v}(r)/r$ that is in gradient balance with the height $\bar{h}(r)$. While our basic state vortex is prescribed to be axisymmetric, the basic state could similarly be obtained by taking an azimuthal mean, i.e.,

$$\bar{v}(r) = \frac{1}{2\pi}\int_0^{2\pi} v(r,\phi)\, d\phi, \tag{12}$$

and similar definitions hold for the other variables. The decomposition for all variables is as follows: $h(r) = \bar{h}(r) + h'(r,\phi,t)$, $v(r) = \bar{v}(r) + v'(r,\phi,t)$, $u(r) = u'(r,\phi,t)$, $\zeta(r) = \bar{\zeta}(r) + \zeta'(r,\phi,t)$, $\delta(r) = \delta'(r,\phi,t)$, $\chi(r,\phi,t) = \chi'(r,\phi,t)$, and $\psi(r,\phi,t) = \psi'(r,\phi,t)$. The perturbations u', v', ζ', and δ' are related to ψ' and χ' as above. Upon substituting the decomposed variables into (3)-(5), enforcing the gradient balance constraint, and neglecting higher order terms involving multiplication of perturbation quantities, we obtain the linearized versions of (3)-(5) which govern small amplitude disturbances about the basic state vortex,

$$\left(\frac{\partial}{\partial t} + \bar{\omega}\frac{\partial}{\partial\phi}\right)\nabla^2\psi' + \bar{\eta}\nabla^2\chi' + \left(\frac{\partial\chi'}{\partial r} - \frac{\partial\psi'}{r\partial\phi}\right)\frac{d\bar{\zeta}}{dr} = 0, \tag{13}$$

$$\left(\frac{\partial}{\partial t}+\bar{\omega}\frac{\partial}{\partial\phi}\right)\nabla^2\chi' - \bar{\eta}\nabla^2\psi' + g\nabla^2 h' - \left(\frac{\partial\chi'}{r\partial\phi}+\frac{\partial\psi'}{\partial r}\right)\frac{d\bar{\zeta}}{dr} = 0, \tag{14}$$

$$\left(\frac{\partial}{\partial t}+\bar{\omega}\frac{\partial}{\partial\phi}\right)h' + \bar{h}\nabla^2\chi' + \left(\frac{\partial\chi'}{\partial r}-\frac{\partial\psi'}{r\partial\phi}\right)\frac{d\bar{h}}{dr} = 0, \tag{15}$$

where $\bar{\eta} = f + \bar{\zeta}$ is the basic state absolute vertical vorticity. We wish to understand the nature and behavior of small disturbances (i.e., waves) on this basic state vortex. Using the Wentzel-Kramers-Brillouin (WKB) approximation, we assume normal mode solutions of the form $\psi'(r,\phi,t) = \hat{\psi}\exp[i(kr+m\phi-\nu t)]$, $\chi'(r,\phi,t) = \hat{\chi}\exp[i(kr+m\phi-\nu t)]$, and $h'(r,\phi,t) = \hat{h}\exp[i(kr+m\phi-\nu t)]$, where k is the radial wavenumber (i.e., the number of wave crests per unit radial distance), m is the azimuthal wave number (i.e., the number of wave crests per 2π radians), ν is the complex frequency, and $\hat{\psi}$, $\hat{\chi}$, and $\hat{h}$ are complex constants. Substituting the assumed solution into (13)-(15), we obtain a matrix eigenvalue problem $\mathbf{Ax} = i(\nu - \bar{\omega}m)\mathbf{Ix}$, where $\mathbf{x} = (\psi', \chi', h')$ and $\mathbf{I}$ is the identity matrix, for the Doppler-shifted frequency $\nu - \bar{\omega}m$. The Doppler-shifted frequency is the frequency of the wave perceived by an observer moving with the hurricane flow, and is always lower than the actual frequency of the wave for positive $\bar{\omega}$. This eigenvalue problem can be written as $\mathbf{Bx} = 0$. By setting the determinant of $\mathbf{B}$ equal to zero, two cubic equations can be obtained that can be solved for the real and imaginary portions (ν_r and ν_i) of the complex Doppler-shifted frequency $\nu = \nu_r + i\nu_i - \bar{\omega}m$. The real portion encapsulates lower frequency vortex Rossby waves, higher frequency inertia-gravity waves, and mixed inertia-gravity-vortex Rossby waves (with inseparable dispersion characteristics) and when combined with a nonzero imaginary portion, exponentially growing and decaying modes exist - the former of which is the Rossby-inertia gravity wave instability (Ford, 1994; Schecter & Montgomery, 2004; Hodyss & Nolan, 2008; Hendricks et al., 2010). It is not our intent in here to describe this more complex dispersion relationship, but rather to isolate the primary modes of variability on shallow water vortices by making certain simplifying assumptions on (13)-(15), foremost of which is neglecting the unstable modes. The local dispersion relationship for the inward and outer propagating inertia-gravity waves can be found by setting $d\bar{\zeta}/dr = d\bar{h}/dr = 0$ in equations (13)-(15) and assuming a constant rotation rate $\bar{\omega}$, yielding

$$(\nu - \bar{\omega}m)^2 = (f+2\bar{\omega})^2 + g\bar{h}\left(k^2 + m^2/r^2\right). \tag{16}$$

The local radial $c_r = \nu/k$ and azimuthal $c_\phi = r\nu/m$ phase velocites are then

$$c_r = \frac{\bar{\omega}m}{k} \pm \frac{1}{k}\left[(f+2\bar{\omega})^2 + g\bar{h}\left(k^2+m^2/r^2\right)\right]^{1/2}, \tag{17}$$

$$c_\phi = \bar{v}(r) \pm \frac{r}{m}\left[(f+2\bar{\omega})^2 + g\bar{h}\left(k^2+m^2/r^2\right)\right]^{1/2}. \tag{18}$$

Note that inertia-gravity waves may propagate both radially outward and inward, and azimuthally prograde and retrograde. Similarly, in the nondivergent limit of (13)-(15) (i.e., $\chi' = 0$ and $h' = 0$) we obtain the vortex Rossby wave dispersion relationship obtained by Montgomery & Kallenbach (1997)

$$\nu - \bar{\omega}m = \frac{m}{r}\frac{d\bar{\zeta}/dr}{k^2+m^2/r^2}. \tag{19}$$

Note that the Doppler-shifted frequency is directly proportial to the basic state vorticity gradient $d\bar{\zeta}/dr$, and inversely proportial to the square of the total wavenumber $k^2 + m^2/r^2$. The local radial and azimuthal phase velocities for vortex Rossby waves are

$$c_r = \frac{\bar{\omega} m}{k} + \frac{m}{kr} \frac{d\bar{\zeta}/dr}{k^2 + m^2/r^2}, \tag{20}$$

$$c_\phi = \bar{v}(r) + \frac{d\bar{\zeta}/dr}{k^2 + m^2/r^2}. \tag{21}$$

As shown by equations (20) and (21), vortex Rossby waves may propagate both radially inward and outward, and azimuthally prograde and retrograde with respect to the local flow, depending on the sign of $d\bar{\zeta}/dr$. For a shallow water vortex with monotonically decreasing basic state vorticity, vortex Rossby waves propagate azimuthally slower than the local flow. Therefore, they are analogous to planetary Rossby waves, with $d\bar{\zeta}/dr$ being the local beta effect. The radial group velocity for vortex Rossby waves is

$$\frac{\partial \nu}{\partial k} = c_{gr} = \frac{-2km}{r} \frac{d\bar{\zeta}/dr}{(k^2 + m^2/r^2)^2}. \tag{22}$$

Equation (22) indicates that the energy from a vortex Rossby wave packet is directed radially outward for $d\bar{\zeta}/dr < 0$ (e.g., a monotonic vortex as above) and radially inward for $d\bar{\zeta}/dr > 0$. For this reason, vortex Rossby waves that are excited locally can affect the vortex in regions away from the location of wave excitation through wave-mean flow interaction. According to the radial phase velocity equation, as the radial wavenumber k increases, the radial phase velocity approaches zero. Since the anticyclonic shear outside the radius of maximum wind in hurricanes has a symmetrizing affect on disturbances (increasing the radial wavenumber), this implies a stagnation radius where the wave can no longer propagate radial outward. In this regard, vortex Rossby waves are quite different from inertia-gravity waves since they cannot radiate outward to infinity. As we will see in section 3, for this reason only inner spiral rainbands can be described by vorted Rossby wave theory, since the outer bands exist in regions of approximately zero basic state PV gradient.

Dispersion curves are plotted for the all three wave types in Figure 1. It is readily apparent that vortex Rossby waves and inertia-gravity waves have very different dispersion characteristics. In particular, the frequencies of the outward and inward propagating inertia-gravity waves increase with increasing radial wavenumber, while for vortex Rossby waves the frequency approaches zero with increasing radial wavenumber. For vortex Rossby waves, the frequency always has the same sign as $d\bar{\zeta}/dr$, resulting in waves that propagate to the local *west* of the radial basic state vorticity gradient. Additionally the absolute value of the frequency is larger for lower azimuthal wavenumbers than for higher azimuthal wavenumbers. When $d\bar{\zeta}/dr < 0$, this indicates that absolute phase speeds for those waves are smallest.

2.4 Energy and angular momentum principles

The energy principle associate with (1),(2) and (5) can be derived by first rewriting (1)-(2) in their rotational forms as

$$\frac{\partial u}{\partial t} - hPv + \frac{\partial}{\partial r}\left[gh + \tfrac{1}{2}(u^2 + v^2)\right] = 0, \tag{23}$$

$$\frac{\partial v}{\partial t} + hPu + \frac{\partial}{r\partial \phi}\left[gh + \tfrac{1}{2}(u^2 + v^2)\right] = 0. \tag{24}$$

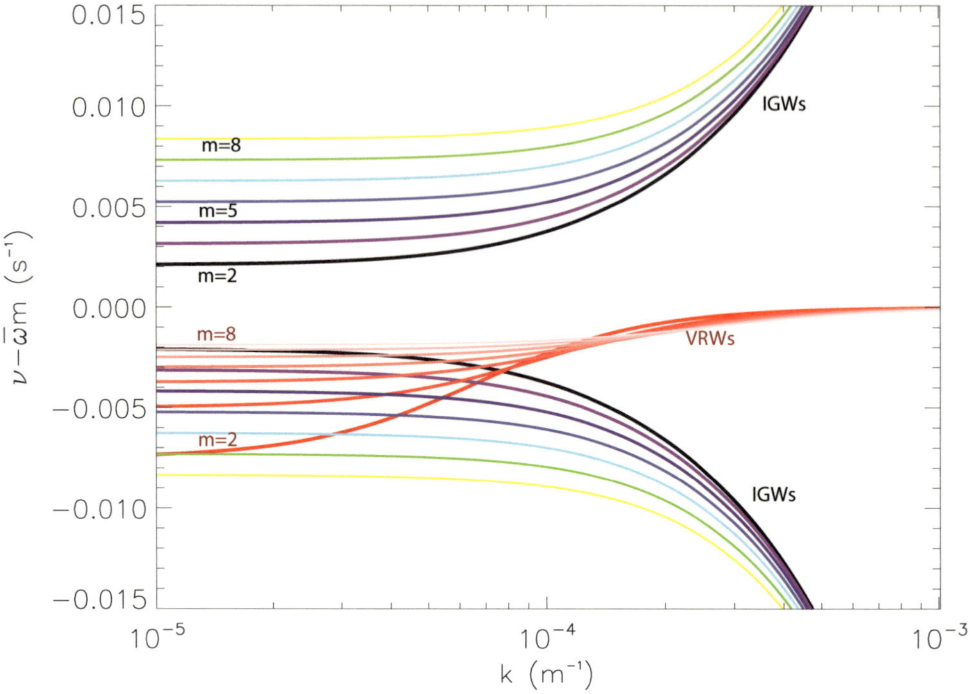

Fig. 1. Plot of the Doppler-shifted frequency $\nu - \bar{\omega}m$ versus radial wavenumber k for the inward and outwardly propagating inertia-gravity waves (IGWs) and vortex Rossby waves (VRWs) for azimuthal wavenumbers $m = 2 \ldots 8$. IGWs are plotted for a basic state vortex with $f = 5 \times 10^{-5}\,\mathrm{s}^{-1}, \bar{\omega} = 1 \times 10^{-4}\,\mathrm{s}^{-1}, \bar{h} = 100$ m, and at $r = 30$ km. The VRWs are plotted for a basic state vortex with $d\bar{\zeta}/dr = -5\times 10^{-7}\,\mathrm{s}^{-1}\,\mathrm{m}^{-1}, \bar{\omega} = 1 \times 10^{-4}\,\mathrm{s}^{-1}$, and at $r = 30$ km.

Multiplying $(hu)\cdot(23)$ and $(hv)\cdot(24)$, adding them together, and combining with the mass conservation equation (5), we obtain

$$\frac{\partial(K+gh)}{\partial t} + \frac{\partial[rhu(K+gh)]}{r\partial r} + \frac{\partial[hv(K+gh)]}{r\partial\phi} = 0, \tag{25}$$

where $K = (u^2 + v^2)/2$ is the kinetic energy and gh is the gravitational potential energy. By integrating (25) over the entire polar domain, we obtain

$$\frac{d}{dt}\iint (K+gh) r dr\, d\phi = 0, \tag{26}$$

which indicates that the total integrated energy is conserved in time. In an integrated sense, changes in gravitational potential energy must be balanced by equal changes in kinetic energy. By multiplying $r\cdot(2)$ and rewriting using the absolute angular momentum $M = rv + fr^2/2$ (or the angular momentum of an air parcel due to both the vortex and Earth's rotation) we obtain the absolute angular momentum principle

$$\frac{DM}{Dt} = -g\frac{\partial h}{\partial\phi}. \tag{27}$$

Equation (27) indicates that absolute angular momentum is materially conserved in the absence of the azimuthal torque $-g\partial h/\partial\phi$. This principle has profound implications for hurricane dynamics. In hurricanes, the radial pressure gradient force $g\partial h/\partial r$ is much larger than the azimuthal pressure gradient force $(g/r)\partial h/\partial\phi$, often M is nearly materially conserved above the boundary layer in real hurricanes. In the boundary layer, there exists friction so that as air parcels spiral inward they lose angular momentum. The near material conservation of M indicates that any mechanism that is able to move air radially inward to smaller radii will also cause the tangential winds to increase, and similarly air that moves radially outward will cause the tangential winds to decrease. This is analogous to the figure skater who spins faster when his or her arms are brought inward and vice versa. Additionally, asymmetric convection can create azimuthal torques to change M. Therefore an air parcel that spirals inward with a favorable azimuthal torque from asymmetric convection can rapidly increase its absolute angular momentum.

2.5 Eddy-mean flow interaction

In the preceding section, perturbations were assumed to be small in comparison to their means, which allowed us to neglect higher order terms. However, in many cases, the PV asymmetries are of sufficiently large magnitude so that a linear model is no longer valid. In this sense, the term "wave" should be replaced with the more general term "eddy", and the eddies may interact in nonlinear ways with the mean vortex. For example, in certain locations the eddies may spin-down the mean vortex while in other locations they may spin-up the mean vortex.

The eddy-mean flow interaction can be elegantly described using the isentropic formulation of Andrews (1983) and Andrews et al. (1987) reduced to our one-layer shallow water fluid. For the radial wind u, the mass-weighted azimuthal average is defined by $\hat{u} = \overline{hu}/\bar{h}$. The deviation from the ordinary average is defined by $v' = v - \bar{v}$, and the deviation from the mass-weighted average is defined by $u^* = u - \hat{u}$. Similar definitions hold for the other variables.

Applying $\overline{(\ )}$ to each term in (5) and noting that $\overline{hu} = \bar{h}\hat{u}$, we can write the azimuthal mean mass continuity equation as

$$\frac{\partial\bar{h}}{\partial t} + \frac{\partial(r\bar{h}\hat{u})}{r\partial r} = 0. \tag{28}$$

Similarly, applying $\overline{(\ )}$ to each term in (24) and noting that $\overline{hPu} = \bar{h}\hat{P}\hat{u} + \overline{hP^*u^*}$ we can write the azimuthal mean v equation as

$$\frac{\partial\bar{v}}{\partial t} + (f + \bar{\zeta})\hat{u} = -\overline{hP^*u^*}. \tag{29}$$

Now consider the radial momentum equation. Since both (28) and (29) contain $\hat{u}$, we would like to transform (23) into a prediction equation for $\hat{u}$. This requires putting (23) into a flux form before taking the azimuthal average. Thus, combining (23) and (5), we obtain the flux form

$$\frac{\partial(hu)}{\partial t} + \frac{\partial(rhuu)}{r\partial r} + \frac{\partial(hvu)}{r\partial\phi} - \left(f + \frac{v}{r}\right)hv + gh\frac{\partial h}{\partial r} = 0. \tag{30}$$

Taking the azimuthal average of (30), we obtain

$$\frac{\partial(\overline{hu})}{\partial t} + \frac{\partial(r\overline{huu})}{r\partial r} - \overline{\left(f + \frac{v}{r}\right)hv} + \overline{gh\frac{\partial h}{\partial r}} = 0. \tag{31}$$

Noting that $\overline{hu} = \bar{h}\hat{u}$, $\overline{huu} = \bar{h}\hat{u}\hat{u} + \overline{hu^*u^*}$, $\overline{h(\partial h/\partial r)} = \bar{h}(\partial\bar{h}/\partial r) + \overline{h'(\partial h'/\partial r)}$, and using the azimuthal mean continuity equation (28), we obtain the advective form (32).
Collecting the above results, we obtain the complete set of azimuthal mean equations

$$\frac{\mathcal{D}\hat{u}}{\mathcal{D}t} - \left(f + \frac{\bar{v}}{r}\right)\bar{v} + g\frac{\partial\bar{h}}{\partial r} = \mathcal{F}, \tag{32}$$

$$\frac{\mathcal{D}\bar{M}}{\mathcal{D}t} = r\mathcal{G}, \tag{33}$$

$$\frac{\mathcal{D}\bar{h}}{\mathcal{D}t} + \bar{h}\frac{\partial(r\hat{u})}{r\partial r} = 0, \tag{34}$$

where

$$\frac{\mathcal{D}}{\mathcal{D}t} = \frac{\partial}{\partial t} + \hat{u}\frac{\partial}{\partial r} \tag{35}$$

is the azimuthal mean material derivative, $\bar{M} = r\bar{v} + \frac{1}{2}fr^2$ is the azimuthal mean absolute angular momentum, and where

$$\mathcal{F} = -\frac{1}{\bar{h}}\left[\frac{\partial(r\overline{hu^*u^*})}{r\partial r} - \left(f + \frac{\bar{v}}{r}\right)\overline{h'v'} - \frac{\overline{(hv)'v'}}{r} + \overline{gh'\frac{\partial h'}{\partial r}}\right] \tag{36}$$

and

$$\mathcal{G} = -\overline{hP^*u^*} \tag{37}$$

are the eddy-induced effective mean radial and azimuthal forces per unit mass. By examining these equations, it is directly clear how the eddies interact with the mean flow. The mean radial momentum can be modified by the eddy-induced force $\mathcal{F}$ and the mean angular momentum can be changed by the eddy-induced torque $r\mathcal{G}$.

2.6 Numerical simulations

Numerical simulations presented in sections 3.1 and 3.2 were obtained using an f-plane pseudospectral shallow water model, with the Cartesian formulation of equations (3)-(5). The numerical solution was obtained on a 600 km $\times$ 600 km doubly periodic domain using 512 $\times$ 512 equally spaced points. After dealiasing of the quadratic advection terms in (3)-(5), the number of retained Fourier modes was 170 $\times$ 170, yielding an effective resolution of 3.52 km. A third order Adams-Bashforth time scheme was used with a time step of 5 s. Numerical diffusion terms were added to the right hand side of the Cartesian forms of (3)-(5), with a coefficient $\nu = 100\ \mathrm{m^2\ s^{-1}}$. The Coriolis parameter $f = 1 \times 10^{-4}\ \mathrm{s^{-1}}$ and the mean fluid depth $\bar{h} = 4077$ m. Numerical simulations presented in section 3.3 were performed using the nondivergent limit of the Cartesian formulations of (3)-(5), i.e. (3) combined with $\nabla^2\psi = \zeta$.

3. Internal dynamical processes

In this section the dynamical model and associated principles are used in conjunction with observations to illustrate barotropic aspects of the important internal dynamical processes responsible for hurricane structure and intensity variability: spiral rainbands, instabilities and mixing between the eye and eyewall, and concentric eyewall cycles.

3.1 Spiral bands

One of the most distiguishable features on tropical cyclones are spiral rain bands. If spiral rainbands are thought of as small amplitude propagating perturbations on a basic state vortex, we might expect that linear wave theory may be able to explain some aspects of their structure and propagation speeds. An earlier view (Kurihara, 1979; Willoughby, 1978) was that spiral bands were inertia-gravity waves. In order to illustrate how spiral bands may be viewed as propagating inertia-gravity waves, we assume $d\bar{\zeta}/dr = d\bar{h}/dr = 0$ and a constant angular velocity $\bar{\omega}$ in (13)-(15) to obtain

$$\left(\frac{\partial}{\partial t} + \bar{\omega}\frac{\partial}{\partial \phi}\right)\nabla^2\psi' + \bar{\eta}\nabla^2\chi' = 0 \tag{38}$$

$$\left(\frac{\partial}{\partial t} + \bar{\omega}\frac{\partial}{\partial \phi}\right)\nabla^2\chi' - \bar{\eta}\nabla^2\psi' + g\nabla^2 h' = 0 \tag{39}$$

$$\left(\frac{\partial}{\partial t} + \bar{\omega}\frac{\partial}{\partial \phi}\right)h' + \bar{h}\nabla^2\chi' = 0, \tag{40}$$

Assuming separable solutions of the form $h'(r,\phi,t) = \hat{h}(kr)\exp[i(m\phi - \nu t)]$ (and similar definitions hold for the other variables), a ordinary differential Bessel equation is obtained which governs the radial structure of the waves. The outward propagating solution to this equation are Hankel functions $H_m^{(1)}(kr)$, which are given in terms of the Bessel function $J_m(kr)$ and the Neumann function $Y_m(kr)$ by $H_m^{(1)}(kr) = J_m(kr) + iY_m(kr)$ as

$$\begin{pmatrix} \chi'(r,\phi,t) \\ \psi'(r,\phi,t) \\ h'(r,\phi,t) \end{pmatrix} = -\frac{h_0}{\bar{h}k^2}\begin{pmatrix} i(\nu - \bar{\omega}m) \\ f + 2\bar{\omega} \\ -\bar{h}k^2/h_0 \end{pmatrix} H_m^{(1)}(kr)e^{i(m\phi - \nu t)}, \tag{41}$$

where h_0 is an arbitrary constant, and where the radial wavenumber k and the frequency ν are related by (16). The relative vorticity and divergence associated with the streamfunction and velocity potential according to (7) are

$$\begin{pmatrix} \delta'(r,\phi,t) \\ \zeta'(r,\phi,t) \end{pmatrix} = h_0\begin{pmatrix} i(\nu - \bar{\omega}m)/\bar{h} \\ (f + 2\bar{\omega})/\bar{h} \end{pmatrix} H_m^{(1)}(kr)e^{i(m\phi - \nu t)}, \tag{42}$$

From (6) we find that the perturbation radial velocity $u' = (\partial\chi'/\partial r) - (\partial\psi'/r\partial\phi)$ and the perturbation azimuthal velocity $v' = (\partial\chi'/r\partial\phi) + (\partial\psi'/\partial r)$ can be written as

$$\begin{pmatrix} u'(r,\varphi,t) \\ iv'(r,\varphi,t) \end{pmatrix} = \frac{ih_0}{\bar{h}k^2}\cdot\begin{pmatrix} (\nu - \bar{\omega}m + f + 2\bar{\omega})(m/r)H_m^{(1)}(kr) \\ -(\nu - \bar{\omega}m)kH_{m-1}^{(1)}(kr) \\ (\nu - \bar{\omega}m + f + 2\bar{\omega})(m/r)H_m^{(1)}(kr) \\ -(f + 2\bar{\omega})kH_{m-1}^{(1)}(kr) \end{pmatrix} e^{i(m\varphi - \nu t)}, \tag{43}$$

where we have made use of the Hankel function derivative relation $dH_m^{(1)}(kr)/dr = kH_{m-1}^{(1)}(kr) - (m/r)H_m^{(1)}(kr)$. Equations (41)–(43) consititute the exact analytic solution to (38)-(40) for outwardly propagating cylindrical waves.

In Figure 2a, a large spiral rainband observed in Hurricane Alex on 1925 UTC is shown, and in Figure 2b the linear solution of $h'(r,\phi)$ is shown with overlayed vectors of $u'(r,\phi)$ and

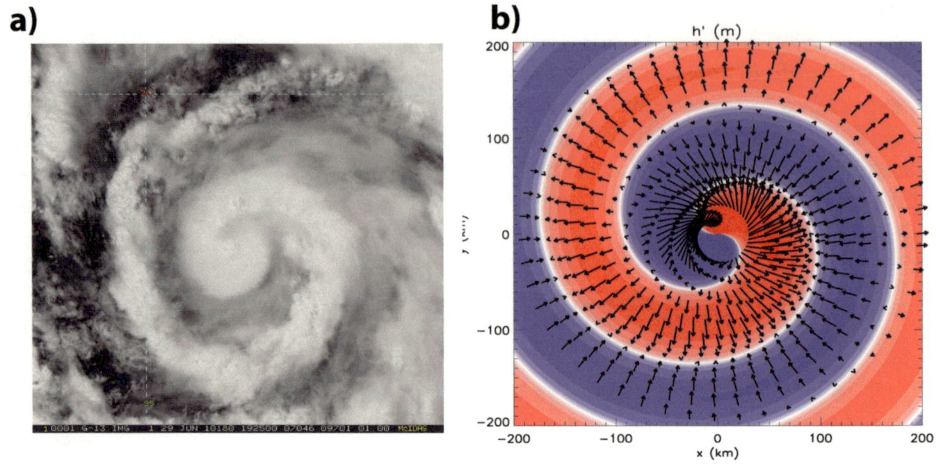

Fig. 2. Panels: a) Observed large spiral rainband in Hurricane Alex on 1925 UTC on June 29, 2010 (left panel), and b) Linear solution for h' with overlayed vectors of u' and v' for the shallow water equations in polar space. The constants used are $\bar{h} = 4288$ m, $\bar{\omega} = 5 \times 10^{-4}$ s^{-1}, $k = 4 \times 10^{-5}$ m^{-1}, $f = 3.7 \times 10^{-5}$ s^{-1}, $h_0 = 10$ m, and $m = 1$. Vectors are plotted in the region 30 km $< r <$ 200 km. Panel a) is courtesy of NOAA/NESDIS/RAMMB in Fort Collins, CO, USA.

$v'(r, \phi)$. Here, the red region denotes positive h' (fluid depth slightly higher) and the blue region denotes negative h' (fluid depth slightly lower). In this particular instant the large spiral rainband of Alex resembles the linear solution to the shallow water equations with azimuthal wavenumber $m = 1$. Note that the perturbation winds are highly divergent and that the orientation and the divergence pattern of the winds with respect to the perturbation height field favors outward propagation of the wave. An propagating spiral region of low level convergence could create shallow or deep convective activity similar to the observations of Alex. However, in order to be sure the frequencies are that of the observed wave, the linear theory would need to be compared.

Another viewpoint is that spiral bands are vortex Rossby waves (MacDonald, 1968; Guinn & Schubert, 1993; Montgomery & Kallenbach, 1997; Chen and Yau, 2001). Their arguments were based off the fact that spiral bands are often observed to propagate azimuthally slower than the mean flow, and the fact that the bands became tighter moving radially outward, similar to PV filaments. Since vortex Rossby waves are confined to live where there exists a radial PV gradient, radial propagation can no longer continue once the waves propagate outward to a certain stagnation radius. At this time, and at earlier times, the symmetrizing effect of the swirling winds tends to shear the waves into tightly wrapped spirals. Eventually the waves decay and axisymmetrization is complete.

Using the numerical model described in section 2.5, a simulation was run with an initially elliptical PV field. Hurricanes are sometimes observed to have distinct elliptical shapes in the inner-core (Kuo et al., 1999; Reasor et al., 2000). In polar coordinates, the initial vorticity field was specified by

$$\zeta(r, \phi, 0) = \zeta_0 \begin{cases} 1 & 0 \le r \le r_i \alpha(\phi) \\ 1 - f_\lambda(r') & r_i \alpha(\phi) \le r \le r_o \alpha(\phi), \\ 0 & r_o \alpha(\phi) \le r \end{cases} \tag{44}$$

where $\alpha(\phi) = [(1-\epsilon^2)/(1-\epsilon^2\cos^2(\phi))]^{1/2}$ is an ellipticity augmentation factor for the ellipse $(x/a)^2 + (y/b)^2 = 1$ (where a is the semi-major axis and b is the semi-minor axis) with eccentricity $\epsilon = (1-(b^2/a^2))^{1/2}$. Here, ζ_0 is the maximum vorticity at the center, $f_\lambda(r') = \exp[-(\lambda/r')\exp(1/(r'-1))]$ is a monotonic shape function with transition steepness parameter λ, $r' = (r - r_i\alpha(\phi))/(r_o\alpha(\phi) - r_i\alpha(\phi))$ is a nondimensional radius, and r_i and r_o are the radii where the vorticity begins to decrease and where it vanishes, respectively. For the special case of $\alpha(\phi) = 1$ the field is axisymmetric. For the experiment conducted, $\zeta_0 = 3.0 \times 10^{-3}\ \mathrm{s}^{-1}$, $\lambda = 2.0$, $\epsilon = 0.60$, and r_i and r_o were set to 10 km and 110 km, respectively. The balanced mass field was determined by solving the nonlinear balance equation, and the initial divergence was set to zero everwhere in the domain.

In Figure 3a, spiral rainbands observed in radar imagery are shown as Hurricane Hugo (1989) made landfall in Charleston, South Carolina. In Figure 3b, spiral rainbands are shown in Hurricane Danielle (2010) in the visible satellite imagery. Note that in both of these panels, the spiral bands are tightly wrapped, or have a larger radial wavenumber. Figure 3c depicts the evolution of $P = (\bar{h}/h)(f+\zeta)$ in the numerical simulation, with brighter colors denoting larger PV. The first panel is the initial condition and each subsequent panel is 1.25 h apart. As the ellipse symmetrizes, two PV filaments are visible. These PV filaments are vortex Rossby waves which are breaking anticyclonically due the swirling winds of the vortex outside the radius of maximum wind, and resemble the tighter spiral bands observed in real hurricanes. Eventually the elliptical PV field symmetrizes into a axisymmetric monopole. In the real atmosphere, external factors such as vertical and horizontal shear as well as internal processes such as convective activity can excite vortex Rossby waves. For monotonic vortices, these waves will generally decay with time, creating spiral PV filaments resembling spiral bands. As stated earlier, while inner spiral rainbands are likely sheared vortex Rossby waves, the outer bands may often be outwardly radiating inertia-gravity waves. Guinn and Schubert (1993) also hypothesized that the outer bands may form during a combined barotropic-baroclinic instability of the intertropical convergence zone. In this sense, the large PV filaments formed from this breakdown are breaking Rossby-like waves, but not vortex Rossby waves.

There is continuing debate on the net effects of spiral rainbands on TC intensity. In a forced context, where the effect of deep convection are parameterized by region of enhanced PV, vortex Rossby wave theory indicates that axisymmetrization of PV anomalies (Carr & Williams, 1989) will spin-up the mean winds as the waves propagate radially outward (Montgomery & Kallenbach, 1997). In an unforced context (as described here), the anticyclonic wave-breaking outside the vortex core actually spins down the vortex winds there. The real atmosphere is much more complicated. Since spiral bands have diabatic heating, there can be associated downdrafts and subsidence in between successive bands. The spiral bands can also cut off inflow into the core of the vortex, resulting in potential weakening (Barnes et al., 1983). Further research is certainly warranted on the net effects of spiral bands, either generated by vortex Rossby or inertia-gravity waves on hurricane intensity.

3.2 Eyewall processes

The eyewall of intense or intensifying hurricanes is often observed to be a annular tower of elevated potential vorticity (Kossin & Eastin, 2001; Möller & Smith, 1994), that extends from the surface to approximately 8-12 km in height. Thus at any vertical level the radial gradient of azimuthal mean PV changes sign moving from the vortex center to the boundaries. This satisfies the Charney-Stern necessary condition for dynamic instability. In the three-dimensional atmosphere, this instability due in part to both barotropic and

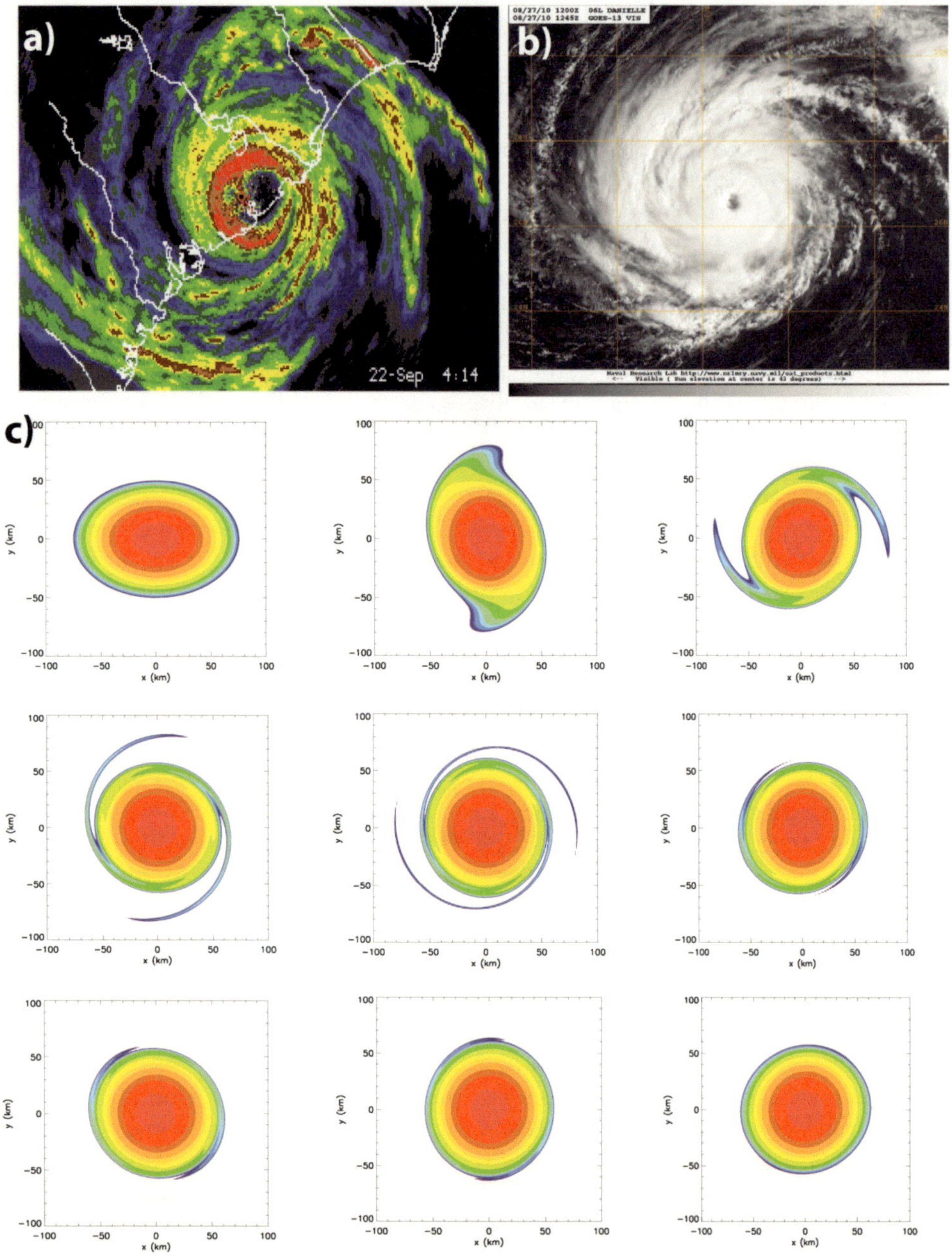

Fig. 3. Panels: a) Spiral rainbands observed in Hurricane Hugo as it made landfall on September 22, 1989 at 0414 UTC, b) Spiral rainbands observed in visible satellite imagery of Hurricane Danielle on August 23, 2010. c) Evolution of the elliptical PV field in a shallow water model. The first figure is the initial condition and each subsequent figure is 1.25 h apart. Panels (a) and (b) are courtesy of NOAA/AOML/Hurricane Research Division in Miami, FL, USA.

baroclinic instabilities, with the former instability being attributed to the large radial shear of the tangential wind as well as the curvature vorticity, and the latter instability being owed to the vertical shear of the vortex tangential wind. We will focus here on barotropic instability and ensuing mixing process between the eyewall and eye.

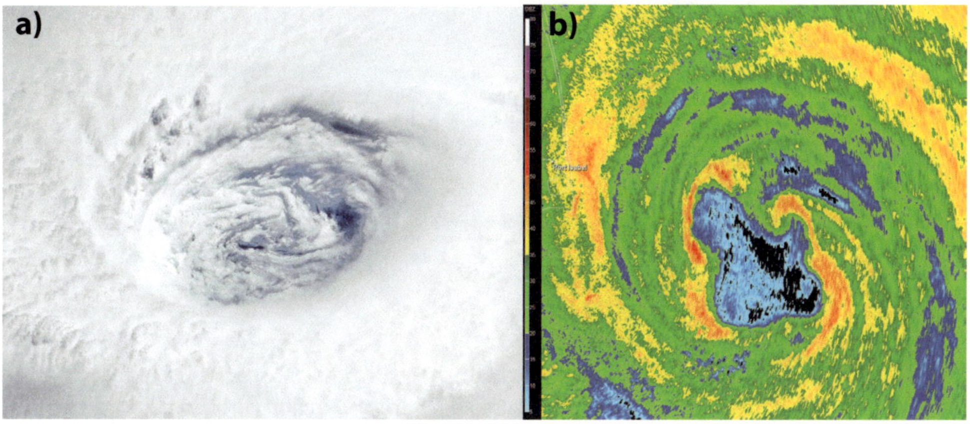

Fig. 4. Panels: a) Vortical swirls in the eye of Hurricane Igor (2010) observed from MODIS, and b) A wavenumber-4 pattern in the radar reflectivity in Hurricane Dolly as it approached the Texas coast at 1052 UTC on July 23, 2008. Panel a) is courtesy of NASA and panel b) is courtesy of the National Weather Service in Brownsville, TX, USA.

In Figure 4, two images are shown to illustrate some observed inner-core asymmetries and eye-eyewall mixing in hurricanes. In Figure 4a, vortical swirls observed at low levels in the eye of Hurricane Igor are shown and in Figure 4b, a wavenumber-4 pattern is shown in Hurricane Dolly (2008) as it approached southern Texas. Polygonal eyewall shapes, and eye mesovortices have been hypothesized by Schubert et al. (1999) to all be part of an intrinsic mixing process occuring between the eye and eyewall of hurricanes resulting from dynamic instability of the eyewall PV hollow tower.
Barotropic instability on a circular vortex is best understood by considering the interaction of two counter-propagating vortex Rossby waves. Consider a ring of elevated vorticity as shown in Figure 5. As shown in section 2, the local azimuthal phase velocity for disturbance vorticity $\zeta'(r,\phi)$ is

$$c_\phi = \bar{v}(r) + \frac{d\bar{\zeta}/dr}{k^2 + m^2/r^2}. \tag{45}$$

The corresponding local wave angular phase velocity is

$$\omega_\phi = \bar{\omega}(r) + \frac{1}{r}\frac{d\bar{\zeta}/dr}{k^2 + m^2/r^2}, \tag{46}$$

Now consider two Rossby waves which exist on the inner and outer edge of the vorticity annulus in Figure 5. For the inner wave, $d\bar{\zeta}/dr$ is positive, while for the outer wave, $d\bar{\zeta}/dr$ is negative. According to (46), the inner wave has a larger angular velocity than the vortex angular velocity, while the outer wave has a smaller angular velocity than the vortex angular velocity. Thus, the inner wave progrades relative to its local flow, and the outer wave retrogrades relative to its local flow. Since the angular velocity profile for this ring has a minimum at the location of the first basic state vorticity gradient, and a

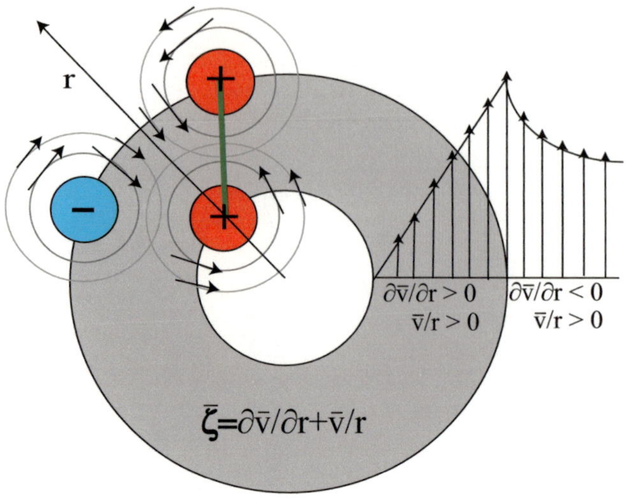

Fig. 5. Conceptual model of the interaction of counter-propagating vortex Rossby waves on a ring of elevated vorticity. Positive sign indicates a cyclonic vorticity anomaly, and negative sign indicates a anti-cyclonic vorticity anomaly. At this time the inner and outer vorticity anomalies are locked in phase (thick green line) with upshear tilt, and the circulation associated with each anomaly is advecting the other anomaly, leading to distortion of the initially axisymmetric ring.

maximum at the location of the second basic state vorticity gradient, it is possible for the two waves to be phase-locked, or have the same absolute absolute angular velocity. If the waves become phase-locked with the disturbance vorticity tilted upshear, as shown in Figure 5, each anomaly is in the optimum phase to interact with the other, eventually leading to distortion of the axisymmetric ring. Note in particular that the cyclonic circulation associated with the outer vorticity anomaly is advecting the inner anomaly inward (enhanced by the anticyclonic anomaly next to it), while the circulation of the inner vorticity anomaly is advecting the outer anomaly radially outward. Clearly, the interaction of vorticity anomalies is ehanced for thinner rings and with rings of larger basic state vorticity. A mathematically rigorous description of the interaction of these counter-propagating vortex Rossby waves in the inner-core is provided by Schubert et al. (1999). Interestingly, this barotropic instability can also exist in the outer region of the tropical cyclone (Peng et al., 2008), affecting its outer structure and intensity.

In the early phase of barotropic instability, both the inner and outer vortex Rossby waves are growing exponentially with time. After a number of *e*-folding times, the infinitesimal perturbations have grown to significant amplitude such that the linearized equations are no longer valid. The flow now is governed by the nonlinear PV equation (8), where the advection terms are of principal importance. In order to understand the resultant vortex structure and intensity change resulting from barotropic instability, we must use a numerical model to simulate the PV advection. In the real atmosphere, there also exists diffusion, which can be thought of as a parameterized eddy viscosity coefficient assuming flux gradient theory. This new PV equation

$$\frac{\partial P}{\partial t} + u\frac{\partial (rP)}{r\partial r} + \frac{v}{r}\frac{\partial P}{\partial \phi} = \nu \nabla^2 P, \tag{47}$$

governs advective mixing (or the combined effects of advection and diffusion) in a barotropic framework, where ν is the eddy viscosity coefficient. Given a PV distribution at some initial time, equation (47) governs the evolution of PV at subsequent times. The advecting winds u and v change each time step, and are determined by the invertibility principle discussed previously. The winds also advect an arbitrary passive tracer $c(r, \phi, t)$, where

$$\frac{\partial c}{\partial t} + u\frac{\partial (rc)}{r\partial r} + \frac{v}{r}\frac{\partial c}{\partial \phi} = \nu \nabla^2 c. \tag{48}$$

In the real atmosphere, this quasi-passive tracer could be water vapor molecules or a derived thermal property such as equivalent potential temperature. It is insightful to examine the interplay of the advection and diffusion in (47) and (48). Advection, or more specifically differential advection (where the advecting winds vary in space and time), acts to create a complicated geometric structure of an arbitrary active or passive tracer. However, true mixing cannot occur unless air parcels exchange properties with one another through diffusion. Thus, advective mixing is due to the combined effects of reversible advection and irreversible diffusion. In the view of Nakamura (1996), differential advection acts to deform tracer contours so that there is more interface for diffusion to act. He provided a derivation of a modified "effective diffusion" equation, where the effects of the differential advection are parameterized by a larger diffusion coefficient. Hendricks & Schubert (2009) applied this effective diffusivity diagnostic to evolving barotropic hurricane-like vortices, and determined the location and magnitude of chaotic mixing and partial barrier regions. Mixing regions were located near breaking vortex Rossby waves, while partial barrier regions were located near tangential jets.

A numerical simulation was performed in the shallow water model for a PV ring similar to that shown in Figure 5. The initial ring was defined by

$$\bar{\zeta}(r,0) = \begin{cases} \zeta_1 & 0 \le r \le r_1 \\ \zeta_1 S(\frac{r-r_1}{r_2-r_1}) + \zeta_2 S(\frac{r_2-r}{r_2-r_1}) & r_1 \le r \le r_2 \\ \zeta_2 & r_2 \le r \le r_3, \\ \zeta_2 S(\frac{r-r_3}{r_4-r_3}) + \zeta_3 S(\frac{r_4-r}{r_4-r_3}) & r_3 \le r \le r_4 \\ \zeta_3 & r_4 \le r \le \infty \end{cases} \tag{49}$$

where $\zeta_1 = 0$, $\zeta_2 = 3 \times 10^{-3}$ s^{-1}, and ζ_3 is set to a minimal negative value so that the circulation vanishes at the boundary. The radii $r_1 = 20$ km, $r_2 = 24$ km, $r_3 = 38$ km, and $r_4 = 42$ km, and $S(s) = 1 - 3s^2 + 2s^3$ is a cubic Hermite shape function that provides smooth transition zones. The eye is defined as the region where $r < r_1$, the eyewall is defined as the region between r_2 and r_3, and the transition zones are defined as the regions between r_1 and r_2, and r_3 and r_4. To initiate the instability process, a broadband perturbation (impulse) was added to the basic state vorticity (6) of the form

$$\zeta'(r,\phi,0) = \zeta_{\text{amp}} \sum_{m=1}^{8} \cos(m\phi + \phi_m) \quad \times \begin{cases} 0 & 0 \le r \le r_1, \\ S(\frac{r_2-r}{r_2-r_1}) & r_1 \le r \le r_2, \\ 1 & r_2 \le r \le r_3, \\ S(\frac{r-r_3}{r_4-r_3}) & r_3 \le r \le r_4, \\ 0 & r_4 \le r < \infty, \end{cases} \tag{50}$$

where $\zeta_{\text{amp}} = 1.0 \times 10^{-5}$ s^{-1} is the amplitude and ϕ_m the phase of azimuthal wavenumber m. For this set of experiments, the phase angles ϕ_m were chosen to be random numbers in the range $0 \leq \phi_m \leq 2\pi$. In real hurricanes, the impulse is expected to develop from a wide spectrum of background convective motions. As in the ellipse, the height $h(r, \phi)$ was determined by solving the nonlinear balance equation using the basic state vorticity and the initial divergence $\delta(r, \phi)$ was set to zero.

In Figure 6, the nonlinear evolution of the PV ring above in the shallow water model is shown. In Figure 6a, the asymmetric PV and asymmetric velocity vectors are shown at $t = 2$ h in this simulation, and in Figure 6b the same asymmetric velocity vectors are overlayed on the perturbation height field. Examining Figure 6a, note that there are two vortex Rossby wave-trains that exist at locations of the the inner and outer PV gradients, respectively. Both wave-trains are phase-locked and propagating at the same angular velocity relative to Earth. Note that the most unstable mode for this PV ring is $m = 3$, and that outer PV anomalies are tilted upshear with respect to the inner PV anomalies. Also note that the perturbation winds are anticyclonic for negative PV anomalies (blue) and cyclonic for positive PV anomalies (red), and that upshear phase-locking is allowing the inner and outer PV anomalies to optimally interact with one another, leading to distortion of the initial axisymmetric ring, similar to the conceptual model in Figure 5. In Figure 6b, note that cyclonic anomalies are associated with lower heights, and anticyclonic anomalies are associated with higher heights. Finally, note that the winds are approximately parallel to the contours of constant perturbation height, which indicates that these vortex Rossby waves are largely balanced. If the center of the vortex is viewed as the North Pole, these vortex Rossby waves are quite similar to large-scale planetary Rossby waves responsible for low frequency variability in weather patterns. The vortex Rossby wave patterns in this barotropic model are similar to full-physics model simulations by Wang (2002a). In their simulation, cyclonic PV perturbations on the vortex are coupled with updrafts and low level convergence, and as such in reality, *convectively coupled* vortex Rossby waves often exist in hurricanes. This coupled wave has been hypothesized by Montgomery & Enagonio (1998) and Möller & Montgomery (2000) to be important for tropical cyclogenesis as well as hurricane intensity change.

In Figure 6c, the evolution of the PV ring is shown for the entire experiment, with each panel being 5 h apart. As the inner and outer vortex Rossby waves grow in tandem, the circular ring is distorted into a polygonal shape with straight line segments. The inner vortex Rossby wave breaks cyclonically drawing low PV air from the eye into the eyewall. Similarly the outer vortex Rossby wave breaks anticyclonically similar to the ellipse, drawing high PV from the eyewall outward in the form of PV filaments. During the evolution of this ring, the eye contracts in an asymmetric fashion, and the vortex slowly symmetrizes into a monopole. Interestingly, Hodyss & Nolan (2008) found that these polygonal shapes can also form when the Rossby-inertia gravity wave instability is the most unstable mode.

During these PV mixing events, significant changes can occur in the vortex structure and intensity. Equation (33) can be rewritten as

$$\frac{\partial \bar{M}}{\partial t} = -(f + \bar{\zeta})r\hat{u} - r\overline{hP^* u^*} \qquad (51)$$

where the left hand side is the local change in azimuthal mean angular momentum, the first term on the right hand side is the change in $\bar{M}$ due to the mean torque, and the second term on the right hand side is the change in $\bar{M}$ due to the eddy torque.

In Figure 7, the angular momentum budget for this simulation was computed from 0–35 h. The initial radius of maximum wind is at the edge of the PV ring, or approximately at $r = 42$

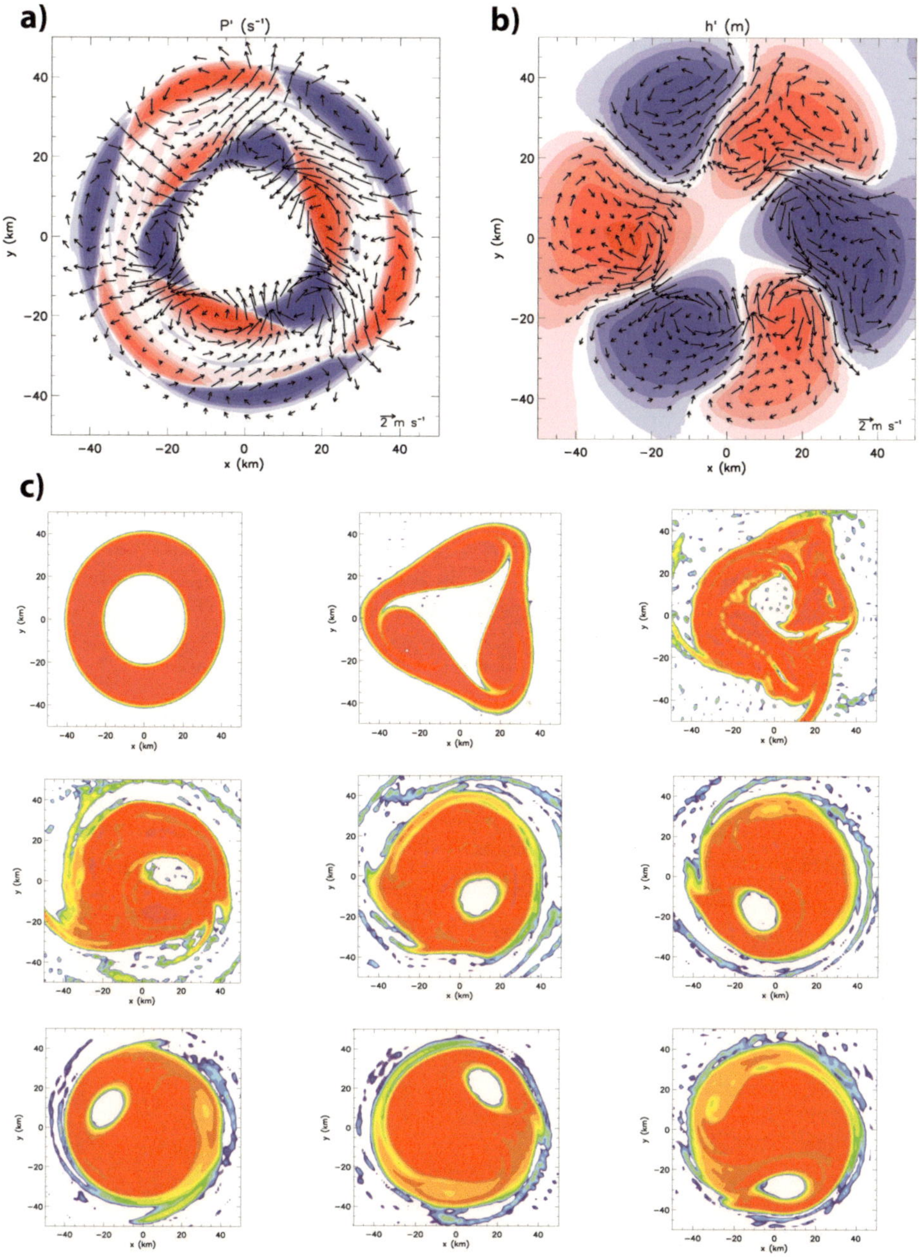

Fig. 6. Panels: a) Asymmetric PV with asymmetric wind vectors showing the most unstable mode of $m = 3$. b) Asymmetric wind vectors overlayed on the perturbation height field. c) The evolution of PV for the ring experiment in the shallow water model. The first figure is the initial condition and each subsequent figure is 5 h apart. In panels a) and b), vectors are plotted in the region 15 km $< r <$ 45 km, and red (blue) shading denotes positive (negative) values.

km. Three curves are plotted. The "ACTUAL" curve depicts the $\Delta\bar{M}$ due to the left hand side of (7), and the "MEAN" and "EDDY" curves depict the integration of the left hand side terms from 0–35 h using the trapezoidal rule on the 15-minute resolution output. Since PV mixing is a highly asymmetric process, the mean term is approximately zero everywhere. This is largely because $\hat{u}$ is nearly zero. However, in real hurricanes with inflow at lower levels, this term would contribute positively to the change in $\bar{M}$ at lower levels. The dominant term is the eddy-induced term. The eddies exert a positive torque in the region $0 < r < 35$ km and a negative torque in the region $35 < r < 70$ km. Viewing this in terms of the mean tangential momentum $\bar{v}$, this corresponds to a $\Delta\ \bar{v} = 25$ m s^{-1} at $r = 20$ km and $\Delta\ \bar{v} = -5$ m s^{-1} at $r = 40$ km. Additionally the radius of maximum wind contracts inward, and the larger winds at smaller radii increase the v^2/r term in the gradient wind balance equation 11, causing the height (or pressure) field to be lower after PV mixing than the initial condition. This is the dual nature of PV mixing with regard to hurricane intensity in a barotropic framework: it simultaneously lowers the central pressure and maximum wind (Kossin & Schubert, 2001; Rozoff et al., 2009; Hendricks et al., 2009). Results from simulation of three-dimensional unstable baroclinic vortices in a primitive equation model support these barotropic results (Hendricks & Schubert, 2010). However since observations and full-physics hurricane simulations rarely show that hurricanes simultaneously lower their central pressure and maximum wind (Wang, 2002b), it is possible that moist processes and boundary layer processes modify these results to some degree in the real atmosphere.

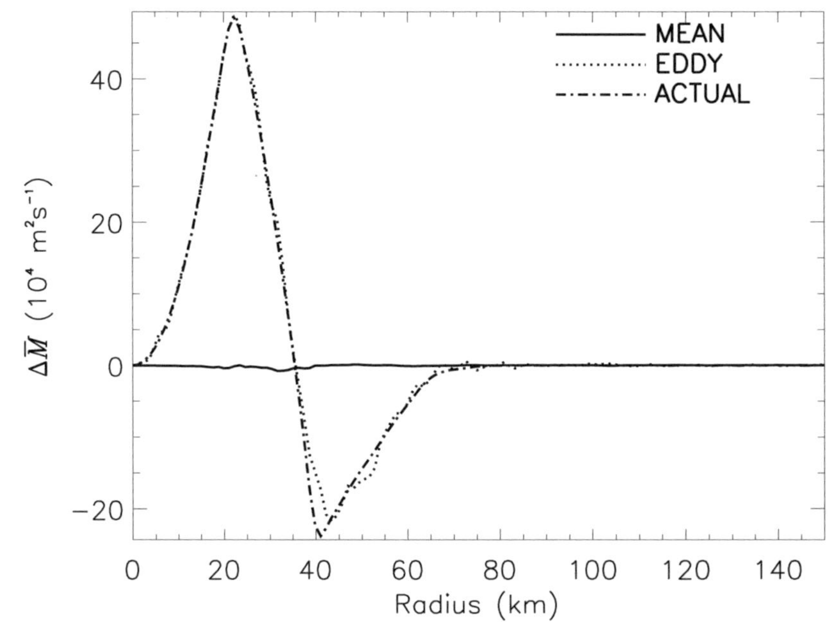

Fig. 7. Absolute angular momentum budget for the ring PV field simulation from 0-35 h.

Another interesting aspect of PV mixing is the horizontal mixing of quasi-passive tracers. Since many intense hurricanes have low level cloud swirls (i.e., airborne condensate) in the eye, it is likely that there is an internal mixing process occuring quite regularly, affecting structure and intensity. At low levels in the eye of hurricanes, there exists a very warm moist air mass (Eastin et al., 2002). In fact, the highest equivalent potential air in the hurricane is typically at low levels in the eye. During the breakdown of the PV ring, the inner breaking vortex Rossby wave can mix this high energy air back into the eyewall, giving the air extra buoyancy as it rises in the deep cumulonimbus convection of the eyewall. This intensification mechanism has been studied in recent work (Cram et al., 2007; Montgomery et al., 2006; Persing & Montgomery, 2003), and also has implications for the maximum intensity that a hurricane may achieve in a given environment.

3.3 Concentric eyewall cycles

Another internal process that can cause significant structural and intensity variability are concentric eyewall (or eyewall replacement) cycles, which encompass the formation of the secondary eyewall, and its eventual contraction and replacement of the inner eyewall. Aircraft observations of Hurricane Gilbert (1988) (Black & Willoughby, 1992) indicated the presence of concentric eyewalls structure in their radar reflectivity. Approximately 12 hours after reaching its minimum sea level pressure of 888 hPa, Hurricane Gilbert displayed concentric eyewalls and quickly followed by the eyewall replacement cycle where the inner eyewall dissipated and storm weakened. The radar observation of typhoon Lekima (2001) near Taiwan (Kuo et al., 2004) indicates a huge area of convection outside the core vortex wraps around the inner eyewall to form concentric eyewalls in about 12 hours. Kuo et al. (2008) also reported cases of passive microwave data that the formation of a concentric eyewall from organization of asymmetric convection outside the primary eyewall into a symmetric band that encircled the eyewalls. Two such concentric eyewalls examples are shown in Fig. 8 for typhoons with for Typhoons Dujuan (2003) and Imbudo (2003). The time interval is approximately 12 h, and estimated maximum winds are indicated at the top of the image.

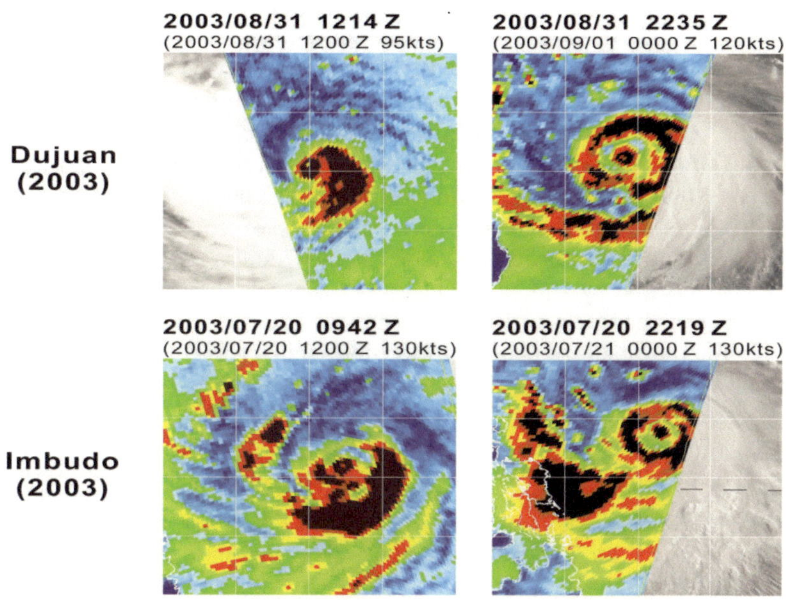

Fig. 8. Passive microwave image sequence for two western Pacific typhoons with concentric eyewalls (Dujuan and Imbudo). For each typhoon the time interval is approximately 12 h, and the estimated maximum winds are indicated at the top of the image (courtesy of Naval Research Laboratory, Monterey, CA, USA)

Dritschel & Waugh (1992) described the general interaction of two barotropic vortices with equal vorticity but with different sizes and separation distances. They performed experiments on the f-plane by varying two parameters, the ratio of the vortex radii and the separation distances normalized by the radius of the larger vortex. The resulting end states can be classified according to the two parameters into elastic interaction, merger, and straining out regimes. The elastic interaction involves distortion to the vortices in a mutual cyclonic rotation. The merger regime involved part of the smaller vortex being removed, and some of it being incorporated into the larger vortex. In the straining out regime, a thin region of filamented vorticity bands surrounding the central vortex with no incorporation into the central vortex. Many observed cases of tropical cyclone interactions resemble the idealized simulation of Dristchel and Waugh (Kuo et al., 2000; Lander & Holland, 1993; Larson, 1975; Prieto et al., 2003). Moreover, the straining out regime appeared to resemble the concentric eyewalls with a moat. However, the outer bands which result from the smaller vortex are much too thin to be identified with the outer eyewall of a tropical cyclone.

An important issue in the formation of concentric eyewalls appears to be the development of the symmetric structure from asymmetric convection. It appears that the formation of concentric eyewall has two important aspects: (i) an existing asymmetric potential vorticity distribution is organized by horizontal advective processes and (ii) the potential vorticity is diabatically enhanced during the organization process. Although both aspects are important, it is useful to isolate (i) through the study of highly idealized barotropic model. The dynamics can be idealized as the interaction of a small and strong vortex (representing the tropical cyclone core) with a large and weak vortex of various spatial scales (representing the vorticity induced by the moist convection outside the core vortex). This interaction of a strong and small vortex with a large and weak vortex was not studied by Dritschel & Waugh (1992) as their vortices are of the same strength, and their larger vortex was always the "victor" and the smaller vortex was the one being partially or totally destroyed. Kuo et al. (2004) introduced a new parameter "vorticity strength ratio" to the binary vortex interaction and adds a new dimension to the Dristchel-Waugh parameter space. As a result, binary interactions manifest a new stable end state: concentric vorticity structure. The axisymmetrization of positive vorticity perturbations around a strong and tight core of vorticity to produce concentric eyewall structure is studied in detail by Kuo et al. (2004; 2008).

To illustrate the barotropic dynamics of concetnric eyewalls formation, we first illustrate the straining out effect of the vortex. The straining out effect is due to the strong differential rotation outside the radius of maximum wind produces vorticity filaments. The strain effect may be quantified with the filamentation time τ_{fil}, in the nondivergent flow

$$\tau_{\text{fil}} = 2(S_1^2 + S_2^2 - \zeta^2)^{-\frac{1}{2}}, \tag{52}$$

provided $S_1^2 + S_2^2 > \zeta^2$, and where S_1 and S_2 are the stretching and shear deformation, respectively (Rozoff et al., 2006). The filamentation dynamics and its role to the moat formation was discussed by Rozoff et al. (2006). The filamentation time decreases as the strain flow strengthens. The presence of local vorticity field may offset the straining flow and lengthen the filamentation time. Figure 9 shows the vorticity fields at t = 0, 6, and 12 h for the straining out experiments. During the first 6 hours of the integration, the neighboring weak vortices become distorted into an approximately elliptical shape. As a result of the motion induced by its own vorticity, this ellipse turns counterclockwise, just as in the idealized Kirchhoff ellips solution (Kuo et al., 1999). Thus, one tip of the ellipse moves closer to the strong core vortex, while the other moves father away. The part of the weaker vorticity region that is closest to the strong core vortex is then rapidly strained out and wound around the core vortex at close radius.

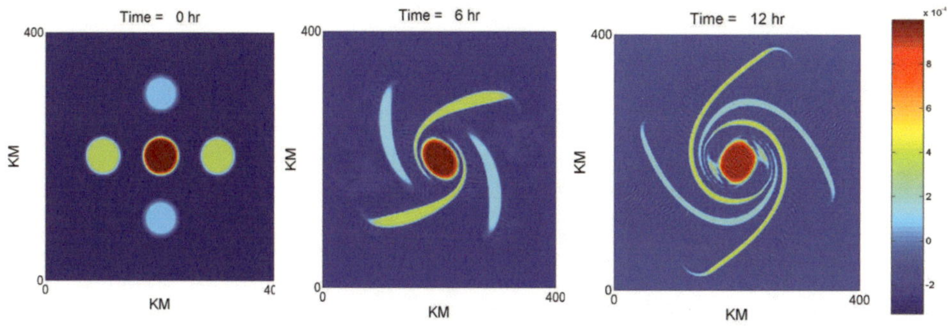

Fig. 9. Vorticity fields at $t = 0, 6, 12$ h in the straining out experiments.

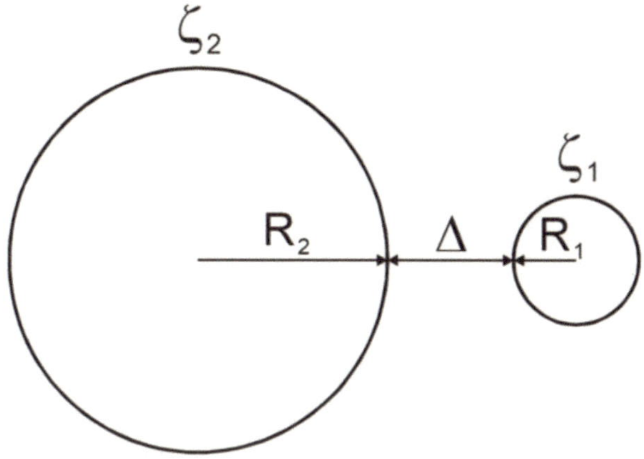

Fig. 10. Initial configuration of two circular vortices with parameters of separation distance, and two different radii and vorticity strengths. The small and strong vortex is the core.

Figure 10 illustrates the initial configuration of the binary vortex interaction for the concentric eyewall experiment, with relevant parameters of separation distance, radii and vorticity strength of the two vortices. Mallen et al. (2005) demonstrated that tropical cyclones are often characterized by a relatively slow decrease of tangential wind outside the radius of maximum wind, and hence by a a corresponding cyclonic vorticity skirt. The vorticity skirt can be easily incorporated into the binary vortex experiments. We consider an experiment with core vortex 7 times stronger in vorticity strength, and radius 4 times smaller in size than the neighboring large and weak vortex. The separation distance is 5 times the core radius and a vorticity skirt is also added to the core vortex outside the radius of maximum wind. Figure 11 gives the vorticity fields for such an experiment at hours 0, 6, and 12. By $t = 6$ h, Fig. 11 indicates the thin vorticity filaments have been wound several times, and by $t = 12$ h, vorticity mixing has created a halo of vorticity around the core vortex. Such a structure at $t = 12$ h is similar to the concentric eyewalls in the tropical cyclone. Noted that if such an event were to occur in a real tropical cyclone, and if it were observed by only a few radial legs of research aircraft,

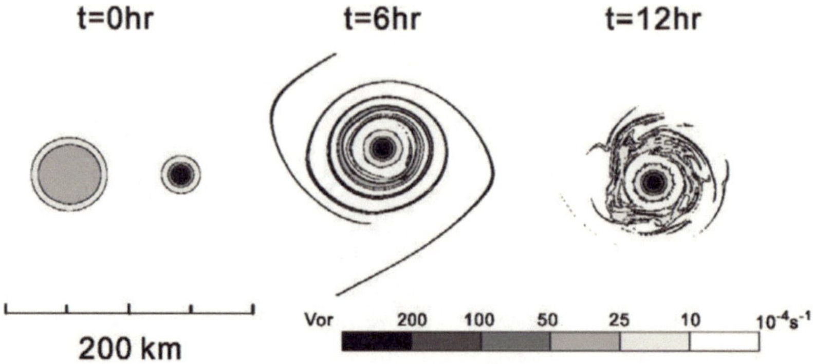

Fig. 11. Vorticity fields at t = 0, 6, and 12 h for an experiment with core vorticity 7 times stronger, size radius 4 times smaller than the large and small vortex nearby. The separation distance is 5 times the core radius. A vorticity skirt is also added to the core vortex outside the radius of maximum wind.

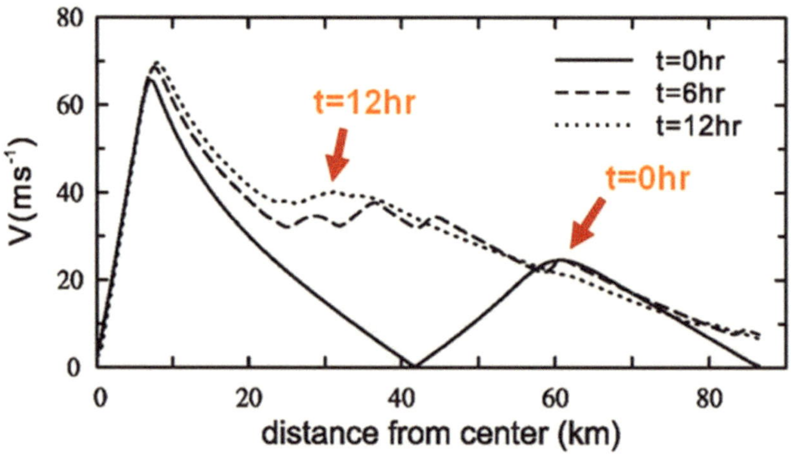

Fig. 12. The same as Fig. 10 except the tangential wind speed along a radial emanating westward from the center of the strong core vortex.

the asymmetric nature of the early revolution maybe lost. Figure 12 shows the tangential wind speed along a radial leg emanating westward from the center of the core vortex. The wind profiles show a secondary wind maximum that apparently contract s over a period of 12 h. During the contracting period, the secondary wind maximum increases from 25 m s^{-1} at 60 km to 40 m s^{-1} at 35 km. The features of the contraction are in general agreement with the observations of Hurricane Gilbert (Black & Willoughby, 1992). While the outer band contracting dynamics is often argued to be an axisymmetric gradient-balanced process involving advection by the divergent part of the flow (Shapiro & Willoughby, 1982), Fig. 12 suggests that vorticity advection by the nondivergent part of the flow can also contribute to the contraction dynamics in the concentric eyewalls formation.

In summary, we present observations indicating area of asymmetric convection outside the core vortex that wraps around the inner eyewall to form the concentric eyewalls in about 12 h. The observation suggests that an important element of concentric eyewalls formation maybe the development of the symmetric structure from asymmetric convection. The dynamics are studied with an idealization of the interaction of a tropical cyclone core with nearby weaker vorticity with nondivergent bartropic dynamics. The results suggest that the contraction of the secondary wind maximum maybe a feature of vorticity dynamics. Kuo et al. (2004) suggests the pivotal role of the vorticity strength of the core vortex in maintaining itself, and in strectching, organizing and stabilizing the outer vorticity field, and the shielding effect of the moat to prevent further merger and enstrophy cascade processes in concentric eyewall dynamics. Kuo et al. (2008) further suggests that double eyewall of different sizes maybe explained by the binary vortex interaction with skirted parameter. The result also suggests that vorticity generation in the core and in the environment (via mesoscale convection) are of great importance in double eyewall dynamics. The vorticity dynamics of concentric eyewalls also serve as a nice example of the two-dimensional turbulence theory with the selective decay principle (McWilliams, 1984). The strong differential rotation outside the core vortex may yield straining out effect and produces vorticity filaments. With microwave satellite data, Kuo et al. (2009) found that the filamentation process tends to make an important contribution to the organization of the moat in strong typhoons with maximum wind greater than 130 kts. Filamentation effects are important for the convection process in TC environment, and likely relevant to the organization of convection, asymmetric structure, and moat formation. Since filamentation increases with increasing TC intensity, it plays a limitation role in convection particularly in the outer spiral band region. While the barotropic dynamics fall well short of a complete theory because of its neglect of vertical motion, the frictional boundary layer and diabatic process, the idealization gives significant insight to the organization and axisymmetrization aspects of concentric eyewalls structures.

4. Conclusion

In this chapter, we have reviewed important barotropic aspects of hurricane structure and intensity variability. First the dynamical model which governs divergent barotropic motions on an f-plane was introduced, and the fundamental dynamical properties were described. The important wave motions (vortex Rossby waves and inertia-gravity waves) of a linearized set of equations governing perturbations on an arbitrary axisymmetric vortex were described. The linear solutions were used to illustrate some barotropic aspects of observed spiral bands in hurricanes. Spiral rainbands close to the eyewall (or inner spiral rainbands), were shown to be most often sheared vortex Rossby waves, while the outer spiral bands that form away from the vortex basic state PV gradient can often be from a different mechanism. They could be inertia-gravity waves generated by flow imbalances from diabatic effects in the inner-core or generated spontaneously by the vortical motion in the hurricane core, or formed during the formative stages of the tropical cyclone via dynamic instability of larger-scale flow features, such as the inter-tropical convergence zone.

The barotropic model was next used to examine eyewall processes: barotropic instability of the hurricane eyewall and the ensuing PV and passive tracer mixing process between the eyewall and the eye. A conceptual model was described to illustrate the interaction of counter-propagating vortex Rossby waves. Numerical simulations of an unstable PV ring were then conducted to understand details of the mixing process. The numerical simulation exhibited polyonal eyewalls with straight line segments in the early part of the mixing process, and then the initially axisymmetric ring slowly symmetrized into a monopole. Observed polygonal eyewall shapes and eye swirls and mesovortices are believed to be evidence of this quasi-barotropic process occuring in hurricanes.

Finally, the dynamical model and observations was used to examine barotropic aspects of concentric eyewall cycles. Observations indicate that concentric eyewalls may form in asymmetric fashion as a result of the interaction of the vortex core with a weaker vorticity anomaly generated by convection. Due the deformation flow outside the vortex core, the outer weaker vorticity anomaly is strained out and elongated during the axisymmetrization process producing a secondary ring of enhanced vorticity. During this process the secondary wind maximum contracts, consistent with observations and full-physics numerical model simulations, indicating that a portion of concentric eyewall dynamics may be explained by potential vorticity advection.

In conclusion, we note that the divergent barotropic framework here is an oversimplification of the real atmosphere. In addition to neglecting vertical variations, we have also used a quasi-conservative framework without the effects of surface friction and diabatic heating. Both of these processes are elemental to hurricanes, and they create the secondary circulation responsible for intensification. Nonetheless, these results provide some fundamental asymmetric dynamics governing hurricane structure and intensity variability, in the absence of these factors.

5. Acknowledgements

We are grateful to Prof. Wayne Schubert and Dr. Scott Fulton for their comments and assistance. This research was performed in part while the first author held a National Research Council Research Associateship Award at the U.S. Naval Research Laboratory in Monterey, CA.

6. References

Andrews, D. G. (1983). A finite-amplitude Eliassen-Palm theorem in isentropic coordinates, *J. Atmos. Sci.* 40: 1877–1883.

Andrews, D. G., Holton, J. R. & Leovy, C. B. (1987). Middle atmosphere dynamics, *Academic Press*. p. 489 pages.

Barnes, G. M., Zipser, E. J., Jorgensen, D. & Marks, F., Jr. (1983). Mesoscale and convective structure of a rainband, *J. Atmos. Sci.* 40: 2125–2137.

Black, M. L. & Willoughby, H. E. (1992). The concentric eyewall cycle of Hurricane Gilbert, *Mon. Wea. Rev.* 120: 947–957.

Carr, L. E. & Williams, R. T. (1989). Barotropic vortex stability to perturbations from axisymmetry, *J. Atmos. Sci.* 46: 3177–3191.

Chen, Y. & Yau, M. K. (2001). Spiral bands in a simulated hurricane. part 1: Vortex Rossby wave verification, *J. Atmos. Sci.* 58: 2128–2145.

Cram, T. A., Persing, J., Montgomery, M. T. & Braun, S. A. (2007). A Lagrangian trajectory view on transport and mixing processes between the eye, eyewall, and environment using a high-resolution simulation of Hurricane Bonnie (1998), *J. Atmos. Sci.* 64: 1835–1856.

Dritschel, D. D. & Waugh, D. W. (1992). Quantification of the inelastic interaction of unequal vortices in two-dimensional vortex dynamics, *Phys. Fluids* 4: 1737–1744.

Eastin, M. D., Black, P. G. & Gray, W. M. (2002). Flight-level thermodynamic instrument wetting errors in hurricanes. Part I: Observations, *Mon. Wea. Rev.* 130: 825–841.

Ford, R. (1994). The instability of an axisymmetric vortex with monotonic potential vorticity in rotating shallow water, *J. Fluid Mech.* 280: 303–334.

Guinn, T. A. & Schubert, W. H. (1993). Hurricane spiral bands, *J. Atmos. Sci.* 50: 3380–3403.

Hendricks, E. A. & Schubert, W. H. (2009). Transport and mixing in idealized barotropic hurricane-like vortices, *Quart. J. Roy. Meteor. Soc.* 135: 1456–1470.
Hendricks, E. A. & Schubert, W. H. (2010). Adiabatic rearrangement of hollow PV towers, *J. Adv. Model. Earth. Sys.* Vol. 2, Art. #8, 19pp., doi:10.3894/JAMES.2010.2.8.
Hendricks, E. A., Schubert, W. H., McNoldy, B. D. & Fulton, S. R. (2010). Spontaneous-adjustment emission of inertia-gravity waves by unsteady vortical motion in the hurricane core, *Quart. J. Roy. Meteor. Soc.* 136: 537–548.
Hendricks, E. A., Schubert, W. H., Taft, R. K., Wang, H. & Kossin, J. P. (2009). Lifecycles of hurricane-like vorticity rings, *J. Atmos. Sci.* 66: 705–722.
Hodyss, D. & Nolan, D. S. (2008). The Rossby-inertia-buoyancy instability in baroclinic vortices, *Phys. Fluids* 20: 096602.
Kossin, J. P. & Eastin, M. D. (2001). Two distinct regimes in the kinematic and thermodynamic structure of the hurricane eye and eyewall, *J. Atmos. Sci.* 58: 1079–1090.
Kossin, J. P. & Schubert, W. H. (2001). Mesovortices, polygonal flow patterns, and rapid pressure falls in hurricane-like vortices, *J. Atmos. Sci.* 58: 2196–2209.
Kuo, H. C., Chang, C. P., Yang, Y. T. & Jiang, H.-J. (2009). Western North Pacific typhoons with concentric eyewalls, *Mon. Wea. Rev.* 137: 3758–3770.
Kuo, H.-C., Chen, G. T.-J. & Lin, C.-H. (2000). Merger of tropical cyclones Zeb and Alex, *Mon. Wea. Rev.* 128: 2967–2975.
Kuo, H. C., Lin, L.-Y., Chang, C.-P. & Williams, R. T. (2004). The formation of concentric vorticity structures in typhoons, *J. Atmos. Sci.* 61: 2722–2734.
Kuo, H. C., Schubert, W. H., Tsai, C. L. & Kuo, Y. F. (2008). Vortex interactions and barotropic aspects of concentric eyewall formation, *Mon. Wea. Rev.* 136: 5183–5198.
Kuo, H.-C., Williams, R. T. & Chen, J.-H. (1999). A possible mechanism for the eye rotation of Typhoon Herb, *J. Atmos. Sci.* 56: 1656–1673.
Kurihara, Y. M. (1979). On the development of spiral bands in tropical cyclones, *J. Atmos. Sci.* 33: 940–958.
Lander, M. A. & Holland, G. (1993). On the interaction of tropical-cyclone-scale vortices. I: Observations, *Quart. J. Roy. Meteor. Soc.* 119: 1347–1361.
Larson, R. N. (1975). Picture of the month–hurricane twins over the eastern North Pacific ocean, *Mon. Wea. Rev.* 103: 262–265.
MacDonald, N. J. (1968). The evidence for the existence of rossby-type waves on the hurricane vortex, *Tellus* 20: 138–150.
Mallen, K. J., Montgomery, M. T. & Wang, B. (2005). Reexamining the near-core radial structure fo the tropical cyclone primary circulation: Implications for vortex resiliency, *J. Atmos. Sci.* 62: 408–425.
McWilliams, J. C. (1984). The emergence of isolated, coherent vortices in turbulent flow, *J. Fluid Mech.* 46: 21–43.
Möller, J. D. & Montgomery, M. T. (2000). Tropical cyclone evolution via potential vorticity anomalies in a three-dimensional balance model, *J. Atmos. Sci.* 57: 3366–3387.
Möller, J. D. & Smith, R. K. (1994). The development of potential vorticity in a hurricane-like vortex, *Quart. J. Roy. Meteor. Soc.* 120: 1255–1265.
Montgomery, M. T., Bell, M. M., Aberson, S. D. & Black, M. L. (2006). Hurricane Isabel (2003). New insights into the physics of intense storms. Part I: Mean vortex structure and maximum intensity estimates, *Bull. Amer. Meteor. Soc.* 87: 1335–1347.

Montgomery, M. T. & Enagonio, J. (1998). Tropical cyclogensis via convectively forced vortex Rossby waves in a three-dimensional quasigeostrophic model, *J. Atmos. Sci.* 55: 3176–3207.

Montgomery, M. T. & Kallenbach, R. J. (1997). A theory for vortex Rossby waves and its application to spiral bands and intensity changes in hurricanes, *Quart. J. Roy. Meteor. Soc.* 123: 435–465.

Nakamura, N. (1996). Two-dimensional mixing, edge formation, and permeability diagnosed in area coordinates, *J. Atmos. Sci.* 53: 1524–1537.

Peng, J., Li, T., Peng, M. S. & Ge, X. (2008). Barotropic instability in the tropical cyclone outer region, *Quart. J. Roy. Meteor. Soc.* 135: 851–864.

Persing, J. & Montgomery, M. T. (2003). Hurricane superintensity, *J. Atmos. Sci.* 60: 2349–2371.

Prieto, R., McNoldy, B. D., Fulton, S. R. & Schubert, W. H. (2003). A classification of binary tropical cyclone-like vortex interactions, *Mon. Wea. Rev.* 131: 2656–2666.

Reasor, P. D., Montgomery, M. T., Marks, F. D. & Gamache, J. F. (2000). Low-wavenumber structure and evolution of the hurricane inner core observed by airborne dual-doppler radar, *Mon. Wea. Rev.* 128: 1653–1680.

Rozoff, C. M., Kossin, J. P., Schubert, W. H. & Mulero, P. J. (2009). Internal control of hurricane intensity: The dual nature of potential vorticity mixing., *J. Atmos. Sci.* 66: 133–147.

Rozoff, C. M., Schubert, W. H., McNoldy, B. D. & Kossin, J. P. (2006). Rapid filamentation zones in intense tropical cyclones, *J. Atmos. Sci.* 63: 325–340.

Schecter, D. A. & Montgomery, M. T. (2004). Damping and pumping of a vortex rossby wave in a monotonic cyclone: Critical layer stirring versus inertia-buoyancy wave emission, *Phys. Fluids* 16: 1334–1348.

Schubert, W. H., Montgomery, M. T., Taft, R. K., Guinn, T. A., Fulton, S. R., Kossin, J. P. & Edwards, J. P. (1999). Polygonal eyewalls, asymmetric eye contraction, and potential vorticity mixing in hurricanes, *J. Atmos. Sci.* 56: 1197–1223.

Shapiro, L. J. & Willoughby, H. E. (1982). The response of balanced hurricanes to local sources of heat and momentum, *J. Atmos. Sci.* 39: 378–394.

Wang, Y. (2002a). Vortex Rossby waves in a numerically simulated tropical cyclone. Part I: Overall structure, potential vorticity, and kinetic energy budgets, *J. Atmos. Sci.* 59: 1213–1238.

Wang, Y. (2002b). Vortex Rossby waves in a numerically simulated tropical cyclone. Part II: The role in tropical cyclone structure and intensity changes, *J. Atmos. Sci.* 59: 1239–1262.

Willoughby, H. E. (1978). A possible mechanism for the formation of hurricane rainbands, *J. Atmos. Sci.* 35: 838–848.

9

Studies of the Hurricane Evolution Based on Modern Thermodynamics

Chongjian Liu[1], Ying Liu[2], Zhexian Luo[3], Xiaotu Lei[4], Donghai Wang[5] and Xiaogang Zhou[6]

[1]*Shanghai Typhoon Institute, China Meteorological Administration*
[1]*State Key Laboratory of Severe Weather, Chinese Academy of Meteorological Sciences*
[2]*State Key Laboratory of Severe Weather, Chinese Academy of Meteorological Sciences*
[3]*Institute of Remote Sense, Nanjing University of Information Science & Technology, Nanjing*
[4]*Shanghai Typhoon Institute, China Meteorological Administration*
[5]*State Key Laboratory of Severe Weather, Chinese Academy of Meteorological Sciences*
[6]*Training Centre, China Meteorological Administration*
China

1. Introduction

The atmosphere has been likened to a giant thermodynamic engine in which disorganized energy is transformed into the organized kinetic energy of the winds and, the general circulation of the atmosphere can be regarded as simply being driven by temperature differences between the polar and equatorial regions. As a result, some concepts and theories in modern thermodynamics should be relevant and crucial to be used to deeper understanding of the mechanism responsible for the hurricane evolution.

Recently, more attention has been paid to the thermodynamics and statistical mechanics with the atmospheric sciences. In addition to the large number of journal articles (Ozawa and Ohmura, 1997; Duane and Curry, 1997; Barsugli and Battisti, 1998; Egger, 1999; Nicolis, 2002; Fraedrich and Blender, 2003; Gade and Gustafson, 2004), several monographs have been published in succession (Curry, 1999; Bohren and Albrecht, 1999; Tsonis, 2002; Zdunkowski and Bott, 2004).

In this chapter, we will introduce a number of recent studies of hurricane evolution based on modern thermodynamics, which are generally distinguished into the following three categories.

The first category of such studies is closely related with the theory of nonlinear nonequilibrium thermodynamics. As well-known to all, entropy flow is a core concept in the modern theory of dissipative structures just as entropy is in classical thermodynamics and statistical physics. According to the second law of thermodynamics, an isolated system will evolve spontaneously into the equilibrium with maximum entropy where the order or organization of the system is at the minimum. Therefore, negative entropy is very important for a system to keep itself far from its equilibrium, which is also true in biological systems (for a biological individual, equilibrium means death as reflected in the statement in *What Is*

Life? by E. Schrödinger that life's existence depends on its continuous gain of 'negentropy' from its surroundings). Based on this cognition, recently the entropy flows of some hurricanes/typhoons have been calculated using the entropy flow formula appropriate for atmospheric dynamical systems with the National Centre for Environmental Prediction/National Centre for Atmospheric Research reanalysis data employed (Liu and Liu, 2004; Liu et al., 2007; Liu and Liu, 2007, 2008, 2009; Xu and Liu, 2008; Liu et al., 2009; Liu et al., 2010). In this chapter, Hurricane Katrina (2005) will be taken as an example to show the main results. This includes that the intensification of net negative (positive) entropy flow entering into Katrina preceded the strengthening (weakening) of its intensity and that the asymmetries of the entropy flow pattern around Katrina's centre in the lower troposphere (e.g. at 850 hPa as shown in Fig. 4 below for entropy flow fields at the two times of 18 UTC 26 and 29 August with the Katrina's track in Fig. 2 for easy reference) contained some significant information on hurricane track prediction. These results imply that classical thermodynamic entropy may serve as an order parameter for an atmospheric system and that entropy flow analysis might provide a new insight into the mechanism responsible for the life cycle of the system.

The second is basically to improve hurricane numerical prediction based on the 2nd law of thermodynamics which implies that if there exists the heterogeneity in a system initially, (say) heat or particles within the system will then be spontaneously transferred (diffused) from part with higher T or higher concentration to that with lower T or lower concentration. The evolutionary direction of any many-body system like the atmosphere or ocean must be controlled by the second law of thermodynamics. As a matter of fact, the evolution of a many-body system like the atmosphere in non-equilibrium involves irreversible physical processes such as diffusion within the system. A proper description of diffusion is of critical importance for a numerical weather prediction (NWP) model. However, for example, the commonly used fourth-order diffusion schemes in a NWP model, as employed in the fifth Penn State/NCAR non-hydrostatic meso-scale model (called simply MM5), would show an inadequate diffusion formulation from the point of view of modern thermodynamics. In the recent works (Liu and Liu, 2005; Liu et al., 2009), the forecast accuracy of the MM5 as well as a global spectrum model is improved by emulating physical dissipation as suggested by the second law of thermodynamics that controls the irreversible evolutionary direction of a many-body system. The ability of the new thermodynamics-based scheme to improve model accuracy has been demonstrated via the case of a hurricane as well as a yearlong continuous simulation.

The third category of study is toward improved understanding of hurricane climate change. In this regard, we would like to particularly mention the seemingly counterintuitive phenomena as shown by the analyses based on the observational data, that is, under global warming the tropical cyclone (TC, among which the hurricane is the one of highest intensity) numbers tend to become decreasing over some oceanic basins in the world during the recent multiple decades despite a rise in their sea surface temperatures (SSTs). From our recent work, a methodology of revealing the mechanism responsible for such events has been suggested based on the second law of thermodynamics and the well-known thermal wind relation in the atmospheric dynamics and thus the causality between the unusually heterogeneous pattern of SSTs in warming environment and TC frequency over, as an example, the western North Pacific (WNP) is explained. This methodology is important in that it provides with an insight into the problem why high SSTs under global warming as one of the necessary conditions for TC genesis should unexpectedly contribute to TC number reduction.

This chapter is arranged as follows: the entropy flows of Hurricane Katrina (2005) are calculated and discussed in the next section; improving hurricane track forecast based on the 2nd law of thermodynamics is presented in the 3rd section; understanding TC frequency variability in terms of thermodynamic principle is given in the 4th section; and further discussions and suggestions are made in the last section.

2. Implication of entropy flow for the development of a hurricane

2.1 A brief introduction

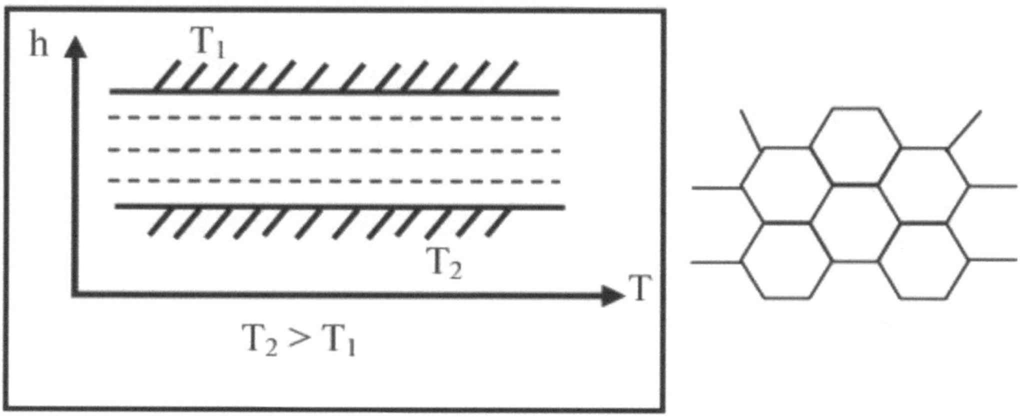

Fig. 1. The schematic diagram for Bénard convection as the prototype of self-organization.

In 1900 Bénard investigated a thin layer of fluid at rest originally, with a free surface at a lower temperature T_1, heated from its bottom at a higher temperature T_2 (see Fig. 1), and noticed that a rather regular cellular pattern of hexagonal convective cells was abruptly organised when the temperature difference (T_2 - T_1) reaching the value of the threshold of primary instability (Prigogine, 1955; De Groot and Mazur, 1984; Glansdorff and Prigogine, 1971; Nitschke, Bestehorn and Thess, 1998), suggesting that every system that obtains heat at a higher temperature but loses heat at a lower temperature will experience net negative entropy flow. In this case the entropy exchange (δs_e) of the system with its environment is as follows:

$$\delta s_e = \frac{Q_2}{T_2} - \frac{Q_1}{T_1} = \frac{T_1 - T_2}{T_1 T_2} Q < 0, \tag{1}$$

where Q_1 and Q_2 are the heat fluxes through the top and bottom of the layer, respectively, and Q is the flux when the system becomes steady so that $Q = Q_1 = Q_2$. Later, in 1940s, Schrödinger stated in his monograph "What Is Life?" that life's existence depends on its continuous gain of "negentropy" from its surroundings (Schrödinger, 1944). This implies that negative entropy flow is something very significant for a system, whether it is living or nonliving (Prigogine, 1955; De Groot and Mazur, 1984; Glansdorff and Prigogine, 1971; Schrödinger, 1944; Olby, 1971; Katchalsky and Curran, 1965). That is, negative entropy flow will cause a system initially at equilibrium or even at rest to be organized, or lead a system already at non-equilibrium to a state further from equilibrium (i.e. cause it to strengthen).

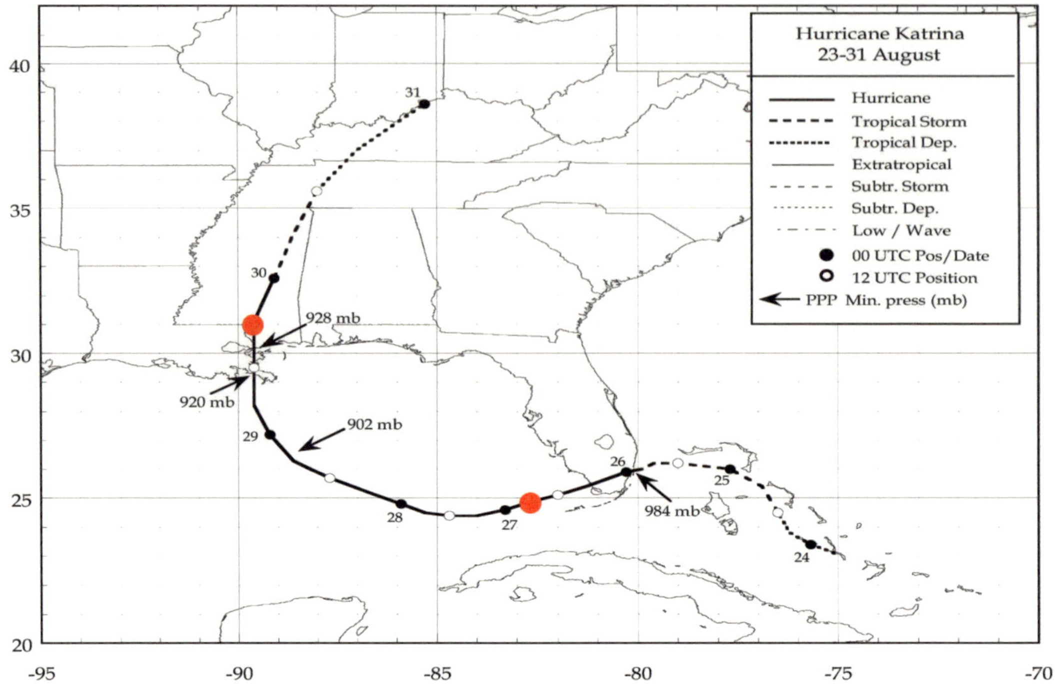

Fig. 2. Best track positions for Hurricane Katrina, 23-30 August 2005. (*Credits:* http://www.nhc.noaa.gov/pdf/TCR-AL122005_Katrina.pdf). The red spot marks are for the corresponding position of the hurricane centre's to that shown in Fig. 4.

The purpose of this section is mainly to use the entropy flow formula, as a logical component of the entropy balance equation (Prigogine, 1955; De Groot and Mazur, 1984; Glansdorff and Prigogine, 1971), to calculate the three-dimensional instantaneous entropy flow field and then discuss the relationship of entropy flow with the development of an atmospheric system (hurricane). The system selected for this study is Hurricane Katrina (2005) (see also Fig. 2) that was extraordinarily powerful and deadly with the total damage cost (about $81 billion) being roughly double that of Hurricane Andrew (1992) and the total number of fatalities over 1800 (Knabb et al., 2005; McTaggart-Cowan et al., 2007; Liu et al., 1997). Considering the scope of its impacts, Katrina was one of the most devastating natural disasters in the history of the United States.

2.2 Theory and the computational formulation

The entropy change in an open system consists of two contributory parts: one is entropy flow that is caused by the positive or negative entropy exchange with its surroundings; and, the other is positive definite entropy production owing to irreversible processes within the system. This can be described by the entropy balance equation which is derived from the Gibbs relation and local equilibrium assumption and, for entropy per unit volume, S, is of the form

$$\frac{\partial S}{\partial t} = -div\ \vec{J}_s + \sigma \tag{2}$$

where

$$\vec{J}_s = S\vec{V} + \frac{1}{T}\vec{J}_q - \sum_k \frac{\mu_k}{T}\vec{J}_k$$

is called the entropy flow vector in which $\vec{J}_q$, T, μ_k and $\vec{J}_k$ are the heat flow, temperature, chemical potential and diffusive flow for component *k*, respectively, with the diffusive flow defined as

$$\vec{J}_k = \rho_k(\vec{V}_k - \vec{V}) \tag{3}$$

where ρ_k and $\vec{V}_k$ are the density and velocity for component *k*, respectively, and $\vec{V}$ the velocity for the mass centre in this context that is related to $\vec{V}_k$ by $\vec{V} = \sum \rho_k \vec{V}_k / \rho$; and, the heat flow expressed approximately, according to the Fourier law (Origogine, 1955; De Groot and Mazur, 1962), as

$$\vec{J}_q = -\rho\tilde{\lambda}\nabla T, \tag{4}$$

where $\tilde{\lambda}$ is the thermal conductivity. In Eq. (2) σ denotes the entropy production, the "*div*"-term is the entropy flow, and, the entropy per unit volume *S* is related to the specific entropy *s* by $S = \rho s$ with ρ being the density (De Groot and Mazur, 1962; Ruelle, 2003). In this chapter the component of liquid water has been omitted owing to the liquid water content (about 5 g m^{-3} on the average over the tropics) being much less than the density of either vapour or dry air (Liu et al., 1997; Mason, 1971) and only two components are taken into account, that is, vapour and dry air with the corresponding density being ρ_v and ρ_d, and fractional mass N_v (= q the specific humidity) and N_d (= 1 - q), respectively. As a consequence, the entropy s per unit mass consists accordingly of s_q for vapour and s_d for dry air so that

$$s = s_q + s_d = q(C_{pv}\ln T - R_v \ln e) + (1-q)[C_{pd}\ln T - R_d \ln(p-e)], \tag{5}$$

where C_{pv} and C_{pd} are the specific heat at constant pressure for vapour and dry air, respectively; e the vapour pressure; and, R_v and R_d are the gas constant for vapour and dry air, respectively. Usually the velocity $\vec{V}_q$ for vapour is assumed to be the same as $\vec{V}_d$ for dry air so that the diffusive flow $\vec{J}_k = 0$ in this case since $\vec{J}_k = \rho_k(\vec{V}_k - \vec{V})$.

In view of that formulas above are derived based on Cartesian coordinates but the NCEP/NCAR reanalysis data used in this letter are on constant pressure layers (Kalnay et al., 1996), it is necessary to transform the expression of entropy flow into that in the *p*-coordinates based on the common transform relations (Haltiner and Martin, 1957).

$$\frac{\partial A}{\partial x} = (\frac{\partial A}{\partial x})_p - \frac{\partial A}{\partial p}\frac{\partial p}{\partial z}(\frac{\partial z}{\partial x})_p, \tag{6}$$

$$\frac{\partial A}{\partial y} = (\frac{\partial A}{\partial y})_p - \frac{\partial A}{\partial p}\frac{\partial p}{\partial z}(\frac{\partial z}{\partial y})_p, \tag{7}$$

$$\frac{\partial A}{\partial t} = (\frac{\partial A}{\partial t})_p - \frac{\partial A}{\partial p}\frac{\partial p}{\partial z}(\frac{\partial z}{\partial t})_p, \tag{8}$$

where A is an arbitrary variable; all derivations made in p-coordinates are denoted by $(\ )_p$ and the others otherwise in z-coordinates. As a result, the entropy flow becomes finally

$$\begin{aligned}
-div\boldsymbol{J_s} = & -\frac{\partial \rho s u}{\partial x} - \frac{\partial \upsilon s v}{\partial y} - \rho\frac{\partial \rho s u}{\partial p}(\frac{\partial \phi}{\partial x}) - \rho\frac{\partial \rho s v}{\partial p}(\frac{\partial \phi}{\partial y}) - \rho\frac{\partial s\omega}{\partial p} \\
& + \frac{\tilde{\lambda}\rho}{T}\{\frac{\partial^2 T}{\partial x^2} + \frac{\partial^2 T}{\partial y^2} + 2\rho\frac{\partial \phi}{\partial x}\frac{\partial^2 T}{\partial p\partial x} + 2\rho\frac{\partial \phi}{\partial y}\frac{\partial^2 T}{\partial p\partial y} + \rho^2[(\frac{\partial \phi}{\partial x})^2 + (\frac{\partial \phi}{\partial y})^2]\frac{\partial^2 T}{\partial p^2} \\
& + \rho(\frac{\partial^2 \phi}{\partial x^2} + \frac{\partial^2 \phi}{\partial y^2} + \rho\frac{\partial \phi}{\partial x}\frac{\partial^2 \phi}{\partial p\partial x} + \rho\frac{\partial \phi}{\partial y}\frac{\partial^2 \phi}{\partial p\partial y})\frac{\partial T}{\partial p} + \rho^2 g^2\frac{\partial^2 T}{\partial p^2} \\
& + (\frac{1}{\rho}\frac{\partial T}{\partial x} + \frac{\partial \phi}{\partial x}\frac{\partial T}{\partial p})(\frac{\partial \rho}{\partial x} + \rho\frac{\partial \rho}{\partial p}\frac{\partial \phi}{\partial x} - \frac{\rho}{T}\frac{\partial T}{\partial x} - \frac{\rho^2}{T}\frac{\partial T}{\partial p}\frac{\partial \phi}{\partial x}) \\
& + (\frac{1}{\rho}\frac{\partial T}{\partial y} + \frac{\partial \phi}{\partial y}\frac{\partial T}{\partial p})(\frac{\partial \rho}{\partial y} + \rho\frac{\partial \rho}{\partial p}\frac{\partial \phi}{\partial y} - \frac{\rho}{T}\frac{\partial T}{\partial y} - \frac{\rho^2}{T}\frac{\partial T}{\partial p}\frac{\partial \phi}{\partial y}) + \frac{\rho^2 g^2}{T}(\frac{\partial T}{\partial p})^2\},
\end{aligned} \tag{9}$$

where u and v are the velocities in the x- and y-direction at constant pressure layers, respectively; ω the vertical velocity in p-coordinates; ϕ the geopotential height with $\phi = gz$ in which g is the gravitational acceleration; and s the specific entropy, expressed mathematically in terms of Eq. (5) in this chapter.

Next we will examine the three-dimensional entropy flow fields, with the computational domain covering the hurricane and its ambient zones, in two ways. One of them is to discuss the vertical cross sections of entropy flow (see fig. 4) based directly on the computational from Eq. (9) and, the other is for the total entropy flow (see Fig. 5) of the hurricane through its lateral boundary, Σ, which is calculated by integrating the divergence of the entropy flow vector $\vec{J}_s$ over the whole system volume V, and a discrete version of this integration is done here by summing over all the gridpoints encompassed by the curved surface. The latter is based on the well-known Gaussian formula (Bronshtein and Semendyayev, 1997):

$$\iint_{\Sigma}(P\cos\alpha + Q\cos\beta + R\cos\gamma)d\Sigma = \iiint_{V}(\frac{\partial P}{\partial x} + \frac{\partial Q}{\partial y} + \frac{\partial R}{\partial z})dxdydz \tag{10}$$

where $cos\alpha$, $cos\beta$ and $cos\gamma$ are the direction cosines of the curved surface Σ at the positive normal direction, and P, Q and R are the three components of the entropy flow vector $\vec{J}_s = P\vec{i} + Q\vec{j} + R\vec{k}$ in the x-, y- and z-directions, respectively.

The total entropy flow calculation for a hurricane is performed by summing the entropy flow over every gridpoint within this system at a constant pressure layer and then over every layer. Therefore, a single figure may stand for the intensity of entropy flow for the system at any given time. The approximate total entropy flow is calculated first on each pressure layer, by summing over all gridpoints within a circle centred on Katrina's centre with its radius of six degrees (thus the horizontal boundaries of the hurricane are defined), and then summed over all layers (thus the top and bottom pressure layer are treated as the vertical boundaries of Katrina).

2.3 Results and discussion

We will first examine changes in the total entropy flow of Hurricane Katrina based on the National Centers for Environmental Prediction/National Center for Atmospheric Research (NCEP/NCAR) 1°×1° (latitude-longitude) resolution reanalysis data (Kalnay et al., 1996).

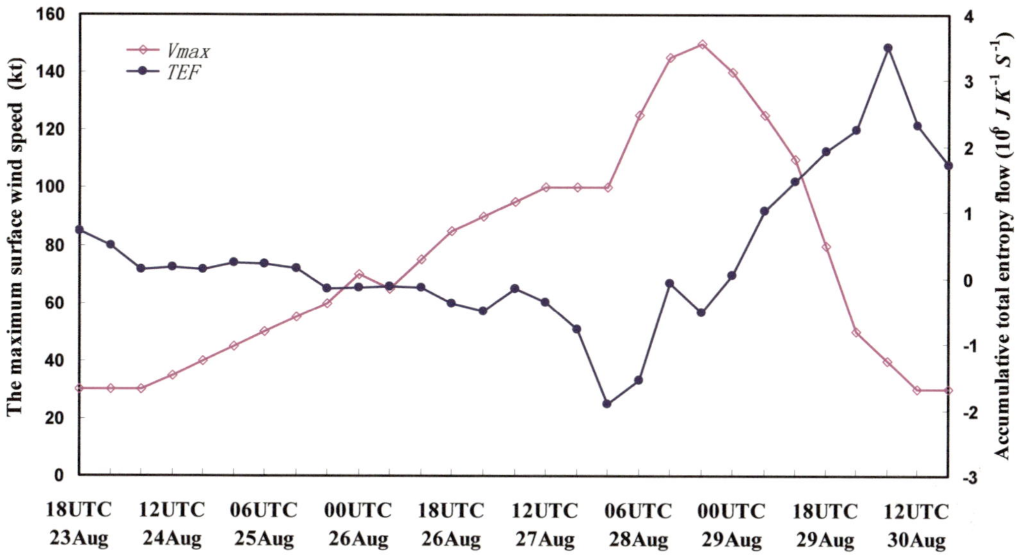

Fig. 3. The evolution of total entropy flow (*TEF*, in 10^6 J K^{-1} s^{-1}) for Hurricane Katrina at 6-hourly intervals, with changes in the maximum sustained wind speed (V_{max}, in *kt*) as a reference.

Figure 3 shows changes, at 6-hourly intervals, in the accumulative total entropy flow for Katrina as well as the corresponding maximum sustained wind speed representative of the intensity of the hurricane. It is seen from Fig. 3 that there exists a general trend of counter-correlation between these two curves with the phase in the total entropy flow preceding that in the maximum wind speed. For example, Katrina experienced more and more increasing net negative entropy flow before 00:00 UTC 28 August 2005 and, which was followed closely by its continuous intensification with very rapid strengthening occurring on the morning of 28 August. On the other hand, the values of negative entropy flow for Katrina decreased distinctly and even turned to positive entropy flow after 00:00 UTC 28 August, which was followed by the gradual weakening of Katrina from its peak intensity of 150 *kt* later at 18:00 UTC 28 August. Especially, Katrina had experienced a dramatic decrease of negative entropy flow and a successive turn to the stage of positive entropy flow during the period from 18:00 UTC 28 to 06:00 UTC 29 August when the very rapid weakening of Katrina followed accordingly 6 hours later starting with 12:00 UTC 29 August. As a result, the total entropy flow of a hurricane may contain some significant information conducive to its intensity forecast.

In addition, it is found that the asymmetries of the entropy flow pattern around Katrina's centre in the lower troposphere (e.g. at 850 hPa as are shown in Fig. 4 for the two times of 18 UTC 26 and 28 August which are marked also with the red spot on the Katrina's track in Fig. 2 for easy reference) contained some significant information on hurricane track prediction.

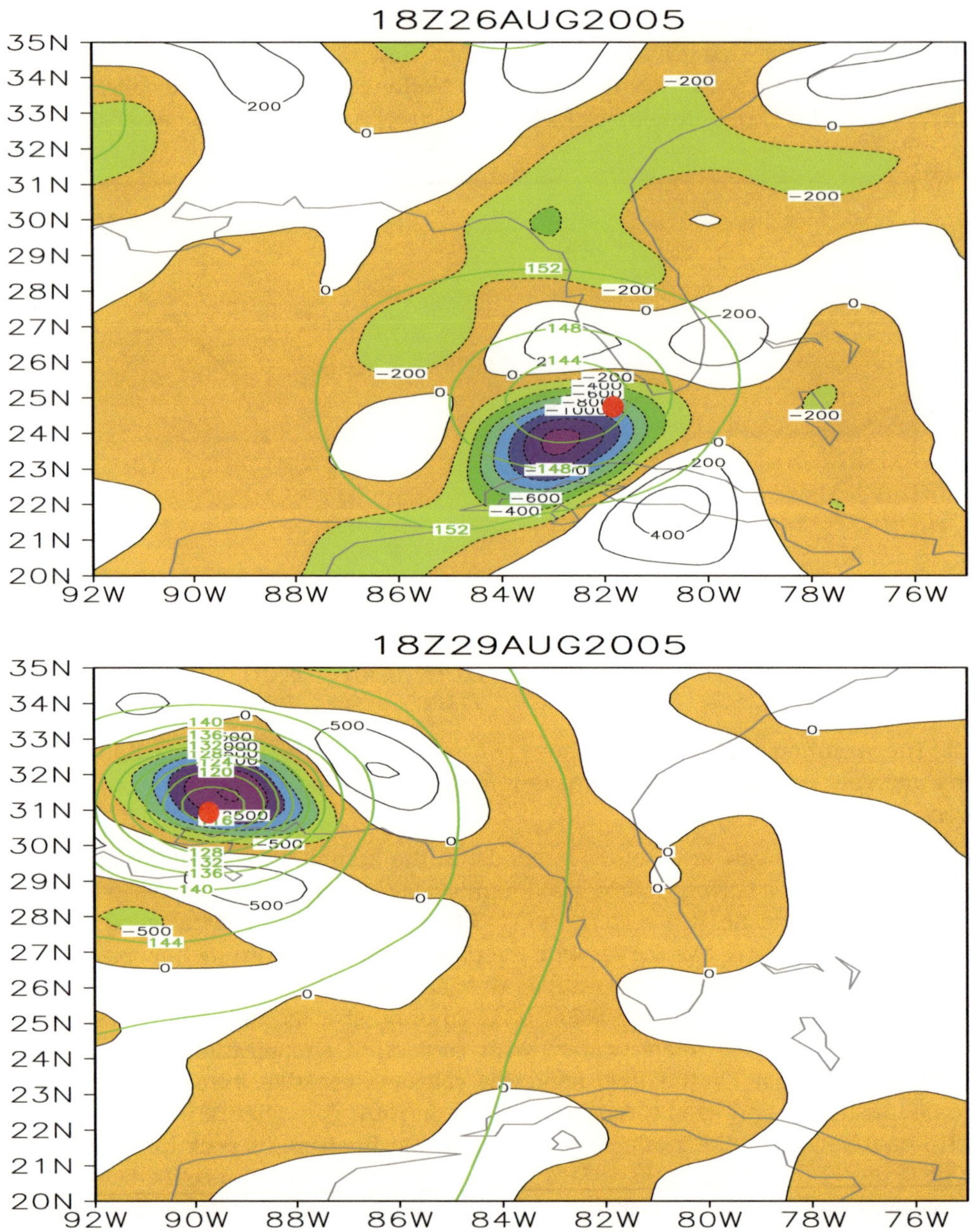

Fig. 4. The entropy flow fields at 850 hPa at 18:00 UTC 26 and at 18:00 UTC 29 August 2005 (the time is marked above the respective panel). The red spot stands for the corresponding position of the hurricane centre's to the respective time.

These results imply that classical thermodynamic entropy may serve as an order parameter for an atmospheric system and that entropy flow analysis might provide a new insight into the mechanism responsible for the life cycle of the system. In view of that Katrina (2005) tends to move towards the side of its centre with stronger negative entropy flows, the entropy flow patterns in the lower levels (e.g. 850 hPa) might act as an indicator of the hurricane track. This is understandable from the theory of dissipative structures (Nitschke et al., 1998): negative entropy flow entering into an open system will potentially enable the system to depart further from equilibrium or cause it to strengthen, and so here a region of stronger negative entropy flow might become a likely area that the hurricane will approach.

3. Improving hurricane track forecast based on the 2nd law of thermodynamics

3.1 A brief introduction

The evolution of a many-body system like the atmosphere in non-equilibrium involves the irreversible physical processes such as diffusion within the system. A proper description of diffusion is of critical importance for a numerical weather prediction (NWP) model (Smagorincky, 1963; Eliasen and Laursen, 1990; Stephenson, 1994; Becher, 2001; Burkhardt and Becher, 2006).

In the early NWP models (Cheng, 1975) the second order diffusion schemes were, indeed, introduced as a dissipative term similar to viscosity into a prognostic equation, which is basically the physically meaningful Fickian diffusion process. However, it is questionable whether the second order scheme can adequately describe mixing processes in the atmosphere. During the 1980s some models progressed to the fourth or higher order diffusion schemes (Anthes *et al.*, 1987; Laursen and Eliasen, 1989). The fourth-order schemes became the most common choice since they have stronger scale-selectivity compared with the second-order schemes (Pielke, 1984). Now the problem arises since a fourth order scheme may create upgradient mass or heat transport, which is spurious and violates the full irreversibility principle of the second law of thermodynamics. In nature, models that use the fourth order diffusion scheme may produce non-physical oscillations like Gibbs phenomenon (Pielke, 1984). Although the new formulation of horizontal momentum diffusion suggested by Becker (2001) is very creative with a firm physical basis, which defines well the local frictional heating rate (dissipation). However, the point might lie in that the dissipation is positive definite only in the case of nabla-square symmetric horizontal diffusion. Therefore, as pointed out in Becker (2001), the physical meaning of higher orders is open to question even if the stress tensor is symmetric. Recently, Burkhardt and Becker (2006) argue even that the new horizontal scheme developed and tested requires the replacement of the hyperdiffusion with a ∇^2-scheme (in order for horizontal dissipation to be positive definite or keep the full irreversibility). To sum up, ∇^2-scheme has been replaced by ∇^4-scheme owing to the latter's merit of better scale-selectivity. Now, we find that ∇^4-scheme can not guarantee being positive definite and then have this chapter of ours to present for suggesting a new scheme that can both remove two-gridlength waves and keep the full irreversibility with the basic theory introduced in section 3.2 below.

The full irreversibility principle exerts an entropy constraint on irreversible processes in a many-body system. The studies (Liu and Liu, 2004, 2005) demonstrated, the discretized atmospheric system must still satisfy the second law of thermodynamics, and improved simulations can be achieved when the irreversibility principle is enforced in a NWP model.

The discretization stencil for the diffusion algorithm in a NWP model presented in this chapter is just based on full irreversibility and, the physics-based "fourth-order" scheme would lead to dramatic improvements in the model outputs as viewed when the new scheme based on the theory described in next section is applied to a meso-scale model (Dudhia, 1993; Grell *et al.*, 1995).

3.2 Theory

The principle of the new diffusion algorithm can be briefly described as follows.

Generally the horizontal diffusion of a prognostic variable, A, is modeled by adding an *ad hoc* term, F_{HA} , to the right-hand side of the prognostic equation for A:

$$\frac{\partial A}{\partial t} = N + F_{HA}, \tag{11}$$

where F_{HA} is just the diffusion-like term added and N denotes all terms other than diffusion.

The commonly used fourth-order diffusion schemes, as employed in the fifth Penn State/NCAR non-hydrostatic meso-scale model (MM5) (Grell et al., 1995), would show an inadequate diffusion formulation from the point of view of the classical thermodynamics.

For simplicity let us consider the central finite difference form of the one-dimensional fourth-order diffusion terms in MM5:

$$\begin{aligned}(F_{HA})_i &= -K_A \nabla^4 A_i \\ &= K_A [4 (A_{i+1}+A_{i-1}) - (A_{i+2}+A_{i-2}) - 6A_i] \\ &= K_A [4 (A_{i+1} - A_i)+4 (A_{i-1} - A_i) - (A_{i+2} - A_i) - (A_{i-2} - A_i)],\end{aligned} \tag{12}$$

where K_A is a positive definite diffusion coefficient and, A_i, $A_{i\pm1}$ and $A_{i\pm2}$ are the values of A at grid points i, $i\pm1$ and $i\pm2$, respectively.

From Eq. (12), it is apparent that the gradients between A_i and A_{i+1} (A_{i-1}) ensure mass or heat downgradient transport, but the gradients between A_i and A_{i+2} (A_{i-2}) cause mass or heat to transport upgradiently, which is not in agreement with the full irreversibility principle. Therefore, the fourth-order diffusion scheme expressed in Eq. (12) is not guaranteed to be a monotonic scheme (Crandall and Majda, 1980).

The horizontal diffusion scheme in MM5 can be reconstructed in accordance with the principle of full irreversibility by changing the sign of the $A_{i\pm2}$ terms in Eq. (12) to enforce an irreversible diffusion process. This new form is:

$$(F_{HA})_i = K_A' \nabla_T A_i, \tag{13}$$

where K_A' is a positive definite diffusion coefficient which might have a different value from K_A but is kept unchanged in the present new scheme, and $\nabla_T A_i$ is given by:

$$\nabla_T A_i = D_1[(A_{i+1} - A_i)+(A_{i-1} - A_i)]+D_2[(A_{i+2} - A_i)+(A_{i-2} - A_i)], \tag{14}$$

where D_1 and D_2 are the diffusive weights. Now all the signs are positive except for those preceding the A_i terms.

It is seen from Eq. (14) that this new scheme is in line with full irreversibility since this scheme ensures the diffusion term must always transport mass or heat downgradiently.

Next we would first verify the merit of this theory with the one-dimensional viscous Burgers equation that has an exact solution when it runs to a steady flow.

The one-dimensional viscous Burgers equation can be written as

$$\frac{\partial u}{\partial t} + u\frac{\partial u}{\partial x} = \frac{1}{R_e}\frac{\partial^2 u}{\partial x^2} \tag{15}$$

where $u \equiv u(x, t)$ is the flow velocity, R_e the Renolds number with R_e = 100.0 set in this chapter. Such a dissipative flow without forcing will eventually tend to the steady flow. If, for example, the initial condition is taken as the linear distribution

$$u = -x \quad \text{at } t = 0, -\infty < x < \infty \tag{16}$$

with the boundary condition as

$$u = \begin{cases} 1 & \text{at} \quad x = -\infty \\ -1 & \text{at} \quad x = \infty \end{cases}$$

there then exists a special solution : $u = -\mathrm{th}[(R_e/2)x]$, that is a steady solution containing no time. In case the central difference scheme is used it will approach, starting from the initial condition Eq. (16) and integrating till a certain time (within 2 *s*), this special solution with a certain accuracy.

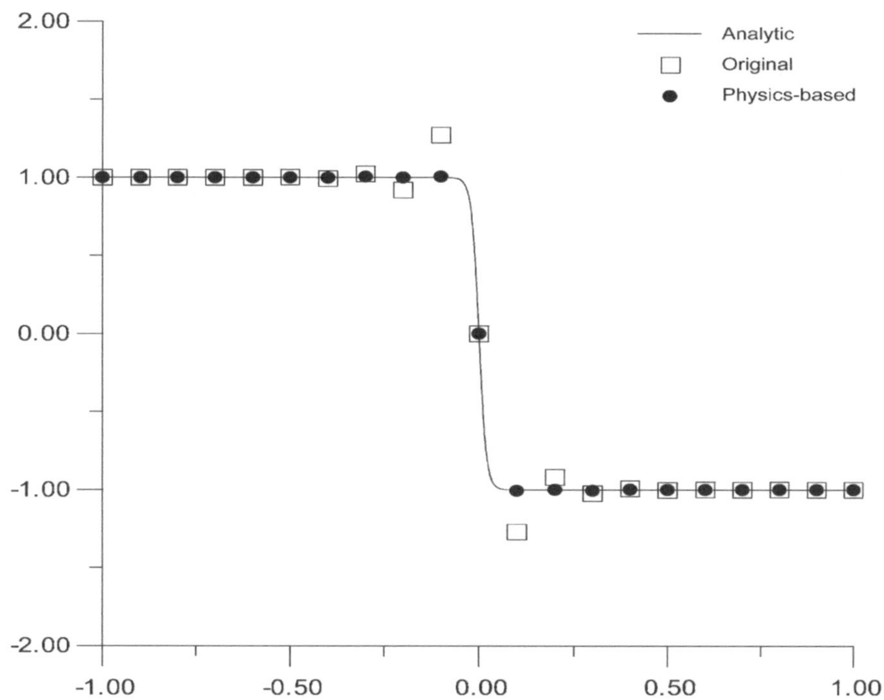

Fig. 5. Comparison of the analytic solution of the one-dimensional viscous Burgers equation that runs to a steady flow with the numerical solutions between the original and physics-based diffusion schemes used (see text for the details).

The computational results show that the final solution with the original diffusion scheme used has a Gibbs oscillation (Pielke, 1984) near the "shock wave" where the velocity gradient augments suddenly (Fig. 5). On the other hand, the numerical solution consistent with the analytic one is reached when the (original fourth order) scheme is changed, based on the theory described in this section, to the physics-based scheme that meets the restriction of the second law of thermodynamics (Fig. 5).

In the next section the effect of the physics-based diffusion scheme will be demonstrated via a series of numerical simulations, especially that of a hurricane in terms of its surface minimum pressure and maximum wind.

3.3 Results and discussion

In this section the diffusion scheme is re-constructed based on the second law of thermodynamics to ensure downgradient transport in the way described above and then applied to the Penn State/NCAR non-hydrostatic mesoscale model (Grell et al., 1995) with Hurricane Andrew (1992) taken as a case to simulate with the emphasis on the evolutions of its surface maximum velocity U_{max} and minimum pressure p_{min}.

Hurricane Andrew in 1992 cost the United States about 25 billion dollars (Wakimoto and Black, 1994; NOAA, 1992). The landfall of the hurricane has been well studied by MM5 simulations (Liu et al., 1997). This hurricane began as a tropical disturbance near the west coast of Africa on 14 August 1992. It developed and then weakened till 1200 UTC 21 August 1992 when the upper level low to its West-North-West started to split from the northern low and became a trough which continued to withdraw and led to the vertical shear decreasing around Hurricane Andrew. It reached hurricane strength (U_{max} > 33 $m\ s^{-1}$ at the surface) on 22 August. From 0000 UTC 21 August to 1800 UTC 23 August, its central pressure decreased from 1014 to 922 *hPa*, a 92 *hPa* drop at a rate of 2.2 *hPa/h*, which meets the criteria for a rapidly deepening hurricane as defined by Holliday and Thompson (1979). After reaching its maximum strength, Hurricane Andrew weakened temporarily, but it strengthened again as it crossed over Florida Strait till landfall. Then Hurricane Andrew weakened as it traversed Florida for the next four hours (from 0830 UTC to 1230 UTC on 24 August).

The MM5 was initialized by the NCEP/NCAR reanalysis, with enhancements by US Navy radiosondes, additional surface observations and sea surface temperature data (Kalnay et al., 1996; Liu *et al.*, 1997), at 1200 UTC 21 August 1992 at which time Hurricane Andrew started to strengthen rapidly. The model was then integrated for 24 hours. The 24 sigma layers in the model are chosen to have high resolution in the planetary boundary layer and a moderate grid, with 18 *km* horizontal resolution and 124×94 grid points, is run nested within a coarse grid, with 54 *km* horizontal resolution and 82×64 grid points, centered on 75W, 25N and 72W, 24N, respectively, and with time-steps of 120 *s* and 40 *s*, respectively.

The boundary conditions of the outer domain are from a linear interpolation of the 12-h NCEP/NCAR reanalysis (Kalnay *et al.*, 1996). The coarser grids provide the finer grids with time-dependent boundary conditions, and the solutions of the finer grids feed back to the coarser grids every time-step due to the two-way nesting.

Two simulations for Hurricane Andrew were conducted. One was with the standard MM5 diffusion scheme, and another was with the new scheme described in section 3.2. Figs. 6 and 7 show the time-series of the surface minimum pressure (central pressure) and surface maximum (sustained) wind velocity, respectively, for the 6hourly observations and simulations for the 24 hour period from 1200 UTC 21 August to 1200 UTC 22 August 1992.

The comparative simulation results show that the new scheme improves noticeably the model accuracy in predicting the intensity of the hurricane. It is seen from Figs. 6 and 7 that the improved outputs are closer to the observations than those from the original simulations.

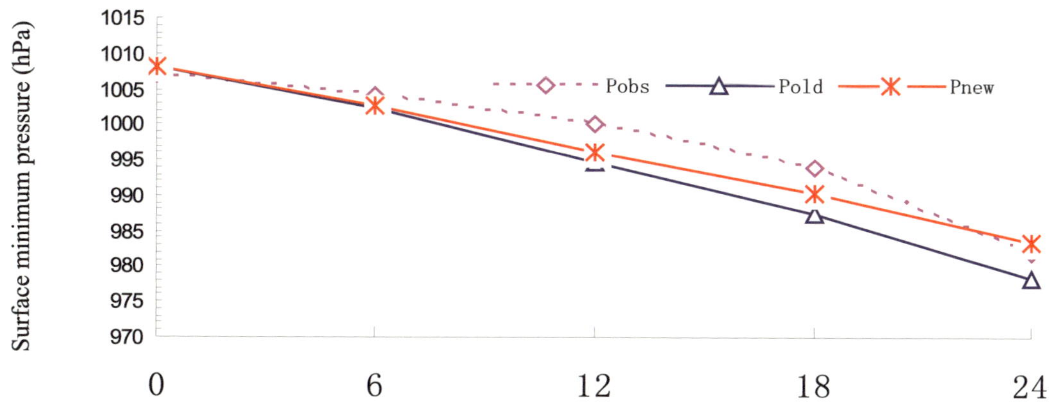

Fig. 6. Comparison of observed changes in the surface minimum pressure, P_{obs}, of the Andrew (1992) with the time from 1200 UTC 21 August to 1200 UTC 22 August, 1992 with the simulated outputs of pressure by the old and new MM5 versions, P_{old} and P_{new}, respectively.

P_{obs} – observed pressure

P_{old} – simulated pressure by the old MM5

P_{new} – simulated pressure by the new MM5

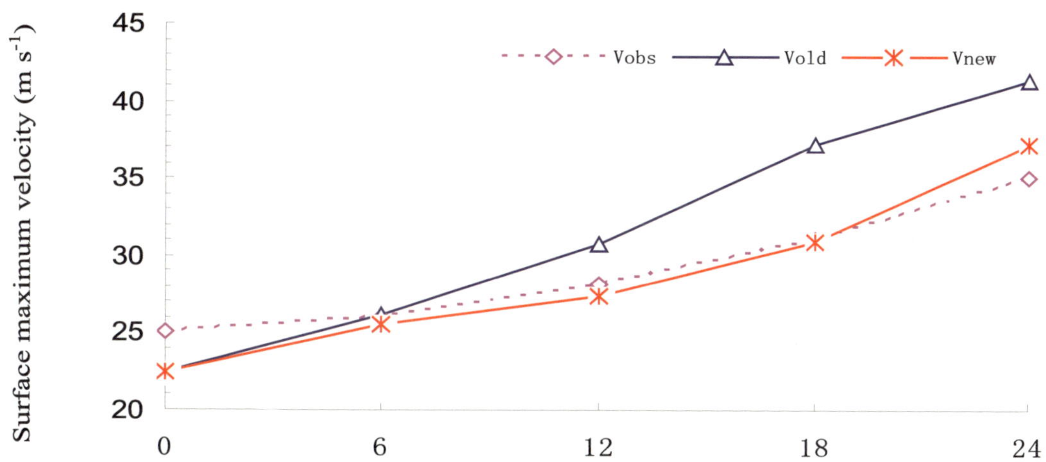

Fig. 7. Same as in Fig. 6 but for the surface maximum wind velocity.

V_{obs} – observed wind velocity

V_{old} – simulated wind velocity by the old MM5

V_{new} – simulated wind velocity by the new MM5

In fact, a yearlong continuous simulation that contains the sum of 365 24 hour runs has been done using the same MM5 model with the same 24 hour simulative settings, from 0000 UTC 1 May 2005 till 0000 UTC 30 April 2006 to validate the reliability of improvement by this physics-based scheme. The one-year averaged (averaged over 365 24 hour simulations) fractional percentage improvement rates (the fractional percentage improvement rate I of the root-mean-square-error (RMSE) for a variable is defined as $I=100^*(E_o-E_n)/E_o$ (%) with E_o and E_n as the errors by the original and new schemes, respectively) are 5.57%, 9.32% and 4.41% for the latitudinal (u), longitudinal (v) velocities and temperature, respectively.

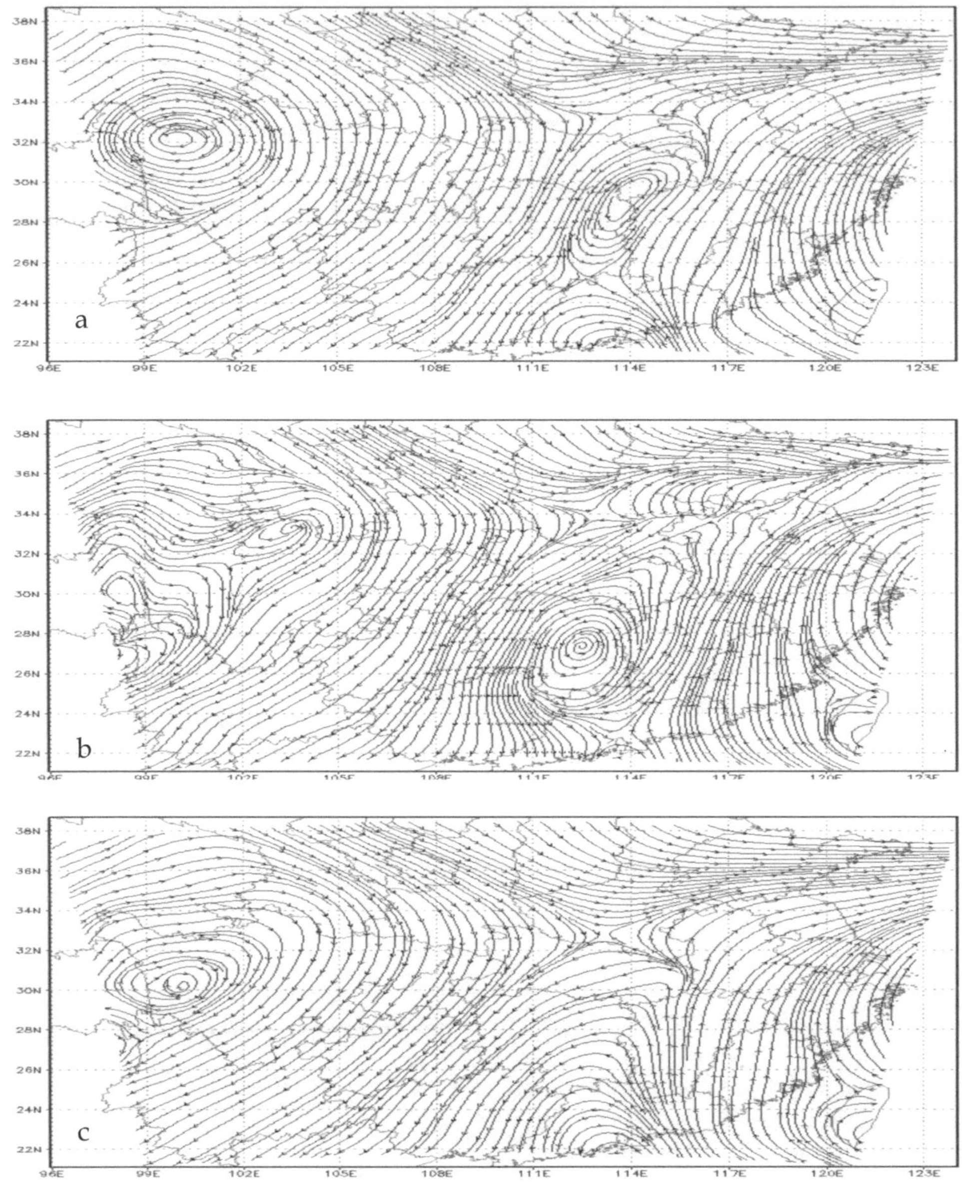

Fig. 8. Comparison of the 500 hPa reanalysis data at 0000 UTC 3 September 2005 (a) with the 24 hour outputs of the horizontal wind field at 0000 UTC 3 September 2005 with the data at 0000 UTC 2 September 2005 as the initial conditions between the old horizontal diffusion scheme (b) and the new scheme (c) used.

As shown in the case of the viscous Burgers equation where there is a Gibbs oscillation near the "shock wave" when the original diffusion scheme is employed, it is prominent that the outputs from the control runs are generally "noisier" than those from the runs using the physics-based scheme. Fig. 8 illustrates an example among the yearlong continuous simulations, which is for comparison of the reanalysis data at 0000 UTC 3 September 2005 with the 24 hour horizontal wind outputs at 0000 UTC 3 September 2005 with the data at 0000 UTC 2 September 2005 as the initial conditions. It is seen from Fig. 8 that, as a whole, the improvement in the longitudinal velocity (v) is larger than that in the latitudinal one (u), but we would like to particularly mention that the huge anticyclone in the western part of the domain is distinctly improved after the new scheme has been employed.

In summary, the present finite difference form of the fourth-order horizontal diffusion scheme employed in a meso-scale model is not consistent with the second law of thermodynamics and should be reformulated by a new approach based on the physical law. The potential of improving the simulative accuracy of a meso-scale numerical weather prediction model via this thermodynamics-based scheme is illustrated through the case of the one-dimensional viscous Burgers equation running to a steady flow as well as a series of simulations of weather events. The new scheme suggested in this chapter resulted in decrease root-mean-square errors and improved forecasts in all the simulations.

4. Understanding tropical cyclone frequency variability in terms of thermodynamic principles

4.1 A brief introduction

The observations show that the tropical cyclone (TC) numbers tend to become decreasing over some oceanic basins during the recent multiple decades in a warming environment. A methodology of understanding the mechanism responsible for such seemingly counterintuitive phenomena is suggested in this chapter.

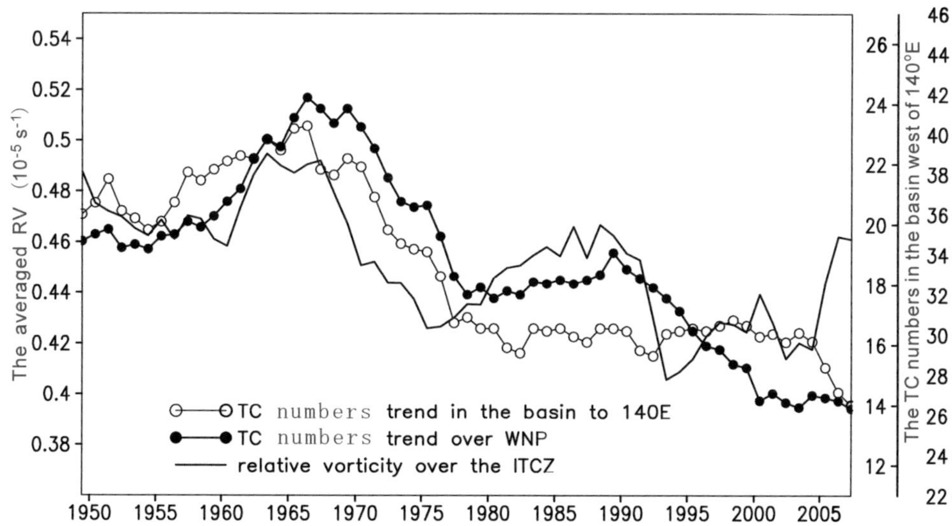

Fig. 9. The long term evolution series for the TC numbers in the WNP and in the basin west of 140°E as well as the averaged RV (in 10^{-5} s^{-1}) in the ITCZ of monsoon trough type during 1949-2007 (see the text for data source).

Here, as mentioned above, the WNP over which the TC occurrence frequencies tended to decrease with minor fluctuations from late 1960s (Fig. 9) is taken as an example to discuss the possible mechanism responsible for the causation.

In Figure 9 the data for TC numbers are from the *Tropical Cyclone Annals* (CMA 1949-2007) and the relative vorticity (RV) is based on the National Centers for Environmental Prediction/National Center for Atmospheric Research (NCEP/NCAR) 2.5° × 2.5° (latitude-longitude grid) reanalysis data.

In view of the fact that the majority of TCs are originated around the intertropical convergence zone (ITCZ) and that the intensity and position of ITCZ can be described reasonably in terms of the RV (see the text below for further details), this study is arranged as follows: the main source of TC genesis and ITCZ of monsoon trough type over the WNP are described in the next section; discussions on the ITCZ variabilities and their possible causes are given in section 4.3 with focus on SSTs changes and the effect of SSTs pattern on the ITCZ migration; the further discussions and conclusive remarks are presented in section 4.4.

4.2 The main source of TC genesis and the ITCZ

According to the previous studies (Gray 1967; Bates 1970; Charney 1971), 80-85% of TCs are originated in the ITCZ or just on its poleward side, and the ITCZ of monsoon trough type located in the basin west of 140°E is the main origin of TCs over the WNP. On the other hand, the necessary conditions of TC genesis and development include higher SSTs, stronger low level vorticity, weaker vertical wind shear, and higher latitudinal position of subtropical anticyclone ridge/ITCZ, and, these conditions are however not equally important. Among them low level vorticity should be the fundamental factor for TC genesis since initial disturbances are the embryo of TCs. In addition, the intensity of the ITCZ as a main system generating TCs can be described in terms of the RV (e.g. use RV at 850 hPa for defining the ITCZ, as is seen in Chan and Evans (2002)). Indeed the TC numbers over the WNP and the basin west of 140°E have a high correlation with the RV around the ITCZ, as is seen in Fig. 9 where these numbers show almost synchronous changes with those of the RV. The corresponding correlation coefficients for the WNP and the basin west of 140°E are 0.6552 and 0.6614 at the 0.001 significance level, respectively. It is noticed, in Ma and Chen (2009), that a 10-year running mean to the annual TC frequency data and NCEP typhoon season mean wind data has been applied to get the long term evolution series of the averaged TC numbers and RV(in $10^{-5}\,s^{-1}$, at 925 hPa) in the ITCZ of monsoon trough type for the period 1949-2007, and, that July-October (JASO hereafter) is defined as the typhoon season since JASO is the most frequent season of TCs over the WNP. It will therefore be a reasonable way to discuss the causality between the TC frequency trend and warming SSTs over the WNP via the ITCZ variability as the medium.

4.3 The ITCZ variabilities and their possible causes

4.3.1 The ITCZ variation

The analyses below are based on the NCEP/NCAR 2.5° × 2.5° resolution reanalysis data for horizontal winds from which the ITCZ fields in terms of RV are figured out via the definition of RV at constant pressure

$$\varsigma_p = \left(\frac{\partial v}{\partial x}\right)_p - \left(\frac{\partial u}{\partial y}\right)_p, \tag{17}$$

where ς_p , u and v are the RV, the latitudinal and longitudinal velocities at constant pressure, respectively, and the 2.0° × 2.0° data for the SST fields (Kalnay et al. 1996). All the means are calculated against the period of JASO as mentioned above.

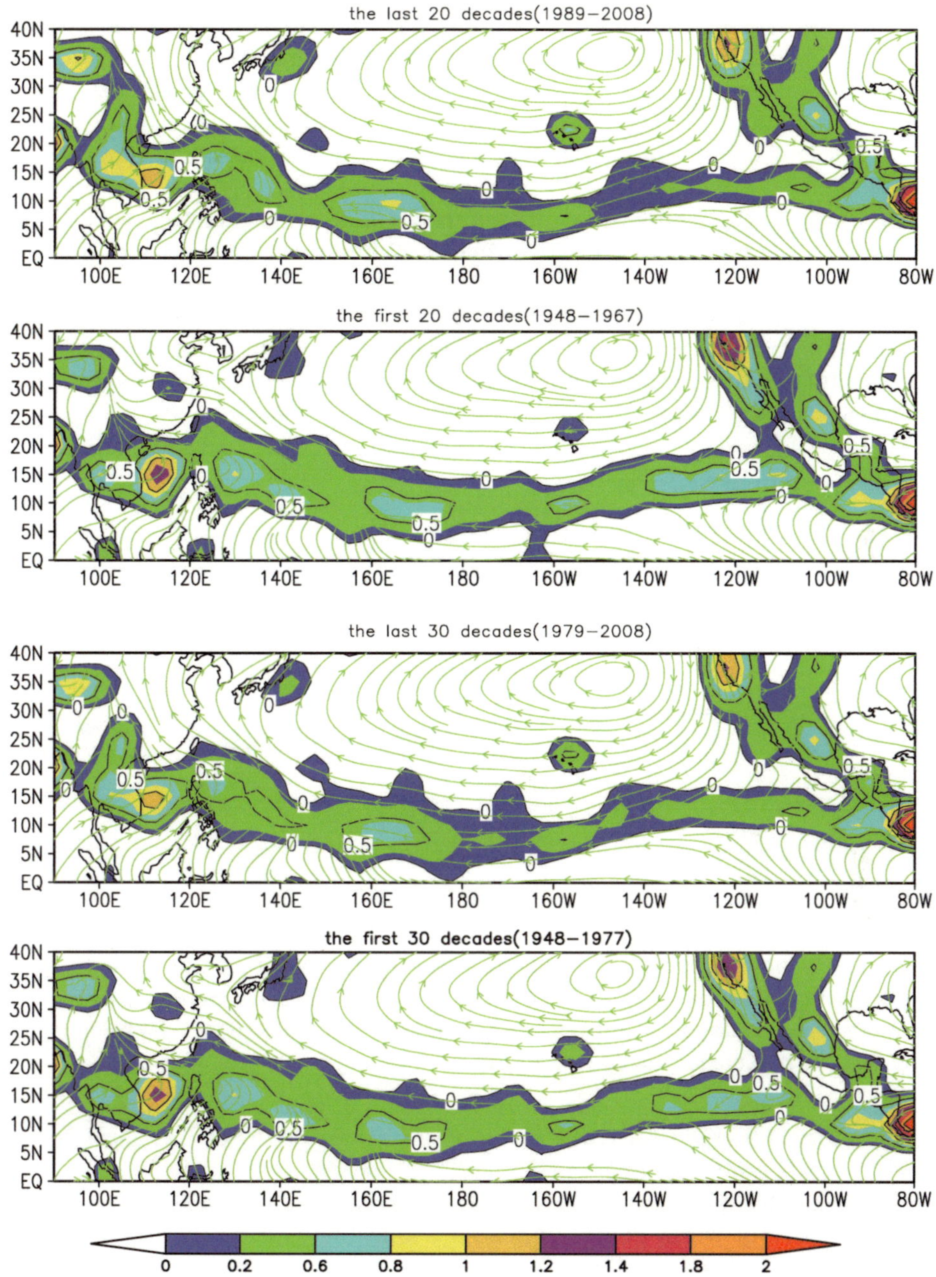

Fig. 10. Comparison of the stream fields superimposed by the positive RV at 925 hPa (isolines are for the positive RV in 10^{-5} s^{-1}) during JASO over the WNP between the first and last 20-year/30-year means (as marked above the respective panel).

Figure 10 shows the stream fields superimposed by the positive RV at 925 hPa during JASO over the WNP. It is found from near the equator in Fig. 10 that the zones of positive RV in the RV field can simply tell the general position of the convergence zones in the stream fields. This convergence zones should be able to represent the ITCZs due to the fact that the ITCZ is mainly formed by the trade winds converging (e.g., Bjerknes et al. 1969; Holton and Wallace 1971). In fact, as mentioned in section 4.2, the attempt using RV at 850 hPa for defining the ITCZ has been done (e.g. Chan and Evans 2002). However, the RV in terms of 850 hPa is not continuous in some sections of ITCZs over the Pacific owing to its weaker intensity (figures not shown here). Therefore the RV at 925 hPa that is more continuous and smoother than that at 850 hPa is chosen instead in this chapter.
It is noticed from Fig. 10 that the variabilities of the ITCZ of monsoon trough type associated with the reduction in the TC numbers over the WNP under warming SSTs during the last several decades are involved with the following two factors: (1) the intensity of the ITCZ of monsoon trough type as the main origin of TCs over the WNP, in terms of RV at 925 hPa, was weakened, and (2) the position of the ITCZ shifted into the latitudes further south (see also Fig. 11), both of which would contribute to the decrease in TC numbers over the WNP. Next we will specifically discuss these two aspects.

4.3.2 The implication of SSTs variation

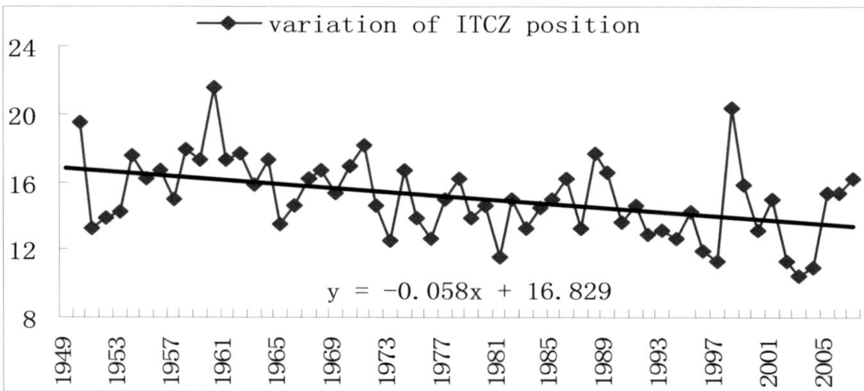

Fig. 11. The illustration of changes in the average position of the ITCZ from 1949 to 2007. Here, the central position of the ITCZ is determined via singling out the points with maximum positive RV along the longitudinal direction within the 100°E – 140°E and then averaging their respective latitudinal position.

Figure 11 shows the every 20-year means of SSTs over the WNP for JASO during 1949-2008. It is seen from Fig. 11 that the SSTs over the North Pacific are gradually increased during the recent 60 years, especially, the areas circled with the isotherm of 28°C that is defined as the critical temperature for the warm pool (Ho et al. 1995) extended eastward prominently. This cross-equatorial area of warm pool was limited within the western Pacific initially which has extended straightforwardly till the eastern Pacific so as to be connected eventually from the western up to eastern Pacific as seen in the panel for the last 20-year period (1989-2008) of Fig. 10. Since there is a close relationship between SSTs and surface wind divergence and convection over the tropical oceans or the ITCZs (Graham and Barnett 1987; Lau et al. 1997;

Lindzen and Nigam 1987) the following questions are then raised: how the SSTs or their gradient influence the intensity of the ITCZ, and why the sea surface warming extent reached around the western Pacific covering some part of the warm pool is not as dramatic as those over the central and eastern Pacific. The latter is relevant to the first factor of the ITCZ variabilities mentioned above. A potential clue would be implied by the modern nonlinear non-equilibrium thermodynamics.

For an isolated thermodynamic system, the state function of the system, entropy s per unit mass, will spontaneously increase with time according to the second law of thermodynamics, which can be expressed by the formula (Prigogine 1955; De Groot and Mazur 1962)

$$\frac{ds}{dt} \geq 0, \tag{18}$$

and be usually called the spontaneous entropy increment principle. As a result, an isolated system will spontaneously tend to homogenization. However, for an open system with the diabatic heating rate Q transferred through its boundaries, Equation (2) should be modified to

$$\frac{ds}{dt} \geq \frac{Q}{T}, \tag{19}$$

where T is the temperature (in ^{o}K) of the system. The nature of the second law of thermodynamics shows that, if there exists difference initially, heat (particles) will be spontaneously transferred (diffused) from part with higher T (concentration) to that with lower T (concentration). Any many-body system like the atmosphere or ocean must be controlled by the second law of thermodynamics and, in fact, the entropy flow properties of the atmospheric systems have been revealed based on this law (Liu and Liu 2005; Liu et al. 2006; Liu and Liu 2007; Xu and Liu 2008).

Specifically, if there exists difference of temperature spatially (say, on sea surface), the original part of warmer sea surface will diffuse its thermal energy (the inner energy, proportional positively to temperature via the formula of $e = C_p T$ where e is the inner energy per unit mass, C_p is the specific heat at constant pressure and T is temperature) to its surrounding parts with lower temperature. As a consequence, compared with the surrounding areas the original warmer area (e.g., the area related with the warm pool) will experience a weaker warming under global warming since it will lose a certain amount of heat via the diffusive process at the same time, and vice versa (Fig. 12 showing the case for the North Pacific is only an example in which it is demonstrated that the SSTs around the ITCZ of monsoon trough type have a smaller increment of temperature while those over the adjacent ITCZ sections near the central and eastern equatorial Pacific have larger increments). In view of that the ITCZ is caused, at least in the initial stages, mainly by the thermodynamic forcings such as the gradient in SST that has been regarded as playing more of a role than the absolute SST value with regard to convection and precipitation (Lindzen and Nigam, 1987) the ITCZ of monsoon trough type should indeed become weaker in response to more uniform SSTs or weakened SST gradient over there.

4.3.3 The effect of SSTs pattern on the ITCZ migration

As described above, in the warming environment the higher SSTs area circled with the isotherm of 28°C related to the warm pool has gradually extended eastward and been connected from the western to eastern Pacific eventually during the last 20 years (Fig. 12) so as to form an apparent zone of higher SSTs with a distinct gradient of nearly north-south direction created on the both northern and southern sides of this zone. Such a SST pattern will cause corresponding changes in the average temperature T_m of an immediate air layer above the sea surface. Thus, in some regions to the north of the zone of higher temperature where the semi-geostrophic relation for the Tropics is valid (Lin and Chao 1998) a western wind component at the higher levels (e.g. at 925 hPa) should be superposed, which will cause a positive increment of the latitudinal velocity u with $\Delta u > 0$ (at the higher level), very similar to the thermal wind relation (Holton 1992; see Eq. (20) below):

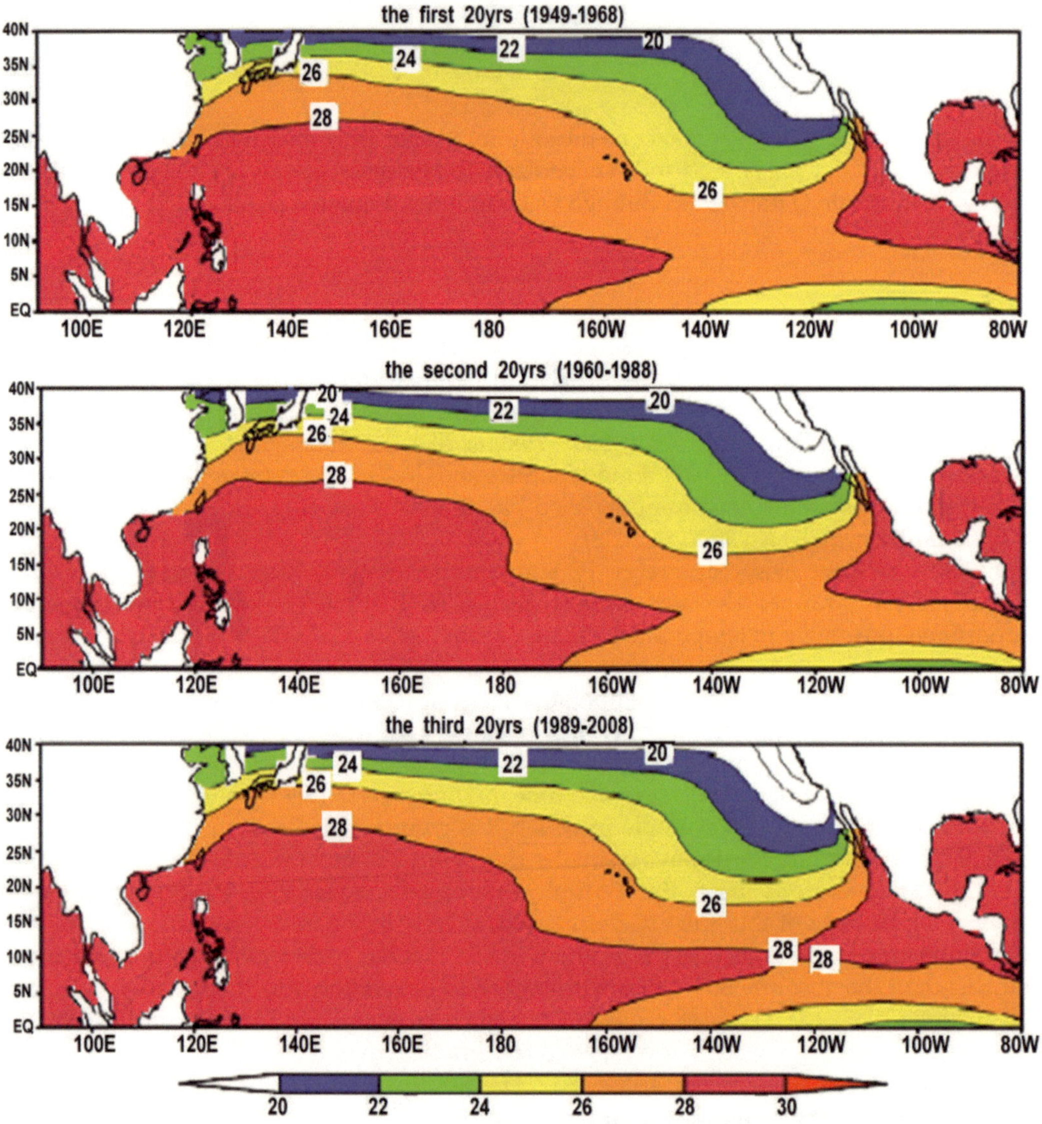

Fig. 12. The every 20-year-averaged SST fields (as marked above the respective panel; unit: °C) for the North Pacific south of 40°N during the period from 1949 to 2008.

$$\vec{V}_T = -\frac{R}{f}\ln\frac{p_1}{p_2}\nabla_p T_m \times \vec{k} \tag{20}$$

where $\vec{V}_T$ denotes the thermal wind between the constant pressure layers p_1 (e.g. 1000 hPa that is near the sea surface) and p_2 (e.g. 925 hPa that is used for defining the ITCZ in terms of RV in this chapter); $\nabla_p T_m$ is the average temperature gradient with T_m *is* the average temperature between p_1 and p_2; $\vec{k}$ the unit vector in the vertical direction; and R and f are the gas constant for dry air and Coriolis parameter, respectively. In contrast, the latitudinal velocity u to the south of the zone should be superposed by an eastern component as the gradient in SSTs in the south regions is reverse to that in the north ones. As a result, the RV, as is expressed by $\varsigma_p = \left(\frac{\partial v}{\partial x}\right)_p - \left(\frac{\partial u}{\partial y}\right)_p$, in the regions to the north of the zone of higher temperature will be decreased (at the higher level) since the term of $-\left(\frac{\partial u}{\partial y}\right)$ becomes smaller with the term of $\left(\frac{\partial v}{\partial x}\right)_p$ having a minor change in this case. At the same time, the RV in the regions to the south of the zone of higher temperature will be increased owing to the reversed gradient in SSTs or T_m there.

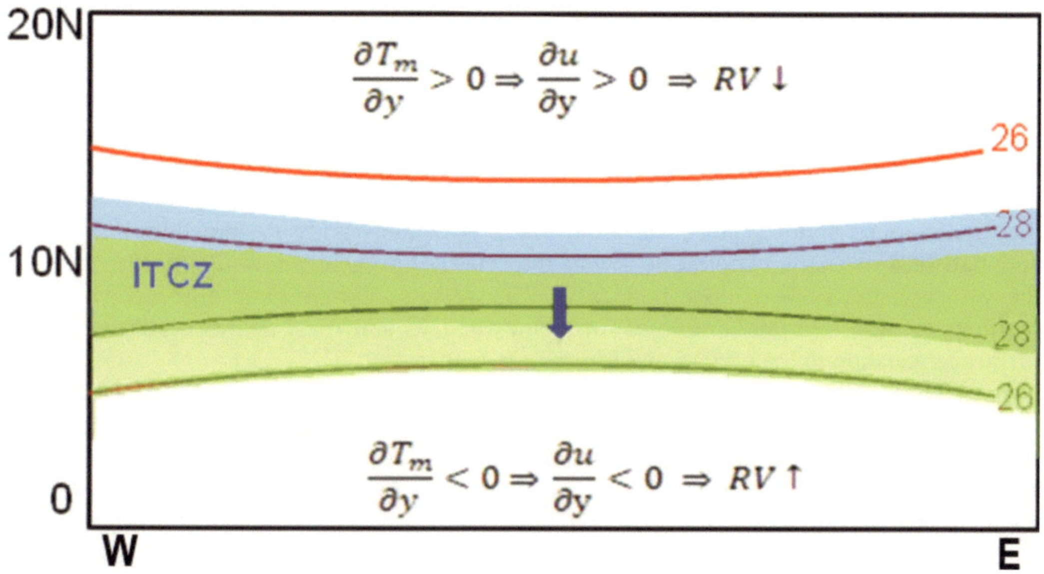

Fig. 13. The diagrammatic sketch for the proposed mechanism for a southern shift of the ITCZ coupled with RV which is contributive to the reduction in TC counts around the WNP (see the text for further details). The red lines represent SSTs showing sketchily the opposite temperate gradients (and the potentially-forced average temperature gradients $\frac{\partial T_m}{\partial y}$ within the immediate layer above the sea surface) in the both sides of the maximum SSTs.

Taking the definition of ITCZ in terms of positive RV into account, we might expect that the part of the ITCZ to the north of the zone of high temperature will tend to weaken and even disappear as the result of reduction in RV there and, on the contrary, the part of the initial ITCZ to the south will enhance and even extend south further (see Fig. 13 for reference). That is able to be used to explain why the ITCZ migrates south apparently, as is relevant to the second factor of the ITCZ variabilities above.

As well-known to all, the necessary conditions for TC genesis are not equally important and, among them dynamical factors such as low level vorticity and vertical wind shear play a more important role than thermodynamic factors such as SST and moist instability as suggested on some simulative results (Sugi et al. 2002; McDonald et al. 2005; Chauvin et al. 2006). This study shows that, warmer SSTs in the WNP even caused fewer TCs, that is, warmer SSTs as one of the necessary conditions do not definitely lead to an increase in TC numbers, which might be attributed to the heterogeneous effects of complicated pattern of SSTs on RV as implied by the second law of thermodynamics with the thermal wind principle taken into account. This study further proves that low level RV associated with ITCZ variations should be a fundamental factor for TC genesis. Based on the analyses in this chapter, a new way of understanding the mechanism responsible for the causality among SSTs, RV and TC occurrence frequency over the WNP is then suggested. As a matter of fact, the WNP is only used as an example in this study and, the principle of the methodology illustrated herein is universal. This analysis can be also applied to the other oceanic basins worldwide than the WNP.

5. Conclusive remarks

In this chapter some crucial concepts and theories in modern thermodynamics have been applied to hurricane researches, which cover the potential mechanism responsible for the hurricane evolution investigated based on the entropy analyses, the simulative improvement of hurricane track forecast using the irreversible thermodynamic operator technique, as well as understanding the TC frequency variability according to the 2nd law of thermodynamics. Indeed, since the atmosphere is a many-body system thermodynamics and statistical mechanics should be a very important approach to revealing the activities of the atmospheric vortices like hurricane. The results shown here are however very preliminary and, a larger sample of hurricanes and its statistical analyses are needed for further application to the operational forecasts in future.

Here we would like to stress that, back to the track and intensity forecasts of Hurricane Katrina with a complex life cycle (McTaggart-Cowan *et al.*, 2007), it is found that some knotty forecast situations during Katrina might have been handled in a proper way using a forecast guidance scheme developed based on the relevant results in this chapter. For example, with regard to the forecast track errors, Knabb *et al.* (2005) particularly mentioned, in its section "d. Forecast and Warning Critique", the difficult forecast scenario associated with Katrina's west-southwestward motion across the southern Florida peninsula and eastern Gulf of Mexico. We notice that all the entropy flow patterns in the vertical cross sections at 6-hourly intervals for Katrina from 00:00 UTC 26 to 00:00 UTC 27 August (figures not presented in this chapter) during which it moves across the southern Florida peninsula and eastern Gulf of Mexico indicate a distinct southerliness in Katrina's track orientation in addition to westering, showing a general agreement with its real west-southwestward motion (see Fig. 2). As for the average official intensity forecast errors during Katrina that are considerably larger than the corresponding Atlantic 10-year (1995-2004) averages and can be partly attributed, as stated in Knabb *et al.* (2005), to the fact that "Katrina was an unusually intense hurricane and underwent two rapid intensification periods, including the very rapid strengthening from Category 3 to 5 on the morning of 28 August" (Knabb *et al.* (2005), we would like to recall the proposition of "negative entropy flow is something very positive for a system to keep far from its equilibrium" and Fig. 3 in this chapter which shows that the increasing continuously strong negative entropy flow entering into Katrina till 00:00 UTC 28 August just precedes the period of very rapid intensification on that morning. In other words, the negative entropy flow analyses in the present chapter suggest that Katrina would meet with a rapid strengthening process immediately after 00:00 UTC 28 August. In view of that "accurately forecasting the timing and magnitude of such events remains an operational challenge, in part because the available intensity guidance models generally have little or no skill in forecasting rapid intensity changes" (Knabb *et al.*, 2005), it is anticipated that a new-generation guidance model might benefit from the scheme based on the entropy flow analysis described in this chapter.

6. Acknowledgments

This work has been jointly supported by 973 Program (2009CB421500), the National Natural Science Foundation of China under Grants 41075048, 40633016, 40875029 and 40975036, as well as the Basic Research Project of the State Key Laboratory of Severe Weather, Chinese Academy of Meteorological Sciences (2008LaSWZI01).

7. References

Anthes, R.A., Hsie, E.-Y. & Kuo, Y.-H.(1987). *Description of the Penn State /NCAR Mesoscale Model Version 4 (MM4). NCAR Tech. Note, NCAR/TN-282+STR,* National Center for Atmospheric Research, Boulder, CO.

Barsugli, J. J. & Battist, D. S. (1998). The Basic Effects of Atmosphere–Ocean Thermal Coupling on Midlatitude Variability. *J. Atmos. Sci.* 55: 477-493.

Bates, J. R., (1970). Dynamics of disturbances on the Intertropical Convergence Zone. *Quart. J. Roy. Meteor. Soc.*, 96: 677-701.

Becker, E. (2001). Symmetric Stress Tensor Formulation of Horizontal Momentum Diffusion in Global Models of Atmospheric Circulation. *Journal of the Atmospheric Sciences* 58: 269-282.

Bohren, C. F. & Albrecht B. A. (1999). *Atmospheric Thermodynamics*, Oxford University Press, London.

Bronshtein, I. N. & Semendyayev K. A. (1997). *Handbook of mathematics, 3rd edition,* Springer-Verlag, London,.

Burkhardt, U. & Bercker E. (2005). A Consistent Diffusion-Dissipation Parameterization in the ECHAM Climate Model. *Mon. Wea. Rev.* 134: 1194-1204.

Cerveny, R. S., Lawrimore, J. Edwards, R. & Landsea, C. (2007). Extreme Weather Records, *Bull. Amer. Meteorol. Soc.* 88: 853-860.

Chan, S.C. & Evans, J. L. (2002). Comparison of the Structure of the ITCZ in the West Pacific during the Boreal Summers of 1989–93 Using AMIP Simulations and ECMWF Reanalysis.*J.Climate.*15: 3459-38.

Charney, J. G. (1971). Tropical cyclonegenesis and the formation of the Intertropical Convergence Zone, in W. H. Reid, (ed.) *Mathematical Problems of Geophysical Fluid Dynamics, Lectures in Applied Mathematics, Vol.13,*. Amer. Math. Soc., 383 pp.

Chauvin, F., Royer J. –F. & Déqué, M. (2006). Response of hurricane-type vortices to global warming as simulated by ARPEGE-climate at high resolution, *Clim. Dyn.* 27: 377 - 399.

Cheng, S. I. (1975). A Critical review of numerical solution of Navier-Stokes equations. *Progress in Numerical Fluid Dynamics, Lecture Note in Physics*, 41: 97-99.

CMA (1949-2007). *Tropical Cyclone Annals*, China Meteorological Press, 150 pp (each).

Crandall, M.G. & Majda, A. (1980). Monotone difference approximations for scalar conservation laws. *Mathermatics of Compatations*, 34: 1-21.

Curry, J. A. (1999). *Thermodynamics of Atmospheres and Oceans*, Academic Press, San Diego.

De Groot, S. R. & Mazur, P. (1962). *Non-equilibrium thermodynamics*, North-Holland Publishing Company.

Duane, G. S. & Curry, J. A. (1997). Entropy of a connecting water-air system and the interpretation of cloud morphogenesis. *Q. J. R. Meteorol. Soc.* 123: 605- 629.

Dudhia, J. (1993). A nonhydrostatic version of the Penn State-NCAR mesoscale model: Validation tests and simulation of an Atlantic cyclone and cold front. *Mon. Wea. Rew.* 121: 1493-1513.

Easterling, D. R., Evana, J. L., Grosman , P. Y. Karl, T. R.,. Kunkel, K. E & Ambenje, P. (2000). Observed Variability and Trends in Extreme Climate Events: A Brief Review, *Bull. Am. Metor. Soc.*, 81: 417-425.

Egger, J. (1999). Numerical Generation of Entropies *Mon. Wea. Rev.* 127: 2211-2216.

Eliasen, E. & Laursen, L. (1990). The effects of horizontal resolution and diffusion in a two-layer general circulation model with a zonally symmetric forcing. *Tellus* 42A: 520-530.

Elsner, J. B. & Jagger, T. H. (2009). *Hurricanes and Climate Change*, Springer.

Fraedrich, K. & Blender, R. (2003). Scaling of atmosphere and ocean temperature correlations in observations and climate models, *Phys. Rev. Lett.* 90 (10) (108501~1-4). DOI: 10.1103/PhysRevLett.90.108501.

Gade, H. G. & Gustafson, K. (2004). Application of classical thermodynamics principle to study of oceanic overturning circulation, *Tellus* 56A(4): 371–387.

Glansdorff, P. & Prigogine, I. (1971). *Thermodynamic theory of structure, stability and fluctuations* , Wiley-Interscience, London,.

Graham, N. E. & Barnett, T. P. (1987). Sea surface temperature, surface wind divergence and convection over tropical oceans, *Science* 238: 657–659.

Gray, W. M. (1967). Global view of the origin of tropical disturbances and storms, *Mon. Wea. Rev.*, 96: 669-700.

Grell, G.A., Dudhia, J. & Stauffer, D.R. (1995). A description of the fifth generation Penn State/NCAR mesoscale model (MM5). NCAR Tech. Note, NCAR/TN-398+STR, National Center for Atmospheric Research, Boulder, CO.

Haken, H. (1983). *Advanced Synergetics*, Springer-Verlag.

Haltiner, G. J. & Martin, F. L. (1957). *Dynamical and Physical Meteorology*, McGraw-Hill, New York,.

Ho, C. -R., Yan, X. -H. & Zheng, Q. (1995). Satellite observations of upper-layer variabilities in the western Pacific warm pool, *Bull. Amer. Meteor. Soc.* 76: 669–679

Holliday, C.R. & Thompson, A.J. (1979). Climatological characteristics of rapidly intensifying typhoons. *Mon. Wea. Rev.* 107: 1022-1034.

Holton, J. R.(1992). *An Introduction to Dynamic Meteorology*, Academic Press.

Kalnay, E. *et al (*1996). The NCEP/NCAR 40-year reanalysis project, *Bull. Amer. Meteor. Soc.*, 77: 437–471.

Karl, T. R., Kukla, G., Razuvayev, V. N., Changery, M. J., Quayle, R. G., Heim R. T., Jr., Easterling, D. R. & Fu C. B. (1991). Global warming: evidence for asymmetric diurnal temperature change, *Geophys. Res. Lett.* 18: 2253-2256.

Katchalsky A. & Curran, P. F. (1965). *Non-equilibrium Thermodynamics in Biophysics* (Harvard University Press, Cambridge, Mass,.

Knabb, R. D., Rhome, J. R & Brown, D. P. (2005). *Tropical Cyclone Report: Hurricane Katrina (23-30 August 2005)* (available online at http://www.nhc.noaa.gov/pdf/TCR-AL122005_Katrina.pdf).

Lau, K. -M., Wu, H. -T. & Bony, S. (1997). The role of large-scale atmospheric circulation in the relationship between tropical convection and sea surface temperature, *J. Climate,* 10: 381–392.

Laursen, L. & Eliasen, E. (1989). On the effects of the damping mechanisms in an atmospheric general circulation model. Tellus A 41: 385-400.

Lin, Y. & Chao, J. P. (1998). The processes of tropical semi-geostrophic adaptation, *Science in China (Series D,* 41: 566-573.

Lindzen, R. S. & Nigam, S. (1987). On the role of sea surface temperature gradients in forcing low level winds and convergence in the tropics, *J. Atmos. Sci.* 44: 2440 -2458.

Liu, C. & Liu, Y. (2004). Negative entropy flow and its effect on the organization of synoptic-scale severe atmospheric systems. *Geophys. Res. Lett.* 31, L01108, doi: 10.1029/2003G018071.

Liu, C. & Liu, Y. (2005). An attempt at improving a global spectral model by incorporating the second law of thermodynamics. Geophys. Res. Lett. 32, L03806, doi:10.1029/2004GL021602.

Liu, C. Liu, Y., & Luo, Z. (2009). Improving numerical simulation of a hurricane by irreversible thermodynamic operators, *Atmos. Res.,* 93: 44–49, doi:10.1016/j.atmosres.2008.10.012.

Liu, C., Liu, Y. & Xu, H. (2006). A physics-based diffusion scheme for numerical models, *Geophys. Res. Lett.* 33, L12805, doi:10.1029/2006GL025781.

Liu, C., Liu, Y. & Xu, H. (2009). A New Diffusion Scheme for Numerical Models Based on Full Irreversibility, *Weather and Forecasting,* 24: 595–600, DOI: 10.1175/2008WAF2222145.1.

Liu, C., Xu, H. & Liu, Y. (2007). Linking entropy flow with typhoon evolution: a case-study, *Journal of Physics D: Applied Physics*, 40: 6694-6704.

Liu, C. et al. (2010). Implication of entropy flow for the development of a system suggested by the life cycle of a hurricane, *Modern Physics Letters B* (*MPLB*) 24 :1747-1757.

Liu, Y. & Liu, C. (2007). On the entropy flow properties of a severe tropical storm, *Appl. Phys. Lett.* 91, 014103. DOI: 10.1063/1.2753538.

Liu, Y. & Liu, C. (2008). Entropy flow and the evolution of a storm, *Entropy*, 10: 430-440, DOI:10.3390/e10040430. 1004-4965(2008)03-0038-07.

Liu, Y. & Liu, C. (2009). Negative entropy flow and the life-cycle of a severe tropical storm, *Atmos. Res.*, 9: 39–43, doi:10.1016/ j.atmosres. 2008. 10.013.

Liu, Y., Zhang, D.-L, & Yau, M. K. (1997). A multiscale numerical study of Hurricane Andrew (1992), Part I: Explicit simulation and verification. *Mon. Wea. Rew.* 125: 3073-3093.

Ma, L. & Chen, L. (2009). The relationship between global warming and the variation in tropical cyclone frequency over the western North Pacific, *J. Trop. Meteor.,* 15: 38-44.

Manton, M. & Eral, J. (2001). Trends in extreme daily rainfall and temperature in Southeast Asia and the South Pacific: 1961-1998, *Int. J Climatol.* 21: 269–284.

Mason, B. J. (1971). *Physics of Clouds* (Oxford University Press, London,.

McBride, J. L. & Ramsay, H. (2009). Relationship between tropical cyclone activity and sea surface temperature in the southern hemisphere, *2ed International Summit on Hurricanes and Climate Change,* May 31 - June 5, 2009, Corfu, Greece,p 23.

McDonald, R.E., Bleaken, D.G., Creswell, D. R., Pope, V. D. & Senior, C.A. (2005). Tropical storms: representation and diagnosis in climate models and the impact of climate change. *Clim. Dyn*. 25: 19-36.

McTaggart-Cowan, R., Bosart, L. F., Gyakum, J. R.& Atallah, E. H. (2007). *Mon. Wea. Rev.* 135: 3905-3926.

Nitschke, K., Bestehorn, M. & Thess, A. (1998). Square cells in surface-tension-driven Bénard convection. *J. Fluid Mech.* 356: 155 - 197.

Nicolis, C. (2002). Irreversible thermodynamics of a simple atmospheric Model.. *Int. J. Bifurcation and Chaos* 12: 2557-2566.

NOAA (1992). *Storm Data*. Department of Commerce Rep.34 (8).

Olby, R. (1971). Schrodinger's problem: What is life?. *J. Hist. Biol.* 4:119-148.

Onsager, L. (1931). Reciprocal relations in irreversible processes. I., *Phys. Rev.* 37: 405–426.

Ozawa, H. & Ohmura, A. *J.* (1997). Thermodynamics of a global-mean state of the atmosphere —a state of maximum entropy increase. J. *Climate* 10(3): 441-445.

Pielke, R. A. (1984). Mesoscale meteorological modeling. Academic Press, Orlando.

Perkey, D. J. & Kreitzberg, W. (1976). A time-dependent lateral boundary scheme for limited area primitive equation models. *Mon. Wea. Rev.* 104: 744-755.

Prigogine I. (1955). *Introduction to thermodynamics of irreversible Processes,* Charles C. Thomas Publisher.

Ruelle, D. P. (2003). Extending the definition of entropy to nonequilibrium steady states. *Proc. Natl. Acad. Sci. USA* 100: 3054-3058.

Schrödinger, E. (1944). *What Is Life?* Cambridge University Press.

Smagorinsky, J. (1963). General circulation experiments with the primitive equations: I. The basic experiment. *Mon. Wea. Rew.* 91: 99-164.

Stephenson, D. B. (1994). The Impact of Changing the Horizontal Diffusion Scheme on the Northern Winter Climatology of a General Circulation Model. *Q. J. R. Meteorol. Soc.* 120: 211-226.

Sugi, M., Noda, A. & Sato, N. (2002). Influence of global warming on tropical cyclone climatology: an experiment with the JMA global model, *J. Meteorol. Soc. Japan,* 80: 249-272.

Tsonis, A. A. (2002). *An introduction to atmospheric thermodynamics,* Cambridge University Press, Cambridge, UK,.

Wakimoto, R.M. & Black, P. G. (1994). Damage survey of Hurricane Andrew and its relationship to the eyewall. Bull. Amer. Meteor. Soc. 75: 189-200.

Wang, C., Li, David S. & Enfield, B. (2008). Atlantic warm pool acting as a link between Atlantic multidecadal oscillation and Atlantic tropical cyclone activity, *Geochem. Geophys. Geosys.* 9: 1-17.

Wang, S. W. & Gong, D.Y. (2000). Enhancement of the Warming Trend in China. *Geophsys.Res. Lett.* 27: 2581-2584.

von Storch, H. (2008). An Attempt to Homogeneously Describe 60 Years Statistics of TC Activity in East Asia, 1948-2007, *2008 Taiwan Climate Workshop,* 18 November 2008, Taipei, Taiwan, China.

Xu, H. & Liu, C. (2008). Entropy flow properties of a typhoon as simulated by a meso-scale model, *Europhys. Lett.,* 83, 18001.doi: 10.1209/0295-5075/83/18001.

Zdunkowski, W. & Bott, A. (2004). *Thermodynamics of the Atmosphere,* Cambridge University Press, Cambridge, UK.

10

Simulation of Tropical Cyclones Using Spectral Bin Microphysics

Alexander Khain and Barry Lynn
Department of Atmospheric Sciences,
The Hebrew University of Jerusalem,
Israel

1. Introduction

1.1 Bulk-parameterization and spectral bin microphysics

Latent heat release is the main energetic source of tropical cyclones (TCs). The intensity of TCs depends on the magnitude of latent heat release (convective heating), as well as on its vertical distribution, and spatial pattern. The latent heat release in clouds depends on the rates of microphysical processes such as: condensational growth/evaporation of droplets, deposition/sublimation of ice, riming and freezing/melting. These microphysical processes are closely related to cloud dynamical properties such as vertical velocity, cloud top height, etc.Droplets and the majority of ice crystals arise on aerosol particles (AP) playing the role of cloud condensational nuclei (CCN) and ice nuclei (IN), respectively. It is well known that droplet size distributions (DSD) and precipitation formation depend on the concentration and size of CCN. Therefore, the convective heating in TCs should depend on the properties of AP in the environment of TCs.

Most equations describing microphysical processes, including those responsible for cloud-aerosol interaction are well known in Cloud Physics. Advanced microphysical schemes are needed in order to describe microphysical processes adequately and to properly take into account microphysical factors (such as aerosols). Yet, advanced microphysical schemes are not generally used in tropical cyclone forecasting models. Rather, until 2006 the current operational TC forecast model developed at the Geophysical Fluid Dynamics Laboratory used large scale convective parameterizations (Kurihara 1973 and a simplified version of the Arakawa and Shubert scheme). Since 2006 this model used a simplified Arakawa-Schubert scheme for cumulus parameterization and a simplified version of the Ferrier bulk-parameterization for large-scale condensation in cases when supersaturation in grid points is reached (Bender et al 2007). Both schemes are insensitive to aerosols.

The development of one and two-moment bulk parameterization schemes and their application in mesoscale models was an important step toward the improvement in the description of convective processes and precipitation in numerical models. These schemes were implemented in mesoscale and/or forecast modles such as the Penn State/NCAR Mesoscale Modeling System Version 5 (MM5) (Dudhia et al., 1997), RAMS (Pielke et al, 1992) and the Weather Research Forecast Model (Skamarock et al, 2005), the operational numerical weather prediction model of the German Weather Service (COSMO) (that is combined with an extended version of the 2-moment bulk scheme by Seifert and Beheng

2006), etc. These models use an *a priori* prescription of the shape of size distribution functions (SDF) of different cloud hydrometeors (such as cloud droplets, rain drops, graupel, aggregates) in the form of exponential Marshal-Palmer distributions or gamma distributions. This reduces the system of equations that describes cloud microphysics to a relatively small number of equations for integral quantities such as mass contents (one moment schemes) and mass contents and number concentrations (two moment schemes). The comparatively small number of prognostic equations makes the schemes computationally efficient, so they are widely utilized in simulating different cloud-related phenomena such as supercell storms, squall lines, etc. These schemes were recently used for simulation of TCs (e.g., Zhang et al, 2007; Fierro et al, 2007).

Note that SDF of different hydrometeors in clouds significantly vary with space and time during cloud evolution. Size distributions of most hydrometeors are bi-modal or multimodal. Sometimes the SDF are very narrow and do not contain small and large hydrometeors. The complicated shape of the SDF is the result of many highly nonlinear microphysical processes. As a result, in many cases SDF cannot be approximated by gamma or exponential distributions. To some extent, the gamma or exponential size distributions can be considered as those obtained by the spatial and time averaging of SDF in clouds. In particular spatial points and particular time instances, the shape of SDF can dramatically deviate from the prescribed shapes. Microphysical phenomena (e.g. precipitation triggering, cloud glaciation, etc.) are determined by local SDF, and not by averaged ones. The utilization of *a priori* assumption of gamma or exponential distributions with a prescribed parameters is only one of limitations of the bulk-parameterization schemes. Microphysical processes are described in most of these schemes using simple semi-empirical relationships with solutions may substantially differ from the solutions of the microphysical equations based on the first principles.

The second approach to simulate microphysical processes is the utilization of spectral bin microphysics (SBM), in which a system of kinetic equations for size distributions of particles of different classes is solved. Each size distribution function is described using several tens of mass (size) bins. The SBM schemes use the microphysical equations based on the first principles trying to avoid utilization of semi-empirical relationships. A brief comparison of the methods used in the SBM and bulk-parameterization approaches is presented in Table 1. One-moment bulk schemes do not use any information about aerosols. Two-moment schemes use some information about aerosols in parameterization of droplet nucleation. The potential source of aerosols is considered to be infinitely large. As a rule, the bulk schemes do not include the transport of aerosols. In contrast, SBM describes aerosol effects on cloud microphysics and dynamics taking into account the aerosol budget, i.e. a limited source of aerosol particles. A special aerosol size distribution is used to take into account the transport and sinks of AP of different size.

As a rule, bulk-parameterization schemes do not solve the equation for diffusion growth of drops. Instead, the saturation adjustment assumption is used according to which all supersaturated water vapor is transferred into cloud water mass, so that the final supersaturation is assumed equal to zero. The assumption of saturation adjustment is not valid in atmospheric clouds, especially in maritime clouds, where supersaturation is high and may exceed 5-10%. The immediate transformation of supersaturated vapor into liquid may lead to overestimation of the rate of latent heat release. In contrast, the SBM schemes solve equation for diffusional droplet growth/evaporation that takes into account effects of aerosols of differential growth rate of droplets.

BULK vs BIN

Properties	BULK –parameterization	Spectral bin microphysics
Main principle	The shape of size distributions is prescribed *a priori*	Model solves system of kinetic equations for size distributions
Aerosols	No aerosol budget, No aerosol transport, no distribution of aerosols with size	Aerosol budget, size distribution of CCN, transport of aerosols, cloud-aerosol interaction
Condensation/evaporation of drops:	No equation for condensational growth ; All supersaturation immediately transfers to cloud water	$r\frac{dr}{dt}=\frac{1}{F}(S-\frac{A}{r}+\frac{Br_N^3}{r^3-r_N^3})$ In maritime clouds S>10%
Collisions	Simplified equations like: $\frac{dq_{rain}}{dt}=k(q_{cloud}-q_{thresh})$	Stochastic collision equation $\frac{\partial f(m,t)}{\partial t}=\int_0^{m/2} f(m',t)K(m',m-m')f(m-m',t)dm' - \int_0^{\infty} f(m,t)K(m,m')f(m',t)dm'$,
Sedimentation	The same fall velocity for particles belonging to the same class	Differential fall velocity depending on particle size , form and air density
Melting , freezing	Assumption that the shape of size distributions remains during the highly non-linear processes	the shape of size distributions changes during these non-linear processes

Table 1. Comparison between bulk-parameterization and spectral bin microphysics

While SBM solves the stochastic equation of collisions to calculate the rate of production of precipitating particles, bulk-parameterization schemes use semi-empirical relationships for autoconversion rates often with hidden or internal assumptions about the production rates of precipitating particles.

Most microphysical processes (advection, settling, collisions, freezing, melting, etc.) are highly non-linear. For instance, the rate of freezing is proportional to drop mass, so that large raindrops should freeze first, small ice particles should melt first. The shape of size distributions changes substantially through these processes. SBM describes these processes explicitly by solving the kinetic equations for size distributions. Bulk-parameterization schemes assume that the size distributions remain gamma or exponential. Therefore, it is not surprising that the SBM and bulk schemes lead to different results (different precipitation, convective heating, cloud type, etc).

Several versions of SBM have been developed at the Department of Atmospheric Sciences of the Hebrew University of Jerusalem. The more detailed version (Khain et al 2008a) contains a larger number of mass bins, higher number of size distribution functions, more accurate description of the processes of big hail formation, melting, etc. This version is used in a 2-D Hebrew University Cloud Model (HUCM) for simulation of single clouds, cloud systems, squall lines.

Another version with a simplified description of some microphysical processes is used in 3-D simulations of supercell storms and TCs (e.g. Lynn et al, 2005a,b; Khain et al 2010) .

In case bulk-parameterizations are used, a simulation of any case study begins with the procedure of tuning of the model parameters. The necessity of the tuning indicates actually

that clouds arising under different conditions have different microphysical structure and obey different precipitation efficiency. The purpose of the tuning is to adjust the results to observations in particular case study. A substantial advantage of the SBM model is that it does not require any tuning of the scheme parameters and can be successfully used *without any tuning of parameters* for simulation of deep maritime convection (Khain et al, 2004, 2008b), continental clouds including pyro-clouds (Khain et al, 2008a), squall lines (Lynn et al, 2005a,b; Tao et al 2007; Li et al 2009a,b; Khain et al, 2009), supercell storms (Khain and Lynn 2009) and arctic stratiform clouds (Fan et al, 2009). All comparisons indicate a substantial advantage of SBM over bulk-parameterization schemes as regards to simulation of precipitation and observed cloud properties. Note, however, that the standard SBM schemes require ~50 times more computer time than standard one-moment bulk-parameterization schemes, which hinders their wide application in operative forecast models, and in particular, in TC forecast models.
Below we describe results of simulations of TC using the SBM.

1.2 Aerosol effects on maritime convection

As suggested above, it is preferred to use SBM schemes when studying aerosol effects on cloud microphysics and dynamics. During the past decade it was found that aerosols (including anthropogenic ones) substantially affect cloud microphysics, and consequently the rate of latent heat release, the dynamics and the precipitation (see, overviews by Levin and Cotton, 2009; Khain 2009; Rosenfeld et al, 2008). In particular, it was found that small aerosols invigorate tropical convection increasing vertical velocities and cloud top heights of deep convective clouds (Khain et al, 2004, 2005, 2008a; Koren et al 2005; Lynn et al, 2005a,b, Wang 2005, Lee et al 2008 ; Khain 2009). Thus, aerosols affect cloud microphysics and dynamics.

An example of aerosol effects on cloud microphysics as follows with simulations using the HUCM is shown in **Figure 1** showing fields of cloud drop and graupel mass contents in deep convective clouds developing in maritime atmospheres under thermodynamic conditions typical of hurricane season. The differences between simulations are only in the concentration of CCN. The left panels show the fields obtained in case of low CCN concentration typical of clean maritime air (100 cm^{-3}). The right panels show the fields in case of high CCN concentration (1500 cm^{-3}). Such concentration is reached when there is penetration of continental aerosols into TC clouds when TCs approach land. One can see that the increase in the CCN concentration dramatically increases the amount of supercooled water aloft, as well as the graupel mass. The physical mechanism of such effect is as follows. An increase in concentration of small aerosols increases droplet concentration and decreases droplet size. The net effect is the decrease in the collision rate, a delay in raindrop formation. As a result, small droplets ascend in cloud updrafts and continue growing by condensation. This leads to an increase in supercooled water content, which intensifies ice-water collisions accompanied by freezing of liquid water. Both processes are accompanied by extra latent heat release leading to an increase in cloud updrafts and sometimes to an increase in cloud top height (Khain 2009). Therefore, clouds developing in polluted air are as a rule stronger and deeper than those developing in clean air. An increase in the mass of condensate means the increase in latent heat release and in the updraft velocity. This effect indicates an aerosol-induced invigoration of convection.

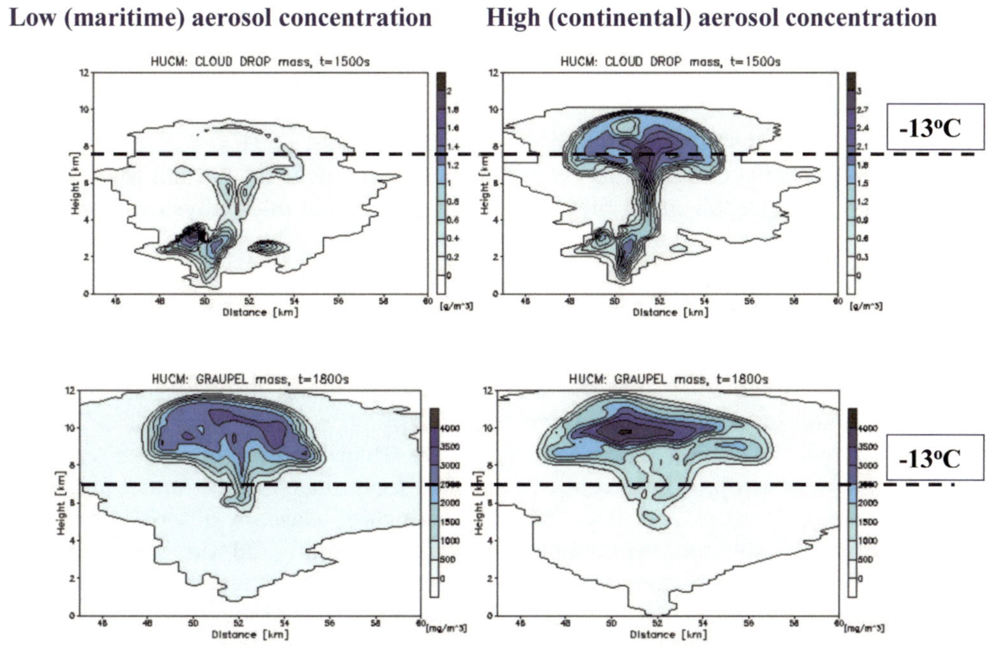

Fig. 1. Liquid water (upper row) and graupel mass content (bottom) fields in simulation of deep martitime clouds using the HUCM. Left panels show the microstructure of cloud developing in clean atmosphere. Right panels show the microstructure of the same cloud but developing in polluted air (after Khain et al 2008b).

Note that an increase in the concentration of CCN creates conditions favorable for lightning. It is known that charge separation in clouds takes place during ice crystals-graupel collisions under the presence of supercooled water. As seen in Fig. 1, an increase in the CCN concentration increased both the supercooled liquid water and graupel content at temperatures colder than -13 ^{o}C . The numerical results show actually, that small CCN are able to transform typical maritime clouds into thunderstorms.

When a TC approaches the land, it imbibes huge masses of continental aerosols into its circulation. This should lead to intensification of convective clouds at the TC periphery. Tropical depressions in the eastern Atlantic Ocean often develop in polluted air (Saharan dust). Therefore, one can expect that aerosols affect the intensity of land falling TCs, as well as TC genesis via their effects on convective clouds.

Indirect evidence of aerosol effects on TC intensity can be derived from enhanced lightning at the periphery of landfalling TCs (Molinari et al 1994; Khain et al 2008b). Khain et al (2008b) simulated the evolution of hurricane Katrina (August 2005) during its movement in the Gulf of Mexico using a two nested grid Weather Research Model (WRF, NCAR version) with the Thompson et al. (2004) one-moment bulk-parameterization. Effects of continental aerosols were simulated by preventing warm rain by shutting off the drop-drop collisions only at the hurricane periphery. A similar approach was used by Rosenfeld et al (2007). Khain et al (2008b) conclude that continental aerosols that penetrated the TC periphery caused enhanced lightning flashes in the areas of penetration. It was also shown that aerosols, invigorating clouds at 250-300 km from the TC center, decrease the convection intensity in the TC eyewall leading to some TC weakening. Similar results were reported by

Rosenfeld et al (2007), who proposed a method of TC mitigation by seeding of clouds at the TC periphery near their cloud base with small aerosol particles of 0.05 μm to 0.1 μm in radius. Simulations of the evolution of an idealized TC using Regional Atmospheric Meteorological System (RAMS) (Zhang et al, 2007) supported the conclusion that aerosols (for instance, Saharan dust) can substantially affect the intensity of TCs.
Below we present results of simulation of hurricane Katrina over the Gulf of Mexico as well as tropical depression Debbie using the WRF with a spectral bin microphysics scheme.

2. Simulation of land-falling TC

2.1 Model and experimental design

2.1.1 Spectral bin microphysics scheme

The SBM scheme implemented into the WRF (Skamarock et al., 2005, version 3) has been described by Khain et al (2004) and Lynn et al (2007). The original scheme is based on solving the kinetic equation system for the size distributions of seven classes of hydrometeors: water drops, three types of crystals (columnar-, plate- and branch-type), aggregates (snow), graupel and hail. Each hydrometeor class is described by a size distribution function defined on mass (size) grid containing 33 doubling bins. The minimum particle mass corresponds to that of the 2 μm radius droplet. Aerosol particles are also described by a size distribution function containing 33 size bins. The size distributions are calculated in the course of the model integration. Using the values of supersaturation, the critical size of aerosol particles to be activated to drops is calculated. Aerosol particles exceeding the critical size are activated and the corresponding mass bins in the aerosol size distribution become empty. The SBM also takes into account possible droplet nucleation during dry air entrainment through the lateral cloud boundaries. An efficient and accurate method of solving the stochastic kinetic equation for collisions (Bott, 1998) was extended to a system of stochastic kinetic equations calculating water-ice and ice-ice collisions. The collision kernels for each pair of particles are calculated using accurate superposition method (Pinsky et al, 2001, Khain et al 2001) and used in the form of lookup tables. The ice nuclei activation is described using an empirical expression suggested by Meyers et al. (1992) and applying a semi-lagrangian approach (Khain et al 2000) to allow the utilization of the proposed diagnostic formulas in a time dependent framework. Secondary ice generation is described according to Hallett and Mossop (1974). The rate of drop freezing follows the observations of immersion nuclei by Vali (1994), and homogeneous freezing according to Pruppacher (1995). Breakup of raindrops is described following Khain et al (2004).
Since the treatment of 8 size distributions requires significant computer time, a Fast-SBM has been developed in which all ice crystals and snow (aggregates) are calculated on one mass grid (one distribution function). The ice particles with sizes below 150 μm are assumed to be crystals, while the larger ones are assigned to aggregates (snow). Similarly, high-density particles (graupel and hail) are also combined into one size distribution (graupel). As a result, the number of size distributions decreases from 8 to 4 (aerosols, water drops, low density ice, high density ice). Note that Fast-SBM keeps the main advantages of SBM: a kinetic equation system is solved using the non-parameterized basic equations, particles of each size have their own settling velocity, particles depending on their mass have different densities, etc. The test simulations showed that Fast-SBM requires less than 20% of the time of the full SBM, which makes it possible to use the Fast-SBM on standard PC-clusters. The comparison of results obtained by the Exact and Fast SBM in simulation of

tropical cloud systems (Khain et al 2009) shows that the Fast SBM produces microphysical and dynamical structure as well as accumulated rain at the surface quite similar to those simulated with Exact SBM.

2.1.2 Experimental design

WRF-SBM simulations were used to study possible aerosol effects on the evolution of Hurricane Katrina (August 2005) in the Gulf of Mexico during three days (beginning with 27 August 00 z) prior to landfall (on about 12z 29 August). A two nested gridded WRF (version 3.01) was used, and the nest moved using a cyclone-following algorithm. The resolution of the finest and the outer grid was 3 km and 9 km, respectively. The number of the vertical levels was 31, with the distances between the levels increasing with the height. The SBM is applied at the finest grid of 400 x 400 km sizes. The initial fields were taken from the Global Forecast System Reanalysis data. The lateral boundary conditions were updated every six hours using the same data source. The Gulf of Mexico surface water temperatures were initialized on 27 August 00 z , and were not updated during the experiments. According to the reanalysis data the SST taken along the TC track reached its maximum near the shore (the place of the TC landfall).

Cloud droplets arise on aerosol particles (AP) playing the role of Cloud Condensational nuclei (CCN). *The initial* (at t=0) CCN size distribution is calculated using the empirical dependence of CCN N_{ccn} concentration on supersaturation with respect to water S_w (in %)

$$N_{ccn} = N_o S_w^k \tag{1}$$

as described by Khain et al (2000). N_o and k are the measured constants for determining the AP concentration and shape of the AP size distribution. At t>0 the the equation for the size distribution of non-activated AP is solved. The initial AP concentration was assumed constant within the lowest 2 km layer and decreased exponentially with height with characteristic scale of 2 km. Aerosols were transported over the entire computational area similarly to other scalars like the mixing ratio.

To investigate aerosol effects on microphysics and the dynamics of TCs, two simulations were carried out: a) in the first "MAR" simulation N_o was set equal to 100cm^3, typical of a maritime atmosphere over the whole computational area b) in the second, semi-continental MAR_CON case the initial CCN concentration over the land N_o was set equal to 1500 cm^3, typical of continents under not very polluted conditions. Initially, over the sea N_o was set equal to 100 cm^3 in all simulations. In all simulations the slope parameter k was set equal to 0.5. When TCs enter the Gulf of Mexico, their circulation transports aerosols from the land to sea, so that some continental aerosols penetrate clouds within TCs and affect their microphysics and dynamics. In all simulations the maximum size of dry AP is equal to 2 μm , which gave rise to droplets of radius 8 μm at cloud base. No ultra giant CCNs (with diameters exceeding 4 μm) that could arise at high winds as a result of spray formation were assumed in the simulations.

2.2 Results of simulations

Figure 2 shows the time dependence of minimum pressure in all numerical simulations and in TC Katrina. One can see that the modeled TC has lower intensity during the first ~50 h of simulations as compared to that of the observed minimum pressure in Katrina. Note in this connection that the WRF initialization used was not a TC forecast initial condition, so that no specific adjustment procedures were used to adopt the TC structure derived from the crude resolution (100 km) reanalysis data to the intensity of the real TC at t=0 (27 Aug 00z). Hence, some relaxation period was required to get the model TC intensity close to the observed one. Yet, the accurate prediction of the Katrina's intensity was not the primary purpose of the study. The main purpose of the simulations was to compare the TC intensity and structure in the simulations with and without aerosol effects on the TC clouds in a strong hurricane, which is able to ingest aerosols from the continent. Figure 2 shows that the TC in the "MAR_CON run turned out substantially weaker, so that at the time instances when the TC reached its maximum intensity the minimum pressure in its center was about ~16 mb higher than in the MAR run. Note that lower (as compared to observed) intensity of the model TC leads likely to an underestimation of aerosol effects, because a weaker TC ingests lower AP amounts into the TC circulation.

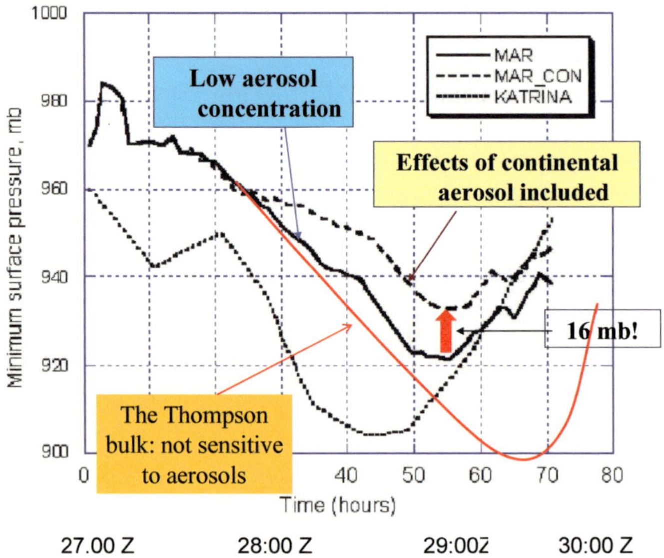

Fig. 2. Time dependence of minimum pressure in numerical experiments and hurricane Katrina (August 2005)

Note that SBM predicts correctly the TC weakening well before landfall. At the same time the bulk-parameterization schemes (for instance, the Thompson scheme) predicts the maximum TC intensity just during landfall (which does not agree with the observations). In the analysis of the aerosol fields we addressed two main questions. The first one was: whether a significant aerosol concentration can enter the TC periphery when it is located at comparatively large distance from coastal line, and second, whether aerosols can penetrate the TC eye. (Note that aerosol fields simulated in the MAR run (not shown) indicate very uniform distribution of AP concentration (which is very low) because the AP concentration over land was assumed equal to that over the sea.)

Figure 3 shows the fields of the AP concentration maximum in MAR_CON run at August 28th, at 23z (left) and at August 29th, at 9 z (right) on the fine grid. The analysis of Figures 3 shows that: a) the AP concentration at the TC periphery approach concentrations similar to those over the continent; b) Over a significant area the aerosol concentration decreases while approaching the TC center partially because of the activation of aerosols to cloud droplets; and c) aerosols can penetrate the TC eyewall clouds along comparatively narrow streams.

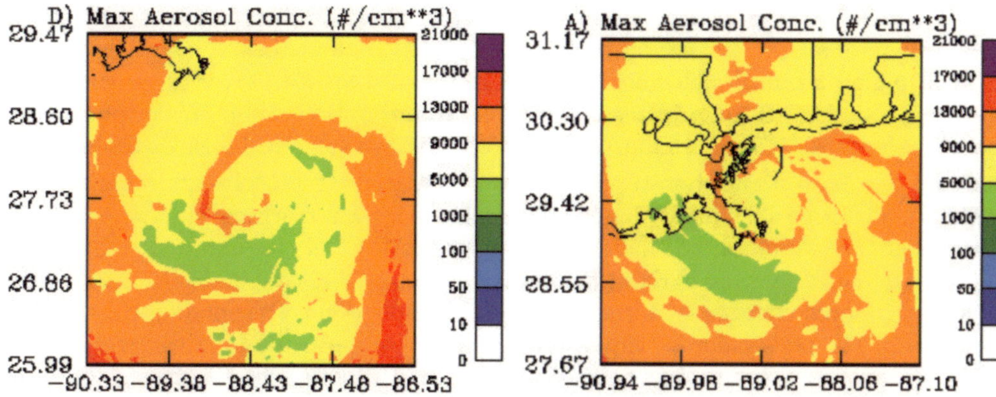

Fig. 3. Fields of maximum AP concentration in MAR_CON simulation at August 28th, at 23z (left) and at August 29th, at 9 z (right) on the fine grid.

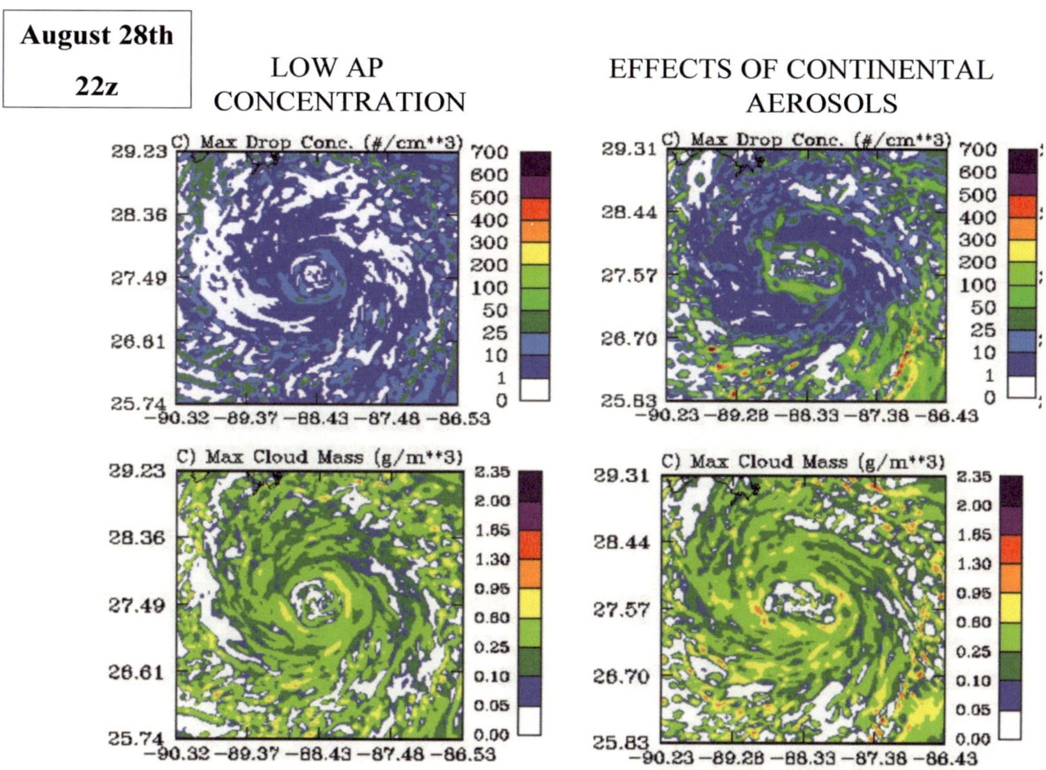

Fig. 4. Fields of the maximum droplet concentrations (upper low) and cloud droplet content (CWC) in simulations MAR (left) and MAR-CON (right) at August 28th 22z at the fine grid.

Figure 4 compares the fields of the column-maximum droplet concentrations (upper row) and cloud water mass content (CWC) (clouds with radii below 40 μm) in clouds in simulations MAR (left) and MAR-CON (right) at August 28th 22z on the fine grid (46h, Figure 2). One can see that in the MAR run droplet concentration does not exceed 50-100 cm^3, which is a typical droplet concentration in clouds arising in clean maritime air. Zones of maximum droplet concentration in the MAR run at the TC periphery indicate zones of higher vertical velocities in rain bands. In MAR-CON, the penetration of continental aerosols led to an increase in droplet concentration at the TC periphery in the zone of high aerosol concentration, as well as in the eyewall. In the MAR-CON run the maximum droplet concentration reached 500 cm^3 (especially high concentrations are at TC periphery), which is substantially higher than those in typical maritime clouds. An increase in droplet concentration within the eyewall in the MAR-CON run indicated that aerosols penetrated to the TC eyewall in the simulations.
The aerosol-induced changes in warm microphysics resulted in corresponding changes in ice microphysics. The penetration of larger amount of drops above the freezing level led to an increase graupel and snow (aggregates) contents at TC periphery (**Figure 5**). Note that convective invigoration of clouds at the TC periphery weakens the updrafts in the eyewall, which immediately resulted in the decrease and even disappearance of graupel and snow in the eyewall. Extra latent heat release caused by droplet condensation and freezing at the

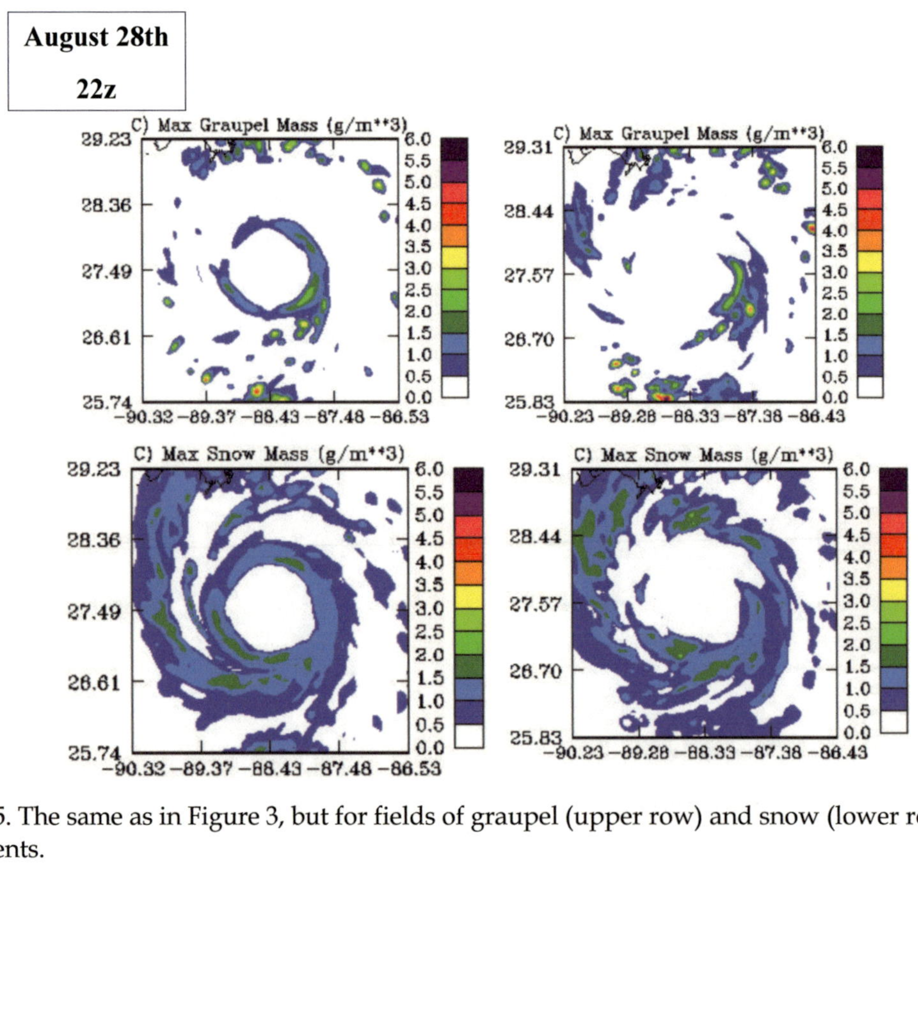

Fig. 5. The same as in Figure 3, but for fields of graupel (upper row) and snow (lower row) contents.

periphery causesd an increase in vertical updrafts velocities and cloud top height (not shown). The values of maximum vertical velocities exceeded 10 m/s which are rare for maritime TC clouds (Jorgensen et al, 1985). Such high velocities are required to form lightning. An increase in cloud top height within polluted air was observed from satellites (Koren et al, 2005) and simulated in many recent studies dedicated to aerosol effects on cloud dynamics (see review by Khain 2009). As it was discussed above, Khain et al (2008b) suggested that the evolution of lightning within TCs approaching land results from the ingestion of continental aerosols iinto the TC periphery. The present study strongly supports this finding.

For instance, **Figure 6** presents the fields of *Lightning Potential Index* (LPI) at 28 Aug. 20 z, and 22 z (t=44 and 46 h in Fig. 1). The LPI was introduced by Yair et al (2009). The LPI is the volume integral of the total mass flux of ice and liquid water within the "charging zone" (0 to -20°C) of the cloud. The LPI has the same meaning as the lightning probability parameter introduced by Khain et al (2008b). Fig. 6 also shows lightning in Katrina (2005) at two different time instances (Shao et al, 2005). The squares show the location of the fine grid approximately corresponding to these time instances. One can see that while in the MAR run the LPI is the highest in the eyewall all the time. Note that simulations of lightning in TC using bulk-parameterization scheme (Fierro et al 2007) (in which aerosol effects were not taken into account) also indicate that lightning was concentrated in the TC central area independently on the stage of TC evolution.

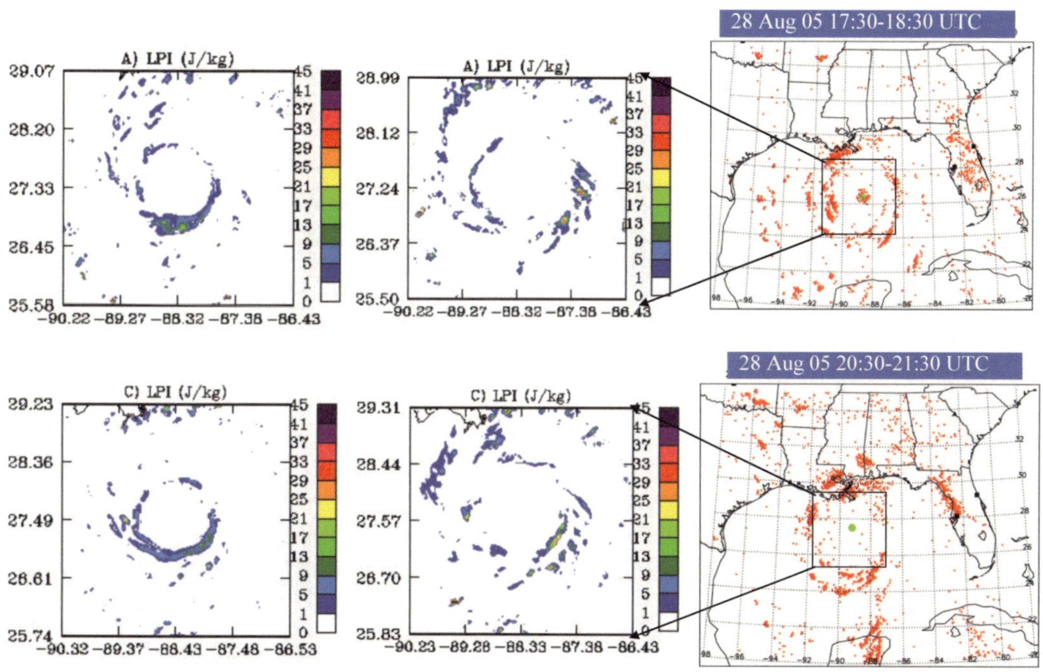

Fig. 6. The fields of Lightning Potential Index (LPI) calculated in MAR and MAR-CON runs at 28 Aug. 20 z, and 22 z. The lightning in Katrina (2005) is also presented (after Shao et al, 2005). Zones of lightning are marked by red dots; the TC eye is marked green. The square shows the location of the fine grid corresponding to these time instances (after Khain et al 2010)

Figure 6 shows the time instance of the dissipation of the internal lightning ring in the course of intensification of convection and lightning at the TC periphery. One can see that the disappearance of lightning in the eyewall in MAR-CON run agrees well with the behavior of lightning in Katrina. The disappearance of lightning in the eyewall in the MAR-CON takes place about 5-6 hours before the TC weakening.

Figure 7 shows the fields of maximum wind speed 28 Aug. 21 z (upper panels), and 22 z in runs MAR and MAR_CON. One can see a significant decrease in the maximum wind speed up to 15 m/s, i.e. by 20-25%. This decrease is substantially stronger than it was reported by Khain et al (2008b). One of the reasons is that Khain et al (2008b) used the bulk-parameterization scheme that is not sensitive to aerosols, as well as the fact that an artificial approach to parameterize aerosol effects by warm rain preventing was performed within the frame of this scheme.

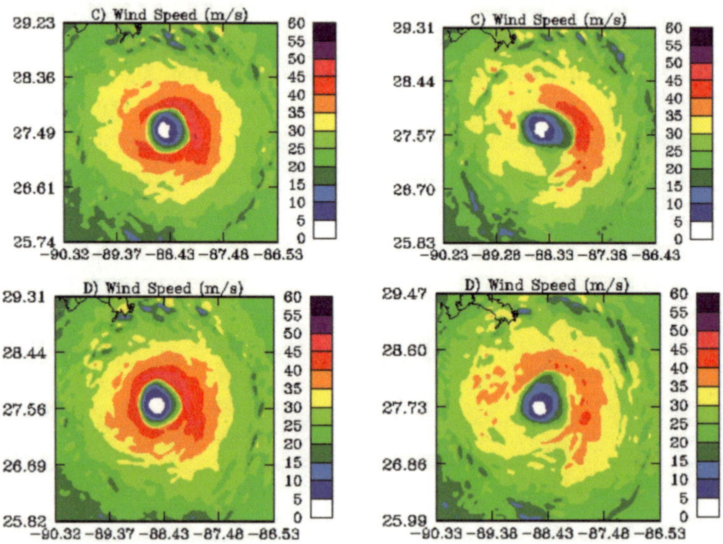

Fig. 7. The fields of maximum wind speed 28 Aug. 21 z (upper panels), and 22 z. in runs MAR (left) and MAR_CON (right).

2.3 Discussion

For the first time, TC evolution was calculated using explicit spectral bin microphysics (SBM). Simulations with grid resolution of 3 km were made with the WRF-SBM. The evolution of Katrina was simulated during 72 hours beginning after it had just bypassed Florida to 12 hours after landfall. In these simulations the effects of continental aerosols ingested into the circulation of the TC on the TC structure and intensity were investigated. It is shown that continental aerosols invigorate convection (largely at the TC periphery), which leads to TC weakening. The TC weakening began ~20 h before landfall, after the instance when the TC intensity reached its maximum. The minimum pressure increased by ~16 mb, and maximum velocity decreased by about 15 m/s. The difference in the intensities remains significant even during the TC landfall. Thus, the results indicate that there is another (in addition to decrease in the surface fluxes) mechanism of weakening of TCs approaching land. This mechanism is related to the effects of continental aerosols ingested into the TC circulation. The mechanism of TC weakening is illustrated in **Figure** 8 showing the vertical cross-section of azimuthally averaged CWC in simulations MAR (left) and MAR-CON (right) at time instance when the maximum difference in the TC intensities took place.

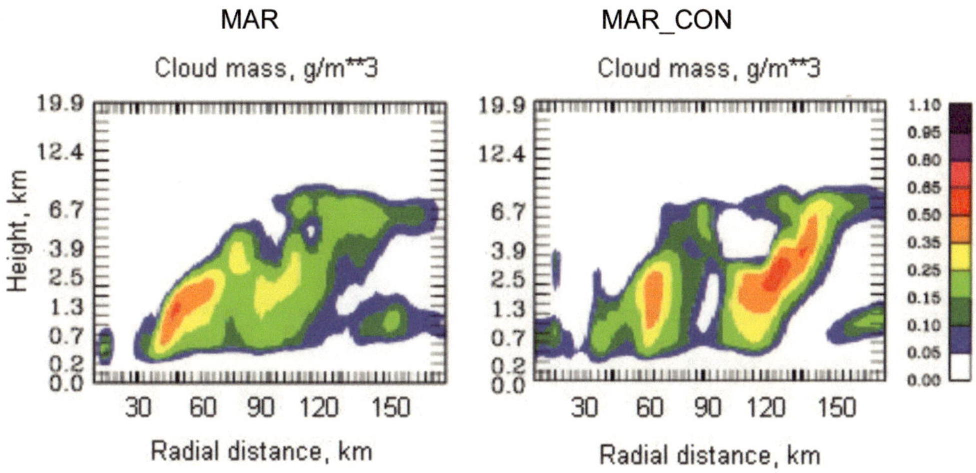

Fig. 8. The vertical cross-section of azimuthally averaged CWC in simulations MAR (left) and MAR-CON (right) at time instance when the maximum difference in the TC intensities took place.

One can see that penetration of aerosols leads to intensification of convection at the TC periphery. Thus, a competition between convection in the TC eyewall and at TC periphery occurs which leads to the weakening of the convection in the TC center. Note, that the weakening of Katrina and the collapse of the inner core occurred in both the observed storm (and in the simulated storm) even though the maximum SSTs were located near the coastal line (and no TC-ocean interaction was taken into account). The results suggest the possibility to mitigate TC intensity by seeding of clouds at the TC periphery with small aerosol particles as it was discussed by Rosenfeld et al (2007) and Khain et al (2008b).

3. Simulation of TC genesis

3.1 Factors affecting TC genesis

Despite many worthy observational and numerical modeling studies in recent decades, our understanding of the detailed physical processes associated with the early stages of tropical cyclone (TC) formation is still inadequate, and, operational forecast skill is still quite low (Gray 1998; Vizy and Cook, 2009).

General parameters affecting TC genesis such as wind shear, air humidity, SST, Coriolis force, etc., were revealed in early studies and led to deriving different expressions for seasonal potential genesis (e.g., Gray 1979; Anthes 1982; Ivanov and Khain 1983; Khainand Sutyrin, 1983; Emanuel 2005). At the same time the seasonal genesis potentials can only present some estimation of the probability of development of tropical depressions (TD) and cannot be used for the purposes of predicting the evolution of particular TDs.

Detailed analysis performed in studies by McBride (1981a,b), McBride and Zehr (1981), Erickson and Gray (1977), Zehr (1992) showed that TDs tend to develop when their vorticity is spatially concentrated and exceeds some critical value. Charney and Eliassen (1964) proposed the mechanism of Conditional Instability of the Second Kind (CISK) according to which convection interacts with a TC-scale vortex via the friction in the boundary layer. The friction-induced radial velocities in the boundary layer concentrate convection (and convective heating) in the central region of a TD. The convective parameterizations based on the CISK allowed developing the first models predicting TC formation. In a set of studies by Montgomery and Farrell (1993), Hendricks et al (2004), and Montgomery et al (2006), the formation of a cloud "coherent" structure in TD and the TD development is attributed to the merger of small scale vortices in conjunction with so-called vortical hot towers (VHT). Tory et al (2007) showed that the exact detail of the vortex interactions was unimportant for qualitative genesis forecast success. Instead the critical ingredients for the genesis were found to be vorticity, comparatively low vertical wind shear, and proper spatial distribution of convective forcing.

The possible sensitivity of TC intensity to the spatial distribution of latent heat release can be derived from lightning data. Lighting indicates zones with the strongest convection. The appearance and intensification of lightning in the eyewall can be a predictor of TC intensification (Lyons and Keen, 1994; Orville and Coyne 1999, Molinari et al 1999; Shao et al 2005; Demetriades and Holle, 2006; Fierro et al 2007, Price et al 2009). Rodgers et al (2000) found that the closer the lightning is to the storm center, the more likely the TC is to intensify.

An increase in computer power has allowed the investigation of TC genesis using models with a resolution of several kilometers (e.g. Wang 2002; Zhu and Zhang 2006; Frisius and Hasselbeck, 2009). Wang (2002) concluded that spiral rain bands forming at the TD periphery have a weakening effect on the tropical cyclone because they introduce low-entropy air by downdrafts into the boundary layer and hinder the boundary-layer inflow towards the eyewall. Sensitivity studies produced by Wang (2002) and Zhu and Zhang (2006) showed that the most intense tropical cyclone resulted when latent cooling processes (evaporation and melting) were switched off. More detailed investigation of spatial distribution of convective heating (latent heat release) and cooling by evaporation and melting was performed by Frisius and Hasselbeck (2009) using axi-symmetric and fully 3-D mesoscale models with bulk-parameterization of microphysical processes. In a set of sensitivity studies, they showed that whether a TD will develop or not, dramatically depends first on the spatial distribution of heating/cooling within an area of a TD. For instance, switching off drop evaporation led to a decrease in the TD eye diameter and convective inhibition outside the eyewall. As a result, the model TD rapidly developed into a TS when evaporation of water droplets or cooling caused by melting were switched off.

The spatial distribution of latent heat within a TC area may depend on the concentration of atmospheric aerosols. Many hurricanes that have reached the US coast have also been found to contain Saharan dust. The transport of Saharan dust may be so strong as to affect precipitation of Florida storms (e.g. Van Heever et al 2006).

In some studies it was shown that the penetration of Saharan dust into TDs hinders their development into tropical storms (e.g. Evan et al 2006). This effect is typically attributed to two factors: a) the penetration of warm air from Africa to the marine layer just or above the inversion, and b) to the radiative effects of dust. In the former, the penetration of Saharan dust layer also increases the stability of the atmosphere and hinders the development of convection. In this case dust plays the "passive" role of tracer of warm air. For instance, Carlson and Benjamin (1980) showed that heating rates associated with Saharan dust can be in excess of 1

^{o}C per day. Using an axisymmetric TC model, Khain and Agrenich (1987) found that solar heating of dust affects TC intensity. Lau and Kim (2007) and Evan et al (2008) suppose that dust intrusions are responsible for chilling the SST in the North Atlantic. It was assumed that the outbreak of Saharan dust led via this mechanism to a decrease in TC activity. In particular, it was assumed that the excess of Sahara dust in the 2006 pre-monsoon season, as compared to 2005, contributed to the big decrease in TC activity from 2005 to 2006.

Substantiality lower attention was paid to the potential effect of aerosols (including Saharan dust) on the intensity and evolution of TD via their effect on microphysics of deep convective clouds. Some observations indicating possible aerosol effects on TD development via their influence on the cloud microphysics of TD clouds have been performed by Jenkins et al (2008) and Jenkins and Pratt (2008). Zhang et al (2007, 2009) investigated the effects of aerosols on the evolution of an idealized TC using RAMS with a 2-moment bulk parameterization of microphysical processes. These simulations suggested that under some conditions aerosols could lead to TC weakening.

As was shown above, continental aerosols penetrating the circulation of a TC approaching the land invigorated convection at the TC periphery and weakened its intensity in the Eye Wall leading to the weakening of the TC at landfall. These results were obtained using the WRF model with spectral bin microphysics (SBM). This microphysics allows the explicit calculation of the effects of aerosols on cloud microphysics through their effect on SDF of droplets and other hydrometeors. In this section, the evolution of TS Debbie (2006) is simulated using the WRF model with SBM.

3.2 Model and experimental design

3.2.1 Case study

TD Debbie (AL042006) 21-26 August 2006 (Zawislak and Zipser 2010) was simulated using the WRF-SBM. Debby formed from a vigorous tropical wave that moved across the west coast of Africa on 20 August. Almost immediately after moving offshore, the wave developed convective banding and a broad closed circulation. The "best track" chart of the tropical cyclone's path is given in **Figure 9**. The depression initially moved west-northwestward to the south of the subtropical ridge. Around 1200 GMT 22 August, the center of the cyclone passed about 100 n mi to the southwest of the southernmost Cape Verde Islands, bringing thunderstorms and gusty winds to the southern islands of Fogo and Brava. The depression strengthened as it moved away from the islands, becoming a tropical storm around 0000 GMT 23 August.

By 1200 GMT that day Debbie's sustained winds reached 45 Knots, and there was little or no change in strength for the next two days as the cyclone moved between west-northwest and northwest at 15-20 Knots over the open waters of the eastern Atlantic. Intensification during this period appeared to be limited by a dry and stable air mass surrounding the cyclone, along with marginal sea-surface temperatures. On 25 August, southerly shear began to increase in association with an upper-level trough, displacing the deep convection to the north of the center. Debbie started to weaken, and became a depression again around 0600 UTC 26 August (Franklin 2006).

In situ measurements in TD Debbie (Zipser et al 2009) showed the existence of strong wind shear in the 5-8 km layer. This shear inhibited convection, so that maximum updraft velocity was quite low (~5m/s). Microphysical measurements showed the dominating contribution of graupel to ice habits. A warm core of 4-5 C was found. The wind maximum was observed on the east side of the storm and exceeded 30 m s^{-1} . Observations also show that

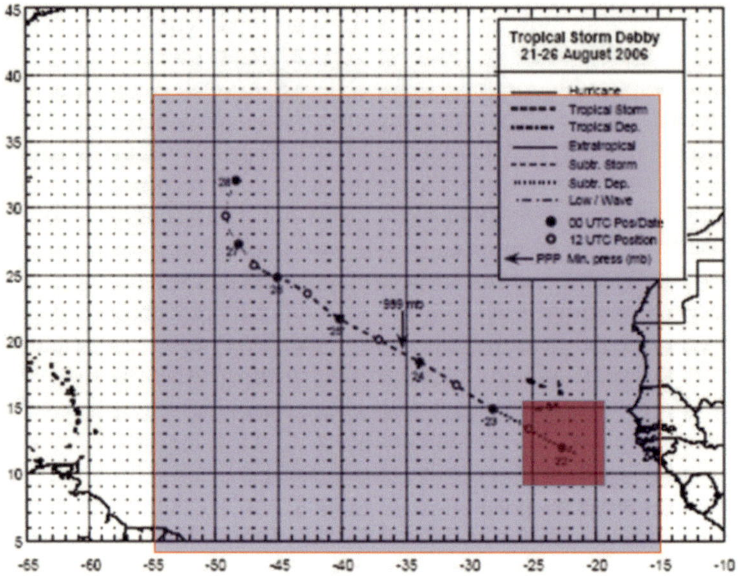

Fig. 9. Best track position for Tropical Storm Debbie, 21-26 Aug. The structure of nested grids is shown.

Debby was completely surrounded by a large Saharan Air Layer (SAL), with a significant aerosol particle layer with depth up to ~ 5 km. According to Zipser et al (2009) the SAL was often adjacent to Debbie's deep convective clouds, sometimes penetrating them and sometimes not. At the same time TD Debbie center region was only weakly affected by continental aerosols (Heymsfield 2010, personal communication).

3.2.2 The simulation design

The WRF model with SBM was similar to that used for simulation of Katrina (2005) (see Section 2). A two-nested gridded WRF (version 3.1) was used, and the nest moved using a cyclone-following algorithm. The spacing of the finest and the outer grid was 3 km and 9 km, respectively. The number of the vertical levels was 31, with the distances between the levels increasing with the height. The SBM was applied at the finest grid of size 720 x 720 km^2. The inner mesh was movable and followed the height minimum at z=850 mb. The outer grid was motionless and was 3240 x 3240 km^2 with spacing of 9 km. The bulk-parameterization of Thompson et al. (2004) was used on the outer grid. The grid structure at t=0 is shown in **Figure 9.** Integral parameters of clouds (mass contents) penetrating from the outer grid into the fine grid were recalculated into size distribution functions assuming a Marshall-Palmer size distribution as defined in the bulk-parameterization scheme. If clouds formed in the internal area penetrated the outermost grid, the size distribution functions of hydrometeors in these clouds were used to calculate mass contents used by the bulk parameterization.

The initial fields were taken from the Global Forecast System Reanalysis data. The lateral boundary conditions were updated every six hours using this data as well. The surface water temperature was initialized at 12 GMT on 20 August, and was not updated during the experiments.

Cloud droplets arise on aerosol particles (AP) playing the role of Cloud Condensational nuclei (CCN). According to observations (Zipser et al 2009) dust in the Saharan air layer

(SAL) is slightly hygroscopic and acts as CCN. The distribution of dust had a maximum within the diameter range 0.1-0.2 μm (Zipser et al 2009). Thus, the high concentration of submicron APs may influence microphysical properties of cloud systems interacting with the SAL. The initial (at t=0) CCN size distribution near the surface is calculated using the empirical dependence of concentration of activated CCN N_{ccn} at supersaturation with respect to water S_w (in %) (eq. 1). The initial AP concentration was assumed constant within the lowest 2 km layer and decreased exponentially with height with a characteristic scale of 2 km. Aerosols were transported over the entire computational area similarly to other scalars like the mixing ratio.

To investigate the possible aerosol effects on the evolution of the TD, two simulations were carried out: a) in the first simulation, N_o was set equal to 100 cm^{-3} typical of maritime atmosphere over the whole computational area (hereafter, referred to as "MAR"); b) in the second, the initial CCN concentration over the land N_o was set equal to 3000 cm^{-3}, typical of continents under not very polluted conditions. Initially, over the sea N_o was set equal to 100 cm^{-3} in all simulations. For the maritime aerosol concentrations, the slope parameter k was set equal to 0.46. It was 0.31 for the continental AP concentrations. The magnitudes of these parameters are typical of maritime and continental conditions, respectively (Pruppacher and Klett 1997). The vertical profiles of the aerosols decreased exponentially with height beyond two kilometers.

The evolution of the background fields was initially simulated for six hours to allow time for the dust to move from over the continent to over the ocean in the experiment MAR-CON. Then, weak temperature sinusoidal heating along the vertical line passing through the point of minimum geopotential at ~800 gPa was used to trigger the TS. The heating zone was moved with the TD. The heating maximum of about 8 °C day^{-1} was applied during 18 hours. This magnitude of the heating is quite small: it is only a few times larger than the radiative forcing. For instance, it is much lower than that in natural clouds, where the heating reaches 100 (and more) °C day^{-1}. The physical meaning of the application of the initial heating is the same as application of bogusing techniques of implementation of initial vortex. However, the implementation of initial heating does not lead to complex processes of mutual adaptation of temperature and pressure fields. The heating does not distort substantially the asymmetric environmental flow as occurs with the implementation of axisymmetric bogus vortex. The advantage of the heating is especially significant in the case of a weak TD, where the initial vortex is very weak and highly asymmetric. Moreover, the further development or weakening of a TD may depend on the rate of asymmetry of the vortex.

The initial triggering was necessary because the crude resolution of reanalysis data does not resolve the TD scales. The heating allowed the system "to recognize" its center, but the further evolution was determined by the cloud formation within the TD, and interaction of the TD with the surroundings, etc.

In addition to simulations with SBM, similar simulations were performed using ten different bulk-parameterization schemes currently implemented into the WRF model.

3.3 Results of simulations

3.3.1 Aerosol effects on microphysics and dynamical structure of simulated TD

Figure 10 shows time dependencies of minimum pressure in Debbie and obtained in numerical simulations using SBM and 10 bulk parameterization schemes. One can see that both SBM simulations reproduced the observed *changes* in surface pressure in good agreement with observations. Toward 72 hr of simulation time, the errors in the prediction of minimum pressure and maximum velocity were relatively low in both simulations when compared to the observed values. The simulation MAR-CON shows, however, better agreement with the observations, especially after 50 hours of simulation time. Trajectories of simulated storms coincided reasonably well with the observed trajectory in both simulations (not shown).

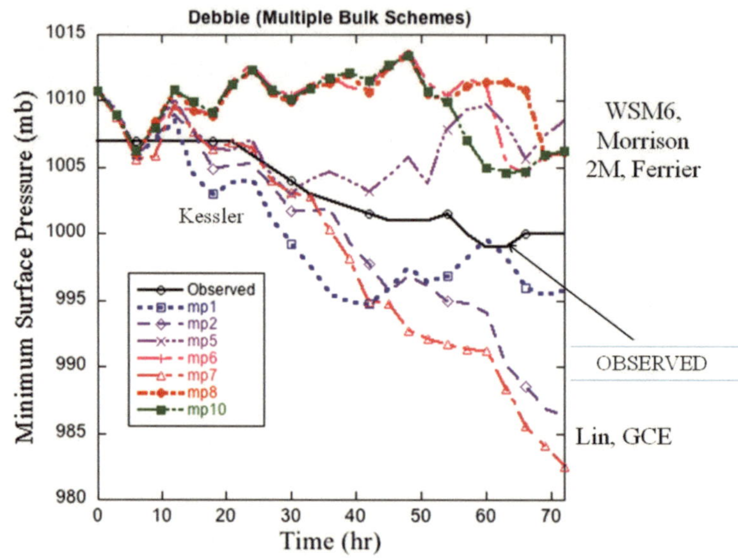

Fig. 10. Time dependencies of minimum pressure in Debbie and obtained in numerical simulations using SBM and 10 bulk parameterization schemes.

One can see that SBM reproduced well the observed evolution of the TD. While the TC intensity was reproduced better in the MAR-CON simulation, the difference between the simulated intensities in simulations with different aerosol concentrations is not large. The better reproduction of the TD microphysics in MAR-CON will be seen by comparison of different microphysical and dynamical fields.

Figure 11 shows that droplet concentration and CWC in CON_MAR are substantially larger than in MAR in agreement with both observations and simulations of maritime clouds developing in clean and polluter air (see Fig. 1). **Figure 12** shows the fields of total liquid water content (QC+QR) in the north-east -south-west vertical diagonal cross -section at t=72 h in MAR and MAR-CON (the line showing the cross-section is shown in Figure 11 by dashed lines. One can see the dramatic effect of APs on microphysics and structure of cloudiness. Droplets in MAR-CON are smaller than in MAR and ascend to higher levels producing supercooled rain drops at ~5 km. While in MAR, the spatial distribution of convection is more or less symmetric with respect to the TD center, it is highly asymmetric in MAR-CON. In MAR-CON droplets penetrate to higher levels and produce stronger supercooled rain that starts at height of 6-7 km.

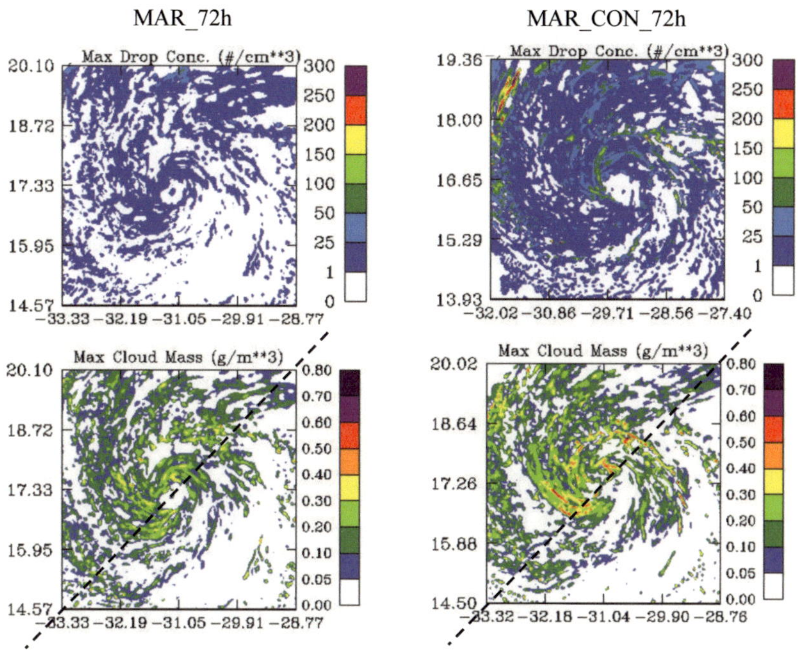

Fig. 11. Fields of maximum values of droplet concentration and cloud water content in simulations MAR (left) and MAR-CON (right) at t=72 h.

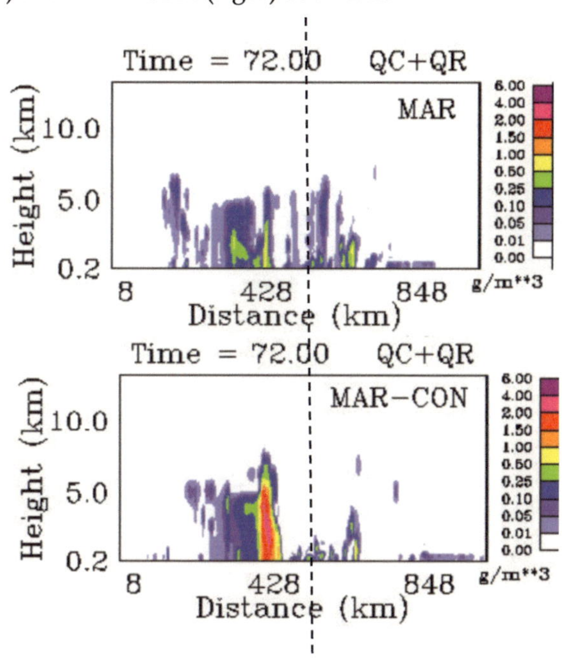

Fig. 12. Fields of total liquid water content (QC+QR) in the north-east -south-west vertical diagonal cross -section at t=72 h in MAR and MAR-CON (the line showing the cross-section is shown in Fig. 11 by dashed lines.

Figure 13 shows fields depicting the maximum updrafts. Prior to the penetration of aerosols into the storm, convection was concentrated in the storm central region in both simulations (not shown). Later the differences in the W fields became significant: W are lower in MAR, but elevated values are more strongly concentrated with respect to the TS center. In MAR_CON, the vertical velocities are higher (up to 7 m/s), but zones with high updrafts are located more that 100 km from the TS center. Convection at the TC periphery is stronger in MAR-CON. A comparison of the two sets of maps reveals the aerosol-induced convection invigoration reported earlier in many observational and numerical studies (see Khain 2009 for detail) and discussed above.

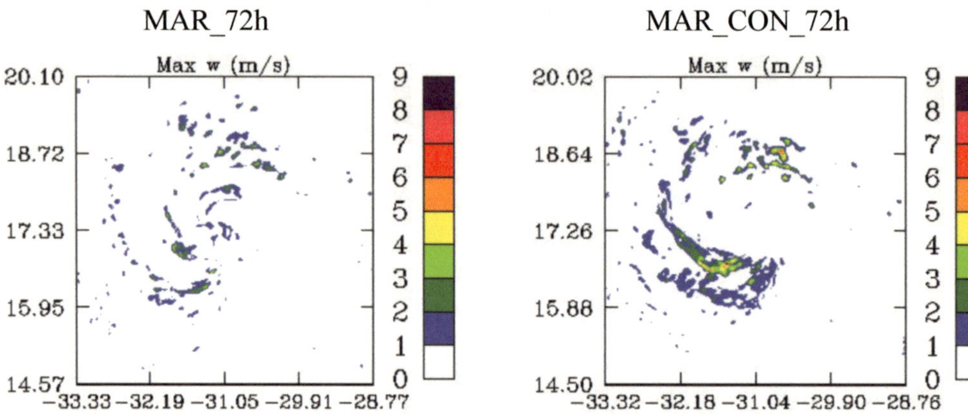

Fig. 13. Fields depicting the maximum updrafts in MAR (left) and MAR_CON (right) at t=72 h.

Differences in warm microphysics and updrafts of clouds in MAR and MAR-CON lead to dramatic difference in ice microphysics. **Figure 14** show fields of maximum values of snow and graupel in MAR and MAR_CON at t=72 h.

Graupel is the dominating solid hydrometeor in MAR-CON, while snow dominates in MAR. The difference in the snow and graupel contents can be attributed to the fact that in MAR-CON supercooled CWC is higher than in MAR that leads to more intense riming of snow, which converts it to graupel. In the MAR experiment there is very little supercooled water and snow does not transfer to graupel as rapidly as in MAR-CON. Again one can see that convection is stronger in the TS center in MAR, while it is stronger in MAR-CON at the TS periphery. Note that according to in-situ observations (Zipser et al 2009), graupel was dominating in Debbie, which corresponds well with the measurements of significant concentrations of AP in the surrounding cloud bands of this TC.

The microphysical and dynamical factors mentioned above lead to significant differences in the wind speed (**Figure 15).** Note that in MAR and MAR-CON maximum wind speed occurs to the east of the TS center which is in agreement with observations (Zipser at al 2009).

In spite of the fact that the maximum values of wind speed are close in both runs, Figure 15 indicates substantial difference in the TS wind structure. In MAR strong winds cover larger area around the TS center than in MAR-CON. At the same time zone of moderate winds covers a much larger area than that in MAR, indicating a much slower decrease of wind speed with the increase in the distance from the TS center as compared to that in MAR. Taking into account that the TC Debbie started weakening at 72 h, results obtained in MAR-CON seem to be closer to observations.

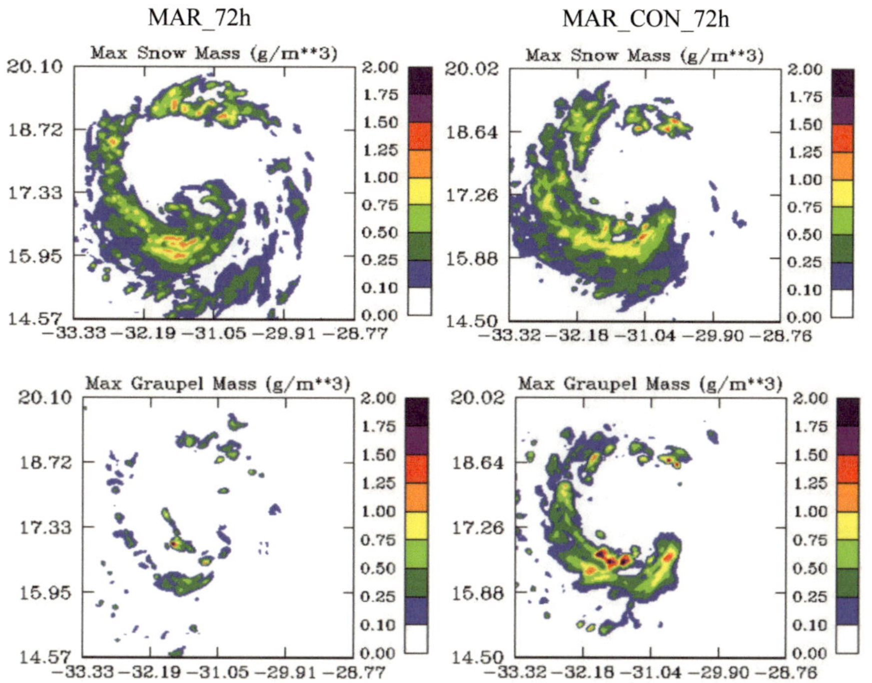

Fig. 14. Fields of maximum of snow and graupel mass contents in MAR and MAR_CON at t=72 h.

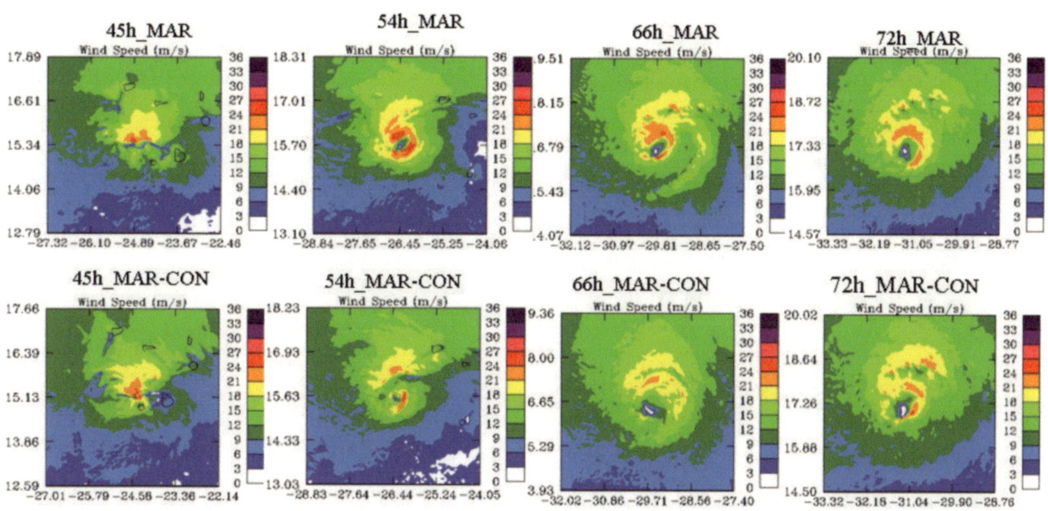

Fig. 15. Fields of the maximum wind in the simulations MAR and MAR_CON at different time instances.

The radar reflectivity calculated in the same cross-section with the observed one (Zipser et al 2009) shows that the structure of convection is substantially asymmetric with the maximum of convection within the south-east quadrant; the calculated maximum radar reflectivity of 55-60 dBZ is reached within the layer from 3 to 5 km which agrees well with the observations.

3.3.2 Comparison of SBM with bulk parameterization schemes

The simulations with bulk-parameterization of convection reveal a wide range of forecasts from non-developing TDs to TCs of hurricane strength (Figure 10), but failed to reproduce the evolution of TD Debbie. Recall that all simulations were performed within the same dynamical framework and the same initial conditions. The difference between the simulations was only in the representation of convection. It is paradoxical that among all bulk-schemes, the best agreement with observations was attained using the simplest one-moment Kessler parameterization (mp1 in Figure 10) which does not include ice processes. Figure 10 stresses the importance of the utilization of correct physics to reproduce the evolution of tropical storms. The second clear conclusion is that the improvement of model resolution will not produce better results without utilization of appropriate microphysics.

Important conclusion that can be derived from Figures 2 and 10 is that in spite of the important role of aerosol effects, the difference between intensities of model TC is determined largely by different representations of physical processes in different microphysical schemes.

A more detailed comparison was performed between the results of the SBM simulations and those using two moment bulk-parameterization scheme, i.e, WDM6 with the prediction for the mixing ratios of six water species (water vapor, cloud droplets, cloud ice, snow, rain, and graupel) (Lim and Hong, 2010). This scheme is one of the most advanced bulk-schemes and predicted the evolution of TD better than most of other bulk schemes. In contrast to most other bulk schemes WDM6 predicts aerosol concentration. This allowed performing simulations with WDM6 similar to those performed with SBM. These simulations will be referred to as WDM6_MAR and WDM6_ CON, respectively. **Figure 16** shows time dependencies of minimum pressure and maximum wind speed in Debbie, as well as in simulations using SBM and WDM6. The TS simulated with WDM6 is weaker than Debbie. Effects of aerosols as it simulated by WDM6 turns out to be opposite to that in SBM, namely, in WDM6 the increase in the aerosol concentration leads to formation of TD of higher intensity.

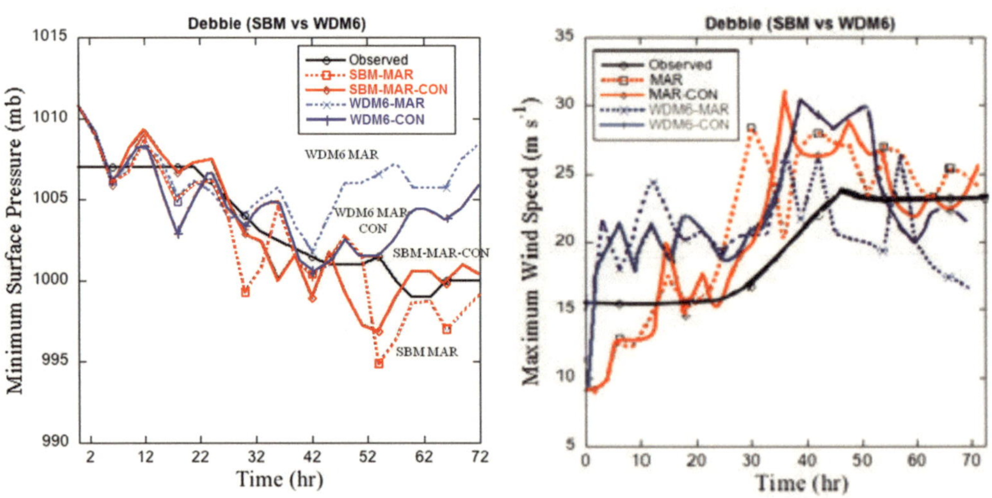

Fig. 16. Time dependencies of minimum pressure and maximum wind speed in Debbie, as well in simulations using SBM and WDM6.

Figures 17 and **18** compare the fields of maximum of cloud and rainwater contents (RWC) in simulations with SBM and WDM6. Comparison of the CWC and RWC shows substantial differences between the fields simulated by SBM and WDM6. Intensity of TD simulated with the SBM is more sensitive to aerosols than in simulations with WDM6. In WDM6 maximum values of CWC are similar in cases of clean and polluted air. This result does not agree with those obtained in many observational and numerical studies (see Figure 1). WDM6 predicts larger maximum values of CWC and RWC than SBM. In WDM6 both CWC and RWC field have a spotted structure indicating very intense clouds. In contrast the SBM simulations show a wide range of CWC and precipitation gradations. The underestimation of low rate precipitation and overestimation of the rate of strong convective precipitation is a typical feature of the bulk schemes.

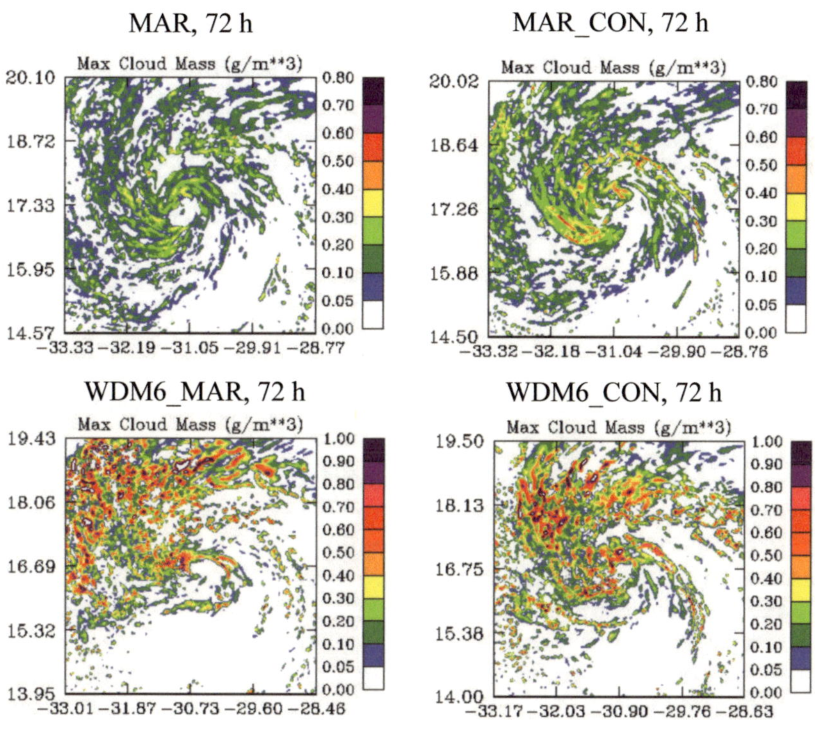

Fig. 17. Fields of the maximum of cloud water contents in simulations with SBM and WDM6 at t=72 h.

The structure of CWC and RWC in the SBM simulations indicates the formation of TC vortex more clearly than that simulated using WDM6.

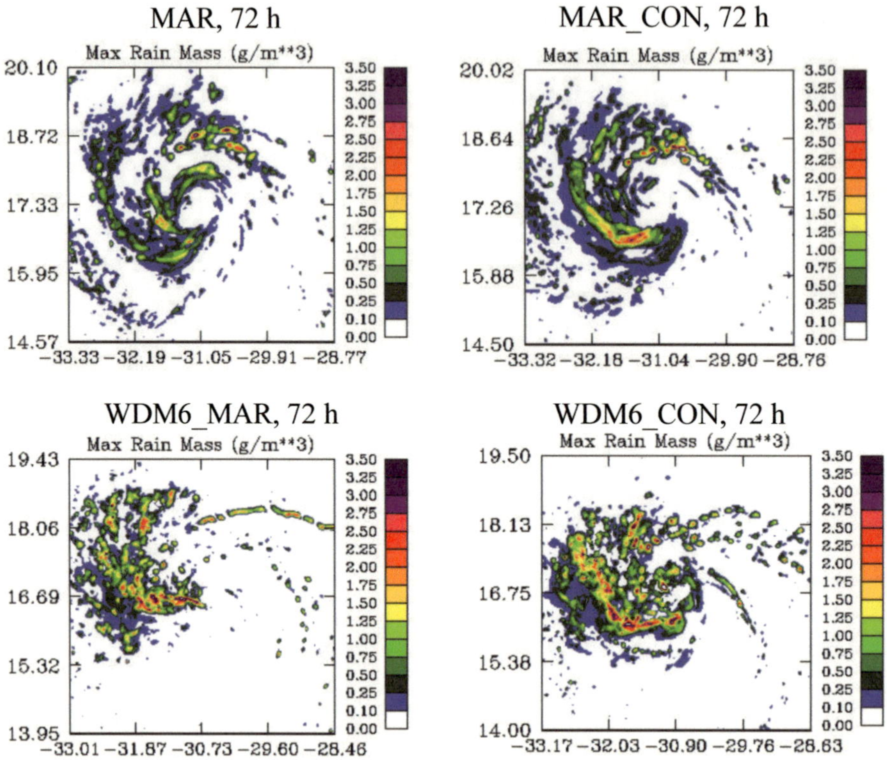

Fig. 18. Fields of the maximum of cloud water contents in simulations with SBM and WDM6 at t=72 h.

Figure 19 shows fields of maximum snow and graupel contents in the WDM6 simulations. The difference between the fields of snow and graupel contents simulated using WDM6 and SBM is even larger than that between fields of liquid water. The maximum values of these ice hydrometeors in WDM6 are substantially lower than those simulated by SBM. Thus, WDM6 indicates stronger warm rain processes (larger CWC and RWC) and much weaker ice processes (lower values of snow and graupel contents) as compared to the SBM. Owing the intense lightning observed in Debbie (lightning is caused by ice-graupel collisions in the presence of supercooled water) the results obtained using the SBM seem to be more realistic. In the WDM6 both graupel and snow contents are larger in the polluted case. In contrast, the SBM predicts lower snow content in the polluted atmosphere. As was discussed above, smaller amounts of snow can be attributed to more intense riming in polluted clouds containing more supercooled water. Other results such as the concentration of the strongest winds in the eastern part of storm, the dominating role of graupel, low vertical velocities (below 5m/s) simulated using SBM also agree with observations better than those obtained with WRM6.

As regards other bulk-parameterizations tested, the microphysical fields produced by these schemes are quite unrealistic, so no comparison is presented here.

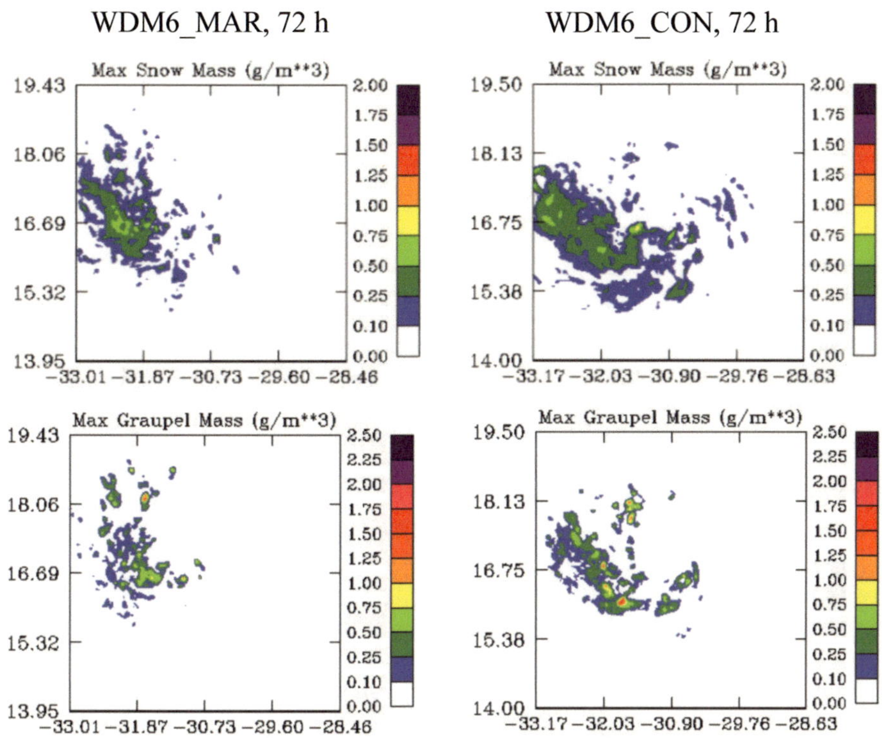

Fig. 19. Fields of maximum snow and graupel contents in the WDM6 simulations at t=72 h.

4. General conclusions

Results obtained using the first simulations of landfalling hurricanes and developing tropical depressions using spectral bin microphysics are quite encouraging. The utilization of SBM based on solving microphysical equations showed significant potential for both further investigation of physical processes in tropical cyclones and for improvement of intensity prediction skill. In all simulations the SBM reproduced the TC intensity (and intensity variations) better than the bulk-parameterization schemes. For instance, the WRF/SBM predicted the TC weakening well before landfall in agreement with observations.

According to Vizy and Cook (2009) "There is currently no parameterization accounting for the influence of the variability of aerosol forcing in the WRF model. These simulations described in this chapter show that such model does now exist. The results show an important effect of aerosols on microphysics and dynamics of TC. The effect of aerosols is strongly related to the spatial distribution of latent heat within the TC area. The aerosol-induced convective invigoration at the TC periphery leads to TC weakening. We suppose that aerosol effects are as important as TC-ocean coupling.

The WRF with spectral bin microphysics seems to be an efficient tool for the investigation of TC genesis and the role of desert dust in this process. The WRF-SBM can serve also as a benchmark model for calibration of bulk parameterization schemes.

The current the WRF/SBM model can be improved in several aspects. In the simulations discussed, the model resolution was 3 km , which is not high enough to resolve comparatively small clouds at the TC periphery. The use of supercomputers should allow simulations of tropical storms with even higher resolution to obtain, quite possibly, even greater sensitivity to aerosols. Other problem to solve is the optimization of the SBM to make it more computationally efficient. Implementation of detailed processes of melting, accurate description of radar signatures from partially melted particles, etc. will be the next step of the SBM improvement in the WRF.

5. Acknowledgements

The study was supported by the Israel Science Foundation (grant 140/07) and scientific project HAMP supported by the US Department of Homeland security (Project FY2008-06-16). The authors express their deep gratitude to Dr. Dudhia, Prof. Baik and to H. Zvi Kruglyak for help in solving many computer problems.

6. References

Anthes, R.A. (1982). *Tropical cyclones-Their evolution, structure , and effects.* Monograph 41, Amer. Meteorol. Soc., 208 pp.

Bender M.; Ginis, I.; Tuleya, R.; Thomas B. and Marchok, T. (2007). The operatuional GFDL coupled hurricane-ocean prediction system and a summary of its performance. *Mon. Wea. Rev.* 135, 3965-3989.

Bott, A., (1998). A flux method for the numerical solution of the stochastic collection equation. *J. Atmos. Sci.*, 55, 2284-2293.

Carlson, T,N.: and Benjamin, S.G. (1980). Radiative heating rates for Saharan dust. *J. Atmos. Sci.* 37, 193-213.

Charney J. G.; Eliassen A. (1964). On the growth of the hurricane depression. *J. Atmos. Sci.* 21: 68–75.

Demetriades N.W.S., and Holle, R.L (2006). Long-range lightning nowcasting applications for tropical cyclones. Preprints, *Conf. Meteorology Application of Lightning Data,* Atlanta, AMS, 9 pp.

Dudhia, J.; Gill, D.; Guo, Y.-R.; Hansen D.; Manning, K.; and Wang, W. (1997): PSU/NCAR mesoscale modeling system tutorial class notes (MM5 modeling system version 2). http://www.mmm.ucar.edu/mm5.

Emanuel K. (2005). *Divine Wind, The History and Science of Hurricanes,* Oxford University Press. 2005, 296pp.

Erickson, S. L.; Gray, W. M., (1977). *Comparison of developing vs. non-developing tropical disturbances.* Dept. of Atmos. Sci. Paper No. 274, Colo. State Univ., Ft. Collins, CO, 80523, 81 pp.

Evan, A.T. ; J.P. Dunion; J.A. Foley; Heidinger, A.K.; and Velden, C.S. (2006). New evidence for a relationship between Atlantic tropical cyclone activity and African dust outbreaks. *J. Geophys. Res.*, 33, L19813; doi: 101029/2006GL026408.

Evan, A.T. and Coauthors, (2008). Ocean temperature forcing by aerosols across the Atlantic tropical cyclone development region. *Geochem. Geophys. Geosyst.*, 9, Q05V04, doi: 101029/2007GC001774.

Fan J., Ovtchinnikov, M.; Comstock, J.; McFarlane, S; and Khain, A. (2009). Modeling Arctic mixed-phase clouds and associated ice formation. J. Geophys. Res. 114, D04205, doi:10.1029/2008JD010782.

Fierro, A.O. ; Leslie, L. ; E. Mansell, Straka, J. ; MacGorman, D. and Ziegler, C. (2007). A high-resolution simulation of microphysics and electrification in an idealized hurricane-like vortex. *Meteorol. Atmos. Phys.* 98, 13-33, Doi: 10.1007/s00703-006-0237-0.

Franklin, J.L. (2006). *Tropical cyclone report: Tropical Storm Debby (AL042006) 21-26 August, 2006*. National Hurricane center, 2 Nov., 2006.

Frisius, T. and Hasselbeck, T. (2009). The effect of latent cooling processes in tropical cyclone simulations, *Quart. J. Roy. Meteorol. Soc.* 135: 1732–1749

Gray, W. M., (1979). *Hurricanes: their formations, structure and likely role in the tropical circulation.* Supplement of Meteorology Over the Tropical Oceans. Published by RMS, James Glaisher House, Grenville Place, Bracknell, Berkshire, RG 12 1BX, D. B. Shaw, (ed.), 155–218.

Gray, W. M. (1998). The Formation of Tropical Cyclones, *Meteorol. Atmos. Phys.* 67, 37-69 Hallett, J., and S. C. Mossop, 1974: Production of secondary ice crystals during the riming process. Nature, 249, 26-28.

Hendricks, E.A, Montgomery M.T, and Davis C. A. (2004). The role of 'vortical' hot towers in the formation of tropical cyclone *Diana* (1984). *J. Atmos. Sci.* 61: 1209–1232.

Ivanov,V. N. and Khain, A. P. (1983). On parameters determining the frequency of tropical cyclone genesis. *Atmospheric and Oceanic Physics*, 19, 787-795.

Jenkins G. S., Pratt A.S.; and A. Heymsfield. (2008). Possible linkages between Saharan dust and tropical cyclone rain band invigoration in the eastern Atlantic during NAMMA-06, *Geophys. Res. Lett.* 35, L08815, doi:10.1029/2008GL034072.2008

Jenkins, G.S. and Pratt, A. (2008). Saharan dust, lightning and tropical cyclones in the eastern tropical Atlantic during NAMMA-06, *Geophys. Res. Lett.*, 35, L 12804, doi: 10.1029/2008GL033979.

Jorgensen, D.P., and LeMone, M.A. (1989). Vertical velocity characteristics of oceanic convection. *J. Atmos. Sci.*, 46, 621-640.

Jorgensen, D. P.; Zipser, E.J. and. LeMone, M.A. (1985). Vertical motions in intense hurricanes. *J. Atmos. Sci.*, 42, 839-856.

Khain, A.P., and Sutyrin, G.G. (1983). *Tropical cyclones and their interaction with the ocean,* Gidrometeoizdat, Leningrad (St. Petersburg), 241p.

Khain, A. P. (2009). Effects of aerosols on precipitation: a review. *Environ. Res. Lett.* 4 (2009) 015004.

Khain, A.P. and Agrenich, E. A. (1987). Possible effect of atmospheric humidity and radiation heating of dusty air on tropical cyclone development. *Proc. Institute Experim. Meteorol.*, 42(127), 77-80.

Khain, A. P., Ovtchinnikov, M.; Pinsky, M. ; Pokrovsky, A. and Krugliak, H. (2000). Notes on the state-of-the-art numerical modeling of cloud microphysics. *Atmos. Res.* 55, 159-224.

Khain A.P., Pinsky, M.B., Shapiro, M. and Pokrovsky, A. (2001). Graupel-drop collision efficiencies. *J. Atmos. Sci.*, 58, 2571-2595.

Khain A.; Pokrovsky; A. Pinsky; M. Seifert, A. and Phillips, V. (2004) Effects of atmospheric aerosols on deep convective clouds as seen from simulations using a spectral microphysics mixed-phase cumulus cloud model Part 1: Model description. *J. Atmos. Sci.*, 61, 2963-2982.

Khain, A.P.; Rosenfeld D. and Pokrovsky, A. (2005). Aerosol impact on the dynamics and microphysics of convective clouds. *Quart. J. Roy. Meteor. Soc.*, 131, 2639-2663.

Khain A.P., Benmoshe, N.; Pokrovsky, A. (2008a). Aerosol effects on microphysics and precipitation in convective clouds with a warm cloud base: an attempt of classification. *J. Atmos. Sci.*, 65, 1721-1748.

Khain, A.; Cohen, N.; Lynn B. and Pokrovsky, A. (2008b) Possible aerosol effects on lightning activity and structure of hurricanes. *J. Atmos. Sci.* 65, 3652-3667.

Khain, A., and Lynn, B. (2009). Simulation of a supercell storm in clean and dirty atmosphere using weather research and forecast model with spectral bin microphysics, *J. Geophys. Res.*, 114, D19209, doi:10.1029/2009JD011827.

Khain, A. P, Leung, L. R.; Lynn, B. and Ghan, S. (2009) Effects of aerosols on the dynamics and microphysics of squall lines simulated by spectral bin and bulk parameterization schemes. *J. Geophys. Res*, 114, D22203, doi:10.1029/2009JD011902

Khain, A. P.; Lynn B. and J. Dudhia (2010). Aerosol effects on intensity of landfalling hurricanes as seen from simulations with WRF model with spectral bin microphysics, *J. Atmos. Sci.* , 67, 365-384

Koren I., Kaufman, Y. J.; Rosenfeld, D.; Remer L. A.; Rudich, Y. (2005) Aerosol invigoration and restructuring of Atlantic convective clouds. *Geophys. Res. Lett.*, 32, L14828, doi:10.1029/2005GL023187.

Kurihara, Y. (1973). A scheme of moist convective adjustment. *Mon. Wea. Rev.*, 101, 547-553.

Lau, W.K.M and Kim, K.-M. (2007) How nature foiled the 2006 hurricane forecasts. *EOS*. Trans. AGU, 88 (9), 105-107.

Lee, S. S., L. J. Donner, V. T. J. Phillips, and Y. Ming, 2008a: The dependence of aerosol effects on clouds and precipitation on cloud-system organization, shear and stability, *J. Geophys. Res.*, 113, D16202, doi:10.1029/2007JD009224.

Levin Z., and Cotton, W.R. (2009) *Aerosol pollution impact on precipitation: A scientific Review*, WMO/IUGG report, Springer, 386 pp.

Li X, Tao, W-K.; Khain, A.P.; Simpson J. and Johnson, D. E. (2009a). Sensitivity of a Cloud-Resolving Model to Bulk and Explicit Bin Microphysical Schemes. Part I: Validation with a PRE-STORM Case. *J. Atmos. Sci.* 66, 3-21.

Li X, Tao, W-K.; Khain, A.P.; Simpson J. and Johnson, D. E. (2009b). Sensitivity of a Cloud-Resolving Model to Bulk and Explicit Bin Microphysical Schemes. Part II: Cloud Microphysics and Storm Dynamics Interactions. *J. Atmos. Sci.* 66, 22-40.

Lim, K-S. S. and Hong, S.-Y. (2010). Development of an Effective Double-Moment Cloud Microphysics Scheme with Prognostic Cloud Condensation Nuclei (CCN) for Weather and Climate Models. *Mon. Wea. Rev.*, 138, 1587- 1612.

Lynn B.; Khain, A. ; Dudhia, J.; Rosenfeld, D.; Pokrovsky, A. and Seifert A. (2005a). Spectral (bin) microphysics coupled with a mesoscale model (MM5). Part 1. Model description and first results. *Mon. Wea. Rev.* 133, 44-58.

Lynn B.; Khain, A.; Dudhia, J.; Rosenfeld, D.; Pokrovsky, A. and Seifert A. (2005b) Spectral (bin) microphysics coupled with a mesoscale model (MM5). Part 2: Simulation of a CaPe rain event with squall line *Mon. Wea. Rev.*, 133, 59-71.

Lynn B. and Khain, A. (2007). Utilization of spectral bin microphysics and bulk parameterization schemes to simulate the cloud structure and precipitation in a mesoscale rain event. *J. Geophys. Res.* 112, D22205, http://dx.doi.org/10.1029/2007JD008475

Lyons, W.A., and Keen, C.S. (1994). Observations of lightning in convective supercells within tropical storms and hurricanes. *Mon. Wea. Rev.*, 122, 1897-1916.

McBride, J. L. (1981a). Observational analysis of tropical cyclone formation, Part I: Basic description of data sets. J. *Atmos. Sci.*, 38, 1117-1131.

McBride, J. L. (1981b). Observational analysis of tropical cyclone formation, Part III: Budget analysis. *J. Atmos. Sci.*, 38, 1152-1166.

McBride J.L.; Zehr R. (1981). Observational analysis of tropical cyclone formation. Part II: Comparison of non-developing versus developing systems. *J. Atmos. Sci.* 38: 1132–1151.

Meyers, M. P.; DeMott, P. J. and Cotton, W. R. (1992). New primary ice-nucleation parameterizations in an explicit cloud model. *J. Appl. Meteor.*, 31, 708-721.

Molinari J., Moore, P. K.; Idone, V.P.; Henderson R.W. and Saljoughy, A.B. (1994). Cloud-to-ground lightning in hurricane Andrew. J. *Geophys. Res.*, 99, 16665-16676.

Molinari J., Moore P., and Idone, V. (1999). Convective structure of hurricanes as revealed by lightning locations, *Mon. Wea. Rev.*, 127, 520-534

Montgomery, M. T., Farrell, B. F. (1993) Tropical cyclone formation. *J. Atmos. Sci.*, 50, 285-310;

Montgomery M.T.; Nicholls M.E.; Cram T.A, Saunders A.B. (2006). A vortical hot tower route to tropical cyclogenesis. *J. Atmos. Sci.* 63: 355–386.

Orville, R.E., and Coyne, J.M. (1999). *Cloud-to-ground lightning in tropical cyclones (1986-1996).* Preprints, 23-rd Conf. on Hurricanes and tropical meteorology, Dallas, Amer. Meteor. Soc., 194pp.

Pielke, R. A., Cotton, W. R.; Walko,R. L.; Tremback, C. J.; Lyons, W. A.; Grasso, L. D.; Nichols M. E.; Moran, M. D.; Wesley, A.; Lee T. J. and Copeland, J. H. (1992) A comprehensive meteorological modeling system-RAMS. *Meteor. Atmos. Phys.*, 49, 69-91.

Pinsky, M., Khain, A. P. and Shapiro, M. (2001). Collision efficiency of drops in a wide range of Reynolds numbers: Effects of pressure on spectrum evolution. *J. Atmos. Sci.* 58, 742-764.

Price C., Asfur M. and Yair Y. (2009). Maximum hurricane intensity precedes by increase in lightning frequency. *Nature Geosceince, Letters*, 6 April, doi: 10,1038/NGEO477.

Pruppacher, H. R. (1995). A new look at homogeneous ice nucleation in supercooled water drops. *J. Atmos. Sci.*, 52, 1924-1933.

Pruppacher, H. R., and J. D. Klett (1997). *Microphysics of clouds and precipitation.* 2-nd edition, Oxford Press, 1997, 963p.

Rodgers E.; Weinman, J.; Pierce, H.; Olson, W. (2000). *Tropical cyclone lightning distribution and its relationship to convection and intensity change.* Preprints, 24th Conf. on Hurricanes and Tropical meteorology, Ft. Lauderdale, Amer. Meteor. Soc. pp. 537-541.

Rosenfeld D., Khain, A.; Lynn, B. and Woodley, W.L. (2007). Simulation of hurricane response to suppression of warm rain by sub-micron aerosols. *Atmos. Chem. Phys. Discuss.*, 7, 5647-5674.

Rosenfeld D., Lohmann, U. Raga, G.B.; O'Dowd, C.D. ; Kulmala, M. ; Fuzzi, S. ; Reissell, A. ; Andreae, M.O. (2008). Flood or Drought: How Do Aerosols Affect Precipitation? *Science*, 321, 1309-1313

Seifert, A. and Beheng, K. (2006). A two-moment cloud microphysics parameterization for mixed-phase clouds. Part 1: Model description, *Meteorol. Atmos. Phys*. 92, 45-66.

Shao X.M., Harlin, J.; Stock, M. ; Stanley, M.; Regan, A.; Wiens, K.; Hamlin, T.; Pongratz, M.; Suszcynsky D. and Light, T. (2005) Los Alamos National Laboratory, Los Alamos, N.M, 18 October 2005, Katrina and Rita were lit up with lightning, *EOS*, 86, No.42, page 398-399.

Skamarock, W.C., Klemp, J.B.; Dudhia, J.; Gill, D.O.; Barker, D.M.; Wang, W.; and Powers, J.G. (2005). A description of the Advanced Research WRF Version 2. *NCAR Tech Notes*-468+STR.

Tao W-K., Li X., Khain, A.; Matsui, T.; Lang, S. and J. Simpson (2007). The role of atmospheric aerosol concentration on deep convective precipitation: Cloud-resolving model simulations. *J. Geophys. Res.* 112, D24S18, doi:10.1029/2007JD008728, 2007

Thompson, G., Rasmussen, R. R. and Manning, K. (2004) Explicit forecasts of winter precipitation using an improved bulk microphysics scheme. Part 1: Description and sensitivity analysis, *Mon. Wea. Rev.*, 132, 519-542.

Tory K. J.; Davidson, N. E. and Montgomery, M. T. (2007) Prediction and Diagnosis of Tropical Cyclone Formation in an NWP System. Part III: Diagnosis of Developing and Nondeveloping Storms. *J. Atmos. Sci.*, 64, 3195–3213.

Yair, Y., Lynn, B., Price, C., Kotroni, V., Lagouvardos, K., Morin, E., Mugnai, A., and Llasat, M. C., 2010: Predicting the potential for lightning activity in Mediterranean storms based on the Weather Research and Forecasting (WRF) model dynamic and microphysical fields, *J. Geophys. Res.*, 115, D04205, doi:10.1029/2008JD010868.

Vali, G. (1994). Freezing rate due to heterogeneous nucleation. *J. Atmos. Sci.*, 51, 1843-1856.

Van den Heever, S. C.; Carrió, G.G..; Cotton, W.R..; Demott, P. J..; Prenni, A. J. (2006). Impacts of Nucleating Aerosol on Florida Storms. Part I: Mesoscale Simulations. *J. Atmos. Sci.*, 63, Issue 7, pp.1752-1775

Vizy E. K. and Cook, K. H. (2009). Tropical Storm Development from African Easterly Waves in the Eastern Atlantic: A Comparison of Two Successive Waves Using a Regional Model as Part of NASA AMMA 2006, *J. Atmos. Sci.* 66, 3313-3334.

Wang Y. (2002). An explicit simulation of tropical cyclones with a triply nested movable mesh primitive equation model: TCM3. Part II: Model refinements and sensitivity to cloud microphysics parameterization. *Mon. Wea. Rev.* 130: 3022–3036.

Wang C., (2005). A modeling study of the response of tropical deep convection to the increase of cloud condensational nuclei concentration: 1. Dynamics and microphysics. *J. Geophys. Res.*, 110; D21211, doi:10.1029/2004JD005720.

Zawislak J. , and Zipser, E. J. (2010). Observations of Seven African Easterly Waves in the East Atlantic during 2006. *J. Atmos. Sci.*, 67, 26-43.

Zehr, R. M., (1976) Tropical cyclone intensification. Dept. of Atmos. Sci. Paper No. 259, Colo. State Univ., Ft. Collins. CO, 80523, 91 pp.

Zehr, R. (1992). Tropical cyclogenesis in the western North Pacific. *NOAA Technical Report NESDIS 16,* 181 pp (available from NESD1S, Washington, DC or CIRA, Colo. State Univ., Ft. Collins, CO).

Zhang H., McFarquhar, G.M. Saleeby, S.M. and Cotton, W.R. (2007). Impacts of Saharan dust as CCN on the evolution of an idealized tropical cyclone. *Geophys. Res. Lett.*, 34, L14812, doi: 10.2029/2007GL029876.

Zhang, H.; McFarquhar, G. M.; Cotton, W. R. and Deng Y. (2009), Direct and indirect impacts of Saharan dust acting as cloud condensation nuclei on tropical cyclone eyewall development, *Geophys. Res. Lett.*, 36, L06802, doi: 10.1029/2009GL037276.

Zhu T. and Zhang D.L. (2006). Numerical simulation of hurricane *Bonnie* (1998). Part II: Sensitivity to varying cloud microphysical processes. *J. Atmos. Sci.* 63: 109–126.

Zipser, E., and co-authors. (2009). The Saharan air layer and the fate of African easterly waves: NASA's AMMA 2006 field program to study tropical cyclogenesis: NAMMA. *Bull. Amer. Meteorol. Soc.*1137-1156.

11

Analysis of Tropical Cyclones in the South China Sea and Bay of Bengal during Monsoon Season

Lei Yang[1], Wei-Wei Li[1], Dongxiao Wang[1] and Yongping Li[2]
[1]*Key Laboratory of Tropical Marine Environmental Dynamics, The South China Sea Institute of Oceanology, Chinese Academy of Science, Guangzhou,*
[2]*Shanghai Typhoon Institute, China Meteorological Administration, Shanghai, China*

1. Introduction

Using reanalysis and satellite data sets, numerical simulation and statistical methods are applied for investigating tropical cyclone (TC) of two ocean basins: the South China Sea (SCS) and Bay of Bengal (BOB). Influenced by Asian monsoon, TCs' feature in these two ocean basins differ from the one of other open oceans. In this chapter, a unique TC case passing through SCS as well as TCs track characteristics in BOB are examined.

The Fifth Pennsylvania State University and National Center for Atmospheric Research Mesoscale Model (MM5) is utilized to study the precipitation and wind speed during Typhoon Chanchu (2006) in SCS. Five model experiments with different physical parameterizations and sea surface temperature (SST) distributions are carried out in the study. Simulations are evaluated using satellite observations. It is found that the control experiment that is configured with the Blakadar boundary scheme, Resiner2 moisture, the Betts-Miller cumulus scheme and daily updated SST has the most reasonable precipitation. The MRF boundary scheme tends to simulate a dryer boundary layer and stronger vertical mixing, which can greatly reduce the intensity of tropical cyclone (TC), resulting in smaller maximum wind speed but larger range of medium wind speed (25-30 m/s). Constant SST through the TC cycle provides more energy from ocean surface, which could cause a significant increase in TC's intensity and therefore result in the largest overestimation on rainfall and maximum wind speed. Longitudinally-uniform SST distribution before the rapid intensification could reduce TC's intensity and heat fluxes, which can partially compensate for the overestimation of precipitation in the control experiment.

Based on a mixture quadratic regression model and the best track dataset 1980-2008, six distinct clusters of TC tracks in the Bay of Bengal (BOB) were identified. For better capturing the background controlling factors, reclassifying is carried out by treating all westward tracks as one type. Thereafter, three track types are obtained, northeastward (fewest in amount), westward (most in amount), and northward. Seasonal variability indicates that northward track type should be divided by two stages (April-May and October-January). After examining the background circulation of each track type, it is found that except for the northeastward one, anticyclonic circulation located in Indo-China Peninsular as well as trough activity in the region of India-Burma played important roles on modulating local wind. These systems assisted steering TC passing through this region. In addition, TCs happened in April and May (pre-summer-monsoon), generally were prone to be affected by the cross-equatorial-flow and move northward.

Results will be given in the following two sections: "South China Sea: Wind and precipitation pattern during typhoon Chanchu (2006) - Comparison between a mesoscale model and remote sensing" and "Bay of Bengal: Cluster Analysis of Tropical Cyclone Tracks".

2. South China Sea: wind and precipitation pattern during Typhoon Chanchu (2006) - Comparison between a mesoscale model and remote sensing

2.1 Introduction

Tropical cyclones (TCs) are one of the most deadly nature hazards to the coastal areas, causing large amounts of casualties and property losses. The South China Sea (SCS) is the largest semi-enclosed marginal sea (~ 3.5x106 km2) in the Northwest Pacific Ocean (WNP), the region of most frequent TC formation in the earth. About 13.2% of TCs in the WNP originates from the SCS (Chen and Ding, 1979). Certain amounts of TCs generated in the WNP also enter the SCS with large inter-annual and decadal variability (Goh and Chan, 2009).

However, compared to the TCs over other ocean basins, there are relatively fewer studies in the literature on TC formation and development over the SCS. The numbers of TCs entering the SCS from the WNP are found below normal during El Nino events, but above normal during La Nina events (Goh and Chan, 2009). It is more complicated for the interannual and decadal variability of the TCs formed in the SCS. The significant atmospheric characteristic in the SCS is that it experiences winter and summer monsoon every year (Liang, 1991). The onset of summer monsoon in the SCS usually starts during mid-May and maintains until September. It is found that during summer (winter) monsoon period, TCs are mainly formed in the northern (southern) part of the SCS (e.g., Wang et al., 2007). The TC activity in the SCS correlates well with the sea surface temperature (SST) and outgoing longwave radiation (OLR) variation (Lee et al., 2006). The formation region of the TCs corresponds with the area where relative vorticity of surface wind (RVSW) is positive, i.e., almost no TCs formed in the negative RVSW area (Lee et al., 2006; Wang et al., 2007). Lee et al. (2006) examined 20 TCs in the SCS during 1972-2002 (May-June), 11 of which are associated with weak baroclinic environment of a mei-yu front and rest of which are more barotopic and possibly intensifies into a stronger TC. The Columnar water vapor, columnar liquid water and the total latent heat release derived from the Special Sensor Microwave/Imager (SSM/I) are found to have significant different in the developing and non-developing tropical disturbances in the SCS (Wang et al., 2008).

The studies of TCs in the SCS, in terms of model simulations and evaluations, are even less. Numerical model simulation can help us to better understand the dynamics of a TC process and therefore help to improve TC forecast skill. Observational data, such as sounding data, Argo floats, satellite data and others have been widely used to improve model simulation (e.g. Soden et al., 2001; Zhao et al., 2005; Chou et al., 2008; Langland et al., 2009; Rakesh et al., 2009). Besides incorporating observations in the simulation to improve forecast skill, evaluation or validation of a model simulation is also possible by comparing the simulations with observational data (e.g. Li et al., 2008; Nolan et al., 2009; Zou et al., 2009). Several simulation studies have been done for the TCs in the SCS. Typhoon Leo (1999) was successfully simulated using two nested domains in relatively coarse resolution (54 km and 18 km) (Lau et al., 2003). The Fifth Pennsylvania State University and National Center for Atmospheric Research Mesoscale Model (MM5) was applied to simulate several characteristics of Typhoon Fitow (2001), including land falling, center position, and precipitation (Li et al., 2004). SST effects on the simulation of the genesis of Typhoon Durian (2001) were investigated using the Weather Research and Forecasting (WRF) model (Wang et al., 2010). MM5 incorporating 4D variational data assimilation system with a full-physics adjoint model was found to greatly improve typhoon forecast in track, intensity, and landfall position (Zhao et al., 2005) .

In this study, we mainly discuss the simulated precipitation and wind pattern during the case of Typhoon Chanchu (2006) using MM5. Previous studies on Typhoon Chanchu (2006) in the SCS mainly focused on analyzing its track and intensity using National Centers for Environmental Prediction (NCEP)/NCAR reanalysis data (Luo et al., 2008). Except for the impact of meteorological factors, ocean effect such as sea surface temperature (SST) on the TC's development has also been studied (Jiang et al., 2008). It is found that considering typhoon-induced cooling in the model simulation can greatly improve the simulation skill of the TC's intensity. Sensitive studies with different SST configurations in Weather Research and Forecasting (WRF) showed that SST variation could cause change in TC's intensity and slightly affect its track as well (Liu et al., 2009).

2.2 Overview of Chanchu (2006)

Typhoon Chanchu (2006) is the first tropical storm in 2006, with a maximum sustained wind speed reaching 46 m/s near its centre during the rapid intensification (RI) on 15 May 2006 in the SCS. It is the most intense typhoon on the Hong Kong Observatory (HKO) record to enter the SCS in May. The TC was originated at the south Philippine Sea on 8 May 2006, when it was declared as a tropical depression. It later strengthened into a typhoon and struck the Philippines twice. It entered the SCS on 13 May 2006 and rapidly intensified into Category 4 (in Saffir-Simpson tropical cyclone scale) on 14 May 2006. During its passage over the SCS, it moved to NW first and turned sharply to NNE while it made its final landfall near Shantou in eastern Guangdong Province, China.

2.3 Model

MM5 is a non-hydrostatic, primitive equation model with a terrain-following sigma-coordinate (Grell et al. 1995). Model simulations of a TC process can vary with many factors, such as model physical parameters, ocean conditions, topography, among others. A total of five experiments are performed to test model sensitivities to some of the above-mentioned factors. All the model runs are initiated at 00 UTC 12 May when the storm was about to enter the SCS. The model version has a horizontal resolution of 15 km and 5 km for Domain 1 and Domain 2, respectively, and 27 vertical levels. Blackdar PBL, Resiner2 moisture and the Betts-Miller cumulus scheme are used in the CTL. Other four experiments are named as PBL_MRF (using MRF PBL), CUM_G (using the Grell cumulus scheme), SST_C (constant SST distribution during TC process, i.e., ignoring the TC induced ocean cooling) and SST_U (longitudinally-uniform SST distribution in the central SCS before the RI).

NCEP Final Analysis (FNL) 6-hourly data are used for the initial and boundary conditions. SST is updated daily with real-time, global SST (RTG_SST) developed at the NCEP/Marine

Modeling and Analysis Branch (MMAB) to represent storm-induced SST changes during the life-cycle of a typhoon. The data has a spatial resolution of 0.5°. Although it is more desirable to use a coupled ocean model to represent storm-induced SST changes with a high resolution, our simulations should have smaller SST errors than those from a coupled model due to technique issues. Regular sounding and synoptic observations were used to refine the NCEP FNL analysis data before the simulations start. Wind bogus is inserted for better results of initial vortex (Zou and Xu 2000). Corresponding winds and locations are obtained from the Joint Typhoon Warning Center (JTWC) of the U.S.

2.4 Data

2.4.1 TRMM

Three-hourly merged rain rate from the Tropical Rainfall Measuring Mission (TRMM) satellite's Precipitation Radar (PR) is used in this study to evaluate the simulations. The PR is a three-dimensional space-borne precipitation radar that, at its nadir point, has horizontal and vertical resolutions of 4.3 by 4.3 km and 250 m, respectively. The PR retrieves reflectivities at a frequency of 13.8 GHz from the surface to 20 km above the Earth ellipsoid. It has a 215-km swath width with a minimum detectable signal of nearly 17dBZ. The 3-hourly rainfall data has a horizontal resolution of 0.25° by 0.25°, and extends from January 1998 to June 2006. The 3-hourly rain rate is then used to calculate 24-houlyr accumulated rainfall in this study.

2.4.2 QuikSCAT

The SeaWinds scatterometer on the QuikSCAT is a microwave radar launched and operated by the U.S. National Aeronautics and Space Administration (NASA) from July 1999 to November 2009. The QuikSCAT measures the back scatter radiation in the Ku-band (~13.402 GHz). The return signal power is proportional to surface stress over the ocean, and the signal can be related to surface wind vector at 10 m height assuming neutral stratification. In addition to wind vector data, the times (minutes of UTC) of measurements are also provided. The data are regridded to 0.25° by 0.25° (latitude by longitude) map for ascending and descending passes. In order to make a better comparison, we choose the model hours that are the closest to the time when the satellite passes the SCS.

2.5 Comparisons between model-simulated and TRMM-derived precipitations

The spatial patterns of 24-hour-accumulated precipitation are shown for three days: 14, 15, and 17 May 2006, which represent before, during and after the RI, respectively. Figure 1 shows 24-hour-accumulated precipitation during the three days simulated by the CTL and derived from TRMM measurements. CTL simulates overall larger rainfall than TRMM, especially in the area surrounding the storm center. The difference in main rainfall area is obviously due to the faster translation speed simulated by CTL, i.e., more northward in CTL than in TRMM. In addition, the rainfall patterns simulated by CTL have more compact structure and clearer spiral rainbands. The maximum accumulated rainfall in CTL is clearly larger than the TRMM measurement, with the largest difference during the RI period. Large rainfall area in the model simulation is mainly located at the center before and after the RI (not shown), but at west or northwest of the rainfall area during the RI period (Fig. 2). After the RI, the TC moves closer to the coast; the formation of maximum rainfall should be related to coastal-enhanced mechanism, which seems much stronger in the model

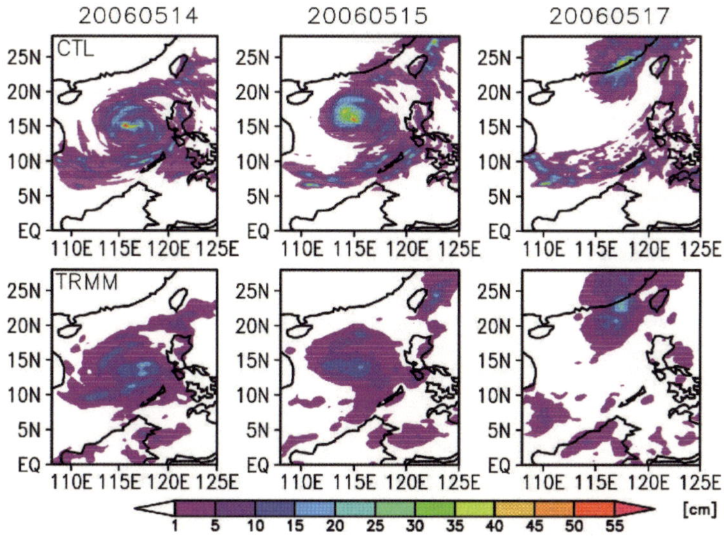

Fig. 1. Comparison of 24-hour-accumulated precipitation derived from CTL (top) and TRMM PR (bottom) during the three days.

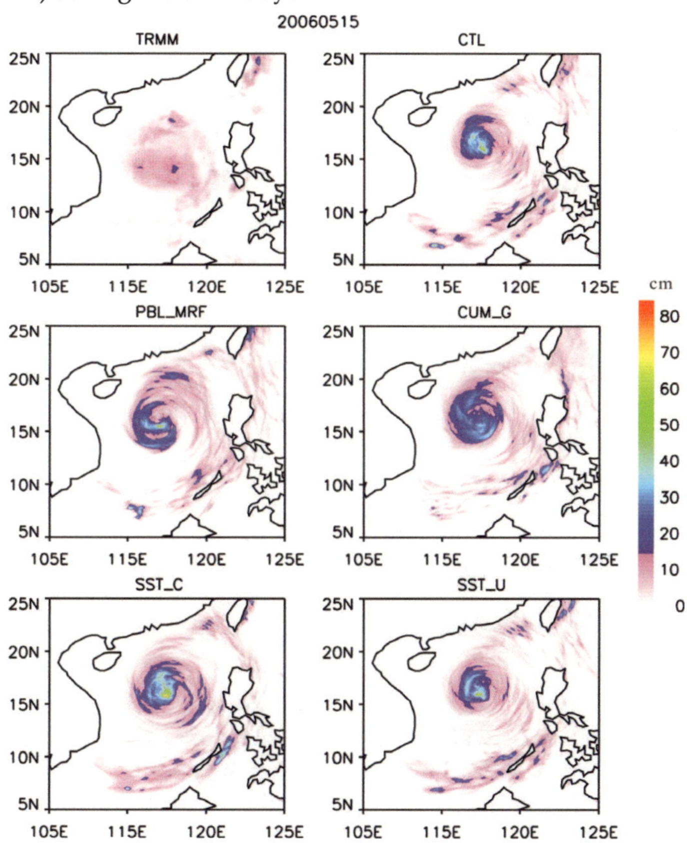

Fig. 2. Comparison of 24-hour-accumulated precipitation derived from TRMM PR and the five model simulations on 15 May 2006 (the RI period).

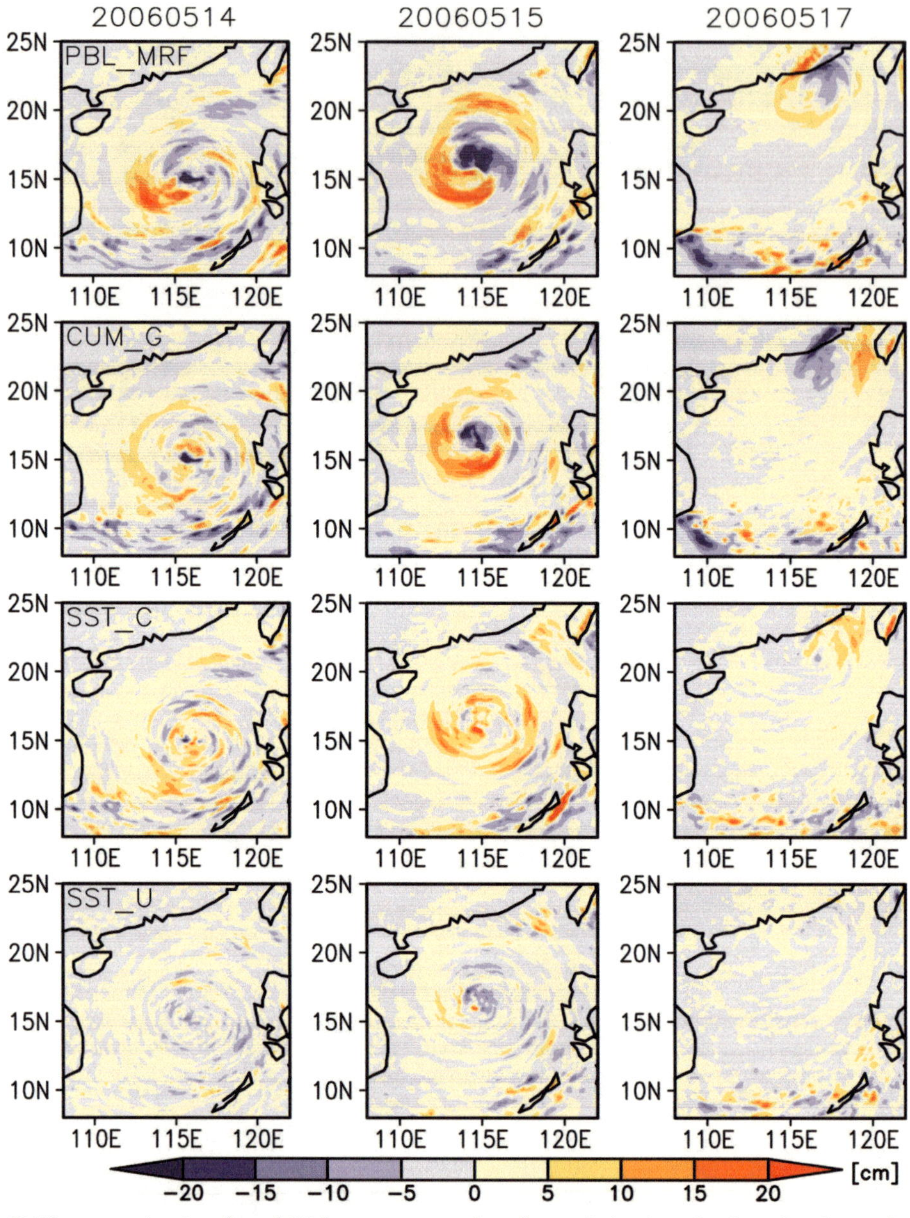

Fig. 3. Differences in simulated 24-hour-accumulated precipitation during the three days: sensitive run minus CTL.

simulation. TRMM gives larger range of the rainfall but weaker strength. The differences of the large rainfall area location between the sensitive runs and CTL can be clearly seen in Fig. 3, in which rainfall simulated by the four sensitive runs are compared with CTL.

The differences between the four sensitivity runs and CTL for the three days are illustrated in Fig. 3. Though having a dryer PBL, PBL_MRF tends to simulate a deeper PBL due to excessive vertical mixing outside of the eye wall (Hong and Pan 1996), which could lead to the overall heavy precipitation in PBL_MRF run (Fig. 3). CUM_G allows more grid-resolved precipitation (Grell 1995). SST_C simulates the largest precipitation (especially during the RI, Fig. 3). In comparison to the CTL, PBL_MRF and CUM_G tend to simulate more precipitation outside the eye wall, especially in the area southwest of the center during the RI. SST_C simulates overall higher precipitation. After the RI, large precipitation area moves to the east or southeast of the center in CUM_G and SST_C, while it still shows large precipitation at the west in PBL_MRF. SST_U has the smallest difference with CTL.

2.6 Comparisons between model-simulated and QuikSCAT-derived wind fields

QuickSCAT-derived and model-simulated wind fields during the three days (phases) are given in Figs. 4 and 5, respectively. QuikSCAT passes over the SCS twice a day (descending and ascending passes), with different spatial coverage of the SCS. We choose the pass when QuikSCAT has the best coverage of the TC process for that day and compare with the simulations at the closest hour. Figure 4 clearly shows the evolution of the TC in the SCS. The wind derived from QuikSCAT is stronger at the inner core than that simulated by the model, especially before and during the RI period. Before the RI, the wind patterns from the five experiments are similar, with slight difference in maximum wind speed. During the RI, PBL_MRF simulates relatively smaller maximum wind speed than the other model runs, but has larger range of medium wind speed (20-25 m/s). SST_U has the weakest wind, including maximum and medium wind speeds. After the RI, maximum wind speed increases significantly in SST_C. Compared to PBL_MRF and CUM_G, the three experiments of CTL, SST_C, and SST_U simulate a more compact storm structure during the RI.

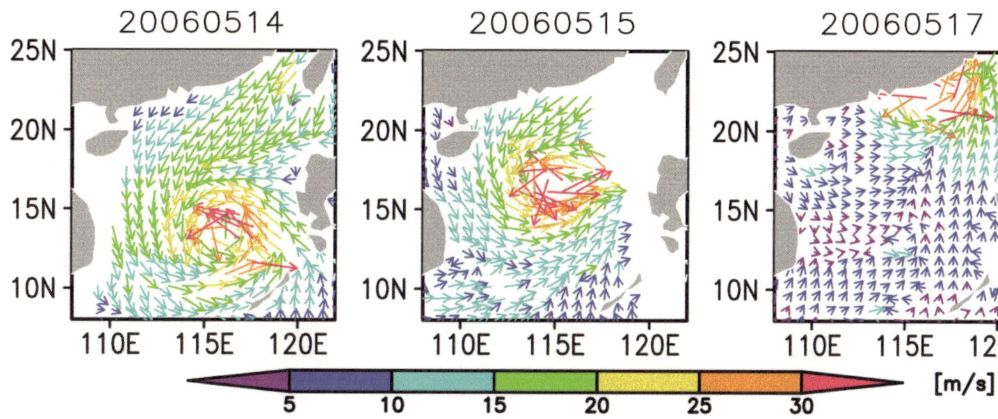

Fig. 4. Wind field derived from QuikSCAT during the three days.

Different atmospheric conditions in PBL_MRF (boundary condition) and CUM_G (cumulus scheme) could cause an increase in latent heat flux and precipitation, but not in storm intensity, such as maximum wind speed and center pressure. However, a warmer SST (in SST_C) could result in an increase in storm intensity in addition to the increase in latent heat flux and precipitation.

2.7 Conclusions and discussions

Model simulated precipitation and wind speed are compared with satellite data, in order to evaluate model performance and to test model sensitivity to several parameters that may impact simulation results. Atmospheric boundary condition, cumulus scheme and SST distribution can have different influences on rainfall and wind patten during different periods of a TC process. All five model runs in this study overestimate precipitation and underestimate maximum wind speed during Chanchu (2006). Large rainfall area mainly occurs on the west or northwest side of the rainfall center. Areas around rainfall and maximum wind center are chosen to quantify the difference between model simulation and satellite observation. The model configured with Blakadar PBL, Resiner2 moisture, the BM cumulus scheme and daily-updated SST has the best simulation of precipitation. Using the MRF PBL scheme would greatly reduce TC's intensity, which can be clearly reflected in the simulation of maximum wind speed. Constant SST through the TC life cycle provides more energy to the TC, which could cause a significant increase in TC's intensity, therefore leading to the largest overestimation on rainfall and maximum wind. Longitudinally-uniform SST distribution before the RI would reduce TC's intensity and heat flux due to less energy from the ocean.

3. Bay of Bengal: cluster analysis of tropical cyclone tracks

3.1 Introduction

The Indian sub-continent is one of the worst areas in the world affected by Tropical Cyclones (TCs), although TCs in this region just account for about 7% of the total number of global TCs (*Gray* 1968). Unlike TCs in the western Pacific mainly occurring after the monsoon onset, TCs in the Bay of Bengal (BOB) have two seasons. The primary season is during the post-monsoon period and the second one is the pre-monsoon season (*Mohanty* 1994). During the monsoon, less TCs or tropical disturbances form because strong tropospheric ventilation produced by the large vertical wind shear inhibits storm development (*Gray* 1968). Compared with TCs in the Pacific and Atlantic, the TC genesis process in the northern Indian Ocean received little attention probably because of lacking in observations. *Kikuchi et al.* (2009) indicated that the incipient disturbances are virtually absent in the northern Indian Ocean and the initiation process of tropical depression is expected to be different from those in the Pacific and Atlantic Oceans.

Murty et al. (2000) examined the Effective Oceanic Layer for Cyclogenesis (EOLC) parameter, which was related with the near-surface stratified layer developed due to the spread of low salinity waters under the influence of freshwater influx. They found that EOLC should be considered for indentifying the zones of cyclogenesis and for better prediction of cyclone tracks in BOB. Their results indicated that zones of cyclogenesis and mean-cyclone tracks fairly coincide with the zones of higher EOLC. Notably, research about Cyclone Heat Potential (CHP) in BOB partly explained the cyclone genesis in this ocean basin. *Sarma et al.* (1990) first revealed the seasonal distribution of CHP. By means of CTD observation data sets collected from five cruises during 1993-1996, *Sadhuram et al.* (2004) found that high value of CHP coincided with anticyclonic gyre (ACG) and vice versa, which emphasized the importance of gyres in the distribution of CHP on the intensification of cyclones/depressions. In terms of decadal variability, *Singh et al.* (2000) have proved that there had been a increase trend in the enhanced cyclogenesis during November and May in BOB which account for the maximum number of severe cyclones over the north Indian Ocean. Ocean responses to TC have also been investigated in BOB, which mainly put concern on the sea surface temperature (SST) change under the passage of TC cases (*Sadhram* 2004). Inertial oscillations signals forced by TC were captured by time-series measurements from a moored data buoy located in BOB set in September 1997 (*Joseph et al.*

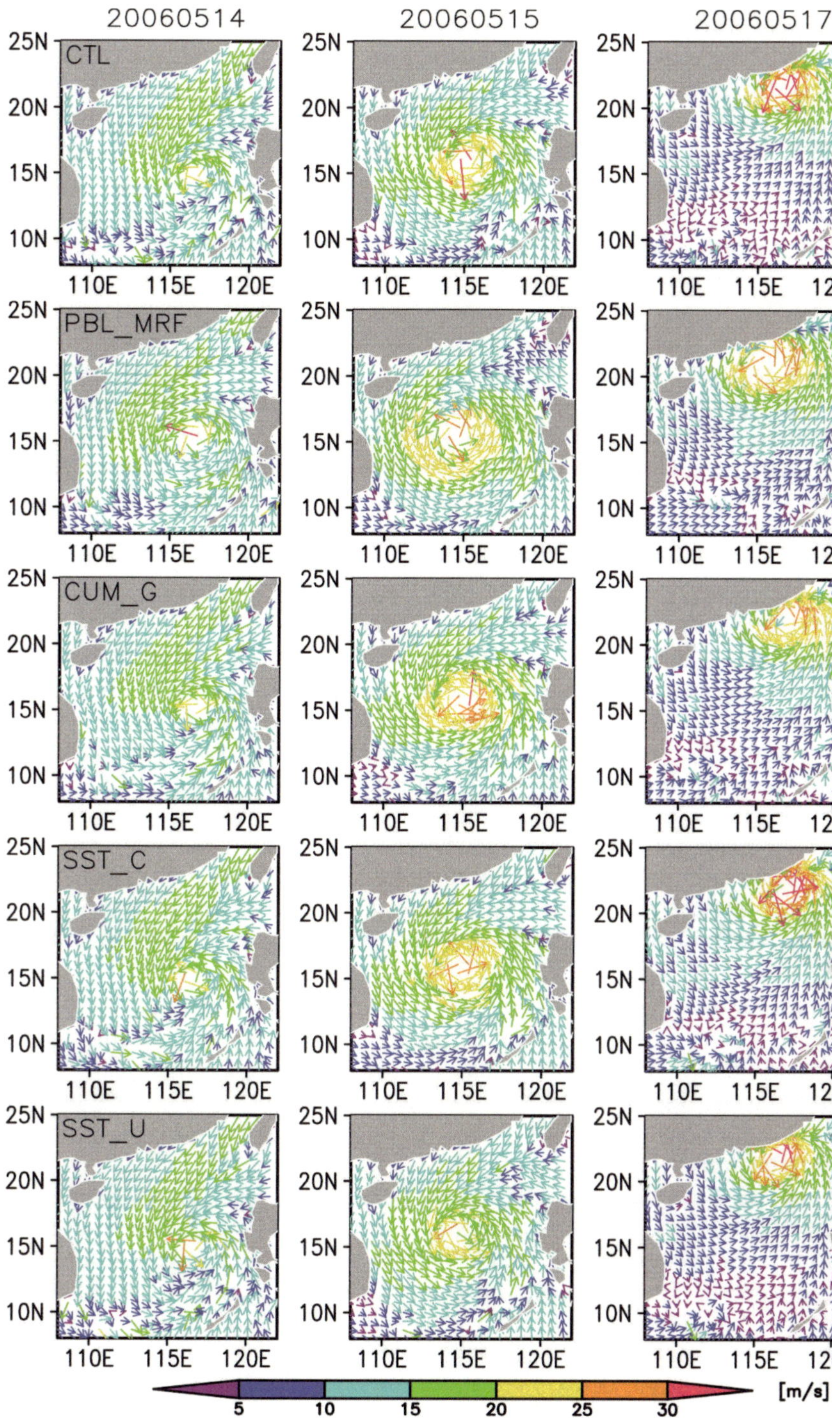

Fig. 5. Wind pattern simulated by the five model runs during the three days.

2007). Add to the above research, TCs case studies were carried out to find the physical explanations under the specific TC conditions which help improve the operational forecast capability, such as the disastrous TC Nargis in 2008 (*Webster* 2008; *Lin et al.* 2009; *Kikuchi et al.* 2009; *Yanase et al.* 2010; *Yokoi et al.* 2010; *Yamada et al.* 2010; *McPhaden et al.* 2009).
The TC's track or movement is to be affected by many internal and external factors (*Mohanty and Gupta* 1997). Steering flow is the prominent external force on TCs, accounting for 70-90% of the motion. When steering flow is weak, TCs tend to move poleward and westward resulting from the internal force (*Chan and Gray* 1982; *Elsberry et al.* 1987). In addition, TCs have a tendency to move toward a warmer ocean surface (*Orlanski* 1998; *Mandal et al.* 2007). Several methods have been used in sorting tracks, which include *K*-means method (*MacQueen* 1967). *K*-means method has been applied in studying North Pacific (*Elsner and Liu* 2003) and North Atlantic (*Elsner* 2003) TCs. Nevertheless, *K*-means method cannot accommodate tracks in different lengths. To solve this issue, the finite mixture polynomial regression model (*Gaffney* 2004) was used to objectively classify the TC tracks, not only based on a few points of trajectory, but on trajectory shape. The technique has been applied to western North Pacific typhoon tracks (*Camargo et al.* 2007a,b), eastern North Pacific hurricane tracks (*Camargo et al.* 2008) and the climate modulation of North Atlantic hurricane tracks (*Kossin et al.* 2010). In recent years, some progresses have been made by using in models for operational track forecast in India, BOB in specific (*Mohanty and Gupta* 1997; *Gupta and Bansal* 1997; *Prasad and Rao* 2003). Sensitive experiments are used to examine the effective factors controlling TC tracks (*Mandal et al.* 2002, 2003).
These previous studies regarding TC tracks in BOB mainly focused on operational model study. However, objective and systematic analysis about TC tracks in this region remain unknown. Clarifying the physical background of TC tracks in BOB will help to minimize the error of operational forecast for similar TCs in the future, and reduce the loss of lives and properties. Therefore, the focal point of the present study is to show the TC track types classified by results from the mixture regression model, and then depict the seasonal variability and circulation field of each track type.
The study is organized as follows. Datasets used in this study are given in Section 2). Characteristic of track types is described in Section 3). Section 4) shows the seasonal variability, which help further sort tracks. Environmental flow is given in section 5).

3.2 Data and methods

TC data is derived from the Joint Typhoon TC tracks (JTWC) for the time interval 1980-2009. The year 1980 is chosen because TC data is more reliable and man-made satellites had already been launched and used for TC observation and forecast since 1980 in BOB. In this study, only TCs with landfall are included in our analysis, for the purpose of examining the environmental flows effectively.
TC track types are detected by a mixture of polynomial regression models which was developed by *Gaffney et al.* (2004). Compared with the previous cluster analysis methods, this model can fit the geographical "shape" of the trajectories which also allow quadratic function by extending the standard multivariate finite mixture model. Maximizing the likelihood of the parameters helps to find the best fitting mean curve. This method can easily accommodate TC tracks with different length.
Variables from the National Centers for Environmental Prediction/ National Centers for Atmospheric Research (NCEP/NCAR) reanalysis (*Kalnay et al.* 1996) since 1980 are used to analyze the background flows of each type of tracks. The entire data records in each case contribute to the composition of their associated track type.

3.3 Characteristics of track types

Under this mixture of polynomial regression models, as measure of goodness of fit, log-likelihood and within-cluster error values are used to obtain the optical mean regression trajectories (track types) of the observed tracks. Within-cluster error is defined as difference in latitude and longitude from the mean regression trajectories squared and summed over all tracks in the cluster for different cluster numbers (*Camargo et al.* 2007). Fig. 6 shows the log-likelihood values and within-cluster error for different number of clusters on different regression order. Larger log-likelihood with smaller within-cluster error indicates good fit. Here, we choose *K*=6 and quadratic regression order to sort tracks into six types. Fig. 7 shows the historical tracks from 1980 to 2009 of each cluster. They are (1) northeastward with recurvature; (2) westward at lower latitude; (3) longest westward; (4) longer northwestward; (5) shorter northwestward; and (6) northward.

To better find the physical relationship between TC activity and seasonal variation, all westward cases (cluster 2, 3, 4, 5) are merged into one category - here as "westward type". Hence, three new track types are listed below: westward (68 cases), northward (31 cases), and northeastward (11 cases) (Fig. 8).

3.4 Seasonal variability

Fig. 9a shows the seasonal variability which indicates total TC count during year 1980-2009. Fig. 9b gives the contribution of each cluster in each month to the total number of TCs for each cluster during 1980-2009. It shows that westward type have the largest number of cases. Most of these TCs (over 85%) occurred after winter monsoon onset (September to January), with a peak in October (over 30%). Northward type has two seasons. The primary season is after October which is similar as the westward type (but delayed for one month), and the second one is pre-summer-monsoon season (April and May). TC numbers in these two seasons are closed. Main peak took place in May, and sub-peak in November (also with one month lag compared to westward type). The northeastward type with recurvature and landing at Myanmar is unique, not only for its fewest number (only 11 cases) but for the two peaks in April and May, which implicit that connection to summer monsoon onset may be considered.

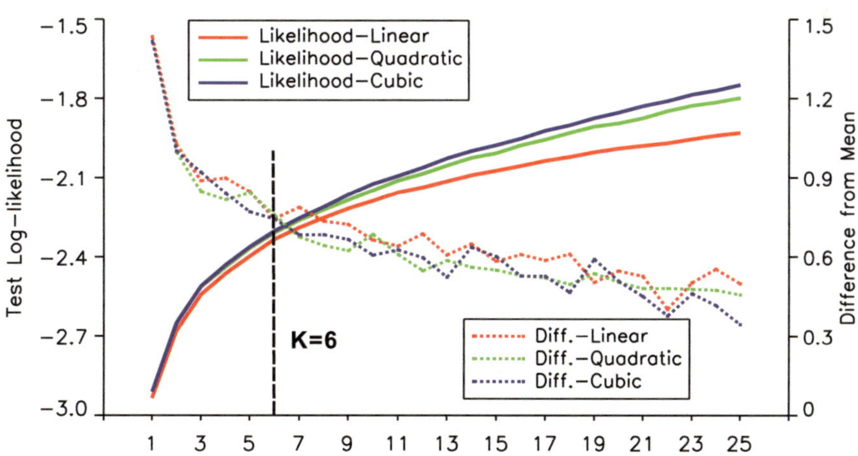

Fig. 6. Log-likelihood values and within-cluster error for different number of clusters on different regression order.

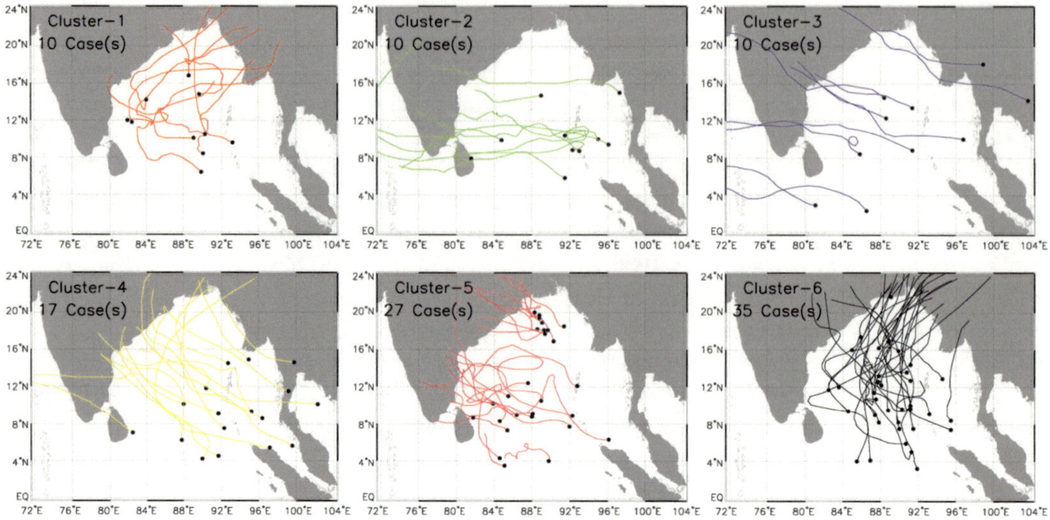

Fig. 7. Historical tracks from 1980 to 2009 of each of the six clusters.

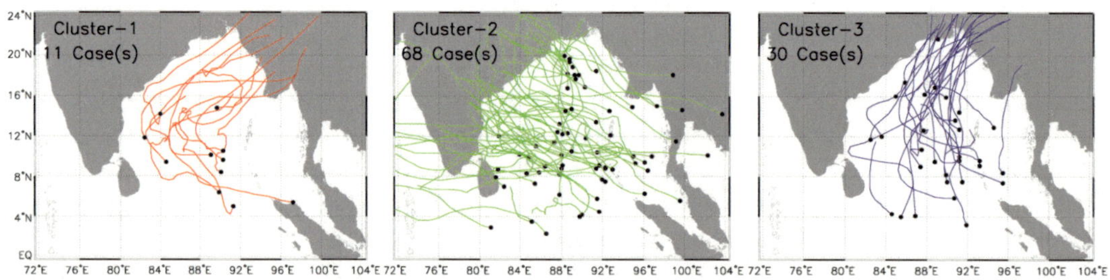

Fig. 8. Historical tracks from 1980 to 2009 of each of the three clusters after reclassifying.

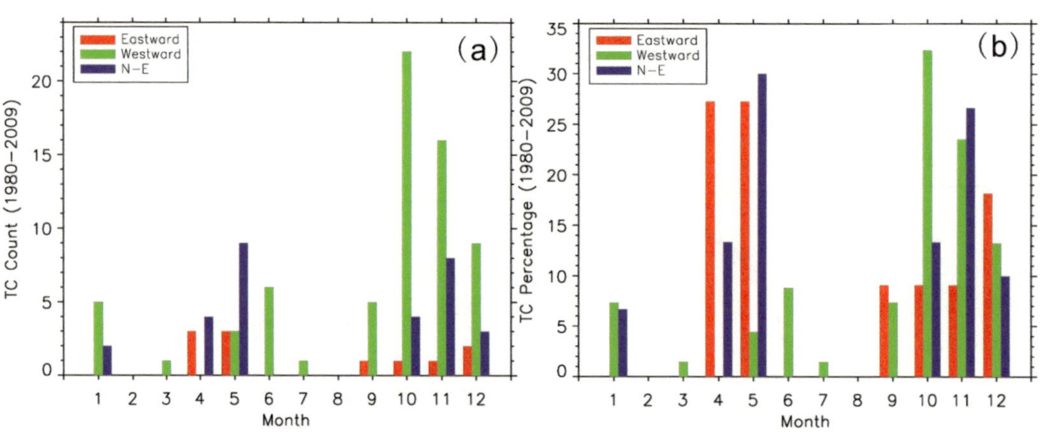

Fig. 9. Seasonal variability of (a) total TC count of year 1980-2009, (b) contribution of each cluster in each month to the total number of TCs for each cluster from 1980-2009

3.5 Environmental flow

Based on the seasonal variability discussed above, northward tracks are divided into two kinds, the one during pre-summer-monsoon (April and May) and the other during post-winter-monsoon (October to January). Figs. 10-13 show the daily vertical wind composition (600-100hPa) for TCs in each cluster, which composites on all days during the lifecycle of each TC. Demonstration of each track type in detail is given below:

Northeastward: Southerly appeared under 500hPa across large area, which seemed to be the low level cross-equatorial-flow. Above 500hPa, an anticyclonic circulation occupied Indo-China Peninsular. This system became stronger with height.

Westward: Easterly spread over wide region from northern to middle part of BOB and extend to the Arab Sea. A significant unclosed long and narrow anticyclonic circulation appeared in Indo-China Peninsular which also extended to the Arab Sea. This circulation pattern which also became stronger with height may be associated with the extension of West Pacific Subtropical High.

Northward (April and May): Remarkable southerly also appeared under 500hPa which was similar to the northeastward type. Nevertheless, this channel of wind was limited to the east of 90E (east BOB), mainly due to northward invasion of the cross-equatorial-flow and southward meander of the northerly in north BOB. These two channels of flow with different properties induced a trough. This trough should belong to the India-Burma Trough which is one of the most frequent-occurring and significant synoptic systems during the pre-monsoon season. Above 500hPa, the anticyclonic circulation still occupied Indo-China Peninsular which is weaker than the one of westward track type.

Northward (October-January): Anticyclonic circulation in Indo-China Peninsular also appeared which is similar to the westward track type but stronger. Difference is easterly could not extend to India let alone the Arab Sea. It is mainly owing to the trough activity in north India, which can constrain such anticyclone system west invasion. Compared with northward tracks in April and May, the one of this post-winter-monsoon season mostly took place in the western part of BOB.

3.6 Conclusion and discussion

By using the mixture quadratic regression model, six clusters of TC tracks in BOB are classified. To better analysis the background physical factors which can affect TC tracks of this region, all westward tracks are sorted as one type. Thereafter, three track types are obtained, northeastward, westward, and northward. After combining the analysis of seasonal variability, it is found that westward TCs which mainly occurred in the post-winter-monsoon period (September-January) were largest in amount. Northward TC were in the next place in amount. Whereas, this type of TCs should be divided by two stages (April-May and October-January), which is more suitable for examining the background controlling systems. Northeastward tracks were fewest in amount. However, this type of TCs generally took place in April and May.

The analysis of background circulation of each track type indicates that except for the northeastward one, anticyclonic circulation located in Indo-China Peninsular as well as the trough activity across the region of India-Burma played important roles on local wind pattern which assisted steering TCs. In addition, TCs happened in April and May (pre-summer-monsoon), were generally affected by the cross-equatorial-flow and prone to be with northward motion.

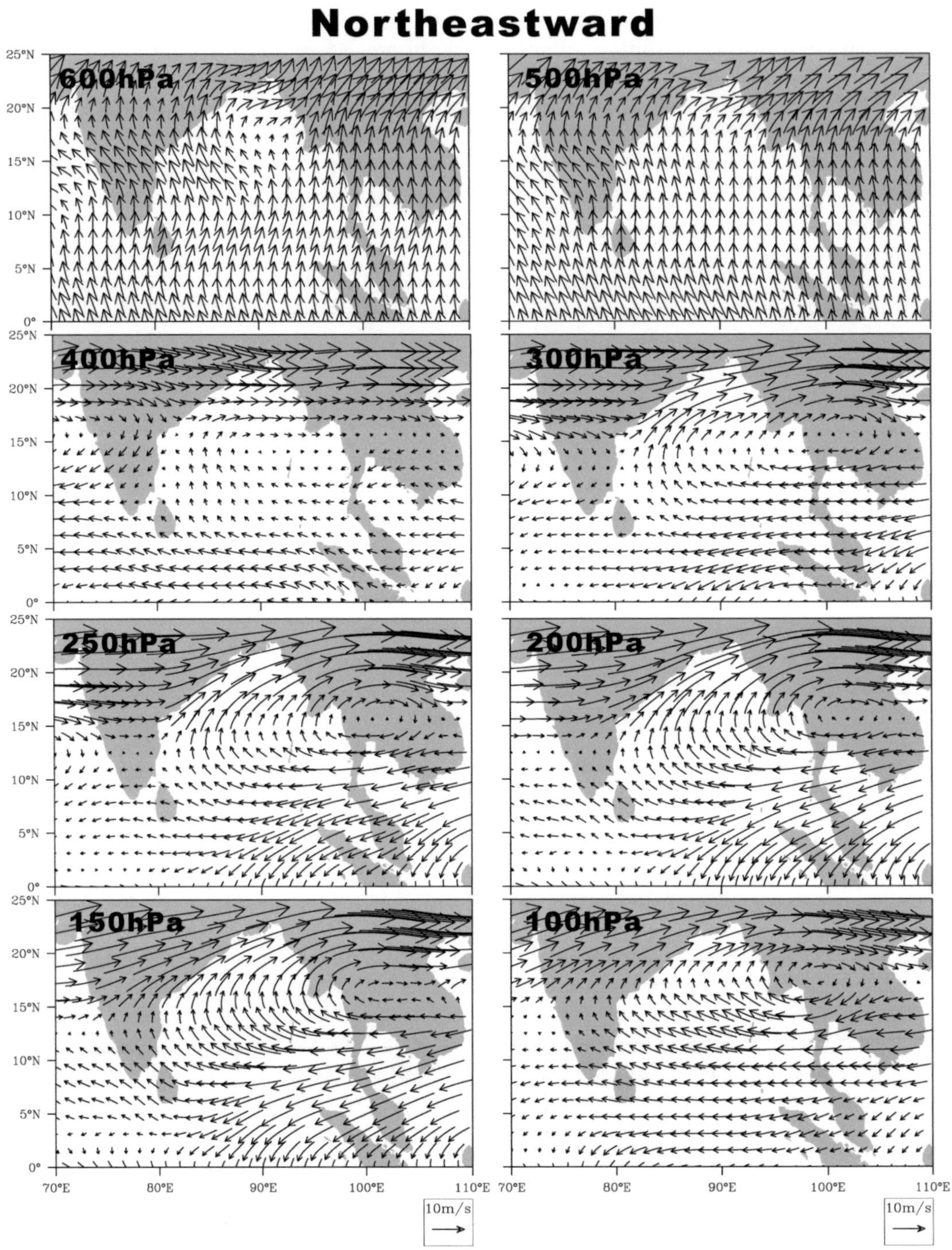

Fig. 10. Daily wind composition (600-100hPa) for TCs with northeastward tracks, which composites on all days during the lifecycle of each TC.

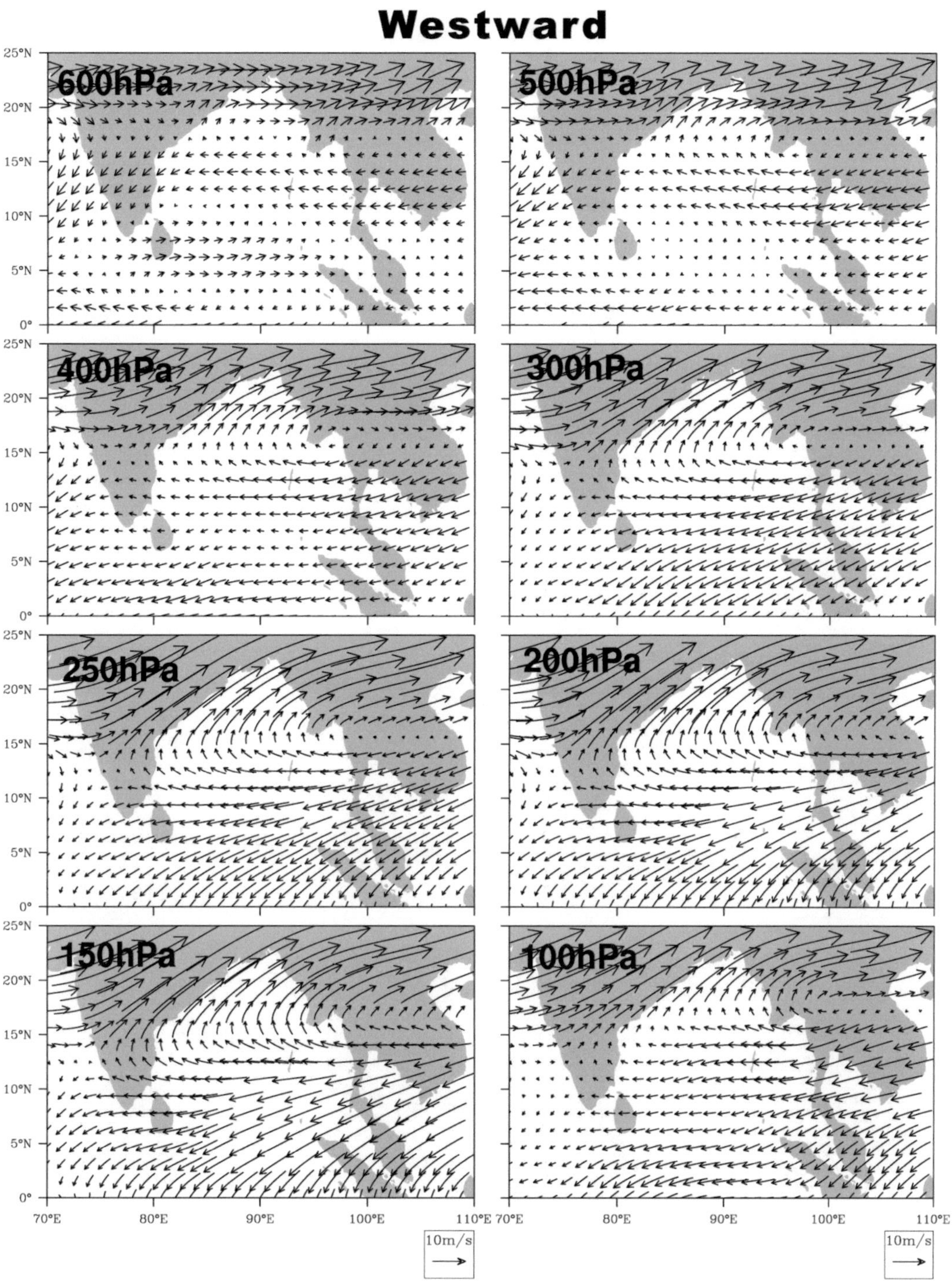

Fig. 11. Same as Fig. 10 but for westward track type.

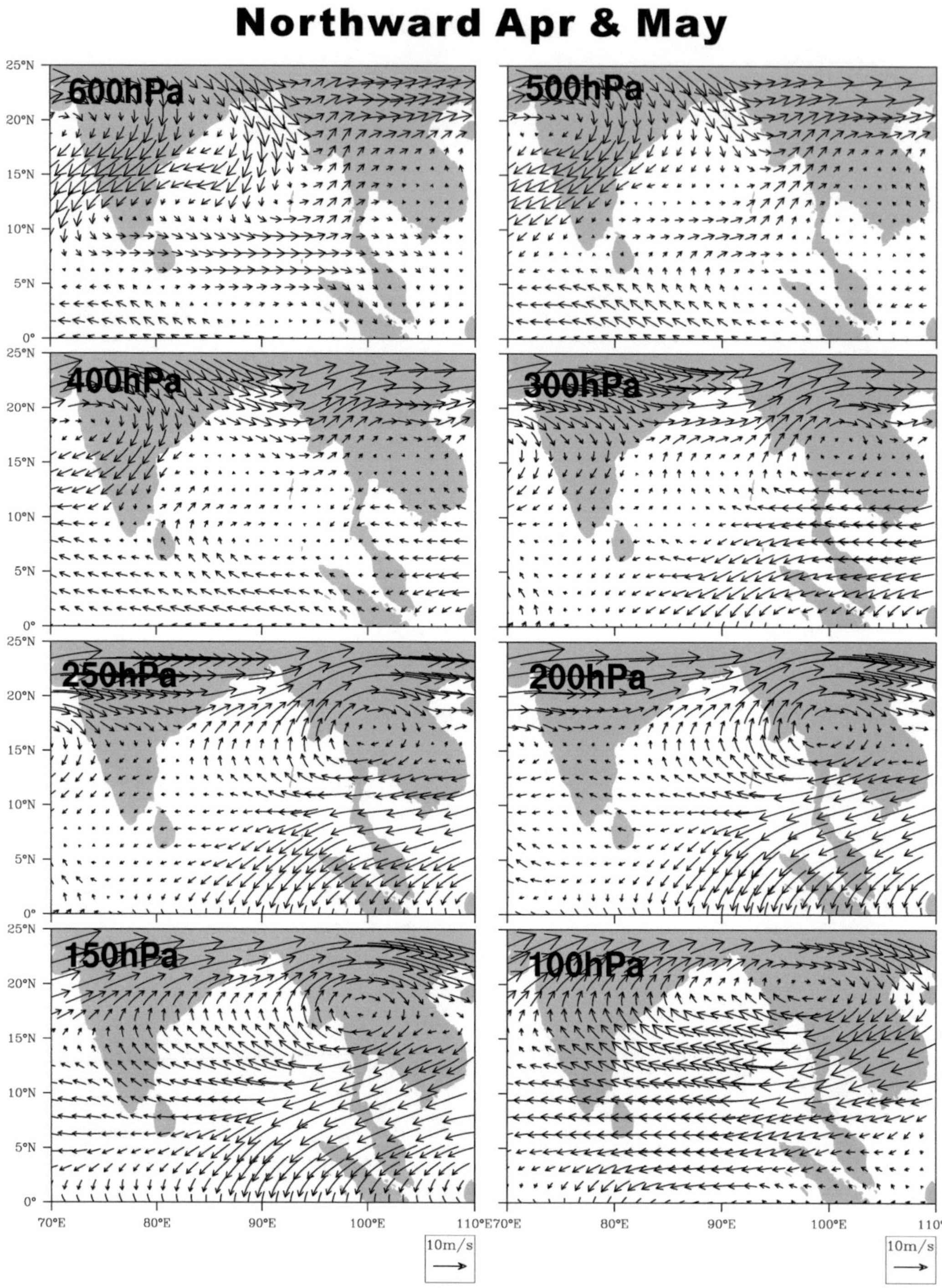

Fig. 12. Same as Fig. 10 but for northward track type in April and May.

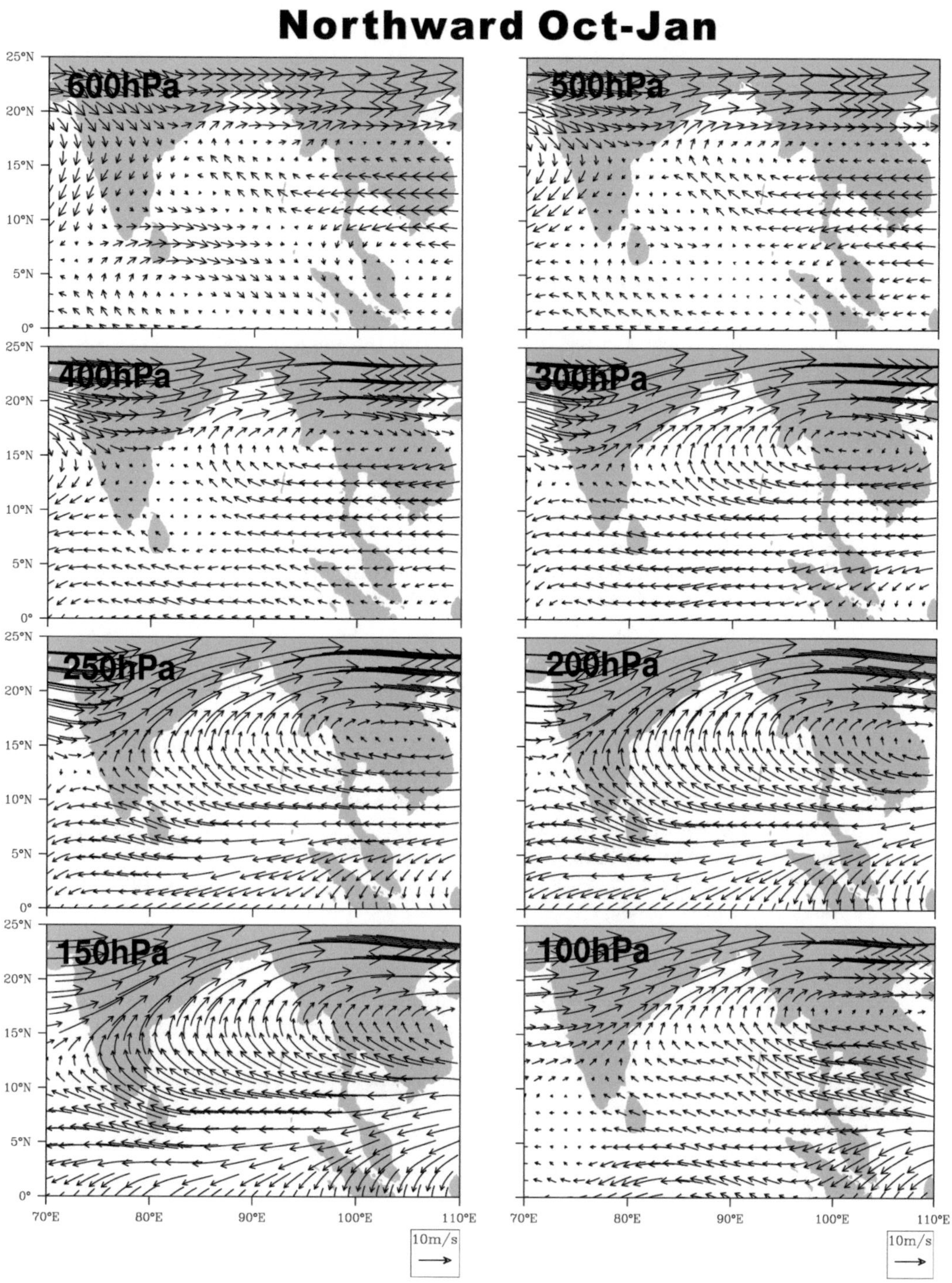

Fig. 13. Same as Fig. 10 but for northward track type from October to January.

The wind fields display the corresponding circulation flow and potential effective local weather systems. In this study, the West Pacific Subtropical High of winter-time may be the critical modulator of TC tracks in this region, specifically during post-winter-monsoon period. Moreover, such strong weather system is associated with Northeast Trade Wind and East Asia Winter Monsoon. How westward this system extends is partly result in whether TC in BOB moves westward or not. Thus, mechanism and causal relationship between the West Pacific Subtropical High as well as possible modulated systems will be investigated in the future.

4. Acknowledgments

This research was supported by National 973 projects (No. 2011CB403505), the Natural Science Foundation of China (Grants U0733002), Natural Science Foundation Guangdong, China (Team Research Project 8351030101000002), and the Postdoctoral Fund of China (No. 20090450911).

5. Reference

Camargo, S. J., A. W. Robertson, S. J. Gaffney, P. Smyth, and M. Ghil, 2007: Cluster analysis of typhoon tracks. Part I: General properties. *J. Climate*, 20, 3635–3653.

— —, A. W. Robertson, S. J. Gaffney, P. Smyth, and M. Ghil, 2007: Cluster analysis of typhoon tracks. Part II: Large-Scale Circulation and ENSO. *J. Climate*, 20, 3654–3676.

— —, A. W. Robertson, A.G. Barnston, and M. Ghil, 2008: Clustering of eastern North Pacific tropical cyclone tracks: ENSO and MJO effects, *Geochemistry, Geophysics and Geosystems*, 9, Q06v05.

Chan, J. C. L. and W.M. Gray, 1982: Tropical cyclone movement and surrounding flow relationships, *Mon. Wea. Rev.*, 110, 1354–1374.

Chen, L. S. and Y. H. Ding, 1979: An Introduction to the West Pacific Ocean Typhoons, *Science Press, Beijing, China*, pp. 400-490 (In Chinese).

Chou, S.-H., Nelkin, E., Ardizzone, J., Atlas, R. M. and Shie, C.-L., 2003: Surface turbulent heat and momentum fluxes over global oceans based on the Goddard satellite retrievals version2 (GSSTF2). *J. Clim.*, 16, 3256-3273.

Elsberry, R. L., W. M. Frank, G. J. Holland, J. D. Jarrell, and R. L. Southern, 1987: A global view of tropical cyclones, *Univ. Chicago Press*, 185 pp.

Gaffney, S. J., 2004: Probabilistic curve-aligned clustering and prediction with regression mixture models. *Ph.D. thesis, University of California, Irvine*, 281.

Goh, A. Z., and J. C. L. Chan, 2009: Interannual and interdecadal variations of tropical cyclone activity in the South China Sea, *Int.J. Climtaolo.*, doi:10.1002/joc.1943.

Grell, G. A., J Dudhia, and F. R. Stauffer, 1995: A description of the fifth-generation Penn State/NCAR Mesoscale Model (MM5). *NCAR Tech. Note NCAR/TN-398 +STR*, 122pp.

Gray, W. M., 1968: Global view of the origin of tropical disturbances and storms, *Mon. Wea. Rev.*, 96, 669-700.

Gupta, A., R. K. Bansal, 1997: Performance of a global spectral model in predicting the track of a hurricane in the Bay of Bengal using synoptic vortex, *NCMRWF, New Delhi Tech. report*.

Jiang, X.-P., Z. Zhong, and C.-X. Liu, 2006: The effect of typhoon-induced SST cooling on typhoon intensity: the case of typhoon Chanchu (2006). *Adv. Atmos. Sci.*, 25 (6), 1062-1072.

Joseph, K. J., A. N. Balchand, P.V. Hareeshkumar, and G. Rajesh, 2007: Inertial oscillation forced by the September 1997 cyclone in the Bay of Bengal, *Current Sci.*, 92, 790-794.

Kikuchi, K., B. Wang, and H. Fudeyasu, 2009: Genesis of tropical cyclone Nargis revealed by multiple satellite observations, *Geophys. Res. Lett.*, 36, L06811, doi:10.1029/2009GL037296, 2009.

Kossin, J. P., S. J. Camargo, M. Sitkowski, 2010: Climate Modulation of North Atlantic Hurricane Tracks. *J. Climate*, 23, 3057–3076.

Langland, R. H., C. Velden., P. M. Pauley and others, 2009: Impact of Satellite-Derived Rapid-Scan Wind Observations on Numerical Model Forecasts of Hurricane Katrina. *Mon Wea Rev.*, 137, 1615-1622.

Lau, K-H., Z. Zhang, H. Lam and others, 2003. Numerical simulation of a South China Sea typhoon Leo (1999), *Meteoro Atmos Phys.*, 83, 147-161.

Lee, C. S., Y. L. Lin, and K. K.W. Cheung, 2006: Tropical cyclone formations in the South China Sea associated with the Mei-Yu front, *Mon. Wea. Rev.*, 134, 2670-2687.

Li, J., A. Wang, E. Hou, G. Li, X. He, T. Peng, and Q. Zeng, 2004: A numerical prediction experiment of track and heavy rainfall about Typhoon Fitow. *J. Tropical Oceanog.*, 23 (1) (In chinese).

Li Q., Y. Duan, H. Yu and others, 2008, A High-Resolution Simulation of Typhoon Rananim (2004) with MM5, Part I: Model Verification, Inner-Core Shear, and Asymmetric Convection, *Mon Wea Rev*, 136(7), 2488-2506.

Liang, B. Q., 1991: Tropical atmospheric circulation system over the South China Sea, *China Meteorology Press, Beijing, China*, Pp. 100-224 (In Chinese).

Lin, I.- I., C.- H. Chen, I.- F. Pun, W. T. Liu, and C.-C. Wu, 2009: Warm ocean anomaly, air sea fluxes, and the rapid intensifi cation of tropical cyclone Nargis (2008), *Geophys. Res. Lett.*, 36, L03817, doi:10.1029/2008GL035815, 2009

Liu, X. G. Jiang, and H. Zhuo, 2009: Numerical experiment for the impact of SST to Typhoon Chanchu. *Marine Forecasts*, 26(3), 1-10. (In Chinese)

Luo, B., C. Zhang, M. Huang, Y. Li, and L. Li, 2008: The analysis of 0601 strong typhoon 'Chanchu, *Marine Forecasts*, 25(1), 95-101. (In Chinese)

Mandal, M., U.C. Mohanty, P. Sinha, and M.M.Ali, 2007: Impact of sea surface temperature in modulating movement and intensity of tropical cyclones, *Nat. Hazards*, 41, 413-427.

– –, U.C. Mohanty, K.V.J. Potty, and A. Sarkar, 2003: Impact of horizontal resolution on prediction of tropical cyclones over Bay of Bengal using a regional weather prediction model, *Proc. Indian Acad. Sci. (Earth Planet. Sci.)*, 112, 79-93.

– –, U.C. Mohanty, S.Raman, 2002: A Study on the Impact of Parameterization of Physical Processes on Prediction of Tropical Cyclones over the Bay of Bengal with NCAR/PSU Mesoscale Model, *Nat. Hazards*, 31, 391-414.

McPhaden, M. J., G. R. Foltz, T. Lee, V. S. N. Murty, M. Ravichandran, G. A. Vecchi, J. Vialard, J. D. Wiggert, and L. Yu, 2009: Ocean-Atmosphere Interactions During Cyclone Nargis, *EOS*, 90(7), 53-60.

Mohanty, U.C., 1994: Tropical cyclones in the Bay of Bengal and deterministic methods for prediction of their trajectories, *Sadhana*, 19, 567-582.

– – and A. Gupta, 1997: Deterministic methods for prediction of tropical cyclone tracks, *Mausam*, 48, 257-272.

Murty, V.S.N, M.S.S. Sarma, and V. Tilvi, 2000: Seasonal cyclogenesis and the role of near-surface stratified layer in the Bay of Bengal, *PORSEC Proceedings*, 1, 453-457.

Nolan, D. S., J. Zhang, and D. P. Stern, 2003: Evaluation of Planetary Boundary Layer Parameterizations in Tropical Cyclones by Comparison of In Situ Observations and High-Resolution Simulations of Hurricane Isabel (2003), Part I: Initialization, Maximum Winds, and the Outer-Core Boundary Layer, *Mon Wea Rev.*, 137 (11): 3651-3674.

Orlanski, I., 1998: Poleward deflection of storm tracks, *J. Atmos. Sci.*, 55, 2577–2602.

Prasad, K. and Y. V. Rama Rao, 2003: Cyclone Track prediction by a quasi-Lagrangian model, *Meteor. Atmos. Phys.*, 83,173-185.

Rakesh, V., R. Singh, P. K. Pal and others, 2009: Impacts of Satellite-Observed Winds and Total Precipitable Water on WRF Short-Range Forecasts over the Indian Region during the 2006 Summer Monsoon , *Wea. Forecast*, 24, 1706-1731.

Sarma, Y.V.B, V.S.B. Murty, and D.P. Rao, 1990: Distribution of cyclone heat potential in the Bay of Bengal, *Indian J. of Marian Sci.*, 19, 102-106.

Sadhuram, Y., B.P. Rao, D.P. Rao, P.N.M. Shastri, and M.V. Subrahmanyam, 2004, *Nature Hazards*, 32, 191-209.

— —, 2004, *Current Sci.*, 86, 383-384.

Singh, O.P., Tariq Masood Ali Khan, and Md. Sazedur Rahman, 2000: Has the frequency of intense tropical cyclones increased in the north Indian Ocean?, *Current Sci.*, 80, 575-580.

Soden, B. J., C. S. Velden, and R. E. Tuleya, 2001: The Impact of Satellite Winds on Experimental GFDL Hurricane Model Forecasts, *Mon Wea Rev.*,129, 835-852.

Wang, G., J. Su, Y. Ding, and D. Chen, 2007, Tropical cyclone genesis over the South China Sea, *J. Marine Sys.*, 68, 318-326.

Wang, L. K-H. Lau, and C-H. Fung, 2010: Numerical simulation of the genesis of Typhoon Durian (2001) over the South China Sea: the effect of sea surface temperature, *J. Ocean Univ. China*, 99-115.

Wang, L., K-H. Lau, C-H. Fung, and J-P. Gan, 2007: The relative vorticity of ocean surface winds from the QuikSCAT satellite and its effects on the geneses of tropical cyclones in the South China Sea, *Tellus*, 59A, 562-569.

Weber, H. C., 2001: Hurricane Track Prediction with a New Barotropic Model, *Mon. Wea. Rev.*, 129, 1834-1858.

Yanase, W., H. Taniguchi, and M. Satoh: 2010: The genesis of tropical cyclone Nargis (2008): environmental modulation and numerical predictability, *J. Meteor. Soc. Japan*, 88, 497-519.

Yamada, H., Q. Moteki, and M. Yoshizaki, 2010: The unusual track and rapid intensification of cyclone Nargis in 2008 under a characteristic environmental flow over the Bay of Bengal, J. *Meteor. Soc. Japan*, 88, 437-453.

Yokoi, S., and Y. N. Takayabu, 2010: Environmental and external factors in the genesis of tropical cyclone Nargis in April 2008 over the Bay of Bengal, *J. Meteor. Soc. Japan*, 88, 425-435.

Zeng, L.-L., P. Shi, W. T. Liu, and D. Wang, 2008: Evaluation of a satellite-derived latent heat flux product in the South China Sea: A comparison with moored buoy data and various products. *Atmos. Res.*, 94 (1), doi:10.16/j.atmosres.2008.12.007.

Zhao, Y., B. Wang, Z.J., X. Liang, G. Deng, and X. Zhang, 2005: Improved track forecasting of a typhoon reaching landfall from four-dimensional variational data assimilation of AMSU-A retrieved data, *J. Geophys. Res.*, 110, D14101, doi:10.1029/2004JD005267.

Zou, X., and Q. Xiao, 2000: Studies on the initialization and simulation of a mature hurricane using a variational bogus data assimilation scheme. *J. Atmos. Sci.*, 57, 836-860.

12

Air-sea Interaction under Hurricane Wind Conditions

Yuliya Troitskaya, Daniil Sergeev,
Alexander Kandaurov and Vasilii Kazakov
Institute of Applied Physics
Russia

1. Introduction

One of the main characteristics appearing in the models of forecasting wind over the sea is the roughness of the sea surface determined by the parameters of the wind waves, quantitatively parameterised by the sea surface drag coefficient C_D. To define it we introduce the turbulent shear stress or turbulent momentum flux far from the sea surface

$$\tau_{turb}(z) = \rho_a u_*^2 , \tag{1}$$

where ρ_a is the air density, u_* is the wind friction velocity. Wind is the turbulent boundary layer with the logarithmic mean velocity profile:

$$U(z) = \frac{u_*}{\kappa} \ln \frac{z}{z_0} , \tag{2}$$

Similar to the resistance law of the wall turbulent flow the sea surface drag coefficient is introduced as follows:

$$C_D = \frac{\tau_{turb}}{\rho_a U_{10}^2} = \frac{u_*^2}{U_{10}^2} , \tag{3}$$

where U_{10} -the wind velocity at a standard meteorological height H_{10}=10 m. which relate this coefficient to U_{10} are obtained either by generalizing empirical data (Garratt, 1977; Large & Pond, 1981, Taylor & Yelland, 2002; Fairall et al., 2003) or by numerical models (see, for example, Janssen, 1989; Janssen, 1991, Makin et.al, 1994; Hara & Belcher, 2004). Numerous field measurements give increasing dependencies of C_D on wind speed, which relates to increasing of wave heights with the wind.

The aerodynamic drag coefficient of the sea surface is a critical parameter in the theory of tropical hurricanes (Emanuel, 1995). To illustrate it we consider here the ideas of theory of energy balance in a tropical cyclone suggested by (Emanuel, 1986; Emanuel, 1995, Emanuel, 2003). According to this theory the mature tropical cyclone may be idealized as a steady, axisymmetric flow whose energy cycle is very similar to that of an ideal Carnot engine,

where the hot reservoir is the ocean with the temperature T_S and the cold reservoir is .the troposphere with the temperature T_0. The details of construction and operation of this heat engine are presented in (Emanuel, 1986; Emanuel, 1995, Emanuel, 2003), but one of the most important characteristics of a tropical cyclone, the maximum surface wind velocity, which determines its category, can be estimated without details from the Carnot theorem.
According to the Carnot theorem, the maximum efficiency of the ideal heat engine is determined by the absolute temperatures of the hot and cold reservoirs:

$$\eta = \frac{W}{Q_s} = \frac{T_s - T_0}{T_0}, \tag{4}$$

where Q_s is the heat energy entering the system from the hot reservoir and W is the mechanical work done by the system. Heat energy support of the tropical cyclone comes from the ocean (heat flux from the sea surface) and mechanical energy dissipated in the marine turbulent boundary layer (Emanuel, 2003), the heat energy entering the system is the surface integral of the heat flux from the sea F_q and mechanical energy dissipation rate F_p:

$$Q_s = \int \left(F_q + F_p \right) ds \tag{5}$$

Mechanical work done by the system compensates mechanical energy dissipation, then

$$W = \int F_p ds \tag{6}$$

The heat flux from the sea and the mechanical energy dissipation rate are determined by the bulk formula:

$$F_q = C_k \rho \left| \vec{V} \right| \left(k_0 - k \right) \tag{7}$$

here k_0, k are enthalpy at the sea level and in marine atmospheric boundary layer.

$$F_p = C_D \rho \left| \vec{V} \right|^3 \tag{8}$$

In (7)-(8) C_k is heat exchange coefficient (or the Stanton number), C_D is surface drag coefficient, defined by equation (3).
Taking into account the Carnot theorem (4) and estimating integrals (5) and (6) yields estimate for the maximum surface wind velocity in a tropical cyclone as a function of ratio C_D/C_k.:

$$\left| \vec{V} \right|_{\max} = \sqrt{\frac{C_k}{C_D} \frac{T_s - T_0}{T_0} \left(k - k_0 \right)} \tag{9}$$

Conventional bulk formulas, derived by generalizing experimental data (Garratt, 1977; Large & Pond, 1981, Taylor & Yelland, 2002; Fairall et al., 2003) obtained at wind velocities less than 30 m/s, overestimate the drag coefficient of the sea surface under hurricane winds. The estimates presented in (Emanuel, 1995) indicate that energy dissipation due to friction proves too high to explain the observed velocity of a hurricane wind for realistic sources of energy.

The problem of explaining high wind velocities during hurricanes can be resolved if the drag coefficient of the sea surface does not increase with increasing wind velocity. To explain high quantities of wind speeds observed in tropical cyclones Emanuel, 1995 suggested that the drag coefficient flattens and even decreases at high wind speed in contradiction with intuition, since, it follows then, that sea surface should be effectively smoothed under the hurricane conditions. However, in the late 90-th these dependencies were observed experimentally in the field and laboratory conditions (Powell et al, 2003).

2. Observations of the sea surface drag reduction

The effect of sea surface drag reduction under hurricane wind was discovered by (Powell et al, 2003) in their experiments on measurements of the wind velocity profiling in the marine atmospheric boundary layer associated with tropical cyclones by 331 Global Positioning System sondes dropped in 15 storms. The wind friction velocity u* can be easily retrieved from equation (2) and sea surface drag can be calculated from its definition (3). Analysis of these measurements in (Powell et al, 2003) showed that that the drag coefficient of the sea surface is much less than the extrapolation of data measured at "usual winds" and even decreases if the wind velocity exceeds 30–35 m/s (see fig.1a). More precisely, according to (Powell, 2007) surface drag depends significantly on the sector of the tropical cyclone, where it is measured.

The similar dependencies of the surface drag coefficient on the wind speed were retrieved from the measurements of the ocean currents driven by the tropical cyclone Andrew (Jarosz et al., 2007). As it was reported in (Jarosz et al., 2007) on 15 September 2004, the centre of Hurricane Ivan passed directly over current and wave/tide gauge moorings on the outer continental shelf in the north-eastern Gulf of Mexico. Analysis of the along-shelf momentum balance in the water column, when the current structure was frictionally dominated was made within the following equation:

$$\frac{\partial U}{\partial t} - fV = \frac{\tau_{sx}}{\rho H} - \frac{rU}{H} \tag{10}$$

(where U and V are depth-integrated along- and cross-shelf current velocity components, f is the Coriolis parameter, ρ is the water density, τ_{sx} is the along-shelf wind stress component, H is the water depth, and r is the constant resistance coefficient at the sea floor). Equation (10) enables one to retrieve the wind stress τ_{sx} and estimate the sea surface grad coefficient by of the equation (3) using independently measured wind velocity. The results produced from evaluation of this procedure presented in (Jarosz et al., 2007) show a decreasing trend of C_D for wind speeds greater than 32 m s^{-1} (see fig.1b).

So field measurements of the wind stress both from the atmospheric and ocean sides of the air-sea interface show that the sea surface drag coefficient is significantly reduced at hurricane wind speeds in comparison with the extrapolation of the experimental data obtained at "normal" wind speeds and even decreases for U_{10} exceeding 35 m/s.

The similar effect was observed in laboratory experiments performed at the Air-Sea Interaction Facility at the University of Miami (Donelan et al., 2004). In that experiment the

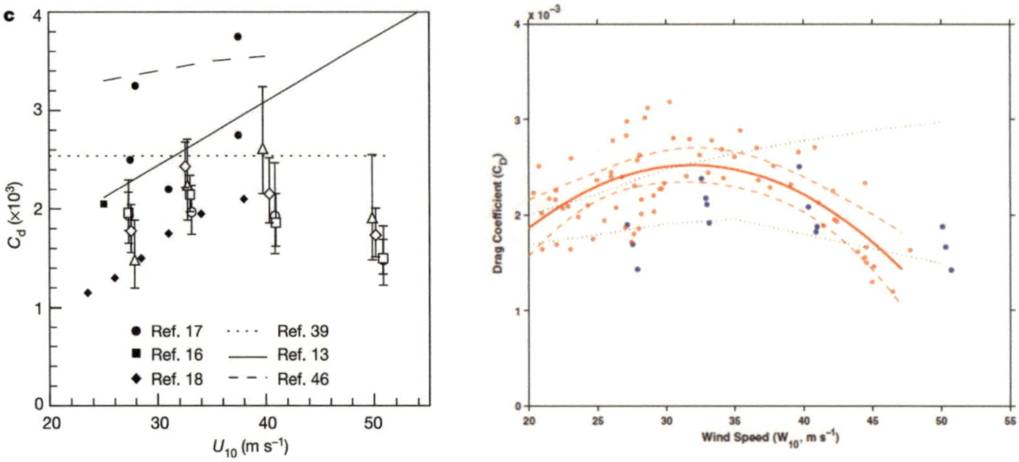

(a) (b)

Fig. 1. Sea surface drag coefficient via 10-m wind speed: (a) – from Powel et.al, 2003, (b) – from science, 2007.

aerodynamic resistance of the water surface was measured by three different methods: using the profile method (in which the vertical gradient of mean horizontal velocity is related to the surface stress), the Reynolds stress method, and the momentum budget method based on analysis of a momentum budget of water column sections of the tank (Donelan et al., 2004). In comparison with two others, the latter method is insensitive to droplets suspended in the airflow at high wind speeds. The wind speed was measured at 30 cm height in the tank and extrapolated to the standard meteorological height of 10 m using the well-established logarithmic dependence on height (Donelan et al., 2004). All methods were in excellent agreement, and the momentum budget method enabled Donelan et al 2004 to measure the wind stress and aerodynamic resistance coefficient of the water surface up to equivalent 10-m wind speeds about 60 m/s. Figure 2 from (Donelan et al., 2004)

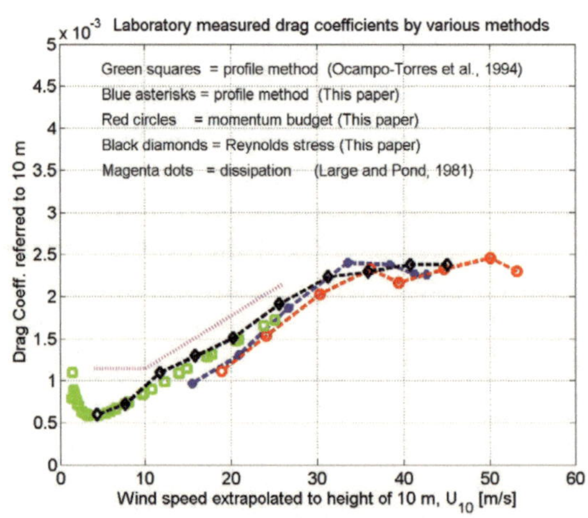

Fig. 2. Laboratory measurements of the neutral stability drag coefficient (reproduced from Donelan etal, 2004)

demonstrates a remarkable levelling of the drag coefficient for the 10-m wind speed exceeding 33 m/s. The difference between C_D dependencies on the wind speed in field and laboratory experiments is discussed in (Donelan et al., 2004). Possibly it is due to strong inhomogenity and non-stationarity of the wind in the hurricane eye walls, where the constant stress concept derives from the boundary layer Reynolds equations is not confirmed.

So it can be concluded both from field and laboratory data, that the growth of the aerodynamic roughness of the water surface with wind speed is significantly reduced at extremely high winds in spite of increasing of surface wave heights. Several theoretical models were suggested for explanation of this empirical fact.

3. Possible mechanisms of the sea surface drag reduction at extreme wind speeds

Among a number of possible theoretical mechanisms suggested for explanation of the effect of the sea surface drag reduction at hurricane winds two groups of the models can be specified. First, (Kudryavtsev & Makin, 2007) and (Kukulka et al., 2007) explain the sea surface drag reduction by peculiarities of the airflow over breaking waves, which determine the form drag of the sea surface. For example, in (Donelan et al., 2004), the stabilization of the drag coefficient during hurricane winds is qualitatively explained by a change in the shape of the surface elevation in dominant waves at wind velocities above 35 m/s, which is accompanied by the occurrence of a steep leading front. In this case, occurrence of flow separation from the crests of the waves is assumed. This assumption is based on the laboratory experiments by (Reul et al., 1999), where airflow separation was observed at the crests of breaking waves by the PIV method. According to hypothesis by Donelan et al, 2004, existence of the airflow trapped in the separation zone skips the portions of the water surface in the troughs of the waves and thus, in conditions of continuous breaking of the largest waves the aerodynamic roughness of the surface is limited. Besides, generation of small-scale roughness within the separation zone is reduced due to sheltering, which can also reduce the surface resistance. This effect is expected to be dominant for the case of young sea (or in laboratory conditions as in (Donelan et al., 2004), when wave breaking events are not rare even for energy containing part of the surface wave spectrum.

Another approach more appropriate for the conditions of developed sea exploits the effect of sea drops and sprays on the wind-wave momentum exchange (Andreas & Emanuel, 2001, Andreas, 2004, Makin, 2005, Kudryavtsev, 2006). (Andreas & Emanuel, 2001) and (Andreas, 2004) estimated the momentum exchange of sea drops and air-flow, while (Makin, 2005) and (Kudryavtsev, 2006) focused on the effect of the sea drops on stratification of the air-sea boundary layer similar to the model of turbulent boundary layer with the suspended particles by (Barenblatt & Golitsyn 1974). Suspended heavy particles (drops) in the marine turbulent boundary layer create stable stratification suppressing the turbulence, and then decreasing the effective viscosity of the turbulent flow and the aerodynamic resistance. In the same time, there is another effect of sea drops, the particles injected from the water surface should be accelerated, and then they consume some momentum flux from the airflow, increasing the surface drag in the turbulent boundary layer.

In the paper by (Troitskaya & Rybushkina, 2008) the sea surface drag reduction at hurricane wind speed is explained by reducing efficiency of wind-wave momentum exchange at hurricane conditions due to sheltering, but sheltering without separation. This assumption is motivated by reports of eye-witnesses of strong ocean storms, who confirmed that the sea at hurricane wind is unexpectedly smooth and wave breaking is a relatively rare event (see references in (Andreas, 2004). Relatively smooth water surface presents at the video-films taken on board of the research vessel ``Viktor Buinitsky", when it passes a polar lo in the Laptev sea and the Kara sea in October 2007 (cruise within the project NABOS - Dr.Irina Repina private communication). These visual observations are also confirmed by the instrumental measurements by (Donnely et al., 1999),, who observed saturation of the C-band and Ku-band normalized radar cross-section (NRCS) for wind speed above 25-30 m/s. Similar reduction of NRCS was observed in the laboratory tank experiments by (Donelan et al., 2004). Microwave power scattered from the water surface is formed by i) the Bragg scattering at short waves and ii) by reflection from wave breakers, i.e. the NRCS reduction supports evidence of smoothing of the sea surface by reducing both short wave roughness and wave breaking events. Mechanisms of unusual smoothness of the sea surface are unknown. (Andreas, 2004) suggested two possible explanations of this effect. One supposes the effect of bubbles on surface tension. Another possible explanation exploits the effect of spume drops torn from the wave crests by wind and then falling back as a kind of strong rain, which causes effective damping of surface waves according to a number of experiments (see ex. (Tsimplis, & Thorpe, 1975).

In spite of a number of theoretical hypotheses the problem of explanation of the effect of surface drag reduction at hurricane winds is not solved mostly due to the lack of experimental data.

4. Laboratory modelling of the air-sea interaction under hurricane wind

In this section we describe the results of new laboratory experiments devoted to modelling of air-sea interaction at extremely strong winds.

4.1 Experimental setup and instruments

The experiments were performed in the wind-wave flume built in the Large Thermostratified Tank of the Institute of Applied Physics. The centrifugal fan equipped with an electronic frequency converter to control the discharge rate of airflow produces the airflow in the flume with the straight part 10 m. The operating cross section of the airflow is 40*40 cm^2, whereas the sidewalls are submerged at a depth of 30 cm. Wave damping beach made of a fine mesh is placed at the airflow outlet at the end of the flume.

Aerodynamic resistance of the water surface was measured by the profile method at a distance of 7 m from the outlet. Wind velocity profiles were measured by the L-shaped Pitot tube intended for measuring flow velocities of up to 20 m/s (the axis velocity in the flume 25 m/s approximately corresponds to U_{10}=50-60 m/s). Simultaneously with the airflow velocity measurements, the 3-channel string wave gauge measured waves at the water surface. The experiment was accompanied by video shooting of the top view of the water surface.

4.2 Peculiarities of the profile method for measuring surface drag coefficient in aerodynamic tunnels

The classical profiling method of measuring surface drag coefficient is based on the property of steady wall turbulent boundary layer to conserve tangential turbulent stress u_*^2, then the average flow velocity is logarithmic and the wind friction velocity u_* can be easily determined from (2), if the velocity profile is measured. However developing turbulent boundary layers are typical for the aerodynamic tubes and wind flumes, then three sub-layers at different distances from the water can be specified: viscous sub-layer, layer of constant fluxes and "wake" part (see fig.3a). The viscous sub-layer, where viscous effects are essential, exists over the hydrodynamically smooth surfaces at the distances less than 20÷30 ν/u_* (ν is the kinematic viscosity), for moderate winds it is about 1 millimetre. The "wake" part is the outer layer of the turbulent boundary layer, where the boundary layer flow transits to the outer flow in the tube. Its thickness δ increases linearly from the outlet of the flume. The layer of constant fluxes is extended from the upper boundary of the viscous sub-layer to approximately 0.15δ. Only in the layer of constant fluxes the flow velocity profile is logarithmic and can be extrapolated to the standard meteorological height H_{10}. Typically in wind flumes the constant layer thickness is less than 10 cm. Measuring of wind velocity profiles at the distance less than 10 cm from the wavy water surface at strong winds is a difficult problem mainly due to the effect of sprays blown from the wave crests. Fortunately, parameters of the layer of the constant fluxes can be retrieved from the measurements in the "wake" part of the turbulent boundary layer, because the velocity profile in the developing turbulent boundary layer is the self-similar "law of wake" (see Hinze, 1959). The self-similar variables for the velocity profile and vertical coordinates are z/δ and $(U_{max}-U(z))/u_*$., where U_{max} is the maximum velocity in the turbulent boundary layer. The self-similar velocity profile can be approximated by the following simple equations (see Hinze, 1959):

- in the layer of constant fluxes

$$U_{\max} - U(z) = u_*\left(-2.5\ln(z/\delta) + \alpha\right), \tag{11}$$

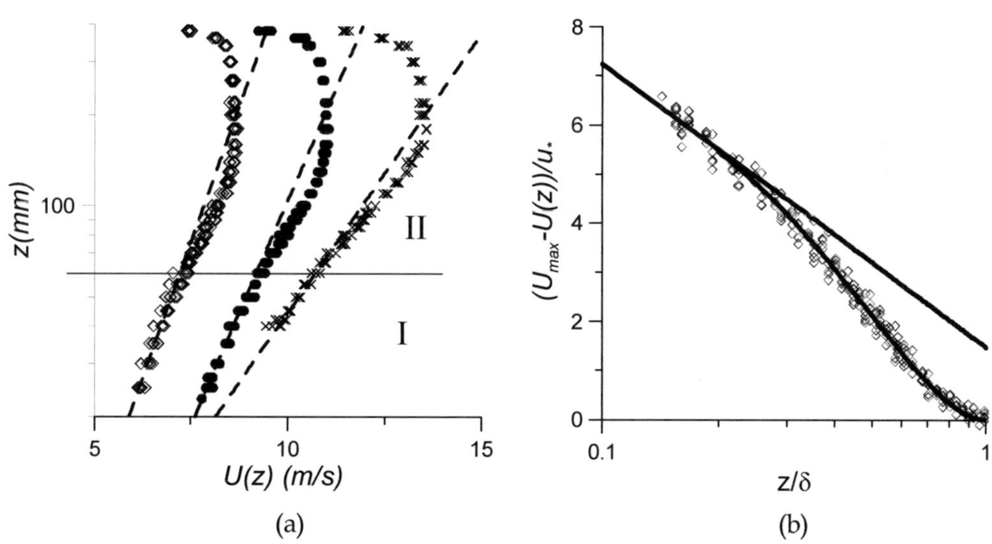

Fig. 3. Airflow velocity profiles in the aerodynamic flume over the waves for different airflow velocities (a); dashed curves are logarithmic approximations in the layer of constant fluxes. I – the layer of constant fluxes, II – the "wake" part. Air-flow velocity profiles measured at different wind speeds over waves in self-similar variables (b).

in the "wake" part

$$U_{\max} - U(z) = \beta u_* (1 - z/\delta)^2 . \quad (12)$$

Collapse of all the experimental points in one curve in self-similar variables occurred in our experiments (see fig.3b). The parameters in equations (11) (12) were obtained by the best fitting of the experimental data: α=1.5, β=8.5.
The parameters of the logarithmic boundary layer can be retrieved from the measurements in the wake part of the turbulent boundary layer, first, retrieving parameters of turbulent boundary layer (U_{max} and δ) by fitting experimental data by equation (12) and then calculating parameters of the logarithmic boundary layer by the following expressions:

$$U(z) = 2.5u_* \ln(z/z_0) , \quad (13)$$

where

$$z_0 = \delta \exp(-\kappa U_{\max} / u_* + \alpha\kappa) . \quad (14)$$

Expression for C_D via measured parameters u_*, U_{max} and δ follows from equations (13-14):

$$C_D = \frac{\kappa^2}{\left(\kappa U_{\max} / u_* - \alpha\kappa + \ln(H_{10} / \delta)\right)} . \quad (15)$$

Wind velocity profiles were measured for 12 values of the centerline velocity from 6 m/s to 24 m/s with the resolution 0.3-0.5 cm. Each point at the velocity profile was determined by averaging over 30 sec. C_D and U_{10} were calculated by equations (15) and (11) respectively. The obtained dependency of the surface drag coefficient on 10-m wind speed is presented in fig. 4a together with the data taken from the paper by Donelan et al, 2004. The data obtained at two different facilities are rather close to each other both at the low and high wind speeds; the difference in C_D is less than 10%. The tendency to saturation of the surface drag coefficient is clearly visible for both data sets, although the thresholds of wind speeds for the saturation are slightly different (33 m/s for the data by Donelan et al, 2004 and 24.5 m/s for our data set). Possibly it is due to differences in the details of data processing.
The values of C_D obtained in laboratory by (Donelan et al., 2004) slightly exceed the data obtained in field conditions (see fig.4b). Besides, decreasing of C_D for 10-m wind speed exceeding 35 m/s reported in (Powel et al., 2003) was not observed by (Donelan et al., 2004). Our laboratory data set is in better agreement with the field data, although, extremely large wind speeds, when C_D is decreasing, were not achieved in our facility. Possibly, the differences between field and lab data are due to two main reasons. First, although similarity between lab and sea conditions[1] can be expected at strong winds, because in both cases the wave phase velocities is much less than the wind speed and then the peculiarities of the air flow over the waves are similar, the fetches in the laboratory facilities are much lower than in the field conditions. Then the waves in the lab are shorter and steeper than in the sea and enhanced aerodynamic resistance of the water surface can be expected. Second

---

[1] The question about similarity between labortory modeling of air-sea interaction and field conditions is not obvious. We will briefly discuss it in section 5.5.

reason was suggested by (Donelan et al., 2004). In laboratory facility we investigate wind-wave interaction in stationary conditions of spatially developing turbulent boundary layer. In the field conditions, the wind in hurricane eye walls is strongly unsteady and inhomogeneous flow.

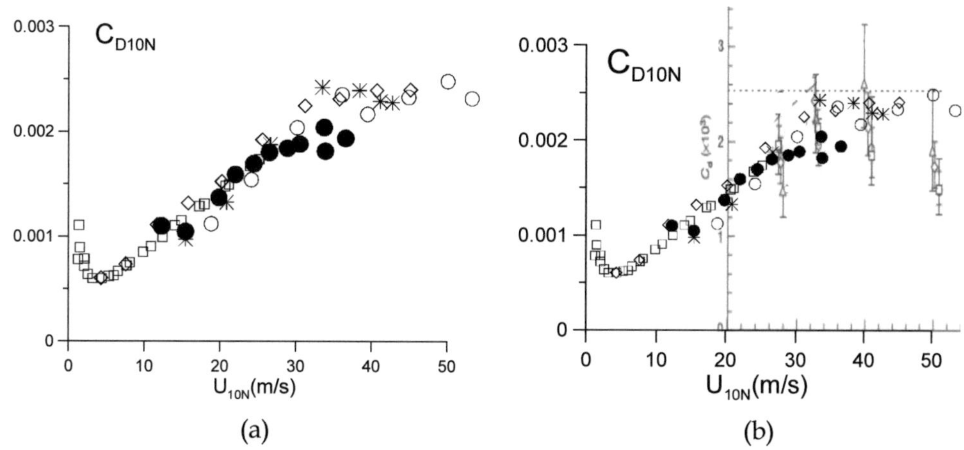

Fig. 4. Surface drag coefficient. (a) – laboratory data, black open symbols (squares, circles, diamonds, asterics) are taken from Donelan et al, 2004, closed circles – measurements at TSWiWaT, (b) – compilation of the field and laboratory data.

4.3 Wave field at strong winds in laboratory conditions

Aerodynamic roughness of the sea surface is conditioned to waves at the water surface including strong wind conditions. According to (Powel, 2007) surface drag depends significantly on the part of the tropical cyclone, where it is measured. The sea surface drag is strongly enhanced in the left front part of the tropical cyclone in comparison with right and rare parts. The data are not sufficient for final conclusions, but it seems, that the aerodynamic drag depends on the wave field, which is significantly different in different sectors of the tropical cyclone.

The wind wave field parameters in the flume was investigated by three wire gauges positioned in corners of an equal-side triangle with 2 cm side, data sampling rate was 100 Hz Three dimensional frequency-wave-number spectra were retrieved from this data by the wavelet directional method (M.Donelan et al., 1996). The wave fields at different wind speeds are characterized by narrow wave-number spectra (fig.5a) with the peak wave-number decreasing with the wind speed. It is clearly visible from fig.5a that the shapes of the spectra tend to saturation from the 10-m wind speed U_{10} exceeding 24.5 m/s. The similar tendency occurs in the dependence of the integral parameters of the wave field on the wind speed. For example, fig.5b clearly shows, that the average slope of the peak wave $S=H_s k_p/4$ (where H_s is the significant wave height, k_p is peak wave number) saturates when U_{10}>25m/s (see fig.5b). It means that at the wind speed about 24.5 m/s changing of the regime of the wave field occurs. Comparing the dependencies of the dominant wave slope on the wind speed with the drag coefficient dependency also shown in fig. 5b shows that the wave field regime changing correlate with saturation of the surface drag dependence of the wind speed.

The photos of top views of the water surface elucidate a possible origin of the changing of the regime of the waves wave field at 10-m wind speeds exceeding 24.5 m/s. Starting from this threshold wave breaking is intensified, because the crests of the waves are blown away by the strong tangential wind stress. It is accompanied with sprays, drops and bubbles near the wave crests, visible at the photos. Blowing away the crests of waves which steepness exceeds a definite threshold lead to the effective smoothening of the waves and the slope of the dominant wave then does not depend on wind speed as it shows fig. 6c. Basing on the theoretical model of wind turbulent boundary layer over wavy water surface, we investigated, whether this wind smoothening of the surface is sufficient for explanation the surface drag reduction.

5. Theoretical model of aerodynamic resistance of the wavy water surface at extreme wind conditions

The first step in the theoretical interpretation of the effect of the sea surface drag reduction at strong winds is calculation of the surface form grad. This part of the total aerodynamic resistance describes influence of the roughness of the surface. We can expect, that smoothening of the water surface by very strong wind significantly reduces the form drag and possibly can explain the experimental results. Then the effect of sprays and drops will be estimated.

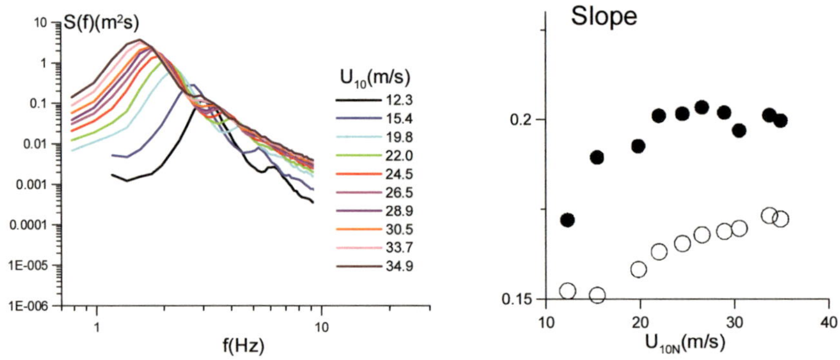

(a) (b)

Fig. 5. Frequency power spectra of the waves for a definite fetch and different wind speeds (a) and dependence of the average slope of the peak wave (b) (open circles are the dependencies of C_D on wind speed).

5.1 The theoretical model of turbulent wind over waved water surface

The wind is regarded as a turbulent boundary layer over the wavy water surface described within the first order semi-empirical model of turbulence based on the set of the Reynolds equations:

$$\frac{\partial \langle u_i \rangle}{\partial t} + \langle u_j \rangle \frac{\partial \langle u_i \rangle}{\partial x_j} + \frac{1}{\rho_a} \frac{\partial \langle p \rangle}{\partial x_i} = \frac{\partial \sigma_{ij}}{\partial x_j}, \tag{16}$$

and the following expressions for the tensor of turbulence stresses:

$$\sigma_{ij} = \langle u_i' u_j' \rangle = \nu \left(\frac{\partial \langle u_i \rangle}{\partial x_j} + \frac{\partial \langle u_j \rangle}{\partial x_i} \right), \tag{17}$$

Here <...> denotes the quantities averaged over turbulent fluctuations, ν is the turbulent viscosity coefficient, a given function of *z*. We use a self-similar expression for the eddy viscosity coefficient in the turbulent boundary layer:

$$\nu = \nu_a f\left(\frac{\eta \sqrt{\tau_{turb}}}{u_*^2} \right), \tag{18}$$

where ν_a is the air molecular viscosity.
We used the approximation for *f* obtained by (Smolyakov, 1974) on the basis of the laboratory experiments on a turbulent boundary layer. Finally, the expression for ν(z) takes the form:

$$\nu = \nu_a \left\{ 1 + \kappa \frac{u_* \eta \sqrt{1 - \tau_{wave} / u_*^2}}{\nu_a} \left[1 - e^{-\frac{1}{L}\left(\frac{u_* \eta}{\nu_a}\right)^2 \left(1 - \frac{\tau_{wave}}{u_*^2}\right)} \right] \right\}, \tag{19}$$

In this expression *L* is a number, which determines the scale of the viscous sublayer of a turbulent boundary layer; it depends on the regime of the flow over the surface. Comparison with the parameters of the velocity profile in the turbulent boundary layer from (Miles, 1959) gives *L*=22.4 for hydrodynamically smooth surface, *L*=13.3 for the transition regime of a flow over surface, and *L*=1.15 for rough surface.
The boundary conditions at the air-sea interface $z=\xi(x,y,t)$ are:

$$\left. \frac{\partial \xi}{\partial t} + \langle u \rangle \frac{\partial \xi}{\partial x} + \langle v \rangle \frac{\partial \xi}{\partial y} \right|_{z=\xi(x,y,t)} = \langle w \rangle |_{z=\xi(x,y,t)}, \tag{20}$$

$$\langle \vec{u}_\tau^w \rangle \big|_{z=\xi(x,y,t)} = \langle \vec{u}_\tau^a \rangle \big|_{z=\xi(x,y,t)}, \tag{21}$$

<*u*>, <*v*> are the *x*- and *y*- components of the velocity field in the air, averaged over turbulent fluctuations, are the tangential velocity components in water and in air.
The random field of the water surface elevation is presented as a Fourier-Stieltjes transform:

$$\zeta(\vec{r},t) = \int dA(\vec{k},\omega)\ e^{i(\vec{k}\vec{r}-\omega t)},$$

here $\vec{k} = (k_x, k_y)$ is a two-dimensional wave vector, ω is the frequency of surface waves.
For a statistically homogeneous and stationary process the wavenumber-frequency spectrum $F(\vec{k},\omega)$ can be introduced as follows

$$\langle dA(\vec{k},\omega) dA(\vec{k}_1,\omega_1) \rangle = F(\vec{k},\omega)\delta(\vec{k}-\vec{k}_1)\delta(\omega-\omega_1) d\vec{k} d\vec{k}_1 d\omega d\omega_1,$$

To avoid strong geometric nonlinearity, the transformation to the wave-following curvilinear coordinates is performed:

$$x = \zeta_1 + \int i \cos\vartheta \; e^{i\left(k(\zeta_1 \cos\vartheta + \zeta_2 \sin\vartheta) - \omega t\right) - k\eta - i\varphi} dA,$$

$$y = \zeta_2 + \int i \sin\vartheta \; e^{i\left(k(\zeta_1 \cos\vartheta + \zeta_2 \sin\vartheta) - \omega t\right) - i\varphi - k\eta} dA, \qquad (22)$$

$$z = \eta + \int e^{i\left(k(\zeta_1 \cos\vartheta + \zeta_2 \sin\vartheta) - \omega t\right) - i\varphi - k\eta} dA,$$

here φ is the angle between the wavenumber wave vector $\vec{k}$ and direction of x-axis. In the linear approximation, the coordinate surface η=0 coincides with the waved water surface. The solution to the set of the Reynolds equations (16) is searched as a superposition of mean wind field $\vec{U}_0(\eta)$ and disturbances induced in the airflow by waves at the water surface. Then, the velocity field is as follows:

$$\langle \vec{u} \rangle = \vec{U}_0(\eta) + \int \vec{u}'(\eta) e^{i\left(k(\zeta_1 \cos\vartheta + \zeta_2 \sin\vartheta) - \omega t\right) - i\varphi - k\eta} kdA$$

The wave-wind interaction is considered here in the quasi-linear approximation similar to the approach developed by (Jenkins, 1992), (Janssen, 1989) and (Reutov & Troitskaya, 1995). Then the wave disturbances induced in the airflow by the waves at the water surface are described in the linear approximation and can be considered independently. The coordinate transformation (22) can be considered as a superposition of formal coordinate transformations for each single harmonic. Nonlinear terms or wave momentum fluxes enter into the equations for the components of mean velocity.

Consider first equations for the disturbances induced by a single harmonic wave at the water surface with the wave vector $\vec{k}$, frequency ω and amplitude dA. We introduce the formal coordinate transformation, where the coordinate line η=0 coincides with the water surface disturbed by this single harmonic wave

$$x = \zeta_1 + i\cos\vartheta e^{i(k(\zeta_1 \cos\vartheta + \zeta_2 \sin\vartheta) - \omega t) - k\eta - i\varphi} dA,$$

$$y = \zeta_2 + i\sin\vartheta e^{i(k(\zeta_1 \cos\vartheta + \zeta_2 \sin\vartheta) - \omega t) - k\eta - i\varphi} dA, \qquad (23)$$

$$z = \eta + dA e^{i(k(\zeta_1 \cos\vartheta + \zeta_2 \sin\vartheta) - \omega t) - i\varphi - k\eta},$$

The linear coordinate transformation

$$\zeta'_1 = \zeta_1 \cos\vartheta + \zeta_2 \sin\vartheta - \frac{\omega}{k} t,$$

$$\zeta'_2 = \zeta_2 \cos\vartheta - \zeta_1 \sin\vartheta = y_2 \cos\vartheta - y_1 \sin\vartheta = y', \qquad (24)$$

defines the reference frame following this harmonic wave, where the wave field does not depend on ς_2' (or Cartesian coordinate y'), i.e. it depends only on two coordinates ς_1' and η. Tangential velocity components are transformed similar to (24), and in the new reference frame

$$u' = u\cos\vartheta + v\sin\vartheta - \frac{\omega}{k},$$

$$v' = -u\sin\vartheta + v\cos\vartheta \tag{25}$$

It means that the stream function Φ can be introduced for the motions in the plane $\varsigma_2'=y'=const$ as follows

$$u' = \frac{\partial\Phi}{\partial\eta},\ w = -\frac{\partial\Phi}{\partial\varsigma'_2},$$

and the Reynolds equations can be formulated in terms of stream function Φ and vorticity χ

$$\frac{\partial\chi}{\partial t} + \frac{1}{I}\frac{\partial\chi}{\partial\zeta_1'}\left(\frac{\partial\Phi}{\partial\eta}\right) - \frac{1}{I}\frac{\partial\chi}{\partial\eta}\left(\frac{\partial\Phi}{\partial\zeta_1'}\right) = \Delta(\nu\chi) - \frac{2}{I^2}\nu_{\eta\eta}\frac{\partial^2\Phi}{\partial\zeta_1'^2} - $$
$$-\frac{I_\eta}{I^3}\left((\Phi_\eta\nu_\eta)_\eta - \nu_\eta\Phi_{\zeta_1'\zeta_1'}\right) - \frac{I_{\zeta_1'}}{I^3}(2\nu_\eta\Phi_{\zeta_1'\eta} - \Phi_{\zeta_1'}\nu_{\eta\eta}) + \Phi_\eta\nu_\eta\frac{I_{\zeta_1'}^2 + I_\eta^2}{I^4}, \tag{26a}$$

$$\Delta\Phi = \chi = \frac{1}{I}\left(\Phi_{\zeta_1'\zeta_1'} + \Phi_{\eta\eta}\right) \tag{26b}$$

here I is the Jacobian of transformation (23). The transversal velocity component v' does not enter the equations (26a,b), and v' obeys the following equation:

$$\frac{\partial v'}{\partial t} + \frac{1}{I}\left(\frac{\partial v'}{\partial\zeta_1'}\frac{\partial\Phi}{\partial\eta} - \frac{\partial v'}{\partial\eta}\frac{\partial\Phi}{\partial\zeta_1'}\right) = \Delta(v'\nu) + \frac{1}{I}v'_\eta\nu_\eta \tag{27}$$

We search the solution to the system (26a,b), (27) as a superposition of the mean field and harmonic wave disturbance:

$$\Phi = \int\left(U_0(\eta)\cos\vartheta + V_0(\eta)\sin\vartheta - \frac{\omega}{k}\right)d\eta + \Phi_1(\eta)dA\,e^{ik\zeta_1'}$$

$$v = V_0(\eta)\cos\vartheta - U_0(\eta)\sin\vartheta + V_1(\eta)dA\,e^{ik\zeta_1'} \tag{28a}$$

$$\chi = U_{0\eta}\cos\vartheta + V_{0\eta}\sin\vartheta + X_1(\eta)dA\,e^{ik\zeta_1'} \tag{28b}$$

Equations for complex amplitudes $\Phi_1(\eta), \chi_1(\eta), V_1(\eta)$ are obtained by linearization of system (26a,b), (27).

$$(\Phi_{0\eta}X_1 - \Phi_1\chi_{0\eta})ik - \left(\frac{d^2}{d\eta^2} - k^2\right)(X_1\nu) = -2\nu_\eta\Phi_1k^2 - 2kAe^{-k\eta}(\Phi_{0\eta}\nu_\eta)_\eta, \tag{29a}$$

$$\frac{d^2\Phi_1}{d\eta^2} - k^2\Phi_1 = X_1 - 2ke^{-k\eta}\Phi_{0\eta\eta} \tag{29b}$$

$$(\Phi_{0\eta}V_1 - \Phi_1\hat{V}_\eta)ik = \nu\left(\frac{d^2}{d\eta^2} - k^2\right)V_1 + \nu_\eta V_{1\eta}k^2 \tag{30}$$

We consider solutions to the system (29a,b), (30) decreasing at large distances from the surface, i.e.

$$\Phi_1\big|_{\eta\to\infty} \to 0; \quad V_1\big|_{\eta\to\infty} \to 0$$

Boundary conditions at the water surface for the system (29a,b) follow from (20) and (21) expressed in curvilinear coordinates (see (Reutov & Troitskaya, 1995)) for details.

$$\Phi_1\big|_{\eta=0} = 0; \quad \Phi_{1\eta}\big|_{\eta=0} = 2\omega; \quad V_1\big|_{\eta\to 0} = 0 \tag{31}$$

The only nonlinear effect taken into account in the quasi-linear approximation is the demodulation of the wave disturbances induced in the air flow by waves at the water surface. Equations for mean velocity profile components $U_0(\eta)$ and $V_0(\eta)$ are obtained by the following steps. Averaging of (26a,b) over ς_1' gives equation for Φ_0 and averaging of (27) yields equation for $v_0(\eta)$. Expressing $U_0(\eta)$ and $V_0(\eta)$ via $\Phi_0(\eta)$ and $v_0(\eta)$ by inversion (28a,b) and integrating over the wind wave spectrum gives:

$$\frac{d}{d\eta}\left(\nu\frac{d(U_0,V_0)}{d\eta}\right) = \iint\left(\tau_{\|}(\eta,k,\varphi,\omega)(\eta)\begin{pmatrix}\cos\varphi\\ \sin\varphi\end{pmatrix} + \tau_{\perp}(\eta,k,\varphi,\omega)(\eta)\begin{pmatrix}-\sin\varphi\\ \cos\varphi\end{pmatrix}\right)k^2F(k,\varphi,\omega)kdkd\varphi d\omega \tag{32}$$

here $\tau_{||}(\eta,k,\theta,\omega)(\eta)$, $\tau_{\perp}(\eta,k,\theta,\omega)(\eta)$ are the components of the wave momentum flux induced by the surface wave with wave number k, frequency ω propagating at the angle θ to the wind.

Expression for $\tau_{||}(\eta,k,\theta,\omega)(\eta)$ follows from (26a,b)

$$\tau_{\|}(\eta,k,\varphi,\omega)(\eta) = k\left[k\nu_\eta \operatorname{Re}\left(\Phi_{1\eta} - k\Phi_1\right)e^{-k\eta} + 2k^2e^{-2k\eta}\nu_\eta U_0\cos\varphi\right]$$

and expression for $\tau_{\perp}(\eta,k,\theta,\omega)(\eta)$ follows from (27)

$$\tau_{\perp}(\eta,k,\omega) = -\frac{1}{2}k\frac{d}{d\eta}\operatorname{Im}\left(\Phi_1^* V_1\right)$$

Equations (32) express the conservation law for the vertical flux of two projections of the horizontal momentum component in the turbulent boundary layer. If the turbulent shear stress at a large distance from the surface is directed along x, the conservation law for the mean momentum components may be written as follows:

$$\tau_{turb}^{(x)}(\eta) + \tau_{||}(\eta) = u_*^2$$

$$\tau_{turb}^{(y)}(\eta) + \tau_{\perp}(\eta) = u_*^2$$

Since according to (32) nonlinear addition to the wind velocity profile is determined by the frequency-wavenumber spectrum $S(\omega,k,\varphi)$ of surface waves, then the spectrum is a critical component of the model. In our laboratory experiment we measured the airflow velocity profile together with the elevation of the water surface in 3 close points. These data are sufficient for retrieving 3-dimensional spectrum $S(\omega,k)$ needed for calculating the form drag and comparing with the experimental data. High frequency part of the spectrum, which can contribute to the surface roughness, was interpolated on the base of available data by the power law with the exponent determined from the experimental spectra.

5.2 Fine structure of turbulent air flow over steep and breaking waves

Let us discuss first applicability of the suggested model for description of the airflow over steep and breaking waves which occurred in the flume at strong winds. The model exploits two main suppositions: the closing hypothesis and the quasi-linear approximation for description of the the wind-wave interaction. The quasi-linear approximation presumes, that wave-induced disturbances in the air flow are considered in the linear approximation, i.e. wave-induced disturbances in the air flow are considered in the linear approximation, but the resistive effect of the wave momentum flux on the mean flow velocity profile is taken into account, i.e. within the model the mean airflow over waves is treated as non-separated.

One can expect existence of strong nonlinear phenomena such as sheltering, flow separation, etc., for the cases of steep and breaking waves. These phenomena were investigated by means of contact methods and smoke visualization in the laboratory experiments by (Banner & Melville, 1976), (Kawamura & Toba, 1979), (Kawai, 1981), (Kawai, 1982), (Hsu & Hsu, 1983), (Hsu et al., 1981). Major difficulties in these experiments are concerned with the measuring of the airflow close to the water surface, especially in the troughs of the waves. These measurements can be performed by means of the wave-following contact technique ((Hsu & Hsu, 1983), (Hsu et al., 1981), (Donelan et al., 2005). Also, the problem of measurement of the wind flow below the crests of the waves was solved by (Kawai, 1981), (Kawai, 1982) by means of seeding the flow by small particles visualized with a strobe source of light and application of special photograph technique. Kawai's experiments demonstrated occurrence of the airflow separation from the crests of steep waves on the instant images of the flow.

Recently, the structure of an airflow over waves was investigated in detail by the method of particle image velocimetry (PIV) (Adrian, 1991), when the flow is seeded with the small particles illuminated by the laser light and then taken with a digital camera. This technique was applied by (Reul et.al., 1999), (Reul et.al., 2008) and (Veron et. al., 2007) and clearly demonstrated the effect of the airflow separation from the crests of the waves and reattachment at the windward face of the wave on the instantaneous patterns of the vector velocity fields.

It should be emphasized that the PIV technique provides an instant picture of the velocity field, but the flow separation in the turbulent boundary layer over a gravity wave is a strongly non-stationary process due to both the stochastic character of the airflow and the brevity of the breaking event, which usually occurs within a small part of the wave period (Duncan et al., 1999). At the same time, the models of the air-sea fluxes and wind wave growth exploit the wind flow parameters averaged over turbulent fluctuations. We combined measurements of the instant airflow velocity fields over surface waves with

statistical averaging (Troitskaya et al., 2010). The statistical ensemble of such vector fields for subsequent averaging was obtained by means of high-speed video filming and processing of the video-films by the PIV algorithm. Individual flow realizations manifested the typical features of flow separation similar to those obtained by (Kawai, 1981), (Kawai, 1982), (Reul et.al., 1999), (Reul et.al., 2008) and (Veron et. al., 2007). The average parameters were retrieved by the phase averaging of the individual vector fields. The averaged flow patterns appear to be smooth and slightly asymmetrical, with the minimum of the horizontal velocity near the water surface shifted to the leeward side of the wave profile (compare fig.6a, b).
The results of these measurements were compared with the calculations within the quasi-linear model of turbulent boundary layer described in section 5.1. The wave parameters (wavelength, celerity, steepness), used in this comparison of theory with experiment, were retrieved from the same video films as those used for the airflow velocity calculations. The model calculations were in a good agreement with the experimentally measured and conditionally averaged mean wind velocity, turbulent stress and also amplitude and phase of the main harmonics of the wave-induced velocity components. (see fig.7a,b).

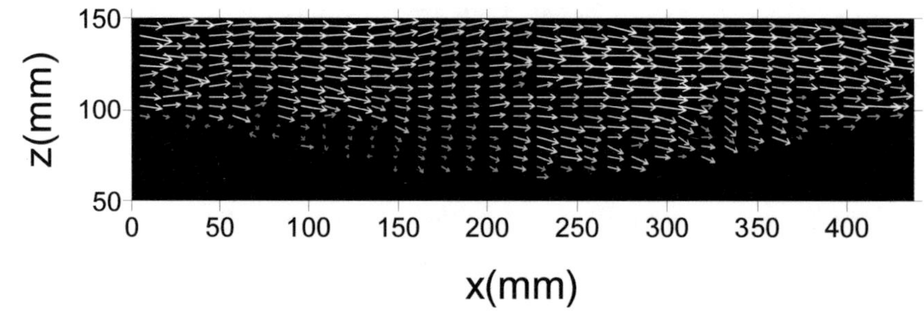

(a)

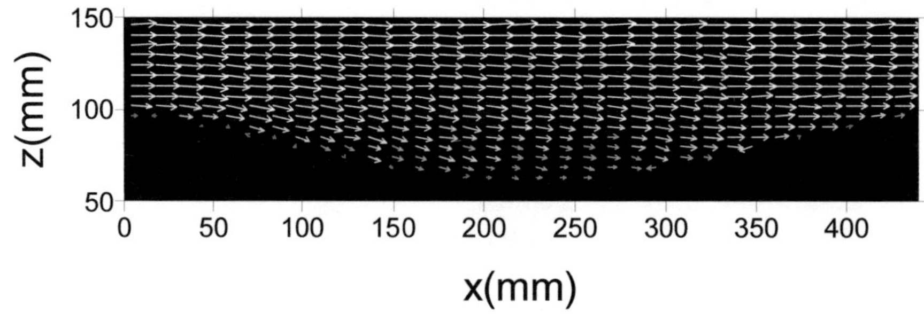

(b)

Fig. 6. The vector velocity field in the airflow over the paddle-generated wave in the wave-following reference frame retrieved from high-speed video filming by digital particle image velosimetry. Wind friction velocity u_*=200 mm/s, wavelength k=0.15 cm-1, slope ka=0.25. (a) – the instantaneous pattern, (b) ensemble averaged pattern.

Applicability of the non-separating quasi-linear theory for description of average fields in the airflow over steep and even breaking waves, when the effect of separation is manifested at the instantaneous flow images, can possibly be explained qualitatively by the strongly non-stationary character of the separation process with the typical time much less than the wave period, and by the small scale of flow heterogeneity in the area of separation. In such a situation small-scale vortices produced within the separation bubble affect the mean flow and wind-induced disturbances as eddy viscosity. Then the turbulence of the flow affects the averaged fields as a very viscous fluid. Then the effective Reynolds number for the average fields determined by the eddy viscosity is not large even for steep waves. It follows from this assumption that strongly non-linear effects such as flow separations should be not expected in the flow averaged over turbulent fluctuations. We were encouraged by these results to apply the quasi-linear model for calculation of the form drag of the water surface at strong winds.

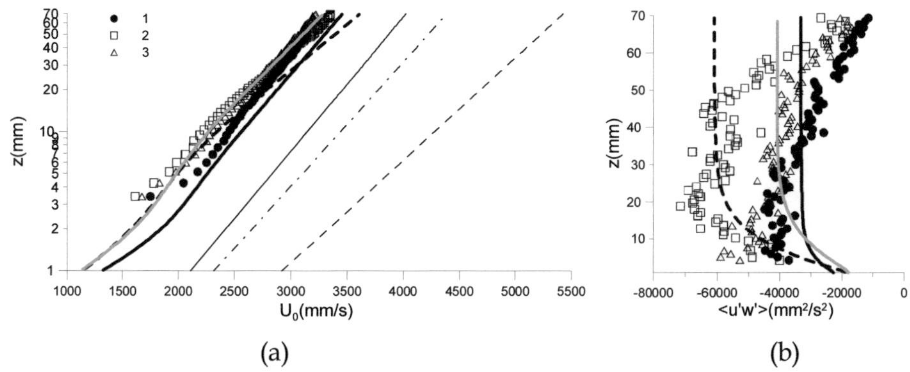

Fig. 7. Comparing theory and experiment: (a) is the mean velocity profile; (b) - the stress profile. Symbols - experiment curves - calculations within the qausi-linear model for the measured parameters of the wave and the wind (see Troitskya et al., 2010).

Dependency of the form drag of the water surface calculated within the model (29-32) for the parameters measured in the flume are presented in fig.8 together with the measured sea surface drag coefficient. It is clear that both at low and moderate winds and at strong winds (exceeding threshold value 24.5 m/s) the model is in good agreement with data. So saturation of the average slope of dominant waves enables one to explain saturation of the surface resistance, both qualitatively and quantitatively.

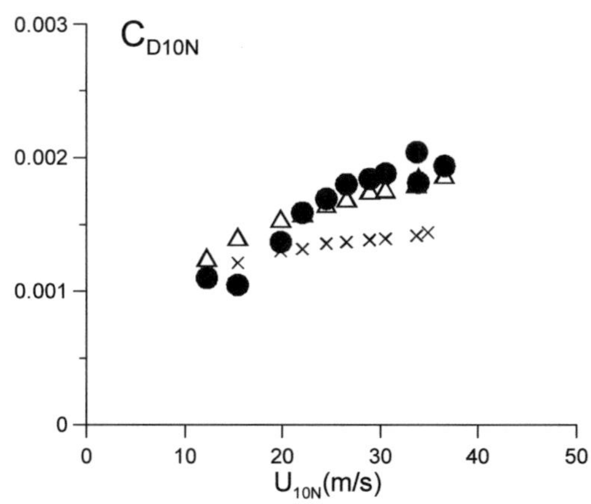

Fig. 8. Dependence of surface drag coefficient on wind speed, comparing theory and laboratory experiment. Measurements - closed circles, theoretical calculations: squares - with short wave spectrum of surface waves, crosses - neglected short wave spectrum of surface waves.

We also investigated sensitivity of the model to the spectrum of surface waves. In calculations shown in fig.8 by crosses the effect of short surface waves was eliminated by cut off surface wave spectrum at the wave number 120 m^{-1}. Comparison presented in fig.8 shows noticeable dependence of C_D on the high wave number part of the spectrum. Unfortunately, measurement of the spectrum of short waves (cm and mm wave length) with high space resolution is a difficult problem especially at strong wings. Optical methods developed by (Jähne et al, 2005), (Rosholtz, 2010) are promising for laboratory conditions.

6. Conclusion and discussion

So at extremely strong wind sea surface drag coefficient in the field and laboratory conditions demonstrates anomalous flattening or even decreasing dependency on the wind speed. Here we suggest a possible explanation of this phenomenon due to smoothening of the water surface by strong wind stress blowing away the crests of steep surface waves. This supposition is confirmed by our laboratory experiments when both wind and wave spectra were measured. Predictions of the theoretical model are in good agreement with the measurements.

It is important to discuss relation of wind-wave interaction in field and laboratory conditions at extremely strong winds. First of all, the laboratory and field conditions strongly differ in fetch. Waves in laboratory flumes are extremely "young" in comparison with nature, i.e. they correspond to the initial stage of the wave development. At this stage the phase velocity of the waves corresponding to the spectral peak c_p is small in comparison with the wind speed U_{10}. , i.e. the age parameter of the waves $c_p/U_{10}<<1$. The wind-wave interaction is the resonant process and interaction is concentrated in the layer where the wind speed is close to the phase velocity of the wave. In the theory by (Miles, 1957), (Miles, 1959) the interaction occurs in the critical layer, where the wind and wave velocities coincide. In more complicated models including the model described in section 4 the interaction occurs in a layer of constant thickness. Since the wind velocity profile is very sharp, this layer is very thin and it is positioned very close to the water surface up to $c_p<U_{10}/2$ (Phillips, 1977) and the approximation of the "young" sea is valid. Then it follows, that the region of wind-wave energy exchange is positioned close to the water surface for "young" waves and slightly depends on the wave age parameter. In other words, since "young" waves propagate much slowly than wind they interact with the wind as almost stationary surface roughness.

The wave age parameter can be easily estimated by the empirical formula suggested by (Donelan, 1985):

$$\frac{U_{10}}{c_p} = 22\left(\frac{xg}{U_{10}^2}\right)^{-1/3}$$

The wind speed distribution in a tropical cyclone is inhomogeneous. Let us first estimate the wave age parameter of waves generated by maximal wind speed, then x approximately equal to the radius of the maximum wind r_0. For a typical r_0=50 km, U_{10}=60 m/s we get U_{10}/c_p=4.3. For the gale force wind area typically x=150 km, U_{10}=20 m/s, then U_{10}/c_p=1.4. These estimates show, that the approximation of the "young sea" can be applied for description of the air-sea interaction near the radius of maximum winds similar to laboratory conditions.

Another governing parameter of the flow over rough surface is the Reynolds number, while a flow averaged over turbulent fluctuations is controlled by the effective Reynolds number Re_t determined by the effective eddy viscosity ν_t. In the case of wind-wave interaction $\nu t = u_*^2/ck$, where c and k are the phase velocity and wave number (see Troitskaya & Rubushkina, 2008). The characteristic scale of the roughness of the wavy water surface is the peak wave amplitude a_p and the airflow velocity is scaled by U_{10}. For the dominant wave the simple algebra gives

$$\mathrm{Re}_t = \frac{c_p}{U_{10}} \frac{k_p a_p}{C_D}$$

The detailed analysis of the dependency of the wave field energy on fetch (see Janssen, 2002)) gives

$$k_p a_p = 0.05 \left(\frac{c_p}{U_{10}} \right)^{-1/2}$$

and

$$\mathrm{Re}_t = \frac{0.05}{C_D} \left(\frac{c_p}{U_{10}} \right)^{1/2}$$

Taking into account that C_D at high wind is saturated at approximately 0.002 yields for the field conditions of hurricane near the region of maximal winds $Re_t \approx 10$ and in the laboratory conditions $Re_t \approx 5$. So, the quantities of the governing parameters in the laboratory and field conditions are close. Then we can expect, that we can model of the air-sea momentum exchange in laboratory flumes.

It should be mentioned, that there are some other effects of air-sea interaction at hurricane wind conditions, which can be specially investigated in laboratory. The hurricane wind waves are strongly affected by swell generated near the hurricane walls, where the wind speed is maximal. This swell is "mixed" with the local wind waves producing complex multi-modal spectra (Young, 2006). Effect of the swell on the sea surface drag was discussed qualitatively in (Powel, 2007). In the flumes the swell can be modelled by artificially generated long waves.

7. References

Adrian, R. J., (1991) Particle Imaging techniques for experimental fluid mechanics. *Annu. Rev. Fluid Mech.*, Vol. 23, p. 261-304.

Andreas, E. L. and K. A. Emanuel, (2001): Effects of sea spray on tropical cyclone intensity. *J. Atmos. Sci.*, Vol. 58, No 24, p. 3741-3751.

Andreas E. L. (2004) Spray stress revised. *J. Phys. Oceanogr.*, , v.34, No 6, p.1429-1440.

Barenblatt G. I. and Golitsyn G. S. (1974) Local structure of mature dust storms *J. Atmos. Sci.*, , Vol.31, No 7, p. 1917- 1933.

Donelan M.A., Haus B.K , Reul N., Plant W.J., Stiassnie M., Graber H. C., Brown O. B., Saltzman E. S. (2004) On the limiting aerodynamic roughness of the ocean in very strong winds. *Geophys. Res. Lett.*, , v.31, L18306.

Donelan M., Hamilton J. , Hui W. H. (1985) Directional spectra of wind generated waves *Philos. Trans. Roy. Soc. London, Ser. A.*, Vol 315, p.509–562.

Donelan, M. A., Babanin A. V., Young I. R., Banner M. L., McCormick C. (2005) Wave follower field measurements of the wind input spectral function. Part I: Measurements and calibrations. *J. Atmos. Oceanic Technol.*, Vol 22, No 7, No 799–813.

Donnely W. J., Carswell J. R., McIntosh R. E. Revised ocean backscatter model at C and Ku band under high-wind conditions. (1999) *J. Geophys. Res.*, v.104, No C5, p.11485–11497.

Duncan, J. H., Qiao H., Philomin V., and Wenz A. (1999) Gentle spilling breakers: crest profile evolution. *J. Fluid Mech.*, Vol. 379, p. 191–222.

Emanuel, K.A. (1986) An air-sea interaction theory for tropical cyclones. Part I: Steady state maintenance. *J. Atmos. Sci.*, Vol. 43, No 6, p. 585-604.

Emanuel, K.A. (1995) Sensitivity of tropical cyclones to surface exchange coefficients and a revised steady-state model incorporating eye dynamics. *J. Atmos. Sci*, Vol.52, No 22, p.3969-3976.

Emanuel, K. (2003) Tropical Cyclones. *Ann Rev. Earth Planet. Sci.*, Vol. 31, p. 75-104.

Fairall C.W., Bradley E.F., Hare J.E., Grachev A.A., Edson J.B. (2003) Bulk parameterization of air–sea fluxes: updates and verification for the COARE algorithm . *J. Climate*, , Vol.16, No 4, p.571–591.

Garratt J.R. (1977) Review of drag coefficients over oceans and continents. *Mon. Weather Rev.*, , Vol.105, No7, p.915-929.

Hara T., Belcher S.E. (2004) Wind profile and drag coefficient over mature ocean surface wave spectra . *J. Phys. Oceanogr*, Vol.34, No 11, p. 2345–2358.

Hinze, J. O. (1959) Turbulence: An Introduction to its Mechanism and Theory. New York: McGraw-Hill. 586 p.

Hsu, C. T., Hsu E. Y. (1983) On the structure of turbulent flow over a progressive water wave: theory and experiment in a transformed wave-following coordinate system. Part 2. *J. Fluid Mech.*, Vol. 131, p.123–153.

Hsu, C. T., Hsu E. Y., Street R. L. (1981) On the structure of turbulent flow over a progressive water wave: theory and experiment in a transformed, wave-following co-ordinate system. *J. Fluid Mech.*, Vol. 105, p. 87–117.

Jarosz E., Mitchell D. A., Wang D.W., Teague W.J. (2007) Bottom-up determination of air-sea momentum exchange under a major tropical cyclone. *Science*, v 315, p. 1707-1709 DOI: 10.1126/science.1136466.Janssen P.A.E.M. (1989) Wave-induced stress and the drag of air flow over sea waves. *J. Phys. Oceanogr.*, , v.19, No 6, p.745-754.

Janssen P.A.E.M. (1991) Quasi-linear theory of wind wave generation applied to wave forecasting *J. Phys. Oceanogr.*, , v.21, No 11, p.1631-1642.

Jenkins A.D. (1992) Quasi-linear eddy-viscosity model for the flux of energy and momentum to wind waves using conservation-law equations in a curvilinear coordinate system. *J. Phys. Oceanogr.*, , Vol.22, No 8, p. 843–858.

Jenkins A.D. (1993) Simplified quasi-linear model for wave generation and air–sea momentum flux Journal of Physical Oceanography, Vol. 23, No 9, p. 2001–2018.

Kudryavtsev V. N. (2006) On the effect of sea drops on the atmospheric boundary layer *J. Geophys. Res.*, , Vol.111, C07020.

Jähne B., Schmidt M., Rocholz R. (2005) Combined optical slope/height measurements of short wind waves: principles and calibration Measurement Science & Technology, Vol. 16, p. 1937—1944

Kawai, S., (1981) Visualisation of air flow separation over wind wave crest under moderate wind. *Boundary Layer Meteorol.*, Vol. 21, p. 93–104.

Kawai, S., (1982) Structure of air flow separation over wind wave crest. *Boundary Layer Meteorol.*, Vol. 23, p. 503–521.

Kudryavtsev V., Makin V. (2007) Aerodynamic roughness of the sea surface at high winds. *Boundary-Layer Meteorol.*, , v.125, p. 289–303.

Kukulka, T., T. Hara, and S. E. Belcher, (2007): A model of the air-sea momentum flux and breaking-wave distribution for strongly forced wind waves. *J. Phys. Oceanogr.*, Vol 37, No 11, p. 1811-1828.

Large W.G., Pond S. (1981) Open ocean momentum flux measurements in moderate to strong winds. *J. Phys. Oceanogr.*, , Vol. 11, No 1, p.324–336.

Miles J.W. (1957) On the generation of surface waves by shear flows. *J.Fluid Mech.*,Vol.3, p.185-204.

Miles J.W. (1959) On the generation of surface waves by shear flows Part 2. *J.Fluid Mech.*, Vol.6, p.568-582.

Makin V. K. (2005) A note on drag of the sea surface at hurricane winds. *Boundary Layer Meteorol.*, , Vol. 115, No1, p.169-176.

Makin V. K., Kudryavtsev V. N., Mastenbroek C. (1995) Drag of the sea surface. *Boundary-Layer Meteorol.*, Vol.79, p.159–182.

Phillips O. M. (1977) *The Dynamics of the Upper Ocean.* Cambridge Monographs on Mechanics Cambridge University Press; 2 edition), ISBN-10: 0521298016, ISBN-13: 978-0521298018

Powell, M.D., Vickery P.J., Reinhold T.A. (2003) Reduced drag coefficient for high wind speeds in tropical cyclones. *Nature* , v.422, p.279-283.

Powell M. D. (2007) Final Report "Drag Coefficient Distribution and Wind Speed Dependence in Tropical Cyclones"Principal In vestigator: Mark D Powell NOAA/AOML, 2005-2007, Taylor P.K., Yelland M.J. (2001) The dependence of sea surface roughness on the height and steepness of the waves. *J. Phys. Oceanogr.*, , v.31, No 2, p.572–590.

Reul, N., Branger H., Giovanangeli G.P. (1999) Air flow separation over unsteady breaking waves. *Phys. Fluids*, , v.11, No 7, p.1959-1961.

Reutov, V.P. and Yu. I. Troitskaya, (1995) On the nonlinear effects in the interaction of gravity waves with turbulent airflow. *Izvestiya, Atmospheric and Oceanic Physics*, Vol. 31, No 6, p. 825–834.

Rocholz R., Jähne B. (2010) Spatio-temporal measurements of short wind water waves EGU General Assembly, Symposium AS2.2, EGU2010-5509,

Smolyakov, A. V., (1973) Spectrum of the quadruple radiation of the plane turbulent boundary layer. Acoust. Phys., Vol 19, No 3, p. 420–425

Yu. I. Troitskaya, G. V. Rybushkina. (2008) Quasi-linear model of interaction of surface waves with strong and hurricane winds. *Izvestiya, Atmospheric and Oceanic Physics* Vol. 44, No 5, p. 621-645

Tsimplis, M. , Thorpe S. A. (1975): Wave damping by rain. *Nature,* Vol.342, p. 893-895.

Veron F., Saxena G., Misra S. K., (2007) Measurements of the viscous tangential stress in the airflow above wind waves. *Geophys. Res. Lett.,* Vol. 34, L19603, doi: 10.1029/2007GL031242.

R. Young (2006) Directional spectra of hurricane wind waves *J. Geophys. Res,* Vol. 111, C08020, doi: 10.1029/2006JC003540,

13

How Vertical Wind Shear Affects Tropical Cyclone Intensity Change: An Overview

Levi Thatcher and Zhaoxia Pu
University of Utah
United States

1. Introduction

Over the last 30 years tropical cyclone (TC) intensity forecasts, for various (yet somewhat puzzling) reasons, have not achieved near the level of improvement of the TC track forecast. Although TCs have been studied intensively throughout the twentieth century, the community has surprisingly little quantitative knowledge as to how these storms interact with their environments, particularly with respect to changes in core structure (Frank & Ritchie, 1999). Rogers et al. (2006) stated that the lack of skill in numerical forecast of TC intensity can be partly attributed to inadequate understanding of the physics of TCs and the way they interact with their environment. In fact, TC structure and intensity changes are affected by a large and complex array of physical processes that govern the inner core structure and the interaction between the storm and both the underlying ocean and its atmospheric environment (Wang & Wu, 2004). Among other issues cited, crude parameterizations, difficulties in treating multiscale interactions, and the uncertainties involved with initializing the model over areas with sparse data coverage have received substantial attention.

In order to predict TC intensity, one of the important questions has been how to first accurately predict a TC's maximum potential intensity (MPI). Despite the fact that various methods for predicting a storm's MPI have been put forth, the failure of the NWP community to realistically forecast TC intensity largely lies in the fact that there are various unexplained processes keeping TCs from reaching their theoretical MPI. While the mechanisms involved are myriad, there are essentially two kinds that have been identified as having the largest impact on TC intensification: 1) internal dynamics and 2) external forcing from environmental flow. Below these two headings fall most TC intensity-related topics: vertical wind shear-induced asymmetries in the core region, the cooling of the sea surface due to oceanic upwelling under the eyewall region, the role of inner and outer rainbands, vortex Rossby waves (VRWs), embedded mesovortices, and eyewall cycles. Tropical cyclones often fail to reach their theoretical MPI because prominent MPI calculations use the basic assumption of TC axisymmetry (Camp & Montgomery, 2001), whereas TC structure is rarely symmetric, even in mature storms. While the tangential wind field and other TC features are axisymmetric, many significant features, such as VRWs, eyewall cycles, rainfall, convection, radial winds, and outer rainbands are often highly asymmetric attributes that impact TC intensity change; it is thus no surprise that nearly all TCs fail to reach their MPI.

The effect of storm asymmetries on TC intensification has been the subject of much debate, and it appears that the relationship is quite fundamental to the TC intensity change question overall. Of the many asymmetry-related topics affecting TC intensity, that of environmental vertical wind shear is one of the most prominently disputed in the TC intensity literature, perhaps partially because of its obvious relationship to storm asymmetry. This, however, is certainly not the only reason, as the basic physical mechanisms behind shear's effect on TC intensification are still not clear. Due to the prominence of the shear-intensification relationship and the questions that still remain, but despite the many other issues related to the TC asymmetry-intensification question, this is the field that we will be discussing for the duration of the chapter. First, early work discussing shear's effect on TC intensity will be overviewed. Following this will be a description of two of the most prominent theories that have been put forth: first, the mid-level warming hypothesis of DeMaria (1996) and then the venting hypothesis of Gray (1968) and Frank and Ritchie (2001). Afterwards a view of recent research will be given and, finally, some conclusions will be made.

2. Early work

While the effects of moderate vertical wind shear on TC intensification are now generally acknowledged to be quite nuanced, the early literature focuses more on their general effect on TC intensification without much debate as to the nature of the relationship; the seeds of the current discussions, however, were sown quite early. Despite the lack of proper atmospheric observations over much of the oceans, by the 1950s the scientific community had reached some general conclusions about TC formation. Besides the need for SSTs above 26°C and latitude greater than 5 to 8°, it was determined that high levels of vertical wind shear are prohibitive to TC intensification (Ramage, 1959). In his paper on the frequency of TCs in the various tropical basins, Ramage went on to describe the dearth of TCs over the China Sea, Bay of Bengal, and the Arabian Sea as being due to vertical shear associated with summer monsoon circulations.

Into the 1960s there was still little consensus as to the general environmental flow features and dynamical processes associated with TC formation (Gray, 1968). Using oceanic upper air data that had just recently become available, Gray published a seminal paper (Gray, 1968) that not only gave a substantive discussion on the atmospheric conditions that accompany TC formation, but also marked the beginning of the "venting" hypothesis (of core moisture and energy out of the top of the TC core) in explaining the method by which vertical shear inhibits TC growth.

Global in its reach, the study looked at over 300 TC development cases, analyzed the prevailing environmental characteristics of the recognized TC basins, and consistently found a strong association between regions of minima in climatological vertical shear and regions of frequent storm formation. Constricting the data by month, Gray found that this relationship held quite well temporally as well as spatially. In explaining these results he said that large levels of vertical shear produce a sizable ventilation of heat away from the developing disturbance. Thinking in terms of the TC core, the condensational heat released by convection in the upper troposphere is advected in a different direction relative to the center at lower levels. This asymmetry, he noted, was deadly for TC growth in that it renders the concentration of heat through the entire troposphere quite difficult (Gray, 1968).

In stark contradiction to the findings of Gray (1968) and the community in general, Tuleya and Kurihara (1981) found that small (nonzero) levels of vertical shear, or moderate easterly shear, were beneficial to TC development. They performed their study using an 11-level primitive equation model in which simple environmental flows were superposed on a shallow perturbation resembling an easterly wave. They explained these results by saying that both the formation of a relatively warm area in the upper levels, and significant moisture convergence in the lower levels, were necessary for TC growth. If these two events were strongly coupled in the vertical then an intensification feedback between the two could occur. They found that surface easterlies were beneficial (along with the easterly shear) for TC development because they promoted a strong coupling between the upper and lower levels. While subsequent studies did not make much mention of the benefits of easterly shear on TC, this early study was notable in that it emphasized the importance of the coupling of the upper and lower levels on the mechanics of intensity change. Later this same topic would command the attention of many as they strove to explain the processes behind the way in which TCs often maintained this coupling despite the influence of vertical shear.

With the increasing interest in TC maximum potential intensity, Merrill (1988) set out to determine the role of shear in keeping a TC from reaching this theoretical upper limit. To do so he used composite National Hurricane Center (NHC) upper-tropospheric observational data from rawinsondes, satellites, and commercial aircraft to study how TC environments differ for developing and nondeveloping storms. His study included 28 hurricanes from the Atlantic basin which were segregated depending on 1) intensity change over a 24 hour period, 2) the proximity of the storm to its maximum potential intensity (or efficiency), and 3) the corresponding climatological SSTs in the storm's vicinity. Considering vertical shear as being the main component of what kept certain storms well below their MPI, Merrill analyzed the data from such storms and concluded, as expected, that vertical shear is lower for intensifying hurricanes; interestingly, he found this to be especially the case at radii of 1000 km or more.

Having sufficiently established the detrimental effect of shear on TC intensification, subsequent studies moved into the mechanisms that lay behind the interaction. These studies took numerous forms and included such topics as the effect of shear on TC movement (Shapiro, 1992; Flatau et al., 1994; Wang & Holland, 1996), the effect of shear on the distribution of convection (Corbosiero & Molinari, 2002; Rodgers et al., 2003; Chen et al., 2006), the way in which TCs with certain characteristics (i.e., a certain strength, horizontal extent, or latitude) are able to better resist shear and how the TC can maintain its vortex nearly vertical in areas of adverse environmental conditions (Jones, 1995; DeMaria, 1996).

Due to disparate research findings, many researchers in the 1990s investigated the way in which vertical shear influences the movement of the storm relative to what the usual environmental steering flow would dictate. Flatau et al. (1994) adeptly began by basing his study on the observation, by Gryanik and Tevs (1989), that, when considering the displacement of the centers of a low-level cyclone and upper-level anticyclone relative to one other, the lower- and upper-level vortices can be thought of as interacting in a manner similar to the interaction of two-dimensional vortices on a horizontal plane.

Using his semispectral, primitive equation model to simulate the movement of a baroclinic vortex, Flatau et al. (1994) found that in westerly linear shear the simulated TC vortex moved toward the north, while in easterly shear the vortex propagated southward. These results contrasted with those of Shapiro (1992), who, when performing similar tests using a three-layer multinested model with a compressible fluid, found southward propagation in westerly sheared environmental flow. Flatau et al. (1994) attributed these differences to 1) the fact that the direction of propagation depends on the position of the upper and lower PV anomalies relative to each other; 2) differences in the two papers' respective wind profiles; and 3) the effects of the background potential vorticity gradient. While these details may seem irrelevant to a discussion on TC intensification, the reasons why Flatau et al. and Shapiro drew their particular conclusions presaged the way in which later researchers explained the methods used by TCs to cope with vertical shears and a tilting vortex. First, Flatau et al.'s motion explanation depends upon the relative movement and interaction between the upper and lower levels of the vortex. In westerly vertical shear the upper-level anticyclone is pushed east of the lower-level anticyclone. The cyclone-anticyclone pair moves toward the north due to the fact that the low-level PV anomaly (by which the TC center is often described) advects the upper-level negative PV anomaly and vice versa.

Despite the contradicting results, Shapiro (1992) and Flatau et al.'s (1994) work both involved TC propagation as caused by a similar interaction between TC layers and the same horizontal transport of potential vorticity. Using a 5-layer primitive equation model on both an f-plane and β-plane, Wang and Holland (1996) also found that under most circumstances the surface vortex was advected to the left of the vertical shear. They too noted that the leftward or rightward enhancement of propagation relative to the vertical shear depended on the relative magnitude of the different motion tendencies. The fact that all three of these motion studies (Shapiro, 1992; Flatau et al., 1994; Wang & Holland, 1996) relied all or partially on the interaction of the upper and lower PV anomalies as a means to explain TC movement lent credence to the importance of vortex tilt and the myriad consequences and complications that arose from this phenomena. The interesting part of the Wang and Holland (1996) paper is the fact that they started to attribute a significant amount of the tilt-reducing coupling (especially for strong and large vortices) to the relative positioning of the upper and lower-level cyclone, not the lower-level cyclone and the upper-level anticyclone, as had the majority of the previous studies had done when looking to explain the way in which shear affected TC structure.

Much of the present day discussion revolves around the behavior and characteristics of the tilt of the cyclonic portion of the TC core. As vertical shear results in the differential advection of PV, this tilt can become quite pronounced and is often used as a proxy for the overall impact vertical shear has on the TC core. The tilt is often an effective way to analyze the shear-intensity relationship not only because of its central role in TC core dynamics, but also because of its ubiquity. For example, Huntley and Diercks (1981), looking at difference between the TC core at the 500 and 700hPa levels, noted that there was evidence of tilt in excess of 100km in 11 of 23 named TCs during the 1979 West Pacific Typhoon season. In addition, they noted that the tilt extended to much higher levels, and, foreshadowing much of the research that was to come, said that the tilt was highly correlated with the magnitude and direction of the vertical shear.

Despite the similarities with other prominent studies of their day, Flatau et al. (1994) in particular pointed the way forward in terms of vortex layer coupling by demonstrating that the vortex tilt is somewhat mitigated by the presence of diabatic heating by upward advection of high PV air at the center and downward advection of low PV at larger radius. Along with Wang and Li (1992), Flatau et al. (1994) considers this the primary means by which a vortex remains vertically coupled, although they do not discuss exactly how the vertical circulation contributes to this coupling (Jones, 1995). This coupling is central to the explanation of how shear affects TC intensification, as it seems to be one of the primary ways in which vertical shear (and environmental influences in general) interacts with TC structure. Explaining and determining the abilities of these vertical coupling mechanisms for the various sizes, intensities, and positions of TCs essentially determines their abilities to resist vertical shear and it thus provides us good reason for including the seemingly tangentially related discussion on TC movement.

This background enables us to discuss the way in which this internal coupling takes place and how exactly it relates to the way in which sustained levels of shear negatively impact TC intensification. While the PV and θe "venting" theories regarding the shear-intensification relationship are not too elaborate, the important midlevel warming theory does require a bit of backdrop.

3. The Midlevel warming hypothesis

As mentioned, Wang and Holland (1996) alluded to the importance of looking at just the tilt of the cyclonic vortex itself, without explicitly considering the impact of the positioning of the upper-level anticyclone and lower-level cyclone. Towards the mid-1990s this new way of analyzing the impact of shear on TC intensity was becoming quite ubiquitous. The literature was now focusing more on the coping mechanisms used solely by the TCs' cyclone (now ignoring the anticyclone) to withstand vertical shear and why these coping mechanisms often broke down for certain TC characteristics; the impact of this shear was now inexorably associated with the tilt of the cyclonic vortex and the interactions between its upper and lower levels.

This refined way of thinking of the shear-intensity relationship was perhaps most importantly spelled out in a landmark paper by Jones (1995). Despite the extensive amount of relevant research done in this field at the time, her paper stands out as a standard which laid much of the groundwork behind the midlevel warming hypothesis. Jones' (1995) paper is based on her research done by looking at the behavior of initially-barotropic vortices in environmental flows. Her calculations were performed using a dry, hydrostatic primitive-equation numerical model on an f-plane. Despite the fact that she was performing barotropic simulations of TC vortices and examining the dynamical and adiabatic mechanisms that countered vertical shear, her explanations of such were so fundamental to the shear-intensity relationship that even later diabatic descriptions, such as the midlevel warming theory, owe much to her research.

Broadly, Jones' work was closely aligned with that of Flatau et al. (and other contemporaries), in that she was using a three-dimensional primitive equation model whose results indicate that the vertical shear of the environmental flow results in differential advection of the storm's potential vorticity. Her approach was unique for her time, however, in that, in order to isolate the vertical coupling mechanism between vortex levels, she performed her studies without the inclusion of diabatic processes, as she believed that none of the TC shear coping mechanisms depended on the presence of the these processes inherent in tropical cyclones. Jones specified the shear such that there was no potential vorticity gradient; the resulting absence of diabatic effects precluded the development of an upper-level anticyclone. Partly due to this technique, future researchers looked almost exclusively to the cyclonic tilt to explain the shear-intensity relationship without dwelling on the complicating implications of the anticyclone.

Jones' (1995) interest in the vertical coupling mechanisms was piqued when, in specifying a westerly flow where a 4m/s wind at the surface decreased to 0 at the model top (at 10km), she found that the vertical tilt of the accompanying vortex was much smaller than that which would be implied by simple advection by the basic flow. As imagined, the lower-level vortex was first displaced to the east of the upper-level vortex. Upon investigating, she found that the subsequent motion of the surface centre has a northward component, while the motion of the upper-level center has a southward component. This resulted in the fact that, while the center at 5km height was moving almost due east, the upper- and lower-level centers revolved cyclonically about it. After 24 hours this resulted in the fact that the vortex at lower levels is located to the west of the vortex at upper levels, which tilt is opposite to that which would be expected from the direction of the vertical shear.

This response of the TC to vertical shear can be explained by first thinking of the initial flow as consisting of two identical potential-vorticity anomalies, which consist of an upper and a lower anomaly that initially line up in the vertical and are under the wind regime specified above. As the lower level anomaly is displaced to the east, because of the shear, the upper level anomaly remains in its prior position. The downward projection of the upper anomaly gives a cyclonic circulation at the surface; since the lower anomaly is displaced to the east, it thus has a southerly component across its center because of the effect of the upper anomaly. This tends to advect the lower anomaly to the north until the line joining the centers of the two anomalies no longer lies in the east-west direction; this leads the downward projection to create an easterly component of the wind onto the lower level anomaly. Thus both the upper and lower anomalies continue to revolve around each other in a cyclonic manner.

Once this cyclonic revolution results in the upper- and lower-level anomalies being aligned in the north-south direction, the downward projection of the upper-level anomaly has a purely easterly component across the lower anomaly; thus, considering that the shear was specified as having a lower-level westerly flow, this particular vertical orientation of the PV anomalies acts as a significant coping mechanism enjoyed by the TC to counter the effects of vertical wind shear. A second, related, coping mechanism is that which comes about after the mutual anomaly revolution has placed the lower-level anomaly west of the upper-level anomaly. With the tilt now leaning into the wind, the environmental shear, in this case, works to reduce the vortex tilt and thus promotes a stable TC structure. Of course, these effects can be, and are, reversed. When the lower-level anomaly is south of the upper, the downward projection ceases to counter the shear and thus ceases to promote vortex alignment; similarly, when the lower-level anomaly is east of the upper, the vertical shear is exacerbating the vertical vortex tilt.

More directly relevant to the central theme of this paper is Jones' (1995) discussion of the vertical circulations that result from the impact of sheared environmental flows. In basic terms, the development of the vertical circulation can be associated with the maintenance of equilibrium. Once the vertical circulation tilts the PV anomaly, a thermal adjustment is required for the flow to be balanced; this is composed of a cooling down-shear and a warming up-shear. This cooling down-shear raises the isentropes there and later leads to adiabatic vertical motion, as parcels moving tangentially around the TC center follow the isentropes upward as they pass the down-shear part of the storm.

Since there are no diabatic processes in her model that could produce thermal anomalies, Jones (1995) concludes that these departures from the initial potential-temperature field can arise only due to the advection of the environmental temperature field by the vortex flow. After analyzing her barotropic cyclone simulations she noted that, after the cyclonically-rotating-upper-and-lower-PV-anomaly mechanism, these vertical motions are the second major way she believed a TC maintained an upright vortex in significant levels of shear. As the strength of the temperature anomalies required to maintain thermal wind balance depends on the size of the vertical tilt, the strength of the vertical velocity will also depend upon the size of the tilt. When modeling the behavior of a vortex, these are very measurable parameters that, at least adiabatically, have a seemingly clear relationship. Analyzing the durability of this relationship would enable the researcher to determine how general TC characteristics alter the storms' ability to withstand vertical shear.

While Jones (1995) discussed the TC adiabatic response, Flatau et al. (1994) had examined the importance of the diabatic secondary circulation to the ability of a TC to resist shear. He explained that this was manifested in the transport of high PV air from below at the center and downward advection of low PV air at larger radius; however, Jones (1995) noted that this did not sufficiently explain how the diabatic effects contributed to vortex coupling.

Both diabatic and adiabatic components of the TC vertical circulation's response to vertical shear play a central role in its relationship with TC intensity change. How this was so was explained in an important paper by DeMaria (1996), who is the principal author of the midlevel warming hypothesis with regard to the way in which vertical shear impacts TC intensity. In essence, his theory comes from an analysis of the consequences of the shear-induced PV tilt. As stated, the vortex tilt is accompanied by a change in thermal structure of the vortex so that the mass and wind fields remain quasi-balanced. His hypothesis is that these tilt related variations in the thermal structure affect atmospheric stability near the storm center, and thus storm intensity.

Essentially, the horizontal displacement of upper and lower PV anomalies causes a thermal shift such that there are colder relative temperatures down-tilt, and warmer temperatures in the direction of the low-level PV anomaly (or up-tilt). While this had been noticed previously (Jones, 1995), DeMaria (1996) made the astute observation that the midlevel temperature increases near the low-level vortex center. He also made the observation that the relative temperature changes associated with the vortex tilt will not only lead to a decrease in convection near the low-level center (because of the warming), but it will also lead to increased convection outside the eyewall which would further disrupt storm symmetry and circulation.

In order to test his hypothesis, DeMaria (1996) used a simple, two-layer barotropic model, as this was determined to be the simplest context in which the effects of vertical shear could be illustrated. For prognostic equations, he used the two horizontal momentum equations and the continuity equation for height. He also employed a PV approach to the model flow analysis. This is often an effective approach for a two-layer model, as the entire flow field can be determined from the PV distribution with suitable boundary conditions and a balance between the mass and wind fields.

The barotropic approach, which, by definition, neglects the upper level anticyclone, can be quite accurate in simulating intense storms, as such TCs often have cyclonic circulations that extend through the depth of the troposphere. Because of this fact, and then-recently published research, DeMaria (1996) decided to focus exclusively on the positive PV anomaly in both model layers. He made this decision partially based on the fact that Shapiro and Franklin (1995), in studying Hurricane Gloria, indicated that there was a core of high PV values within 150km radius of the storm center that extended up to at least 200hPa. The fact that this important study from DeMaria (1996) and that that from Jones (1995) both produced their vortex simulations using a at least initially, barotropic vortex, and thus essentially ignored the upper-level anticyclone, demonstrates the way in which the community had changed its way of thinking about TC tilt. Considering that only a decade earlier the widely accepted method of viewing the TC response to storm tilt was related to the interaction between the upper-level anticyclone and lower-level cyclone, this represents a notable sea change in the analysis of the TC shear-intensity relationship.

In order to simulate the effects of TC tilt on midlevel warming, DeMaria (1996) performed several experiments in which he placed the low-level PV anomaly at the domain center with the upper-level anomaly displaced to the east. After using upper-level eastward displacements of 0, 20, 40, and 60km, he observed that the midlevel temperature consistently increased near the low-level vortex center in order to maintain balance. He found that the maximum midlevel temperature anomaly had increased by roughly 3K when the upper-level PV anomaly was displaced 60km (at radius of maximum wind) to the east of the low-level vortex center. This level of midlevel warming, if experienced in a real TC, would be sufficient to inhibit convection and thus weaken TC intensity. Essentially, at 60km of horizontal displacement, the meridional component of the winds from each level have their maxima in terms of positioning vis-à-vis the center of the opposite layer. DeMaria (1996) hypothesized that not only would the tilt-induced midlevel warming prove inhibitive of convection and TC intensification, but the increase in asymmetry due to the increased convection down-tilt of the storm's low-level axis of rotation also appeared to hinder coherent storm structure and intensification.

As mentioned, for a purely eastward displacement, the overlapping winds at the centers of the PV anomalies would be increasingly meridional; in DeMaria's (1996) work, this maximum meridional wind (and maximum PV anomaly interaction) occurs at a displacement of 60km. DeMaria (1996) echoes Jones (1995) in explaining that the resulting cyclonic revolution of upper- and lower-level PV anomalies works to resist the effects of vertical wind shear. He then goes on to perform a regression analysis for all named TCs from the Atlantic basin to determine the relationship between shear and intensification for various types of TCs, the reasons for which will be discussed below.

While examining the mechanism by which a TC weakens in vertical shear is the topic of interest in this chapter, it is also of interest to know why these processes occasionally break down, as seen in TCs which appear to be able to strengthen in adverse environments. This interest has mainly been manifested by those in the DeMaria camp, because of the effect of various environmental parameters on the extent of TC vertical coupling, and thus it will now be discussed as an extension of the midlevel warming hypothesis.

The resistance of a vortex to vertical shear is a function of the Rossby penetration depth (D), which is a measure of the amount of vertical coupling in a TC. This parameter can explain the way in which many different factors can affect the tilt and rotation rate of a vortex. A few of these major influences include the Coriolis parameter, the strength of the vortex, static stability, the height of the vortex, and the width of the vortex. Accordingly, DeMaria (1996), in his TC regression analysis, used TC latitude, intensity, and size when analyzing the predictability of the shear-intensity relationship. He posited that TCs were more able to mitigate shear's impact when they were strong, large, and at higher latitudes.

Just the year before, Jones (1995) performed several penetration depth-related tests using her dry, primitive equation model on an f-plane while simulating initially barotropic vortices. Specifically, she used her model to look at how the magnitude of the vertical penetration depth depended on the Coriolis parameter, vortex strength, static stability, and horizontal length scale. She found that a weaker vortex (with a max tangential wind of 20m/s instead of 30m/s) led to a slower upper- and lower-level relative revolution rate and greater vertical tilt. Worth noting, however, is her explanation for why this took place; despite it making intuitive sense, she renders the mechanisms involved especially lucidly. Seeing that, in a weaker vortex, the tangential flow at the level of the upper-level PV anomaly is reduced, the flow due to the downward projection is also weaker. This results in reduced advection at a given level due to anomalies at other levels (Jones, 1995). The weaker vortex cannot counter the shear because its upper-level and lower-level PV anomalies cannot revolve (cyclonically) sufficiently for the resulting tilt to counter the prevailing shear.

When similarly conducting experiments by changing the Coriolis parameter she found that by increasing the latitude, from 12.5° N to 20° N, at which the f-plane is situated, the vortex manifested significantly less tilt, with a greater revolution rate. This was attributed to the fact that, everything else being equal, a higher Coriolis parameter would cause the isentropes to be more strongly tilted; since the TC's method to counter shear depends on the dynamic and thermodynamic processes having to adjust to these tilted isentropes, this increased revolution rate is expected. Jones (1995) also performed analogous experiments related to the length scale and height of the vortex and found that an increasing TC width led to a smaller vertical tilt and a greater revolution rate, whereas a higher vortex top led to a larger vortex tilt and smaller revolution rate, as the vertical coupling would obviously be reduced as vertical scale increases.

While Jones' (1995) simulations and DeMaria's (1996) six-year case study confirmed what had been hypothesized about the effects of penetration depth, their work did go a step further in that they quantified these effects through their model runs, multiyear analyses, and development of penetration depth equations. Jones (1995) first attempted to use quasi-geostrophic theory to define penetration depth, where f is the Coriolis parameter, L is the length scale, and N is the static stability. Using typical scales for her particular model, she came up with a depth of 930m. Admitting that that was unreasonable, largely because of the fact that TCs experience significant ageostrophic vertical motions, she then turned to the work of Hoskins et al. (1985) and Shapiro and Montgomery (1993), who formulated an alternate expression for penetration depth of an axisymmetric vortex as being

$$(f_{loc}(f+\zeta))^{1/2}L/N \text{, where } f_{loc}=f+2v_T/r \quad (1)$$

and ζ is the vertical component of vorticity. Noting the fact that the vorticity profile of her vortex was strongly peaked, she expressed concern as to how to use the above formula, but ultimately used average values of relative vorticity and tangential winds and found a reasonable Rossby penetration depth of 14km. Nevertheless, she still was not satisfied with these methods and conceded that, for the parameters involved in her experiments, it was not entirely clear how a penetration depth should be defined. DeMaria (1996) made attempts similar to those of Jones (1995) in an effort to define penetration depth for his particular model setup and appears to have enjoyed more success. To begin, he referenced the work of Shapiro and Montgomery (1993) as he looked to define Rossby penetration depth for the general atmosphere; this is done by solving the definition of the local Rossby radius for an axisymmetric vortex for the vertical scale, which gives

$$D \sim \frac{IL_R}{N}, \quad (2)$$

where L_R is the horizontal scale, N is the static stability and I is the inertial stability parameter defined by

$$I=\left[(f+\zeta)\left(f+\frac{2V}{r}\right)\right]^{1/2} \tag{3}$$

where V is the tangential wind and r is the radius. Equations 2 and 3 plainly show that D increases with latitude, length scale, and intensity (V and ζ) and decreases with static stability; DeMaria (1996) then went on to define D for his two-layer model using quasi-geostrophic arguments, which, while clever, are not relevant here. Equations 2 and 3 will suffice in giving an idea of what type of resistance to vertical shear a particular tropical cyclone will exhibit.

4. The venting hypothesis

While the inner-core θe venting hypothesis of Gray (1968) is somewhat simple, its effect, nonetheless, is not very well understood, as some would see the cooling in the upper-levels (upon which the theory is based) as a destabilizing factor beneficial to TC development (DeMaria, 1996). While both modeling (Frank & Ritchie, 2001) and observational studies (Knaff et al., 2004) have convincingly shown that TCs weaken from the top down through a vertical lowering of the upper-level warm core and accompanying vortex, it is also still not clear how this core erosion occurs. Thus, a bit of background regarding the theory and its beginnings, development, and treatment will be beneficial in order to understand the current state of research.

Gray (1968), in his global observational study of atmospheric conditions surrounding TC development, is noted as being the first to elucidate the venting hypothesis as a method by which to explain the effect of shear on TC intensity. He concluded that the dynamics of TC development are best viewed as a hydrostatic problem of temperature-pressure-wind adjustment. In his opinion, to form and maintain a TC, the mean tropospheric temperature must be increased and concentrated. This is a typical view of those in the venting camp; the strength of a TC is seen as being concentrated and maintained in the upper levels of the core, primarily manifested in high levels of equivalent potential temperature and potential vorticity. Gray (1968) explained shear's impact as producing a large ventilation of heat away from the developing disturbance. He found that the condensational heat released by the cumulus convection to the upper troposphere was advected in a different direction relative to the heat released at lower levels. The loss of heat in the core would lead to a hydrostatic adjustment to a higher minimum pressure and thus a weaker TC.

Through the following decades, there was little talk in the literature about the venting hypothesis as most TC researchers were more concerned with shear's effects on TC track (Shapiro, 1992; Flatau et al., 1994; Wang & Holland, 1996), TC structure (Bender, 1997), and with specifying the effect of different types of shear on TC intensity (Tuleya & Kurihara, 1981; Merrill, 1998), rather than how exactly shear produced these effects. However, as mentioned, general explanations as to how these processes occurred generally focused on dynamical processes (Jones, 1995), midlevel warming (DeMaria, 1996), and asymmetries between surface fluxes and moisture convergence (Peng et al., 1999).

Around the year 2000, there was renewed interest in the venting hypothesis because of the interesting results produced by Frank and Ritchie (1999 & 2001). In their discussions regarding this method of explaining TCs' reaction to shear, the arguments center on a few general tenets. First is the fact that the flux of θe and PV out of the core is responsible for TC weakening; the second is the positioning of the PV and θe maxima relative to the core; and the third is based on the fact that the TCs are seen as weakening from the top down, which ties in closely with the first two.

In their 2001 paper, using the MM5 with three embedded meshes, a 5km grid spacing, and a fully explicit representation of the moist convective processes, they examined the response of mature, idealized hurricane-like vortices to various types and magnitudes of vertical wind shear. The MM5 solves the nonlinear, primitive equations using Cartesian coordinates in the horizontal and a terrain-following sigma coordinate in the vertical. This new type of model was crucial to the goals of their study as numerical simulations of tropical cyclone-shear interactions require a model with a very large domain that is capable of resolving the storm's environment and external circulation features, as well as a high-resolution core region capable of resolving the small, intense core processes of the storm (Frank & Ritchie, 1999). Their simulations were performed on an f-plane centered on 15 degrees north, so that complications would not arise from the interaction between the storm flow and the planetary meridional absolute vorticity gradient, as these can often alter the mean flow in the storm's core region (Frank & Ritchie, 2001).

Looking at wind regimes of 1) no large-scale flow, 2) easterly flow, 3) 5m/s shear, 4) 10m/s shear, and 5) 15m/s shear (whose sheared winds varied from the surface to the upper-troposphere), they found that, in the case of a vortex in shear, as opposed to a vortex embedded in uniform zonal flow, the weakening of the storms took place through a well defined, multistep process. First, the storm core develops a strong, wavenumber one asymmetry with regard to vertical motion, rainfall, and cloud water at most levels, shortly after the shear is applied. The asymmetry is then great enough so that upper-levels of θe and PV lose their concentration in the eye of the storm and become concentrated solely within portions of the eyewall and other rainbands; this occurs primarily through outward eddy fluxes. This loss of core warmth leads to an increase in surface pressure, which decreases the circulation at all levels. This asymmetric pattern then moves down with time, accompanying the tilt's progression, such that the structure continues to weaken. At some point this is halted because the stronger circulation at lower levels somewhat preserves the symmetric vortex, such that equilibrium is reached at an intensity well below the storm's MPI.

As to how well these storms were able to avoid this process and withstand shear, Frank and Ritchie (2001) report that the vortex in 5m/s shear was able to maintain its strong, vertically aligned structure for about a day and a half, while the vortex in 10m/s shear continues to strengthen for 18-24 hr before weakening. While shear of 15m/s tore an intense, idealized storm apart within about one day, the delayed response of the storm under light shear suggests that the processes that tend to keep the storm axisymmetric are able to temporarily overcome shear of less than 10m/s.

DeMaria subsequently worked with Knaff et al. (2004) to perform a composite analysis of TC warm cores as they relate to various levels of shear. What was unique about this study is that they used temperature soundings from the Advanced Microwave Sounding Unit (AMSU) A instruments to analyze the characteristics of 186 TCs in various levels of shear from 1999-2002 TC seasons in the Atlantic and east Pacific basins and from May-December 2002 for the western North Pacific. Defining shear as the 24 hr averaged vector difference of the horizontal wind between 200 and 850hPa, they created two TC composites. One was for intense storms (46-52m/s maximum surface winds) that displayed favorable shear conditions (<7.5m/s) and the other included intense storms that were experiencing significant shear (>7.5m/s); comparing the two composites, they analyzed the inner core temperature anomalies in hopes of finding a consistent change in the TC axisymmetric thermal structure due to shear variations.

Knaff et al. (2004) calculated the TC temperature anomalies relative to the azimuthally and radially averaged temperature from a 500-600km radius and found that, as the shear increases, the height of the balanced hurricane vortex decreases. Confirming the results of Frank and Ritchie (2001), Knaff et al. (2004) concluded that the general effect of vertical shear on a mature TC is the top-down erosion of its warm-core structure due to the downward propagating horizontal fluxes of potential temperature, although they admitted they weren't sure how this occurred.

5. Recent developments

Much of the recent work in this field, besides directly addressing or confirming the results above (Jones 2000; Wong and Chan 2004), looks into which large-scale parameters accompany TC intensification (Paterson et al., 2005; Zeng et al., 2007,2008; Garner et al., 2009; Hendricks et al., 2009), discussed how TC asymmetries affect storm intensity (Yang et al., 2007; Sang et al. 2008), examines the role of turbulent fluxes on intensification (Zhu, 2008; Bryan and Rotunno, 2009; Rotunno, 2009), and proposes several new mechanisms and frameworks to describe the way TCs are detrimentally affected by shear (Reasor et al., 2004; Riemer et al., 2009; Tang and Emanuel 2010). Using a primitive equation model to study baroclinic vortices on an f-plane, Jones (2000) found, contrary to the work of DeMaria (1996), that shear-induced tilt in the TC core region may actually reduce stability in certain parts of the vortex, although the implications for TC intensity weren't entirely clear. Similar to the studies of Frank and Ritchie (1999,2001), Wang and Chan (2004) used the MM5 at 4km resolution to analyze the response of idealized TCs on an f-plane to various levels of vertical shear. What they found was that while TCs under shears of 6-8m/s are not as intense as TCs under no shear, it's only under shears above 10m/s when a TC weakens significantly. Confirming the results of Jones (1995), they found that TCs resist tilt by a mutual rotation between the various levels; they also found that there was significant warming in the lower half of the troposphere near the center of the TC, which may corroborate part of what DeMaria (1996) was seeing happen in the midlevels of his TCs.

In the last five years there have been several attempts to distinguish the environmental characteristics most favorable for TC intensification. Using NCEP-NCAR reanalyses, Paterson et al. (2005) found that, in the Australian region, shears of 2-4m/s actually favor significant intensification and shears above 12m/s favored rapid weakening; using the same data, Zeng et al. (2007) examined the effect of transitional speed and vertical shear on TC intensification in the North Pacific and North Atlantic (Zeng et al., 2008) from 1981-2003. They found that few TCs intensified when their environmental wind shear was above 20m/s and they developed a new empirical maximum potential intensity (MPI) which includes the combined negative effect of translational speed and vertical shear. Hendricks et al. (2009) similarly used the NOGAPS global analysis and the TRMM Microwave Imager (TMI) to examine environmental and climatological characteristics of tropical cyclones undergoing different intensity changes in the western North Pacific and North Atlantic basins. Interestingly, they found that the environment of rapidly intensifying TCs and moderately intensifying TCs is quite similar. The exceptions to this was the fact that in the Atlantic rapid intensification (RI) events occurred in environments with weaker deep-layer shear than moderately intensifying evens; an important finding was that the rate of TC intensification was not critically dependent on SSTs. Using a compressible, hydrostatic, dynamical model nested in global analysis fields to investigate the relative effects of shear and thermal stratification on TC variability, Garner at al. (2009) found that the sensitivity of the model's tropical storm activity to projected changes in vertical shear was substantial; projecting through the 21st century according to the IPCC's A1B scenario, it was found that the model's reduction in Atlantic tropical cyclone activity was largely driven by the increased seasonal-mean vertical shear in the western Atlantic and Caribbean (Garner et al., 2009). While all of this serves to confirm the importance of vertical shear in TC intensification, there have also been many recent, forward-looking papers helping to explain the nature of the relationship.

In the last ten years one of the most unique studies was that of Reasor et al. (2004), who, employing a Boussinesq PE model to simulate barotropic vortices on an f-plane, found that VRWs, rather than the diabatic secondary circulation, are primarily responsible for a TC's resistance to shear. Notably, this resistance was sustained for shears above 10m/s without the direct aid of cumulus convection. While their work explained several neglected topics (e.g., the presence of quasi-stationary tilted states for a TC in shear and the prominence of radially and azimuthally propagating VRWs that arise from a TC forced by shear), they admit that the much-debated role of diabatic processes on TC resiliency was still unclear. Subsequently, Riemer et al. (2009) developed a new explanation of the vertical shear-intensity relationship. He emphasized the role of downdrafts and their transport of low θe air from the mid-troposphere into the boundary layer and its effect on eyewall energetics from a Carnot perspective. Relatedly, Tang and Emanuel (2010) established an idealized framework based on steadiness, axisymmetry, and slantwise neutrality to assess how shear-related ventilation affects TC intensity via downdrafts outside the eyewall and through midlevel eddy fluxes directly into the eyewall. As did Riemer et al. (2009), Tang and Emanuel (2010) explain mechanisms' detrimental effect on TC intensity by considering the storm as a Carnot heat engine; when these shear-related processes reduce the maximum entropy present in the core of the TC, and thus decrease the thermodynamic efficiency, the result is a decrease in the total amount of work that can be performed by the TC in combating frictional dissipation. As mentioned by many previous papers, TCs with higher potential intensities (PIs) are better able to withstand wind shear as the steady state intensity increases while the minimum intensity needed for sustainability decreases with increasing potential intensity (Tang and Emanuel 2010). It thus is well established that vertical shear leads to a decrease in the amount of energy available to the TC to be used as work, however, exactly where and how this occurs is still being debated.

As mentioned in the research of Frank and Ritchie (2001), it appears that turbulent eddies play a key, yet little understood role in the shear-related venting that works to reduce a TC's thermodynamic efficiency. The importance of this turbulent flux on TC intensification has recently come to the fore due to 2 papers involving Richard Rotunno. First, in Bryan and Rotunno (2009), they touched on the subject indirectly when using an axisymmetric numerical model to evaluate the maximum PIs of tropical cyclones. They showed that turbulence in the radial direction limits maximum intensity by weakening the radial gradients of angular momentum and of entropy; maximum wind speed turned out to be very dependent on horizontal mixing length (Bryan and Rotunno, 2009). Echoing Frank and Ritchie (2001), they subsequently declared that the difficulty in real time forecasts of intensity may be partly related to the specification of turbulence in NWP models and/or the general lack of understanding of turbulence effects in hurricanes (Bryan and Rotunno, 2009). Considering the typical mesoscale model has a resolution of 1-3km and much of the critical venting occurs through turbulent eddies on the order 100m the importance of the models' ability to resolve or approximate turbulence is essential.

Building on this fact, Rotunno et al. (2009) published a study which examined the importance of turbulent transfer on TC modeling using a large-eddy simulation of a tropical cyclone using the Advanced Research WRF. Employing six nested grids with a horizontal resolution of 62m, they analyzed the impact of the resolution of turbulent eddies on maximum TC wind speed. They found that the storm strength increased significantly when the grid spacing was reduced while the turbulence was parameterized (from 1.67km to 185m), but the mean winds actually decreased when increasing the resolution from 185 to 62m, as turbulent eddies are now being resolved (Rotunno et al., 2009). Especially interesting is the fact that the study confirmed that there is strong turbulence along the inner edge of the eyewall of intense hurricanes; considering the essential role turbulence plays in several hypothesis concerning venting in this part of the TC (Frank and Ritchie, 2001; Tang and Emanuel, 2010), this area of research is key to being able to explain how TCs interact with their environment. Further, the clear dependence of mean TC intensity and radius of the TC eyewall on the effects of resolved turbulence demonstrates the importance of correctly approximating turbulent diffusion when modeling TC intensification.

Recently, research conducted by Thatcher (2010) attempted to address whether a TC is negatively impacted by vertical shear through a midlevel warming (DeMaria, 1996) or an upper-level entropy venting (Frank & Ritchie, 2001). Using the WRF-ARW at a 3km grid spacing while simulating Typhoon Jangmi (2008), it was found that the TC's intensification ended while under moderate shear (~7m/s) due to the establishment of a theta-e maxima from 200-300hPa in the eyewall; prior to this, the concentration had been relatively uniform across the core of the TC at upper-levels. It was found that Jangmi's mean sea-level pressure (MSLP) was most sensitive to theta-e values from 200-300hPa in the inner part of the eyewall (30-50km). The end of the intensification appears to be due to two factors related to upper-level core venting: 1) the MSLP is affected through a hydrostatic adjustment to reduced inner eyewall energy values and 2) the upper part of the eyewall stabilizes due to outward energy flux. However, due to the 3km resolution of the model, it was difficult to determine the role venting-related turbulence played in the end of Jangmi's intensification.

6. Conclusion

Research over the past 4 decades has established that environmental forces at large radii have a significant impact on tropical cyclone intensification. It has also been established that environmental vertical wind shear has a detrimental effect on TC strength. This has been confirmed repeatedly even in the last 5 years (Paterson et al., 2005; Zeng et al., 2007, 2008; Garner et al., 2009; Hendricks et al., 2009). While small amounts of vertical shear have been seen as beneficial to TC development (Tuleya & Kurihara, 1981; Paterson et al., 2005), shears above 8-12m/s (DeMaria Kaplan, 1999; Wong and Chan, 2004; Paterson et al., 2005) have proven deleterious to TC intensity and structure. The question is how this works.

Generally accepted is the fact that, overall, the TC approximates a Carnot heat engine. These storms convert heat into work. Any interaction that inhibits this (e.g., internal dynamics, large-scale environmental flow, or ocean welling) will keep a TC from reaching its maximum potential intensity. Theories as to how vertical shear keeps a TC from its MPI are myriad, yet they generally are seen as either relating to the dynamics or thermodynamics. The theories based on a fluxing of low-entropy air into the core of a TC have found support (Frank and Ritchie, 2001; Tang and Emanuel, 2010), as have the theories stating that down drafts of low entropy-air into the near-core boundary layer lessen eyewall buoyancy and thus TC intensity (Riemer et al., 2009; Tang & Emanuel, 2010). Works on the dynamical aspect have mostly concentrated on the ways by which TCs are able to cope with high levels of vertical shear (Jones, 1995; Schecter et al. 2002; Reasor, et al., 2004) rather than how the TC's intensification is detrimentally affected; it thus appears that a TC's intensification is ultimately inhibited through mostly thermodynamic processes.

Nevertheless, despite decades of effort, the true nature of relationship between vertical wind shear and TC intensification is not properly understood. Along with a lack of proper model initialization, the perpetual mystery behind the TC's interaction with its environment is one of the major issues behind the community's lack of ability in forecasting TC intensity change. The frustrations voiced in solving this mystery have been myriad. For example, although he performed a much-cited modeling study, DeMaria (1996) expressed the need for detailed observations to determine the true effect of shear on TC midlevel temperature perturbations. He went on to clearly state the need for three-dimensional model simulations with explicitly resolved convection in order to discern the feedback process between the vertical temperature structure and convection near the TC center, which process is key to understanding the shear-intensity relationship. Despite the confident pronouncements that resulted from their TC composite analysis, Knaff et al. (2004) made the concession that their symmetric analysis could not determine how the warm-core erosion occurs, partially due to the coarse (50km) resolution of the AMSU. Even Rotunno et al. (2009) expressed the need for still higher resolution and computing resources, yet lamented the computational cost of the then-current experiments.

One of the ways forward appears to be the proper parameterization, or explicit resolution of the turbulent eddies in the near core region of TCs, especially as they relate to the transport of entropy into and out of the TC eye (Frank and Ritchie, 2001; Bryan and Rotunno, 2009; Rotunno et al., 2009). The impact of this turbulence is large, as, when parameterized, it looks to consistently affect both the radius of the eyewall and the TC's maximum tangential velocity (Rotunno et al., 2009). It appears that a significant amount of these eddies are occurring on the order of 100m (Rotunno et al., 2009), however, the computational cost of routinely simulating a TC at that resolution is currently too expensive. Thus, large-eddy simulations (LES) will be primarily used to determine the eddy-diffusion coefficients necessary to properly parameterize these motions when using a grid spacing on the order of a kilometer.

Going forward, increasing wind observations and the assimilation of these wind measurements into the models would improve model initial conditions in terms of the accurate representation of the wind shears and thus enhance the numerical model's ability to produce accurate TC intensity forecasts (e.g., Pu et al. 2008). The future in the shear-intensity area of research looks particularly dependent on increases in computing power, which will enable scientists to examine the above questions without as much of a reliance on parameterizations. It appears that many of the above issues ultimately won't be resolved until simulations are able to explicitly resolve the small scale convection and turbulent eddies which populate (and affect) much of the TC inner vortex, while also properly simulating environmental effects at 500km radius from the TC center.

7. References

Bender, M.A., 1997: The effect of relative flow on the asymmetric structure in the interior of hurricanes. *J. Atmos. Sci.*, 54, 703-724.

Bryan, G.H., & R. Rotunno, 2009: The maximum intensity of tropical cyclones in axisymmetric numerical model simulations. *Mon. Wea. Rev.*, 137, 1770-1789.

Camp, J.P., & M.T. Montgomery, 2001: Hurricane Maximum Intensity: Past and Present. *Mon. Wea. Rev.*, 129, 1704-1717.

Chen, S.S, J.A. Knaff, & F.D. Marks Jr., 2006: Effects of vertical wind shear and storm motion on tropical cyclone rainfall asymmetries deduced from TRMM. *Mon. Wea. Rev., 134, 3190-3208.*

Corbosiero, K.L., & J. Molinari, 2002: The effects of vertical wind shear on the distribution of convection in tropical cyclones. *Mon. Wea. Rev.*, 130, 2110-2123.

DeMaria, M., 1996: The effect of vertical shear on tropical cyclone intensity change. *J. Atmos. Sci.*, 53, 2076-2087.

DeMaria, M., & J. Kaplan, 1999: An updated Statistical Hurricane Intensity Prediction Scheme (SHIPS) for the Atlantic and eastern North Pacific basins. *Wea. Forecasting*, 14, 326-337.

Flatau, M., W.H. Schubert, & D.E. Stevens, 1994: The role of baroclinic processes in tropical cyclone motion: the influence of vertical tilt. *J. Atmos. Sci.*, 51, 2589-2601.

Frank, W.M., and E.A., Ritchie, 1999: Effects of environmental flow upon tropical cyclone structure. *Mon. Wea. Rev.*, 127, 2044-2061.

Frank, W.M., & E.A., Ritchie, 2001: Effects of vertical wind shear on the intensity and structure of numerically simulated hurricanes. *Mon. Wea. Rev.*, 129, 2249-2269.

Garner, S.T., I.M. Held, T. Knutson, & J. Sirutis, 2009: The roles of wind shear and thermal stratification in past and projected changes of Atlantic tropical cyclone activity. *J. of Clim.*, 22, 4723-4734.

Gray, W.M., 1968: Global view of the origin of tropical disturbances and storms. *Mon. Wea. Rev.*, 96, 969-700.

Gryanik, V.M., & M.V. Tevs, 1989: Dynamics of singular vortices in a N-layer model of the atmosphere. *Bull. USSR Acad. Sci. Atmos. Oceanic Phys.*, 25, 179-188.

Hendricks, E.A., M.S. Peng, B. Fu, & T. Li, 2010: Quantifying environmental control on tropical cyclone intensity change. *Mon. Wea. Rev.*, 138, 3243-3271.

Holland, G.C., & R.T. Merrill, 1984: On the dynamics of tropical cyclone structural changes. *Q.J.R. Met. Soc.*, 110, 723-745.

Hoskins, B.J., M.E. McIntyre, & A.W. Robertson, 1985: On the use and significance of isentropic potential vorticity maps. *Q.J.R. Met. Soc.*, 111, 877-946.

Huntley, J.E., & J.W. Diercks, 1981: The occurrence of vertical tilt in tropical cyclones, *Mon. Wea. Rev.*, 109, 1689-1700.

Jones, S.C., 1995: The evolution of vortices in vertical shear. I: Initially barotropic vortices. *Q.J.R. Met. Soc.*, 121, 821-851.

Jones, S.C., 2000: The evolution of vortices in vertical shear. III: Baroclinic vortices. *Q.J.R. Met. Soc.*, 126, 3161-3185.

Knaff, J.A., S.A. Seseske, M. DeMaria, & J.L. Demuth, 2004: On the influences of vertical wind shear on symmetric tropical cyclone structure derived from AMSU. *Mon. Wea. Rev.*, 132, 2503-2510.

Merrill, R.T., 1988: Environmental influences on hurricane intensification. *J. Atmos. Sci.,* 45, 1678-1687.

Paterson, L.A., B.N. Hanstrum, N.E. Davidson, & H.C. Weber, 2005: Influence of environmental vertical wind shear on the intensity of hurricane-strength tropical cyclones in the Australian Region. *Mon. Wea. Rev.,* 133, 3644-3660.

Peng, M.S., B.-F. Jeng, & R.T. Williams, 1999: A numerical study on tropical cyclone intensification. Part I: Beta effect and Mean Flow Effect. *J. Atmos. Sci.,* 56, 1404-1423.

Pu, Z.; Li, X; Velden, C.; Aberson, S. & Liu, T. W, 2008: Impact of aircraft dropsonde and satellite wind data on the numerical simulation of two landfalling tropical storms during TCSP. *Wea. Forecasting.,* 23, 62-79.

Ramage, C.S., 1959: Hurricane Development. *J. of Meteorology,* 16, 227-237.

Reasor, P.D., M.T. Montgomery, L.D. Grasso, 2004: A new look at the problem of tropical cyclones in vertical shear flow: Vortex resiliency. *J. Atmos Sci.,* 61, 3-22.

Riemer, M., M.T. Montgomery, M.E. Nicholls, 2009: A new paradigm for intensity modification of tropical cyclones: thermodynamic impact of vertical wind shear on the inflow layer. *Atmos. Chem. Phys. Discuss..,* 9, 10711-10775.

Rodgers, R., S. Chen, J. Tenerelli, H. Willoughby, 2003: A Numerical study of the impact of vertical shear on the distribution of rainfall in Hurricane Bonnie (1998). *Mon. Wea. Rev.,* 131, 1577-1599.

Rogers, R., Aberson, S., Black, M., Black, P., Cione, J., Dodge, P., Dunion, J., Gamache, J., Kaplan, J., Powell, M., Shay, N., Surgi, N., & Uhlhorn, E., 2006: The Intensity Forecasting Experiment: A NOAA multiyear field program for improving tropical cyclone intensity forecasts. B. Am. Meteorol. Soc., 87, 1523–1537.

Rotunno, R., Y. Chen, W. Wang, C. Davis, J. Dudhia, & G.J. Holland, 2009: Large-eddy simulation of an idealized tropical cyclones. *Bull. Amer. Meteor. Soc.,* 90, 1783-1788.

Sang, N.V., R.K. Smith, M.T. Mongomery, 2008: Tropical-cyclone intensification and predictability in three dimensions. *Q.J.R. Meterorol. Soc.* 134, 563-582.

Schecter, D.A., M.T. Montgomery, & P.D. Reasor, 2002: A theory for the vertical alignment of a quasigeostrophic vortex. *Mon. Wea. Rev.,* 59, 150-168.

Shapiro, L.J., 1992: Hurricane vortex motion and evolution in a three-layer model. *J. Atmos. Sci.,* 49, 140-153.

Shapiro, L.J., & Franklin, J.L, 1995: Potential vorticity in Hurricane Gloria. *Mon. Wea. Rev.,* 123, 1465-1475.

Shapiro, L.J., & M.T. Montgomery, 1993: A three-dimensional balance theory for rapidly rotating vortices. *J. Atmos. Sci.,* 50, 3322-3335.

Tang, B. & K. Emanuel, 2010: Midlevel ventilation's constraint on tropical cyclone intenisty. *J. Atmos. Sci.,* 67, 1817-1830.

Thatcher, L. 2010. How vertical wind shear affected the rapid intensification of Typhoon Jangmi (2008). Masters thesis, University of Utah. 165 pp.

Tuleya, R.E., & Y. Kurihara, 1981: A numerical study on the effects of environmental flow on tropical storm genesis. *Mon. Wea. Rev.,* 109, 2487-2506.

Wang, Y., & G.J. Holland, 1996: Tropical cyclone motion and evolution in vertical shear. *J. Atmos. Sci.,* 53, 3313-3332.

Wang, B., & X. Li, 1992: The beta drift of three-dimensional vortices: a numerical study. *Mon. Wea. Rev.,* 120, 579-593.

Wang, Y., & C.-C. Wu, 2004: Current understanding of tropical cyclone structure and intensity changes—a review. *Meteo. And Atmos. Physics,* 87, 257-278.

Wong, M.L.M, & J.C.L. Chan, 2004: Tropical cyclone intensity in vertical wind shear. *J. Atmos Sci.,* 61, 1859-1876.

Yang, B., Y. Wang, B. Wang, 2007: The effect of internally generated inner-core asymmetries on tropical cyclone potential intensity. *J. Atmos. Sci.,* 64, 1165-1188.

Xiao, Q., X. Zou, & B. Wang, 2000: Initialization and simulation of a landfalling hurricane using a variational bogus data assimilation scheme. *Mon. Wea. Rev.,* 128, 2252-2269.

Zeng, Z., Y. Wang, C.-C. Wu, 2007: Environmental dynamical control of tropical cyclone intensity—An observational study. *Mon. Wea. Rev.,* 135, 38-59.

Zeng, Z., L. Chen, Y. Wang, 2008: An observational study of environmental dynamical control of tropical cyclone intensity in the Atlantic. *Mon. Wea. Rev.,* 136, 3307-3322.

Zhu, P., 2008: Simulation and parameterization of the turbulent transport in the hurricane boundary layer by large eddies. *J. Geophys. Res.,* 113, D17104.

14

Formation of Gravitational Hurricanes Simulated by Numerical Gravitational Gas Dynamics Model

A. Pavlov
Kepler Simulation Lab, St Andrews House, 90 St Andrews Road, Cambridge CB4 1DL, UK

1. Introduction

The formation and development of hurricanes is of great importance to the forecast centers. Computer simulations essentially improved the forecasts of hurricanes and allow to model their tracks (Dudhia 1993, Grell *et al.* 1994, Zhang *et al.* 2000). The forecasts rely on monitoring real data of satellite atmospheric images (Chelton 2000), Doppler winds (Marks *et al.* 1992), baroclinic disturbances (Davis 2001) and vortices (Kurihara 1998) to detect the hurricanes at the beginning of their formation. However, the simulation models are very complicated (Liu 1999, Tong 2004) including complex parametric schemes for initializing the simulation programs. A more important problem is that there exists no simple physical model for hurricanes and therefore the estimation of their intensity (DeMaria 1994, DeMaria 1999, Camp 2001, Bister and Emanuel 1998) is inadequate and prediction times are relatively short. In this paper we show that density variations in the atmosphere may produce gravitational instability that can lead to the formation of a hurricane. This is a result of a self-gravitational contraction of a slowly rotating gaseous cloud. We developed a gravitational N-body model to simulate the formation and dynamics of a hurricane. The simulation program enables us to predict hurricanes and their intensities at early stages using only two initial parameters, density and velocity. Hurricanes (Holland 1997) are very large sustainable rotating clouds of atmospheric gas with typical size of 1000 km having as usual an eye in the center and spiral arms, whose extreme velocities exceed 30 m/s (Simpson and Riehl 1981). They form above the warm oceans in tropical and subtropical regions (Whitney and Hobgood 1997) and the general conditions favoring their formation have been known for a long time (Gray 1968, McBride and Zehr 1981). Hurricanes are a consequence of the processes in the atmosphere under strong influence of the ocean and the Sun (Merrill 1988, DeMaria 1996). The earth's atmosphere is a very dynamical system. It has been observed that atmospheric pressure, temperature, precipitations and winds including upper winds vary with time (Miller 1958, Malkus and Riehl 1960). Also, variations in angular momentum of the atmosphere can affect the Earth's angular momentum, and consequently, the length of day (Rosen and Salstein 1983, Brzezinski *et al.* 2002). The original model of the hurricane proposed in this paper uses variation of density within the atmosphere resulting in gravitational instability and development of cloud's vorticity. Not only the gravitation of the Earth but also the gravitational interaction between gas molecules plays a very important role in global climate dynamics, particularly, in formation of hurricanes and tropical cyclones.

2. Materials and methods

We consider a numerical modeling of a gaseous cloud using a many-body (N-body) model. The model was used to simulate gravitational systems such as spiral galaxies (Pavlov and Pavlova 2003a) and Saturn (Pavlov and Pavlova 2003b). That model was modified to simulate a self-gravitation of relatively thin disk-like gaseous clouds. The model uses two physical principles - the central gravitational field originating from the center of mass of the cloud and conservation of the angular momentum. In the N-body system, the gravitational force exerting on a *i*th body from all the other bodies is equivalent to a gravitational force from a virtual mass M_i placed in the center of mass. The value M_i depends on the positions of all the bodies relative to the position of the *i*th body. Thus, a whole system is characterized by a set of virtual masses M_i whose values are continuously changing during the simulation due to the change in the relative positions of the bodies.

We simulated the formation of a hurricane as a result of gravitational contraction of the gaseous cloud. The size of the initial cloud is about 2000 km. The algorithm aims for future practical applications. Therefore, we introduced into the simulation 2500 bodies. This number corresponds to a mesh 50x50 with grid spacing 40 km corresponding to a coarse-mesh domain size used for monitoring and computer simulations of the tracks of the hurricanes (Rosenthal 1970). To further simplify the procedure of obtaining the initial conditions, we use randomly distributed positions of the bodies allowing monitoring measurements at random places instead of equally spacing rectangular domains in traditional meshes.

The rotation of the cloud is initiated by the difference in linear movement of the rotating earth's surface when the atmosphere at the equator moves faster than at larger latitudes. There are two observations that help us to understand this process. It is observed that hurricanes are formed outside the equatorial line of around 200 km and rotate counter clockwise in the northern hemisphere and clockwise in the southern hemisphere. If the cloud is formed north from the equator then its northern part moves slower than its southern part and due to the west-to-east general movement of the earth's surface a counter clockwise rotation is generated. Consequently, in the southern hemisphere a clockwise rotation is analogously generated. If the cloud is situated exactly over the equator, then two opposite torques compensate each other. This explains the existence of the calm 200 km-wide equatorial strip.

3. Results

Figure 1 shows the time-evolution of the cloud due to internal gravitational forces only with no influence from the surrounding atmosphere. The cloud has a shape of an elliptical disk. During the gravitational contraction, the disk transforms into a spherical central part and a two-armed spiral. After 2.34×10^5 seconds (65 hours) of the contraction process the eye starts to develop. The eye is a region where the centripetal forces are larger than gravitational ones. The density of the cloud near the eye edge reaches its maximum. Because the density increases faster in the center, this part of the cloud starts to fall downwards the earth's surface (ocean surface) due to the earth's gravitation. This will result in formation of the eyewall. Figure 2 shows the development of the eye in more detail. It is seen that prior to the eye the density in the center increases. This feature can be used as additional indication of hurricanes. During the contraction process, the cloud intensifies reaching the category 1 after 2.50×10^5 seconds. This stage corresponds to the appearance of the eye. With further contraction the hurricane approaches category 2 after 2.7×10^5 seconds, category 4 (3.5×10^5 s.) and category 5 (4.0×10^5 s.) after around 5 days.

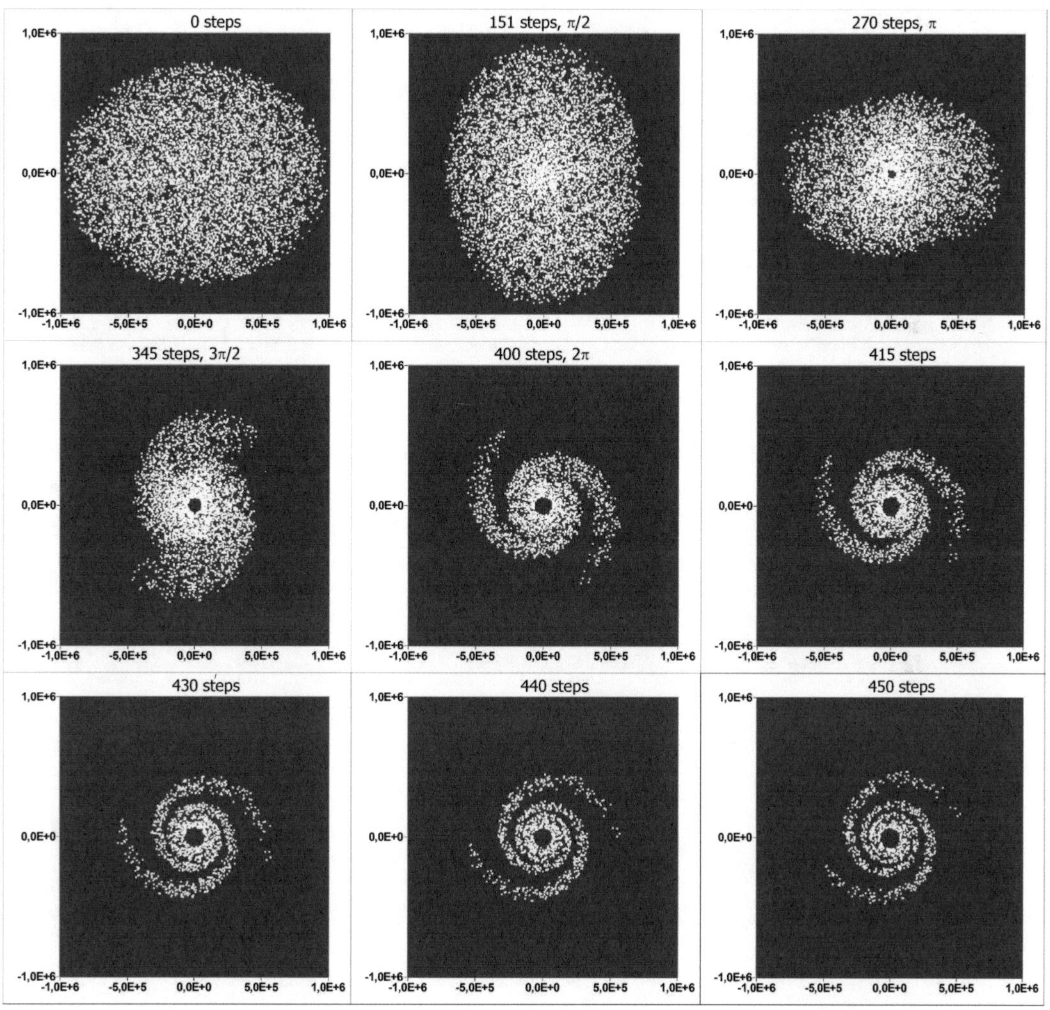

Fig. 1. Time development of the slowly rotating elliptical cloud during self-gravitational contraction. One time step is equal to 1000 seconds. The cloud rotates counter clockwise. The initial angular velocity is 1.0×10^{-5} rad/s. The turning angle shows an average rotation of the cloud relative to the initial orientation of the cloud.

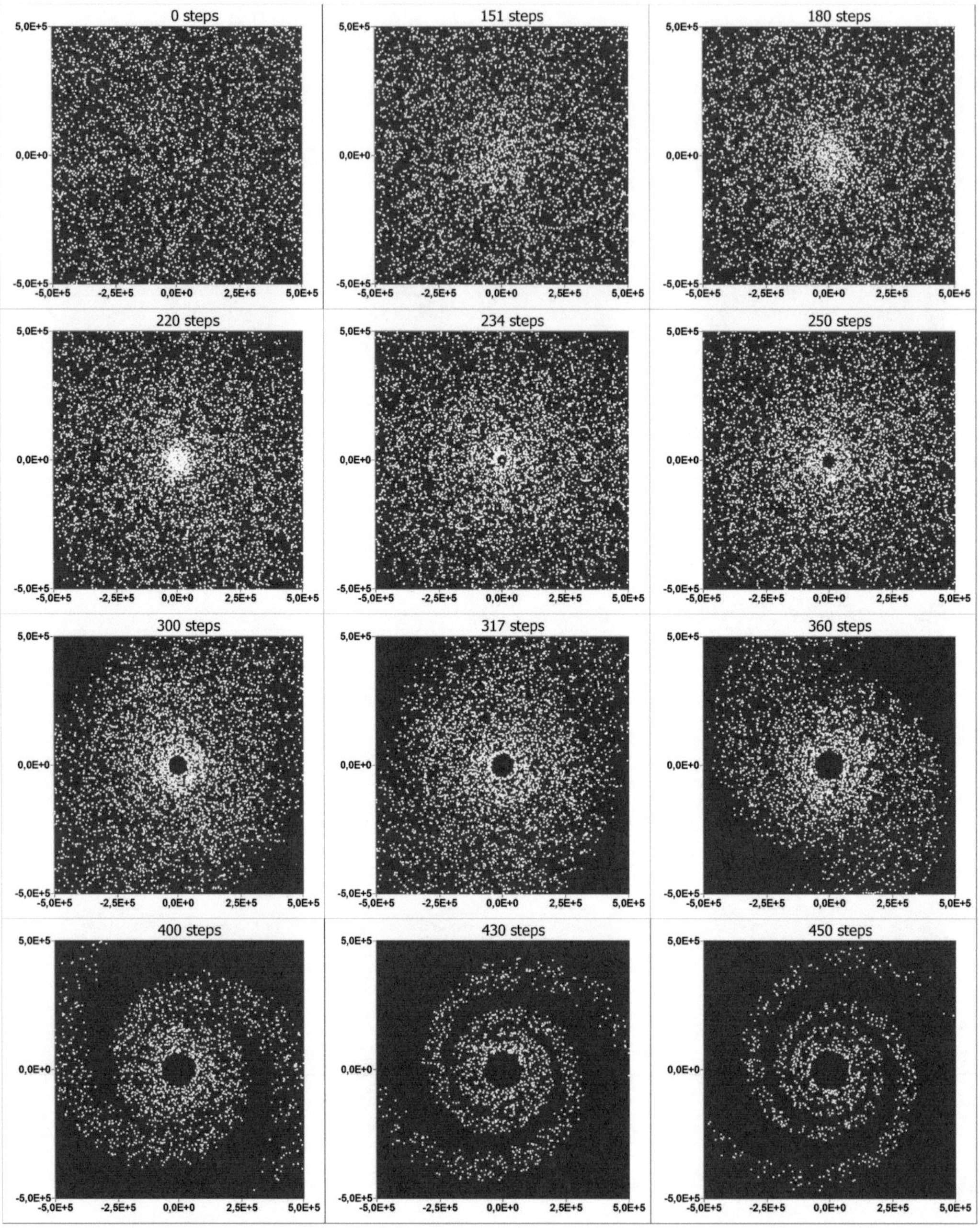

Fig. 2. The formation of the eye of the hurricane is shown in more detail. It starts to form in a middle stage of the process. Density near the eye's edge varies with time and periodic formation of the dense rings is observed as can be seen in the 317 steps graph.

Figure 3 shows the tangential and radial velocities of the cloud for different stages. At the beginning, the cloud rotates with angular velocity of $1x10^{-5}$ rad/s and the tangential velocities of the bodies form a linear-like radial dependency with a maximum speed of around 10 m/s at distances 1000 km from the center. This distance corresponds to double the radius of the final hurricane. During the contraction process, the linear dependency transforms dramatically. The bodies situated closer to the center start to rotate faster reaching typical extreme hurricane's velocities near the eye edge. The velocity dependency forms a smooth curve revealing that the process is a self-clocking in which every body contributes into the dynamics of the whole system. Inside the eye region, the density is very low and velocities are almost zero corresponding to calm weather and clean sky. These features are usually observed in the eye's regions of real hurricanes. The increase of the maximum velocity during the formation process correlates well with the extreme wind dependency proposed by Emmanuel (1999).

To simulate the contraction of the cloud in the presence of the surrounding atmosphere, we introduced a stationary external gravitational field by applying a repulsive force. The external field used in this simulation is symmetric about the axis of rotation. This associates with the homogeneous surrounding atmosphere. Figure 4 shows the result of the gravitational contraction of the same initial cloud as in Figure 1 but in different external gravitational fields. The intensity of the hurricane decreases with the increase of the external field. The less the difference between the environmental atmosphere and the cloud the slower-rotating and less intensive system is formed. The equalizing of the density of the cloud with the density of the surrounding atmosphere will result in no tropical cyclone.

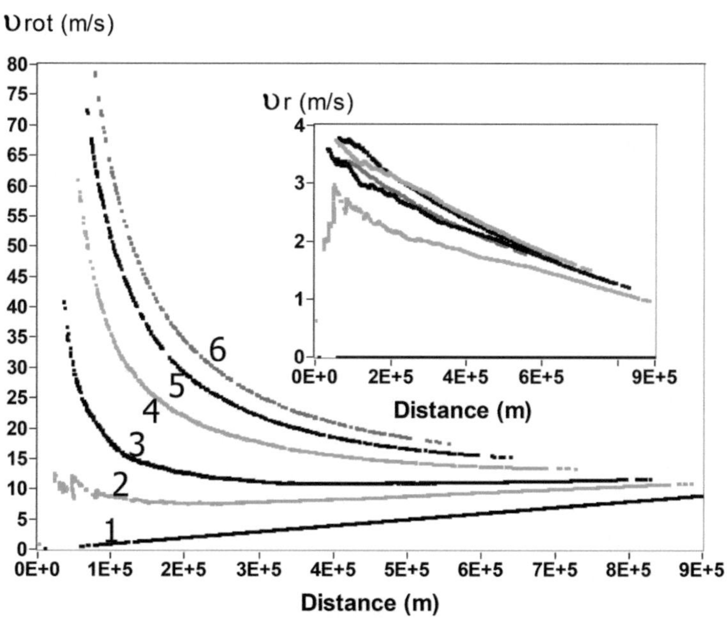

Fig. 3. Tangential v_{rot} and radial v_r velocities vs. distance from the centre of the cloud are shown for different stages of the formation of the hurricane.

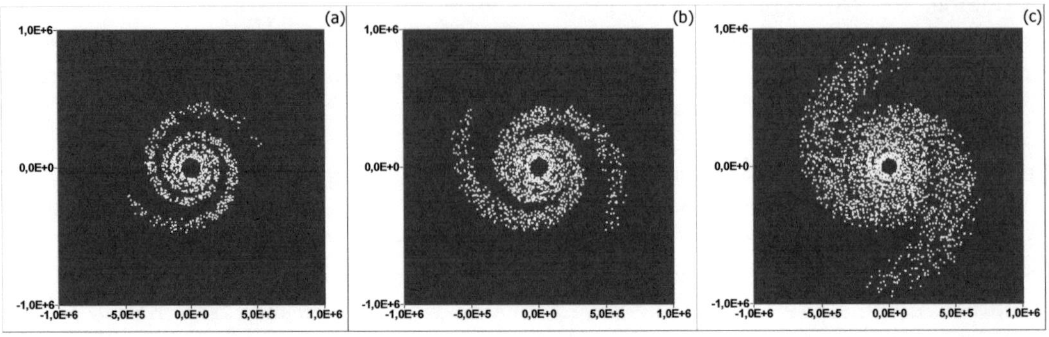

Fig. 4. The result of simulation of the same initial cloud as in Figure 1 but in the presence of surrounding gas of different density. The external gravitational field has central symmetry and is given by an effective density that is zero for graph (a), 2×10^{-2} kg m^{-3} (b) and 5×10^{-2} kg m^{-3} (c). All three hurricanes are shown after 450 steps of simulation. The graph (a) corresponds to the hurricane shown in Figure 1 (last graph). Apparently, the maximum velocities of the hurricane decrease with increase of the external density. The maximum velocity of the hurricane (a) is around 80 m/s whereas the maximum velocity of the hurricane (c) is 65 m/s. The diameter of the eye of the hurricane (c) decreased as well.

4. Discussion

In real conditions, the circulation process and surface winds produce a feedback mechanism when the loss of mass due to precipitations is compensated by upwards-rising water vapor flow along the eye wall in the middle of the hurricane. The difference between the falling-down mass and the rising mass leads to decreasing the intensity of the hurricane.

The disturbance of the balance in the atmosphere may result in more tropical storms. There are natural factors such as solar activity and local season weather that may influence the creation of the hurricanes. The variations of density are observed in other planets as well. A sharp density gradient (called the ionopause) in the atmosphere of Venus correlates with solar activity because this planet has no intrinsic magnetic field and therefore the solar wind interacts directly with ionosphere. Thus, magnetosphere of the Earth is the main natural obstacle for the hurricanes.

There is also a human factor that can be dominant in future global climate. The change of chemical composition of the atmosphere can distract its balance (Henderson-Sellers *et al.* 1998). For example, heavy elements, like CO_2, result of the industrial activity and disappearing the forests, may result in more frequent hurricanes (Knutson and Tulea 1999).

We can conclude that the gravitational model explains the origin of the hurricanes and can help to predict their formation at early stages. Our model uses minimum parameters making feasible the practical applications of the simulation program. The practical simulation can be performed based on initial data obtained from measuring the atmospheric gas density and wind velocity. The model for the hurricane can be generally applied for any tropical cyclone. Less dynamic tropical cyclones are characterized by less mass and smaller cloud's size and consequently weaker extreme winds.

5. References

Bister M, Emanuel KA (1998) Dissipative heating and hurricane intensity. Meteorol. Atmos. Phys. 65, 233.

Brzezinski A, Bizouard C, Petrov SD (2002), Influence of the atmosphere on Earth rotation: what new can be learned from the recent atmospheric angular momentum estimates? Surv. Geophys. 23: 33.

Camp JP (2001), Hurricane maximum intensity: past and present. Mon. Weath. Rev. 129: 1704.

Chelton DB, Wentz FJ, Gentemann CL, de Azoeke RA, Schlax MG (2000) Satellite microwave SST observations of transequatorial tropical instability waves. Geophys. Res. Lett. 27: 1239.

Davis CA, Bosart FL (2001) Numerical Simulations of the Genesis of Hurricane Diana (1984). Part I: Control Simulation. Mon. Weath. Rev. 129: 1859.

DeMaria M, Kaplan J (1994) A statistical hurricane intensity prediction scheme (SHIPS) for the Atlantic basin. Weath. Forcast. 13: 209.

DeMaria M (1996) The effect of vertical shear on tropical cyclone intensity change. J. Atmos. Sci. 53: 2076.

DeMaria M, Kaplan J (1999) An updated statistical hurricane intensity prediction scheme (SHIPS) for the Atlantic and Eastern North Pacific basins. Wea. Forecasting 14: 326.

Dudhia JA (1993) A nonhydrostatic version of the Penn State-NCAR mesoscale model: Validation tests and simulation of an Atlantic cyclone and cold front. Mon. Wea. Rev., 121: 1493.

Emanuel KA (1999) Thermodynamic control of hurricane intensity. Nature 401: 665.

Holland GJ (1997) The maximum potential intensity of tropical cyclones. J. Atmos. Sci. 54: 2519. The terms "hurricane" and "typhoon" are regionally specific names for a strong "tropical cyclone". A tropical cyclone is the generic term for a non-frontal synoptic scale low-pressure system over tropical or sub-tropical waters with organized convection (i.e. thunderstorm activity) and definite cyclonic surface wind circulation.

Henderson-Sellers A, Zhang H, Berz G, Emanuel K, Gray W, Landsea C, Holland G, Lighthill J, Shieh S-L, Webster P, McGuffie K (1998) Tropical cyclones and global climate change: a post-IPCC assessment. Bull. Am. Meteorol. Soc. 79: 19.

Gray WM (1968) Global view of the origin of tropical disturbances and storms. Mon. Weath. Rev. 96: 669.

Grell GA., Dudhia J, Stauffer DR (1994) A description of the fifth-generation Penn State/NCAR Mesoscale Model (MM5). NCAR Tech. Note: 398.

Knutson TR, Tulea RE (1999) Increased intensities with CO_2 -induced warming as simulated using the GFDL hurricane prediction system. Climate Dinamics 15: 503.

Marks Jr. FD, Jr. Houze RA, Gamache JF (1992) Dual-aircraft investigation of the inner core of Hurricane Norbert. Part I: Kinematic structure. J. Atmos. Sci. 49: 919.

Kurihara Y, Tuleya RE, Bender MA (1998) The GFDL hurricane prediction system and its performance in the 1995 hurricane season. Mon. Weath. Rev. 126: 1306.

Liu Y, Zhang D-L, Yau MK (1999) A multiscale numerical study of Hurricane Andrew (1992). Part II: kinematics and inner-core structures. Mon. Weath. Rev. 127: 2597.

Malkus JS, Riehl H (1960) On the dynamics and energy transformations in steady-state hurricanes. Tellus 12: 1.

McBride JL, Zehr R (1981) Observational analysis of tropical cyclone formation. Part II: Comparison of nondeveloping versus developing systems. J. Atmos.Sci.38: 1132.

Merrill RE (1988) Environmental influences on hurricane intensification. J. Atmos.Sci.45: 1678.

Miller BI (1958) On the maximum intensity of hurricanes. J. Meteor. 15: 184.

Pavlov A, Pavlova Y (2003) Evolution of elliptical galaxies and mechanism of formation of spiral galaxies. Mod. Phys. Lett. A 18: 2265.

Pavlov A, Pavlova Y (2003) Formation and dynamics of Saturn and its disk simulated by using a new N-body model, Central European Journal of Physics 1: 634.

Rosen RD, Salstein DA (1983) Variations in atmospheric angular momentum on global and regional scales and the length of day. J. Geophys. Res. 88: 5451.

Rosenthal SL (1970) Experiments with a numerical model of tropical cyclone development: some effects of radial resolution. Mon. Weath. Rev. 98: 106.

Simpson RH, Riehl H (1981) The Hurricane and Its Impact. Louisiana State Univ. Press, Baton Rouge (IBSN 0-8071-0688-7).

Tong Z., Zhang D-L, Weng F (2004) Numerical simulations of hurricane bonnie (1998). Part I: Eye wall evolution and intensity changes. Mon. Weath. Rev. 132: 225.

Whitney LD, Hobgood JS (1997) The relationship between sea surface temperatures and maximum intensities of tropical cyclones in the eastern North pacific Ocean (1997). J. Clim. 10: 2922.

Zhang D-L, Liu Y, and Yau MK (2000) A multiscale numerical study of hurricane Andrew (1992). Part III: dynamically induced vertical motion. Mon. Weath. Rev.128: 3772.

Part 3

Forecasting of Tropical Cyclones

15

Improvements of Hurricane Forecast with Vortex Initialization using WRF Variational (WRF-Var) Data Assimilation

Qingnong Xiao
College of Marine Science, University of South Florida
St. Petersburg, FL 33701,
USA

1. Introduction

Although the forecasts of hurricane track have steadily improved during the past two decades, intensity forecasts remain unsatisfactory (Elsberry, 2005). Previous studies have shown that vortex structures can significantly affect the behavior of hurricane intensity (Ross & Kurihara, 1995; Willoughby & Black, 1996; Xiao et al., 2000). The initialization of hurricane vortex as well as its environment using an advanced data assimilation technique is a key procedure to improve the accuracy of hurricane forecasts and to extend lead-time for hurricane forecasts with increased certainty. Particularly, assimilation of the data in the vortex region deserves more scientific research and technical development.

Assimilation of Doppler radar observations from coastal and aircraft Doppler radars in hurricane vortex initialization is of great interest to the weather service and research communities but with a lot of challenges (Liu et al., 1997; Marks et al., 1998; Xiao et al., 2000; 2006). The complexity comes from how we should establish the data assimilation system for hurricanes that are usually in the low latitude, compounded by a lack of data over the ocean and inadequate computer resources to resolve the inner core. From the several-year real-time hurricane forecasts using the Advanced-research Hurricane WRF (AHW) model (Davis et al., 2008; Xiao et al., 2009a; b), however, one conclusion is that an advanced analysis scheme has to be implemented for improved vortex initialization.

Data assimilation is a process that incorporates observations into numerical model with consideration of both observational data and model background information. There are several data assimilation techniques that can be used for hurricane initialization. The four-dimensional variational (4D-Var) data assimilation (Courtier et al., 1994) and Ensemble Kalman filters (EnKF, Evensen, 1994) are two of the most advanced in algorithm formulation and technique design. 4D-Var employs a forecast model as a strong constraint in a least-squares fit problem (Lewis & Derber, 1985, Le Dimet & Talagrant, 1986, Thepaut & Courtier, 1991, Navon et al., 1992). It has an implicit update of the flow-dependent background field and the capability to assimilate data at the exact observation time. The 4D-Var adjoint approach is an attractive assimilation technique. As a retrospective assimilation algorithm, it can derive the optimal time-trajectory fit to observational data, including non-synoptic data (Xiao et al., 2002, Simmons & Hollingsworth, 2002). 4D-Var has been implemented for operational numerical weather prediction at many operational centers (e.g. ECMWF, Metero France, UK Met Office, & JMA). As we all know in the data assimilation community, however, the development of adjoint model is tedious, labor-intensive, and often subject to errors (Xiao et al., 2008a).

Another advanced data assimilation technique, EnKF, has gained much attention in recent years in the data assimilation community. EnKF can use flow-dependent background error covariance (**B** matrix) calculated from ensemble forecasts and can be easily implemented without tangent linear and adjoint models. Several formulations of EnKF have been developed, such as double EnKF (Houtekamer & Mitchell, 1998), ensemble square root filter (EnSRF, Whitaker & Hamill, 2001), ensemble adjust Kalman filter (EAKF, Anderson, 2001), ensemble transform Kalman filter (ETKF, Bishop et al., 2001), and local ensemble transform Kalman filter (Ott et al., 2004, Hunt et al., 2004). The advantages and weaknesses of 4D-Var and EnKF are discussed in Lorenc (2003), Gustafsson (2007) and Kalnay et al. (2007). The combination of advantages in 4D-Var and EnKF has attracted a lot of efforts in recent years (Evensen & van Leeuwen, 2000; Zupanski, 2005; Hunt et al., 2004; Fertig et al., 2007). Along this route, we proposed an ensemble-based four-dimensional variational (En4D-Var) algorithm (Liu et al., 2008; 2009). The new data assimilation technique, En4D-Var, uses the flow-dependent background error covariance matrix (**B** matrix) from the WRF ensemble, and performs 4D-Var optimization. It adopts the variational approach but avoids the tangent linear model and its adjoint, the component that is difficult to develop and that make 4D-Var minimization computationally expensive.

Since the establishment of the WRF-Var data assimilation system (Barker et al., 2004, Skamarock et al., 2005), we have contributed considerable efforts to develop the WRF adjoint (Xiao et al., 2008a), WRF 4D-Var (Huang et al., 2009), and WRF En4D-Var (Liu et al., 2009) under WRF-Var framework. However, most of the scientific researches, including hurricane initialization, still rely on the WRF 3D-Var system. During the past decade, remote sensing technology using Doppler radars has advanced rapidly to the point where synthetic dual Doppler observations are now available in real time from the extensive network of coastal radars and from reconnaissance aircraft. For example, Hurricane Rita (2005) was observed by the Key West WSR-88D (KBYX) while it was still a minimum hurricane. When Rita rapidly intensified in Gulf of Mexico into a Category 5 hurricane with a double eyewall, two NOAA P3s, NRL P3, Air Force C-130, and NOAA G-IV observed its structures in three consecutive days. During the TC08 campaign, the NRL P3 had several days of flight penetrating Typhoon Sinlaku. Obviously, hurricane initialization with these data can result in much better definition of the vortex structures.

Although storms approaching the coastline of North America and some typhoon cases in the western North Pacific are well sampled by both airborne and coastal radars, these data are currently being used primarily in the form of images for qualitative interpretation by forecasters. Several initialization techniques using Doppler radar data have been demonstrated to be effective. The mesoscale vorticity method (MVM, Lee et al., 2003, 2006) achieves improved initialization from high temporal resolution radar observations to compute the vertical velocity and radial flow from the better-observed rotational winds by Doppler radars. WRF 3D-Var can directly assimilate the Doppler radial winds and reflectivity data into analysis, which includes increments of vertical velocity and hydrometeors in the vortex initialization (Xiao et al., 2005; 2007; Xiao & Sun 2007; Xiao et al., 2008b). One of our recent efforts on hurricane vortex initialization has been shown that airborne Doppler radar (ADR) data assimilation can significantly enhance the hurricane vortex (Xiao et al., 2009a). With the improved vortex, we are able to study the relationship of hurricane intensity change associated with its initial state and vortex structures, scale interaction between vortex and environment.

Hurricane intensity and intensity change are highly related to vortex structures. In order to have an accurate vortex structure in model forecasts, the initial conditions should be improved using observations in the vortex's inner core region. Numerical weather forecasting is an initial value problem. The enhanced analyses of vortex structure should provide sufficient detail and accuracy in support of the subsequent intensity forecasting. Unfortunately hurricanes are over oceans and there are usually no Doppler radar data available in the vortex region. Satellite data are not reliable due to cloud and precipitation contamination in the vortex. To define a hurricane structure in the model initial conditions, Xiao et al. (2000) and Zou & Xiao (2000) proposed the bogus data assimilation (BDA), a technique that assimilates a synthetic vortex using variational data assimilation to initialize the hurricane structure. With the optimization procedure, the synthetic vortex structures are gradually incorporated into the hurricane initial conditions. While BDA is initially applied to 4D-Var, it is now tested in 3D-Var systems (Xiao et al., 2006; 2009b). Computationally, BDA using 3D-Var is much faster than using 4D-Var. Because integrations of forward model and backward adjoint are involved in the iterative minimization procedure in 4D-Var, extensive experiments or real-time application of hurricane BDA with the 4D-Var framework have their difficulties. On the contrary, 3D-Var with BDA can be easily applied for large numbers of experiments. In addition, the WRF 3D-Var with BDA allows the hurricane structure to be generated by 3D-Var balance transform and background error covariance. Since the background error covariance is produced using WRF, the hurricane initialization can still achieve balance to some extent with the WRF model.

The BDA in WRF 3D-Var has been implemented based on the characteristics of WRF model (Skamarock et al., 2005). The preconditioned control variables are streamfunction, unbalanced velocity potential, unbalanced temperature, pseudo relative humidity, and unbalanced surface pressure. The NMC-method (Parrish & Derber, 1992) is used to construct correlations among the control variables for multivariate analysis. An asymmetric vortex perturbation is extracted from the previous forecast and relocated to the observed position. Summation of the symmetric vortex and the asymmetric perturbation produces a synthetic vortex structure. Assimilation of the synthetic sea-level pressure (SLP) and wind profiles are performed via the minimization procedure in the proposed BDA scheme in WRF 3D-Var. The radius of 34-kt wind in the hurricane advisory is used to calculate the bogus vortex size in BDA implementation (Xiao et al., 2006).

In the next section, a brief description of the WRF variational (WRF-Var) data assimilation technique is provided. The hurricane initialization experiments with Doppler radar data assimilation and BDA are discussed in Sections 3 and 4. In section 5 we summarize our work and draw some conclusions.

2. WRF Variational (WRF-Var) data assimilation

Hurricane initialization can be achieved through data assimilation approach, which blends observations with numerical model. Variational data assimilation achieves this through the iterative minimization of a prescribed cost (or penalty) function, so as to optimally synthesize observations in a model-consistent framework. Efforts have been made to perform vortex initialization for the Advanced-research Hurricane WRF (AHW, Davis et al.,

2008; Xiao et al., 2009b) using WRF-Var. The observations we used to better define the vortex structure are Doppler radar data and synthetic observations.

WRF-Var is a single software that combines 3 and 4-dimensional variational data assimilation systems for the WRF community. It includes WRF 3D-Var and WRF 4D-Var. The variational adjoint technique was introduced in 1980s (Lewis & Derber, 1985; Le Dimet & Talagrand, 1986), and it has played a more and more important role in numerical weather prediction (Daley, 1991; Rabier et al., 2000; Gustaffson, 2007; Kalnay et al., 2007; and Huang et al., 2009). Mathematically, variational data assimilation produces an optimal initial condition $\mathbf{x}_a$ by minimizing the cost function $J(\mathbf{x})$ in the assimilation window,

$$J(\mathbf{x}) = \frac{1}{2}(\mathbf{x} - \mathbf{x}_b)^T \mathbf{B}^{-1}(\mathbf{x} - \mathbf{x}_b) + \frac{1}{2}\sum_{i=0}^{I}(HM\mathbf{x} - \mathbf{y}_i)^T \mathrm{O}^{-1}(HM\mathbf{x} - \mathbf{y}_i). \tag{1}$$

In the equation, $\mathbf{x}_b$ is background state vector at t_0, H is observation operator, M is forecast model and $\mathbf{y}_i$ is observation vector at t_i, $\mathbf{B}$ is the background error covariance matrix and $\mathbf{O}$ is the observation error covariance matrix, and I is the total number of time levels at which observations are available. If all the observations are at one time and no forecast model is involved in the formulation, the problem is 3D-Var. If more than one-time observations are used and the model is involved in the formulation of (1), the problem becomes 4D-Var.

In the incremental approach (Courtier et al., 1994), the analysis increment $\mathbf{x}'$ instead of $\mathbf{x}$, is used in minimizing the cost function $J(\mathbf{x}')$ defined by

$$J(\mathbf{x}') = \frac{1}{2}(\mathbf{x}')^T \mathbf{B}^{-1}(\mathbf{x}') + \frac{1}{2}\sum_{i=0}^{I}(\mathbf{HMx}' + \mathbf{d}_i)^T \mathbf{O}^{-1}(\mathbf{HMx}' + \mathbf{d}_i), \tag{2}$$

where $\mathbf{x}' = \mathbf{x} - \mathbf{x}_b$ is the analysis increment vector, $\mathbf{H}$ is tangent linear observation operator, $\mathbf{M}$ is tangent linear forecast model, and $\mathbf{d}_i$ is the innovations at different time (with subscript i). $\mathbf{d}_i$ is calculated by

$$\mathbf{d}_i = HM(\mathbf{x}_b) - \mathbf{y}_i. \tag{3}$$

With the preconditioning (Gilbert et al., 1989) and introducing the preconditioning matrix $\mathbf{U}$ defined by

$$\mathbf{B} = \mathbf{U}\mathbf{U}^T, \tag{4}$$

we obtain the cost function in control variable ($\mathbf{w}$) space as,

$$J(\mathbf{w}) = \frac{1}{2}\mathbf{w}^T\mathbf{w} + \frac{1}{2}\sum_{i=0}^{I}(\mathbf{HMUw} + \mathbf{d}_i)^T \mathbf{O}^{-1}(\mathbf{HMUw} + \mathbf{d}_i). \tag{5}$$

The gradient of the cost function with respect to the control variables is

$$\nabla_{\mathbf{w}} J = \mathbf{w} + \sum_{i=0}^{I}\mathbf{U}^T\mathrm{M}^T\mathbf{H}^T\mathbf{O}^{-1}(\mathbf{HMUw} + \mathbf{d}_i). \tag{6}$$

And the final analysis is

$$\mathbf{x}_a = \mathbf{x}_b + \mathbf{U}\mathbf{w}. \quad (7)$$

The WRF 3D-Var was developed from MM5 3D-Var (Barker et al., 2004), and has been adopted in real-time forecasting in several operational applications including Air Force Weather Agency (AFWA), Korean Meteorological Administration (KMA) and Taiwan Central Weather Bureau (CWB). It includes a background error covariance statistics package using either NMC method (Parrish & Derber, 1992) or ensemble method (Fisher, 1999). The cost function minimization is performed in terms of analysis increments. The analysis space is grid-point in the horizontal and a projection onto eigenvectors of the background error covariance matrix in the vertical. This projection allows efficient data compression at the expense of some spatial/temporal averaging of vertical error correlations. The WRF 4D-Var has recently developed based on the WRF model and 3D-Var as its basic components (Huang et al., 2009). It used the WRF 3D-Var framework that shares the same software engineering infrastructure as WRF. However, it possesses a number of advantages over 3D-Var including the ability to: a) use observations at the exact times that they are observed, which suits most asynoptic data; b) implicitly use flow-dependent background errors, which ensures the analysis quality for fast developing weather systems; and c) use a forecast model as a constraint, which enhances the dynamic balance of the final analysis. Almost all observations can be assimilated into WRF 3/4D-Var analysis, including the observations from GTS data stream, satellite (Liu & Barker, 2006) and Doppler radar (Xiao et al., 2005; 2007; Xiao & Sun 2007; Xiao et al., 2008b).

Recently, Liu et al. (2008; 2009) expanded WRF-Var to include the ensemble-based perturbation matrix in preconditioning. The advantage of WRF En4D-Var is that we can reformulate the algorithm to avoid tangent linear and adjoint models in calculating the gradient of cost function. However, the analysis is subject to the so-called sampling error problem in ensemble-based data assimilation. Liu et al. (2009) applied the Schur operator for localization in the WRF En4D-Var analysis and demonstrated that the localization technique can effectively alleviate the impacts of sampling errors upon analysis. The experimental results show that the WRF En4D-Var produced a better overall analysis than the WRF En3D-Var cycling. It suggests that the ensemble-based retrospective assimilation, WRF En4D-Var, has more robust analysis ability than ensemble-based sequential algorithms such as WRF En3D-Var cycling approach.

3. Hurricane initialization with Doppler radar data

3.1 Radial velocity

In order to include a capability to assimilate Doppler radial velocity data, the WRF 3D-Var was modified to include vertical velocity increments. Based on Richardson (1922) and White (2000), a balance equation that combines the continuity equation, adiabatic thermodynamic equation, and hydrostatic relation is derived and expressed as:

$$\gamma p \frac{\partial w}{\partial z} = -\gamma p \nabla \cdot \boldsymbol{V_h} - \boldsymbol{V_h} \cdot \nabla p + g \int_z^{\infty} \nabla \cdot (\rho \boldsymbol{V_h}) dz \quad (8)$$

where w is vertical velocity, $\boldsymbol{V_h}$ is the vector of horizontal velocity (components u and v), γ the ratio of specific heat capacities of air at constant pressure/volume, p pressure, ρ density, T temperature, c_p specific heat capacity of air at constant pressure, z height, and g the acceleration due to gravity. For simplicity, hereafter Eq. (8) will be referred to as Richardson's equation. Linearizing Eq. (8) by writing each variable in terms of a basic state (overbar) plus a small increment (prime) gives

$$\gamma\overline{p}\frac{\partial w'}{\partial z} = -\gamma p'\frac{\partial \overline{w}}{\partial z} - \gamma\overline{p}\nabla\cdot\boldsymbol{V}_{\boldsymbol{h}}' - \gamma p'\,\nabla\cdot\overline{\boldsymbol{V}_{\boldsymbol{h}}} - \overline{\boldsymbol{V}_{\boldsymbol{h}}}\cdot\nabla p' \\ -\boldsymbol{V}_{\boldsymbol{h}}'\cdot\nabla\overline{p} + g\int_z^{\infty}\nabla\cdot(\overline{\rho}\boldsymbol{V}_{\boldsymbol{h}}')dz + g\int_z^{\infty}\nabla\cdot(\rho'\overline{\boldsymbol{V}_{\boldsymbol{h}}})dz \quad . \tag{9}$$

The basic state (overbar) variables satisfy Eq. (8). They also satisfy the continuity equation, adiabatic equation and hydrostatic equation. The linear equation (9) is discretized, and its adjoint code is developed according to the code of the linearized equation (Xiao et al., 2005). The observation operator for Doppler radial velocity is

$$V_r = u\frac{x - x_i}{r_i} + v\frac{y - y_i}{r_i} + (w - v_T)\frac{z - z_i}{r_i}, \tag{10}$$

where (u,v,w) are the wind components, (x,y,z) are the radar location, (x_i,y_i,z_i) are the location of the radar observation, r_i is the distance between the radar and the observation, and v_T is terminal velocity. For radar scans at nonzero elevation angles, the fall speed of precipitation particles has to be taken into account. There are different ways to calculate terminal velocity. Here, we use the algorithm of Sun & Crook (1998) to calculate terminal velocity V_t (m/s),

$$v_T = 5.40a\cdot q_r^{0.125}, \tag{11}$$

where q_r is rainwater mixing ratio (g/kg). The quantity a is a correction factor defined by

$$a = (p_0\,/\,\overline{p})^{0.4}, \tag{12}$$

where $\overline{p}$ is the base-state pressure and p_0 is the pressure at the ground.

3.2 Reflectivity

Radar reflectivity measures the radar's signal reflected by precipitation hydrometeors. To assimilate radar reflectivity directly, the WRF 3D-Var system should be able to produce the increments of the hydrometeors (at least, the rainwater mixing ratio). We chose total water mixing ratio q_t as a control variable for reflectivity assimilation. Because total water mixing ratio q_t is used as a control variable, partitioning of the moisture and hydrometeor increments is necessary in the 3D-Var system (Xiao et al., 2007). We introduced the warm rain process of Dudhia (1989) for the partitioning, which includes condensation of water vapor into cloud (P_{CON}), accretion of cloud by rain (P_{RA}), automatic conversion of cloud to rain (P_{RC}), and evaporation of rain to water vapor (P_{RE}). These are the major processes of hydrometeor cycle in the summer season.

The tangent linear and its adjoint of the scheme were developed in the WRF 3D-Var system. Although the control variable is q_t, the q_v, q_c and q_r increments are produced through the partitioning procedure during the 3D-Var minimization. The warm rain scheme builds a constraint: the relation among rainwater, cloud water, moisture and temperature. When rainwater information (from reflectivity) enters the minimization procedure, the forward warm rain process and its backward adjoint distribute this information to the increments of other variables (under the constraint of the warm rain scheme).

Once the 3D-Var system can produce q_c and q_r increments, the setup of the observation operator for assimilation of reflectivity is straightforward. We adopted the observation operator from Sun & Crook (1997):

$$Z = 43.1 + 17.5\log(\rho q_r), \tag{13}$$

where Z is reflectivity in the unit of dBZ and q_r is the rainwater mixing ratio. The relation (13) is derived analytically by assuming the Marshal-Palmer distribution of raindrop size.

3.3 Hurricane initialization with airborne Doppler radar data

We take Hurricane Jeanne (2004) as an example to illustrate the impact of the airborne Doppler radar (ADR) data assimilation on the vortex structure in the initial conditions as well as subsequent forecast. We conducted four sets of experiments: the control run CTL which uses the NCEP/GFS as the initial condition; the experiment GTS which assimilates only the conventional GTS data; GRV which assimilates GTS plus ADR wind data; and GVZ which is the same as GRV except both the ADR wind and reflectivity are assimilated.

Figure 1 shows the sea-level pressure (SLP) and 10-m winds with and without ADR wind assimilation for Hurricane Jeanne (2004) at 1800 UTC 24 September (the initialization time of the forecast). The cyclonic circulation is strengthened with a maximum surface wind (MSW) speed increase to 38 ms^{-1} (Fig. 1c), compared with 23.4 ms^{-1} in GTS (Fig. 1 b) and 23.7 ms^{-1} in CTL (Fig. 1a). The observed MSW speed is 44 ms^{-1}. Its CSLP is decreased in GRV after assimilating ADR wind data. Along with the intensity enhancement in wind speeds, the CSLP is correspondingly decreased about 5 hPa through the multivariate constraint in WRF 3D-Var. The GRV wind speed distribution in Figure 1c is much closer to the H*WIND (figure omitted) than those in Figures 1a and 1b.

WRF 3D-Var also produces temperature increments using ADR data due to the multivariate incremental structure in the system (Fig. 2). At 300 hPa, both temperature and wind vector increments in the vortex region in GTS are small (Fig. 2a), because no GTS observations exist in the vortex at high levels. The small increments for both the wind vector and temperature in the vortex region are due to the covariance with the environment. However, once assimilating the ADR winds in the vortex region, positive temperature increments occur associated with the strong wind vector increments and the maximum positive temperature increment is 1.117°C at 300 hPa (Fig. 2b). In the cross-sections above the hurricane vortex, GTS shows only slight increments of temperature above 850 hPa (Fig. 2c). On the contrary, GRV produces a vertical incremental structure with the largest positive temperature increments in the high troposphere around 300 hPa. In the middle troposphere around 500 hPa, there is a negative layer of temperature increments. Notice that both GTS and GRV produce negative temperature increments above the vortex near the surface, due to the buoys and ships data assimilation. It is clear that the ADR data assimilation not only improves the 3-dimensional inner vortex wind structure, but also contributes to the vortex's warm core structure as well. The temperature response to the wind increments is rather small, due to the "climatological" error covariance from one-month forecasts as opposed to more flow-dependant error covariance that would know about the presence of the hurricane. Vortex region is highly imbalanced; the "climatological" error covariance can only construct a small part of the warm core, a highly unbalanced structure.

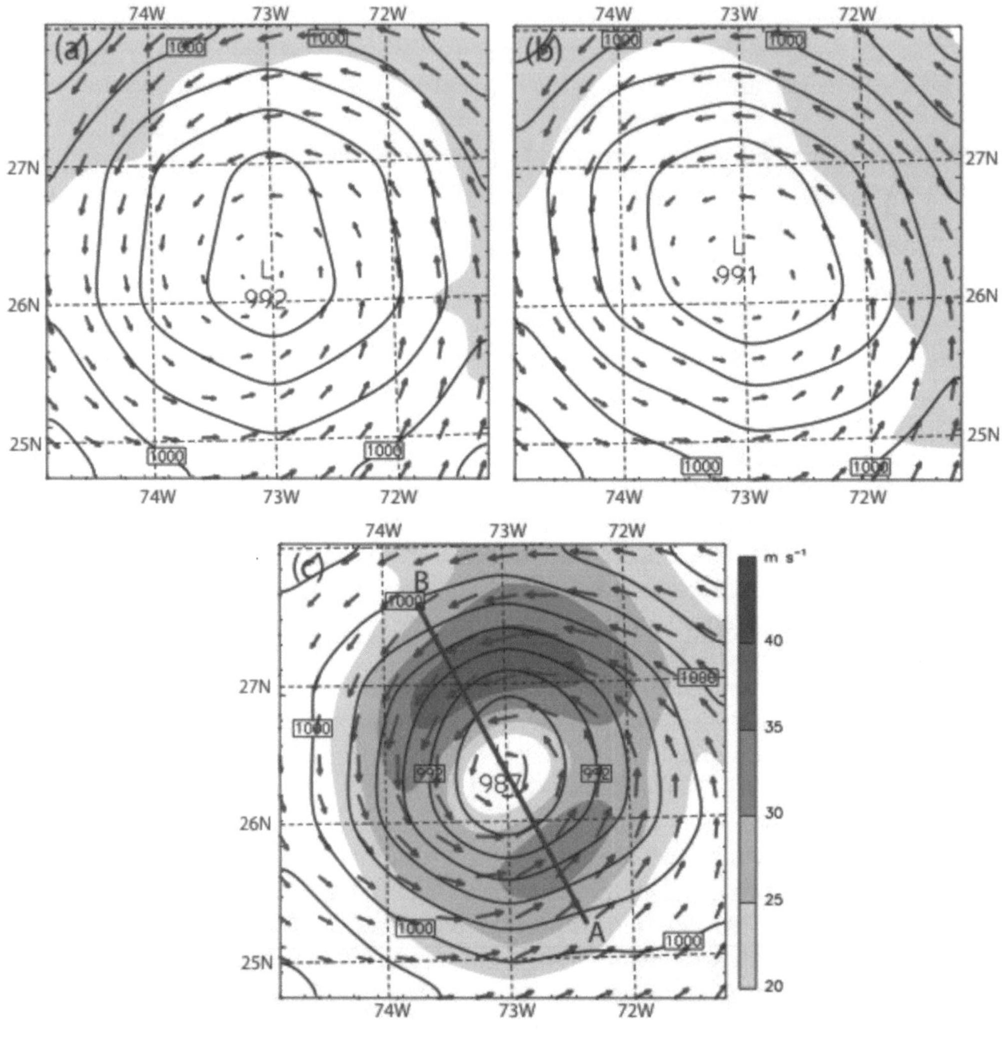

Fig. 1. SLP (thick solid isolines), and surface (10 m) wind vector and speed (shading with thin isolines) for Hurricane Jeanne at 1800 UTC 24 Sep 2004 by experiments (a) CTL, (b) GTS, and (c) GRV. The shading scale for surface wind speed is on the lower right.

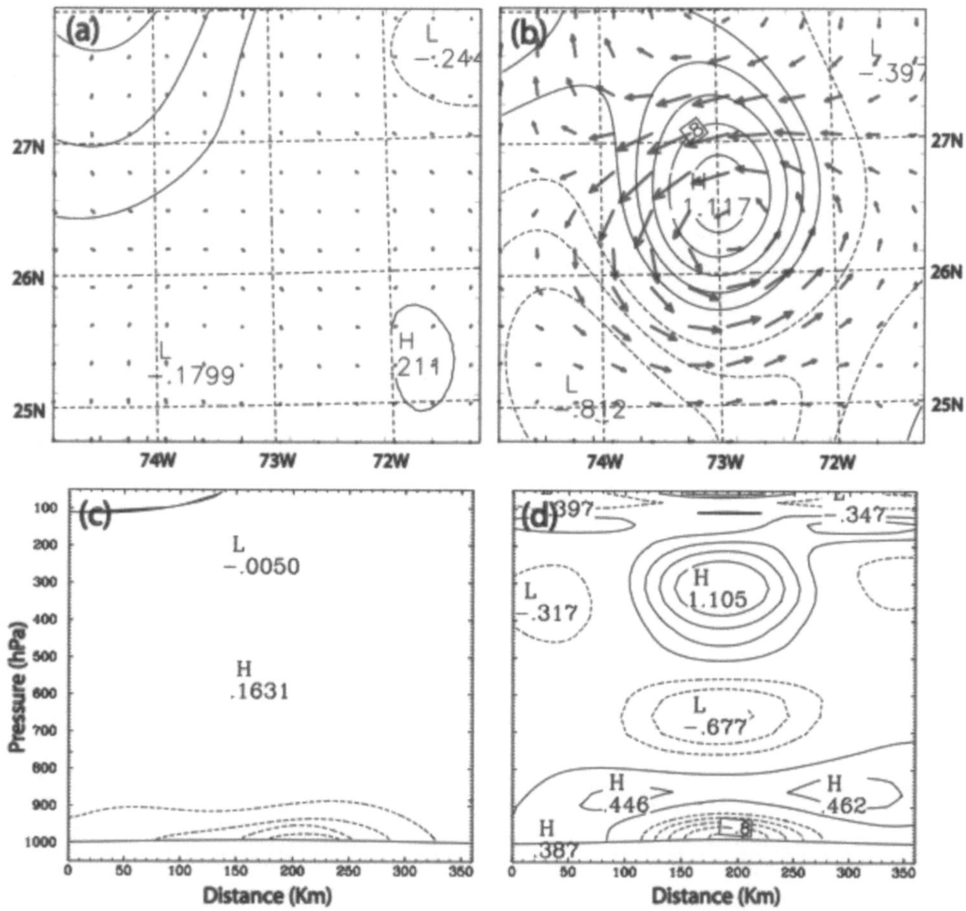

Fig. 2. 300-hPa analytical increments of wind vector (maximum vector represents 29.7 m s-1) and the temperature (isolines with contour interval of 0.2 K, the negative value dashed) by experiments (a) GTS and (b) GRV, and cross sections of temperature increments (0.2 K interval) across the vortex from west to east in (c) GTS and (d) GRV.

Figure 3 shows the ADR reflectivity assimilation results of Hurricane Jeanne at 1800 UTC 24 September 2004 in the GVZ experiment. The CSLP (987 hPa) and 10-m wind analysis in GVZ (Fig. 3a) are similar to the results of GRV (Fig. 1c). However, GVZ produces cloud water and rainwater analyses after assimilating reflectivity data. Because the microphysics involved in the reflectivity assimilation has many "on-off" switches, the linearity assumption used in the WRF 3D-Var for the reflectivity assimilation is compromised. In addition, the warm-rain scheme in WRF 3D-Var does not represent well the microphysical process above the melting level (~ 500 hPa) for hurricanes. Nevertheless, GVZ contains some hydrometeor representation in the initialization, compared to no hydrometeors in GRV.

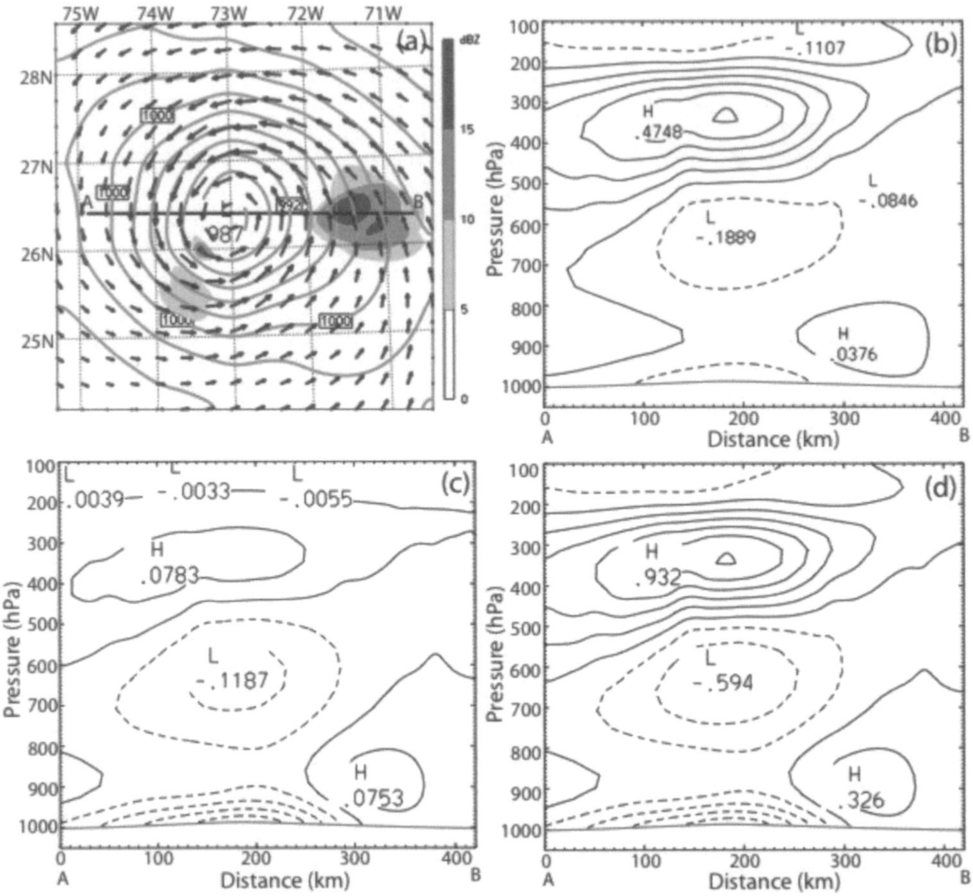

Fig. 3. Initialization with ADR reflectivity assimilation for Hurricane Jeanne at 1800 UTC 24 Sep 2004 by GVZ experiment: a) surface analysis of SLP (gray isolines with the interval of 2 hPa), 10-m wind (arrows with the maximum vector representing 38.4 ms^{-1}), and recovered composite reflectivity (dBZ) in the analysis (shading with the scales on the right), and cross-sections of b) temperature difference (with 0.1°K interval), c) water vapor mixing ratio difference (with 0.05 g kg^{-1} interval), and d) equivalent potential temperature (q_e) difference (with 0.2°K interval) between GVZ and GRV (GVZ-GRV) above the hurricane vortex along the line AB in a).

Figure 3b-d illustrates the changes occurring in variables not directly assimilated owing to the addition of reflectivity data. These changes arise from the multivariate structure in the analysis. The temperature and water vapor mixing ratio in the mid-to-lower levels above hurricane vortex are decreased, while in upper levels are increased (Figs. 3b and 3c). The storm center's stability $\frac{\partial \theta_e}{\partial z}$ increases (Fig. 3d), while at about 120 km radius, the stability between 900 and 500 hPa decreases, with a maximum difference of q_e at around 900 hPa. GVZ also produces a warmer upper level core. In general, ADR reflectivity assimilation increases the atmospheric stability in the vortex center, while decreasing stability outside the eye region. Figure 4 shows the composite reflectivity at 24- and 36-h forecasts for GTS, GRV and GVZ experiments. Experiment GTS does not produce an eye wall, indicating that the vortex is not well organized (Fig. 4a & d). However, GRV and GVZ produce well-organized hurricane structures in reflectivity with compact eyewalls embedded in the vortex. Comparison of the observed reflectivity at landfall (figure omitted) with these forecasts suggests that both GRV and GVZ produce a realistic distribution of reflectivity, but the heavier rainband over the east coast of Florida and the suggestion of a break in reflectivity on the east side of the eye wall in GVZ matches the observations somewhat better.

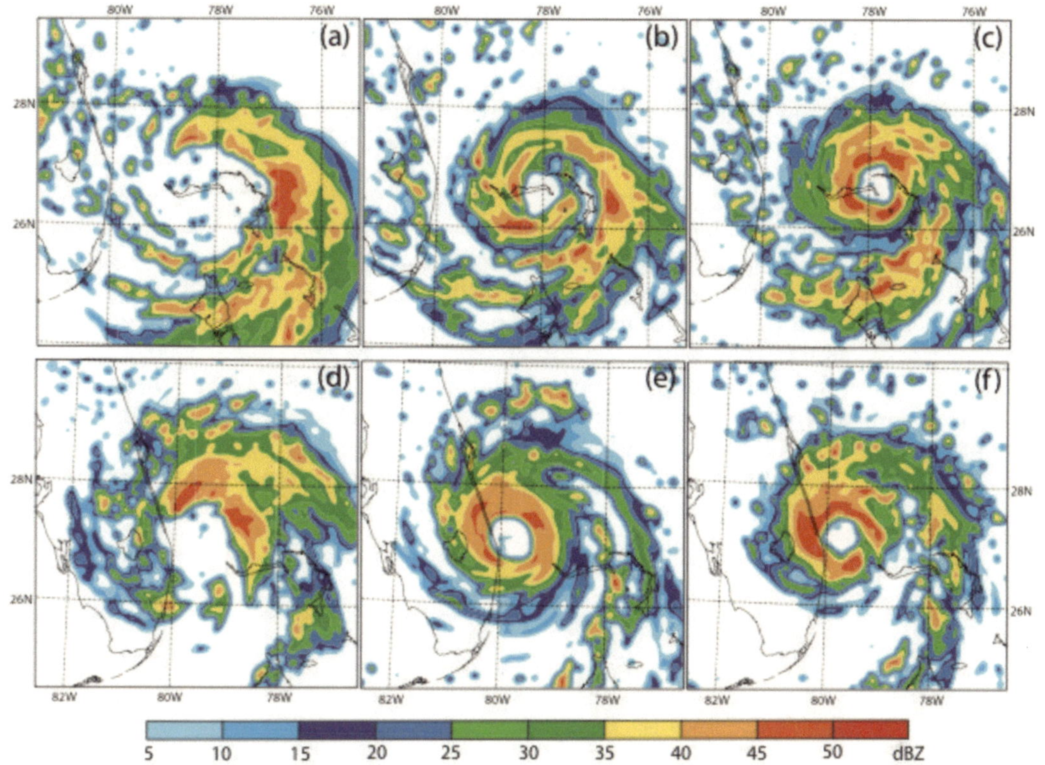

Fig. 4. The column maximum radar reflectivity (dBZ) at 1800 UTC 25 Sep (24-h forecast) for experiments a) GTS, b) GRV, and c) GVZ, and at 0600 UTC 26 September (36-h forecast) for experiments d) GTS, e) GRV, and f) GVZ.

We also carried out ADR data assimilation experiments for Hurricanes Katrina and Rita (2005). Verification of track and intensity for Hurricanes Jeanne (2004), Katrina and Rita (2005) indicates that ADR data assimilation improves the intensity forecast for all three hurricanes. The forecast improvement in track is not as significant as in intensity, but is noticeable. It is consistent with that the hurricane track mostly influenced by the environment, instead of the inner structure of hurricane. Hurricane intensity forecast is more significantly improved by ADR data assimilation than track. While GTS data shows no universal benefits for the intensity forecast, the CSLP average mean absolute errors with ADR data initialization are reduced at each forecast lead time within 48 h. The improvement in MSW is maintained for roughly 30 h, echoing the results for Jeanne alone. Significant reduction of MSW mean absolute errors occurs at the initial time. On average, the error reduction is nearly 29 (25) kt by the ADR data initialization in GRV (GVZ). In response to the vortex wind correction, the average decrease of CSLP from the three cases is about 7 hPa, which is not as significant as the MSW increase. The results indicate that the current correlation between CSLP and MSW is relatively weak in WRF 3D-Var system. The increments in vortex dynamical fields obtained by assimilating ADR wind do not result in a correspondingly large pressure response. Because the background error statistics used in this study were based on statistics averaged over an entire month, they are not flow-dependent and therefore do not reflect the vortex structure among variables in the background covariance. We anticipate further improvements by using specific error covariance that recognizes the hurricane vortex structure.

Nevertheless, the hurricane intensity forecasts are improved with ADR data assimilation using the relatively computationally inexpensive 3D-Var approach. The largest error reduction of CSLP and MSW occurs at 24 h. At 48 h, the CSLPs of both GRV and GVZ experiments still show less error than CTL and GTS.

4. Bogus data assimilation

To better define the vortex structure in hurricane initial conditions, a bogus data assimilation (BDA) scheme is developed in the WRF-Var. The BDA algorithm includes: a) bogus vortex construction and error specification, and b) assimilation of the bogus data with the WRF-Var system. In our studies (Xiao et al., 2006; 2009b; Zhang et al., 2007), the bogus vortex is constructed according to the method of Ueno (1989; 1995). There are two components (symmetric and asymmetric) in the bogus observations. The asymmetric component comes from 3D-Var background (previous forecast).

The bogus vortex fields include SLP and wind profiles at various levels. The distribution of SLP within the bogus area is calculated by the following equation based on the Fujita (1952) formula,

$$P(r) = P_{\max} - \frac{\Delta P}{\left(1 + \frac{1}{2}\left(\frac{r}{R_0}\right)^2\right)^{1/2}}, \tag{14}$$

where

$$\Delta P = P_{\max} - P_C, \tag{15}$$

$$P_{\max} = P_C + \frac{P_B - P_C}{1 - \dfrac{1}{1 + (R_B / R_0)^2}}. \tag{16}$$

In the equations of (14)-(16), r is the distance from the hurricane center (km), R_0 is the radius of maximum wind, P_C is the reported central SLP, and P_B is the average SLP within R_B. R_B is a predefined radius of the bogus area.

The bogus symmetric wind is based on the gradient wind relation,

$$V(r) = \left(\frac{r}{\rho}\frac{\partial P}{\partial r} + \frac{f^2 r^2}{4}\right)^{1/2} - \frac{r|f|}{2}. \tag{17}$$

Empirically, we include the effect of surface friction near the boundary layer by specifying smaller weightings. There are 7 levels (sea-level, 1000, 925, 850, 700, 600, 500 hPa) in the bogus wind profile. The weightings of the symmetric wind speed for each layer are 0.7, 0.8, 0.9, 1.0, 1.0, 1.0, 1.0, respectively. The divergence of upper air is difficult to determine. Therefore, the bogus symmetric winds are assigned 0 in the upper levels above 400 hPa.

The asymmetric component is extracted from the WRF-Var background fields. When the background fields come from the WRF forecast, the position of the hurricane in the background fields can be different from observation. The asymmetric component is the difference between the background and the background typhoon symmetric component. Such asymmetric components are relocated to the right position and added to the symmetric bogus fields.

The bogus observation error is empirically specified. We assume that the error is linearly increased with respect to the distance from the typhoon center. Empirically, the specified errors of the bogus SLP and wind profiles are specified as follows:

$$E_{SLP}(r) = 1 + \frac{3}{R_B} r, \tag{18}$$

$$E_V(r,z) = E_{V0}(z) + \frac{4}{R_B} r, \tag{19}$$

where r is the distance from the hurricane center, z is the height, and R_B is the calculated radius of the bogus area. The errors of the bogus SLP range from 1 hPa in the center to 4 hPa around the outmost edge of the bogus area, compared to the constant surface observation error (2 hPa) for SYNOP. $E_{V0}(z)$ represents the errors of the bogus wind at the center. Its values range from 1 m/s at the sea level to 3.3 m/s at 250 hPa, the same as the other wind errors (e.g. TEMP, SATOB, PILOT, etc.). The specified errors of the bogus wind at sea level range from 1 m/s in the center to 5 m/s around the outmost edge of the bogus area.

The observation error specification is crucial in defining how the observation is used in a data assimilation system (Hollingsworth & Lonnberg, 1986). It determines how much of the observation is contributed to the analysis. However, the true values of the vortex bogus observation error are difficult to establish. The smaller error specified near the hurricane center implies more contribution from the bogus data near the center in the analysis. At the outmost edge of the bogus area, the bogus data have the smallest contribution. This also ensures a smooth transition of the analysis between the bogus area and the surrounding environment.

In the BDA, the hurricane bogus fields are treated as supplemental observations and are assimilated during the WRF-Var analysis. The cost function in Eq. (1) includes the contributions of the bogus data in the observation term. The observational operators for the SLP and wind profiles in the WRF-Var system are just an interpolation scheme because SLP and winds are direct variable of the model.

To verify the capability of BDA, we selected twenty-one cases from seven hurricanes in the 2004 and 2005 seasons to conduct parallel experiments. They are Hurricanes Charley, Frances, Ivan, and Jeanne in 2004, and Katrina, Rita, and Wilma in 2005. We selected three cases for each hurricane before its landfall. These cases are very famous due to their striking effects to Florida in 2004 and devastating impacts in the Gulf of Mexico in 2005. Two parallel sets of experiments are carried out. The first set of experiments (CT) for all cases use GFS (Global Forecast System) analysis as background and assimilate only the conventional GTS. The second set of experiments (GB) is the same as CT but including BDA. The WRF 3D-Var experiments use the same background error covariance calculated from one-month statistics in September 2004 using the NMC method (Parish & Derber, 1992).

Figure 5 shows the mean absolute errors of the forecasts (position, CSLP, and MSW) against the best track observations at 24, 48, and 72 hours. The errors from the BDA experiments (GB) are smaller than that from control experiments (CT) for all of the verification parameters (position, CSLP, and MSW). Bogus data is able to remedy the data sparse issue in the vortex region and BDA improves the hurricane forecasting skill. We calculated the error reduction percentage by BDA from the statistics in Figure 5. The largest reduction of average error is in the forecast of CSLP, with 26.4% reduction of average error by BDA. The improvement in hurricane MSW is also significant; the average error is reduced by 24.0%. The track has the smallest but evident improvement among the three verification parameters.

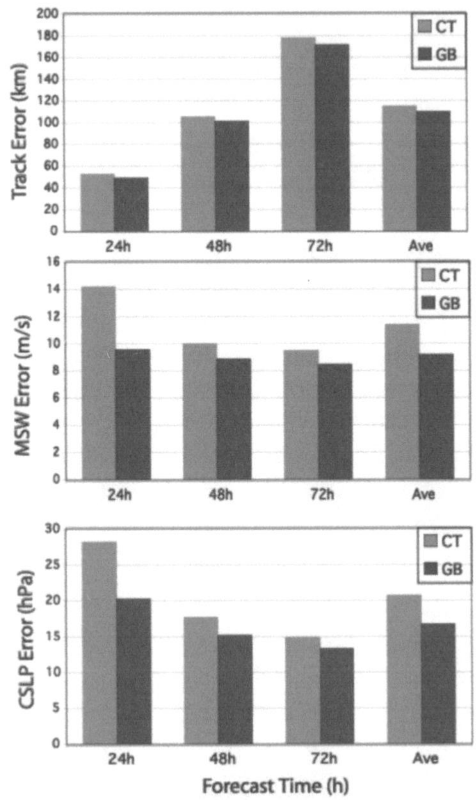

Fig. 5. The mean absolute errors of the hurricane track (top), MSW (middle) and CSLP (bottom) for the forecasts from statistics of 21 cases in 2004 and 2005.

Due to model spin-up problem, both sets of experiments (CT and GB) present smaller CSLP and MSW errors at 48 and 72 hours than at 24 hour. However, the spin-up problem in GB is much less than in CT. BDA alleviates the spin-up problem, and produces larger hurricane intensity improvement at 24 hour than at 48 and 74 hours. The benefit of BDA in hurricane intensity forecasts becomes less with the increase of the forecast time (Fig. 5). The initial intensity using BDA in GB experiments is much closer to the observed than CT. With the model runs, the difference of intensity between GB and CT decreases, reflecting the model forecast at longer time is less sensitive to initial conditions than at shorter time. The improvement of hurricane CSLP and MSW in GB compared to CT at 24 hours is much more remarkable than that at 48 and 72 hours. In the track forecasts, however, experiment GB has the most significant improvement at 72 hours over the experiment CT.

Statistically, the improvement of hurricane intensity using BDA is more significant than that of hurricane track. It is further verified that the large-scale environment influences hurricane track, but that intensity is mainly impacted by hurricane's internal, dynamical and thermo-dynamical vortex structures. The BDA technique, which mainly improves the hurricane vortex structure according to the hurricane concept model, results in significant improvement of the hurricane intensity forecast. To support the assertion, we take Hurricane Katrina at 0000 UTC 26 August 2005 as an example to compare the vortex structures in CT and GB (Fig. 6). The CSLP and MSW errors are both reduced by BDA for Katrina initialized at 0000 UTC 26 August 2005. As shown in Figure 6, the hurricane positions of CT and GB are both very close to the observation. However, GB produces a vortex with lower CSLP and larger MSW than CT. BDA enhances the cyclone circulation and make the vortex much more compact. The area of the circular isobar with 1010hPa in GB is much reduced compared with that in CT. Note that the GFS analysis has its bogussing/relocation procedure, but it is apparently not sufficient for a good hurricane intensity forecast in the WRF ARW model. We verified that the initialization with BDA could improve the hurricane intensity forecast compared with the forecast from a simple GFS analysis.

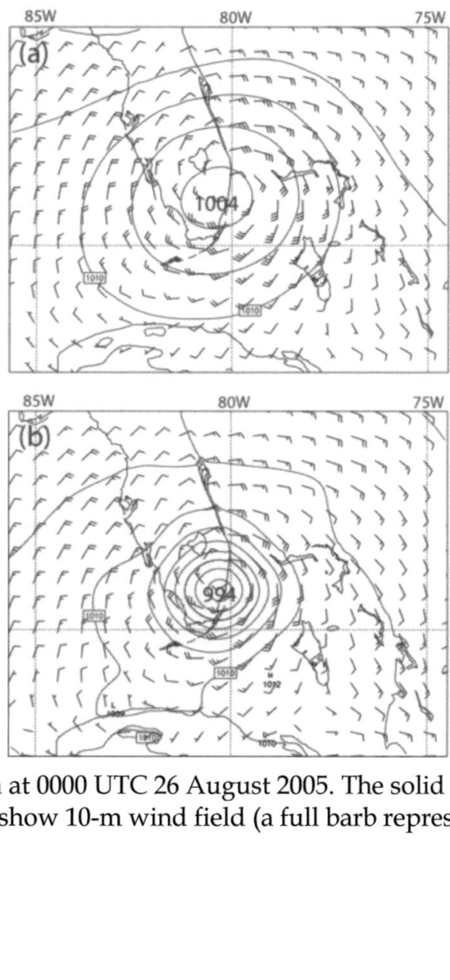

Fig. 6. Hurricane Katrina at 0000 UTC 26 August 2005. The solid lines are SLP (with interval of 2.5 hPa and the barbs show 10-m wind field (a full barb represents 5 m/s) for (a) CT and (b) GB experiments.

The GFDL bogus scheme (Kurihara et al., 1993) has been applied in hurricane initialization for over a decade. To further evaluate the performance of the BDA scheme in WRF 3D-Var, two parallel experiments on Hurricane Humberto at 1200 UTC 12 September 2007 are compared. The first experiment (CT) uses the WRF preprocessing system (WPS) to interpolate the GFDL analysis as initial conditions; and the second experiment (GB) uses the WRF 3D-Var procedure to initialize the hurricane vortex. The experiment GB uses GFS analysis as the first guess in WRF 3D-Var and assimilates the bogussing vortex plus conventional data. The forecasts for both experiments (CT and GB) are executed on 3 domains with the moving nested domains 2 and 3. The grid-spacings of the three domains are 12, 4, and 1.333 km, respectively.

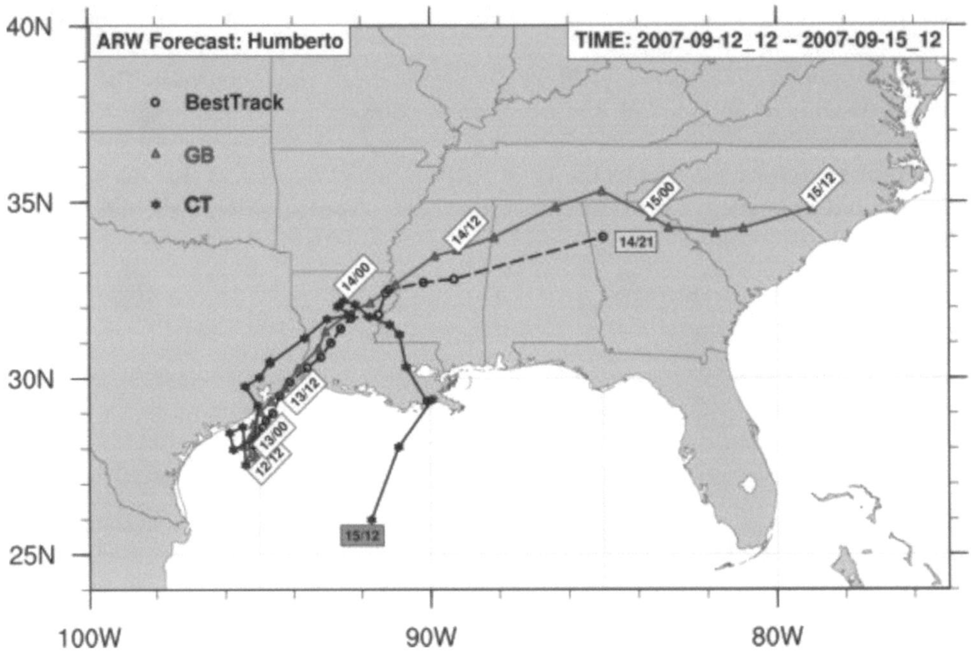

Fig. 7. 72-h track forecast for Hurricane Humberto starting from 1200 UTC 12 to 1200 UTC 15 September 2007. Dashed line with "o" is the best track; the solid line with "*" is the forecast from CT experiment; and the grey line with "Δ" is the forecast from GB experiment. The date/hour are shown in boxes.

Figure 7 shows the forecasted tracks from experiments CT and GB as well as the best track for the period of 1200 UTC 12 - 1200 UTC 15 September 2007. The experiment CT, which has the vortex not well organized at the initial time, fails predicting the storm's inland movement. It lingers along the coastal area of Texas and Louisiana for two days, and then turns back to the Gulf of Mexico. Compared with the best track observation, CT fails to predict the storm's track. On the contrary, the experiment GB successfully predicts the storm's landfall and inland movement. Its predicted track follows the best track observation. Note that in Figure 7 the best track from the National Hurricane Center extends to 2100 UTC 14 September, while the prediction extends to 1200 UTC 15 September.

For the intensity prediction, CT also fails to predict the storm's intensification before landfall. Since it fails to predict the storms' intensification before landfall and fails to predict the storm's inland movement, its overall intensity forecast is not successful at all and therefore omitted in the comparison. In Figure 8, we thus only analyze the storm's intensity in GB and compare it with observation. The trends of CSLP (Fig. 8a) and MSW (Fig. 8b) from 1200 UTC 12 to 1200 UTC 15 September show good agreement between the forecast and observation. GB successfully predicts Humberto's intensification from a tropical storm (1200 UTC 12 September) to a Category-1 hurricane (0600 UTC 13 September) before its landfall over the Texas coast. The observation indicates the maximum intensity of Humberto at 0915 UTC 13 September with a CSLP of 986 hPa and a MSW of 85 kt (44 m/s). GB predicts the maximum intensity at 0945 UTC 13 September with a CSLP of 989 hPa and a MSW of 82 kt (43 m/s). However, it over-predicts Humberto's strength inland. At 0000 UTC 14 September, for example, GB predicts a CSLP of 997 hPa and a MSW of 37 kt (19 m/s), while the observations are 1006 hPa and 25 kt (13 m/s).

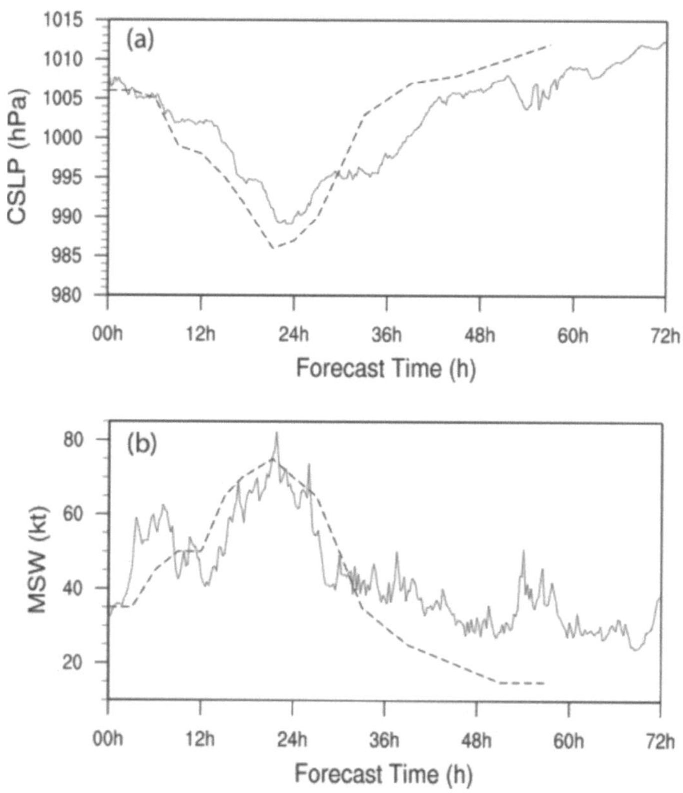

Fig. 8. 72-h intensity forecast for Hurricane Humberto starting from 1200 UTC 12 to 1200 UTC 15 September 2007: (a) Central sea level pressure (CSLP), and (b) Maximum surface wind (MSW). The solid line is the forecast from the GB experiment and dashed line is from the best track observation.

5. Summary and conclusions

Hurricane initialization using the data in the vortex region is important for the intensity forecast. Doppler radar data and synthetic bogus data assimilations (BDA) have been implemented in WRF variational data assimilation (WRF-Var) system, and positive impacts on the analysis and forecast of hurricane structure and intensity have been obtained (Xiao et al., 2009a; b). The capability of airborne Doppler radar (ADR) data assimilation to improve hurricane initialization using WRF 3D-Var is examined for Hurricanes Jeanne (2004), Katrina (2005), and Rita (2005). The BDA technique is tested using 21 cases from 7 hurricanes in the 2004 and 2005 seasons. We also conducted one case study for Hurricane Humberto (2007), and compared the results between BDA and GFDL analyses for the WRF runs. The followings are highlights of our findings from the experiments:

- Assimilation of Doppler radial wind data markedly improve the representation of the hurricane vortex structure both at the initial time and in the forecast out to about 36 h. The ADR wind assimilation makes important contributions to improving hurricane intensity and structure forecasts. Hurricane track forecasts also benefited from assimilation of ADR wind data.
- The ADR reflectivity data assimilation in WRF 3D-Var system retrieves portion of the three-dimensional rainwater and cloud water fields of hurricane vortex at initialization. The multivariate responses in other variables are also reasonable. The addition of ADR data produces a realistic eye wall and associated strong convection. Rain bands are also favorably reorganized and appear more realistic.
- WRF hurricane forecasting using the BDA technique shows an improved forecast skill in hurricane track and intensity compared to initialization from just GFS analysis. Using the WRF 3D-Var system, the bogus SLP and wind profile data can be efficiently assimilated to recover the initial hurricane structure under 3D-Var statistical and physical balances. The forecasts of hurricane track and intensity are therefore improved.
- The enhancement of the hurricane forecast skill using BDA technique reflects in all forecast periods. With BDA, the largest improvement is in hurricane central pressure. The improvement in hurricane maximum surface wind is also statistically significant. The track has the smallest improvement among the three verification parameters.
- BDA with WRF 3D-Var performs well in a case study of Hurricane Humberto (2007), whereas WRF with the initial conditions interpolated from the GFDL analysis failed in the hurricane's landfall. More cases studies and real-time forecasts are necessary to further verify its performance. However, BDA with WRF 3D-Var for hurricane initialization has the potential to improve WRF hurricane forecasts.

One of the most challenges for the hurricane forecaster and researcher is to define the vortex structure in light of insufficient observations over the ocean. The Doppler radar data (from coastal radar or airborne Doppler radar) are valuable data source for hurricane structure. When there is no data at all in the vortex region, synthetic bogus data would help in defining the vortex structure. The BDA has shown promise in WRF hurricane forecasts. Usually, satellite data are less useful within the vortex due to cloud and rainfall contamination, limited spatial resolution, or suboptimal timing of observations (from polar orbiting platforms, for instance). Assimilating Doppler radar data or sometimes synthetic bogus vortex data is feasible and should be considered in forecasting of landfalling hurricanes so as to reduce the loss of life and property in coastal regions.

In terms of WRF 3D-Var for hurricane vortex initialization, some limitations also exist. Firstly, a specific background error covariance for hurricanes should be developed and used in hurricane initialization. The background error statistics used in our studies are from the traditional NMC technique (Parrish & Derber, 1992). It is not totally suitable for the correlations in the hurricane vortex. The correlation of wind and pressure only presents large-scale feature. Secondly, reflectivity assimilation in WRF 3D-Var uses warm-rain process to bridge rainwater with other model variables in the analysis. At high levels above melting layer, however, ice-phase hydrometeors contribute to the most of reflectivity measurement. In this regard, a sophisticated microphysics that builds relationships among the whole hydrometeors and other dynamical and thermo-dynamical variables should be developed in WRF 3D-Var for radar reflectivity data assimilation. Finally, observation error statistics for radar observations and bogus data are only crudely represented at present. In addition, it should also be noted that most of our studies are based on the WRF 3D-Var that does not take into account the time differences but instead ingests data at one instant in time. 4D-Var should be a future direction for hurricane vortex initialization in order to better initialize the time dependence of the vortex necessary to accurately capture rapid intensity change as it is occurring.

6. References

Anderson, J. L. (2001). An ensemble adjustment Kalman filter for data assimilation. *Mon. Wea. Rev.*, **129**, 2884–2903.

Barker, D. M.; Huang, W.; Guo, Y.-R.; Bourgeois, A. J. & Xiao, Q. (2004). A three-dimensional (3DVAR) variational data assimilation system for MM5: implementation and initial results. *Mon. Wea. Rev.*, 132, 897-914.

Bishop, C.H.; Etherton, B.J. & Majumdar, J. S. (2001). Adaptive Sampling with the Ensemble Transform Kalman Filter. Part I: Theoretical Aspects. *Mon. Wea. Rev.*, 129, 420-436.

Courtier, P.; Thepaut, J.-N. & Hollingsworth, A. (1994). A strategy for operational implementation of 4D-Var using an incremental approach. *Quart. J. Roy. Meteor. Soc.*, 120, 1367–1387.

Daley, R. (1991). *Atmospheric data analysis*, Cambridge University Press. Cambridge UK.

Davis, C. A.; Wang, W.; Chen, S. S.; Chen, Y.; Corbosiero, K.; DeMaria, M.; Dudhia, J.; Holland, G. J.; Klemp, J.; Michalakes, J.; Reeves, H.; Rotunno, R. & Xiao, Q. (2008). Prediction of landfalling hurricanes with the advanced hurricane WRF model. *Mon. Wea. Rev.*, 136, 1990-2005..

Dudhia, J. (1989). Numerical study of convection observed during the winter monsoon experiment using a mesoscale two-dimensional model. *J. Atmos. Sci.*, 46, 3077-3107.

Elsberry, R. L. (2005). Achievement of USWRP hurricane landfall research goal. *Bull. Amer. Meteor. Soc*, 86, 643-645.

Evensen, G. (1994). Sequential data assimilation with a nonlinear quasi-geostrophic model using Montre Carlo methods to forecast error statistics. *J. Geophys. Res.*, 99, 10, 143-10,162.

Evensen, G. & van Leeuwen, P. J. (2000). An ensemble Kalman smoother for nonlinear dynamics. *Mon. Wea. Rev.*, 128, 1852–1867.

Fertig, E. J.; Harlim, J.; & Hunt, B. R. (2007). A comparative study of 4D-VAR and a 4D Ensemble Filter: perfect model simulations with Lorenz-96. *Tellus*, 59A, 96-100.

Fisher, M. (1999). Background error statistics derived from an ensemble of analyses. *ECMWF Research Department Tech. Memo.*, 79, 12pp.

Gustafsson, N. (2007). Discussion on "4D-Var or EnKF?", *Tellus*, 59A, 774-777.

Hollingsworth, A. & Lo¨nnberg, P. (1986). The statistical structure of short-range forecast errors as determined from radiosonde data. Part I: The wind field. *Tellus*, 38A, 111–136.

Houtekamer, P. L. & Mitchell, H. L. (2001). A Sequential Ensemble Kalman Filter for Atmospheric Data Assimilation. *Mon. Wea. Rev.*, 129, 123-137.

Huang, X.-Y.; Xiao, Q.; Barker, D. M.; Zhang, Xin; Michalakes, J.; Huang, W.; Hendersen, T.; Bray, J.; Chen, Y.-S.; Ma, Z.; Dudhia, J.; Guo, Y.-R.; Zhang, X.; Won, D.-J.; Lin, H.-C. & Kuo, Y.-H. (2008). Four-dimensional variational data assimilation for WRF: Formulation and preliminary results. *Mon. Wea. Rev.*, 137, 299-314.

Hunt, B. R.; Kalnay, E.; Kostelich, E. J.; Ott, E.; Patil, D. J.; Sauer, T.; Szunyogh, I.; Yorke, J. A. & Zimin, A. V. (2004). Four-dimensional ensemble Kalman filtering. *Tellus*, 56A, 273-277.

Kalnay, E.,; Li, H.; Miyoshi, T.; Yang, S.-C. & Ballabrera-Poy, J. (2007). 4D-Var or Ensemble Kalman filter? *Tellus*, 59A, 758-773.

Kessler, E. (1969). *On the Distribution and Continuity of Water Substance in Atmospheric Circulation. Meteor. Monogr.*, 32, Amer. Meteor. Soc., 84 pp.

Kurihara, Y.; Bender, M. A. & Ross, R. J. (1993). An initialization scheme of hurricane models by vortex specification. *Mon. Wea. Rev.*, 121, 2030–2045.

Le Dimet, F.-X. & Talagrand, O. (1986). Variational algorithms for analysis and assimilation of meteorological observations: Theoretical aspects, *Tellus*, 38A, 97-110.

Lee, J.-L.; Kuo, Y.-H. & MacDonald, A. E. (2003). The vorticity method: Extension to mesoscale vertical velocity abd validation for tropical storms. *Quart. J. Roy. Meteor., Soc.*, 129, 1029-1050.

Lee, J.-L.; Lee, W.-C. & MacDonald, A. E. (2006). Estimating vertical velocity and radial flow from Doppler radar observations of tropical cyclones. *Quart. J. Roy. Meteor. Soc.*, 132, 125-145.

Lewis, J. M. & Derber, J. C. (1985). The use of adjoint equations to solve a variational adjustment problem with advective constraints. *Tellus*, 37A, 309–322.

Liu, C.; Xiao, Q. & Wang, B. (2008). An ensmble-based four-dimensonal variational data assimilation scheme. Part I: technical formulation and preliminary test. *Mon. Wea. Rev.*, 136, 3363-3373.

Liu, C.; Xiao, Q. & Wang, B. (2009). An ensemble-based four-dimensional variational data assimilation scheme. Part II: Observing syetem simulation experiments with advanced research WRF (ARW). *Mon. Wea. Rev.*, 137, 1687-1704..

Liu, Y.; Zhang, D.-L. & Yau, M. K. (1997). A multiscale numerical study of Hurricane Andrew (1992). Pat I: Explicit simulation and veridication. *Mon. Wea. Rev.*, 125, 3073-3093.

Liu, Z. & Barker, D. M. (2006). Radiance assimilation in WRF-Var: Implementation and initial results. *Preprint of the 7th WRF user's workshop*, Boulder, Colorado, 19-22 June 2006.

Lorenc, A. (2003). The potential of the Ensemble Kalman Filter for NWP: a comparison with 4DVar. *Quart. J. Roy. Meteor. Soc.*, 129, 3183–3203.

Marks, F. D.; Shay, L. K.; Barnes, G.; Black, P.; DeMaria, M.; McCaul, B. & Molinari, J. (1998). Landfallinhg tropical cyclones: forecast problems and associated research opportunities. *Bull. Amer. Met. Soc.*, 79, 305-323.

Navon, I. M.; Zou, X.; Derber, J. & Sela, J. (1992). Variational Data Assimilation with an Adiabatic Version of the NMC Spectral Model. *Mon. Wea. Rev.*, 120, 1381–1393.

Ott, E; Hunt, B. R.; Szunyogh, I. ; Zimin, A. V. ; Kostelich, E. J. ; Corazza, M.; Kalnay, E.; Patil, D. J. & Yorke, J. A. (2004). A local ensemble Kalman filter for atmospheric data assimilation. *Tellus*, 56A, 415-528.

Parrish, D. F. & Derber, J. C. (1992). The National Meteorological Center's Spectral Statistical Interpolation analysis system. *Mon. Wea. Rev.*, 120, 1747–1763.

Rabier, F.; Jarvinen, H.; Klinker, E.; Mahfouf, J.-F. & Simmons, A. (2000). The ECMWF operational implementation of four-dimensional variational assimilation. Part I: Experimental results with simplified physics. *Quart. J. Roy. Meteor. Soc.*, 126, 1143-1170.

Richardson, L. F. (1922). *Weather Prediction by Numerical Process*. Cambridge University Press, 236pp.

Ross, R. J. & Kurihara, Y. (1995). A numerical study on influences of Hurricane Gloria (1985) on the environment. *Mon. Wea. Rev.*, 123, 332-346.

Simmons, A. J. & Hollingsworth A. (2002). Some aspects of the improvement in skill of numerical weather prediction. *Quart. J. Roy. Meteor. Soc.*, 128, 647-677.

Skamarock, W. C.; Klemp, J. B.; Dudhia, J.; Gill, D. O.; Barker, D. M.; Wang, W. & Powers, J. (2005). A Description of the Advanced Research WRF Version 2. *NCAR Technical Note*, TN-468+STR. 88 pp.

Sun, J. & Crook, N. A. (1997). Dynamical and microphysical retrieval from Doppler radar observations using a cloud model and its adjoint. Part I: Model development and simulated data experiments. *J. Atmos. Sci.*, 54, 1642–1661.

Sun, J. & Crook, N. A. (1998). Dynamical and microphysical retrieval from Doppler radar observations using a cloud model and its adjoint. Part II: Retrieval experiments of an observed Florida convective storm. *J. Atmos. Sci.*, 55, 835–852.

Thépaut, J.-N. & Courtier, P. (1991). Four-dimensional data assimilation using the adjoint of a multilevel primitive equation model. *Quart. J. Roy. Meteor. Soc.*, 117, 1225–1254.

Ueno, M. (1989). Operational bogussing and numerical prediction of typhoon in JMA. *JMA/NPD Tech. Rep.* 28, 48pp. [Available from Japan Meteorological Agency, Numerical Prediction Division, 1-3-4; Ote-Machi, Chiyodaku, Tokyo, 100, Japan.]

Ueno, M. (1995). A study of the impact of asymmetric components around tropical cyclone center on the accuracy of bogus data and the track forecast. Meteor. *Atmos. Phys.*, 56, 125–134.

Whitaker, J. S. & Hamill, T. M. (2001). Ensemble Data Assimilation without perturbed observations. *Mon. Wea. Rev.*, 130, 1913–1924.

White, A. (2000). A view of the equations of meteorological dynamics and various approximations, *Met Office Forecasting Research Scientific Paper*, 58, 88 pp.

Willoughby, H. E. & Black, P. G. (1996). Hurricane Andrew in Florida: Dynamics of a disaster. *Bull. Amer. Meteor. Soc.*, 77, 543-549.

Xiao, Q.; Zou, X. & Wang, B. (2000). Initialization and simulation of a landfalling hurricane using a variational bogus data assimilation scheme. *Mon. Wea. Rev.*, 128, 2252-2269.

Xiao, Q.; Zou, X.; Pondeca, M.; Shapiro, M. A.; & Velden, C. S. (2002). Impact of GMS-5 and GOES-9 satellite-derived winds on the prediction of a NORPEX extratropical cyclone. *Mon. Wea. Rev.*, 130, 507-528.

Xiao, Q.; Kuo, Y.-H.; Sun, J.; Lee, W.-C.; Lim, E.; Guo, Y.-R. & Barker, D. M. (2005). Assimilation of Doppler radar observations with a regional 3D-Var system: Impact of Doppler velocities on forecasts of a heavy rainfall case. *J. Appl. Meteor*, 44, 768-788.

Xiao, Q. & Sun, J. (2007). Multiple-radar data assimilation and short-range quantitative precipitation forecasting of a squall line observed during IHOP_2002. *Mon. Wea. Rev.*, 135, 3381-3404.

Xiao, Q.; Kuo, Y.-H.; Zhang, Y.; Barker, D. M. & Won, D.-J. (2006). A tropical cyclone bogus data assimilation scheme in the MM5 3D-Var system and numerical experiments with Typhoon Rusa (2002) near landfall. *J. Met. Soc. Japan*, 84, 671-689.

Xiao, Q.; Kuo, Y.-H.; Sun, J.; Lee, W.-C.; Barker, D. M. & Lim, E. (2007). An approach of radar reflectivity data assimilation and its assessment with the inland QPF of Typhoon Rusa (2002) at landfall. *J. Appl. Meteor. Climat.*, 46, 14-22.

Xiao, Q. & coauthors (2008a). Application of an adiabatic WRF adjoint to the investigations of the May 2004 McMurdo Antarctica severe wind event. *Mon. Wea. Rev.*, 136, 3696-3713.

Xiao, Q. & coauthors (2008b). Doppler radar data assimilation in KMA's operational forecasting. *Bull. Amer. Meteor. Soc.*, 89, 39-43.

Xiao, Q.; Zhang, X.; Davis, C. A.; Tuttle, J. D.; Holland, G. J. & Fitzpatrick, P. J. (2009a). Experiments of hurricane initialization with airborne Doppler radar data for the Advanced-research Hurricane WRF (AHW) model. *Mon. Wea. Rev.*, 137, 2758-2777.

Xiao, Q.; Chen, L. & Zhang, X. (2009b). Evaluations of BDA scheme using the Advanced Research WRF (ARW) model. *J. Appl. Meteor. Climat.*, 48, 680-689.

Zhang, X.; Xiao, Q. & Fitzpatrick, F. J. (2007). The impact of multi-satellite data on the initialization and simulation of Hurricane Lili's (2002) rapid weakening phase. *Mon. Wea. Rev.*, 135, 526-548.

Zou, X. & Xiao, Q. (2000). Studies on the initialization and simulation of a mature hurricane using a variational bogus data assimilation scheme. *J. Atmos. Sci.*, 57, 836-860.

Zupanski, M. (2005). Maximum likelihood ensemble filter: theoretical aspects. *Mon. Wea. Rev.*, 133, 1710-1726.

16

Improving Hurricane Intensity Forecasting through Data Assimilation: Environmental Conditions Versus the Vortex Initialization

Zhaoxia Pu
Department of Atmospheric Sciences
University of Utah
USA

1. Introduction

Tropical cyclones (TCs) are one of nature's most intense phenomena and one of all coastal residents' greatest fears. They threaten the maritime industry, devastate coastal regions and cause floods and erosion inland through torrential rainfall, high winds and severe storm surges. Through coastal development and growth, the United States has become more vulnerable to the impact of hurricanes now than at any time in the recent past (Pielke & Pielke, 1997). An extreme example was Hurricane Katrina during the summer of 2005. Owing to the great social and economic impact of hurricanes, it is of great importance that the track and intensity of these hurricanes can be accurately predicted many hours in advance.

Over the last two decades, computer modeling and data assimilation techniques have advanced rapidly. Along with the development of high-resolution numerical modeling and advanced data assimilation capabilities, the skill of numerical weather prediction has improved significantly. As a result of these efforts, TC track forecasts have improved substantially. However, the intensity forecast remains a great challenge in the operational and research communities. According to reports from Rogers et al. (2006) and the National Hurricane Center (NHC, http://www.nhc.noaa.gov/verification), the official 48 h TC track forecast error has been reduced by 45% in the past 15 years, while the intensity forecast error has decreased by only 17%. The rapid intensification of TCs has been especially poorly predicted (Titley & Elsberry, 2000; Houze et al., 2006).

The lack of skill in numerical forecasts of TC intensity can be attributed to three factors (Rogers et al., 2006): 1) inaccurate initial conditions in the storm vortex and environment in numerical models; 2) limitations in numerical models such as imperfect physical parameterizations; and 3) inadequate understanding of the physics of TCs and their development. Among other factors, the uncertainties in hurricane initialization are some of the fundamental reasons for the limited skill in hurricane forecasts.

The processes associated with TC intensifications have been investigated in numerous previous studies (*e.g.*, Malkus, 1958; Willoughby, 1988; Frank & Ritchie, 1999; Montgomery et al., 2006). It has been recognized that the large scale environmental conditions such as vertical wind shear, pre-existing upper level troughs, mesoscale convections, storm-scale inner core structures, and air-sea interactions, such as the ocean surface fluxes all play important roles (*e.g.*, Davis & Emanuel, 1988; DeMaria & Pickle, 1988; Kuo et al., 1991; Merrill & Velden, 1996; Willoughby & Black, 1996; Bosart et al., 2000; Zhu et al., 2002; Montgomery et al., 2006). Therefore, in order to improve TC forecasting, accurate specifications of each of the large scale, mesoscale, and small-scale atmospheric conditions is very important.

Since TCs form over the oceans, where conventional observations are sparse, significant progress has been made with the use of satellite and airborne in-situ and remote sensing data in TC forecasting. It has been shown that these additional observations have a positive impact on TC forecasts. For instance, the U.S. National Oceanic and Atmospheric Administration (NOAA) synoptic surveillance missions are very useful in improving individual TC track forecasts (e.g., Franklin & DeMaria, 1992; Franklin et al., 1993; Burpee et al., 1996; Aberson, 2002; Aberson et al., 2003; Aberson & Etherton, 2006). Other research programs such as the "Dropwindsonde Observations for Typhoon Surveillance near the Taiwan Region" (DOTSTAR) have also demonstrated the positive impact of the targeted observations on the track forecasts of the typhoons over the Western Pacific Ocean (Wu et al., 2007).

Along with the advancements in remote sensing techniques, the amount of usable satellite data has increased rapidly in the last two decades. Many satellite data products have become useful sources for TC analyses and forecasts (e.g., Velden et al., 1992; Leslie et al., 1998; Velden et al., 1998; Pu et al., 2002; Zhu et al., 2002; Pu & Tao, 2004; Hou et al., 2004, Wu et al., 2006; Chen, 2007; Zhang et al., 2007, Pu et al., 2008). However, due to cloud and rain effects, few observations of TC inner core regions are assimilated into the model. Satellite data assimilation has mainly contributed to the improvement in hurricane environmental conditions. Thus, even in a high-resolution model, the initial conditions may not resolve the realistic inner core structure and intensity of TCs. The use of so-called "bogus vortices" is often adopted in research numerical simulations and operational forecasting (Kurihara *et al.*, 1993; Leslie & Holland, 1995; Pu & Braun, 2001; Braun et al., 2006). Due to both the lack of conventional observations and the difficulties in using rain- and cloud-affected satellite observations, the initialization of hurricane inner core structures with realistic dynamic and thermodynamic observations has been a challenging problem in NWP. Fortunately, the amount of available Doppler radar data and recent developments in radar data assimilation have brought hopes of improving TC inner-core thermodynamic and hydrometeor conditions. Several studies (Zhao et al., 2008; Zhao & Jin, 2008; Pu et al., 2009b) have proved that radar data assimilation resulted in a significant impact on TC track and intensity forecasting.

From all the aformentioned studies, improved hurricane environmental conditions and the vortex initialization both play important roles in the accurate forecasting of hurricane rapid intensification and intensity changes. Specifically, through the assimilation of satellite, radar and in-situ aircraft observations, it has been established that the hurricane intensity forecast can be indeed improved by data assimilation. However, a general conclusion regarding how (and to what extent) environmental conditions and initial vortex structure can contribute to significant improvement in hurricane rapid intensification is yet to be established. With the results from author's recent studies, this topic will be addressed in this current chapter.

2. Improving TC environmental conditions

Among all the environmental conditions that have an impact on TC intensity changes, it has been well recognized that the environmental vertical wind shear is an important factor that influences the tropical cyclone structure and development. Rogers et al. (2003) suggested that tropical cyclone convection and precipitation structures are closely related to the magnitude of environmental vertical wind shear. Merriell (1988) and Frank and Ritchie (1999) concluded that weak environmental vertical wind shear is a necessary condition for tropical cyclone deepening.

With a mesoscale community weather research and forecasting (WRF) model (Skamarock et al., 2008) and its three-dimensional variational (3DVAR) data assimilation system (Barker et al., 2004), Figure 1 compares the impact of assimilating various wind data, including these from the National Aeronautics and Space Administration (NASA) QuikSCAT satellite measured ocean surface vectors, NOAA GOES-11 rapid scan atmospheric motion vector, and aircraft dropwindsondes, on the environmental vertical wind shear for Tropical Storm Gert (2005) over the Atlantic Ocean. The significant impact of wind data assimilation on TC environmental conditions has been shown; as a result, the assimilation of aircraft dropsonde data and satellite wind data help the model to reproduce more accurate storm landfall forecasts. About 50% of the track error was reduced. The landfall time and locations are well captured by the forecasts with data assimilation (see Pu et al., 2008 for details).

A further study with Hurricane Cindy (2005) shows that the assimilation of the aforementioned wind data has a great impact on the quantitative precipitation forecast (QPF) during Cindy's evolution (Fig. 2), implying the significant effects of environmental wind shear on the mesoscale convective system associated with Cindy (See Pu et al., 2008 for details).

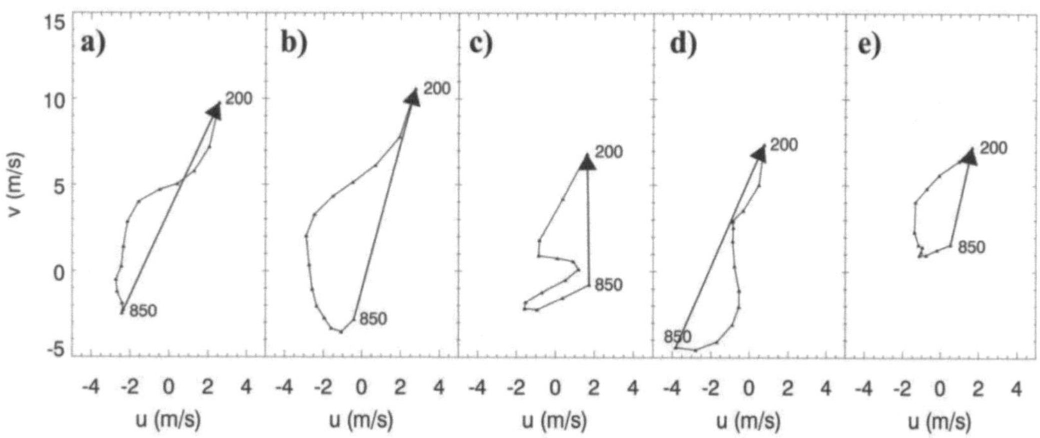

Fig. 1. The hodograph of environmental winds (m s^{-1}) between 850 and 200 hPa in the vicinity of the vortex of tropical storm Gert (2005) at 1200 UTC 24 July 2005. Different panels represent the wind shears obtained from the experiments without the assimilation of wind data (a), and with assimilation of b) NASA QuikSCAT ocean surface vectors, c) NOAA GOES-11 rapid scan atmospheric motion vector, d) Dropwindsondes, and e) all of above wind data mentioned. [Figure from Pu et al. (2008)].

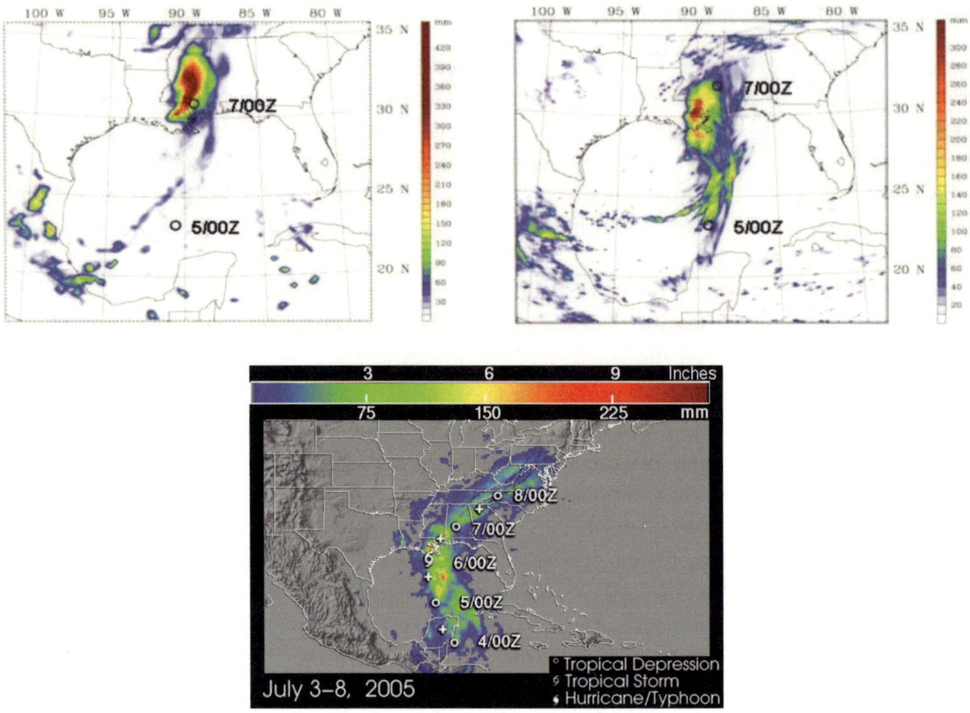

Fig. 2. The weather research and forecasting (WRF) model forecasted rainfall amount over 48 hours from Hurricane Cindy, July 5-7, 2005 (top panels) without (left) and with (right) data assimilation. The bottom panel shows the rainfall accumulation along the track during Cindy's life cycle [courtesy of NASA Tropical Rainfall Measuring Mission (TRMM) project]. Without data assimilation, the forecast shows unrealistically large amounts of rain after Cindy made landfall. It also shows that the forecast missed a large amount of rainfall that occurred before Cindy made landfall. However, the image with data assimilation more accurately portrays the rain that actually fell.

In addition to wind information, accurate atmospheric profile information was generally lacking before the Atmosphere Infrared Sounder (AIRS) was launched into orbit onboard the NASA Aqua satellite in May 2002. In order to evaluate the AIRS data impact on TCs forecast, Pu and Zhang (2010) performed a series of data assimilation experiments with the mesoscale community weather research and forecasting (WRF) model and its data assimilation system. Results show that the assimilation of the AIRS retrieved temperature and moisture profiles has a significant impact on the numerical simulations of TCs, although the overall impacts of the data on numerical simulations of TCs are very sensitive to the data quality. Specifically, it is found that the forecast of the formation of tropical storm Debby (2006, over Atlantic Ocean) is very sensitive to the assimilation of moisture profiles. Compared with the moisture profiles, temperature profiles show a larger impact on Debby's track forecasting, mainly because the wind field responds largely to the analysis increment of the temperature field. When both moisture and temperature profiles are adjusted through bias correction, the data assimilation leads to improved track and intensity forecasts for Debby (Pu & Zhang, 2010). Additional experiment with a Supertyphoon Jangmi (2008) also indicated that the AIRS data has a positive impact on the storm's track forecast of Jangmi.

3. Does accurate initial intensity lead to an accurate TC intensity forecast?

The aforementioned data assimilation experiments have indicated that assimilation of atmospheric temperature, wind and moisture profiles could contribute to accurate specification of TC environmental conditions in the numerical model and thus result in improved TC forecasts. In fact, with the assimilation of available satellite and in-situ data, we are often able to capture the initial intensity of TCs through assimilation of the data around the TC environment. However, the accurate initial intensity of the TC vortices does not always lead to perfect intensity forecasts.

In a recent study by Li and Pu (2008), numerical simulations focus on the early rapid intensification period of Hurricane Emily (2005) from 0600 UTC 14 July to 0600 UTC 15 July 2005, when the observed minimum central sea level pressure (MSLP) changed from 991 to 952 hPa. The WRF model initial conditions were generated using a WRF 3DVAR data assimilation system. The GOES-11 rapid scan atmospheric motion vectors, QuikSCAT ocean surface vector winds, and aircraft dropwindsonde data, collected during NASA Tropical Cloud Systems and Processes (TCSP) experiment (Halverson et al., 2007), were assimilated into the WRF model with the available conventional data in a 6 hourly cycling data

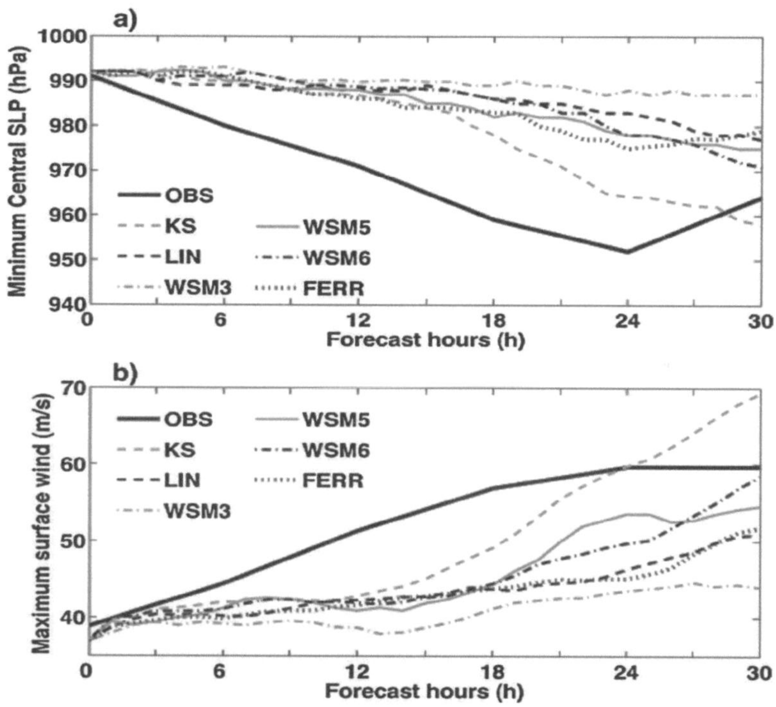

Fig. 3. Time series of a) minimum central sea level pressure (hPa) and b) maximum surface wind speed (m s^{-1}) from the National Hurricane Center best track data (OBS) and the numerical simulations of Hurricane Emily using WRF model during 0600 UTC 14 July to 1200 UTC 15 July 2005 with different microphysical schemes: Kessler warm-rain scheme (KS); Purdue Lin scheme (LIN); WRF single moment (WSM); three-class simple ice scheme (WSM3); WSM five-class mixed phase scheme (WSM5); WSM six-class graupel scheme (WSM6); Eta Ferrier scheme (FERR). [Figure from Li and Pu (2008)].

assimilation within the 12 h assimilation window (1800 UTC 13 to 0600 UTC 14 July 2005). At the end of data assimilation, the intensity (in terms of minimum sea-level pressure and maximum wind) of the initial vortex matches the intensity specified by the National Hurricane Center's best track data. After the data assimilation, a 30-h numerical simulation was conducted at a 3 km grid spacing. Numerical simulation results show significant sensitivity of the forecast of Hurricane Emily to varying cloud microphysical schemes in WRF model. However, even with different combinations of the various physical schemes, none of the experiments were able to capture the real intensification of Hurricane Emily during the simulation period, although all of the simulations start from the same initial conditions with the same storm intensity that matches the best track intensity of the storm in terms of both minimum central sea-level pressure and maximum surface wind.

A diagnostic study (Pu et al., 2009a) is then conducted to examine the accompanying initial and forecast errors. The initial conditions and the simulated hurricane vortices are compared with flight level data acquired from the United States Air Force C-130J aircraft. Unrealistic thermal and convective structures of the storm eyewall are found in the initial conditions (Fig. 4). In addition, the simulated (forecast) eyewall does not contract rapidly enough during the model simulation. Although the possible factors that cause the failure in the intensity forecast are complicated, and the diagnostic study from Pu et al. (2009a) can only explain some of the reasons linked with this failure, overall *results from the study suggest that a more accurate representation of the hurricane vortex, especially the inner core structures in the initial conditions, is necessary for a more accurate forecast of hurricane rapid intensification* (See Pu et al., 2009a for details).

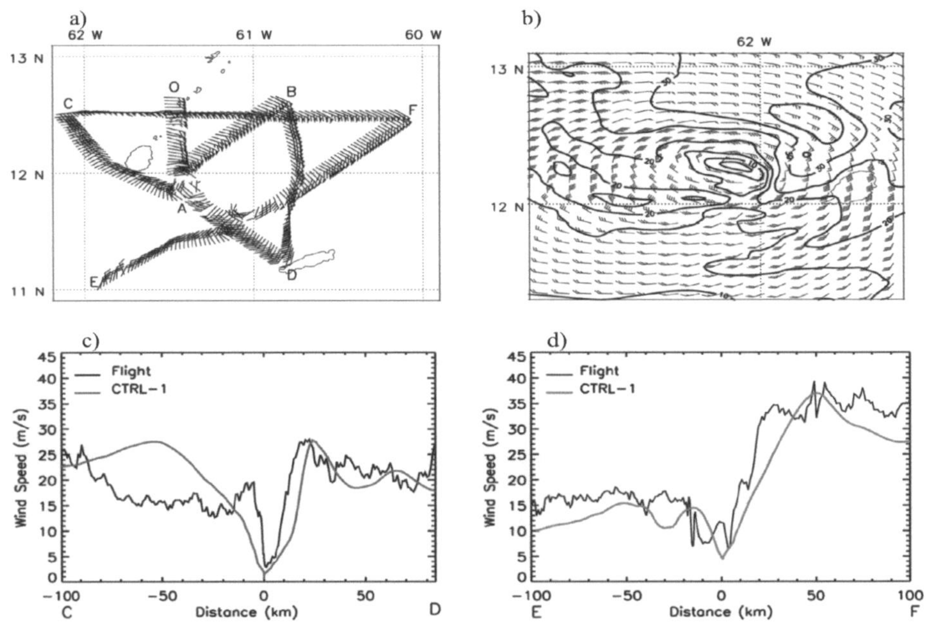

Fig. 4. Comparison of the wind structure of Hurricane Emily from a) US Air Force flight level data; b) the model initial condition at 0600 UTC 14 July 2005 at 1-km resolution (Exp. CTRL-1); and wind speed comparison from the model initial condition and the US air force flight level data at legs CD (c) and EF (d). [Figure from Pu et al. 2009a)]

3. Impact of assimilation of the airborne Doppler radar on TC forecast

As mentioned above and in previous studies, accurate forecasts of TC structure and intensity changes are closely related to the storm inner-core thermal and dynamic structures and their evolution (Jordan 1961; Franklin et al., 1988; Kossin & Eastin, 2001; Houze et al., 2006; Rogers et al., 2006). Unfortunately, most available satellite data over the TC inner-core region are contaminated by heavy precipitation. This produces uncertainties in representing TC structures in model initial conditions. Recent operations of Doppler radar have brought opportunities to sample the thermodynamics, microphysics and dynamic characteristics of TCs (Marks 2003). With data at high spatial and temporal resolution, airborne Doppler radar can reveal detailed structural features of the TCs, and studies have demonstrated that these radar data are useful in studying hurricanes (Xiao et al., 2007; Zhao et al., 2008; Zhao & Jin 2008, Pu et al., 2009b).

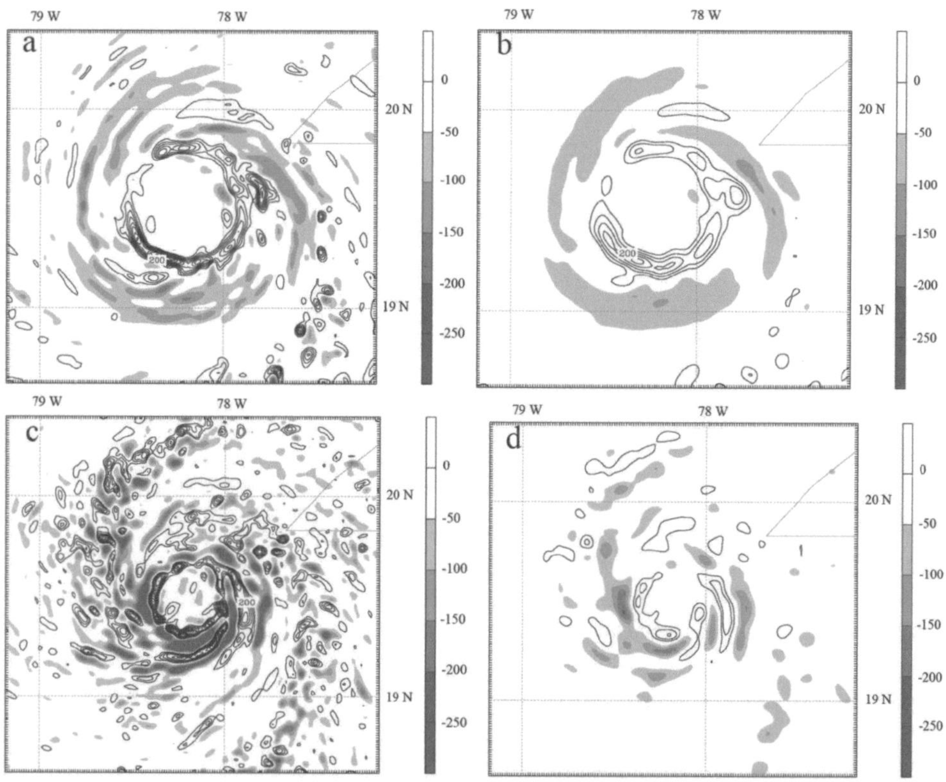

Fig. 5. Divergence fields at 250hPa (top) and 850hPa (bottom) pressure levels. The control experiment (left; without assimilation of radar data), compared with a experiment that assimilates the radar radial velocity data (right) at 0600UTC 8 July 2005. The line contours denote positive values of the divergence, while the shaded contours represent negative values. The contour interval is 50 $x10^{-5}$ s^{-1}. (from Pu et al. 2009b).

With very high resolution (about 1-2 km in the horizontal and 0.5 km in the vertical direction), airborne Doppler radars mostly represent the wind, moisture, and hydrometeor structure within the hurricane eyewall. In a recent study (Pu et al., 2009b), the usefulness of airborne Doppler radar reflectivity and radial velocity in better representing the hurricane inner core structure and improving hurricane intensity forecasts has been investigated. A series of numerical experiments is conducted for Hurricane Dennis (2005) to study its intensity changes near landfall. Both radar reflectivity and radial velocity derived wind fields are assimilated into the WRF model with its 3DVAR system. Numerical results indicate that the radar data assimilation has greatly improved the simulated structure and intensity changes of Hurricane Dennis. The hurricane landfall, intensification and weakening events during the simulation period are well captured by assimilating both radar reflectivity and radial velocity data. As shown in Figure 5, the assimilation of radar radial velocity data improved the divergence/convergence structure over Hurricane Dennis inner-core area. Figure 6 and 7 illustrates the impact of assimilation of airborne Doppler radar data on the numerical simulation of the track intensity changes of Dennis during the simulation period; significant improvement in Dennis's track and intensity forecast is clearly evidenced.

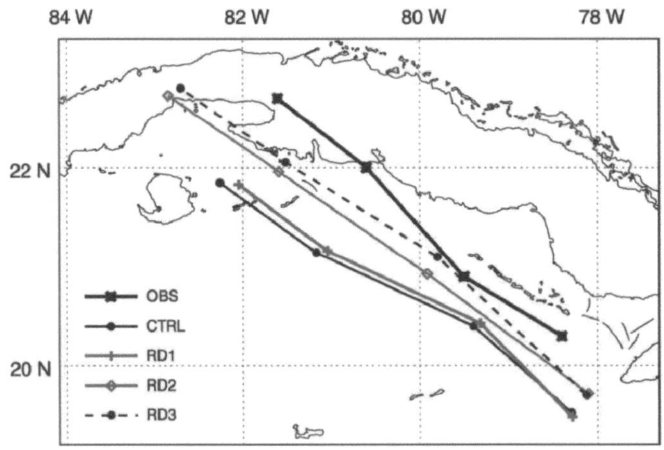

Fig. 6. Storm tracks of Hurricane Dennis (in 6-h interval) between 0600 UTC 8 and 0000 UTC 9 July 2005. National Hurricane Center best track data (OBS) compared with the experimental forecast with assimilation of radar reflectivity (RD1), radial velocity (RD2) and both radar reflectivity and radial velocity (RD3). (from Pu et al., 2009b)

4. Summary and concluding remarks

Forecasting hurricane rapid intensification is a challenging problem for operational weather prediction. Our limited forecasting ability can be mainly attributed to inaccurate initial conditions. Recent studies demonstrated that improved hurricane environmental conditions and the vortex initialization both play important roles in the accurate forecasting of hurricane rapid intensification and intensity changes. Specifically, through the assimilation of satellite, radar and in-situ aircraft observations, it has been established that the hurricane intensity forecast can indeed be improved by data assimilation. However, a general conclusion regarding to how and to what extent the environmental conditions and initial vortex structure can contribute to the significant improvement of hurricane rapid intensification has not yet been established. With the results from recent studies, this book chapter discussed the impact of accurate environmental conditions and improved vortex initialization on hurricane rapid intensification forecasts. It is obvious that both factors are very important for improving hurricane intensity forecasting. In order to improve the current limited skill of hurricane intensity forecasting, an integrated data assimilation effort with the assimilation of observations in TC enviroment and the inner-core region is necessary.

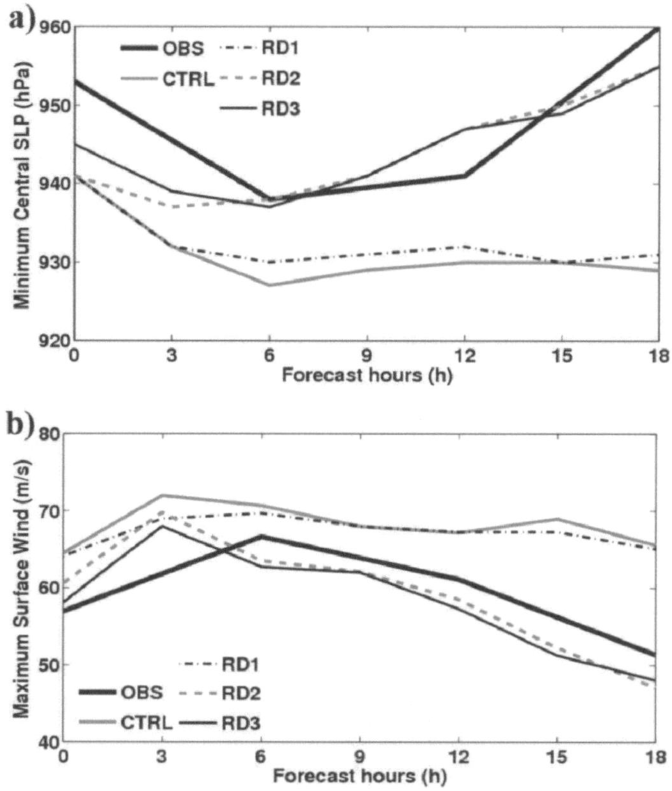

Fig. 7. Time series of observed and simulation minimum Sea Level Pressure (left) and storm track (right) between 0600 UTC 8 and 0000 UTC 9 July 2005. National Hurricane Center best track data (OBS) compared with the experimental forecast with assimilation of radar reflectivity (RD1), radial velocity (RD2) and both radar reflectivity and radial velocity (RD3). (from Pu et al., 2009b)

In addition, since uncertainties in the model initial conditions are only one of the many factors that could influence the accuracy of TC intensity forecasting, studies should also be conducted to improve computer models and our understanding of the processes that control hurricane intensity change.

5. Reference

Aberson, S. D. (2002). Two years of operational hurricane synoptic surveillance. *Wea. Forecasting*, 17, 1101–1110.

Aberson, S. D. & Sampson C. R. (2003). On the predictability of tropical cyclone tracks in the Northwest Pacific Basin. *Mon. Wea. Rev.* ,131, 1491–1497.

Aberson, S. D. & Etherton, B. J. (2006). Targeting and data assimilation studies during Hurricane Humberto (2001). *J. Atmos. Sci.*, 63, 175–186.

Barker, D. M.; Lee, M.-S.; Guo, Y.-R.; Huang, W.; Xiao Q.-N. & Rizvi, R. (2004) WRF variational data assimilation development at NCAR. June 22-25, 2004. *WRF/MM5 Users' Workshop*, Boulder, CO.

Bosart, L. F.; Velden, C. S.; Bracken, W. E.; Molinari, J & Black, P. G. (2000). Evironmental influences on the rapid intensification of Hurricane Opal (1995) over the Gulf of Mexico, *Mon. Wea. Rev.*, 128, 322-352.

Braun, S. A.; Montgomery; M. T. & Pu, Z. (2006). High-resolution simulation of Hurricane Bonnie (1998). Part I: The organization of eyewall vertical motion. *J. Atmos. Sci.*, 63, 19-42.

Burpee, R. W.; Aberson, S. D.; Franklin, J. L.; Lord, S. J. & Tuleya, R. E. (1996). The impact of Omega dropwindsondes on operational hurricane track forecast models. *Bull. Amer. Meteor. Soc.*, 77, 925–933.

Chen, S. H. (2007). The Impact of assimilating SSM/I and QuikSCAT satellite winds on Hurricane Isidore simulation. *Mon. Wea. Rev.*, 135, 549–566

Davis, C. A. & Emanuel, K. A. (1988). Observational evidence for the influence of surface heat fluxes on rapid maritime cyclogenesis. *Mon. Wea. Rev.*, 116, 2649-2659.

DeMaria, M. & Pickle, J. D. (1988). A simplified systems of equations for simulation of tropical cyclones, *J. Atmos. Sci.*, 45, 1542-1554.

Frank, W. M. & Ritchie, E. A. (2001). Effects of vertical wind shear on the intensity and structure of numerically simulated hurricanes. *Mon. Wea. Rev.*, 129, 2249-2269.

Franklin, J. L. & DeMaria, M., (1992). The impact of omega dropwindsonde observations on barotropic hurricane track forecasts. *Mon. Wea. Rev.*, 120, 381–391.

Franklin, J. L.; Lord, S. J.; Feuer, S. E. & Marks, F. D. (1993). The kinematic structure of Hurricane Gloria (1985) determined from nested analyses of dropwindsonde and Doppler radar data. *Mon. Wea. Rev.*, 121, 2433–2451.

Halverson, J. B. & co-authors (2007). NASA's Tropical Cloud Systems and Processes (TCSP) experiment. *Bull. Amer. Meteor. Soc.*, 88, 867-882.

Hou, A. Y.; Zhang, S. Q. & Reale, O. (2004). Variational continuous assimilation of TMI and SSM/I rain rates: Impact on GEOS-3 hurricane analyses and forecasts. *Mon. Wea. Rev.*, 132, 2094-2109.

Houze, R. A.; Chen, S. S.; Lee, W. C.; Rogers, R. F.; Moore, J. A.; Stossmeister, G. J.; Bell, M. M.; Ceterone, J.; Zhao, W. & Brodzik, S. R. (2006). The Hurricane Rainband and Intensity Change Experiment: Observations and modeling of Hurricanes Katrina, Ophelia, and Rita. *Bull. Amer. Meteor. Soc.*, 87, 1503-1521.

Kuo, Y.-H.; Shapiro, M. A. & Donall, E. G. (1991). The interaction between baroclinicand diabatic processes in a numerical simulation of a rapidly intensifying extratropical marine cyclone. *Mon. Wea. Rev.*, 119, 368-384.

Kurihara, Y.; Bender, M. A. & Ross, R. J. (1993). An initialization scheme of hurricane models by vortex specification. *Mon. Wea. Rev.*, 121, 2030 –2045.

Leslie, L. M. & Holland, G. J. (1995). On the bogussing of tropical cyclones in Numerical models: A comparison of vortex profiles. *Meteor. Atmos. Phys.*, 56, 101–110.

Leslie, L. M.; LeMarshall, J. F.; Morison, R. P.; Spinoso, C.; Purser, R. J.; Pescod, N. & Seecamp, R. (1998). Improved hurricane track forecasting from the continuous assimilation of high quality satellite wind data. *Mon. Wea. Rev.*,126, 1248-1258.

Li, X. & Pu, Z. (2008). Sensitivity of numerical simulation of early rapid intensification of Hurricane Emily (2005) to cloud microphysical and planetary boundary layer parameterizations. *Mon. Wea. Rev.*, 136, 4819-4838

Malkus, J. S. (1958). On the structure and maintenance of the mature hurricane eye. *J. Meteor.*, 15, 337-349.

Marks, F. D. (2003). State of the Science: Radar View of Tropical Cyclones. Radar and Atmospheric Science: A Collection of Essays in Honor of David Atlas, Edited by R. M Wakimoto and R. C. Srivastava, *Meteorological Monographs*,

Merrill, R. T. (1988). Environnemental influences on hurricane intensification. *J. Atmos. Sci.*, 45, 1678-1687.

Merrill, R. T. & Velden, C. S. (1996). A three-dimensional analysis of the outflow layer of Supertyphoon Flo (1990). *Mon. Wea. Rev.*, 127, 47-63.

Montgomery, M.T.; Nicholls, M. E.; Cram, T. A. & Saunders, A. B. (2006). A "vortical" hot tower route to tropical cyclogenesis. *J. Atmos. Sci.* (CAMEX Special Issue), 355-386.

Pielke, R.A., Jr. & R.A. Pielke, Sr. (1997). Hurricanes: Their nature and impacts on society. *John Wiley and Sons*, England

Pu, Z. & Braun, S. A. (2001). Evaluation of bogus vortex techniques with four-dimensional variational data assimilation. *Mon. Wea. Rev.*, 129, 2023-2039.

Pu, Z; Tao, W. K.; Braun, S. A.; Simpson, J.; Jia, Y.; Halverson; J., Hou, A. & Olson, W. (2002). The impact of TRMM data on mesoscale numerical simulation of supertyphoon Paka. *Mon. Wea. Rev.*, 130, 2248-2258.

Pu, Z. & Tao, W. K. (2004). Mesoscale assimilation of TMI data with 4DVAR: Sensitivity study. *J. Meteor. Soc. Japan.* 82, 1389-1397.

Pu, Z.; Li, X.; Velden, C.; Aberson, S. & Liu, T. W. (2008). Impact of aircraft dropsonde and satellite wind data on the numerical simulation of two landfalling tropical storms during TCSP. *Wea. Forecasting*,23, 62-79.

Pu, Z., Li, X. & Zipser, E. (2009)a. Diagnosis of the initial and forecast errors in the numerical simulation of the rapid intensification of Hurricane Emily (2005). *Wea. Forecasting*, 24, 1236-1251.

Pu, Z., Li, X. & Sun, J. (2009)b. Impact of airborne Doppler radar data assimilation on the numerical simulation of intensity changes of Hurricane Dennis near a landfall. *J. Atmos. Sci.*,66,3351-3365.

Pu, Z. & Zhang, L. (2010). Validation of AIRS temperature and moisture profiles over tropical oceans and their impact on numerical simulations of tropical cyclones. *Journal of Geophysical Research – Atmospheres*, 115, D24114, doi: 10.129/210JD-014258

Rogers, R.; Chen, S.; Tenerelli, J. & Willoughby, H. (2003). A numerical study of the impact of vertical shear on the distribution of rainfall in Hurricane Bonnie (1998). *Mon. Wea. Rev.*, 131, 1577–159.

Rogers, R.; Aberson, S.; Black, M.; Black, P.; Cione, J.; Dodge, P.; Dunion, J.; Gamache, J.; Kaplan, J.; Powell, M.; Shay, N.; Surgi, N. & Uhlhorn, E. (2006). The intensity forecasting experiment: A NOAA multiyear field program for improving tropical cyclone intensity forecasts. *Bull. Amer. Meteor. Soc.*, 87, 1523-1537.

Skamarock, W. C.; Klemp, J. B.; Dudhia, J.; Gill, D. O.; Barker, D. M.; Duda, M. G.; Huang, X.-Y.; Wang, W. & Powers, J. G. (2008). A Description of the Advanced Research WRF Version 2. NCAR Technical Note, NCAR/TN-468+STR. [available at NCAR, Boulder, CO 80300]

Titley, D. W. & Elsberry, R. L. (2000). Large intensity changes in tropical cyclones: A case study of Supertyphoon Flo during TCM-90. *Mon. Wea. Rev.*, 128, 3556 -3573.

Trinh, V. T. & Krishnamuti, T. N. (1992). Vortex initialization for typhoon track prediction. *Meteor. Atmos. Phys.*, 47, 117–126.

Velden, C. S.; Hayden, C. M.; Franklin; J. L. & Lynch, J. S. (1992). The impact of satellite-derived winds on numerical hurricane track forecasting. *Wea. Forecasting*, 7, 107–118

Velden, C. S.; Olander, T. L. & Wanzong, S. (1998). The impact of multispectral GOES-8 wind information on Atlantic tropical cyclone track forecasts in 1995. Part 1: Dataset methodology, description and case analysis. *Mon. Wea. Rev.*, 126, 1202-1218.

Willoughby, H. E (1988). The dynamics of the tropical cyclone core. *Aust. Meteoro. Mag.* 36, 183-191.

Willoughby, H. E & Black, P. G. (1996). Hurricane Andrew in Florida: Dynamics of a disaster. *Bull. Amer. Meteor. Soc.*, 77, 543–549.

Wu, C. C.; Chou, K. H.; Lin, P. H.; Aberson, S.; Peng, M. S. & Nakazawa, T. (2007). The impact of dropwindsonde data on typhoon track forecasts in DOTSTAR. *Wea. Forecasting*. 22, 1157–1176.

Zhao, Q.; Cook, J.; Xu, Q. & Harasti, P. R. (2006). Using radar wind observations to improve mesoscale numerical weather prediction. *Wea. Forecasting*, 21, 502-522.

Zhao, Q. & Jin, Y. (2008). High-resolution radar data assimilation for Hurricane Isabel (2003) at landfall. *Bull. Amer. Meteor. Soc.*, 89, 1355-1372

Zhu, T.; Zhang, D. L. & Weng, F. (2002). Impact of the Advanced Microwave Sounding Unit measurements on hurricane prediction. *Mon. Wea. Rev.*, 130, 2416–2432.

Zhang, X.; Xiao, Q. & Fitzpatrick, P. J. (2007). The impact of multi-satellite data on the initialization and simulation of Hurricane Lili's (2002) rapid weakening phase. *Mon. Wea. Rev.*, 135, 526-548.

17

The Impact of Oceanic Heat Content on the Rapid Intensification of Atlantic Hurricanes

Kevin Law
Marshall University
United States of America

1. Introduction

With the increased infrastructure and amount of people living along the United States coastline, it is imperative to improve the accuracy of Atlantic hurricane intensity forecasts. Over the last 10 years, there have been many Atlantic hurricanes, including Hurricanes Katrina and Charley that surprised many forecasters with their rapid intensification and power. The rapid intensification of tropical cyclones is the most serious aspect, when it comes to forecasting. It is generally accepted that sufficient surface ocean temperatures (approximately 26°C) are needed to produce and sustain tropical cyclone formation. However, the sea-surface temperature (SST) has shown not to be critical in intensity forecasting by itself, particularly with rapid intensification (Schade & Emmanuel, 1999; Law & Hobgood 2007). Tropical cyclones derive much of their energy from warm, deep ocean water. Therefore, a quantified measure of the amount of this warm, deep water is a better way to measure the amount of energy available to the storm. The oceanic heat content (OHC) is such a variable to measure the amount of warm water available for the tropical cyclone to convert into energy and has been shown to be a much better predictor than SST alone (Zebiak, 1989; McDougall, 2003; Wada & Usui 2007; Palmer & Haines, 2009; Shay & Brewster, 2010).
The accuracy of predicting tropical cyclone tracks currently exceeds the accuracy of intensity prediction. This is largely because track forecasts are largely governed by large-scale processes, such as upper-wind conditions and the positioning of ridges and troughs. These large-scale processes are more easily captured by dynamic numerical models therefore making track forecasts more precise. However, the same dynamic numerical models have difficulty capturing the sub-grid scale features inside of the tropical cyclone. Since the sub-grid scale features, such as the temperature/moisture profiles and heat fluxes, tend to govern the intensity of the systems, the dynamic numerical model accuracy of the intensity forecasts lag behind their track counterparts.
While dynamic numerical models tend to lag behind in forecasting intensity, statistical models have been shown to outperform the performance of dynamic models. Many statistical and statistical-dynamic models are used by the National Hurricane Center in Coral Gables, Florida, USA. These would include such models as the Statistical Hurricane Intensity Prediction Scheme or SHIPS and SHIFOR5 which is a 5-day statistical hurricane intensity forecast model. The SHIPS is a statistical-dynamic model uses a variety of dynamic variables such as the maximum sustained surface wind speed, 12-hour intensity change, 850-200 hPa vertical wind shear, the average temperature at 200 hPa, and the 850

hPa relative vorticity (DeMaria & Kaplan 1999; DeMaria et al., 2005). Many of these climatological and synoptic sub-grid scale variables are averaged over the distance of the storm, to produce intensity forecasts based upon the climatological characteristics of previous tropical cyclones. The SHIFOR5 model uses climatology and persistence as a purely statistical baseline.

Regardless of statistical-dynamic or statistical models, rapid intensification processes still remain to be problematic when it comes to forecasting. The purpose of this paper is to present a statistical-dynamic intensity model that utilizes OHC and show how critical the variable is to forecast the rapid intensification of Atlantic hurricanes.

2. Oceanic heat content background

While the SST may provide the conditions of the sea surface, it does not reveal the conditions of the upper 50 m of the ocean. Strong relationships have been known to exist between the ocean and tropical cyclone development and intensification (Leipper & Volgenau, 1972, Shay et al., 2000, Law & Hobgood, 2007, Maineli et al. 2008). This area is critical to tropical cyclone development and intensification as deep water upwelling provides the warm water necessary to eventually be converted into latent heat energy. The OHC which is sometimes referred to as Tropical Cyclone Heat Potential (TCHP) is defined as the integrated vertical temperature from the sea surface to the depth of the 26°C isotherm. The 26°C isotherm corresponds to the threshold necessary for tropical cyclone development (Gray 1968). However despite being a key variable, OHC fields cannot be created by hydrographic observations alone. These fields have to be produced from those observations by integrating the temperature profile down to the 26°C isotherm. OHC values typically range over 100 kJ cm^{-2} and were found primarily for rapid major tropical cyclone intensification (Law & Hobgood 2007).

OHC is calculated globally and daily from estimated altimeter-derived vertical temperature profiles (Shay et al. 2000). There have been three different methods designed to estimate this parameter. The first version was a reduced gravity model, used prior to 2005. This method used three variables (a) SSTs derived from the Tropical Rainfall Measuring Mission (TRMM) Microwave Imager (TMI), (b) altimeter estimates of the 20°C isotherm chosen because it represents the thermocline and the upper level flow in the Gulf of Mexico and Atlantic (Goni et al, 1996), and (c) the depth of the 26°C isotherm based upon a climatological relationship between the 26°C and 20°C isotherms.

The second version used from January 2005 to October 2008 was based upon a linear regression model between the depth of the 26°C and 28°C isotherms acquired from temperature profiles and topography estimated by Archiving, Validation and Interpretation of Satellite Oceanographic (AVISO) data. The OHC was then computed by integrating the temperature profile from the surface down to the 26°C isotherm. The third version used from October 2008 to present provides better agreement with observations in the Gulf of Mexico. It estimates the isotherm depth based upon sea height anomalies from satellite observations.

3. Statistical and statistical-dynamic intensity models

Statistical models such as SHIFOR5 and Statistical-Dynamic models such as SHIPS use a variety of synoptic and climatological variables to forecast the short-term intensification change of tropical cyclones. Regression equations are created from the climatological tropical cyclone database using each of these variables. However, statistical models primarily have used a single set of regression equations to be used uniformly and throughout the entire life-cycle of a particular tropical cyclone. It has been theorized that different sets of regression equations need to be created because parameters change in importance throughout the storm's life cycle. Law & Hobgood (2007) first developed such a model by incorporating Discriminant Function Analysis (DFA) with the regression equations. By utilizing a DFA, it could be determined how close the tropical cyclone is to its rapid intensification period (RIP). Therefore a specific set of regression equations could be applied based upon how close the tropical cyclone is to the RIP. This particular study expands upon that method of statistical-dynamic intensification model.

3.1 Dataset and variables

The investigation consisted of all tropical cyclones in the Atlantic basin from 2002-2007. There were 96 tropical cyclones including 1 unnamed subtropical storm in 2005 and 1 unnamed system in 2006. A series of 12 variables were gathered for all of the storms in 6-hr intervals (Table 1).

Latitude (LAT)	6-hr Change in Wind Speed (ΔWS)	850 hPa Relative Humidity (RH)
Longitude (LONG)	6-hr Change in Pressure (ΔP)	Sea Surface Temperature (SST)
Wind Speed (WS)	850-200 hPa Wind Sheer Magnitude (ShrMag)	Inception Time (Incep)
Central Pressure (CP)	850-200 hPa Wind Shear Direction (ShrDir)	**Oceanic Heat Content (OHC)**

Table 1. List of 12 variables used in the statistical intensity model.

The latitude (LAT), longitude (LONG), wind speed (WS), and central pressure (CP) are available in the Atlantic Best Track dataset from the National Hurricane Center archives. This file contains the 6-hr locational coordinates as well as the maximum sustained winds and the central pressure. The 6-hr changes in wind speed (ΔWS) and pressure (ΔP) were easily computed from this dataset. The 850-200 hPa wind sheer magnitude (ShrMag) and direction (ShrDir), 850 hPa relative humidity (RH), and sea surface temperature (SST) were obtained from the National Centers for Environmental Prediction-National Center for Atmospheric Research (NCEP-NCAR) reanalysis dataset (Kalnay et al. 1996). This 6-hourly reanalysis dataset is on a 2.5° latitude-longitude grid or approximately a 210 km resolution. To smooth any extremes in the variables, a 10° latitude-longitude "box" was created and positioned over the storm center indicated by the Best Track dataset. This was to create an average value of the variable over the entire area of the storm.

However, the SST data were available on a weekly time scale on a 1° latitude-longitude grid. The grid boxes were adjusted to fit the same area used for the relative humidity and the wind shear variables. The SSTs were derived from a variety of in-situ ship and buoy data, satellites, and simulated SSTs for sea ice. Prior to calculating the database, these parameters were adjusted for biases using simulated SSTs described by Reynolds (1988) and Reynolds and Marsico (1993) and more information describing the ocean interaction analysis can be found in Reynolds and Smith (1994).

The Oceanic Heat Content (OHC) was obtained from a database constructed by the Cooperative Institute for Research in the Atmosphere (CIRA) at Colorado State University. This database utilized techniques described by Mainelli et al. (2008) and had the same 1° latitude-longitude resolution as the SST reanalysis data. These data were similarly averaged over the storm center to give an average OHC over the entire area of the tropical cyclone.

3.2 Methodology

The hypothesis of this study suggests that more than one regression model, rather than just one uniform model, is necessary to accurately forecast hurricane intensity. By using the multiple regression models, it attempts to show the importance of OHC upon rapid hurricane intensification. The statistical-dynamic intensity model used in this study used a variety of multiple regression equations to forecast the 24-hour change in the maximum sustained winds. To form the equations, multiple regression analysis was used from a climatological database that was formed from the 96 tropical cyclones in the sample. This database was created from the all 12 variables listed in Table 1. In total, there were 3348 individual cases.

Every 6-h time interval for each storm was categorized based upon the current intensity of the system: tropical storm, minor hurricane, or major hurricane. Of the 3348 total cases, 2633 were tropical storm status, 445 were minor hurricanes (Category 1 or 2 on the Saffir-Simpson Scale), and 270 major hurricanes (Category 3, 4, or 5). In addition, each time interval was divided based upon the future 6-h intensity change: weakening (>5 kts), little change (±5 kts), strengthening (>5 kts). Of the 3348 total cases, 697 exhibited future weakening intensity, 1857 cases showed little change, and 794 exhibited future strengthening characteristics. Therefore there were nine different categories and sets of multiple regression equations to develop. After each was classified, the multiple regression analysis was completed using future 24-hour maximum sustained wind speed change as the dependent variable and the 12 variables listed on Table 1 as the independents. Figure 1 shows a schematic of the combination of regression equations developed.

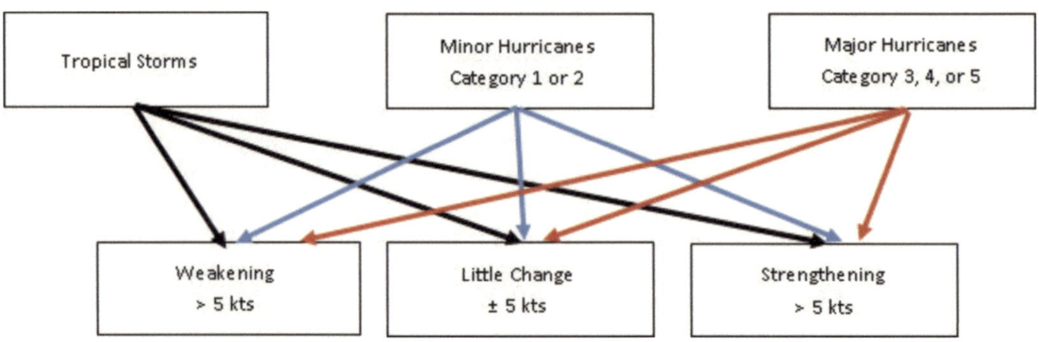

Fig. 1. Schematic of the nine categories of regression equations

When constructing the multiple regression equations and climatology database, the future 6-hour is known. However, when using as a real-time prediction model, obviously the future 6-hour intensity change will not be known and a prediction technique must be used. Based upon the current characteristics of the variables used in the model, the DFA is then used as a prediction method to forecast the future 6-hour intensity change (weakening, little change, strengthening).

The DFA equations were then applied to the same storms in the database to verify their selection accuracy. The most discriminating variables between the 9 categories as well as the most significant variables could be identified. Sample forecasts were created from the regression equations and also applied to the storms in the sample. Mean error statistics were produced so the model could be compared to other statistical models and other official forecasts.

3.3 Revisions

The previous method used by Law and Hobgood (2007) used a two-step DFA. That version was used to predict how close the storm was to its rapid intensification period and then predict the magnitude of the intensification change. The first step of the DFA was used to predict what status the tropical would eventually become (i.e. tropical storm, minor hurricane, or major hurricane). The second step of the DFA was then used to predict how close it was to its RIP (i.e. the 24-h period where the maximum pressure decrease occurred). However, the problem with this version was that it could not be used over the entire life cycle of the storm.

The method in this study used a one-step DFA approach, since it used the current status of tropical cyclone rather than trying to predict what it "would eventually become". The DFA was only used to predict the 6-h future intensity increase. By predicting this parameter, it could be applied over the entire life cycle of the storm. In addition, the other major revision was the inclusion of OHC which was not considered in the previous edition.

4. Regression equations

As mentioned in Section 3, there were 3348 individual cases. Table 2 shows the number of cases for each of the nine types of regression equations which were created. A stepwise-multiple regression technique was performed to create the nine equations in each category. Variables that were not significant at the 95% confidence interval were omitted from the equation. For the tropical storm cases, there were 494, 1621, and 518 cases for the future weakening, little change, and future strengthening cases, respectively. For the minor hurricane cases, there were 122, 151, and 172 cases for the future weakening, little change, and future strengthening cases, respectively. Finally, for the major hurricane cases, there were 81, 85, and 104 for the future weakening, little change and future strengthening cases, respectively.

	Tropical Storms	Minor Hurricanes	Major Hurricanes	
Weakening	494	122	81	**697**
Little Change	1621	151	85	**1857**
Strengthening	518	172	104	**794**
	2633	**445**	**270**	**3348**

Table 2. Number of cases for each of the nine types of regression equations

For all 9 equations, the most significant variable is the closest to the y-intercept and decreases in significance. The first three equations are for **tropical storms** (Equations 1-3). Equations 1, 2, and 3 are for *future weakening, little change,* and *future strengthening* cases, respectively. The next set of equations (Equations 4-6) is for **minor hurricanes**. Likewise, Equations 4, 5, and 6 are for *future weakening, little change* and *future strengthening* cases, respectively. The last set of equations (Equations 7-9) is for **major hurricanes**. Again, Equations 7, 8, and 9 are for *future weakening, little change,* and *future strengthening* cases, respectively.

$$-.219\ SST - .155\ LAT - .075\ LONG - .360\ WS + 75.478 \quad (1)$$

$$.10\ LONG - .193\ \Delta P + .036\ RH - .054\ ShrMag - .047\ WS + .106\ \Delta WS - .005\ Incep - .090 \quad (2)$$

$$-.296\ SST - .146\ ShrMag + .071\ LONG - .178\ CP - .282\ LAT - .834\ \Delta CP + 286.731 \quad (3)$$

For the tropical storm cases, wind speed, latitude/longitude, pressure, and sea surface temperature were significant variables in predicting the intensification. OHC was not significant for the tropical storms. This suggests that while the storms are weak, it is not necessary to have warm, deep ocean water and that using sea surface temperatures are sufficient.

$$.338\ LAT - .428\ RH - .512\ \Delta CP - .244\ LONG + 9.409 \quad (4)$$

$$.107\ CP - 104.760 \quad (5)$$

$$-.654\ ShrMag - .072\ Incep + .317\ LONG - .698\ LAT + 33.955 \quad (6)$$

Similarly, minor hurricanes did not incorporate OHC. Minor hurricane intensification was strongly dependent upon the location (latitude/longitude), either central pressure or the change in the pressure, relative humidity, and if it happens to ben strengthening vertical wind shear. Like tropical storms, it suggests that perhaps OHC is not significant when forecasting the intensity of minor hurricanes.

$$.272\ OHC - 1.157\ LONG + 35.930 \quad (7)$$

$$-.325\ LAT + 5.981 \quad (8)$$

$$-.713\ ShrMag - .133\ OHC - .094\ Incep + .643\ LONG + 1.860 \quad (9)$$

However, with major hurricanes OHC was a very influential variable. In both the weakening and strengthening cases OHC was significant. Latitude and longitude were significant as well as vertical wind shear in strengthening cases. Since OHC was significant rather than SST, it suggested that warm, deep ocean water is important and OHC better represents this parameter.

5. Discriminant function analysis classifications

The DFA was run on each of the three categories (tropical storms, minor hurricanes, and major hurricanes). This was performed to analyze which variables are the most significant in differentiating between groups. The standardized discriminant function coefficients were determined and are listed in Tables 3, 4, and 5 (tropical storms, minor hurricanes, and major hurricanes, respectively. These tables show the relative significance of each variable to discriminate among each group.

OHC was 10th (Table 3) and SST was 3rd showing that SST was more important than OHC when discriminating among tropical storms. Tropical cyclones that tend to remain weak are influenced by smaller changes in central pressure and wind speed. Since tropical storms are relatively weak, they do not have deep water upwelling and therefore are not influenced as much by OHC.

However, with minor and major hurricanes (Tables 4 and 5) OHC was more significant than SST. For minor hurricanes (Table 4), OHC was 5th on the list while SST was 11th. Wind

VARIABLE	COEFFICIENT
Wind Speed	0.999
6-h Change in Central Pressure	0.377
Sea Surface Temperature	-0.327
6-h Change in Wind Speed	-0.281
Central Pressure	0.279
Longitude	0.226
Wind Shear Magnitude	0.153
Latitude	0.121
Inception Time	0.065
Oceanic Heat Content	-0.023
Relative Humidity	0.018
Wind Shear Direction	0.004

Table 3. Standardized discriminant function coefficients for tropical storms

shear, latitude, pressure, and wind speed were the most significant but since OHC was more significant than SST, it suggests that deep ocean water becomes more important for these stronger systems. Likewise, the most significant variables for major hurricanes were wind speed, pressure and latitude (Table 5). OHC was again 5th on the list, while SST was 8th in significance suggesting that deep, water upwelling occurs and OHC indeed captures this parameter better than SST. Another interesting facet was that wind shear magnitude was the most important for minor hurricanes but noticeably lower for major hurricanes. Since major hurricanes exhibit little, if any, vertical wind shear, perhaps this explains the lower significance. On the other hand, minor hurricanes may be experiencing a larger magnitude of wind shear, which is what is preventing it from becoming a major hurricane.

6. Results

6.1 Classification accuracy and mean values

After the DFA was run and significant variables were found, the DFA was then applied to all 3348 cases to examine the accuracy. The overall classification accuracy for tropical storms was 55.4%. Assuming there was an equal possibility of each of nine scenarios, the DFA added skill to the selection (e.g. 55.4% versus 11.1%). Tropical storm (Table 6) weakening cases were classified correctly 69.2% of the time while little change and strengthening cases were accurately classified 45.3% and 65.1%, respectively. Minor hurricane overall classification accuracy was 59.3%. The weakening cases for minor hurricanes were correctly classified 51.9%, little change 49.1%, and strengthening cases 77.4% (Table 7). Overall major hurricane overall accuracy was 68.1%. The weakening cases for major hurricanes were correctly classified 72.5%, little change 60.3%, and strengthening cases 68.0% (Table 8). Since the classification accuracy of major hurricanes was better than tropical storms and minor hurricanes, it suggests that the characteristics of major hurricanes are more discriminating. OHC was one of these key discriminating variables (Table 5) and plays a significant role in determining whether major hurricanes will weaken or strengthen.

VARIABLE	COEFFICIENTS
Wind Shear Magnitude	0.631
Latitude	0.429
Central Pressure	0.392
Wind Speed	0.296
Oceanic Heat Content	-0.270
6-h Change in Wind Speed	-0.249
Longitude	0.240
Inception Time	0.165
6-h Change in Pressure	0.122
Relative Humidity	-0.117
Sea Surface Temperature	0.053
Wind Shear Direction	-0.045

Table 4. Standardized discriminant function coefficients for minor hurricanes

VARIABLE	COEFFICIENTS
Wind Speed	1.107
Central Pressure	0.621
Longitude	0.607
Latitude	0.586
6-h Change in Wind Speed	-0.572
6-h Change in Pressure	-0.357
Oceanic Heat Content	-0.346
Relative Humidity	0.307
Wind Shear Magnitude	0.268
Sea Surface Temperature	0.059
Wind Shear Direction	-0.025
Inception Time	-0.007

Table 5. Standardized discriminant function coefficients for major hurricanes

However, OHC was not a key discriminating variable for tropical storms and minor hurricanes. It is widely known that as the strength of the hurricane increases, warmer SSTs are necessary to support it as the amount of energy extracted from the ocean increases (Anthes & Chang, 1978). Conversely, there is an inverse relationship between vertical wind shear and hurricane intensity. Table 9 shows the mean OHC and wind shear for all 3348 cases and the different categories of tropical cyclones. The mean values for OHC and wind shear were 27.0 kJ cm^{-2} and 12.3 m s^{-1} for tropical storms. OHC slightly decreased and wind shear increased slightly however for Category 1 hurricanes. Then as the hurricane intensity increased, the OHC increased and wind shear decreased as expected. There was a noticeable increase in OHC from Category 3 to Category 4, where it increased from 45.0 kJ cm^{-2} to 66.0 kJ cm^{-2}. Similarly another large increase took place from Category 4 to Category 5 (66.0 kJ cm^{-2} to 81 kJ cm^{-2}). The large OHC values of major hurricanes (Category 3, 4, and 5) were dramatically different from the tropical storms and minor hurricanes. This helps explain why OHC was a key discriminator for major hurricanes but was not as significant for tropical storms and minor hurricanes.

Observed	Predicted Group		
	Weakening (%)	Little Change (%)	Strengthening (%)
Weakening	**69.2**	14.4	16.4
Little Change	20.7	**45.3**	34.0
Strengthening	11.2	23.7	**65.1**

Table 6. Classification table showing % classification for tropical storms

Observed	Predicted Group		
	Weakening (%)	Little Change (%)	Strengthening (%)
Weakening	**51.9**	30.5	17.6
Little Change	34.0	**49.1**	17.0
Strengthening	5.8	16.8	**77.4**

Table 7. Classification table showing % classification for minor hurricanes

Observed	Predicted Group		
	Weakening (%)	Little Change (%)	Strengthening (%)
Weakening	**72.5**	16.9	10.6
Little Change	14.1	**60.3**	25.6
Strengthening	8.0	24.0	**68.0**

Table 8. Classification table showing % classification for major hurricanes

	Mean OHC (kJ cm-2)	Mean Wind Shear (m s-1)
Tropical Storm	27.0	12.3
Category 1	24.0	12.9
Category 2	40.0	10.4
Category 3	45.0	9.2
Category 4	66.0	8.0
Category 5	81.0	6.1

Table 9. Mean OHC and Wind Shear for different categories of tropical cyclones

6.2 Oceanic heat content case studies

Two of the most powerful, infamous hurricanes that rapidly intensified were Hurricanes Ivan (2004) and Katrina (2005). Both hurricanes achieved Category 5 status and underwent a considerable Rapid Intensification Period (RIP). The OHC was extraordinarily high for both events and both hurricanes traveled over the regions of the highest OHC. Figures 2-9 show maps of the OHC for Hurricane Katrina. On Aug. 23, 2005 (Figure 2), Katrina was tropical storm status and located near the Bahamas (Green dot). The OHC near the center of the storm was 78 kJ cm^{-2}, which was much higher than usual for tropical storms. It was natural to expect Katrina to intensify considering the conditions. On Aug. 24, 2005 (Figure 3), Katrina moved toward the northwest over slightly warmer water (99 kJ cm^{-2}) while maximum sustained winds increased from 30 to 40 kts and the pressure dropped from 1008 hPa to 1003 hPa. By the 25th, (Figure 4) Katrina turned west toward Florida, still located near high OHC (104 kJ cm^{-2}), and was near hurricane status with winds of 60 kts. But on the 26th (Figure 5), Katrina passed over the Everglades in Florida and then into the Gulf where slightly cooler waters existed. The OHC dropped to 55 kJ cm^{-2}, nevertheless Katrina still intensified and achieved Category 1 hurricane status due to the nearby extraordinarily high OHC. By the 27th (Figure 6), Katrina was beginning to enter the warm core ring in the Gulf of Mexico. The OHC increased again to 88 kJ cm^{-2} and winds speeds were 100 kts achieving major hurricane (Category 3) status. Over the next 24 hours (Figure 7), Katrina would undergo rapid intensification as it entered the center of the warm core ring with OHC values of 123 kJ cm^{-2}. Maximum sustained winds were 150 kts and was a very strong Category 5 hurricane. It was not until Katrina exited the warm core ring where much lower OHC existed that the intensity lowered and the eventual landfall near New Orleans, Louisiana to further dissipate (Figures 8 - 10).

Hurricane Ivan started much farther to the east near the African coast (green dot) on Sep. 2, 2004. This area typically has much lower OHC as indicated by the value of 21 kJ cm^{-2} (Figure 11). As Ivan progressed to the west, winds gradually increased to tropical storm status, but the OHC was still rather low (22 kJ cm^{-2}) (Figure 12). The relatively low OHC was part of the reason Ivan was slow to initially intensify through the 4th (Figure 13). Over the next 24 hours (Figure 14), Ivan underwent the first rapid intensification period as winds suddenly increased to 110 kts (Category 3) and the OHC increased to 52 kJ cm^{-2}. The winds would gradually decrease as Ivan continued toward the west in the Caribbean on the 6th (Figure 15) despite high OHC values. This temporary weakening was likely due to an eyewall replacement cycle as it would soon start to re-intensify. Over the next 4 days, Ivan was over relatively high OHC ranging from 58 kJ cm^{-2} to 91 kJ cm^{-2} (Figures 16 – 19) as the winds would approach Category 4 status. Then on Sep. 11th (Figure 20), Ivan was located just south of Cuba and underwent a second rapid intensification. Ivan entered extremely high OHC values of 127 kJ cm^{-2} with Category 5 maximum sustained winds of 145 kts. The OHC remained extremely high near Ivan over the 12th and 13th (Figures 21 and 22) which would help maintain the intensity at 130 and 140 kts, respectively. Subsequently on the 14th, Ivan exited the high area of OHC into a much cooler area with values of 47 kJ cm^{-2} (Figure 23). As a result the maximum sustained winds gradually decreased to 120 kts. Ivan passed over a small warm core ring in the central Gulf of Mexico on the 15th (Figure 24) and the OHC of 96 kJ cm^{-2} allowed the winds to maintain major hurricane status. Hurricane Ivan would ultimately exit the warm core ring making landfall along the Gulf Coast dissipating rapidly on the 16th (Figure 25).

6.3 Average error

The regression equations to forecast the 24 hour maximum wind speeds were applied to selected powerful major hurricanes during the 2004-2005 seasons. Based upon the DFA classification, the appropriate regression was applied to forecast the wind speed. The average error for the entire storm track forecast was computed for these tropical cyclones and then compared with the National Hurricane Center 24-h official forecast and another statistical model SHIFOR5 (Table 10) (Franklin 2005; Franklin 2006). The statistical model exhibited in this study exhibited lower error than SHIFOR5 for all seven selected major

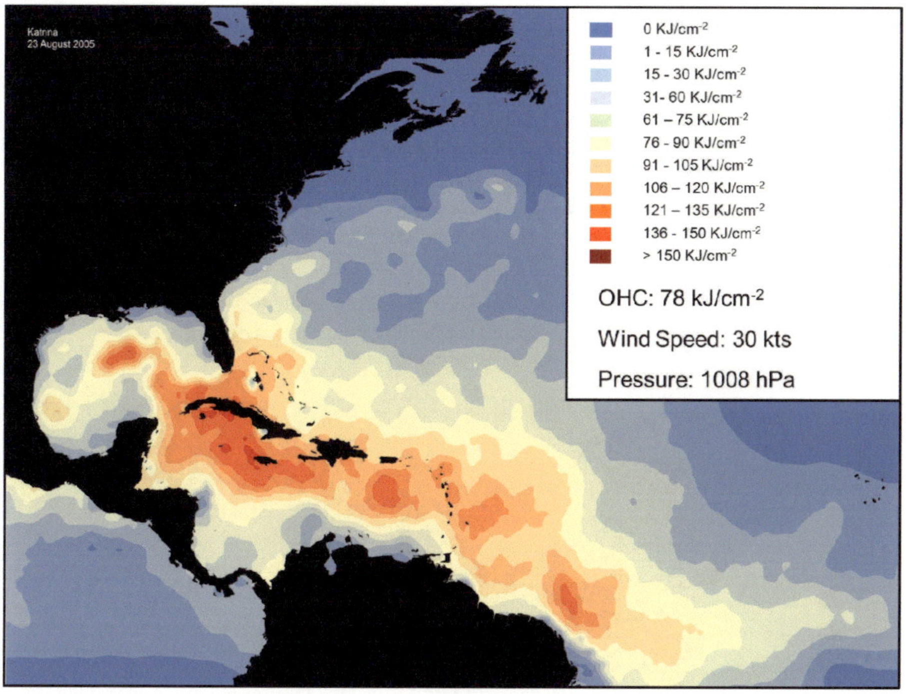

Fig. 2. Oceanic Heat Content of Hurricane Katrina, August 23, 2005 at 1800 UTC.

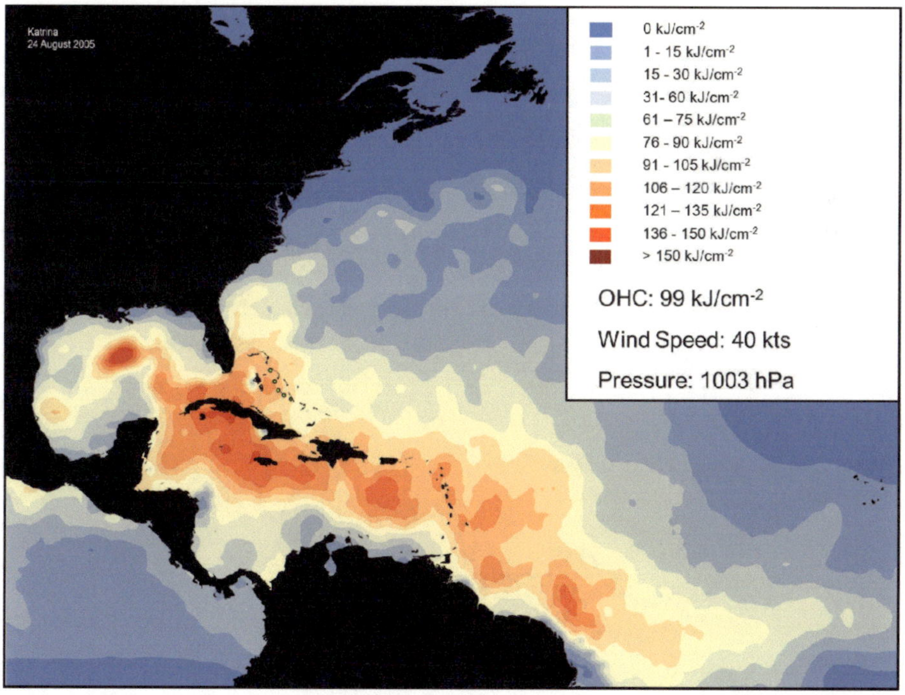

Fig. 3. Oceanic Heat Content of Hurricane Katrina, August 24, 2005 at 1800 UTC.

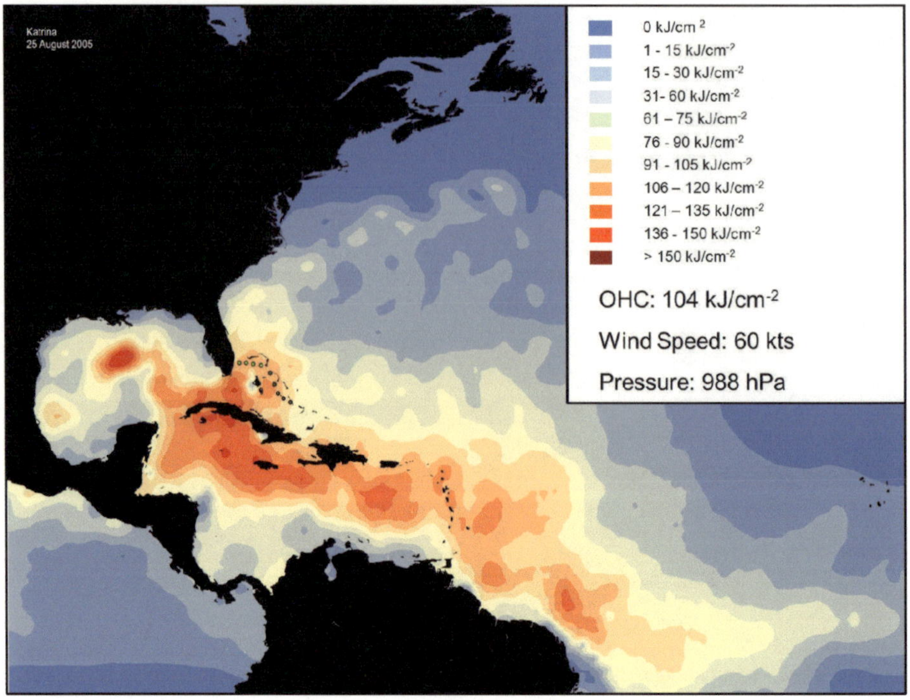

Fig. 4. Oceanic Heat Content of Hurricane Katrina, August 25, 2005 at 1800 UTC.

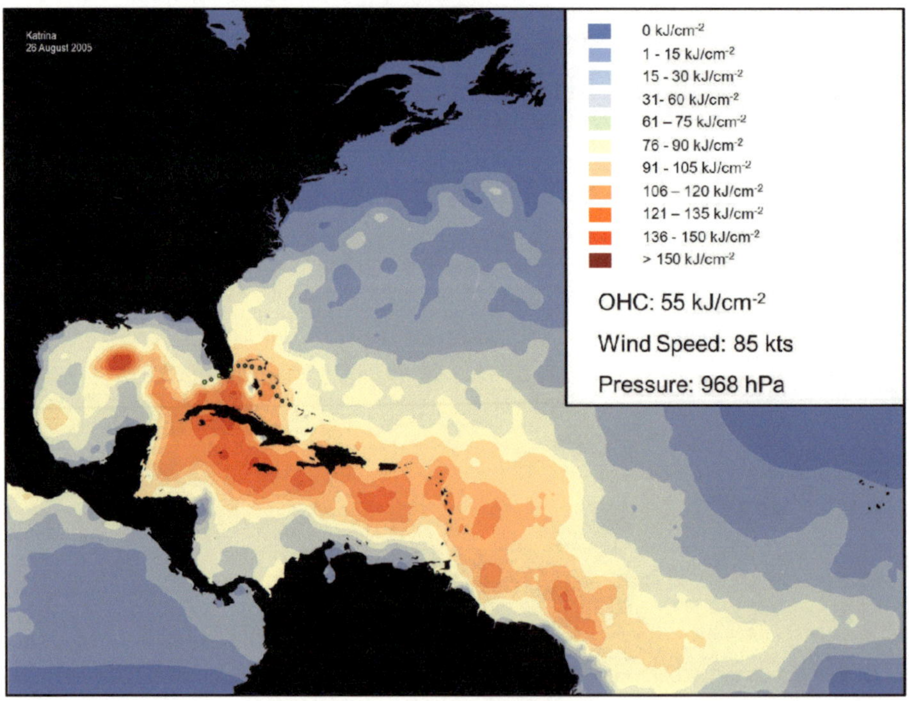

Fig. 5. Oceanic Heat Content of Hurricane Katrina, August 26, 2005 at 1800 UTC.

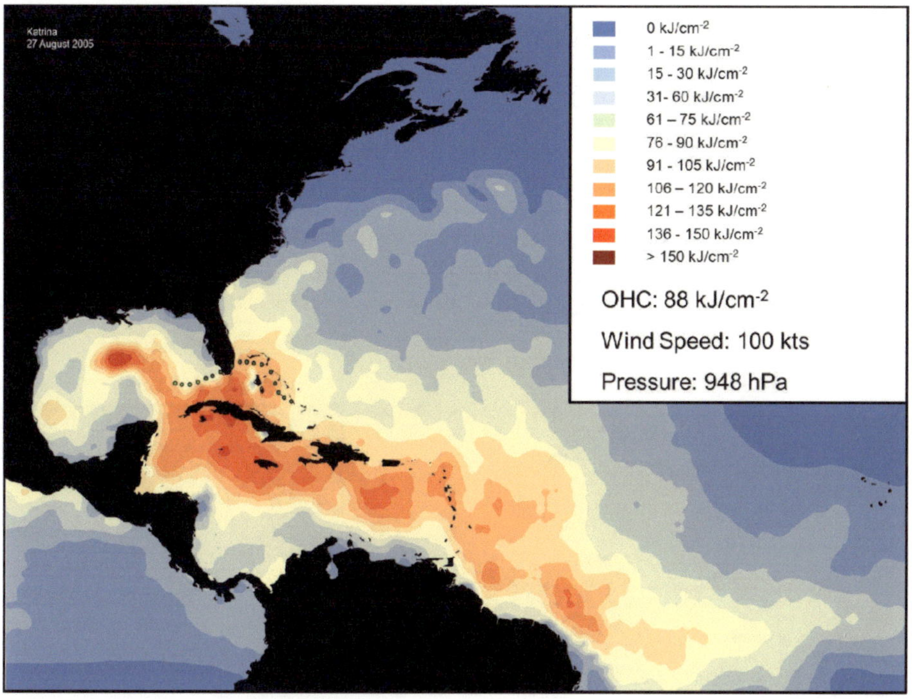

Fig. 6. Oceanic Heat Content of Hurricane Katrina, August 27, 2005 at 1800 UTC.

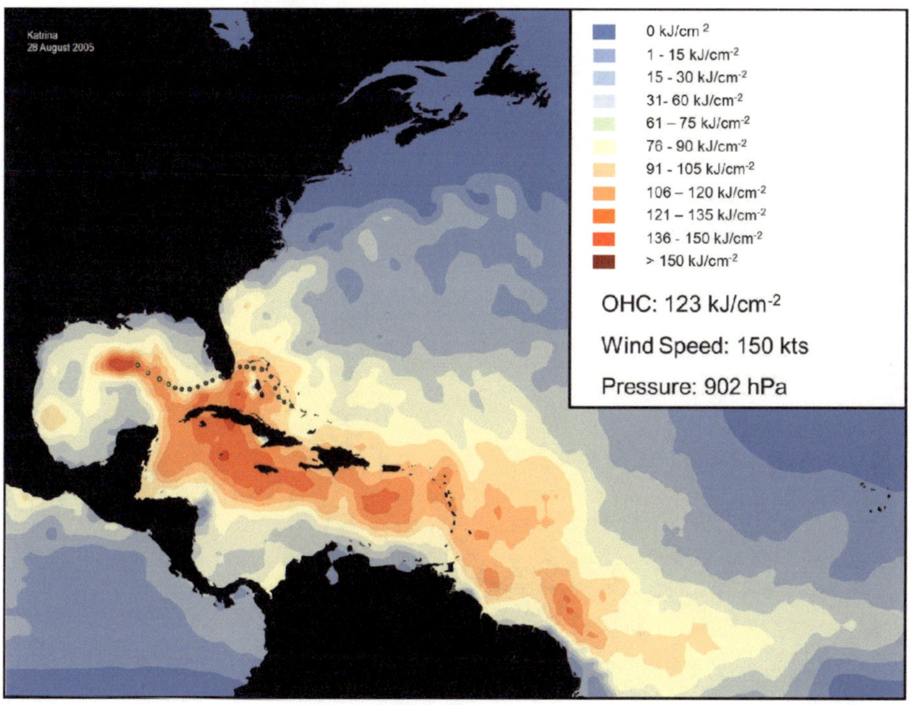

Fig. 7. Oceanic Heat Content of Hurricane Katrina, August 28, 2005 at 1800 UTC.

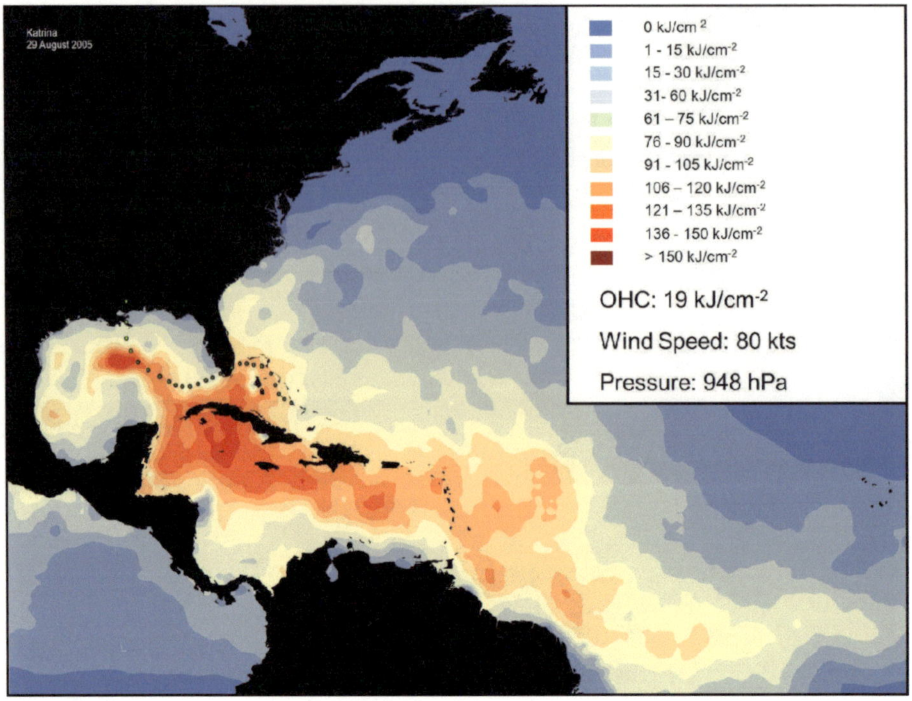

Fig. 8. Oceanic Heat Content of Hurricane Katrina, August 29, 2005 at 1800 UTC.

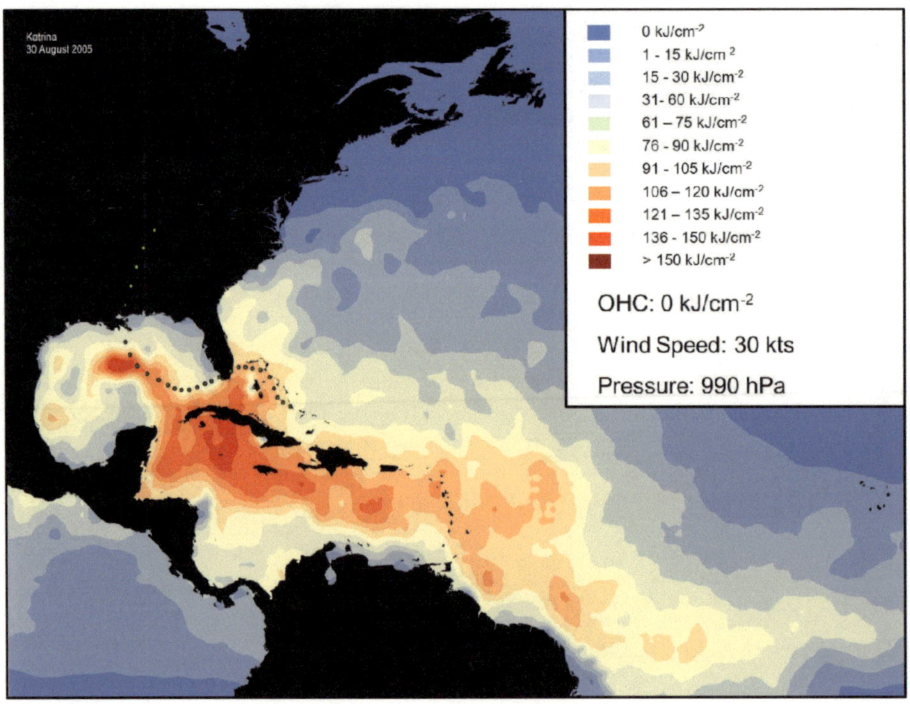

Fig. 9. Oceanic Heat Content of Hurricane Katrina, August 30, 2005 at 1800 UTC.

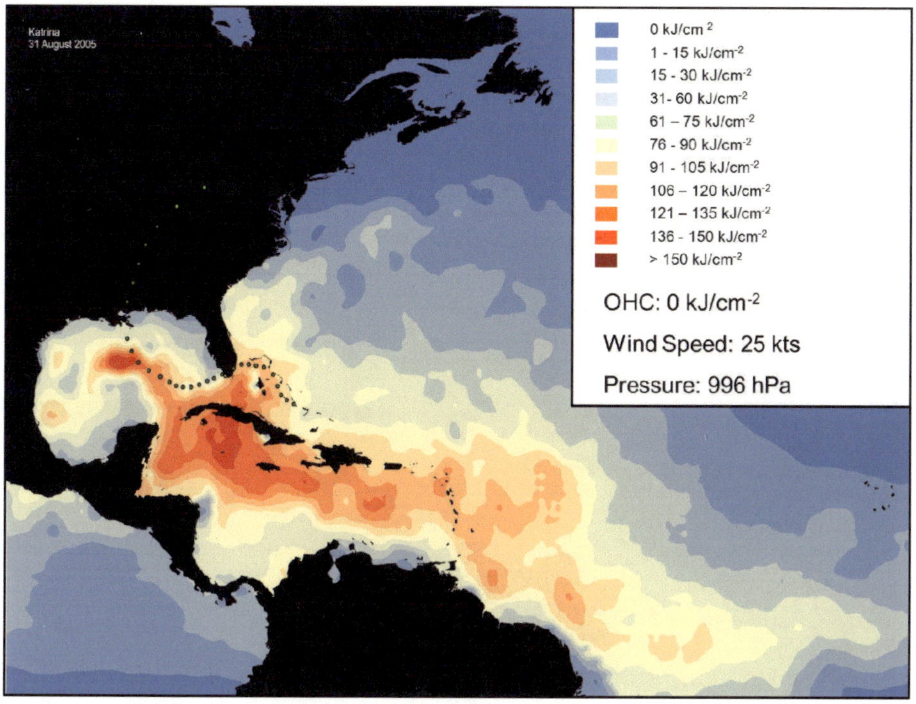

Fig. 10. Oceanic Heat Content of Hurricane Katrina, August 31, 2005 at 1800 UTC.

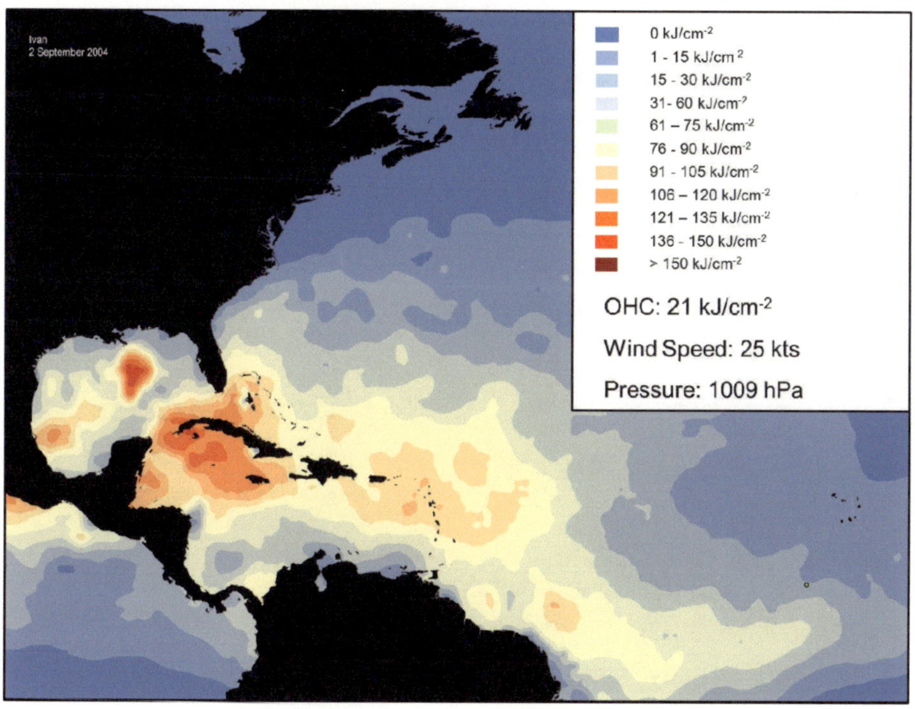

Fig. 11. Oceanic Heat Content of Hurricane Ivan, September 2, 2005 at 1800 UTC.

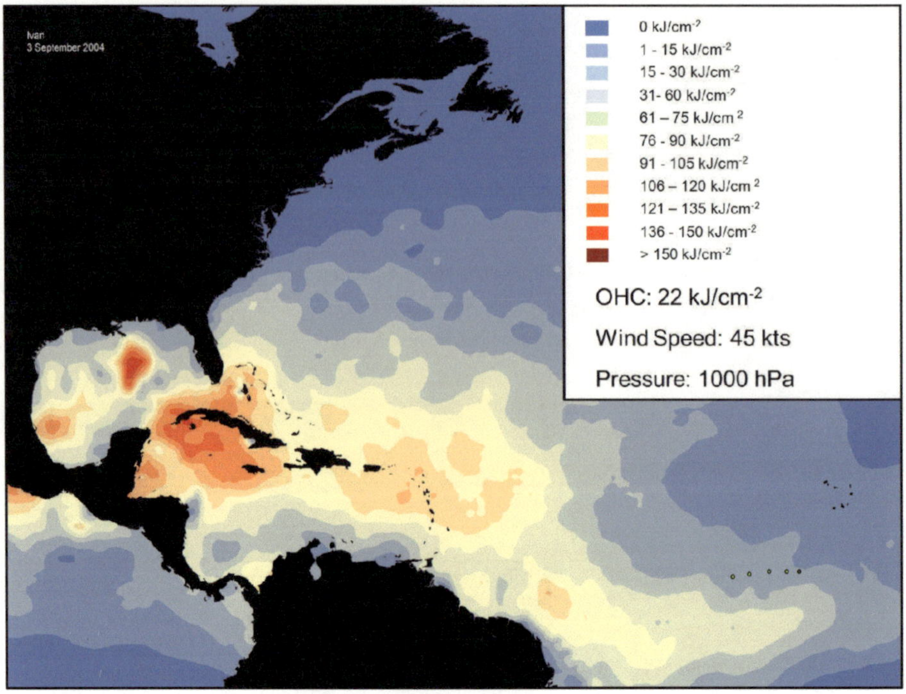

Fig. 12. Oceanic Heat Content of Hurricane Ivan, September 3, 2005 at 1800 UTC.

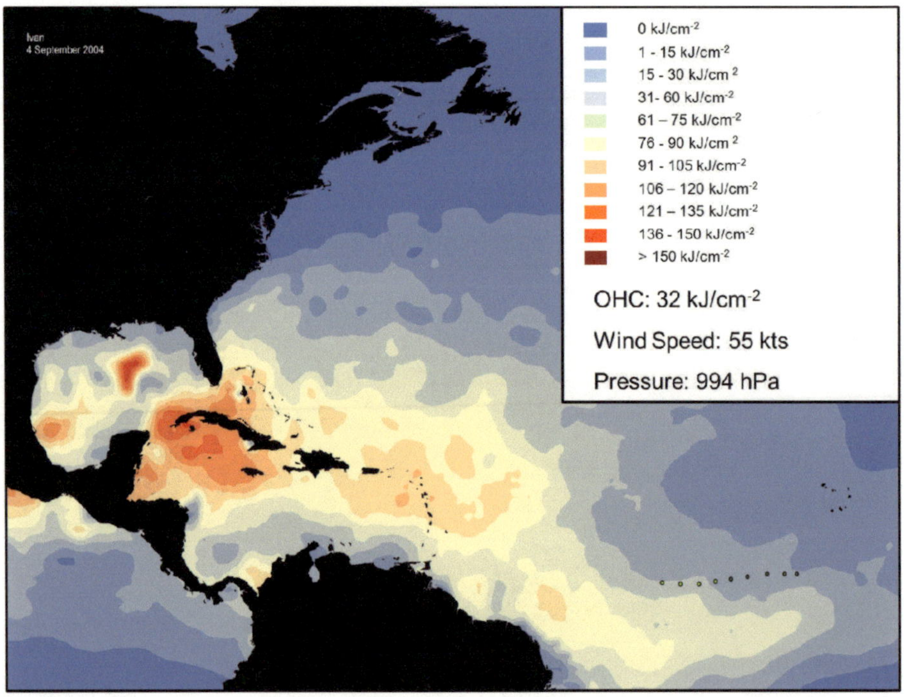

Fig. 13. Oceanic Heat Content of Hurricane Ivan, September 4, 2005 at 1800 UTC.

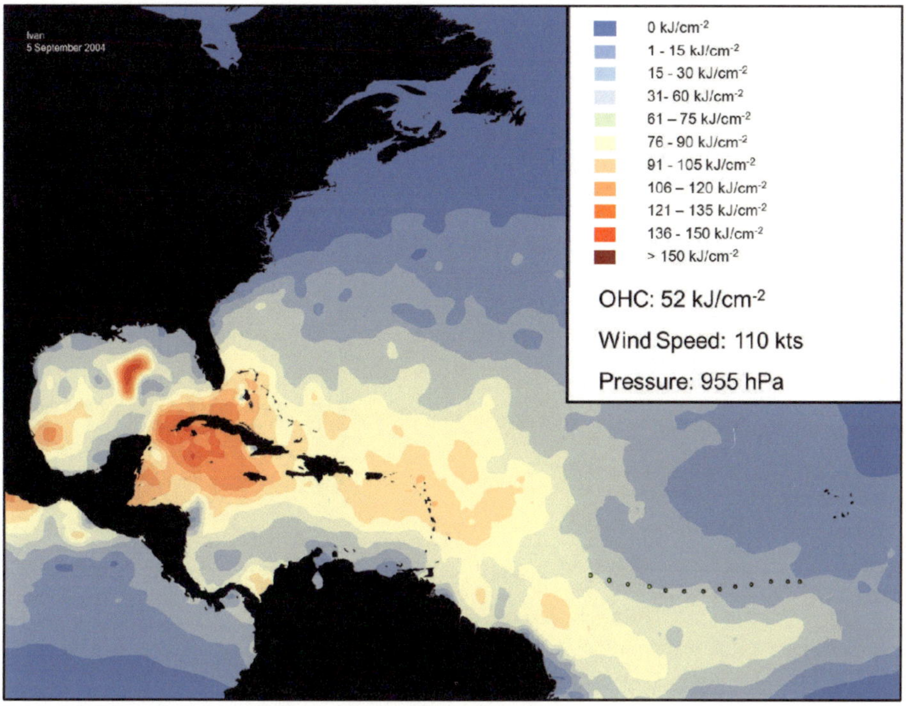

Fig. 14. Oceanic Heat Content of Hurricane Ivan, September 5, 2005 at 1800 UTC.

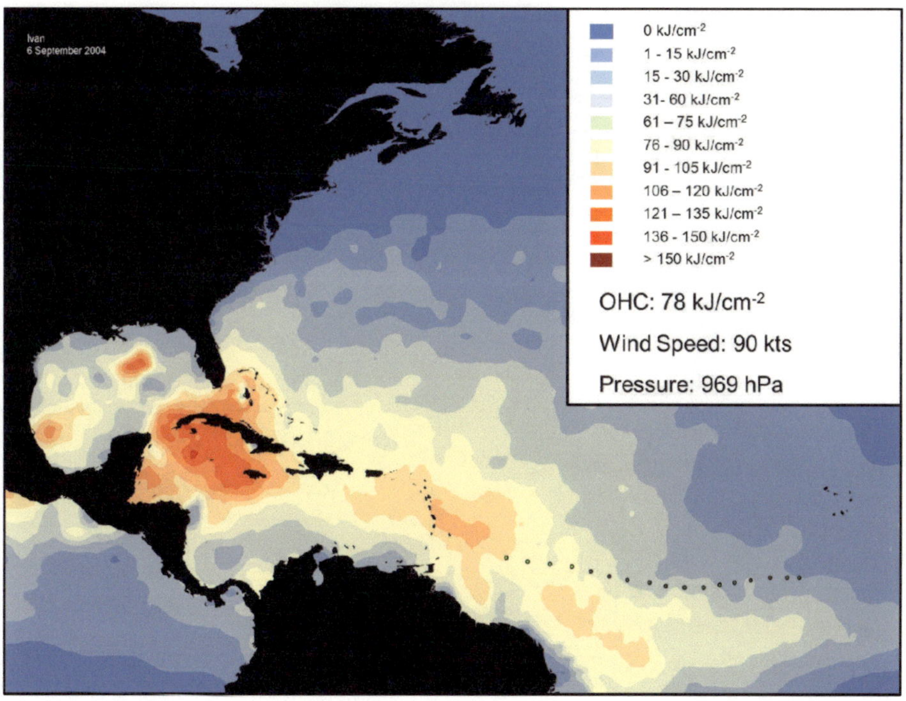

Fig. 15. Oceanic Heat Content of Hurricane Ivan, September 6, 2005 at 1800 UTC.

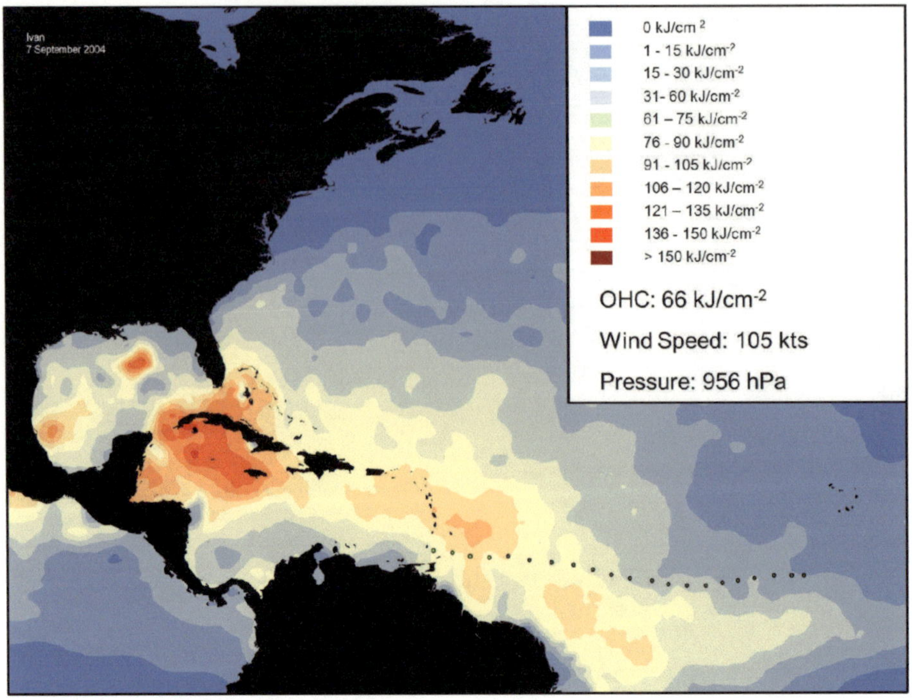

Fig. 16. Oceanic Heat Content of Hurricane Ivan, September 7, 2005 at 1800 UTC.

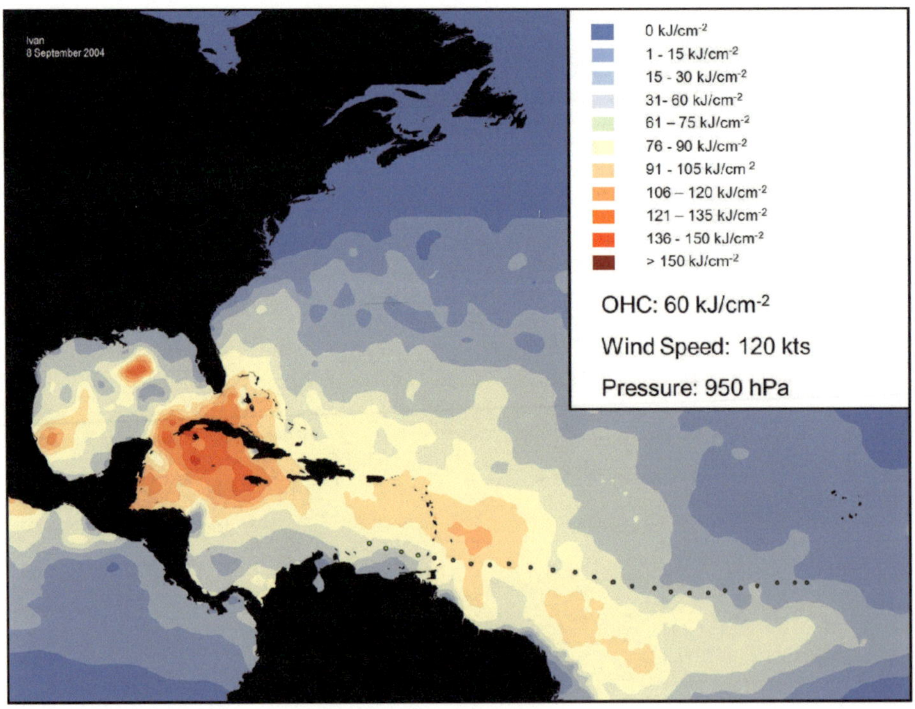

Fig. 17. Oceanic Heat Content of Hurricane Ivan, September 8, 2005 at 1800 UTC.

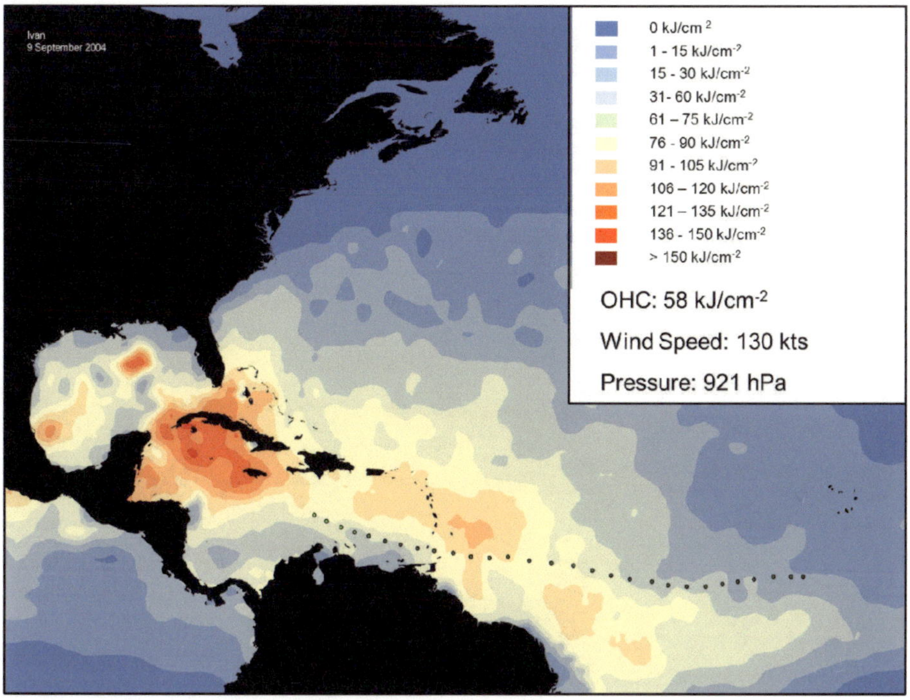

Fig. 18. Oceanic Heat Content of Hurricane Ivan, September 9, 2005 at 1800 UTC.

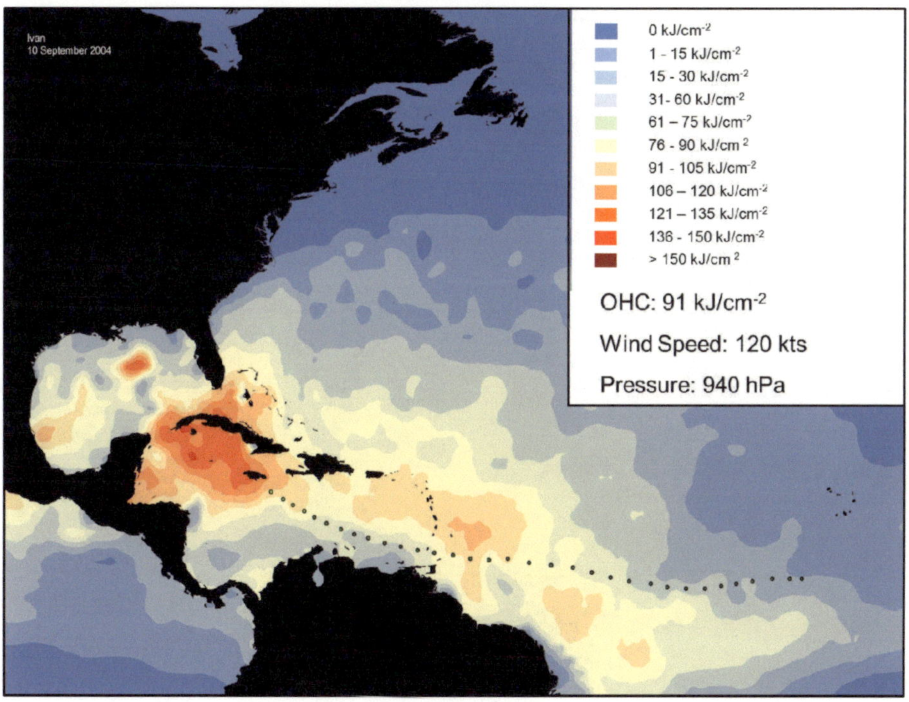

Fig. 19. Oceanic Heat Content of Hurricane Ivan, September 10, 2005 at 1800 UTC.

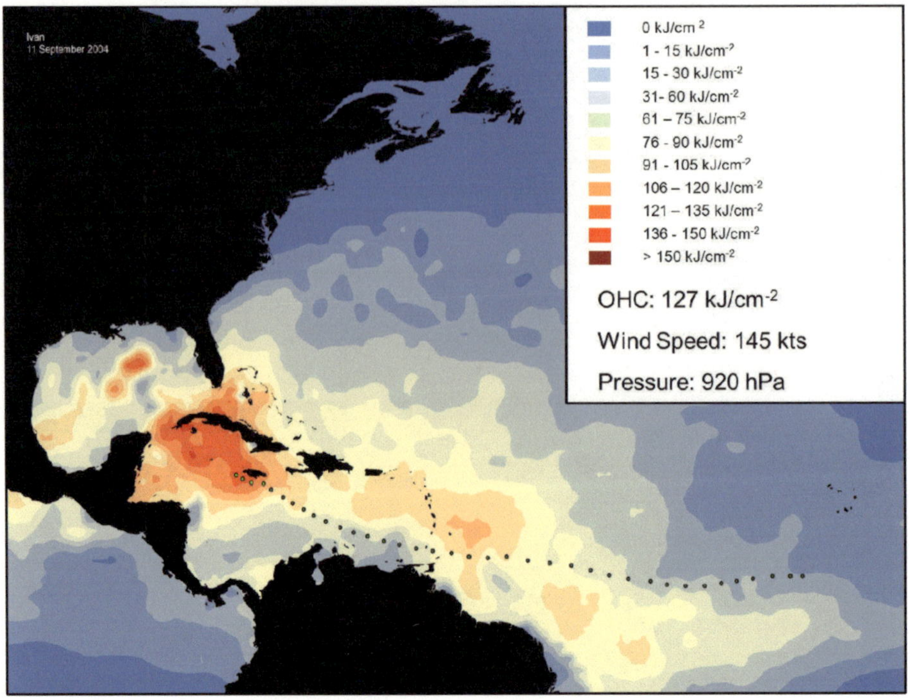

Fig. 20. Oceanic Heat Content of Hurricane Ivan, September 11, 2005 at 1800 UTC.

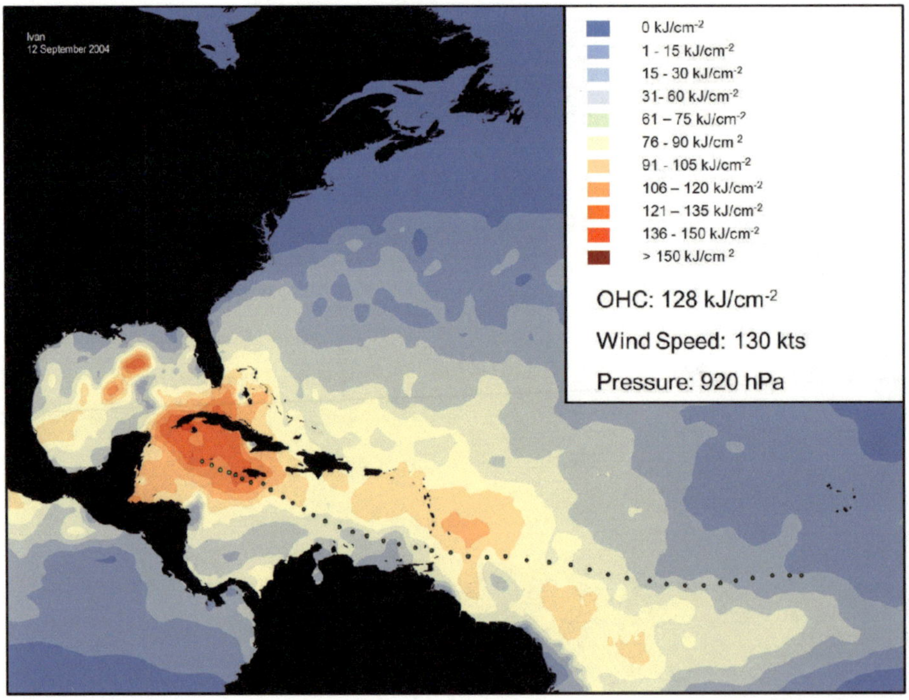

Fig. 21. Oceanic Heat Content of Hurricane Ivan, September 12, 2005 at 1800 UTC.

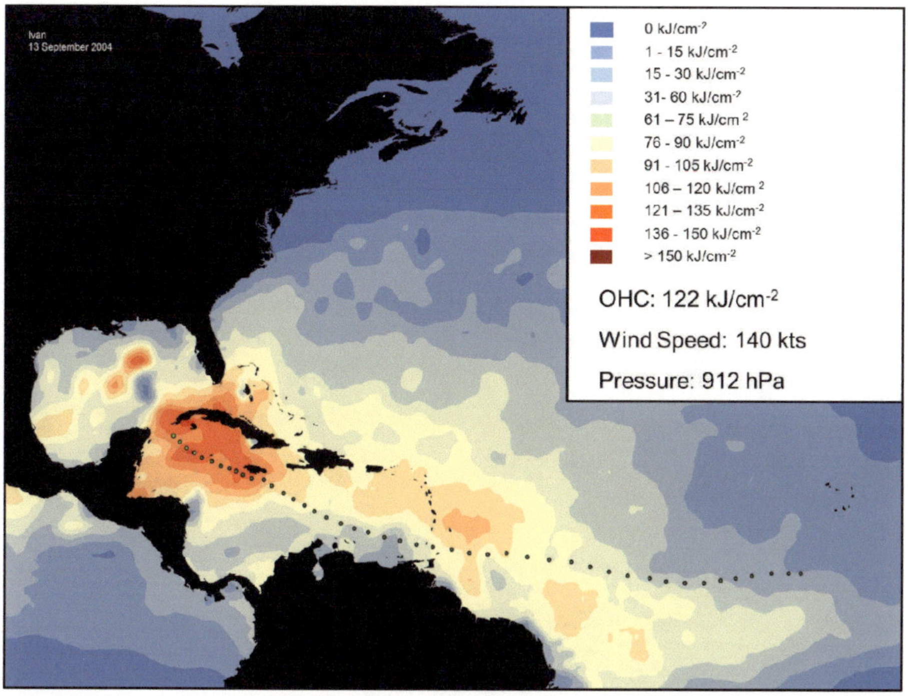

Fig. 22. Oceanic Heat Content of Hurricane Ivan, September 13, 2005 at 1800 UTC.

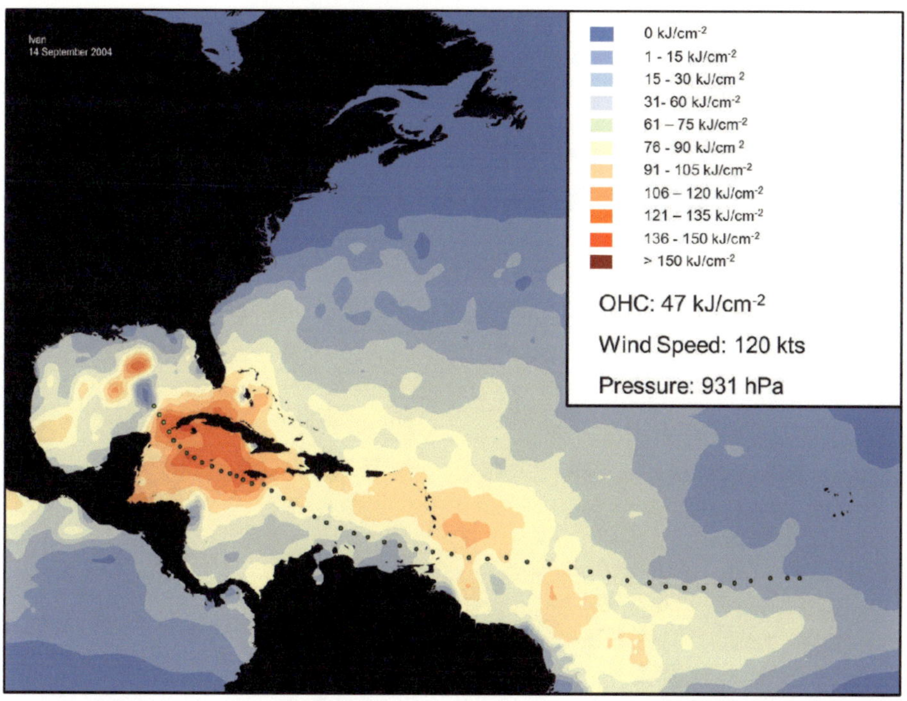

Fig. 23. Oceanic Heat Content of Hurricane Ivan, September 14, 2005 at 1800 UTC.

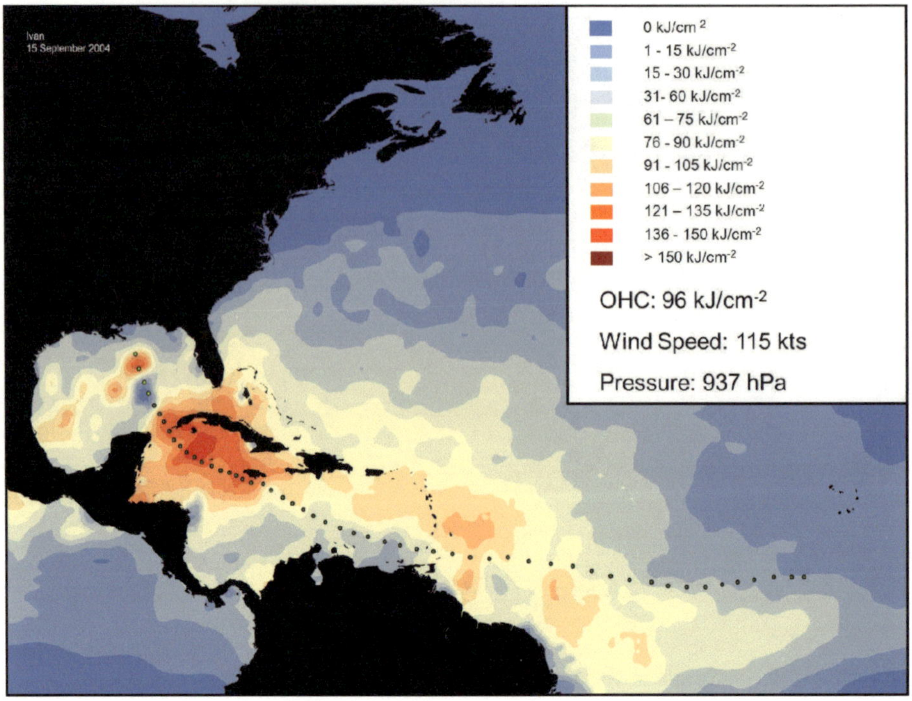

Fig. 24. Oceanic Heat Content of Hurricane Ivan, September 15, 2005 at 1800 UTC.

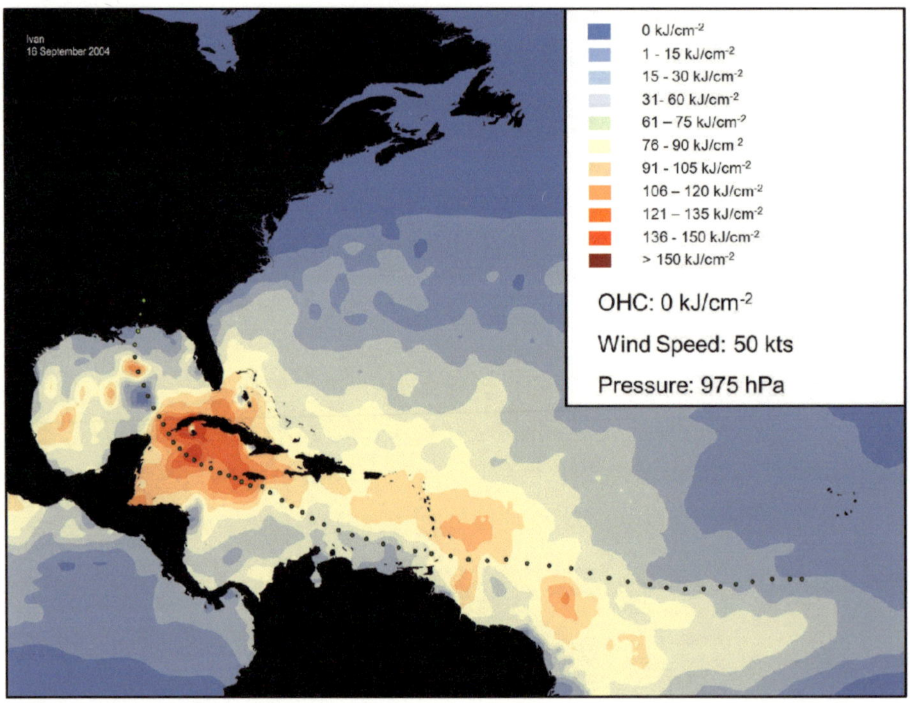

Fig. 25. Oceanic Heat Content of Hurricane Ivan, September 16, 2005 at 1800 UTC.

hurricanes. However, the official intensity forecast from the National Hurricane Center (OFCL) did produce more accurate forecasts for Charley, Ivan, Jeanne, and Rita. It was noteworthy that this particular model produced much more accurate intensity forecasts for Katrina. This is likely due to the inclusion of OHC in the major hurricane strengthening model.

Selected Storm	Average Error (kts)	OFCL error (kts)	SHF5 (kts)
Hurricane Charley (2004)	14.5	9.2	14.7
Hurricane Frances (2004)	9.2	10.4	13.0
Hurricane Ivan (2004)	11.9	11.7	16.5
Hurricane Jeanne (2004)	13.6	8.8	14.1
Hurricane Katrina (2005)	10.8	17.8	28.0
Hurricane Rita (2005)	18.3	16.2	28.1
Hurricane Wilma (2005)	17.5	18.2	19.0

Table 10. 24-h average error for selected tropical cyclones by the statistical intensity model

7. Conclusion

This study showed the importance of OHC in the role of rapid intensification of Atlantic Hurricanes. By utilizing a different type of statistical-dynamic model which applies a regression model in a particular stage during the life of a hurricane, rather than applying one uniform equation, significant variables can be identified. OHC was identified as such during weakening and strengthening cases for major hurricanes. OHC was a better discriminating variable than SST for minor and major hurricanes however SST was better for tropical storms. This is likely due to the fact that minor and major hurricanes churn deep water and need to extract the energy required to maintain their substantial strength.

By including OHC in the model and applying it during the weakening and strengthening cases for major hurricanes, it produced more accurate 24-h forecasts. The model was able to produce more accurate forecasts than SHIFOR5 for seven major hurricanes during 2004-2005, most notably Hurricane Katrina. However, it still did not produce better forecasts than all of the official National Hurricane Center forecasts.

More research needs to be conducted regarding the variables used in the equations to improve the accuracy of the regression equations and DFA classification. Additional forecast intervals need to be investigated besides short-term 24-h forecasts. In the future, other tropical cyclone basins also need to be examined to see how OHC and other parameters perhaps vary and influence hurricane intensification.

8. References

Anthes, R. A., & Chang, S. W., (1978). Response of the Hurricane Boundary Layer to Changes of Sea Surface Temperature in a Numerical Model. *J. Atmos. Sci.*, 35, (1240-1255)

DeMaria M., & Kaplan J., (1999). An Updated Statistical Hurricane Intensity Prediction Scheme (SHIPS) for the Atlantic and Eastern North Pacific Basins. *Wea. Forecasting*, 14, (326-337)

DeMaria M., Mainelli M., Shay, L. K., Knaff, J. A., & Kaplan J. (2005). Further Improvements to the Statistical Hurricane Intensity Prediction Scheme (SHIPS). *Wea. Forecasting,* 20, (531-543)

Franklin J. L, (2005). 2004 National Hurricane Center Forecast Verification Report. National Hurricane Center, NOAA/NWS/NCEP/Tropical Prediction Center, (1-52)

Franklin J. L, (2006). 2005 National Hurricane Center Forecast Verification Report. National Hurricane Center, NOAA/NWS/NCEP/Tropical Prediction Center, (1-52)

Goni, G., Kamholz, S., Garzoli, S., &Olson, D., (1996). Dynamics of the Brazil-Malvinas Confluence Based on Inverted Echo Sounders and Altimetry, J. Geophys. Res., 101(C7), (16,273-16,289)

Gray, W, M., (1968). Global View of the Origin of Tropical Disturbances and Storms. *Mon. Wea. Rev.*, 96, (669-700)

Kalnay, E., and Coauthors, (1996). The NCEP/NCAR 40-Year Reanalysis Project. *Bull. Amer. Meteor. Soc.*, 77, (437-471)

Law, K. T., & Hobgood, J. S, (2007). A Statistical Model to Forecast Short-Term Atlantic Hurricane Intensity. *Wea. Forecasting.*, 22, (967-980)

Leipper, D., & Volgenau, D., (1972). Hurricane Heat Potential of the Gulf of Mexico. *J. Phys. Oceanogr.*, 2, (218-224)

Mainelli, M., DeMaria, M, Shay, L., & Goni, G., (2008). Application of Oceanic Heat Content Estimation to Operational Forecasting of Recent Atlantic Category 5 Hurricanes. *Wea. Forecasting*, 23, (3-16)

McDougall, T., J., (2003). Potential Enthalpy: A Conservative Oceanic Variable for Evaluating Heat Content and Heat Fluxes. *J. Phys. Oceanogr.*, 33, (945-963)

Palmer, M. D., & Haines, K., (2009). Estimating Oceanic Heat Content Change Using Isotherms. *J. Climate*, 22, (4953-4969)

Reynolds R. W., (1988). A real-time global sea surface temperature analysis. *J. Climate,* 1, 75-86.

Reynolds, R. W., & Marsico, D. C., (1993). An improved real-time global seas surface temperature analysis. *J. Climate*, 6, (114-119)

Reynolds, R. W., & Smith, T. M., (1994). Improved global sea surface temperature analyses. *J. Climate*, 7, (929-948)

Schade, L. K., & Emmanuel, K. A., (1999). The Ocean's Effect on the Intensity of Tropical Cyclones: Results from a Simple Coupled Atmosphere-Ocean Model. *J. Atmos. Sci.*, 56, (642-651)

Shay L. K., Goni G. J., & Black P. G., (2000). Effect of a Warm Ocean Ring on Hurricane Opal. *Mon Wea. Rev.*, 128, (1366-1383)

Shay, L. K., & Brewster, J. K., (2010). Oceanic Heat Content Variability in the Eastern Pacific Ocean for Hurricane Intensity Forecasting. *Mon. Wea. Rev.*, 138, (2110-2131)

Wada, A., & Usui, N., (2007). Importance of Tropical Cyclone Heat Potential for Tropical Cyclone Intensification in the Western North Pacific. *J. of Oceanography*, 63, 3, (427-447)

Zebiak, S. E., (1989). Oceanic Heat Content Variability and El Niño Cycles. *J. Phys. Oceanogr.*, 19, (475-486)

Part 4

Societal Impacts: Risk and Damage Assessment

18

Modelling Hurricane Related Hazards and Risk through GIS for Early Warning Systems

Andrea Taramelli[1], Massimiliano Pasqui[2], Laura Melelli[3], Monia Santini[4] and Alessandro Sorichetta[5]
[1]ISPRA, Institute for Environmental Research, via di Casalotti, 300, Roma,
[2]Institute of Biometeorology – National Research Council, Roma,
[3]Department of Earth Science, University of Perugia, Perugia,
[4]euroMediterranean Center for Climate Changes, Lecce,
[5]Dipartimento di Scienze della Terra "Ardito Desio", Università degli Studi di Milano, Milano,
Italy

1. Introduction

In the last decade, developing hazard models for hurricane impact studies, also integrating GIS and remote sensed data, has become a major topic of research. However hurricane hazard areal identification and risk assessment remain largely unsolved problems (Kok & Winograd 2002). Indeed, despite the disastrous effects of hurricanes on coastal and inland communities are well known (O'Hare 2001, Pielke et al. 2003, Watson & Johnson 2005), there is still a need to better understand the contributions to hazards of the different mechanisms related to hurricanes strike like storm surge, floods and high winds. In fact, it is well known that hurricane hazard is 'controlled by' or 'dependent on' a large and complex set of natural and human induced environmental factors (Howard et al. 2003, Shen et al. 2005, Pielke et al. 2008), but to complicate matters further, hurricane related events like storm surges, floods and high winds, require forecasting appraisal that is often founded upon different methods, techniques and tools (Jiang et al. 2003, Bao et al. 2006). All that calls for a multidisciplinary and integrated approach: in this sense, technologies such as GIS and Remote Sensing (RS) have raised great expectations as potential means of coping with natural disasters like hurricanes. GIS is actually the best instrument to compare and analyze spatial data. Coarse and free digital data are now available also for geographic areas traditionally poor of spatial information. Moreover remote sensing provides updated images at high resolution useful to observe remote areas pre and post an hurricane event. Also Digital Elevation Models derived from satellite data are the only possibility to obtain topographic information for developing countries.

In this context a basic approach relying on a Multi-Hazard GIS-based modelling method was described (Taramelli et al. 2010) to assess hurricane related hazards and elements at risk. Understanding both the morphological/atmospheric parameters and the elements at risk is essential for reproducing the response to hurricane hazards. First, the asymmetric hurricane wind model proposed by Holland (1980, 2008) proved to fit the real hurricane's wind profile. In particular, comparing model estimates with observed data it is highlighted an evident benefit in a region far from the maximum speed close to the hurricane's eye with respect to other proposed models (Emanuel, 2004, Willoughby et al 2006, Fujita 1952). In turn, areas potentially affected by high winds, are identified. Second, morphology (e.g. beach height, wetness index) is assumed influencing the available on set zonation. The Multi-Hazard GIS model estimates the morphological response to variations in the balance between the forcing factors (wind velocity) and the related different single hazards (storm surge, floods and high winds).

Earth surface systems in Central America area provided a good study site for developing this understanding, most appropriately at both coastal and inland landform and to related morphological and atmospheric parameters. The applied model is not only conceptual. It is one of the first attempts to evaluate quantitatively at a regional scale relying on a physically based model.

The main advantage of the developed approach consists in its suitability also for hazard/risk future projections supporting early warning systems.

Indeed, there is a general agreement in the scientific community that the climatic tendency is toward a warmer world, with no increase in the number of days with rain, but with more intense rainfall (Saunders 1998, Russel et al. 2000). This could generate extreme rainstorms that often can be related to hurricane events (Kerry 2005, Pielke et al. 2005, Trenberth & Shea 2006). Furthermore in a recent review paper an extensive analysis of detection, attribution and projection assessments of tropical cyclone changes has been proposed by Knutson et al. (2010). This research reviews the existent limitations on the possible attributions of past changes but also the consistent shift, within the future climate scenarios, of globally averaged intensity of tropical cyclones toward stronger storms. Whether tropical cyclone frequency it is likely expected to decrease or remain unchanged, even with a large heterogeneity among basins, intensity and rainfall are expected to increase (Knutson et al. 2010). Coastal vulnerability is thus generally expected to increase in a future climate. Unfortunately there is still a low level confidence on future projections of specific storm characteristics which cause hurricane hazards such as rainfall, high winds and storm - surge floods. This is the reason why developing a robust and comprehensive approach is essential to allow an easy repeatability of risk predictions considering a wide range of hurricane projected characteristic, responses and thus their uncertainties.

2. Description of the Caribbean area

2.1 The West Indies and Central America

The West Indies (Anguilla, Antigua and Barbuda, Aruba, Barbados, British Virgin Islands, Bonaire, Cayman Island, Cuba, Dominica, Curacao, Grenada, Guadeloupe, Dominican Republic, Haiti, Jamaica, Martinique, Montserrat, Netherlands Antilles, Puerto Rico, Saint Kitts and Nevis, Saint Lucia, St. Marteen, Saint Vincent and the Grenadines, Trinidad and Tobago, United States Virgin Islands) and Central America (Guatemala, El Salvador, Honduras, Nicaragua, Costa Rica, and Panama) selected as study area are roughly located lat 0°S and 30°N and long 60°E and 110°W (Figure 1).

The study area is a subplate presently attached to the South American plate (Freeland & Dietz 1971) where natural hazards are related to hurricanes, especially in the northern and eastern parts of the region due to the geographic location (Palmieri et al. 2006). Although largely similar in climatic and biophysical characteristics (Schumann & Partridge 1989), the study area displays large socio-economic (World Bank 1998), environmental and political (Pelupessy 1991) differences. Climate in the Caribbean basin can be classified as dry-winter

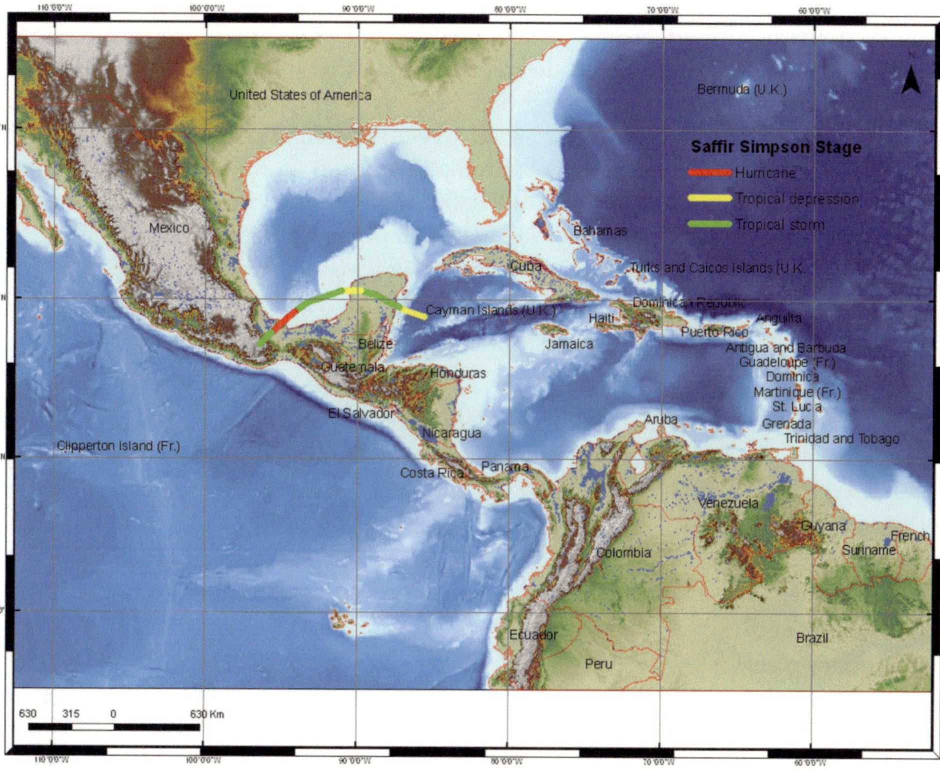

Fig. 1. Location map of the study area showing country boundaries, GEBCO bathymetry, and SRTM topography. The track of Stan event is also reported.

tropical, with significant subregional variations in rainfall annual totals, length of the rainy season, and timing of rainfall maxima. Three rainfall regimes can be related to the geography of the Caribbean-Central American region. From May to October the rainfall regime is typical of the Central America. A regime characterized by a pronounced midsummer break in rainfall accumulations is typical of the interior of the basin (southern coasts of Jamaica and Hispaniola). A regime characterized by a late-fall peak in rainfall is typical of the Caribbean coast of Honduras, of the northern coasts of Jamaica and Hispaniola, of Puerto Rico and of the Lesser Antilles. In this context rainfall-bearing disturbances, known as African easterly waves (Riehl 1954, Burpee 1972), propagate across the Atlantic Ocean into the Caribbean basin from mid June to early October generating hurricanes.

2.2 The stan event on the Caribbean area

A tropical wave that moved off the coast of Africa on September 17th, 2005 was the likely precursor to Stan event (Figure 1). Cloudiness and showers associated with the system began to increase as the wave neared 50° W longitudes on September 22nd but north-north-easterly shear created an environment that was not favourable for tropical cyclone formation. The wave moved into the eastern Caribbean Sea on September 25th, while shear over the system diminished. By September 27th, deep convection associated with the wave became more consolidated over the central Caribbean Sea. Based on the extent and organization of deep convection as well as surface observations, it is estimated that a tropical depression formed around 12.00 UTC on October 1st centred about 115 nmi southeast of Cozumel. Lower to middle-tropospheric ridging to the north and northeast of the tropical cyclone resulted in a west-north-westward steering current and an upper-tropospheric anticyclone became established over the area. The depression strengthened into a tropical storm shortly before its centre made landfall on the east coast of the Yucatán peninsula, just south of Tulum, around 10.00 UTC on October 2nd. Stan crossed the peninsula in about 18 hrs while weakening back to a depression. It quickly regained tropical storm strength, however, after it moved back over water (Pasch & Roberts, 2005).

3. Data source

3.1 Topography dataset

The Shuttle Radar Topography Mission (SRTM; http://www2.jpl.nasa.gov/srtm/) data are characterized by a recent and extensive literature (Farr et al. 2007 and references within). In this analysis the SRTM data version 2 was used. SRTM data in their original format have a resolution of 3-arc-seconds, corresponding approximately to 90 m x 90 m over the study area. Assemblage and local interpolation of the SRTM was performed importing tiles into ArcInfo 9.x (©ESRI) using an Arc Macro Language procedure (Taramelli & Barbour 2006).

3.2 Bathymetry dataset

The General Bathymetric Map of the Oceans (GEBCO; http://www.gebco.net/) One Minute Grid (Jones 2003) is global and also includes land elevations from the IGBP GLOBE database (Figure 1). A medium stage bathymetry datasets exported from GEBCO with 1km horizontal resolution were examined in Interactive Visualization Systems software for further cleaning, geomorphic analysis, and exporting grids to ArcInfo. Resolution of the bathymetry data was such that landscape features and differences at a horizontal distance on the order of 1km were clearly discernable. Data were rigorously edited for spurious points and smoothed and gridded to 1 km interval to minimize data gaps in the final 3D export. The net vertical resolution was multiplied by the single pixel area (90x90 m^2) and re-interpolated to arrive at the net value in all areas of the bathymetry.

3.3 Rain dataset

The Climate Prediction Center Morphing Method (CMORPH; http://www.cpc.noaa.gov/products/janowiak/cmorph_description.html) uses motion vectors derived from half-hourly interval geostationary satellite InfraRed imagery to propagate the relatively high quality precipitation estimates derived from passive microwave data. In addition, the shape and intensity of the precipitation features are modified (morphed) during the time between microwave sensor scans by performing a time-weighted linear interpolation (Levizzani & Mugnai 2004, Joyce & Ferraro 2005).

The hourly analyses of CMORPH at a grid resolution of 1 km have been produced using the INGRID program at the IRI/LDEO Climate data library website (http://ingrid.ldeo.columbia.edu/). The INGRID is an alternative mesh generator for finite element modelling, which is principally used as a fairly complete mesh generator with a wide range of geometric capabilities.

3.4 Wind dataset

The National Hurricane Center's (NHC) Tropical Cyclone Reports (http://www.nhc.noaa.gov/pastall.shtml) contain comprehensive information on each tropical cyclone, including synoptic history, meteorological statistics, casualties and damage, and the post-analysis best track (six-hourly positions and intensities). These data are of key-importance in vulnerability assessment. The lesson learnt in past events can really help to strengthen prospective scenarios. Tropical cyclones include depressions, storms and hurricanes (Abraham et al. 2004). In particular the report was used to calculate the standard temperature and pressure R0 following the asymmetric hurricane wind model from Holland (1980). It assumes that for a generic tropical cyclone, surface pressure field follows a modified rectangular hyperbola, as a function of radius in cyclostrophic balance. Even if the asymmetric is rarely, it is possible to introduce deviation from that geometry in a simple way: for example dividing in quadrants the wind fields. Following the idea proposed in Bao et al. (2006) following the reviewed wind model by Holland (2008) we computed the wind field, for each single quadrants, from the maximum sustained wind observed (Xie et al. 2006) and reported in the NHC website.

3.5 Land cover dataset

The Global Land Cover database (http://edc2.usgs.gov/glcc/glcc.php) is being produced by an international partnership (Hansen et al. 2000). The database contains a global product that combines all regional classes in one consistent legend (Mucher & Badts 2002). As some of classes could be clustered within a same class (e.g. shrub cover with herbaceous cover), to create the final land cover dataset of the study area a reclassification of the 21 different land cover classes was carried out through the decision trees approach. Decision trees provide a more rational approach to land cover classification than traditional statistical supervised classification, allowing the user to specify the exact logical basis of class assignment in the form of a Boolean conditional of arbitrary complexity. So 8 classes were finally obtained: tree cover, regularly flooded shrub, cultivated and managed areas, cropland, bare areas, water, artificial surfaces and associated areas and irrigated agriculture.

3.6 Population dataset

Affected population was assessed on a 2.5' x 2.5' latitude-longitude grid of global population, the Gridded Population of the World, version 3 (GPWv3; http://sedac.ciesin.columbia.edu/gpw/). The GPWv3 depicts the distribution of human population across the globe transforming population census data (corresponding to irregularly vector census block and block group boundaries), which most countries collected for subnational administrative units, into a regular raster-grid. Each cell contains an estimate of total population and population density on land, based on the overlap between the irregular boundaries of administrative units and the regular boundaries of the grid (Taramelli et al., 2010). In this analysis a preliminary version of GPWv3 was used, which contains population estimates for 1990, 1995, and 2000 for approximately 375 000 sub-national administrative units (Center for International Earth Science Information Network et al. 2004).

4. Methodology

For hurricane hazard/risk mapping and predictions at regional scales, it is reasonable to work on a geographic basis given the size of involved areas (Dilley et al., 2005). An accurate computation of the cell area on a geographic grid-based map is necessary to appropriately assess the extent of critical conditions up to detecting hazard/risk distribution (Santini et al., 2010). For this reason all available datasets were first projected in geographic coordinates. Based on the above described datasets, the hurricane hazard/risk estimate in the Caribbean is here described for the Hurricane Stan event affecting these regions on September 2005 (Pasch and Roberts, 2005). The steps to build the structure of the hurricane multi-hazard GIS model presented in this research are (Figure 2):

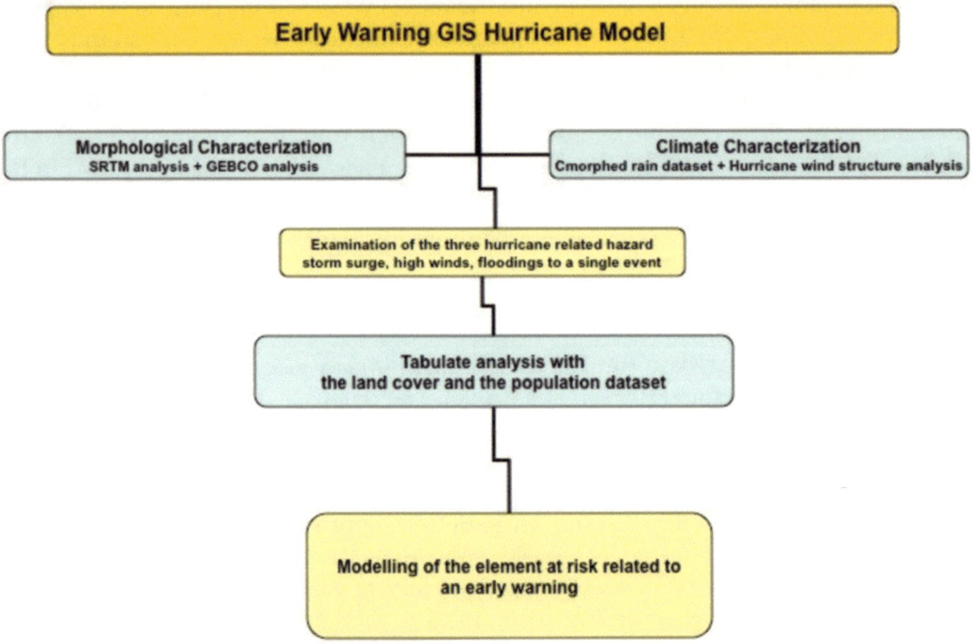

Fig. 2. Hurricane multi-hazard GIS model.

- homogenization of dataset reference systems and resolution;
- modelling of the hurricane hazard related to three different components such as storm surges, high winds and floods;
- examination of the spatial distribution of the hurricane hazard using the Jenks classification (Jenks, 1967). Classes are selected on natural groupings contained in the data with similar values. The break points are identified where huge jumps in data sequence are present (Goodchild et al. 1992, Osaragi 2002).
- assessment of the elements at risk (in terms of areal extent and involved population) to be potentially affected by the components (overlay analysis with the land cover and population dataset) using the ASPHAA (Santini et al., 2010) to calculate areas on a geographically projected grid.

4.1 Hazard assessment

4.1.1 Storm surge

The model implemented in this project calculates the proportional height of the bathymetry near the coast line and consequently the hazard degree value onshore related to the slope angle of the topography (Taramelli et al., 2010). The storm surge was calculated using SRTM and GEBCO Dataset using ArcInfo 9.x. The analysis was cast in different sreps:

- the coastline was modelled as a polyline for the study area using the NOAA/NOS Medium Resolution Coastline designed for 1:70.000 (available at: http://www.ngdc.noaa.gov/mgg/shorelines/shorelines.html);
- based on literature data (Blain et al. 1994, 1998, Zenger et al. 2002) a 5 km width buffer was calculated around the coastline for the far reachable distance value onshore;
- for both onshore and offshore potentially affected area slope grids were calculated respectively from SRTM and GEBCO datasets.
- both slope grids were reclassified in three increasing storm surge hazard classes (from value 1 minimum hazard to value 3, maximum hazard) based on the proportional height of the bathymetry near the coastline related to the slope angle of the topography (Taramelli et al. 2010).
- finally, a final storm surge hazard assessment was produced identifying every grid node, within the 5 km coastline buffer area, having the three different hazard values. The resulting grid (Figure 3) shows values that vary from 1 to 3, indicating respectively the greatest, the medium and the lowest likelihood of hazard signatures.

For Stan event the potentially affected area is around 36 km^2 whereas the potentially affected population is near 22,524 persons. Figure 3a and b show, respectively, the percent distribution of the population and the areas in the three storm surge hazard classes.

Fig. 3. Storm-surge hazard assessment related to the Stan hurricane event. a) diagram: storm surge affected population of the study area, b) diagram: storm surge affected area.

4.1.2 Floods

In this research two groups of data are considered to estimate flood hazard. The first set is about topographic and morphometric attributes of the catchment basins struck by the hurricane. The second one is related to meteorological data: rainfall depths and intensities, magnitude and frequency of rain peak. Due to spatial data integration issues within GIS such as geographical scales of the study area, the geologic factors as permeability and soil type are not taken into account. The final flood hazard analysis was cast in different steps using ArcInfo 9.x:

- A hydrological modelling of the test area is calculated. The delineation of flow direction grid is carried out from DEM exploiting the eight-direction pour point model (Puecker & Douglas 1975).
- The method of Jenson and Domingue (1988) is used to determine the flow accumulation grid and the stream network in a grid structure is then derived. The the stream network is classified according to Strahler method (Strahler 1980) assigning a numeric order to links in a stream network based upon their number of tributaries. For the study area according to previous studies (Correia et al. 1998, Colby et al. 2000, Taramelli & Melelli 2007) the fourth order is the maximum value assigned.
- The river network grid is converted in a vector layer (polyline) and then a buffering is made in order to link the stream order to a potential affected flooded area considering a linear proportional relation between the two variables. Based on literatures (Penning-Rowsell & Fordham 1994, Penning-Rowsell 1996) a buffering width equal to 200 m is measured for the first order, 1000 m for the second, 2000 m for the third and 4000 m for the fourth one. Then the buffer vector layer is converted in a grid format assigning to each pixel the river order value.
- The Wetness Index (WI) is calculated using the Terrain Analysis Using Digital Elevation Models (TauDEM) plug-in (Tarboton 1998, Tarboton & Ames 2001) in order to consider into the hydrologic model the topographic parameter of the flooding areas. This calculation estimates the ratio Slope/Specific Catchment Area, where a specific catchment area is the ratio between a contributing area concerning a specific unit contour length along the slope. This is algebraically related to the more common WI, with the contributing area at the denominator to avoid errors dividing by 0 when the slope value is 0° (Costa-Cabral & Burges 1994, Tarboton 1998).
- The WI grid is reclassified in three increasing hazard classes (Speight 1984, Taramelli et al. 2008). This grid relies on topographic variables only and it is still independent from specific rain values.
- The CMORPH rain dataset for the specific Stan hurricane event (between the 1st and 5th of October, 2005) is calculated in the INGRID mesh generator. It is interesting to notice that high rain values are present not only near by the hurricane track but also far away from the eye of the event due to the topography of the test area. The total rain values grid is then overlaid to the WI one achieving the final flooding hazard grid. Each area is zoned in terms of degree of hazard and each zone is low flooding, medium flooding and high flooding and allocated the numeric values 1, 2 and 3 respectively. A final flooding hazard assessment was produced identifying every grid node within the zones having only three different hazard values that fall within the sum of evidence criteria using the 'reclassify' operator in a GIS environment. The resulting grid shows values that vary from 1 to 3, indicating respectively the lowest, the medium and the greatest likelihood of hazard signatures (Figure 4).

For Stan event the potentially affected area is around 321 km^2 whereas the potentially affected population is near 8500 persons. Figure 4a and b show, respectively, the percent distribution of the population and the areas in the three flooding hazard classes.

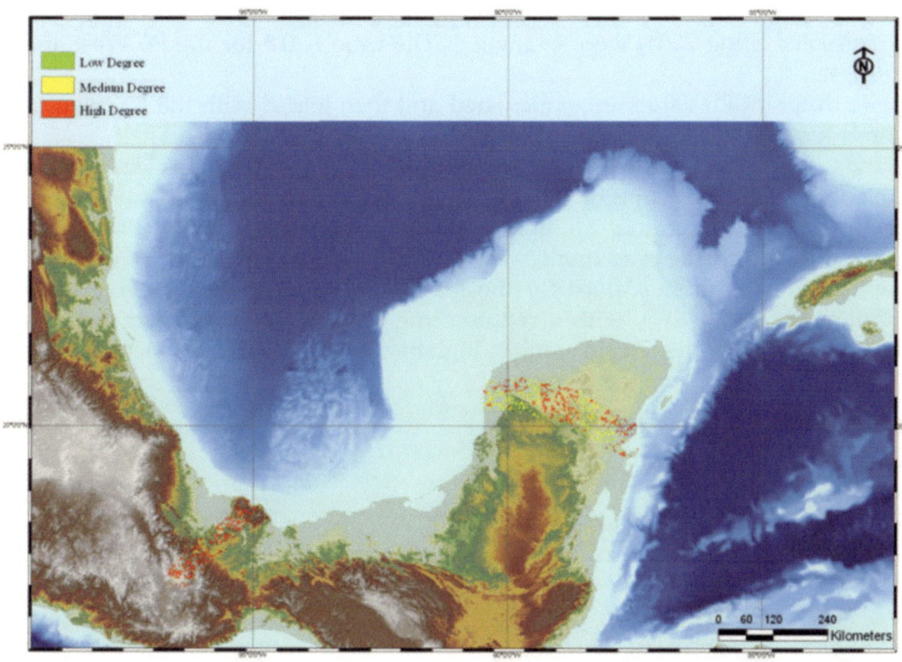

Fig. 4. Flooding hazard assessment related to the Stan hurricane event. The assessment of the rain-rate is within the watershed area using cumulate values. a) diagram: flooding affected population of the study area, b) diagram: flooding affected area.

4.1.3 High winds

Another key variable in the hurricane hazard is the area struck by the high winds. In the absence of detailed instrument observations it was assumed that wind velocity increases linearly from the centre to the outer side (Holland 1980; Holweg 2000) and thereafter decreases exponentially moving outwards. Moreover, the horizontal wind field is asymmetrical and, in northern hemisphere, the strongest winds are found in the right-hand quadrants of the storm (relative to the direction of movement) due to the Coriolis force.

Based on the aforesaid basic model, in this study the NOAA preliminary hurricane report (Pasch & Roberts 2005, Xie et al. 2006) was used in order to gain all the comprehensive information on each hurricane, including synoptic history, meteorological statistics and the post-analysis best track (six-hourly positions and intensities). The track and the intensity evolution (from NHC) of the hurricane Stan, shown in Figure 1, are the first input of the model as lat-long point features. In order to estimate the potentially affected areas by high winds the following sequence was respected:

- the lat-lon *.txt* file from the NOAA preliminary report was imported as event theme,
- the cyclone's mean sustained surface wind was assessed, based on the radius maximum winds speed and on the pressure within the same radius. The wind at each level of the hurricane has been normalized by the wind speed of the different quadrant based on the asymmetric modelling of the hurricane itself and the velocity was calculated based on the Holland model for each quadrants and computing all the model parameters from the hurricane report in the NHC web page (Taramelli et al., 2010),

- the ratios of the R0 East and R0 West related to the stage of the Saffir-Simpson scale of the Hurricane (Blog 2004) were evaluated. The ratio is 0.8 for the R0 West and 1.2 for the R0 East,
- the two different R0 value were calculated and then joined with the XY event theme to generate a polyline vector file,
- a rounded buffer file was created using the two different R0 values leading to polygon vector file within the value of R0 intensities that represent intensity evolution of the hurricane event referred to the temporal variation of the R0,
- the polygon vector file was finally converted to an integer grid file representing the different hazard values. Values for the resulting dataset, named degree values, vary continuously from 1 to 3, with 1 representing terrain with low hazard degree and 3 indicating that all of the terrain exhibits high hazard degree (Figure 5).

Fig. 5. High winds hazard assessment related to the Stan event. a) diagram: High winds affected population of the study area, b) diagram: high winds affected area.

As a preliminary result, it can be noticed that high values of the signature are represented with red, medium values with yellow and low values with light green. As can be seen, in addition to the relatively homogeneous values near the coastal area, there is a wide variety of different signal showing out throughout the ridge region and a widespread boundary region areas of strongly high value composition.
For Stan event the potentially affected area is around 36000 km^2 whereas the potentially affected population is near 1407 people. Figure 5a and b show, respectively, the distribution of the population and the areas in the three high winds hazard classes.

4.2 Risk assessment

The risk assessment requires that critical facilities are identified and data on past physical effects, in terms of structural and functional damages, are collected. These data are to be integrated into the hazard GIS database. In this research the hazard results were overlaid with population and land cover datasets. The approach was then validated on a regional basis using the Stan event report on elements affected by damages (OCHA 2005a, 2005b) over an area that covers both developed and developing countries in the Caribbean Region. The report allows to know the effects of Stan event in the study area and to compare the numbers obtained from the model (in terms of affected area and population) to the real effects summarized in the regional reports. The key layers in the GIS model are the hazard maps of high wind, storm surges and floods with the spatial distribution of elements at risk such as agricultural and managed areas and number of population at a pixel scale. The final risk assessment as set out in this research, after identifying the hazards, consists of the following steps:

- creation of the specific maps showing the zones with elements at risk,
- calculation of a total score of population and total area affected for each single hazard and for the final hazard using ASPHAA (Santini et al., 2010).The choice of using ASPHAA for calculating areas in lat/lon coordinates rather than projecting layers was justified starting from results of Santini et al. (2010), reporting not negligible differences in total affected area when switching from geographic to projected coordinate systems.

The underestimation of the total affected area within the Caribbean by projected coordinate system propagates the error to the final calculation of elements at risk in terms of affected population. Using population data from the GPWv3 dataset the 1,418,634 affected people estimated using ASPHAA on the geographic grid is reduced to 1,010,210 affected people when projecting the grid (Taramelli et al., 2010), even if maintaining the source Datum (i.e. WGS 1984) and choosing the most appropriate linear mesh spacing (0.0008333333°). Based on ASPHAA, for each pixel and then for the whole Caribbean area, the total area covered by the three hazard classes was computed.

Results are shown in Figure 6, in terms of number of pixels belonging to the three hazard severity classes, producing a total affected area (in km²).

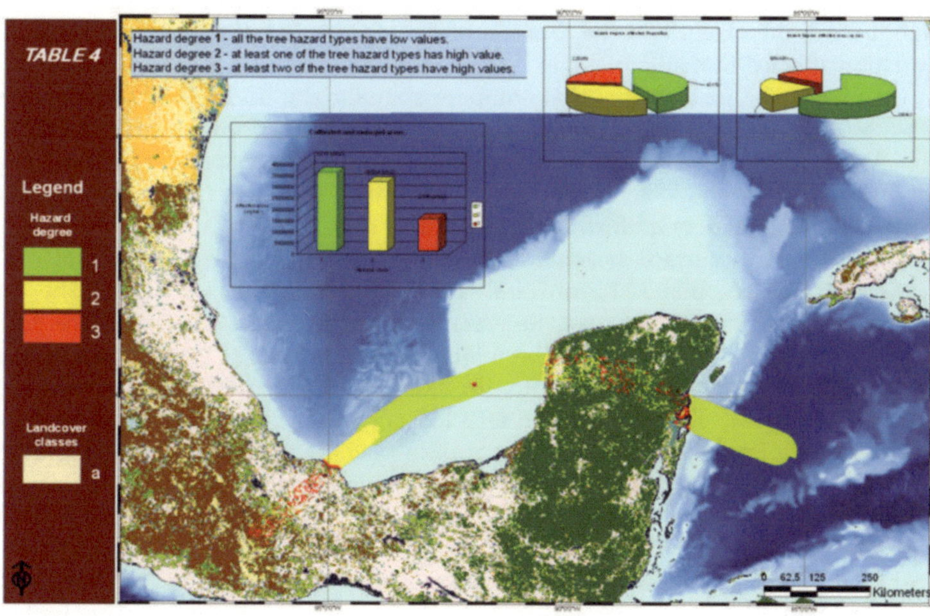

Fig. 6. Total element at risk affected area estimation.

The final affected population and total affected area related to the hurricane Stan appeared to be embedded within the western portion of a broader-scale low-level cyclonic circulation. This larger system produced extensive very heavy rains over portions of extreme eastern Mexico and Central America that resulted in disastrous floods. Estimates of the total number of lives lost in Mexico and Central America are mostly in the range of 1000 to 2000, some even higher (OCHA 2005a, 2005b). Guatemala was hit particularly hard and over 1000 persons may have perished only in that country (UNEP-OCHA 2005). As can be seen, in addition to the relatively homogeneous values of hazard signature near the coastal area, there is a wide variety of different signal showing out throughout the ridge region and a widespread boundary region areas of strongly high value composition far away from the hurricane track. This widespread signature confirms the high correlation between the affected population score and the flooding environmental variables, while shows an overestimation score between the affected population and the storm surge hazard, probably due to some weak in the storm surge modelling such as the non consideration of the impact angle effect.

In order to finally calculate the affected population, people living in the potentially affected areas were considered, comparing the data with the land cover use (in this case the cultivated-managed and artificial surfaces). To this end, a final rough estimate of the affected population was evinced based on the information available. The number of the potentially affected population is estimated based on the hazard evaluation. As a main result the multi-hazard GIS model highlights estimation scores for critical land cover. It can be noticed that topography highlights several high score values thought-out areas well far from the point where the hurricane makes the landfill.

5. Conclusions

In this work the hazard hurricane evaluation is closely related to the possibility of point out an early warning model. The aim is to have a valid instrument to estimate, in case of hurricane hazard, the extent and the location of damaged areas, the economic impact and the population affected immediately after the event and as long-term response.

GIS are the most valid instruments that compare spatial data, useful to assess the hazard zonation, with vulnerability parameters. Moreover, hurricanes strike particular underdeveloped countries with low or no spatial data available. Nowadays several digital data are disposable with a global coverage. Although this data have not an high resolution they are updated and overlay uniformly large lacking areas.

For these reasons earth surface systems in Central America provided the framework for developing the hurricane related hazard and risk estimation GIS based model. The model describes the morphological response to variations in the balance between the forcing factors (e.g. wind velocity) and one of the related single hazards (e.g. storm surge wave). The model could be extended to generate, for example, vector grids of square polygons storing in each feature (cell) the estimated affected population value to be then visualized and used in Google Earth or in other web mapping systems. Anyway authors highlighted how such modelling tools has an increasing usefulness for scales ranging from regional to semi-regional, while for local applications often treated with local mapping the modelling system is not such highly reliable.

As a final remark the GIS hurricane model approach shown in this study could be used for different term simulations and it should be considered as a potential monitoring tool in an integrated management approach to hurricane hazard mitigation and control. It is a practical tool for building possible intervention scenarios both for small and large-scale areas, providing, also, quantitative evaluations of the elements at risk and inter-linkages between the different landforms involved (coastal or in land).

6. References

Arnold, M., Chen, R.B., & Deichmann, U. (2006). *Natural Disaster Hotspots Case Studies.* World Bank Publications, 184pp.

Bao, S., Xie, L., & Pietrafesa, L.J. (2006). An asymmetric hurricane wind model for storm surge and wave forecasting. *AMS Annual Meeting,* Atlanta. 2006.

Blain, C.A. (1997). Modelling methodologies for the prediction of hurricane storm surge. *Recent Advances In Marine Science and Technology,* 96, pp.177-189.

Blain, C.A., Westerink, J.J., & Luettich, R.A. (1994) The influence of domain size on the response characteristics of a hurricane storm surge model. *Journal of Geophysical Research,* 99(18), pp.467-479.

Blog, R. (2004). A Review of Damage Intensity Scales. *Natural Hazards,* 29 (1), pp.57-76.

Burpee, R.W. (1972). The origin and structure of easterly waves in the lower troposphere of North Africa. *J. Atmos. Sci., 29, pp.77– 90.*

Center for International Earth Science Information Network (CIESIN) Columbia University and Centro Internacional de Agricultura Tropical (CIAT). (2004). Gridded Population of the World (GPW), Version 3. Palisades, NY: Columbia University. Available at http://beta.sedac.ciesin.columbia.edu/gpw.

Colby, J., Mulcahy, K., & Wang, Y. (2000). Modelling Flooding Extent from Hurricane Floyd in the Coastal Plains of North Carolina. *Environmental Hazards,* 2, pp.157-168.

Correia, F.N., Rego, F.C., Saraiva, M.G., & Ramos, I. (1998). Coupling GIS with hydrologic and hydraulic flood modelling. *Water Resources Management,* 12, pp.229-249.

Costa-Cabral, M., & Burges, S.J. (1994). Digital Elevation Model Networks (DEMON): A Model of Flow Over Hillslopes for Computation of Contributing and Dispersal Areas. *Water Resources Research,* 30(6), pp.1681-1692.

Dilley, M., Chen, R.B., Deichmann, U., Lerner-Lam, A.L., & Arnold, M. (2005). *Natural Disaster Hotspots: A Global Risk Analysis.* World Bank Publications, 132 pp.

Emanuel, K.A. (2004) Tropical cyclone energetics and structure. Atmospheric Turbulence and Mesoscale Meteorology, E. Federovich, R. Rotunno, and B. Stevens, Eds., Cambridge University Press, 165–192.

Farr, T.G., Rosen, P.A., Caro, E., Crippen, R., Duren, R., Hensley, S., Kobrick, M., Werner, M., Paller, M., Rodirguez, E., Roth, L., Seal, D., Shaffer, S., Shimada, J., Umland, J. Wermer M., Oskin, M., Burbank, D., & Alsdorf, D. (2007). The Shuttle Radar Topography Mission. *Rev. Geophys* 45 RG2004; doi: 10.1029/2005RG000183

Freeland, G.L., & Dietz, R.S. (1971). Plate Tectonic Evolution of Caribbean-Gulf of Mexico Region. *Nature,* 232, pp.20-23.

Fujita, T. 1952: Pressure distribution within a typhoon. Geophys. Mag., 23, 437-451.

Goodchild ,M.F, Guoqing, S., & Shiren, Y. (1992). *Development and test of error model for categorical data,* Int. J. Geographical Information Systems, 6(2), pp.87-104.

Hansen, M., Defries, R., Townshend., & J.G. Sohlberg, R. (2000). Global land cover classification at 1km resolution using a decision tree classifier. *International Journal of Remote Sensing,* 21, pp.1331-1365.

Holland, G.J. (1980). An analytic model of the wind and pressure profiles in hurricanes. *Monthly Weather Review*, 108, pp.1212–1218.

Holland, G. J., (2008) A Revised Hurricane Pressure-Wind Model. Mon. Wea. Rev., 136, 3432–3445.

Holweg, E.J. (2000). Mariner's guide for hurricane awareness in the North Atlantic Basin. National Weather Service, United States.

Howard, R., Dodge, P., Doggett, A., Finney, J., Gurley, K., Levitan, M., Reinhold, T., Schroeder, J., & Stone, G. (2003). The Landfall of Hurricane Lili in Louisiana: A Summary of Cooperative Data Collection Efforts. *International Conference on Wind Engineering*, Lubbock, Texas, 2003.

Jenks G.F. (1967). The Data Model Concept in Statistical Mapping *International Yearbook of Cartography*, 7, pp.186-190.

Jenson, S.K., & Domingue, J.O. (1988). Extracting topographic structure from digital elevation data for geographic information system analysis. *Photogramm. Eng. Remote Sens.*, 54(11), pp.1593-1600.

Jiang, H., Halverson, J.B., Simpson, J. (2003). Difference of rainfall distribution for tropical cyclones over land and ocean and rainfall potential derived from satellite observations and its implication on hurricane landfall flooding prediction. *J. Hydrometeor*, 12 B.1.

Jones, M.T. (2003). User guide to the Centenary Edition of the GEBCO Digital Atlas and its data sets, Available at: http://www.ngdc.noaa.gov/mgg/gebco/.

Joyce, R.J., & Ferraro, R. (2005). Improvements of CMORPH Resulting from Limb Adjustments and Normalization of AMSU-B Rainfall. *J. Hydrometeor*, P1.22.

Kerry, E. (2005). Increasing destructiveness of Tropical Cyclones over the past 30 Years. *Nature*, 436, pp.686-688.

Kok, K., & Winograd, M. (2002). Modelling land-use change for Central America, with spatial reference to the impact of hurricane Mitch. *Ecological Modelling*, 149, pp.53–69.

Knutson, T. R., McBride, J. L. Chan, J. , Emanuel, K., Holland, G., Landsea, C., Held, I., Kossin, J.P., Srivastava, A.K., & Sugi, M. (2010). Tropical Cyclones and Climate Change, *Nature Geoscience*, Review Article, DOI: 10.1038/NGEO779.

Levizzani, V., & Mugnai, A. (2004). Rainfall measurements from space: Where are we? *Proc. 14th Int. Conf. on Clouds and Precipitation*, Bologna, Italy, 18-23 July, 2004.

Mucher, C.A., & Badts, D.E.P.J. (2002). Global Land Cover 2000: Evaluation of the SPOT VEGETATION sensor for land use mapping. Wageningen, Alterra, Green World Research pp. 49.

OCHA: OCHA Situation Report n.2, web source: http://www.reliefweb.int/rw/RWB.NSF/db900SID/SNAO-6GZSTG?OpenDocument, 2005a.

OCHA: OCHA Situation Report n.3, web source: http://www.reliefweb.int/rw/RWB.NSF/db900SID/EVOD-6H3DTK?OpenDocument&rc=2&emid=TC-2005-000173-SLV, 2005b.

O'Hare, G. (2001). Hurricane 07B in the Godavari delta, Andhra Pradesh, India: vulnerability, mitigation and the spatial impact. *The Geographical Journal*, 167(1), pp.23-38.

Osaragi, T. (2002). Classification methods for spatial data representation. *Centre for Advanced Spatial Analysis. Working Paper Series,* 40, pp.1-19.

Palmieri, S., Teodonico, L., Siani, A.M., & Casale, G.R. (2006). Tropical storm impact in Central America. *Meteorol. Appl.,* 13, pp.21-28.

Pasch, R.J., & Roberts, D.P. (2006). *Tropical Cyclone Report - Hurricane Stan,* 1-5 October 2005, National Hurricane Center, web source: http://www.nhc.noaa.gov/2005atlan.shtml

Pelupessy, W., (Ed.) (1991). Perspectives on the Agro-export Economy in Central America, Macmillan, Washington, 175.

Penning-Rowsell, E. (Ed.) (1996). *Improving Flood Hazard Management Across Europe,* Middlesex University Press, London.

Penning-Rowsell, E., Fordham, M., (Eds.) (1994). *Floods across Europe. Flood Hazard Assessment, Modelling and Management,* Middlesex University Press, London.

Pielke, R.A., Jr., Gratz, J., Landsea, C.W., Collins, D., Saunders, M.A., & Musulin, R. (2008). Normalized Hurricane Damages in the United States: 1900-2005. *Natural Hazards Review,* 9(1), pp.29-42.

Pielke, R.A., Jr., Landsea C., Mayfield M., Laver J., & Pasch R. (2005). Hurricanes and Global Warming, *Bulletin of the American Meteorological Society,* 86:11, pp.1571-1575

Pielke, R.A., Jr., Rubeira, J., Landsea, C., Fernandez, M.L., & Klein, R. (2003). Hurricane vulnerability in Latin America and the Caribbean: normalised damage and loss potentials. *Natural Hazards Review* ASCE, Aug., pp. 101–114.

Puecker, T.K., & Douglas, D.H. (1975). Detection of surface-specific points by local parallel processing of discrete terrain elevation data. *Computer Graphics and Image Processing,* 4, pp.375-387.

Riehl, H. (1954). *Tropical Meteorology.* McGraw-Hill, 392 pp.

Russel, G.L., Miller, J.R., Rind, D., Ruedy, R.A., Schmidt, G.A., & Sheth, S. (2000). Comparison of model and observed regional temperature changes during the past 40 years. *Jou. of Geoph. Resea.,* 105(14), pp.891-898.

Santini, M., Taramelli, A., & Sorichetta, A. (2010). ASPHAA: a GIS-based algorithm to calculate cell area on a latitude-longitude (geographic) regular grid. *Transcation in GIS,* 14(3); pp.351-377, doi: 10.1111/j.1467-9671.2010.01200.x

Saunders, M. (1998). *Global warming: the view in 1998,* Benefield Greig Hazard Research Centre, University College-London, 120 pp.

Schumann, D.A., & Partridge, W.L. (Eds.) (1989). The human ecology of tropical land settlement in Latin America. Westview Special Studies on Latin America and the Caribbean. Westview Press, Boulder Arizona, 470 pp.

Shen, B.W., Atlas, R., Reale, O., Lin, S.J., Chern, J.D., Chang, J., Henze, C., & Li, J.L. (2005). Hurricane forecasts with a global mesoscale-resolving model: Preliminary results with Hurricane Katrina (2005). *Geophysical Research Letter,* 33, L13813. doi:10.1029/2006GL026143.

Speight, J.G. (1984). The role of topography in controlling through-flow generation: a discussion. *Earth Surface Processes Landforms,* 5, pp.187-191.

Strahler, A.N. (1980). System theory in general geography. *Phys. Geogr.,* 1, pp.1-27.

Taramelli, A., & Barbour, J. (2006). A new DEM of Italy using SRTM data, *Italian Journal of Remote Sensing,* 36, pp.3-15.

Taramelli, A., & Melelli, L. (2007). Strategy to delineate potentially affected areas by hurricane using GIS approach. *Rend. Soc. Geol. It.,* 4, pp.27-34.

Taramelli, A., Melelli, L., Cattuto, C., & Gregori, G. (2008). A case of geomorphic parametrization from SRTM data. *Mem. Descr. Carta Geol. d'It.*, LXXVIII, pp. 289-298.

Taramelli, A., Melelli, L., Pasqui, M., & Sorichetta, A. (2010). Modelling hurricane element at risk in potentially affected areas by GIS system, *Geomatics, Natural Hazards and Risk,* volume 1, Issue 4 December 2010, pages 349 - 373.

Tarboton, D.G. (1998). A new method for the determination of flow directions and upslope areas in grid digital elevation models. *Water Resources Research,* 33(2), pp.309–319.

Tarboton, D.G., & Ames, D.P. (2001). Advances in the mapping of flow networks from digital elevation data. *World Water and Environmental Resources Congress, May 20- 24,* Orlando, Florida, 2001.

Trenberth, K.E., & Shea, D.J. (2006). Atlantic Hurricanes and Natural Variability in 2005. *Geophysical Research Letters,* 33(12).

UNEP-OCHA: Hurricane Stan: Environmental Impacts from Floods and Mudslides in Guatemala - Results from a rapid environmental assessment in Guatemala, http://ochaonline.un.org/OchaLinkClick.aspx?link=ocha&DocId=100394, 2005.

Watson, C.C., & Johnson, M.E. (2005). Hurricane loss estimation models, *American Meteorological Society,* BAMS - nov.2004, pp.1713-1726.

Willoughby, H.E., R.W.R. Darling, and M.E. Rahn, 2006: Parametric Representation of the PrimaryHurricane Vortex. Part II: A New Family of Sectionally Continuous Profiles. Mon. Wea. Rev., 134, 1102–1120.

World Bank. (1998). *World Development Indicators,* 1997, World Bank CD-ROM. World Bank, Washington.

Xie, L., Bao, S., Pietrafesa, L.J., Foley, K., & Fuentesa, M. (2006). Real-Time Hurricane Surface Wind Forecasting Model: Formulation and Verification. *Monthly Weather Review,* 134, pp.1355–1370.

Zenger, A., Smith, D.I., Hunter, G.J., Jones, S.D. (2002). Riding the storm: a comparison of uncertainty modelling techniques for storm surge risk management. *Applied Geography,* 22(3), pp.307-330.

19

Geospatial Technique for Rapid Delineation of Potential Hurricane Damage

William H. Cooke III[1], John C. Rodgers[1] and Katarzyna Grala[1]
[1]*Department of Geosciences, Mississippi State University*
USA

1. Introduction

Rapid delineation of potential damage from hurricanes to the natural and human-modified landscape is critical for disaster management and for assessment of ecological and economic impacts. Geospatial technologies including Remote Sensing (RS), Geographic Information Systems/Science (GIS), and Global Positioning Systems (GPS) provide new tools for delineating damage potential and for assessing destruction after natural disasters. Historically, airborne and space RS missions have been used for post-hurricane impact assessment, but these data are limited in a variety of ways. Space-based RS data have limitations including differing specifications for spatial, spectral, and temporal resolution. As an example, predicting damage or assessing damage from land-falling hurricanes via satellite data is often hampered by temporal resolution (re-visit frequency) that can result in unavailability of cloud-free images over impacted areas. Airborne RS data are not temporally limited to a specific window of ground coverage and can even acquire data during cloudy conditions. Airborne RS instruments may provide data of higher spectral and spatial resolution than space-borne instruments; however complete 'wall-to-wall' coverage of landscapes impacted by tropical systems as large as hurricane Katrina can be impractical and prohibitively expensive. Consequently, many geospatial analysts first use space-borne data and geospatial modeling at a relatively coarse-resolution to provide an overview of hurricane impacted areas, and then they use this information to pinpoint locations where they need to acquire airborne RS data for higher-resolution subsets of damaged areas. Therefore, there is a need to develop more accurate geospatial models that incorporate readily available data at a regional scale to assist emergency and disaster management agencies with their decision-support.

According to Hodgson et al., (2010), agencies with first responder responsibilities have expressed the critical need for image acquisitions over disaster impacted areas within 24 hours, or within three days of the event at a minimum. Given the limitations for rapid acquisition of RS imagery over hurricane-impacted areas, researchers at Mississippi State University in coordination with Mississippi Emergency Management Agency (MEMA) advisors have developed a geospatial modeling-based tool that enables rapid depiction (within 24 - 48 hours after landfall) of a damage probability statistical grid which incorporates readily available data sources such as storm surge predictions and precipitation and wind forecasts. This grid can be overlain on a variety of image and GIS base layers to help plan for evacuation procedures prior to landfall and to help orient and manage the effective allocation of resources to impacted areas.

2. Background

2.1 Review of geospatial technologies

Geospatial technologies including GIS, RS, and GPS are increasingly used in a variety of fields where data gathering for and information about places, objects, and processes are displayed and analyzed. Each of these technologies comprises vast fields of study and research, but the integration of these technologies provides considerable descriptive, predictive, and analytical power.

GIS can be described as a system comprised of software, hardware, and georeferenced geographic data used to describe, and/or analyze patterns and processes that occur at a variety of scales. GIS provides advanced spatial analyses targeting location, attributes, and relationships of geographic features that exhibit spatial relations including proximity, direction, and topology (Theobald, 2001). Proximity may be understood in terms of relative and absolute distances of features from each other. Directional information is associated with relative and absolute directions from one feature to another. Topology describes numerical relationships between geographic features as encoded by adjacency, linkage, inclusion, or proximity (Clarke, 2003).

Remote Sensing has been defined as "the science and art of obtaining information about an object, area, or phenomenon through the analysis of data acquired by a device that is not in contact with the object, area, or phenomenon under investigation" (Lillesand et al., 2008). Sophisticated passive recording devices flown aboard aircraft and satellite platforms obtain information about feature reflection, transmissivity, and absorption of energy from electromagnetic spectra. Other active devices like radar transmit microwave energy from an antenna then record information about the intensity of the radiation 'backskattered' by the surface (Lillesand et al., 2008).

GIS and RS spatial data structures can be categorized into vector and raster data types. Vector data are comprised of points, lines, and polygons while raster data is comprised of grid cells. In general, vector data are the data of choice for accurate mapping purposes and raster data are the data of choice for mathematical and probabilistic analyses. Integration of GIS and RS can occur on a variety of levels, but an obvious unifying characteristic of both disciplines is raster spatial data. Raster GIS data is generally created by the tessellation of geographic space via interpolation of points to a statistical grid surface or by simply subdividing geographic space into a regular grid. Figure 1 illustrates a polygon that has been tessellated into a square grid to create a raster data type. Remotely sensed data is often the product of information about reflected portions of the electromagnetic spectrum captured on a charge coupled device. The size of pixels generated via a digital to analog conversion process is predicated on several factors including the amount of land area covered by the Instantaneous Field of View and the height of the remote sensing platform above the earth. Both vector and raster data can be displayed, integrated, and analyzed using software-hardware platforms that can accommodate a wide variety of proprietary and open source data formats.

The U.S. Global Positioning System includes a group of at least 24 satellites rotating the earth used to determine ground positions via 'satellite ranging'. Originally developed for defense purposes, locations of features or field observations can be determined very accurately using GPS methods. GPS has resulted in dramatic improvements in the ability to accurately locate features on earth and more effectively orient response and recovery personnel to specific sites, or accurately designate hazardous areas to be avoided.

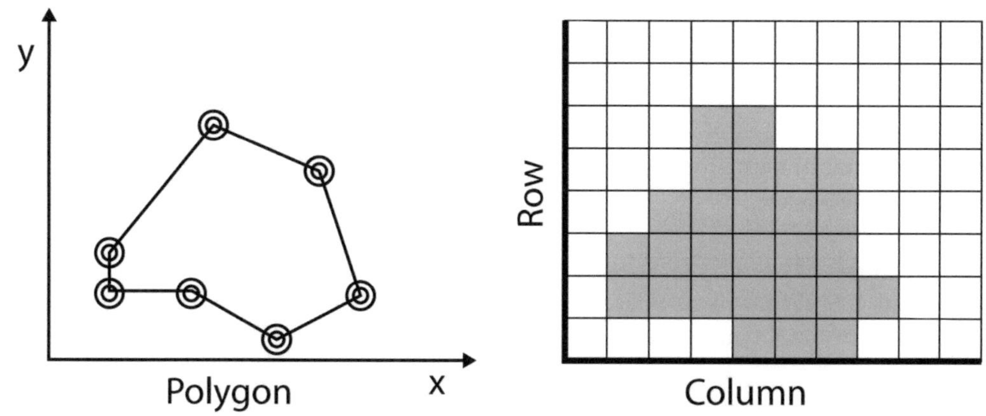

Fig. 1. Tessalation of a polygon to create a raster data type

2.2 Emergency management applications of geospatial technologies

Applications of geospatial technologies for disaster management include emergency preparedness, response, recovery, and mitigation. Geospatial technologies could have improved the characterization of geophysical, topographical, geological, and sociological factors important for mitigating the disastrous effects of the 7.0 magnitude earthquake that struck Haiti on January 12, 2010. Theilen-Willige (2010) used a GIS weighted overlay approach to produce maps representative of areas where factors affecting surface-near earthquake shock waves that aggregate and interfere with each other resulting in increasing vulnerability to soil amplification. This information in turn was used to create landslide and flooding susceptibility maps. By integrating the predictive products developed by Theilen-Willige with GIS base layers that spatially describe population distributions and the built environment, mitigation strategies could have been developed that could have saved many lives and could have minimized the disastrous effects of this seismic event.

Geospatial technologies can be instrumental for determination of post-disaster impacts. For example, they have been used in recent years to model tsunamis (Walsh et al., 2000; Papathoma et al., 2003; and Keating et al., 2004). Chandrasekar et al., (2007) used GIS techniques to assess the impact of the Indian Ocean Tsunami that caused severe damage to coastal zones on December 26, 2004. Their methods enabled accurate mapping of inundation along the coastal villages of Kanyakumari District, India. The area of inundation was mapped by a field survey, superimposed on satellite imagery, and interpreted using GIS techniques. The results showed a variation in percentage of inundation from 7% to 39%. Accurate depiction of the range of inundation provides valuable information that can be combined with demographic data and tsunami early warning systems to minimize loss of life. Similarly, once a tsunami impacts land areas, more precise inundation information can help guide post-disaster rescue and response activities.

Post-disaster response begins with initial rescue and relief efforts, followed by recovery activities that can initiate mitigation procedures including development of series of pre-impact preparedness actions, in the literature often referred as "emergency response cycle" (Cutter, 2003). Cutter (2003) describes 'GI Science' applications citing examples of the use of GIS in the September 11 terrorist event and essential activities in the Coalition War in Iraq where spatial decision support systems helped identify command and control structures and movements of opposing forces and other activities. While Cutter (2003) maintains that "GI Science can, and should make a difference in emergency preparedness, response, recovery,and mitigation activities", she also warns that effective utilization of GI Science requires a more complete understanding by the GI Science community of the limitations and constraints of the practitioner community, and the lack of fundamental data in some of the most hazardous places.

2.3 Applications of geospatial technologies for hurricane disaster management

A significant U.S. Department of Homeland Security (DHS)-funded study was undertaken following Hurricane Katrina to identify successes and failures of geospatial technologies for response and recovery along the Mississippi Gulf Coast. The study entitled 'Capturing Hurricane Katrina Data for Analysis and Lessons-learned Research, in tandem with the report to the President entitled 'The Federal Response to Hurricane Katrina: Lessons Learned, provide insight into the usefulness of geospatial technologies for hurricane disaster management. Findings from the DHS-funded study revealed the need to: develop and maintain centralized geospatial database comprised of locally accurate data, develop and improve geospatial capabilities at local/state/federal levels, identify response culture similarities and differences that require standardized or customized geospatial products, develop input data criteria for analytical models, and develop tools that rapidly delineate damage severity and extent on post-disaster RS imagery is critical for effective disaster management (Lessons Learned Final Report, 2008). A similar finding is presented in the report to the President as a 'Lesson Learned'. The 'Lesson' specifies the need for coordination between the Department of Homeland Security (DHS) and the Environmental Protection Agency to oversee efforts aimed at the Federal government's capability to quickly gather environmental data and to provide the public and emergency responders the most accurate information available. The rapid dissemination of information can help determine whether it is safe to operate in a disaster environment or to return to an area after evacuation. Further, the DHS is encouraged to work more closely with its State and local homeland security partners to plan and to coordinate an integrated approach to debris removal during and following a disaster.

The importance of geospatial technologies for hurricane disaster response was further illustrated in the report to the President in APPENDIX B – WHAT WENT RIGHT by the story of National Guard member Ronnie Davis. Davis, also an employee of the USDA National Resources Conservation Service (NRCS), combined the NRCS's digital data that are collected to develop conservation plans and generated in Texas, with the NRCS Digital Togographic Support System to create much needed maps of the affected regions of Mississippi. The result was that over 800 maps were delivered to support sector operations, needs of local police and law enforcement officers arriving from other states, and FEMA. Similarly, NVision Solutions Inc., the map producing entity at the local Emergency Operations Center (EOC) in Hancock County, MS, was a rich source of information on the development and distribution of geospatial products and tools in the aftermath of Katrina at a location generally considered to be the epicenter of the storm.

There are a variety of geospatial tools available for hurricane emergency management purposes. Among the most widely used geospatial tools are the Sea, Lake, and Overland Surge (SLOSH) model developed by the Federal Emergency Management Agency (FEMA),

United States Army Corps of Engineers (USACE), and the National Weather Service (NWS) used to predict storm surge heights; FEMA's Hazards U.S. Multi-Hazard (HAZUS-MH) model that estimates potential losses from earthquakes, hurricane winds, and floods; and the HURREVAC computer program which is used to track hurricanes and assist in evacuation planning.

The SLOSH model provides information on the potential for flooding both at the coast and inland by computing water height over a geographical areas covered by a network of grid points (Jarvinen & Neumann, 1985). The authors (Jarvinen and Neumann, 1985) state that the SLOSH model's primary use is to define flood-prone areas for evacuation planning, as evacuation of the flood plain is the planned response of many coastal communities to the threat of a hurricane landfall. There is some evidence that the coarse resolution of the SLOSH polar grids with cell sizes of 500-7000 m results in too much uncertainty in storm surge flooding estimates. Zhang et al., (2008) present results of the Coastal and Estuarine Storm Tide (CEST) model for Hurricanes Andrew (1992), Hugo (1989) and Camille (1969) that uses a fine-resolution grid with cell sizes of 100-200 m which reduced uncertainty when compared with field-measured elevations of high water marks and the locations of debris lines. Nevertheless, the SLOSH model is widely used and accepted for modeling both historic and predicted storms.

Tran et al., (2009) showed the HAZUS-MH model to be a valuable planning tool when assessing implications of the combined effects of storm surge with hurricane force winds Using HAZUS analysis for estimating disaster associated losses, storm surge, residual flooding, and wind damage associated with hurricane categories 1 thru 5 enabled planning for health care contingencies for predicted damage and flooding that would occur at various levels of storm intensity. Vickery et al., (2006) compared modeled and observed losses for Hurricanes Andrew, Hugo, Erin, and Opal; and they concluded that there was overall agreement at the zip code level but suggested that the damage and loss models may underestimate the small losses that occur at lower wind speeds (less than 100 mph). This underestimation is less significant for regional assessments but could become problematic for loss estimation in smaller areas.

HURREVAC is a U.S. federally-funded program supported by FEMA and USACE. State emergency management agencies in coastal areas often implement the HURREVAC model for hurricane storm tracking and decision processes associated with evacuation. Combining the National Hurricane Center Forecast/Advisory product with data from various state Hurricane Evacuation Studies (HES) helps local emergency managers determine proper evacuation decision time and the arrival time of storm effects including wind and storm surge (HURREVAC, 2010). There are many forecast features available including hourly wind ranges, track predictions, and hourly error ellipses, and the display shoes storm position and size at certain hours of the forecast. Concentric rings represent the extent of tropical storm force winds and output of many features is available in GIS format.

These and many other models designed to aid emergency managers require significant investments of hardware, software, and analyst training to optimize their usefulness. Rapidly produced products available in a web-based environment are becoming more available and necessary to emergency managers. Hodgson et al., (2010) describe the need for rapid identification and acquisition of RS products immediately following a disaster. The Coastal Response Research Center, a partnership between the University of New Hampshire (UNH) and the National Oceanic and Atmospheric Administration (NOAA) Office of

Response and Restoration (ORR) have created the Environmental Response Management Application (ERMA®) is a web-based GIS tool designed to assist both emergency responders and environmental resource managers who deal with incidents that may adversely impact the environment (ERMA). The Southeast Region Research Initiative (SERRI) of the Department of Homeland Security (DHS) facilitated and funded research at Mississippi State University designed to integrate geospatial technologies for the rapid characterization of potential damage from land-falling hurricanes. A variety of geospatial models have been developed using damage data collected following hurricane Katrina that estimate forest damage, damage to infrastructure, and damage extent and severity. The damage extent and severity models are available within 24 hours of hurricane landfall and require three model factors: estimates of surge, wind, and rain. All models have been reviewed by a State advisory committee that comprises the Mississippi Emergency Management Agency (MEMA), the Mississippi Department of Environmental Quality (MDEQ) and the Mississippi Forestry Commission (MFC) and are currently being transitioned to the State agencies.

2.4 Delineating potential damage from Hurricane Katrina: An example of geospatial applications to disaster management

Hurricane Katrina provided an unparalleled opportunity to study a variety of damage conditions from hurricane force wind and rain over a wide area and range of landscape conditions. According to a National Hurricane Center (NHC) (Knabb et al., 2005) report, Hurricane Katrina was likely at Saffir-Simpson Category 4 storm a couple of hours before making landfall. Comparisons have been made between Hurricane Camille (1969) and Hurricane Katrina (2005). Fritz et al., (2007) mention that Hurricane Camille made more powerful landfall, but was considerably smaller, with hurricane wind gusts reaching only 100 km to the east of the eyewall while Katrina hurricane force winds extended 140 km to the east of the center. The resulting Katrina storm surge extent was also much wider and likely exceeded the 6.9 m high water mark recorded in the aftermath of hurricane Camille. The heaviest rainfall occurred in southeast Lousisana with over 30 reporting stations in seven states recording rainfall in excess of 15.24 cm (Graumann et al., 2005).

According to NOAA (NOAA, 1999), accurate measurements of surge height, wind speed, and rainfall totals depend on a combination of direct measurement and indirect (observational) methods. Direct measurements are obtained primarily from reconnaissance aircraft, ships, and buoys and Automated Surface Observation Systems (ASOS) once a hurricane is near and/or on land. Indirect observational methods include satellite imagery and Doppler radar. In particular satellites have greatly improved the ability to monitor and understand hurricanes. Radar data are important after landfall for forecasting hurricane-related weather. Identification of direct and indirect measurement data useful for delineating potential damage from hurricanes based on information collected for Hurricane Katrina was a primary goal of the SERRI project mentioned in section 2.3.

2.5 Damage extent and severity model

Hurricane damage potential concepts were reviewed by researchers at the Institute for Clean Energy Technology (ICET), the Forest Products Laboratory (FPL) the Geosystems Research Institute (GRI), the Geosciences Department, the Signal Research Center at the University of Southern Mississippi (USM) and the Savannah River National Laboratory (SRNL). Identification of specific measurements used as input variables for modeling damage potential was primarily a collaborative effort among researchers at ICET, GRI, and SRNL. The modeling system that was developed by the research team uses readily available, easily obtainable, and GIS-suitable data to provide map products within 24 hours that are useful for hurricane emergency response.

The research staff in the Department of Geosciences at Mississippi State University includes the State Meteorologist, experts in satellite and radar meteorology, synoptic meteorology, remote sensing, and geographic information systems and extensive computing and simulation resources. The GIS-based model enables rapid depiction of a continuum of damage severity probabilities across large landscape areas. The model input factors (surge, wind, and rain) weights can be modified by the user, and the continuous output data histogram can be categorized using a variety of methods to construct damage severity output classes that support multiple emergency management needs.

Figure 2 shows the continuous output grid from the 'Damage Extent and Severity Model' (DESM). The output grid is based on a simple additive weighted GIS model that integrates surge predictions from the Sea, Lake and Overland Surges from Hurricanes (SLOSH) model with H*Wind estimates from the National Hurricane Center (NHC), and Multi-Sensor Precipitation Estimate (MPE) data from the National Weather Service (NWS). The DESM output product enhances the effectiveness of disaster management by providing a rapid (within 24 hours) depiction of areas where damage is likely, and provides users the opportunity to categorize the output into three or five levels of potential damage. In addition, the user can specify a combination of surge, wind, and rain as model factors; or specify a reduced set of factors that include wind and rain only. Factor weightings can be modified over a restricted range to emphasize different hurricane features and a number of output formats are supported.

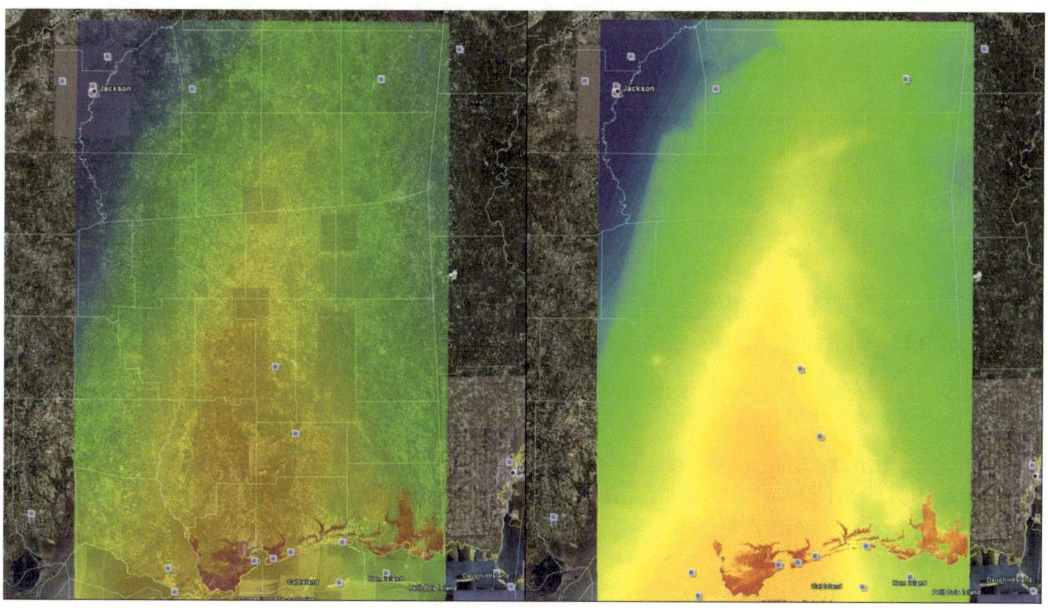

Fig. 2. DESM model output as a KML output file overlaid on Google Earth™

2.6 Factors

Surge

Storm surge is frequently one of the most damaging features of a hurricane. As mentioned previously, The SLOSH model provides widely used information on the potential for flooding both at the coast and inland by computing water height over a geographical areas covered by a network of grid points (Jarvinen and Neumann, 1985). The National Hurricane Center (NHC) publishes data on probability of storm surge exceeding incremental heights above mean sea level from 2 to 25 feet. Model output is available beginning when the NHC issues a hurricane watch or warning for the continental U.S. and updates are available every six hours from the NHC (NOAA, 2010). The DESM model uses the Maximum of the Maximum Envelopes of Water (MEOW) designated as 'MOM' and modifies the data to fit a locally-accurate map projection (Albers Equal Area). Then it calculates the difference between the maximum potential surge height and the terrain height maintained as a 10 m cell-size resolution from a Digital Elevation Model (DEM). The resulting 10 m output grid that represents potential surge height above the terrain (Figure 3).

Fig. 3. DESM model output illustrating area of surge using height above terrain as the 'surge' modeled variable

Wind

Wind is the primary factor responsible for damage outside the limits of the surge. Powell and Houston (1996) describe the importance of wind as a damaging force in tropical cyclone events. Their study emphasizes the importance of maximum sustained wind speed and using damage data from Hurricane Andrew, graphically illustrate that a hyperbolic tangent model has the desired characteristics of an exponential increase of damage followed by a leveling out at the most extreme wind speeds. Burpee et al., (1993) describe the basic process used by the NHC to combine airborne and land-based radar reflectivity and

Doppler velocities used to estimate surface winds (H*Wind) products that are available from the NHC Hurricane Research Division (HRD) in real-time prior to and continuing following hurricane landfall. Following landfall the HRD seeks supplemental wind observations and documentation of instruments and site exposures (Powell et al., 1998). The DESM model uses downloaded 3-hour interval GIS-formatted H*Wind files for the wind factor. The downloaded data are reprojected to a locally-accurate map projection and interpolated. Due to the density of the 2-degree latitude-longitude point grid associated with the surface wind estimates, a simple Inverse Distance Weighting interpolator is implemented to create a 'down-sampled' 10 m output grid.

Rain

Rain is a contributing factor to damage from hurricanes, particularly with respect to wetting of soils that can interact with wind resulting in tree blowdown. Blowdown is associated with both direct and indirect damage to infrastructure and the built environment. In addition, roof damage from wind and subsequent rainfall can result in considerable damage to built structures. Next Generation Weather Radar (NEXRAD) Multisensor Precipitation Estimate (MPE) data are readily available from the National Weather Service (NWS) and are used for rainfall estimates in DESM. The MPE rainfall estimates are generated by combining NEXRAD radar-derived precipitation fields with hourly surface-based observations from the hydrometeorological data system (HADS), leading to a nominal 4x4km dataset with hourly temporal resolution. Wang et al., (2008) examined the performance of NEXRAD MPE and observed that there are limitations in the accuracy of radar estimates when compared to gauge networks, but also observed that the radar-based products do a better job of spatial representation of rainfall. The geospatial nature of GIS-based modeling using factors that provide good spatial representation of rainfall amounts was the deciding factor in the choice of NEXRAD MPE data for the rain component of DESM. The data are downloaded, reprojected, and interpolated to create a 10 m output grid.

2.7 DESM model

The DESM model is a Simple Additive Weighted (SAW), commonly implemented in the GIS raster environment (Malczewski, 1999). It is particularly suited to modifications by the user and is the most commonly used technique for multiattribute decision making. Malczewski (1999) describes the process whereby the decision maker directly assigns weights of relative importance to each attribute (factor) and a total score is obtained by multiplying the importance weight assigned for each factor by the scaled (standardized) value given to each factor at each grid cell location and summed for each factor. Assignment of weights was determined heuristically and weights were normalized according to the following equation:

$$W_j = \frac{n - r_j + 1}{\sum(n - r_k + 1)}$$

Where:

w_j is normalized weight for the jth criterion

n is the number of criteria under consideration (k = 1,2,...n)

r_j is the rank position of the criterion

A number of linear scale transformations were considered and for DESM modeling functions the 'maximum score' standardization technique was used. The 'maximum score' technique simple uses the maximum value in a data range as the divisor for all values and insures on a range of 0 to 1 that each factor's original intervallic relationships are maintained following rescaling. It should be noted that in areas beyond the limits of surge inundation, that values associated with the surge grid default to 0.

As mentioned previously, model runs can be modified to meet the objectives of the users including modification of weights and modification of the continuous output histogram into 3 or 5 levels based on three partitioning criteria: Jenk's Natural Breaks, Quantile divisions, or Equal Interval divisions. Natural breaks as described by Jenks and Caspall (1971) is a form of variance-minimization classifications where an iterative process is used to define break values for a pre-determined number of classes *n*. The individual values in each class are systematically assigned to adjacent class by adjusting the class boundaries and a solution is reached when the total sum of squared deviations around each class mean is a minimum. Quantile intervals are simply determined so the number of observations (grid cells) in each interval is the same (de Smith et al., 2007). Equal Interval divisions are determined by dividing the range of values by the number of classes *n*. Each class interval has the same width, often called 'slice' when used with raster maps.

Grid processing for the DEMS model is accomplished within the GIS environment and is currently automated using the Arc Macro Language (AML) to control mathematical processes within the ArcInfo™ GRID program. Due to the number and size of grids required for modeling, database and processing steps are carried out in the Unix environment while the web interface programming and data formatting and display activities are carried out in the PC environment.

3. Conclusions

Model results have been presented to the State advisory team and the model results have been corroborated by various agency representatives. The models compared well with Mississippi Forestry Commission post-hurricane data acquired for a selected damage types including pine shear (breakage) and hardwood wind-throw (blowdown). Model usefulness to State agencies is further evidenced by letters of support and continued SERRI/DHS funding aimed at transitioning the model to the MEMA operations center.

The research team believes that DESM can lead to better disaster preparedness and disaster recovery. Prior to landfall, DESM output can be overlain with existing residential geospatial datasets, and notices of evacuations can be targeted or monitored for specific locations. Officials may also be able to acquire pre-landfall lists of residents and property owners, which may help determine survivorship. Additionally, overlaying DESM with a road network layer can help managers predict which roads to use or not to use in an evacuation and where to set up disaster management headquarters. After landfall, a DESM reanalysis can provide decision support for rescue operations and to determine areas in need of reconnaissance.

Thus, the DESM offers emergency managers a quick, readily available, and inexpensive means to prepare, plan, and recover from land-falling hurricanes. The success of geospatial models such as this illustrates how Geospatial Techniques have the potential to mitigate damage and ultimately save lives.

4. References

Burpee, R. W., Aberson, S. D., Black, P. G., Demaria, M., Franklin, J. L., Griffin, J. S., (1993). Real-time guidance provided by NOAA's Hurricane Research Division to forecasters during Emily of 1993. *Bulletin of the American Meteorological Society, 75*(10), 1765-1784.

Chandrasekar, N., Immanuel, J. L., Sahayam, J. D., Rajamanickam, M., & Saravanan, S. (2007). Appraisal of tsunami inundation and run-up along the coast of Kanyakumari district, India - GIS analysis. *Oceanologia, 49*(3), 397-412.

Clarke, K. C. (2003). *Getting started with Geographic Information Systems*. Upper Saddle River, New Jersey: Prentice Hall.

Cutter, S. L. (2003). GI science, disasters, and emergency management. *Transactions in GIS, 7*(4), 8p.

ERMA. Environmental Response Management Application. Coastal Response Research Center at the University of New Hampshire. Available at: http://www.crrc.unh.edu/erma/

Fritz, H. M., Blount, C., Sokoloski, R., Singleton, J., Fuggle, A., McAdoo, B. G., et al. (2007). Hurricane Katrina storm surge distribution and field observations on the Mississippi Barrier Islands. *Estuarine Coastal and Shelf Science, 74*(1-2), 12-20.

Graumann, A., Houston, T., Lawrimore, J., Levinson, D., Lott, N., McCown, S., et al. (2005). Hurricane Katrina, a climatological perspective. *Technical Report 2005-01*. Ashville, NC.: NOAA National Climatic Data Center.

Hodgson, M. E., Davis, B. A., Cheng, Y., & Miller, J. (2010). Modeling remote sensing satellite collection opportunity likelihood for hurricane disaster response. *Cartography and Geographic Information Science, 37*(1), 7-15.

HURREVAC (2010). Hurrevac2010 user's manual. (pp. 116): FEMA, US Army Corps of Engineers, NOAA/NWS. Available at: http://www.hurrevac.com/documents/Hurrevac2010_Users_Manual.pdf

Jarvinen, B. J., & Neumann, C. J. (1985). An evaluation of the SLOSH storm surge model. American Meteorological Society Bulletin, 66(11), 1408-1411.

Jenks, G.F. & Caspall F. C. (1971) Error on choroplethic maps: definition, measurement, reduction. Annals of the Association of American Geographers, 61(2), 217-244

Keating, B., Whelan, F., & Brock, J. B. (2004). Tsunami deposits at Queen's beach, Hawaii-initial results and wave modeling. *Science of Tsunami Hazards, 22*(1), 23-43.

Knabb, R. D., Rhome, J. R., & Brown, D. P. (2005). Tropical cyclone report, Hurricane Katrina, 23-30, August 2005. National Hurricane Center. Available at: http://www.nhc.noaa.gov/2005atlan.shtml

Lessons Learned Final Report (2008). Capturing Hurricane Katrina data for analysis and lessons-learned research, Final Report. (pp. 155). Mississippi State: Mississippi State University. Available at: http://www.gri.msstate.edu/research/katrinalessons/Documents/Final_report_BC_New1.pdf

Lillesand, T. M., Kiefer, R. W., & Chipman, J. W. (2008). Remote sensing and imageinterpretation. New York: John Wiley & Sons, Inc.

Malczewski, J. (1999). *GIS and multicriteria decision analysis*. New York, NY: John Wiley & Sons.

NOAA (1999). Hurricane basics. National Oceanic and Atmospheric Administration, National Hurricane Center, National Weather Service. Available at: http://hurricanes.noaa.gov/pdf/hurricanebook.pdf

NOAA (2010). Tropical cyclone storm surge probabilities. National Oceanic and Atmospheric Administration, National Weather Service, National Hurricane Center. Available at: http://www.nhc.noaa.gov/aboutpsurge.shtml.

Papathoma, M., Dominey-Howes, D., Zong, Y., & Smith, D. (2003). Assessing tsunami vulnerability, an example from Herakleio, Crete. *Natural Hazards and Earth System Sciences, 3*(5), 377-389.

Powell, M. D., & Houston, S. H. (1996). Hurricane Andrew's landfall in south Florida. Part ii: Surface wind fields and potential real-time applications. *Wea. Forecasting, 11*, 329-349.

Powell, M. D., Houston, S. H., Amat, L. R., & Morisseau-Leroy, N. (1998). The HRD real-time hurricane wind analysis system. *Journal of Wind Engineering and Industrial Aerodynamics, 77&78*, 53-64.

De Smith, M. J., Goodchild, M. F., & Longley, P.A. (2007) Geospatial Analysis: A comprehensive guide to principles, techniques, and software tools (2nd Edition). Leicester, UK.. Matador Publishing Ltd.

Theilen-Willige, B. (2010). Detection of local site conditions influencing earthquake shaking and secondary effects in Southwest-Haiti using remote sensing and GIS-methods. *Natural Hazards and Earth System Sciences, 10*(6), 1183-1196.

Theobald, D. M. (2001). Topology revisited: Representing spatial relations. *International Journal of Geographical Science, 15*(8), 689-705.

Tran, P., Weireter, L., Sokolowski, W., Lawsure, K., & Sokolowski, J. (2009). HAZUS modeling for hurricane effect on a healthcare campus: Implications for health care planning. *American Surgeon, 75*(11), 1059-1064.

Vickery, P. J., Skerlj, J. L., Lin, J., Twisdale, L. A., Young, M. A., & Lavelle, F. M. (2006). HAZUS-MH hurricane model methodology. Ii: Damage and loss estimation. *Natural Hazards Review, 7*(2), 94-103.

Walsh, T. J., Caruthers, C. G., Heinitz, A. C., Myers, E. P., Baptista, A. M., Erdakos, G. B., et al. (2000). Tsunami hazard map of the southern Washington coast: Modeled tsunami inundation from a Cascadian Subduction Zone earthquake. *Washington Div. Geol. Earth Res. Rep, GM-49*, 12.

Wang, X. W., Xie, H. J., Sharif, H., & Zeitler, J. (2008). Validating NEXRAD MPE and stage iii precipitation products for uniform rainfall on the Upper Guadalupe River Basin of the Texas Hill Country. *Journal of Hydrology, 348*(1-2), 73-86.

Zhang, K. Q., Xiao, C. Y., & She, J. (2008). Comparison of the CEST and SLOSH models for storm surge flooding. *Journal of Coastal Research, 24*(2), 489-+.

20

Storm Surge: Physical Processes and an Impact Scale

Hal Needham and Barry D. Keim
Department of Geography and Anthropology Louisiana State University U.S.A.

1. Introduction

Tropical cyclone-generated storm surges create natural disasters that are among the most deadly and costly global catastrophes. Individual disasters have inflicted hundreds of thousands of fatalities and billions of dollars in damage. In 1970, a tropical cyclone in the Bay of Bengal generated a 9.1-meter surge which killed approximately 300,000 people in Bangladesh (Frank and Husain 1971; Dube et al. 1997; De et al. 2005). More recently, and well into the age of satellite meteorology, a storm surge in 1991 killed approximately 140,000 people in Bangladesh (Dube et al. 1997). Although the magnitude of storm surge heights and loss of life are highest along the shores of Bangladesh and India, such disasters are not limited to countries with developing economies. The 1900 Galveston Hurricane generated a 6.1-meter surge (Garriott 1900), which killed between 6,000 and 8,000 people in Galveston, Texas (Rappaport and Fernandez-Partagas 1995), producing the most deadly natural disaster in United States history (National Oceanic and Atmospheric Administration 1999). More recently, Hurricane Katrina (2005) generated an 8.47-meter surge (Knabb et al. 2006), which claimed more than 1,800 lives along the coasts of Louisiana and Mississippi, and inflicted $81 billion dollars in damage (McTaggart-Cowan et al. 2008).

While tropical cyclone-generated storm surge is a deadly and costly hazard, it is also scientifically complex, because meteorological, oceanographic and geographic factors influence the height, extent and duration of storm surge flooding. Such factors include maximum sustained hurricane wind speed at landfall and offshore, hurricane size, hurricane forward speed, the angle of hurricane approach to the coastline, bathymetry of coastal waters, coastline shape and the presence of barriers or obstructions to surge waters on land. Although relationships between some of these factors and resultant surge heights may seem intuitive (e.g. the assumption that stronger hurricanes always generate higher surges), storm surge observations reveal that a combination of physical factors influence surge characteristics. As a result, weaker hurricanes sometimes generate higher surges than stronger hurricanes. For example, Hurricane Ike (2008) generated a 5.33-meter surge along the Upper Texas Coast, although maximum sustained winds at landfall were only 175 km/ hour (Berg 2009), whereas Hurricane Charley in Western Florida had maximum sustained winds of 240 km/ hour at landfall, but only generated a 2.13-meter surge, partly because the storm rapidly intensified just before making landfall (Pasch et al. 2004).

The complex nature of storm surge makes this phenomenon difficult to forecast and difficult for coastal populations to understand. In cases where storm surge heights and extents are accurately forecast, coastal populations still commonly misunderstand storm surge impacts on coastal buildings, marine infrastructure, utilities, and transportation infrastructure, such as coastal roads and bridges. This chapter addresses these issues by discussing the physical processes that generate storm surges, while providing some insight into impacts in low-lying coastal communities along the United States' Gulf of Mexico Coast. Such information is likely helpful to professionals in the fields of emergency management and law enforcement, planning, insurance, construction, urban planning, health care, science and engineering, as well as to coastal populations who live in regions vulnerable to tropical cyclone-generated storm surges.

2. Physical processes that generate storm surge

A storm surge is defined as a dome of increased sea level height in association with the approach of an intense cyclone. Persistent onshore winds, in conjunction with reduced air pressure, force water levels to rise, especially in shallow water of impacted coasts. Although tropical or extratropical cyclones can theoretically produce storm surges in any ocean basin, tropical cyclones produce the most destructive surges in terms of extreme water-level heights and human impacts.

An understanding of tropical cyclone-generated storm surge climatology requires a brief examination of the physical processes that shape tropical cyclone development. These parameters include sea surface temperatures exceeding 26 degrees C (Ali 1996; Holland 1997; Gray 1998), and proximity to the Intertropical Convergence Zone (ITCZ), or its regional manifestation, such as the South Pacific Convergence Zone (SPCZ) (De Scally 2008). Proximity to this global band of low pressure is necessary to provide the atmospheric lift for cyclonic development (Ali 1996; Dube et al. 1997; De Scally 2008). Furthermore, tropical cyclones tend to form between 5°-25° of latitude because at these latitudes the Coriolis parameter is sufficient to generate the dynamic potential of an area of convection (Gray 1998). As the Coriolis parameter is a function of latitude, with no Coriolis forcing at the equator and higher forcing found at higher latitudes, tropical cyclone development is generally suppressed near the Equator (Gray 1975). Although the Coriolis parameter is more favorable for cyclonic development in higher latitudes where forcing is greatest, cooler ocean waters and detachment from the ITCZ in latitudes poleward of approximately 25 degrees latitude suppress the development of tropical cyclones. Figure 1 and Table 1 reveal the most favorable regions for tropical cyclogenesis, and the average annual number of tropical cyclones observed in each ocean basin (Landsea 2007).

Once a tropical cyclone has developed, atmospheric circulation will usually steer the cyclone from the point of origin, sometimes transporting the disturbance thousands of kilometers from the point of origin. The atmospheric circulation of the specific ocean basin will govern the exact path of the storm. Typical circulation patterns exist for each basin, depending on several physical factors. For example, the clockwise circulation around the Bermuda and Hawaiian Highs, the dominant mid-oceanic features in the North Atlantic and North Pacific Oceans, respectively, typically produce easterly trade winds between 10 and 25 degrees latitude, the region of highest tropical cyclogenesis. This typical circulation pattern causes most tropical cyclones in these basins to initially travel westward in the easterly trade winds, before making a northerly turn as they approach North America or East Asia. Figure 2 depicts this pattern with the 270 August tropical cyclone tracks that took place in the north Atlantic from 1886 - 2006.

Spatial patterns of tropical cyclone landfalls develop as the processes of tropical cyclone development and movement are considered within the context of physical geography. Exposed islands, peninsulas and capes often experience the highest landfall frequencies. Coasts that curve in a convex manner, such as the North Carolina coast, along the eastern seaboard of the United States, protrude from the continent and experience higher frequencies of tropical cyclone strikes than concave-bending coasts, such as the Georgia coast, 650 kilometers to the southwest (Keim et al. 2007).

As tropical cyclones approach a coastline, several additional factors determine the characteristics of the resultant storm surges. These factors include maximum sustained cyclonic winds at landfall and offshore, minimum central pressure, cyclone size, forward movement and angle of approach to the coast, shape and bathymetry of the coastline, and presence of obstructions to flowing water.

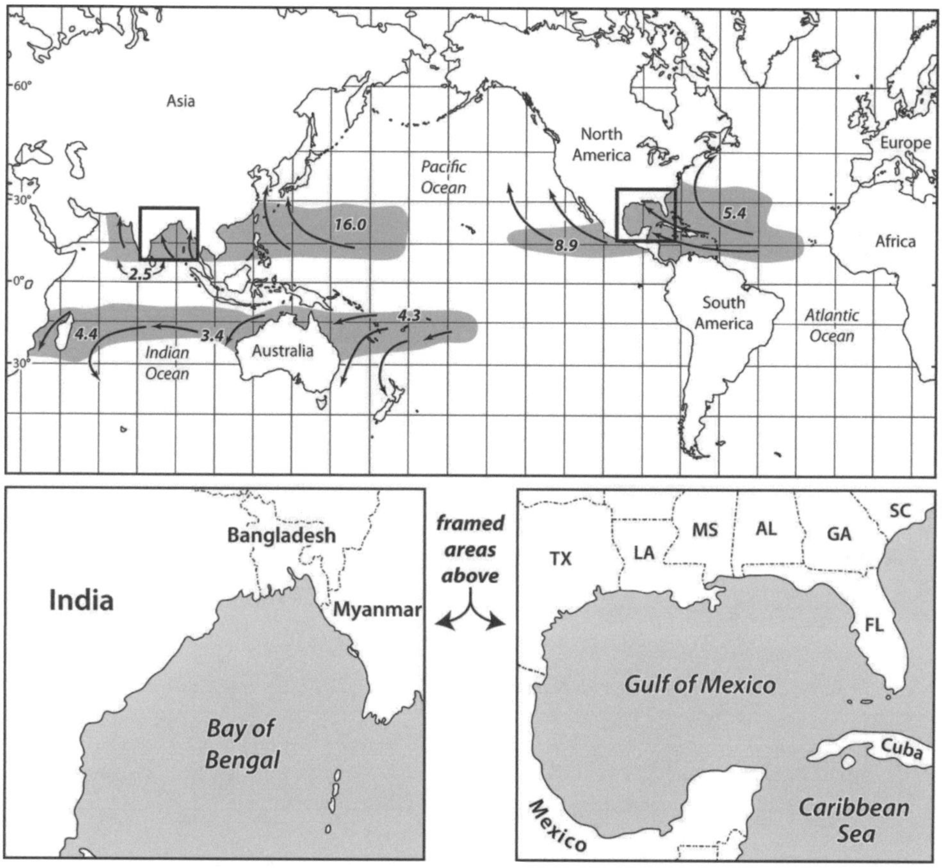

Fig. 1. Regions of global tropical cyclogenesis, and the average annual number of hurricane-strength systems per basin.

Wind stress is the predominant factor forcing storm surge and destructive waves generated by hurricanes (Harris 1963). A quantitative relationship exists between wind speed and water heights, as wind stress exerts a force on water that increases exponentially as wind speeds increase (Ali 1996). Kurian et al. (2009) state that wind stress accounts for 80-85% of

	Tropical storm or stronger (sustained winds greater than 17 m/s)		Hurricane/ Typhoon (sustained winds greater than 33 m/s)	
Basin	Average	Percentage	Average	Percentage
Atlantic	9.7	11.6	5.4	12.0
NE Pacific	16.5	19.7	8.9	19.8
NW Pacific	25.7	30.7	16.0	35.6
N Indian	5.4	6.5	2.5	5.6
SW Indian	10.4	12.4	4.4	9.8
Aus SE Indian	6.9	8.2	3.4	7.6
Aus SW Pacific	9.0	10.8	4.3	9.6
Globally	83.7	100	44.9	100

Table 1. Average annual number of tropical cyclones that develop in selected ocean basins, from Landsea (2007).

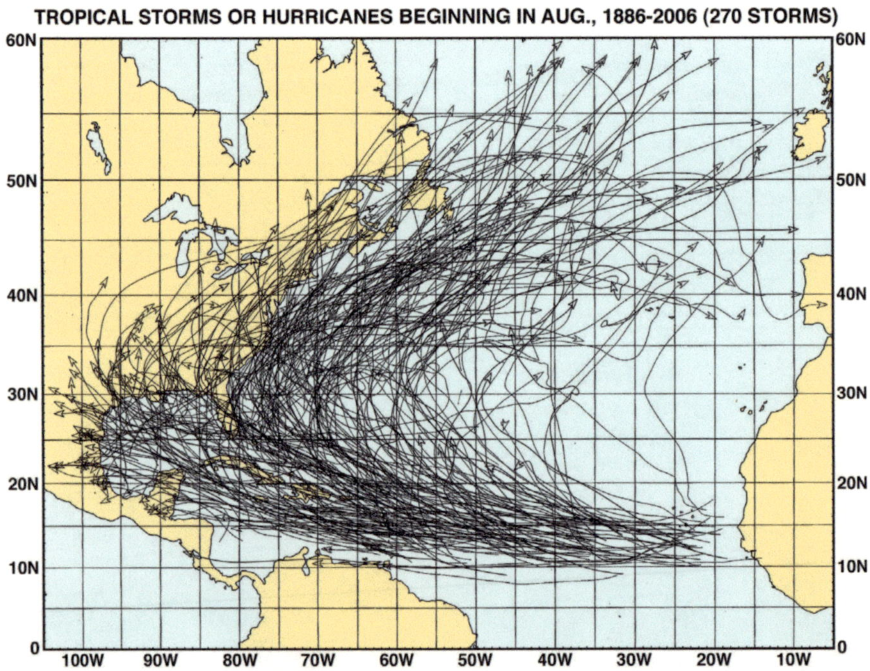

Fig. 2. August tracks of North Atlantic Basin hurricanes, 1886-2006, from Keim and Muller (2008), adapted from Neumann et al. (1993).

the generated surge, however, as recent research has discovered the importance of other cyclonic variables on storm surge heights, some scientists may argue that these figures provided by Kurian et al. (2009) overestimate the influence of wind speed.

It should be noted that pre-landfall wind speeds likely influence maximum storm surge heights to a greater extent than wind speeds at landfall, especially for hurricanes that rapidly strengthen or weaken as they approach the coast. Jordan and Clayson (2008) discovered that a scaled 24-hour, pre-landfall intensity average and a 12-hour pre-landfall instantaneous wind speed seem to be the best predictors of storm surge height. This may seem counterintuitive, as many meteorologists, as well as the media, focus attention on predicted wind speeds at landfall. However, the transfer of momentum from the atmosphere to the water, as strong winds blow over the water surface, requires time, therefore enabling hurricanes that rapidly weaken (strengthen) as they approach the coast to often generate higher (lower) surges than anticipated. Hurricanes Katrina and Wilma (2005) exemplify this phenomenon. The National Hurricane Center downgraded Katrina's surge forecast as the storm rapidly weakened while approaching the Mississippi Coast, but upgraded Hurricane Wilma's surge forecast as the hurricane rapidly intensified before making landfall in Southwest Florida. In hindsight, both of these changes increased forecasting errors, as the surge levels predicted using offshore intensities were more accurate than the forecasts that accounted for sudden intensity changes before landfall (Jordan and Clayson 2008).

An inverse relationship exists between air pressure and storm surge heights. This relationship is sometimes called the inverted barometer effect, or law of the inverted barometer (Welander 1961). As barometers measure the amount of atmospheric pressure exerting downward force on a given location at Earth's surface, reduced barometric readings mean less atmosphere is pressing down from above, thus, enabling sea levels to rise slightly. The sea level rises approximately one centimeter for every millibar of air pressure reduction (Welander 1961). This indicates that differences in air pressure account for a relatively small component of storm surge rise. For example, consider Hurricane Camille, an unusually intense hurricane that struck the Mississippi Coast in the United States Gulf of Mexico in 1969, produced one of the lowest air pressure observations ever recorded in the United States. The minimum central pressure in this hurricane was 905 millibars (Simpson et al. 1970), or approximately 108 millibars less than the atmospheric average pressure of 1013 millibars (National Aeronautics and Space Administration 2010). The inverted barometer effect indicates that reduced air pressure in this cyclone accounted for only 1.08 meters in storm surge rise, or about 14 percent of the maximum storm surge height of 7.5 meters at Pass Christian, Mississippi (Simpson et al. 1970).

Cyclone size is an important factor that has been underestimated until recently, after very large hurricanes Katrina (2005) and Ike (2008) generated larger surges than anticipated (Irish et al. 2008). The relationship between storm size and surge levels has likely been underestimated because research conducted on storms from the 1950s through 1970s lacked examples of massive storms like Katrina (Irish et al. 2008). Hurricane Katrina's size likely enabled the storm, a category-3 storm on the Saffir-Simpson Scale at landfall, to generate larger surge heights than category-5 Camille along the same coastline. Katrina's Category-5 winds, one day before landfall, enabled the storm to generate such a massive surge. It is believed that the massive size of Katrina then enabled it to maintain this enormous surge even as it weakened before striking the coastline. As such, both Hurricanes Charley (2004) and Katrina (2005) also indicate that the lifecycle of hurricanes before landfall can play some role in producing surge. For example, Hurricane Charley's rapid intensification before landfall did not allow for a large build-up of surge over time like Hurricane Katrina, which weakened before landfall.

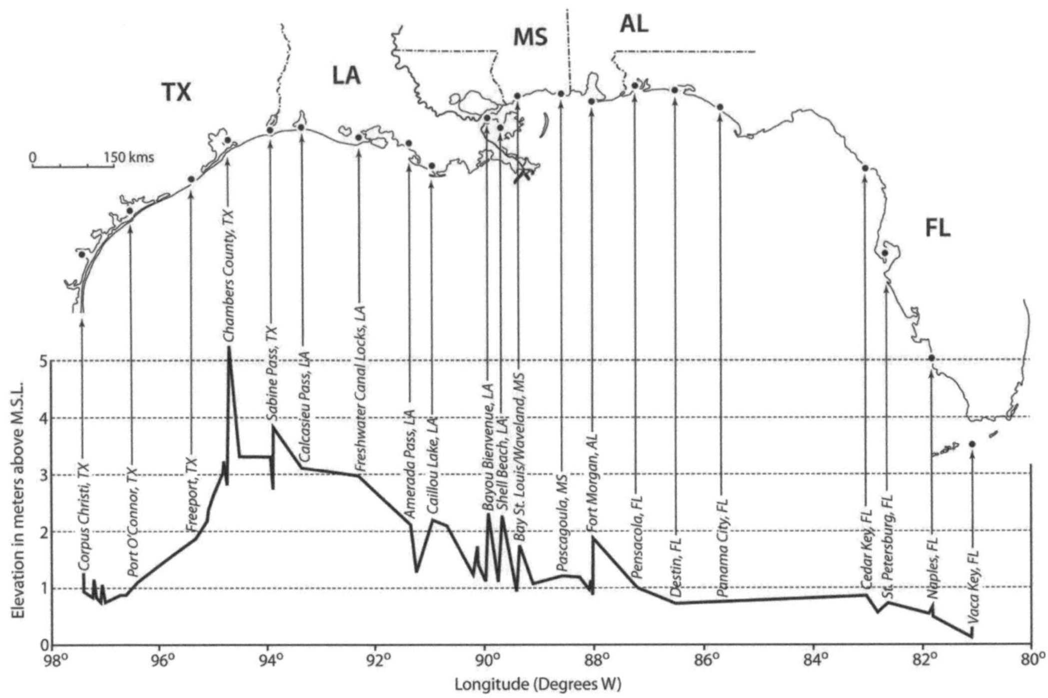

Fig. 3. High water profile for Hurricane Ike in the Gulf of Mexico, created from selected storm surge observations provided by Berg (2009).

In addition to producing higher surges, larger cyclones also inundate more coastline. In a modeling experiment, Irish et al. (2009) discovered that a 10 km increase in the hurricane pressure radius, a measure of hurricane size, produced a 20-100% increase in the extent of coastline inundated by high surge, at least 2 meters high in this particular experiment. This reasoning partially explains the reason why Hurricane Ike (2008) inundated more than 350 km of Texas and Louisiana Coastline with at least 2 meters of storm surge (See Figure 3).

Forward cyclonic movement is a factor often overlooked in surge research (Rego and Li 2009). Weisberg and Zheng (2006) resolved that slower-moving cyclones generate higher surges than faster-moving cyclones, because slower-moving storms have more time to redistribute water. It is also reasonable that slower-moving storms would elevate water levels for longer time periods, therefore, increasing the amount of time that coastal communities are susceptible to damage from surge and waves.

Coastline shape and the presence of natural and artificial obstructions to flowing water, such as coastal forests and levees, have profound effects on storm surge, sometimes producing extreme localized water level differences (Jarvinen and Neumann 2005). The highest surges usually occur where water approaches the coast in a perpendicular direction, especially when funneled into inlets and bays. The Advanced CIRCulation Model for Coastal Open Hydrodynamics (ADCIRC) of Hurricane Katrina's surge clearly depicts these localized differences along the Mississippi River south of New Orleans (Interagency Performance Evaluation Taskforce Report 2006). Both natural and artificial levees along the river inhibited surge from flowing across the river from east to west, therefore enhancing surge levels on the east side of the river. In some locations surge levels on the east side of the river were more than three meters higher than the west side, less than two kilometers away

(Figure 4). Storm surge levels were also enhanced along the Mississippi River Gulf Outlet Reach 2 (MRGO Reach 2), as strong winds blew from the northeast, nearly perpendicular to the levees in this section (Ebersole et al. 2010).

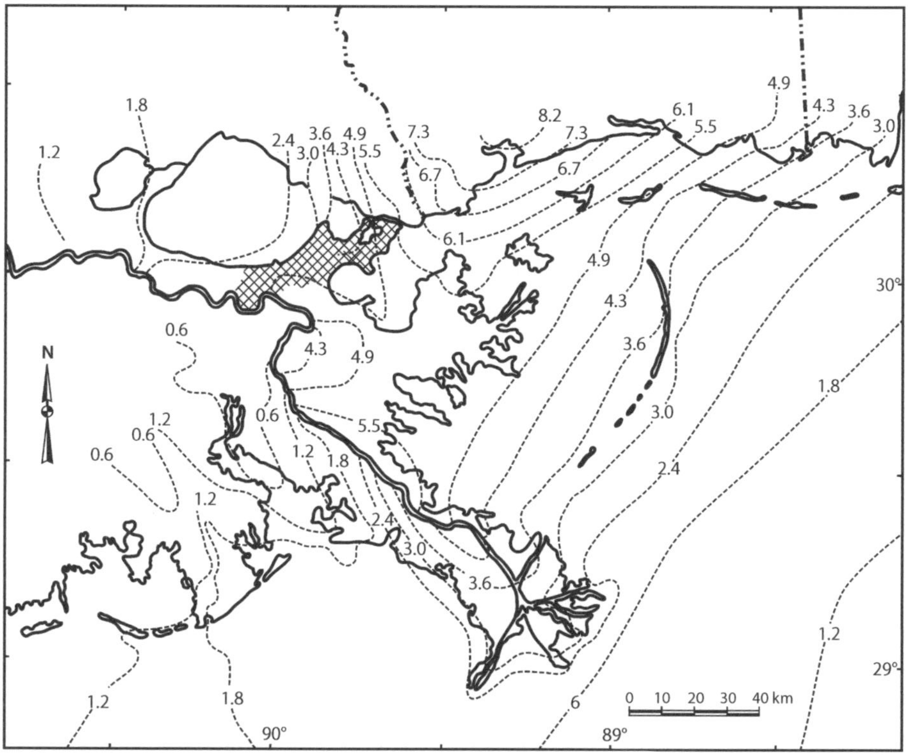

Fig. 4. The Advanced CIRCulation Model for Coastal Open Hydrodynamics (ADCIRC) map of maximum storm surge generated by Hurricane Katrina along the Louisiana and Mississippi coasts in the Gulf of Mexico. Original source is Interagency Performance Evaluation Taskforce (IPET) Report, 2006, although this graphic is adapted from Keim and Muller (2009).

Notes: 1. Water levels in this figure are given in meters above NAVD88. 2. The ADCIRC surge model reveals dramatic localized difference in storm surge along the Mississippi River levee, south of New Orleans. Surge levels on the east side of the river are more than 3 meters higher than the west side of the river in some locations.

Shallow bathymetry enhances storm surge because deeper water currents cannot carry away excess water, and as a result, water accumulates in shallow areas (Rappaport and Fernandez-Partagas 1995). Bays and gulfs, especially associated with large river deltas, therefore, generally experience larger surges than shorelines adjacent to open ocean or steeper continental shelves. For example, the extremely shallow waters near the mouth of the Mississippi River, created by extensive silt deposits, enhance the surge levels of hurricanes that track near the delta. These shallow waters enhanced the surge heights of Hurricanes Camille (1969) and Katrina (2005), helping these cyclones generate the two largest storm surges in United States history. Pass Christian, Mississippi, on the eastern shore of Bay St. Louis, observed the location of peak storm surge in both of these events.

Chen et al. (2008) hypothesize that the shallow waters south of the Mississippi Coast, produced by deltaic lobes from the Mississippi River, were the primary reason that Hurricane Katrina's surge was extraordinarily high, and that the surge may have been four meters lower had Katrina tracked over a wider and more steeply sloping continental shelf, similar to the bathymetry south of Mobile Bay, Alabama (Figure 5).

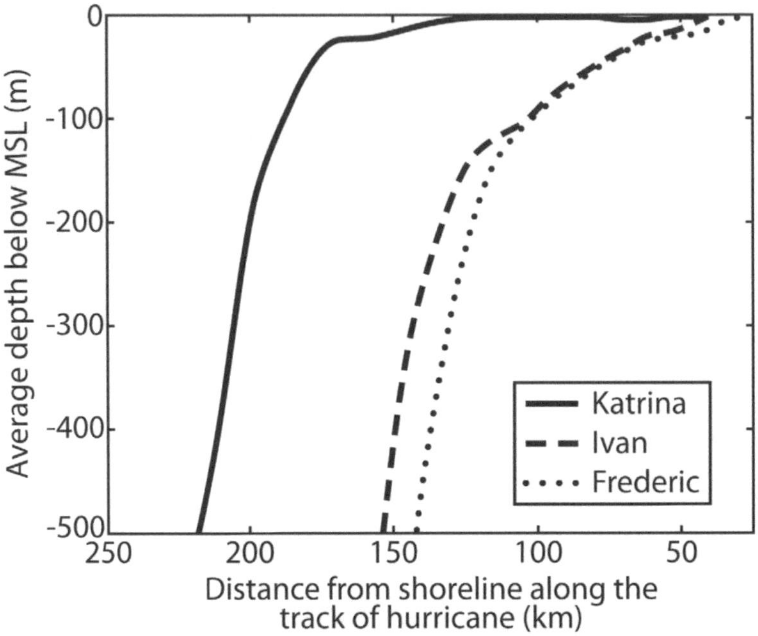

Fig. 5. Average bathymetry, or water depth, located 90 km east of the tracks of Hurricanes Katrina, Ivan and Frederic in the Gulf of Mexico. The bottom profiles indicate that the onshore winds of Hurricane Katrina blew over much shallower water than Ivan and Frederic, as the storm approached shore. From Chen et al. (2008).

Storm surges sometimes inundate entire regions; widespread surge events can flood more than 1,000 kilometers of coastline (Berg 2009). Larger cyclones generally inundate more extensive areas than smaller cyclones with similar maximum wind velocities, while cyclones that track parallel or at an oblique angle to the coastline will usually flood longer stretches of shoreline than cyclones that make a perpendicular approach to the coast at landfall. The high water profiles depicted in Figures 3 and 7 reveal that Hurricane Ike inundated a much larger expanse of coastline than Hurricane Camille, even though Hurricane Ike made landfall as a Category-2 storm on the Saffir Simpson Scale, compared with Camille's extraordinary Category-5 sustained winds and 190-mile-per-hour gusts (National Weather Bureau 1969) along the Mississippi coast. Ike's much larger geographic size and oblique path relative to the coastline enabled the cyclone to inundate a wider swath of the coast.

Generally, the portion of coastline that observes onshore winds will usually experience the greatest surge levels, with water heights increasing closer to the path of the cyclone. In contrast, the region that observes offshore winds will often experience much lower surge levels. In the northern hemisphere, onshore winds are observed to the right of the path of the cyclone, while offshore winds are experienced to the left (Figure 6).

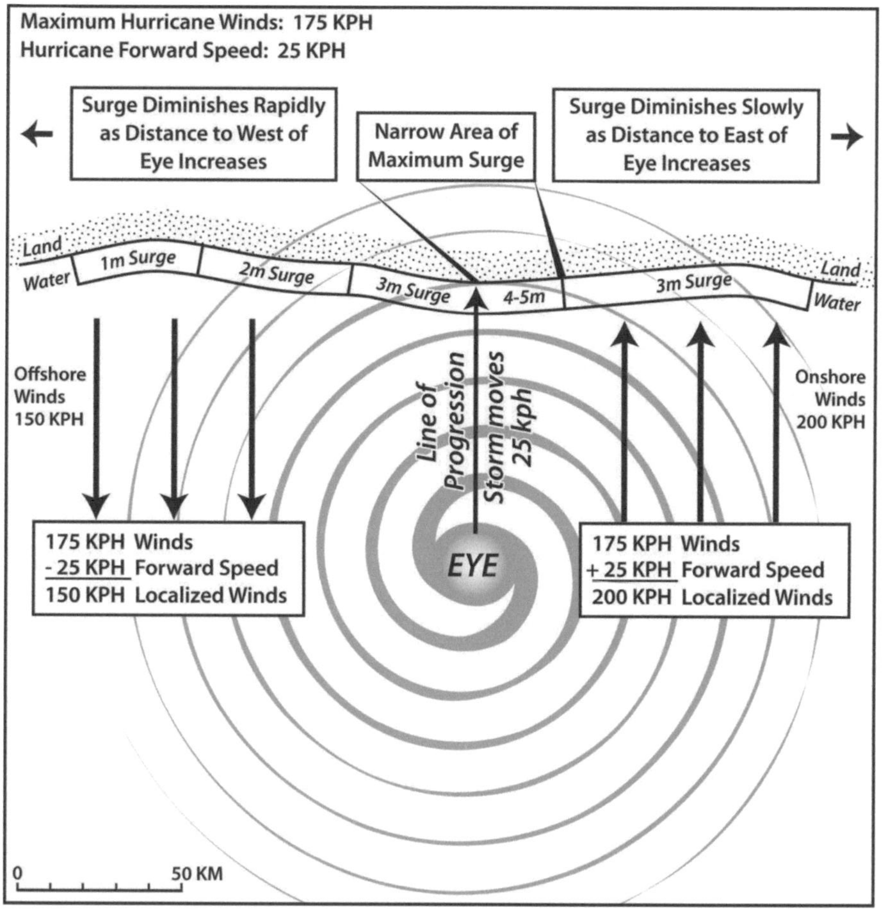

Fig. 6. Generalized schematic of storm surge heights relative to position, path and wind speeds of a tropical cyclone.

In certain cases, offshore winds create negative surges, in which winds push enough water away from land that water levels fall below normal levels. This phenomenon is usually experienced when offshore winds are observed on the leeward side of a peninsula, island or isthmus. These geographic features experience the majority of negative surges because they are wide enough to impede the surge that strikes the windward coastline, yet thin enough to not significantly weaken the cyclone. The west coast of Florida, for example, observes negative surges because cyclones commonly track from east to west across the peninsula, producing strong offshore winds to the north of the cyclonic path. Examples of this phenomenon include the negative surges of minus 2.01 meters at Tampa in October 1910 (Cline 1926), minus 1.83 meters at Tampa in September 1926 (Harris 1963) and minus 1.37 meters at Cedar Key in September 2004 (Lawrence and Cobb 2005). Sometimes positive surges immediately follow negative surges, such as the 1.07 foot positive surges that followed both the 1910 and 2004 negative surges at Tampa and Cedar Key, respectively (Cline 1926; Lawrence and Cobb 2005). Negative surges can damage property or infrastructure that are dependent upon the presence of water, such as marinas or industries that require water as a cooling agent. They can also endanger people, who are enticed by the novel opportunity to walk on the bed of a bay or river, unaware that a sudden change in wind direction could bury them under a torrent of rushing water. Bevan (2001), of the National Hurricane Center, reports that curious people did this very thing during Hurricane Keith (2000) in Belize.

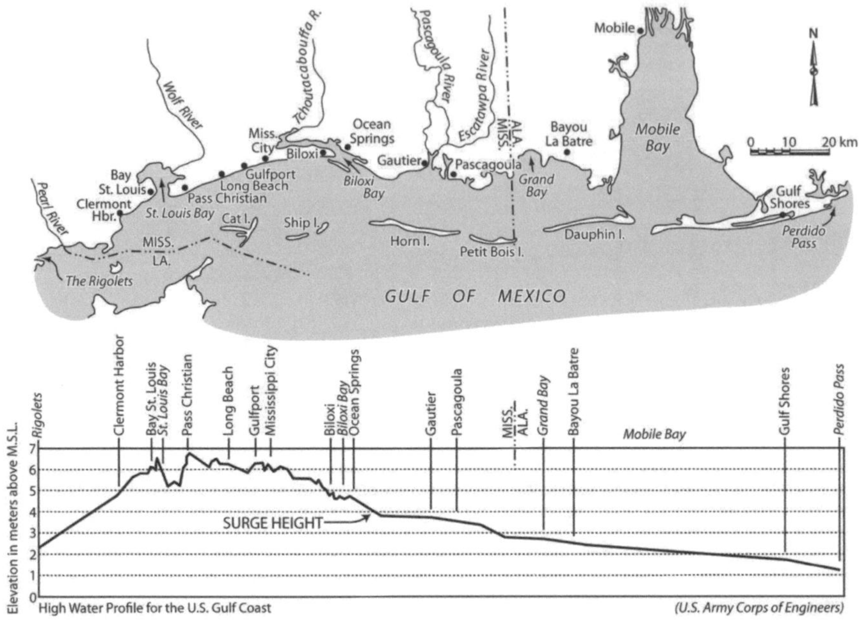

Fig. 7. High water profile for Hurricane Camille in the northern Gulf of Mexico, from U.S. Army Corps of Engineers (1970).

3. Storm surge impact scale for coastal communities

Scientific literature contains little information that describes the incrementally destructive nature of storm surge inundations in coastal communities as the magnitude of storm surges increase. In this regard, coastal communities lack critical information that could help them improve long-and short-term decisions in preparation for storm surge inundation. With regard to long-term decision-making, such information could improve coastal planning, thereby reducing both human and economic losses in surge events. Coastal planning would likely improve, as people understand the major, moderate and minor impacts of storm surge in their community, as well as the frequency with which those impacts are likely to occur. For example, beach erosion was a relatively minor impact of Hurricane Katrina along the Mississippi Coast, at least compared to the loss of human life and total economic losses. However, the Mississippi beaches are important for the local economy, as tourism along the Mississippi Gulf Coast generates $1.3 billion in revenue and employs around 27,500 people (www.gulfcoast.org). Beach restoration projects have attempted to restore the sand, and return beaches to pre-Katrina conditions; an effort aimed at enhancing the quality of life in this area and boosting the local economy.

However, communities along the Mississippi Coast will likely benefit from making informed decisions on such projects, which involve knowledge of storm surge impacts and frequencies. Relevant questions may address the critical storm surge level that erodes most of the sand and/or how often can we expect that surge level to be repeated. As these questions are answered, local and regional governments can make informed decisions that account for localized surge risk.

With regard to short-term decision-making, such information could help in the evacuation process, as well as localized emergency preparedness. As communities understand impacts of various surge heights, they are more likely to take appropriate precautions, but less likely to over-react and take unnecessary precautions. Overreaction of local governments to storm surge threat may have negative long-term implications, as coastal populations become more skeptical of future warnings. For example, many people who evacuated the Upper Texas Coast as Hurricane Rita (2005) approached did not evacuate before Hurricane Ike (2008), because they became skeptical after Rita did not produce the anticipated surge in their specific region. Unfortunately, Ike generated a catastrophic 5.33-meter surge, which directly claimed 7 lives and was indirectly responsible for 55 other fatalities, including several people dying from carbon monoxide poisoning, a bus crash during the evacuation, and heat exhaustion. Ike also caused an estimated $10 billion in damages (Knabb et al. 2006).

In view of this gap in the scientific literature, this chapter introduces the Storm Surge Impact Scale for Coastal Communities. This index provides information on the incremental impacts of surge inundation in coastal communities as surge levels increase, providing four categories of increasingly destructive surge impacts, listed as Impact Level 1 through Impact Level 4. These levels were defined from storm surge height and damage descriptions provided by anecdotal hurricane literature for the United States Gulf Coast. This research is based on anecdotal descriptions of over 230 surge events along the United States Gulf of Mexico from 1880-2009 which were utilized to help develop this classification system; the U.S. Government provided 15 of these quotes, newspapers provided eight, scientific journals provided five and books provided three. These quotes provide information on the surge height as well as the damage description for each surge event.

As the creation of this index was based upon anecdotal storm accounts from the U.S. Gulf Coast, specific storm surge heights associated with each increment may not apply to other water basins, although the increasingly destructive impacts of each increment would likely relate to most locations susceptible to tropical cyclone-generated storm surges. Even along the U.S. Gulf Coast, some localized differences in surge heights associated with each increment may exist, although the surge height generalizations presented in this index hold true for most communities, as this coastal area is remarkably flat, with most coastal communities residing between 1.0 and 1.5 meters above mean sea level. The lowest-impact category, Impact Level 1, describes storm surges that are less than or equal to 1.22 meters. The highest category, Impact Level 4, describes storm damage for surges of 2.44 meters or greater. Although surges on the Gulf of Mexico sometimes substantially exceed this level, the obvious visible thresholds, such as flooding of roads and coastal structures, are surpassed, making anecdotal distinctions between surges of various magnitudes in this category more difficult. Also, coastal populations should likely take similar precautions for all destructive storm surges (Impact Level 4), regardless of the storm surge magnitude within this class. It should also be noted that this index provides a valuable tool for researchers investigating physical and social history. In most locations, anecdotal accounts of extreme weather phenomena pre-date information recorded by trained scientists or instruments. Therefore, anecdotal storm descriptions are often more prevalent in historical literature than are exact measurements. This index provides a method for researchers to estimate historic storm surge heights, a practice that may be directly valuable in the fields of meteorology, climatology, oceanography, and geomorphology, as well as indirectly useful in fields such as engineering, health care, and history.

This new index is similar to the Storm Impact Scale for Barrier Islands, created by Sallenger, Jr. (2000), which also defines four levels of storm impacts along the coast, while focusing on the physical impacts to barrier islands, such as beach erosion and sand transport. The Storm Surge Impact Scale for Coastal Communities includes some information about beach erosion, but also includes information about impacts to coastal communities and infrastructure, such as roads and bridges.

Figure 8 displays a summarized version of this impact scale in the form of a schematic that associates surge levels with damage descriptions. The generalized patterns found in this index are subject to localized differences in topography, as well as differences in the physical landscape, such as the presence of levees, dunes or other natural or artificial flood-control devices.

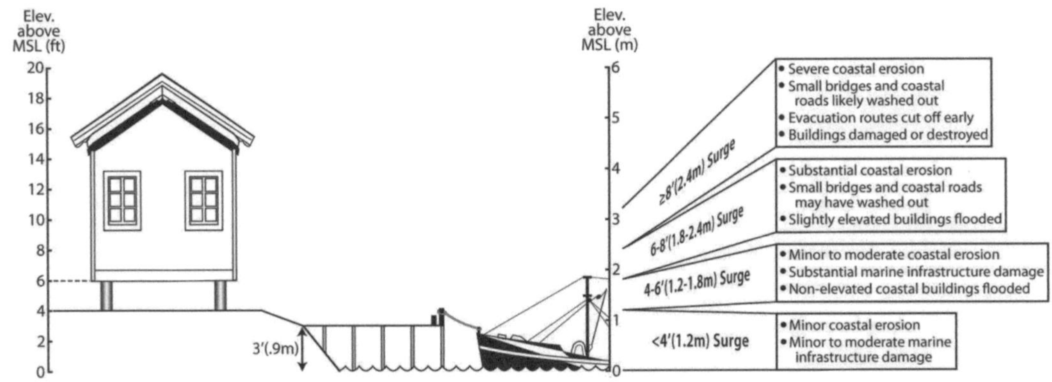

Fig. 8. Generalized schematic of storm surge impacts in communities along the U.S. Gulf coast. Marine infrastructure is estimated at .9 meters (3 feet), ground level of communities is estimated at 1.22 meters (4 feet), and slightly elevated buildings in these communities is estimated at 1.83 meters (6 feet) above MSL. Generalized damage descriptions associated with surge heights are listed in the right portion of diagram.

4. Storm surge impact scale for coastal communities

4.1 Impact level 1: "Marine impacts"

U.S. gulf coast surge height: < 1.22 meters (< 4 feet)

Damage Description: Water levels increase and are perhaps accompanied by waves, but relatively minor damage is reported. Marine infrastructure, such as docks, wharves, piers, marinas, or bath houses may sustain some damage. Non-elevated industrial equipment located outside the protection of seawalls or levees may be damaged as well. Minor beach erosion may occur. If accompanied by wave action, water may wash up against, or even over, smaller seawalls, and may reach coastal roads.

Impact Level 1 Generalization: Water is high, but mostly remains in marine areas; water does not usually reach buildings and barely crossed coastal roads if it reached them at all. See Table 2 for a list of anecdotal quotes that relate this surge height with the damage description.

4.2 Impact level 2: "Overwash impacts"

U.S. Gulf Coast Surge Height: ≥ 1.22 meters and < 1.83 meters (≥ 4 feet and < 6 feet)

Damage Description: Water washes over, and may damage coastal roads and smaller sea walls. Beach erosion occurs and is substantial if accompanied by large waves. Non-elevated structures in the coastal zone that are not protected by levees or seawalls are likely flooded. Marine infrastructure, such as docks, wharves, marinas, bath houses, as well as other structures located adjacent to the water, may sustain substantial damage or wash away, especially if accompanied by wave action.

Impact Level 2 Generalization: Water damages marine areas, crossed coastal roads and flooded non-elevated coastal buildings. See Table 3 for a list of anecdotal quotes that relate this surge height with the damage description.

Storm Name	Year	Location	Source	Quote
Ella	1958	Florida Keys	Kurtsweil, J.P., 1958	*"Tides were two feet above normal. There was no flooding, but great quantities of seaweed, sand and small rocks were washed onto the boulevard at the southeast portion of the island."*
Beulah	1959	Texas	National Weather Bureau, 1959	*"Tides 2 to 3 feet above normal and rough seas were reported along the lower Texas coast. No reports of damage or loss of life have been received."*
Judith	1959	Florida	*Tampa Tribune*, October 19, 1959, Pg. 4	*"A 75-foot long section of boardwalk next to Naples' city fishing pier was washed away in the pounding surf – three feet above normal."* *"The Fort Myers-to-Naples stretch, south of the storm's center, felt the main force of the blow but there was comparatively little damage."*
Tropical Storm #1	1960	Texas	Mozeney, R.P., 1960	*"One shrimp boat beached. One shrimp boat and two crewmen missing. Three private fishing piers wrecked at Bayside on Copano Bay."*
Danielle	1980	Louisiana	National Hurricane Center, 1980	*"Tides were no more than 2 to 3 feet above normal on the southwest Louisiana and upper Texas coasts. Beach erosion was minor."*

Table 2. Anecdotal quotes for Impact Level 1: "Marine Impacts"

Storm Name	Year	Location	Source	Quote
Great Miami Hurricane	1926	Florida Panhandle	Mitchell, C.L., 1926, Pg. 413	*"On the 21st the tide was estimated at 4.2 feet at 6:30 a. m.; it was then overflowing low ground along the water front with highest waves running possibly 5.0 feet, portions of Water Street being then from 6 to 8 inches under water."* Note: The previous quote refers to Apalachicola, Florida.
Unnamed	1936	Florida	Barnes, 2007, pgs. 160-161	*"Just after the storm center passed, an 'abnormally high tide' occurred, which measured 5.5 feet above normal and left water 18 inches deep in the streets of Everglades City."*
Unnamed	1941	Florida	Sumner, H.C., 1941, pg. 303	*"The lowest pressure, at Everglades City, 995.6 millibars (29.40 inches), was accompanied by winds exceeding 65 miles per hour and a tide of 4.1 feet which flooded the town and surrounded low country to a depth of about 1 foot."*
Baker	1950	Florida Panhandle	*Dothan Eagle*, September 1, 1950, pg. 1	*"Some beach property and wharves were damaged when water rose five and half feet above normal."*
Esther	1957	Mississippi	*Times-Picayune*, September 19, 1957, pg. 5	*"US Highway 90 along the beach was pronounced unsafe for travel by the state police which sent extra troopers to detour motorists as five-foot tides smashed over the seawall."*
Ella	1958	Texas and Louisiana	Lichtblau, S., 1958	*"Highest tides were between 2 and 4 feet along the Texas and Louisiana coasts. Some coastal roads were closed due to tide and waves."*
Candy	1968	Texas	Sugg and Hebert, 1969, pg. 233	*"Tides ranged up to 4 ft in San Antonio Bay and Corpus Christi Bay and were 2 to 3 ft elsewhere on the central and upper Texas coasts. Damage was confined mainly to the formation of cuts along Padre Island and coastal oil industry equipment."*
Bob	1979	Louisiana	National	*"Tides were generally 3 to 5 feet above*

			Hurricane Center, 1979, pg. 1	*normal and rainfall totals between 3 and 6 inches...No serious flooding was reported in Louisiana, Mississippi, or Alabama." "A number of boats were sunk or damaged and there was considerable pier damage due to high tides and rough seas."*
Jeanne	1980	Texas	National Hurricane Center, 1980, pg. 2	*"Tides of 2 to 4 feet above normal occurred along the Texas coast accompanied by a prolonged period of rough seas and heavy swells. However, only minor beach erosion was reported."*
Gilbert	1988	Texas	National Hurricane Center, 1988, pg. 2	*"Tides of 3 to 5 feet above normal were reported along the Texas coast with a number of low-lying roads under water. There was considerable beach erosion on Padre Island."*
Alberto	1994	Florida Panhandle	*Intelligencer*, July 4, 1994, pg. 41	*"A 5-mile stretch of U.S. Highway 98 from Destin west to Fort Walton Beach was closed for two hours because a sea wall protecting the highway had eroded."*
			Rappaport, 1994, pg. 3	*"A storm tide of 5 feet was estimated near Destin. A 3 foot storm surge (NGVD) occurred at Panama City, Panama City Beach, Turkey Point and Apalachicola."*
Josephine	1996	Louisiana	National Weather Service, Lake Charles, LA, 2009	*"The highest tide noted was 5.5 feet at Bayou Bienvenue, near Lake Borgne. Highway 1 was under a foot of water. A few homes and roads were also flooded in Orleans and St. Bernard Parishes, outside the flood control levees."*

Table 3. Anecdotal quotes for Impact Level 2: "Overwash Impacts"

4.3 Impact level 3: "Community inundation impacts"

U.S. Gulf Coast Surge Height: ≥ 1.83 meters and < 2.44 meters (≥ 6 feet and < 8 feet)

Storm Name	Year	Location	Source	Quote
Florence	1953	Florida	*The Panama City News-Herald*, September 27, 1953, pg. 1	*"For a short time Port St. Joe was isolated from the east and west but roads to the north remained open. A washed- out bridge east of the city interrupted travel between Port St. Joe and Apalachicola while high tides on the west at Hiland View, a Port St. Joe suburb,*

				cut the route between this city and Panama City."
			Barnes, 2007, pg. 194	*"Tides were five feet above normal at Apalachicola and Carrabelle and "six or seven feet" at Panacea."*
				Note: These quotes associate a surge height of 6 to 7 feet with washed out bridges. However, Barnes reports the surge at Apalachicola was only 5 feet. As Apalachicola was located closer than Panacea to the washed out bridges, it is possible that the surge height was around 5 feet where the bridges washed out.
Flossy	1956	Florida	Barnes, 2007, pg. 195	*"Flossy's highest tides in Florida ranged from five to six feet and caused minor damages to piers and small craft. Beachfront erosion was severe, and several homes were destroyed."*
Gladys	1968	Florida	Sugg and Hebert, 1969, pg. 236	*"Highest tides were estimated at 6.5 feet, causing considerable beach erosion and flooding of coastal areas."*
Delia	1973	Texas	Hebert and Frank, 1974, pg. 286	*"Tides of 4 to 6 ft MSL in Galveston Bay caused flooding of the Baytown area and an estimated $3 million in losses for homeowners."*
			Baytown Sun, Sept. 5, 1973, pg. 1	*"At least 500 people fled their Brownwood homes Tuesday during a threat of tides five to seven feet above normal."*
			Baytown Sun, Sept. 6, 1973, pg. 1	*"More than 1,000 Brownwood residents were evacuated early Thursday as tidal waters gushed over the elevated perimeter roadway in the worst flood since Hurricane Carla.* *At least 140 homes were flooded…"* *"A maximum 6.4 foot-tide at 4:24 a.m. Thursday was 1.5 feet above the Valentine's Day flood of 1969 and five feet below the devastating Carla flood in September, 1961."*
Chantal	1989	Texas	Gerrish, H.P., 1989, pg. 3	*"Tides at High Island were 7.0 feet MSL…There was extensive beach erosion from High Island to Sea Rim State Park."*

Table 4. Anecdotal quotes for Impact Level 3: "Inundation Impacts"

Damage Description: Widespread damage or even destruction of marine infrastructure, as well as industrial infrastructure located outside flood protection levees. Coastal communities are flooded, as water enters non-elevated and slightly elevated (one meter or less) structures in the coastal zone. Coastal roads are flooded. Small bridges as well as portions of roadway may wash out. Substantial coastal erosion is observed. Storm surge impacts the coast and areas slightly inland.

Impact Level 3 Generalization: Water damages coastal buildings and infrastructure, such as roads, bridges and causeways. See Table 4 for a list of anecdotal quotes that relate this surge height with the damage description.

4.4 Impact level 4: "Destructive impacts"

U.S. Gulf Coast Surge Height: ≥ 2.44 meters (≥ 8 feet)

Storm Name	Year	Location	Source	Quote
Danny	1985	Louisiana	National Weather Service, Lake Charles, Louisiana, 2009	*"Storm surges of 8 feet were seen along the coast of South Central Louisiana. Highway 46 near Hopedale in St. Bernard Parish was impassable due to the high waters. The pier at Grand Isle State Park was damaged, while a pier near Slidell was demolished. Coastal erosion was greatest in Terrebonne and Lower Jefferson Parishes."*
Juan	1985	Louisiana	National Weather Service, Lake Charles, Louisiana, 2009	*"Storm surges were 8 feet at Cocodrie...LA 1 south of Leeville and LA 3090 near Fourchon were destroyed. Three bridges were washed out near Lacombe on LA 434. Levees were overtopped in Lockport, Marrero, Oswego, and Myrtle Grove; this added to the already serious flooding. Two hundred cattle were drowned in Terrebonne Parish. Grand Isle was under 4 feet of sea water; 1200 residents were trapped on the island as the storm surge cut off any evacuation attempts early on."*
Allison	1995	Florida	Pasch, R.J., 1997	*"Maximum storm surge heights were estimated at 6 to 8 ft from Wakulla through Dixie counties..."* *"Damage was greatest in the coastal sections of Dixie, Levy, Taylor and Wakulla counties, mainly from storm surge effects, with 60 houses and businesses damaged. A house collapsed at Bald Point in Franklin County. About 5000 people evacuated from the coast. Other costal effects included mostly minor beach erosion, damage to sea walls and coastal roadways, and the sinking of several small boats."*
			New Mexican, June 6, 1995, pg. 4	*"More than 65 coastal homes, three hotels and at least one restaurant were flooded as the morning storm caused the ocean to surge 8 feet along a 150-mile stretch of Florida's Big Bend, where the Panhandle meets the Peninsula."*
Josephine	1996	Florida	Pasch, R.J., 1997	*"County officials estimated storm tides (storm surge plus astronomical tide) ranged from up to 9 feet in Levy County...These tides produced widespread flooding of roads, dwellings, and businesses."*

Table 5. Anecdotal quotes for Impact Level 4: "Destructive Impacts"

Damage Description: Severe coastal erosion is observed. Severe damage is inflicted on marine infrastructure, as well as any industrial equipment, residential or commercial structures located outside flood protection levees. Structures in coastal zone, even if elevated or protected by levees, may be damaged or completely destroyed. Small bridges and coastal roads are likely washed out. Evacuation routes are cut off hours before the peak surge level arrives. Water may overtop levees. Surge likely penetrates inland.
Impact Level 4 Generalization: Water greatly damages coastal structures, washing some away, or creating complete loss. Water also penetrates inland. See Table 5 for a list of anecdotal quotes that relate this surge height with the damage description.

5. Summary and conclusion

Several physical processes are responsible for generating storm surges in association with landfalling tropical storms and hurricanes. Although wind stress is the predominant factor forcing surge heights, several other processes often enhance or diminish maximum surge heights. Low air pressure clearly contributes to storm surge, but its overall impact is modest. Cyclonic size is an important factor that has been underestimated until recently, after very large hurricanes Katrina (2005) and Ike (2008) generated larger surges than anticipated. Forward cyclonic movement is another factor which influences surge heights, as slower moving storms often produce higher surges, and inundate the coastline for longer periods of time. Coastline shape and the presence of natural and artificial obstructions to flowing water, such as coastal forests and levees, sometimes have profound localized effects on storm surge height, especially where water approaches the coast or a barrier in a perpendicular direction, and is forced to rise. Shallow bathymetry is another factor which enhances surge heights, as underwater currents cannot distribute water unless the bathymetry is relatively deep.
After recognizing these contributing factors leading up to storm surge, it should come as no surprise that Hurricane Category alone cannot accurately estimate surge at a specific coastline. As a result, this chapter also presented a classification system for surge based on anecdotal information mostly from U.S. government sources, newspapers and books. The classification system has four categories of increasingly destructive surge impacts, ranging from Impact Level 1, with minor "Marine Impacts" to Impact Level 4, with "Destructive Impacts" that often penetrate far inland. The classification should be of some benefit to coastal zone planners, emergency managers, and researchers.

6. References

Ali, A., 1996: Vulnerability of Bangladesh to climate change and sea level rise through tropical cyclones and storm surges. *Water, Air, and Soil Pollution*, 92, 171-179.

Barnes, J., 2007: *Florida's Hurricane History*. 2nd edition. University of North Carolina Press, 407 pp.

Baytown Sun, Baytown, Texas, Wednesday, September 5, 1973, Page 1. Author unknown.

Baytown Sun, Baytown, Texas, Thursday, September 6, 1973, Page 1. Author unknown.

Berg, R., 2009: National Hurricane Center Tropical Cyclone Report on Hurricane Ike (AL092008), updated January 23, 2009. Published on the Web at http://www.nhc.noaa.gov/pdf/TCR-AL092008_Ike.pdf.

Bevan, J., 2001: National Hurricane Center Tropical Cyclone Report on Hurricane Keith, updated January 29, 2001. Published on the Web at http://www.nhc.noaa.gov/2000keith.html.

Chen, Q., L. Wang, R. Tawes, 2008: Hydrodynamic response of northeastern Gulf of Mexico to hurricanes. *Estuaries and Coasts*, 31, 1098-1116.

Cline, I, 1926: *Tropical Cyclones*. Macmillon Company, 301 pp.

De, U.S., R.K. Dube, G.S. Prakasa Rao, 2005: Extreme weather events over India in the last 100 years. *Journal of Indian Geophysical Union*, 9, 173-187.

De Scally, F.A., 2008: Historical tropical cyclone activity and impacts in the Cook Islands. *Pacific Science*, 62, 443-459.

Dothan Eagle, Dothan, Alabama, Friday afternoon, September 1, 1950. Author unknown.

Dube, S.K., A.D. Rao, P.C. Sinha, T.S. Murty, N. Bahulayan, 1997: Storm surge in the Bay of Bengal and Arabian Sea: The problem and its prediction. *Mausam*, 48, 283-304.

Ebersole, B.A., J.J. Westerlink, S. Bunya, J.C. Dietrich, and M.A. Cialone, 2010: Development of storm surge which led to flooding in St. Bernard Polder during Hurricane Katrina. *Ocean Engineering*, 37, 91-103.

Frank, N.L., and S.A. Husain, 1971: Deadliest tropical cyclone in history. *Bulletin of the American Meteorological Society*, 52, 438-&.

Garriott, E.B., 1900: West Indian hurricane of September 1-12, 1900. *Monthly Weather Review*, 28, 371-378.

Gerrish, H.P., 1989: Preliminary Report, Hurricane Chantal, 30 July to 3 August 1989. Report published at the National Hurricane Center, Miami, Florida. Available on the Web at: http://www.nhc.noaa.gov/archive/storm_wallets/atlantic/atl1989-prelim/chantal/prelim03.gif.

Gray, W.M., 1975: Tropical cyclone genesis. Atmospheric science paper No. 234, Department of Atmospheric Science, Colorado State University, Fort Collins, CO.

Gray, W.M., 1975: Global view of tropical cyclone genesis. *Bulletin of the American Meteorological Society*, 56, 322.

Gray, W. M., 1998: The formation of tropical cyclones. *Meteorology and Atmospheric Physics*, 67, 37-69.

Harris, L.D., 1963: Characteristics of the Hurricane Storm Surge. U.S. Weather Bureau, *Technical Paper No. 48.*

Hebert, P.J. and N.L. Frank, 1974: Atlantic Hurricane Season of 1973. *Monthly Weather Review*, 102, 280-289.

Holland, G.J., 1997: The maximum potential intensity of tropical cyclones. *Journal of the Atmospheric Sciences*, 54, 2519-2541.

Intelligencer, Doylestown, Pennsylvania, Monday, July 4, 1994, Page 41.

Interagency Performance Evaluation Taskforce Report (IPET), 2006: Performance Evaluation of the New Orleans and Southeast Louisiana Hurricane Protection System, Draft Final Report of the Interagency Performance Evaluation Task Force. *U.S. Army Corps of Engineers*, Volume 1- Executive Summary and Overview, June 1, 2006.

Irish, J.L., D.T. Resio, J.J. Ratcliff, 2008: The influence of storm size on hurricane surge. *Journal of Physical Oceanography*, 38, 2003-2013.

Irish, J.L., D.T. Resio, and M.A. Cialone, 2009: A surge response function approach to coastal hazard assessment. Part 2: Quantification of spatial attributes response functions. *Natural Hazards*, 51, 183-205.

Jarvinen, B.J., and C.J. Neumann, 1985: An evaluation of the SLOSH storm surge model. *Bulletin of the American Meteorological Society*, 66, 1408-1411.

Jordan II, M.R. 2008: Development of a new storm surge index for global prediction of tropical cyclone generated storm surge. Dissertation, Florida State University, Tallahassee, FL, 82 pp.

Jordan II, M.R., and C.A. Clayson, 2008: Evaluating the usefulness of a new set of hurricane classification indices. *Monthly Weather Review*, 136, 5234-5238.

Jordan II, M.R., and C.A. Clayson, 2008: A new approach to using wind speed for prediction of tropical cyclone generated storm surge. *Geophysical Research Letters*, 35, article number: L13802.

Keim, B., R. Muller, G. Stone, 2007: Spatiotemporal Patterns and Return Periods of Tropical Storm and Hurricane Strikes from Texas to Maine. *Journal of Climate*, 20, 3498-3509.

Keim, B.D., and R.A. Muller, 2008: Overview of Atlantic Basin Hurricanes. Chapter 4, pp. 79-89, in: Walsh, P.J., S.L. Smith, L.E. Fleming, H.M. Solo-Gabrielle, W.H. Gerwick, *Oceans and Human Health*. Academic Press, Burlington, Massachusetts, 644 pp.

Keim, B.D., and R.A. Muller, 2009: *Hurricanes of the Gulf of Mexico*. Louisiana State University Press, 216 pp.

Knabb, R.D., J.R. Rhome, D.P. Brown, 2006: National Hurricane Center Tropical Cyclone Report on Hurricane Katrina. Published on the Web at: http://www.nhc.noaa.gov/pdf/TCR-AL122005_Katrina.pdf.

Knabb, R.D., D.P. Brown, J.R. Rhome, 2006: National Hurricane Center Tropical Cyclone Report on Hurricane Rita. Published on the Web at: http://www.nhc.noaa.gov/pdf/TCR-AL182005_Rita.pdf.

Kurian, N.P., N. Nirupama, M. Baba, K.V. Thomas, 2009: Coastal flooding due to synoptic scale, meso-scale and remote forcings. *Natural Hazards*, 48, 259-273.

Kurtsweil, J.P., 1958: Narrative Report of Hurricane Ella. Report published as U.S. Weather Bureau Office Memorandum, Key West, Florida. Available on the Web at: http://www.nhc.noaa.gov/archive/storm_wallets/atlantic/atl1958/ella/preloc/psheyw.gif.

Landsea, C., 2007: Record number of storms by basin. Table published on the Web through the Atlantic Oceanographic and Meteorological Laboratory, Hurricane Research Division at: http://www.aoml.noaa.gov/hrd/tcfaq/E10.html.

Lawrence, M, and H. Cobb, 2005: National Hurricane Center Tropical Cyclone Report on Hurricane Jeanne, updated January 7, 2005. Published on the Web at http://www.nhc.noaa.gov/pdf/TCR-AL112004_Jeanne.pdf.

Lichtblau, S., 1958: Report on Tropical Storm Ella Sept. 3 to 6, 1958, in the Gulf of Mexico. Report published by the National Weather Bureau. Available on the Web at: http://www.nhc.noaa.gov/archive/storm_wallets/atlantic/atl1958/ella/preloc/pshnew.gif.

McTaggart-Cowan R., G.D. Deane, L.F. Bosart, C.A. Davis, T.J. Galarneau, Jr., 2008: Climatology of tropical cyclogenesis in the North Atlantic (1948-2004). *Monthly Weather Review*, 136, 1284-1304.

Mitchell, C.L., 1926: The West Indian Hurricane of September 14-22, 1926. *Monthly Weather Review*, 54 (10). Available on the Web at: http://www.aoml.noaa.gov/general/lib/lib1/nhclib/mwreviews/1926.pdf.

Mozeney, R.P., 1960: Tropical Low Report for June 23-24. Report published by the National Weather Bureau. Available on the Web at: http://www.nhc.noaa.gov/archive/storm_wallets/atlantic/atl1960/td1/preloc/pshcrp.gif.

National Aeronautics and Space Administration, 2010: NASA Quest- an Educational Web Site. Available on the Web at: http://quest.nasa.gov/aero/planetary/mars.html

National Hurricane Center, Miami, Florida, 1979: Preliminary Report, Hurricane Bob, 9-16 July, 1979. Author only identified by initials, "GBC." Available on the Web at: http://www.nhc.noaa.gov/archive/storm_wallets/atlantic/atl1979-prelim/bob/prelim01.gif.

National Hurricane Center, Miami, Florida, 1980: Report on Tropical Storm Danielle, 4-7 September 1980. Author unknown. Available on the Web at: http://www.nhc.noaa.gov/archive/storm_wallets/atlantic/atl1980-prelim/danielle/prelim02.gif.

National Hurricane Center, Miami, Florida, 1980: Preliminary Report, Hurricane Jeanne, November 7-16, 1980. Author only identified by initials "JMP." Available on the Web at: http://www.nhc.noaa.gov/archive/storm_wallets/atlantic/atl1980-prelim/jeanne/prelim02.gif.

National Hurricane Center, Miami, Florida, 1988: Preliminary Report, Hurricane Gilbert, 08-19 September 1988. Author only identified by initials, "GBC." Available on the Web at: http://www.nhc.noaa.gov/archive/storm_wallets/atlantic/atl1988-prelim/gilbert/prelim02.gif.

National Oceanic and Atmospheric Administration, 1999: NOAA's Top U.S. Weather, Water and Climate Events of the 20th Century. Available on the Web at: http://www.noaanews.noaa.gov/stories/s334c.htm

National Weather Bureau, 1959: Preliminary report on Tropical Storm Beulah, June 15-18, 1959. Author unknown. Available on the Web at: http://www.nhc.noaa.gov/archive/storm_wallets/atlantic/atl1959/beulah/prenhc/prelim1.gif.

National Weather Bureau, September 1969: Preliminary Report on Hurricane Camille. Author unknown, document published on the Web at: http://www.nhc.noaa.gov/archive/storm_wallets/atlantic/atl1969-prelim/camille/TCR-1969Camille.pdf.

National Weather Service, Lake Charles, Louisiana, 2009: Louisiana Hurricane History. Author unknown. Available on the Web at: http://www.srh.noaa.gov/lch/research/lalate20hur2.php.

National Weather Service, Lake Charles, LA, 2009: Louisiana Hurricane History. Author unknown. Available on the Web at: http://www.srh.noaa.gov/lch/research/lalate20hur3.php.

Neumann, C.J., B.R. Jarvinen, C.J. McAdie, J.D. Elms, 1993: *Tropical cyclones of the North Atlantic Ocean*. U.S. Department of Commerce, prepared by the National Climate Data Center, Asheville, North Carolina, 193 pp.

New Mexican, Santa Fe, New Mexico, Tuesday, June 6, 1995. Author unknown.

Panama City News-Herald, Panama City, Florida, September 27, 1953. Author unknown.

Pasch, R.J., 1996: Preliminary Report, Hurricane Allison, 3-6 June 1995. Report published at the National Hurricane Center, Miami, Florida. Available on the Web at: http://www.nhc.noaa.gov/1995allison.html.

Pasch, R.J., Florida, 1997: Preliminary Report, Tropical Storm Josephine, 4-8 October 1996. Report published at the National Hurricane Center, Miami, Florida. Available on the Web at: http://www.nhc.noaa.gov/1996josephin.html.

Pasch, R.J., D.P. Brown, and E.S. Blake, 2004: Tropical Cyclone Report, Hurricane Charley, 9-14 August 2004. Available on the Web at: http://www.nhc.noaa.gov/pdf/TCR-AL032004_Charley.pdf.

Rappaport, E.N., 1994: Preliminary Report, Tropical Storm Alberto, 30 June – 7 July 1994. Report published at the National Hurricane Center, Miami, Florida. Available on the Web at: http://www.nhc.noaa.gov/archive/storm_wallets/atlantic/atl1994/alberto/prenhc/prelim03.gif.

Rappaport, E.N, and J.J. Fernandez-Partagas, 1995: The deadliest Atlantic tropical cyclones, 1492-1994. *NOAA Technical Memorandum NWS NHC-47*, National Hurricane Center, 41 pp.

Rego, J.L., and C. Li, 2009: On the importance of the forward speed of hurricanes in storm surge forecasting: A numerical study. *Geophysical Research Letters*, 36, L07609.

Sallenger, Jr., A.H., 2000: Storm Impact Scale for Barrier Islands. *Journal of Coastal Research*, 16, 890-895.

Simpson, R.H., A. L. Sugg, and Staff, 1970: The Atlantic Hurricane Season of 1969. *Monthly Weather Review*, 98, 293-306.

Sugg, A.L. and P.J. Hebert, 1969: The Atlantic Hurricane Season of 1968. *Monthly Weather Review*, 97 (3), pp. 225-255.

Sumner, H.C., 1941: Hurricane of October 3-12 and Tropical Disturbance of October 18-21, 1941. *Monthly Weather Review*, October 1941. Available on the Web at: http://www.aoml.noaa.gov/general/lib/lib1/nhclib/mwreviews/1941.pdf.

Tampa Tribune, Tampa, Florida, October 19, 1959. Author unknown.

Times-Picayune, New Orleans, Louisiana, September 19, 1957. Author unknown.

United States Army Corps of Engineers, 1970: Hurricane *Camille, 14-22 August 1969*. Mobile, Alabama.

Welander, Pierre, 1961: *Numerical Prediction of Storm Surges,* in *Advances in Geophysics*, 8. Edited by H.E. Landsberg and J. van Mieghem. Published by Academic Press, Inc., New York, USA, 316 pp.

Weisberg, R.H., and L. Zheng, 2006: Hurricane Storm Surge Simulations for Tampa Bay. *Estuaries and Coasts*, 29, 899-913.

21

Estimating the Vulnerability of U.S. Coastal Areas to Hurricane Damage

Jeffrey Pompe and Jennifer Haluska
Francis Marion University
U.S.A.

1. Introduction

Although hurricanes can strike most U.S. coastal states and cause damage in noncoastal states as well, some coastal areas are much more likely than others to experience the most costly storms. In the U. S., 40 percent of all major hurricanes have battered Florida, and 83 percent of category 4 or 5 hurricane strikes have pummeled either Florida or Texas (Blake et al., 2007). It is not surprising, therefore, that the Texas and Florida coastlines have experienced much more damage from hurricanes than other southeastern coastal areas. Of the 10 most costly hurricanes, 9 have struck the Texas or Florida coasts (Table 1).

Hurricanes have become much more costly in recent years; 8 of the 10 most damaging hurricanes have occurred since 2004. In total, the seven hurricanes in 2004 and 2005 caused \$79.3 billion in insured losses. An increase in hurricane activity may explain some of the increased cost. After a period of infrequent hurricane activity between 1971 and 1994, hurricane activity has increased in recent years. The five most intense consecutive storm seasons on record occurred between 1995 and 2000. In 2004 an unprecedented four hurricanes, viz. Charley, Frances, Ivan and Jeanne, damaged Florida communities. And the 2005 hurricane season was the busiest and most costly in United States history, with 28 named storms, 15 of which were hurricanes, including Katrina (South Carolina Department of Insurance, 2007, 14).

Although a period of increased hurricane activity may have contributed to increased hurricane damage, other factors such as population growth and increased property values have contributed to the higher costs. In order to examine factors that influence hurricane damage we consider eight southeastern states: Alabama, Florida, Georgia, Louisiana, Mississippi, North Carolina, South Carolina, and Texas. We focus on this area because the greatest likelihood of severe damage from hurricanes is along the coastlines of the southeastern Atlantic and Gulf of Mexico states where 112 major hurricanes have struck between 1851 and 2006 (Blake et al., 2007). Understanding more about factors that increase coastal area vulnerability will help mitigate other coastal threats such as northeasters. Northeasters, which are powerful storms that have winds that blow from the northeast, can generate storm surge and waves that cause even greater damage than hurricanes. The Ash Wednesday storm of 1962, which was one of the most damaging northeasters, created waves more than 30 feet high and caused millions of dollars in damage along the mid-Atlantic coast.

In this study, we consider the various factors that increase an area's vulnerability to hurricane damage. We focus on the risk to the built environment not the loss of life from hurricanes, which has been greatly reduced in recent years, generally. However, the

Rank	Date	Area	Name	Dollars when Occurred (millions)	2008 dollars (millions)
1	Aug. 25-29, 2005	LA, MS, Fl	Katrina	$40,600	$45,309
2	Aug. 23-24, 25-26, 1992	SE FL, SELA	Andrew	15,500	23,786
3	Sept. 12-14, 2008	TX, LA	Ike	11,700	11,700
4	Oct. 24, 2005	SE & SW FL	Wilma	10,300	11,355
5	Aug. 13-15, 2004	SW FL	Charley	7,475	8,520
6	Sep. 16-21, 2004	NW FL, AL	Ivan	7,110	8,104
7	Sep. 17-18, 21-22, 1989	SC	Hugo	4,195	7,284
8	Sep. 20-26, 2005	LA, TX, FL	Rita	5,000	6,203
9	Sep. 5, 2004	SE FL	Frances	4,595	5,237
10	Sep. 15-25, 2004	SE FL	Jeanne	3,440	4,166

Table 1. The Ten Costliest U.S. hurricanes

Source: Insurance Information Institute

information we provide can increase the safety of coastal area residents and reduce the disruption to social order. In order to measure the degree of risk to coastal areas we calculate an index value that is composed of factors that affect hurricane damage costs. A better understanding of the factors that increase an area's vulnerability may help create polices that mitigate hurricane damage more effectively. In addition, we consider what effect increased vulnerability may have on insurance premiums for property owners.

2. Vulnerability analysis of coastal areas

Given the risks to which coastal communities are subject, understanding what affects the degree of risk for residents is important. Various authors have considered factors that influence the vulnerability of coastal areas to hurricanes and other threats.

Davidson and Lambert (2001) develop a Hurricane Disaster Risk Index in which they use factors such as hurricane frequency and emergency evacuation routes to examine the relative risk of economic and life loss in coastal counties. Jain et al. (2005) produce a case study of two North Carolina counties in which they find that building inventory changes affect losses significantly. Zandberger (2009), who examined the historic record of hurricanes for the period 1851 to 2003, finds that the top 10 counties in terms of cumulative exposure are in coastal Florida (6), North Carolina (3), and Louisiana (1). The factors that describe the exposure to hurricanes include distance to coast, latitude, longitude, size, and shape of the counties.

Thieler and Hammer-Klose (1999) create a physical vulnerability index to examine the vulnerability of U.S. coastal areas to sea level rise using mean tidal range, coastal slope, rate of relative sea level rise, shoreline erosion and accretion rates, mean wave height, and erodability. They find that sections of the mid-Atlantic coast – from Maryland south to North Carolina and northern Florida - are some of the most vulnerable areas to sea level rise. The most vulnerable areas generally are high-energy coastlines where the regional coastal slope is low and where the major landform type is a barrier island. Although they consider an areas vulnerability to sea level rise, similar physical conditions may contribute to increased flood damage from hurricanes. Boruff et al. (2005) use socioeconomic and physical characteristics to examine the vulnerability of coastal area residents to erosion. They find that physical characteristics (such as coastal slope, sea level rise, and mean wave height) determine Atlantic and Pacific Ocean vulnerability and social characteristics (such as persons age 65 and over) determine Gulf of Mexico vulnerability.

3. Factors that influence hurricane damage costs

Most damage from hurricanes is caused by high winds and storm surge, which is the water that a hurricane's wind pushes toward shore. A major hurricane, which would be a category 3, 4, or 5 on the Saffir/Sampson Hurricane Scale, has winds that exceed 110 miles per hour. A storm surge of up to twenty-eight feet above average sea level was recorded in Mississippi during Katrina (Masters, 2010). If the storm surge is accompanied with high tide, the damage is even greater.

Three interacting elements contribute to the amount of damage caused by a hurricane: the amount and type of property at risk, the degree of vulnerability of the property, and the likelihood of the natural disaster. We examine the contribution of the three elements to property damage for the coastal counties of the eight southeastern states. We use only the coastal counties because hurricane damage is greatest, generally, in coastal areas where the damage from wind and storm surge is most severe. We define a coastal county as one that has some land area exposed to the Atlantic Ocean or Gulf of Mexico, which is the definition used by the United States Geological Survey. There are 99 coastal counties in the southeastern U.S. states that meet this criterion.

Population growth and increased property values in coastal areas are major contributors to the increased damage costs from hurricanes. Although the 10 most costly hurricanes have occurred since 1989, earlier hurricanes could be more damaging but not as costly because population and development was much less. Pielke et al. (2008) normalize hurricane damage by adjusting for inflation, wealth, and changes in population and housing units. After adjustments they find that the decade with the greatest cumulative damage was 1926 to 1935, and the most damaging single storm was the 1926 Great Miami storm, which caused $140 to $157 billion of normalized damage.

Coastal population has increased much more rapidly than the nation's population. Population in the U.S. and southeastern coastal counties has increased by 27 and 40 percent, respectively, from 1970 to 2000. In addition, population density is much higher in coastal areas. In 2000, average population density in the U.S. and southeastern counties was 80 and 290 people per square mile, respectively (Table 2). In many coastal areas housing growth has exceeded population growth as property owners have built second homes and rental properties. The number of houses in southeastern coastal counties has increased by 51.9 percent from 1970 to 2000. The density of houses in coastal counties was 72.3 houses per square mile in 2000 which was an increase of 59.3 percent from 1970 (National Oceanic Economics Program, 2009).

In addition to the increased number of houses, property values in coastal areas have increased in recent decades. The AIR Worldwide Corporation (2008) estimates that insured property value in the coastal U.S. was $8,891 billion in 2007. The insured value of coastal area properties grew at a compound annual growth rate of just over 7 percent from 2004 to 2007; that growth rate will lead to a doubling of the total value every decade (AIR 2008). Several factors such as inflation, higher building costs, and larger homes cause higher housing prices. In the U.S. the average size of a single family house has increased from 983 to 2434 square feet between 1950 and 2005 (National Home Builders Association, 2007). In

Coastal Counties	Population 1970	Population 2000	Population Growth Rate	Average Population Density (per square mile of land)
Alabama	376,690	540,258	43.42%	206
Florida	5,388,655	12,285,697	127.99%	410
Georgia	281,157	439,154	56.20%	159
Louisiana	1,415,649	1,658,507	17.16%	431
Mississippi	239,944	363,988	51.70%	199
North Carolina	477,404	792,902	66.09%	112
South Carolina	497,984	884,925	77.70%	140
Texas	2,830,354	5,041,960	78.14%	262
United States (Total)	203,211,926	281,421,906	38.49%	80

Table 2. Population Growth in Southeastern Coastal Counties

Source: U.S. Census (author's calculations)

coastal areas, homes as large as 5000 square feet are not uncommon and rental homes as large as 10,000 square feet with 16 bedrooms have been built. Large oceanfront homes that provide seasonal rentals have replaced smaller single-family cottages that were once a popular form of accommodation for visitors.

Strengthening buildings with hurricane resistant construction reduces damage costs. Fronstin and Holtman (1994) found that Hurricane Andrew, which struck Florida in 1992, caused less damage in subdivisions with higher average home prices. Although houses were more expensive in higher income areas, because the houses were more storm-resistant, nearby houses suffered less damage from wind-blown debris. States enforce building codes that have been designed to create more hurricane resistant housing. However, agencies may not do an adequate job of enforcement. Areas with a larger number of older houses, which are less hurricane resistant, generally, may suffer greater damage costs.

Although it is not possible to predict where or when hurricanes will strike, the likelihood of a major hurricane striking an area can be estimated using climatology data. The climatological probability of a major hurricane strike in the southeastern states is highest in Florida (21 percent) and lowest in Georgia (1 percent) (U.S. Landfalling Hurricane Probability Project, 2010). In addition, the past history of hurricane strikes provides some indication of which areas are likely to be struck by a hurricane. The highest and lowest expected return period of a major hurricane is 9 and 34 years for Miami-Dade County (Florida) and Chatham County (Georgia), respectively (Blake et al., 2007).

The geology of a coastal area may increase the likelihood of hurricane damage. For example, storm surge may be more damaging if the shoreline has a shallow beach slope or a bay that funnels storm surge. The rate of shoreline erosion (which averages two to three feet per year along the Atlantic coast and six feet per year on the Gulf coasts) may affect hurricane damage because beaches absorb and dissipate wave force. The force of waves causes much of the storm related damage in coastal areas. Mean wave height and sea level rise are two other physical factors that may impact hurricane damage.

4. The hurricane vulnerability index

In order to rank the susceptibility of coastal areas to hurricane damage we calculate a Hurricane Vulnerability Index (HVI) which is a composite of the elements that contribute to hurricane damage. Three elements contribute to the degree to which communities experience hurricane damage: (1) the level of exposure, (2) physical susceptibility to the hurricane, and (3) the hurricane's frequency and intensity. Each element is comprised of several indicators that represent characteristics of risk. The seven indicators that we employ are: population, number of housing units, house value, probability of hurricane strike, building code effectiveness, building age, and vulnerability to sea-level rise.[1] The rationale for choosing these indicators was discussed in the previous section.

We use multiattribute utility theory to formulate the HVI (Keeney and Raiffa 1976). We standardize the indicators using a range from 1 to 10 because the indicators are measured in different units (dollars and percentage, e.g.). The value 1 is equal to the minimum risk and 10 is the maximum risk. The formula used to standardize the indicators is:

$$R_{ij} = (x_{ij} - min_i) * 10 / (max_i - min_i) \quad (1)$$

where R_{ij} and x_{ij} are the scaled and unscaled value of risk indicator i for county j; and min_i and max_i are the minimum and maximum values for the ith indicator, respectively.

The HVI is calculated with the following equation:

$$HVI = (E)(S)(H) \quad (2)$$

where E and S are the exposure and susceptibility to the hurricane, and H is likelihood of the hazard. We use a multiplicative model because risk is a product of exposure, susceptibility, and hazard. We weight the scaled values in order to adjust for the relative importance of the indicators. We use Saaty's (1980) analytic hierarchy process, which creates a pair wise comparison of the indicators, to guide in the selection of the appropriate weights. The three elements are calculated with the following equations:

$$E = w_{E1}R_{E1} + w_{E2}R_{E2} + w_{E3}R_{E3} \quad (3)$$

$$S = w_{S1}R_{S1} + w_{S2}R_{S2} + w_{S3}R_{S3} \quad (4)$$

$$H = w_{H1}R_{H1} \quad (5)$$

where R_{E1}, R_{E2}, and R_{E3} are population, housing units, and housing value; R_{S1}, R_{S2}, and R_{S3} are building code effectiveness, average building age, and vulnerability to sea-level rise ; R_{H1} is hurricane probability; and w is the appropriate weight for each indicator

5. Case study

We estimate the HVI for a sample of southeastern counties in order to demonstrate the estimation process and method to interpret the results. This sample of counties provides at least one county from each of the eight states; we chose two counties from three states (Texas, Florida, and North Carolina) that have large coastlines and a high probability of hurricane strikes. The 11 counties are: Cameron (TX), Galveston (TX), Orleans (LA), Harrison (MS), Mobile (AL), Miami-Dade (FL), Bay (FL), Chatham (GA), Charleston (SC), New Hanover (NC), and Dare (NC). The indicator definitions, and minimum and maximum values, are listed in Table 3. There is significant variation between the indicators for the

[1] We do not include indicators for non-residential construction but assume that it would be proportional to residential construction.

sample counties. Most of these coastal areas have experienced rapid population growth and development growth in recent decades despite the risk of hurricane damage.

Indicator	*Minimum Risk*	*Maximum Risk*
Resident Population	0	2253362
Number of Housing Units	0	852278
Median value of owner-occupied housing	0	137200
Climatological probability of hurricane strike (percent)	0	100
Average Building Code Enforcement Grade	1	10
Percentage of homes built after 1990	0	100
Vulnerability to sea-level rise	1	4

Table 3. Hurricane Vulnerability Index Indicators

We use values for population and housing data from the 2000 U.S. Census to estimate the four demographic HVI indicators. We use the coastal vulnerability index from Thieler and Hammer-Klose (1999) for a measure of the physical shoreline characteristics that increase the susceptibility to hurricane damage. We use the Insurance Services Office (ISO) rating on building code enforcement. The ISO's Building Code Effectiveness Grading Schedule evaluates the effectiveness of a municipality's enforcement of building codes. For the climatological probability, we use U.S. Landfalling Hurricane Probability Project estimations of the likelihood that a hurricane will strike a particular county.

The HVI value for the 11 sample counties is presented in Table 4. A higher HVI value indicates a county that where the risk of hurricane damage is greater. The maximum and minimum risk possible would be indicated by a value of 10 and 0, respectively. The most and least vulnerable county to hurricane damage from our sample is Miami (5.52) and Chatham (0.11), respectively, based on the HVI values. Examining the individual indicators can explain much of the ranking for a particular county. For example, of the 11 sample counties, Miami is the most at risk for three of the indicators - population, number of houses, and probability of a strike - and among the most risky for the other four indicators. Chatham County is the least at risk for two variables - probability and vulnerability - and near the lowest risk for two variables - population and the number of houses.

County	HVI Value
Miami-Dade, FL	5.52
Dare, NC	1.10
Charleston, SC	1.34
New Hanover, NC	0.42
Bay, FL	0.19
Harrison, MS	0.69
Galveston, TX	0.71
Orleans, LA	0.78
Cameron, TX	0.68
Chatham, GA	0.11
Mobile, AL	0.65

Table 4. HVI for sample counties

The hurricane return period, probability of a strike, and hurricane damage cost for each sample county is listed in Table 5. Comparing the HVI value to the historical hurricane damage cost provides some validation of the HVI values. The county with the most and least cumulative damage is Miami-Dade and Chatham, respectively. The two counties have the highest and lowest HVI, respectively. Between 1851 and 2006, the greatest number of major hurricane strikes occurred in Florida (37), and the fewest strikes occurred in Georgia (3) (Blake et al., 2007). Coastal areas that are predicted to have a higher probability of a hurricane strike - Miami-Dade, Charleston, Cameron, and Dare. - have the highest HVIs, generally. Although Cameron County has the third highest strike probability, its HVI is ranked sixth. This is partially the result of Cameron having the lowest housing values of the sample.

Examining individual indicators can indicate which factors contribute the most to risk for a particular county, and therefore which factors should receive the most attention. For example, Bay County's HVI is slightly higher than Chatham County's HVI. The principle difference between the indicators for the two counties is that Bay County has a much higher vulnerability indicator than Chatham. This difference may account for some of the higher historical damage costs for Bay County. The higher vulnerability for Bay County would suggest that hurricane mitigation efforts and resources should be focused on creating less vulnerable communities in Bay County. Building hurricane-resistant housing or protecting shorelines would be productive activities.

Counties	Return Period Hurricane (years)	Hurricane Strike Probability[1] (climatological)	Hurricane Damage Costs(1950-2008) (2008 $)
Cameron, TX	25	2.7	47,301,684
Galveston, TX	18	1.9	869,886,903
Orleans, LA	19	1.7	988,065,664
Harrison, MS	18	1.9	1,322,042,826
Mobile, AL	23	1.9	622,166,867
Bay, FL	17	0.5	164,097,087
Miami-Dade, FL	9	5.0	3,523,941,209
Chatham, GA	34	0.4	3,596,486
Charleston, SC	15	2.9	878,583,973
New Hanover, NC	16	1.1	399,547,814
Dare, NC	11	2.4	88,484,892

[1]Probability of intense hurricane striking the county

Sources: Blake et al., 2007; U.S. Landfalling Hurricane Probability Project; Hazards & Vulnerability Research Institute

Table 5. Hurricane Statistics for Sample Counties

The HVI can be helpful in creating equitable rates for property insurance. With the recent frequency of storms and hurricanes, insurance companies and state agencies have reassessed their insurance policies in coastal markets. Insurance companies began changing their coastal policy writing practices shortly after 1992's Hurricane Andrew, which struck south Florida and caused an estimated $32 billion in property damages. Insurers immediately attempted to limit coverage and raise rates where coverage was provided in areas subject to hurricane impacts. Additional, heavy losses in the years following Andrew have caused coastal insurance for wind and hail to become much more expensive and even difficult to obtain in many areas as insurance companies attempt to reduce risk. For example, following the 2004 hurricane season, premiums doubled for windstorm insurance in many parts of the Florida with owners of 1,500 square foot homes facing premiums of $10,000 for wind damage alone with total insurance costs of $13,000 and deductibles of up to $18,000 (Mortgage News Daily 2007).

As insurance companies have reduced coverage and raised rates, consumers have become increasingly concerned and groups such as the real estate industry have called for increased government intervention. State insurance agencies, which provide oversight of insurance rates, can use the HVI in order to more accurately regulate insurance companies. For example, Miami-Dade's HVI value of 5.52 is much higher than Charleston County's of 1.34, which is the second highest HVI. This indicates that the risk of damage, and therefore insurance rates should be much higher in Miami-Dade County. However, examining individual indicators is important in order to assess rates correctly. Dare County's HVI, for example, is ranked third; this is due in part to the modest population and housing totals. Other indicators, such as hurricane probability and vulnerability are very high for Dare, which would suggest that rates should be adjusted accordingly.

In addition, the HVI can be used to inform residents about areas of greater risk. Property owners moving to counties with high HVI values can be informed of the potential risk. Although we provide an illustration of the HVI calculation using a sample of coastal counties, expanding the analysis to include all counties from the southeastern U.S. or adding other indicators would follow this methodology.

6. Mitigating hurricane risk

The HVI provides a guide to areas that are the most vulnerable coastal communities. Although hurricane risk cannot be eliminated, various actions can help to minimize the damage to communities. For example, natural systems can provide some protection against storm damage. Protecting coastal wetlands, which buffer shorelines against erosion and absorb floodwaters from storms will reduce storm damage. Costanza et al., (1989) estimated that coastal Louisiana wetlands provided $452/acre/year (in 2009 dollars) in storm reduction benefits. Although the Mississippi deltaic plain bordering Louisiana provides valuable storm-damage reduction benefits, numerous human activities have destroyed large areas of the coastal marshes. Since 1900, about 4900 km^2 of wetlands in coastal Louisiana have been lost at rates as high as 100 km^2/year, most of the loss is the result of human activities. A large-scale effort is now considering ways to protect and enhance these coastal areas.

Engineering solutions such as levees to protect against flooding may provide some protection from hurricane damage. The British and Dutch have built large, expensive barriers that provide some protection against North Sea storm surge. However, engineering solutions may fail as was the case with New Orleans. An independent study by the American Society of Civil Engineers (2007) explains that engineering and planning failures caused much of the catastrophic flooding from Katrina. Two engineering failures were responsible for the levee breaches in New Orleans: (1) the I-walls were poorly designed, which caused their collapse and (2) overtopping resulted because the levees were not protected from erosion. In addition, the existing pump stations, which should have removed the water during and after the storm, were inoperable.

Storm resistant construction as defined by the International Building and Residential Codes (IBRC) can effectively reduce hurricane damage. In order for homes built along the coast to withstand winds of 130 to 150 miles per hour (the guideline set by the IRBC), it is necessary to use more steel and concrete, hurricane-resistant windows, metal strapping from the foundation to the roof, and plywood to wrap around the house (McLeister, 2007). Recent changes in state building codes require such construction. Following Katrina, state agencies in Louisiana, Mississippi, and Alabama, recognizing the importance of hurricane resistant construction, adopted better building standards.Unfortunately, building codes are not always enforced. Building codes were poorly enforced at the time of the Gulf coast hurricanes in 2005 for the states of Alabama, Louisiana, Mississippi, and Texas (Burby, 2006). Insurance companies offer lower insurance rates for actions such as storm resistant construction. Lower insurance rates could encourage property owners to pay the added cost for the construction.

As shown earlier, in recent decades population and housing growth in coastal areas has escalated; this has been a major contributor to the increased damage from hurricanes. Ensuring that those who choose to live in high risk areas bear the cost of potential damage will decrease development and help to lower damage costs. Market signals, such as higher property insurance rates in high hazard areas, may discourage development in areas that are most susceptible to storm damage. Following the increase in hurricane damage in recent years, property insurance rates in coastal areas have increased, in some cases dramatically. Hairston (2007) reported on the waves of transplants from Florida coastal communities that relocated to Atlanta because of increased property insurance costs.

Government can direct development away from hazardous areas which would limit the potential damage from storms. However, government efforts to discourage development of hazardous areas through the National Flood Insurance Program (NFIP) have been ineffective. The NFIP, begun in 1968, was to guide development away from flood-prone areas and to enforce building standards, in exchange for subsidized flood insurance for property owners. Due to the influence of special interest groups such as land developers, the policy has not been effective. Burby (2001) found that more than a third of the 6.6 million buildings located in the 100-year floodplains of participating communities were built after the start of the NFIP floodplain management plan.

Communities in certain high-risk areas, which are included in the 1982 Coastal Barrier Resource Act (CBRA), are restricted from receiving NFIP insurance. CBRA prohibits federal financial assistance, post-storm reconstruction, and erosion control in undeveloped areas of barrier islands designated by the Department of Interior. Barrier islands parallel mainland areas thus providing a buffer against storms and offering a valuable habitat for fish and wildlife. The goal of CBRA, which was amended by the Coastal Improvement Act in 1990, is to protect barrier island ecosystems by not encouraging development with federal subsidies. The original 186 CBRA locations were expanded to 590 in 1990 (Pasternick 1998, 146).

Unfortunately, the NFIP and other government policies have encouraged development in high hazard areas. Lowering the risk of damage costs to property owners increases coastal development and the damage costs from hurricanes. State wind pools, for example, provide wind insurance at below market prices in some high risk areas. The Florida state government, for example, has created the Citizens Property Insurance Corporation (CPIC) which is a state-run insurance company. The CPIC, which was intended to be an insurer of last resort, is a tax-exempt entity that charges premiums, issues policies, and pays claims, just as a private insurance company does. The CPIC competes with private companies and has the advantage of financing its deficits by levying a surcharge on its competitors. Limiting policies that lower costs of development in coastal areas will be more equitable and will reduce hurricane damage costs.

7. Conclusion

Coastal areas can be hazardous locations in which to reside and additional challenges may increase the risk of damage to coastal communities. For example, land subsidence and melting polar ice caps are causing sea levels to increase, which will cause coastal flooding. Geologists have concluded that large portions of North Carolina's Outer Banks could disappear within the next several decades if sea level continues to increase at the current level or if one or more major hurricanes directly hit the barrier islands (Riggs et al., 2008).

Some project that climate change will cause an increase in global temperatures that could melt enough ice cover in Greenland and Antarctica to increase sea level by 2 to 3 feet in the current century. Because coastal processes occur over so narrow a range (all wave and current energy that affects the shore occurs within a vertical range of 65 feet on the shoreline), small changes in sea level can have significant impacts. The increased flooding caused by sea level rise will exacerbate hurricane damage. In addition, some evidence suggests that climate change may also increase hurricane activity and intensity.

Hurricane strikes will continue to put coastal community residents and their property in harms way. Although we focused on property damage in this study, the information provided here can aid in the protection of individuals and help create communities that can better adapt to the threats from natural disasters such as hurricanes. Despite this risk of severe property damage from hurricanes and other events, all evidence suggests that increasing numbers of people will choose to live in coastal areas. The HVI formulated here can better inform the public about areas that are located in high hazard locations. If individuals are given accurate information about risk, they should be willing to accept the cost of the potential damage.

The HVI can provide valuable information to the insurance industry, which is important to coastal area residents because insurance helps to minimize the financial risk of living in hazardous area. For insurance policies to be feasible, risks must be estimable and manageable for the company, and the rates must be affordable to consumers. A better understanding of the factors that affect hurricane damage can create more equitable rates for policyholders and decrease uncertainty for insurance companies. Better information on the areas that are most likely to suffer from damaging storms can guide policy-makers in creating plans that mitigate storm risk. The factors that contribute most to risk can be addressed and funds can be allocated to the most vulnerable areas.

Although we cannot control the path of a hurricane we can undertake actions to minimize the risks to coastal communities. Improving our understanding of the factors that increase those risks will lead to adaptation and avoidance practices that can help make coastal communities less vulnerable to the vagaries of nature.

8. Acknowledgements

We would like to thank the Francis Marion University Quality Enhancement Plan for providing funding for this project.

9. References

AIR Worldwide Corporation. (2008). The Coastline at Risk: Estimated Insured Value of Coastal Properties. Available at: http://www.air-worldwide.com/PublicationsItem.aspx?id=14604

American Society of Civil Engineers Hurricane Katrina External Review Panel. (2007). The New Orleans Hurricane Protection System: What Went Wrong and Why. Available at:
http://www.asce.org/uploadedFiles/Publications/ASCE_News/2009/04_April/ERPreport.pdf

Blake, E.; Rappaport, E. & Landsea, C. (2007). The Deadliest, Costliest, and Most Intense United States Tropical Cyclones from 1851 to 2006 (and Other Frequently Requested Hurricane Facts). National Oceanic and Atmospheric Administration Technical Memorandum NWS TPC-5. Miami: National Weather Service.

Boruff, B.; Emrich, C. & Cutter, S. (*2005*). Erosion Hazard Vulnerability of US Coastal Counties. *Journal of Coastal Research*, September, Vol. 21, No. 5, pp. 932-942.

Burby, R. (2006). Hurricane Katrina and the Paradoxes of Government Disaster Policy: Bringing about Wise Governmental Decisions for Hazardous Areas. *Annals of the American Academy of Political and Social Science*,3 vol. 604: 171 - 191.

Burby, R. (2001). Flood Insurance and Floodplain Management: The U.S. Experience. *Environmental Hazards* 3, no. 3: 111-122.

Costanza, R.; Farber, S. & Maxwell, J. (1989). The valuation and management of wetland ecosystems. *Ecological Economics*, 1, 335.

Davidson, R., & Lambert. (2001). Comparing the Hurricane Disaster Risk of U.S. Coastal Counties. *Natural Hazards Review*, Volume 2, Issue 3, pp. 132-142.

Fronstin, P. & Holtman, A. (1994). The determinants of residential property damage caused by Hurricane Andrew. *Southern Economic Journal*, 61, 387-97.

Hairston, J. (2007). Insurance-scarred Florida Émigrés Stream to Atlanta:
Wave of Transplants Affects New Architecture, Real Estate Marketing Campaigns, Schools. *The Atlanta Journal-Constitution*, Published on June 23. Available at:
http://www.ajc.com/metro/content/printedition/2007/06/23/biztransplants0623a.html

Hazards & Vulnerability Research Institute (2009). The Spatial Hazard Events and Losses Database for the United States, Version 7.0 [Online Database]. Columbia, SC: University of South Carolina. Available at http://www.sheldus.org

Hewlitt, M. & Railey, J. (2003). Built on Sand: Damage Renews Development Debate, September 22, 2003, *Winston Salem Journal*, Winston-Salem, NC, p. 1A.

Keeney, R. & Raiffa, H. (1993). Decisions with Multiple Objectives: Preferences and Value Trade-Offs. Cambridge University Press.

Insurance Information Institute. (2007). The National Flood Insurance Program. Available at: http://www.iii.org/media/facts/statsbyissue/flood/.
Insurance Services Office (2009). The Building Code Effectiveness Grading Schedule. Jersey City, NJ.

Jain, V.; Davidson, R. & Rosowsky, D. (2005). Modeling Changes in Hurricane Risk over Time. *Natural Hazards Review,* May 2005, 88-96.

Masters, J. U.S. Storm Surge Records. Available at:
http://www.wunderground.com/hurricane/surge_us_records.asp

McLeister, D. (2007). Lessons from Katrina: better building, codes, materials. July 7: 2, 2007.
www.hgtvpro.com/hpro/nws_dstr_huric_torndo/article/0,2624,HPRO_265224503255,00.html

Mortgage News Daily. (2007). Florida Tornados and Scientific Reports Bad News For Homeowners. Thursday April 19. Available at: http://www.mortgagenewsdaily.com/262007_Homeowners_Insurance.asp

National Oceanic Economics Program. (2009). Population and Housing Data. Available at: http://www.oceaneconomics.org/Demographics/demogSearch.asp

National Association of Home Builders. (2007). Housing Facts, Figures, and Trends. February 2007, NAHB Public Affairs and NAHB Economics, Washington, DC.

Pasterick, E. T. (1998). The National Flood Insurance Program. In Howard Kunreuther and Richard J. Roth, Sr., eds., *Paying the Price: The Status and Role of Insurance Against Natural Disasters in the United States*. Washington, DC: Joseph Henry Press.

Pielke, R.; Gratz, J. & Landsea, C. & Saunders, D. & Musulin, R. (2008). Normalized Hurricane Damage in the United States: 1900–2005. *Natural Hazards Review*, Volume 9, Issue 1, pp. 29-42, February.

Riggs, S.; Culver, D. & Ames, D. & Mallison, D. & Corbett, R. & Walsh, J. (2008). North Carolina Coasts in Crisis: A Vision for the Future. North Carolina Coastal Geology Cooperative Research Program, 2008.

Saaty, T. (1980). The Analytic Hierarchy Process: Planning, Priority Setting, Resource Allocation. McGraw-Hill International Book Co.; New York.

South Carolina Department of Insurance. 2007. *Coastal Property Insurance Issues in South Carolina*. Columbia, South Carolina.

Thieler, E.R., and Hammar-Klose, E.S. (1999). A Preliminary Database for the U.S. Atlantic, Pacific and Gulf of Mexico Coasts. U.S. Geological Survey Digital Data Series - 68. Available at: http://pubs.usgs.gov/dds/dds68/htmldocs/ofreport.htm

U.S. Landfalling Hurricane Probability Project. Landfall Probability Table. Available at: http://www.e-transit.org/hurricane/welcome.html

U. S.Census Bureau. (2000). County population. U.S. Census Bureau, Population Division, Washington, D.C., Available at: http://www.census.gov/main/www/cen2000.html

Zandberger, P. (2009). Exposure of US counties to Atlantic tropical storms and hurricanes, 1851–2003. Natural Hazard, Jan. 2009.48:83–99.

Part 5

Societal Impacts: Evacuation Risks and Hazards

22

Factors Associated with Hurricane Evacuation in North Carolina

Horney, JA[1], MacDonald, PDM[1], Berke, P[1],
Van Willigen[2], M and Kaufman, JS[3]
[1]*University of North Carolina at Chapel Hill*
[2]*East Carolina University*
[3]*McGill University*
[1,2]*USA*
[3]*Canada*

1. Introduction

On September 18, 2003, Hurricane Isabel made landfall as a Category 2 storm between Ocracoke Island, North Carolina, and Cape Lookout, North Carolina. The storm entered the Albemarle Sound where strong winds of up to 105 miles per hour and storm surge of 4–6 feet caused extensive flooding and downed trees and power lines. One death and over $450 million in property damage were directly attributed to Hurricane Isabel.

Pasquotank, Perquimans, and Chowan counties are located in northeastern North Carolina between the Albemarle Sound and the Virginia-North Carolina border (Figure 1). In a 2005

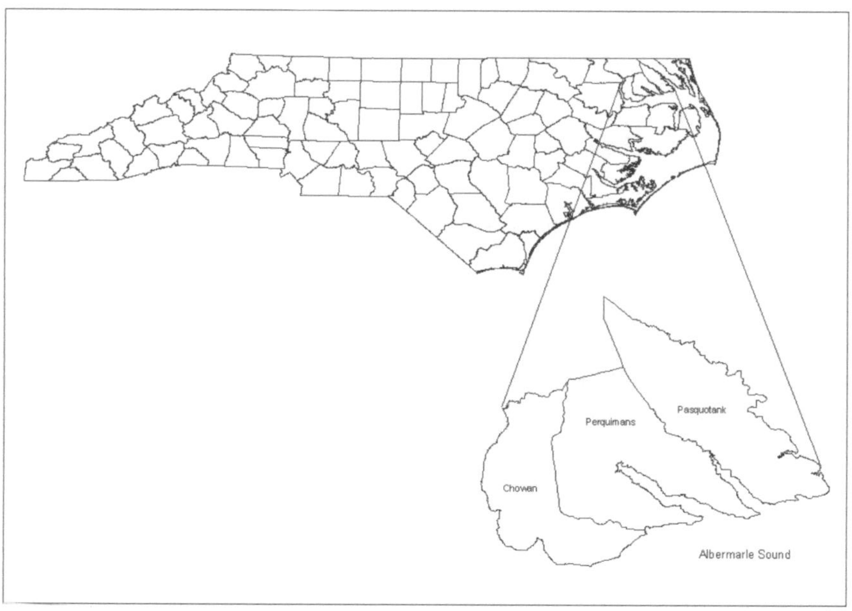

Fig. 1. Map of North Carolina with 3-county study area expanded

report, all 3 counties were classified by the U.S. Army Corps of Engineers as Category 1–3 storm surge areas for Hurricane Isabel, although prior to landfall only Pasquotank County was under a mandatory evacuation order, which replaced a voluntary order approximately 24 hours before landfall. A voluntary order was issued in Perquimans County and no evacuation order was issued in Chowan County. The three counties are part of the 7-county Albemarle Regional Health District and share staff, emergency preparedness plans and other area-level public health and emergency response resources.

2. Background

Existing research on evacuation behavior during hurricanes and flooding has focused primarily on individual demographic characteristics to understand why some households evacuate at higher rates than others. However, the inconsistency of published results and the inability of public health and safety officials to address these factors make it difficult to develop effective interventions or to draw conclusions that will hold true over multiple storms. Social factors such as access to social capital, levels of social control, and the extent of social cohesion also play a role in evacuation behavior. While social factors are generally considered to encourage evacuation, particularly for those with access to large networks and stocks of social capital, the potential for negative effects among certain groups are relatively unexamined.

Previous research has focused extensively on the role of prior disaster experience in evacuation decision making (Aguirre 1994; Dash and Gladwin 2007; Moore et al. 2004; Riad, Norris and Ruback 1999; Strope, Devaney and Nehnevajsa 1977; Wilkinson and Ross 1970). Individual stories of storms that did not make landfall as strongly as or in the location predicted are common, since relatively few areas have experienced direct hits by major hurricanes. Based on reports in the literature, this judgment of the risk of an approaching storm based on the last one that affected the area may lead to successful evacuation or failure to evacuate. Hurricane experience predicted evacuation for residents of Charleston, South Carolina, for Hurricane Emily, which made landfall just 4 years after Hurricane Hugo devastated that city, but hurricane experience was not associated with preparation for or evacuation from Hurricane Fran, which struck the area in 1999, 11 years after Hugo (Sattler 2001).

There is near consensus in the existing literature that people take action regarding evacuation on the basis of their perception of risk (Lindell and Hwang 2008; Lindell and Perry 2004; Riad and Norris 1998). However, how they develop this perception is unclear. In order to accurately assess risk, residents must feel that they are in danger and that leaving the area will be beneficial (Arlikatti et al. 2006; Fitzpatrick and Mileti 1991). Official watches, warnings, and evacuation orders are generally related to evacuation (Baker 2000; Drabek 1969; Edwards et al. 2001; Gladwin and Peacock 1997; Moore et al. 1963; Whitehead et al. 2000; Wilkinson and Ross 1970). However, several studies indicate that personal communications with family, friends, and co-workers and first-hand assessments of the dangers are even more important to the evacuation decision than official warnings (Drabek and Boggs 1968; Killian 1954; Windham, Ross and Spencer 1977).

Recent studies have reported significant associations between evacuation and gender (Bateman and Edwards 2002; Gladwin 2005; Lindell, Lu and Prater 2005; Whitehead et al. 2001), race (Riad et al. 1999; Van Willigen et al. 2005), having children at home (Lindell et al. 2005) and special medical needs (Maiolo et al. 2001; Van Willigen et al. 2002), although overall, associations between personal characteristics and hurricane evacuation have been inconsistent in the published literature (Baker 1991).

Gender differences in evacuation have generally been attributed to variations in socio-economic status, care-giving responsibilities, and perception of risk (Bateman and Edwards 2002). Studies of Hurricane Bonnie (1998) found that female gender of the head of household and lower education levels were significant predictors of evacuation (Bateman and Edwards 2002; Whitehead et al. 2001). However, neither Whitehead (2001) nor Bateman (2002) found an association between gender and evacuation from Hurricanes Dennis (1999) or Floyd (1999), even though Bonnie, Dennis, and Floyd all made landfall in south-eastern North Carolina as Category 2 hurricanes within a 1-year period. In Gladwin's study of Hurricane Ivan evacuation, male gender of the respondent was significantly associated with increased evacuation (Gladwin 2005).

The majority of the literature finds no difference in evacuation based on race or ethnicity, although race and ethnicity may be more strongly correlated with vulnerability to property damage from hurricanes due to differential quality of housing than to evacuation (Van Willigen et al. 2005) and susceptibility of housing locations, particularly to flooding. Race may also be associated with differential access to information and services necessary for successful evacuation.

Quarantelli (1980) reported that evacuation from all types of events was positively associated with having children under age 18 living in the home. However, Baker's (1991) later review of evacuation studies of 12 hurricanes that made landfall between 1961 and 1989 did not find a consistent relationship between households with children under age 18 living at home and evacuation. In North Carolina, the effect of children on the evacuation decision may be explained by the fact that having children in the household typically increases the likelihood of living in a mobile home, generally a predictor of evacuation, by nearly 50% (Edwards et al. 2001).

Age is frequently included in studies of evacuation because of concerns about the limited mobility and special health needs of the elderly. Most studies have failed to find an association between age and evacuation. Those over age 60 were reportedly less likely to evacuate after Hurricanes Carla (Moore et al. 1963) and Andrew (Gladwin and Peacock 1997) made landfall in Florida, and each 5-year increase in age decreased the odds of evacuation by 10% when Hurricane Floyd made landfall in North Carolina (Van Willigen et al. 2005). However, in Perry and Lindell's (1997) review of nine disasters, those older than age 65 were no less likely to comply with disaster warnings than younger residents.

Previous research has been inconsistent with regard to the evacuation of those with special medical needs or disabilities. In a survey conducted following Hurricane Bonnie (1998), households that reported a special medical need were more likely to evacuate, while those with reported physical or mental disabilities were less likely to evacuate than households that did not report a medical need or disability (Whitehead et al. 2001). After Hurricane Floyd, households that included someone who was disabled reported the lowest rates of evacuation of any population sub-group (Van Willigen et al. 2002). Special medical needs may include conditions that require electricity, medical equipment, or home health care, all of which are likely to be disrupted during a hurricane. On the other hand, disabilities may make transportation and accommodation more difficult due to the need for handicap accessible accommodations and personal care facilities.

In terms of social factors, the evacuation literature has generally emphasized a positive association between social capital, social cohesion, and social control and evacuation

(Aguirre 1994; Bland et al. 1997; Dash and Gladwin 2007; Moore et al. 2004; Morrow 1999; Riad et al. 1999; Van Heerden and Streva 2004). However, a number of studies have found negative associations between high levels of social capital, social cohesion, or social control and evacuation (Buckland and Rahman 1999; Cordasco 2006; Gladwin, Gladwin and Peacock 2001; Solomon 1986).

Individuals look to others during an emergency for tangible assistance with evacuation, such as transportation, as well as for emotional support. When warnings or evacuation orders are issued, families tend to gather at home to reach consensus about what action to take (Drabek and Boggs 1968; Drabek and Stephenson 1971; Moore et al. 2004) and prefer to go to the homes of friends or relatives rather than public shelters (Aguirre 1994; Mileti, Sorensen and O'Brien 1992; Moore et al. 2004). For example, studies completed 1 year after Hurricane Hugo in 1989 and 6 months after Hurricane Andrew in 1992 found that perceived social support was a strong predictor of evacuation (Riad et al. 1999). In areas with high levels of social capital and social cohesion, community members may be able to act together to assist those who otherwise may not evacuate. For example, after Hurricane Floyd, residents of eastern North Carolina who believed their property to be safe from flooding reported being awoken by neighbors knocking on their doors during the night warning them of rising water (Moore et al. 2004).

Some research indicates that families with relatives outside the affected area are more likely to evacuate (Drabek and Boggs 1968; Drabek and Stephenson 1971). Called "evacuation by invitation," family members in safe areas provide accommodations for those in affected areas (Quarantelli 1980). This type of evacuation by invitation may be due to the strength of weak ties (Granovetter 1983). Weak ties with those outside ones primary social network may allow for access to information and organizational capacity that encourage evacuation but may not be available to those with a narrower and more local network.

Other research has examined the role of social capital in the receipt and provision of financial, information, and emotional support following major life events. After a disaster, social capital has been shown to be a strong predictor of the help that is received, but a weaker predictor of help that is provided (Kaniasty and Norris 1995). After other major life events (e.g. divorce) the most effective assistance generally comes from strong ties rather than weak ones (Lin, Woelfel and Light 1985). Stronger ties may be a more effective buffer against stress and other impacts of emergencies. However, during a disaster such as a hurricane, which affects entire communities, small dense network ties may not be able to function efficiently if all the members of the group have been negatively impacted by the storm. Individuals who are most strongly embedded in dense, homogeneous, or family dominated networks receive and expect more social support during normal times and during an emergency (Haines, Hulbert and Beggs 1996); whether or not this support encourages or discourages evacuation should be further explored.

Few previous studies have explored the influence of group memberships, such as churches, community organizations, or voluntary associations on hurricane evacuation. Buckland and Rahman's study of floods in several communities in Canada found that the more community organizations a person was a member of, the less likely they were to evacuate. Two studies have explored the role of religious organizations on the evacuation process in areas where the church was central to the community, including research around the Teton Dam floods, where the majority of the population was Mormon (Golec 1980) and the Toccoa Falls Dam flood in Toccoa, Georgia, which killed 39 on the campus of a fundamentalist Christian university.

Several studies have pointed to potential negative impacts of strong social factors on evacuation. Qualitative data collected following Hurricane Katrina from those evacuated to the Houston Astrodome indicated that even when respondents had access to transportation and the financial means to evacuate, some decided not to leave. In addition to financial considerations, "shared norms, local culture and traditions, responsibilities to social networks, and a collective history leading to trusting one's network rather than the authorities" all contributed to the decision not to evacuate (Cordasco 2006). Similar findings were reported by Buckland and Rahman (1999) regarding community preparedness for the 1997 Red River floods. Rosenort, the community with the highest level of social capital as measured by civic involvement, experienced the most conflict in decision making around evacuation. Although their social capital seemed to facilitate better preparation before the floods, peer pressure from residents who chose not to evacuate led many other residents to ignore the mandatory evacuation order.

For a theoretical framework to help understand the potential associations between social factors and evacuation, we turned to Weber's *Economy and Society*. Weber (1922) defined status groups as communities who share the same lifestyle and social restrictions. Certain groups set themselves apart as a status group by mobilizing and investing social resources for returns of wealth, status, or power (Lin 1999). During times of need, such as a disaster, these group ties may be more important than political or governmental structures. These group ties may evolve in such a way as to require "submission to the fashion that is dominant at a given time in society" (Gerth and Mills 1946), in this case, the refusal of certain residents to evacuate. They may also precipitate responsibility for the actions of many others, including friends, family, extended family, and pets, so it becomes less stressful to do nothing and avoid evacuation. Finally, the actions of these groups may be influenced by the "common experience of adversity" (Portes 1998) that are a result of their experiences with past storms or distrust of authorities.

Based on this theoretical framework and the existing literature, we hypothesized that social factors, such as high levels of social capital, social cohesion, and social control, may be more appropriate for explaining a social action such as evacuation than personal or household demographic characteristics. In addition, social factors could mediate the associations between demographic factors and evacuation, helping to explain some of the inconsistency of previous results. Examining associations between social factors and evacuation behavior may help to better understand differential patterns of evacuation and provide opportunities for interventions that could increase rates of evacuation among certain groups or improve preparedness of groups that have been identified as unlikely to evacuate.

3. Methods

3.1 Data sources

Flood insurance rate maps for the 3 counties were obtained from the North Carolina Floodplain Mapping Program. To ensure that each flood zone was represented in the study, census blocks were first stratified by flood zone based on the designation of the block's physical center. Thirty census blocks in each stratum were then randomly selected based on probability proportionate to population size. Within each selected block, 7 interview locations were chosen from a simple random sample of all existing parcels using a geographic information systems-based survey site selection toolkit developed by the North Carolina Division of Public Health in ESRI ArcMap 9.2 (Redlands, CA).

Data were collected between March 15, 2008, and August 23, 2008, using global positioning systems-equipped Trimble Recon Field Data Collectors via in-person interviews with one adult member of each selected household. Data were electronically recorded at the time of interview. Interviewers were routed to each location with a map generated with ESRI ArcPad 6.0.3 Street Map USA (Redlands, CA). Selected households were approached by an interviewer or interview team and gave informed consent. In order to qualify for inclusion, the resident had to be living in the same place as they did when Hurricane Isabel made landfall and all survey questions referred to the respondent's situation at the time of landfall. This research received approval by the Institutional Review Board of the UNC-Chapel Hill School of Public Health (Public Health IRB #06-0426).

3.2 Study variables

Evacuation from Hurricane Isabel was defined as self-reported relocation of a household or any household members to any location other than their primary residence prior to landfall of Hurricane Isabel on September 18, 2003. Social capital, social cohesion, and social control among the study sample were determined using established measures for social cohesion (Sampson, Raudenbush and Earls 1997), social control (Sampson et al. 1997), and social capital (Coleman 1990; Kawachi and Berkman 2000; Putnam 2000). Additional social control measures included markers of territoriality (Riad et al. 1999) and property preparation (Baker 1991; Buckland and Rahman 1999).

Social cohesion was represented by 5 survey questions that asked respondents about their willingness to help neighbors, how close-knit they felt their neighborhood was, whether they trusted their neighbors, how they got along with their neighbors, and whether neighbors shared their values. Responses to each question were on a 5 point Likert scale and had a possible total score of 4 (0 = strongly disagree, 4 = strongly agree) with higher values reflecting greater social cohesion. To determine the consistency of the 5 questions in measuring the single construct of social cohesion, a Cronbach's alpha statistic was calculated in SAS 9.1.3 (Cary, NC).

Social control was represented by 5 survey questions that asked respondents about their likelihood of taking action if they saw children from their neighborhood destroying property, skipping school, fighting, or being disrespectful to an adult. Respondents were also asked about the likelihood that they would write a letter or attend a community meeting if they heard that budget cuts were likely to eliminate a program that was important to them, such as a local fire station. Responses to each question were on a 5 point Likert scale and had a total possible score of 4 (0 = highly unlikely; 4 = highly likely) with higher values reflecting greater social control. Similar to social cohesion, a Cronbach's alpha statistic was calculated in SAS 9.1.3 (Cary, NC) to determine the consistency of responses to the 5 questions.

Social capital was measured in two ways. Following Putnam's (2000) model of civic involvement, respondents were asked to report any memberships in business, civic, community, and religions organizations (0 = no; 1 = yes). To measure engagement, respondents reported the number of meetings they attended each month. Information on organizational social capital specifically related to Hurricane Isabel was also collected, including dichotomous questions for whether the organization provided information or assistance to area residents affected by Hurricane Isabel and whether the respondent themselves volunteered through these organizations to provide assistance to anyone impacted by Hurricane Isabel (0 = no; 1 = yes). The density of friendship and kinship ties was also examined by having respondents report the number of local and non-local friends and family, whether or not the respondent received assistance after Hurricane Isabel from local or non-local friends and family (0 = no; 1 = yes), and whether or not local friends and family evacuated from Hurricane Isabel (0 = no; 1 = yes). Respondents were also asked to report whether how many of their neighbors evacuated from Hurricane Isabel (0 = none; 1 = some; 2 = most or all). To determine if length of residence or hurricane experience was associated with evacuation, respondents were asked to report the number of years they had lived in their current home and in the county, as well as how may hurricanes they had experience in their lifetime. Tenure in the home and county were divided at the median value of 8 years for home and 22 years for county for analysis. Hurricane experience was divided at the median value of hurricanes that respondents reported they had experienced, which was 4.

Several other social factors were measured. Prior to starting an interview, interviewers recorded the presence of markers of territoriality at the residence, including names on mailboxes, no trespassing signs, beware of dog signs, and fenced in yards (0 = no; 1 = yes for each) (Riad et al. 1999). To measure the extent of property preparation and the potential for residents to fail to evacuate in order to monitor their property, residents were asked whether they prepared their property in advance of Hurricane Isabel by putting plywood on windows or taking other measures to protect their property (0 = no; 1 = yes). If residents reported making preparations, they were asked whether keeping an eye on those preparations was part of their reason for failing to evacuate (0 = no; 1 = yes) (Baker 1991; Buckland and Rahman 1999). To measure confidence in local governments' ability to provide evacuation-related services, residents were also asked if they agreed that their county or city provided services they needed in general, such as healthcare, after school and recreation programs for children and other municipal services and how likely they would be to intervene if they saw looters stealing from a neighbor after a hurricane. Responses to both of these questions were on a 5 point Likert scale and had a total possible score of 4 (0 = strongly disagree; 4 = strongly agree).

The interviewer assessed and recorded the type of home prior to the start of the interview. Homes were categorized by the interviewer as a stick built, mobile home or multi-unit dwelling (0 = stick-built; 1 = mobile home; 2 = multi-unit). Respondents were asked whether they owned or rented their homes (0 = rent; 1 = own). Additional demographic covariates measured included age (a bivariate variable was created around the median age of 50 years and used in analysis), race (0 = African-American or other; 1 = white), gender (0 = female; 1 = male), marital status (0 = widowed / divorced / never married; 1 = married), having children under age 18 living at home, having pets, or having a special medical need (0 = no; 1 = yes).

Respondents were asked whether they believed that their home was under an evacuation order prior to Hurricane Isabel's landfall (0 = no; 1 = yes; 2 = don't know). Respondents also were asked to report whether they had an evacuation plan and a disaster supply kit with at least three days of food and water for every member of the household and each pet (0 = no; 1 = yes). Perceived risk was measured by asking respondents to separately characterize the risk of flood and wind damage to their home during a hurricane similar to Hurricane Isabel. (0 = low; 1 = medium; 2 = high).

3.3 Data analysis

Bivariate analyses were performed using generalized linear models to identify any associations between hurricane evacuation and demographic, storm related, and social factor variables. Crude risk differences and 95% confidence intervals (CI) were estimated. CIs that did not include the null value were interpreted as indicating a statistically significant difference in the absolute risk of evacuation between the referent group and the exposed group. For both dichotomous and multilevel exposures, reference categories were selected because they were considered to be most similar to a logical zero. All statistical analyses were conducted in SAS 9.1.3 (Cary, NC).

Multivariate analyses were used to create a parsimonious model of evacuation behavior and adjust for potential confounding. A full multivariable model was developed based on a review of published studies. Since the outcome of interest was common, generalized linear modeling was used to produce risk differences. In order to construct the full model, variables were removed from the full model 1 at a time except indicator variables, which were removed as a group. Based on the χ^2 values from Likelihood Ratio Tests (LRT), variables with a p-value of ≤ 0.20 were retained in the final model (Kleinbaum and Klein 2005). Confounding was assessed by removing variables one at a time from the full model, except indicator variables which were removed as a group. Variables that resulted in a change in the risk difference of greater than or equal to 10% were considered to be confounders and retained in the final model (Maldonado and Greenland 1993).

4. Results

Of those eligible to participate, 86.8% responded to the survey. In the study sample, 28% (n= 162) of the residents interviewed reported evacuating prior to Hurricane Isabel landfall, while 72% (n=408) did not evacuate.

Residents 50 years or older were 10% (95% CI: 2%, 18%) less likely to have evacuated when compared with younger residents. Households with children under the age of 18 living at home were 13% (5%, 22%) more likely to have evacuated than those without children. Race, gender and marital status were not significantly associated with evacuation status. Respondents who had lived in their home for more than the sample median of 8 years were 10% (95% CI: 3%, 18%) less likely to have evacuated. Those who had experienced more than the median number of 4 hurricanes were also 10% (95% CI: 3%, 18%) less likely to have evacuated (Table 1). These findings were consistent with previous studies that have reported lower rates of evacuation for older persons and higher rates for families with children. The finding that long-term residents and those with more hurricane experience were less likely to evacuate was consistent with the hypothesis, assuming that long-term residents generally have higher levels of social capital, social control, and social cohesion.

Those living in mobile homes were 36% (95% CI: 27%, 45%) more likely to have evacuated from Hurricane Isabel than those living in stick-built homes. Homeowners were 10% (95% CI: 0%, 21%) less likely to have evacuated compared to those who rented their homes. Having pets or having a special medical need was not significantly associated with evacuation status (Table 2). These finding were consistent with previous studies that have reported higher rates of evacuation for residents of mobile homes and renters. However, they were inconsistent with recent research reporting lower evacuation rates among pet owners.

Responses for the five social cohesion questions were closely associated (Cronbach's alpha = 0.92) and were therefore aggregated. The range for total social cohesion was 0 to 20 with a median of 15. In the crude analysis, a 1-unit increase in social cohesion was associated with a 1% (95% CI: 0%, 2%) decrease in evacuation. When the social cohesion factors were examined separately, those who strongly agreed they were willing to help neighbors were 6% (95% CI: 1%, 11%) less likely to have evacuated, those who characterized their neighborhood as close knit were 5% (95% CI: 1%, 9%) less likely to have evacuated, those who strongly agreed that they trusted their neighbors were 6% (95% CI: 2%, 10%) less likely to have evacuated, and those who strongly agreed that their neighbors got along well were 5% (95% CI: 1%, 9%) less likely to have evacuated. These findings were consistent with the hypothesis that higher levels of reported social cohesion would be associated with lower rates of evacuation.

Responses for the five social control questions were closely associated (Cronbach's alpha = 0.87) and were therefore aggregated. The range for total social control was 0 to 20 with a median of 17. In the crude analysis, there was no change in evacuation for a 1-unit increase in social control. When the social control factors were examined separately, those who

Variable Description	Evacuated (n=162)		Did not evacuate (n=408)		Risk differences (95% CI)
	n	%	n	%	
Age					
Less than 50	91	33.83	178	66.17	REF
50 Years or Older	71	23.75	228	76.25	-0.10 (-0.18, -0.02)
Race					
African-American or Other	48	32.00	102	68.00	REF
White	114	27.14	306	72.86	-0.05 (-0.14, 0.05)
Gender					
Female	100	31.65	216	68.35	REF
Male	62	24.41	192	75.59	-0.07 (-0.15, 0.01)
Marital Status					
Widowed, Never Married, or Divorced	50	26.88	136	73.12	REF
Married	112	29.17	272	70.83	0.02 (-0.11, 0.06)
Children in Household					
No	77	36.84	132	63.16	REF
Yes	85	23.55	276	76.45	0.13 (0.05, 0.22)
Tenure in Home					
≤8 Years	95	33.57	188	66.43	REF
> 8 Years	67	23.34	220	76.66	-0.10 (-0.18, -0.03)
Hurricanes Experienced					
≤4	89	34.10	172	65.90	REF
> 4	73	23.62	236	76.38	-0.10 (-0.18, -0.03)

Table 1. Distribution, crude risk differences and 95% confidence intervals (95% CI) for demographic factors potentially associated with evacuation from Hurricane Isabel, 2003 (n=570)

strongly agreed they were willing to confront or report children skipping school were 3% (95% CI: 0%, 6%) less likely to have evacuated, while those who strongly agreed they were willing to confront or report children showing disrespect to elders were 3% (95% CI: 0%, 7%) less likely to have evacuated. The other individual social control variables had no effect on evacuation.

Variable Description		Evacuated (n=162)		Did not evacuate (n=408)		Risk differences (95% CI)
		n	%	n	%	
Home Type						
	Stick Built	78	19.50	322	80.50	REF
	Mobile Home	82	55.03	67	44.97	0.36 (0.27, 0.45)
	Multi-Unit	2	9.52	19	90.48	-0.20 (-0.35, -0.04)
Homeownership						
	Rent	44	36.67	76	63.33	REF
	Own	118	26.22	332	73.78	-0.10 (-0.21, 0.00)
Pets						
	No	93	29.43	223	70.57	REF
	Yes	69	27.17	185	72.83	0.02 (-0.05, 0.10)
Special Medical Needs						
	No	15	24.59	46	75.41	REF
	Yes	147	28.88	362	71.12	-0.04 (-0.17, 0.08)

Table 2. Distribution, crude risk differences and 95% confidence intervals (95% CI) for contextual factors potentially associated with evacuation from Hurricane Isabel, 2003 (n=570)

Variable Description		Evacuated (n=162)		Did not evacuate (n=408)		Risk differences (95% CI)
		n	%	n	%	
Markers of Territoriality						
	No	132	31.35	289	68.65	REF
	Yes	30	20.13	119	79.87	-0.11 (-0.19, -0.03)
Prepared Property						
	No	61	28.91	150	71.09	REF
	Yes	101	28.13	258	71.87	-0.01 (-0.09, 0.07)

Table 3. Distribution, crude risk differences and 95% confidence intervals (95% CI) for social control measures potentially associated with evacuation from Hurricane Isabel, 2003 (n=570)

Contrary to our hypothesis, higher levels of social control had little to no effect on evacuation. However, other variables that may also measure social control were supportive of our hypothesis. Residents who had markers of territoriality at their homes, including names on mailboxes, no trespassing or beware of dog signs, or fenced-in yards were 11% (95% CI: 3%, 19%) less likely to have evacuated. Respondents who indicated that they spent time preparing their property prior to the storm were no more or less likely to have evacuated (Table 3).

Social capital was measured in two ways, including organizational participation and the number and location of friends and family. Organizational participation variables were supportive of our hypothesis. Respondents who reported that they were members of a church were 11% (95% CI: 3%, 19%) less likely to have evacuated compared with those who were not church members. Members of business or civic organization (e.g., Rotary, Ruritan, or the American Legion) were 16% (95% CI: 5%, 28%) less likely to have evacuated when compared with those who did not report membership this type of organization. Those who attended more church services or organizational meetings per month were no more or less likely to have evacuated than those who attended fewer meetings. There was also no difference in evacuation for respondents if the organizations they participated in provided relief services to those affected by Hurricane Isabel or if the organizations provided information about Hurricane Isabel to the respondent. However, if the respondent reported being a volunteer through one of these organizations following Hurricane Isabel they were 12% (95% CI: 3%, 21%) less likely to have evacuated (Table 4).

Variable Description		Evacuated (n=162)		Did not evacuate (n=408)		Risk differences (95% CI)
		n	%	n	%	
Attend Church						
	No	69	35.75	124	64.25	REF
	Yes	93	24.73	283	75.27	-0.11 (-0.19, -0.03)
Member of a Club						
	No	155	29.92	363	70.08	REF
	Yes	7	13.46	45	86.54	-0.16 (-0.28, -0.05)
Provided Hurricane Relief						
	No	103	27.83	267	72.17	REF
	Yes	42	24.85	127	75.15	0.03 (-0.05, 0.11)
Provided Hurricane Information						
	No	145	29.29	350	70.71	REF
	Yes	17	23.29	56	76.71	0.06 (-0.05, 0.17)
Volunteered						
	No	144	30.51	328	69.49	REF
	Yes	18	18.37	80	81.63	-0.12 (-0.21,-0.03)

Table 4. Distribution, crude risk differences and 95% confidence intervals (95% CI) for social capital measures potentially associated with evacuation from Hurricane Isabel, 2003 (n=570)

When considering their neighbors' behavior, respondents who indicated that some neighbors evacuated were 21% (95% CI: 11%, 31%) more likely to have evacuated, while those who indicated that most or all neighbors evacuated were 65% (95% CI: 53%, 78%) more likely to have evacuated. Residents who believed that their home was under an evacuation order issued by local authorities were 34% (95% CI: 18%, 50%) more likely to have evacuated when compared with those who believed that an evacuation order did not cover their home. Those who reported that they did not know whether or not an evacuation order covered their home were also 21% (95% CI: 6%, 37%) more likely to have evacuated when compared with those who believed that an evacuation order did not cover their home. Having an evacuation plan was important for successful evacuation, with those who had a

Variable Description		Evacuated (n=162)		Did not evacuate (n=408)		Risk differences (95% CI)
		n	%	N	%	
Neighbors' Evacuation						
	None	71	18.07	322	81.93	REF
	Some	48	39.02	75	60.98	0.21 (0.11, 0.31)
	Most or All	41	83.67	8	16.33	0.65 (0.53, 0.78)
Believed Home Under Evacuation Order						
	No	113	23.84	361	76.16	REF
	Yes	26	57.78	19	42.22	0.34 (0.18, 0.50)
	Don't Know	23	45.10	28	54.90	0.21 (0.06, 0.37)
Evacuation Plan						
	No	31	19.25	130	80.25	REF
	Yes	128	33.86	250	66.14	-0.15 (-0.23, -0.06)
Disaster Supply Kit						
	No	67	30.73	151	69.27	REF
	Yes	95	26.99	257	73.01	-0.04 (-0.12, 0.04)
Perceived Flood Risk						
	Low	105	27.78	273	72.22	REF
	Medium	32	25.81	92	74.19	-0.02 (-0.11, 0.07)
	High	25	36.76	43	63.24	0.09 (-0.04, 0.22)
Perceived Wind Risk						
	Low	35	22.15	123	77.85	REF
	Medium	73	31.60	158	68.40	0.09 (0.00, 0.19)
	High	54	29.83	127	70.17	0.08 (-0.02, 0.18)

Table 5. Distribution, crude risk differences and 95% confidence intervals (95% CI) for storm related measures potentially associated with evacuation from Hurricane Isabel, 2003 (n=570)

plan being 15% (95% CI: 6%, 23%) more likely to have evacuated when compared to those without an evacuation plan. On the other hand, having a disaster supply kit was not significantly associated with evacuation. Neither perceived risk for flood or wind damage was significantly related to hurricane evacuation (Table 5).

Since the associations between demographic, social, and storm-related variables may be confounded by other variables, a multivariable analysis approach was also used. Backward elimination modeling resulted in a final model that included home type, having an evacuation plan, and neighbor's evacuation status. Assessment of confounding resulted in retaining all of these variables in the model due to a change in the RD of more than 10% when each variable was removed from the model. Among survey respondents, those living in mobile homes were 14% (95% CI: 6%, 21%) more likely to have evacuated controlling for having an evacuation plan and neighbor's evacuation. Those with an evacuation plan were 9% (95% CI: 1%, 18%) more likely to have evacuated controlling for home type and neighbor's evacuation and those who reported that some, most, or all of their neighbor's evacuated were 46% (95% CI: 31%, 61%) more likely to have evacuated controlling for home type and having an evacuation plan.

5. Discussion

The associations between hurricane evacuation and individual and household demographic factors have been somewhat inconsistent in published studies. In this study, there were no significant associations between demographic variables and evacuation failure except age and having children under age 18 at home. The finding that older residents were less likely to evacuate may be due to actual or perceived difficulties in evacuation or based on experience with previous storms. Since the last major hurricane to affect this area was Hurricane Hazel in 1954, older residents may have believed that they were not at risk. The fact that there are few differences between demographic groups in this study leads us to question the construction of social difference in this region and suggests that further research focusing on other factors related to the evacuation decision is warranted.

The type of home and whether the respondent rented or owned the home were strongly predictive of evacuation. Those who live in mobile homes are clearly aware of added dangers of failing to evacuate during severe weather when compared to those who live in single family homes. It is also reasonable that renters have less at stake in terms of the damage that may occur to their homes. Renters are unlikely to have any financial or other responsibility for damages that may occur to a landlord's property, and therefore have little interest in staying through a storm to see how the property fares, although they do have their own contents such as furniture or clothing at risk of damage or loss.

The belief that their property was covered by an evacuation order issued by local government officials was an important factor in residents' decision to evacuate from Hurricane Isabel, indicating that the issuance of evacuation orders is effective in encouraging evacuation, or at least in shaping risk perception. In addition, those who reported that they did not know whether or not an evacuation order covered their home were also more likely to chose to evacuate, perhaps feeling that it was better to be safe than sorry. A closer examination of the covariate pattern for those reporting that they did not know whether their home was covered by an evacuation order showed that they were more than twice as likely to live in a mobile home and about three times as likely to have children under 18 years old living at home compared to the overall study sample.

Although it has been a consistent predictor of evacuation in the published literature (Drabek 1969; Drabek and Boggs 1968; Edwards et al. 2001; Gladwin and Peacock 1997; Killian 1954; Lindell and Hwang 2008; Lindell and Perry 2004; Moore et al. 1963; Riad and Norris 1998; Wilkinson and Ross 1970; Windham et al. 1977; Whitehead et al. 2001), the perception of risk of damage from either flooding or high winds was not associated with evacuation in this study (Horney 2010). This leads us to ask: How bad must respondents perceive conditions are before they decide to evacuate? Perceived risk includes not only the official or personal assessment of the severity of the threat (e.g., the issuance of the evacuation order) but also the individuals' perceived susceptibility (Houts et al. 1984; Perry et al. 1981; Riad and Norris 1998). While residents' perceived susceptibility to flooding or wind damage was not enough to spur evacuation from Hurricane Isabel, perceived severity as determined by an evacuation order was. Those who live in an area where they feel the risks for flood and wind damage are severe may not see a way to avoid the anticipated negative effects of a strong storm and decide to take no action. The issuance of an evacuation order removes perceived barriers by providing information on open shelters and evacuation routes as part of the issuance of the evacuation order. Having an evacuation plan may also provide a cue to action for evacuation similar to that of an evacuation order.

Higher levels of social cohesion were associated with an increase in hurricane evacuation failure. There is much scientific and anecdotal evidence that communities come together in the face of a disaster. It makes sense that neighbors who trust each another, get along well, and are willing to help each another may feel more comfortable remaining in their homes and neighborhoods rather than evacuating. The social resources available through direct ties to neighbors can provide access to the temporary support necessary for coping with storm impact and dealing with the initial phases of recovery (Lin 1999). These findings are consistent with the importance that neighbors' evacuation status had on the respondent's evacuation. Neighbors who do not evacuate may contribute to a downward leveling of norms which encourages those they know and trust not to evacuate. However, these results may be unrelated to social factors. Neighbors are likely to have the same information about a storm's anticipated severity, either due to location (e.g., areas near water or low-lying areas) or housing quality (e.g., trailer parks or suburban developments); therefore, their decision to evacuate may be unrelated to the influence of their neighbors.

Higher levels of overall social control were not associated with evacuation failure. However, indicators of social control such as markers of territoriality (Riad et al. 1999) were important. Posting no trespassing signs may indicate an unwillingness to follow government-issued evacuation orders or a lack of interest in taking part in the social action of an evacuation. Markers of territoriality may demonstrate a type of "territorial defense" (Riad et al. 1999) which makes residents who choose to utilize them more likely to avoid evacuation in order to protect their property from flooding, a storm surge, or looting. Personalization (names of mailboxes), signs (no trespassing) and barriers (fences) may also be markers of long-term ownership or territorial behavior (Riad et al. 1999). Since the presence of markers of territoriality were noted by the interviewer prior to making contact with the respondent, this measure may be a more unbiased indicator of whether or not a respondent would trust their neighbors as some respondents may be reluctant to report distrust of neighbors who they believe may also be approached by the interviewer.

Civic involvement was an important factor in hurricane evacuation failure. Members of churches and other community or civic groups were less likely to evacuate, as were volunteers. Clearly, civic engagement engenders ties to the community that may inhibit evacuation, either through peer pressure or the anticipated need for assistance through volunteerism. Friendship and kinship ties were not associated with an increased risk of evacuation failure. Additional analyses of the density of relationships with friends and family using splines or other methods to account for outliers (e.g., some respondents who reported hundreds of local friends and relatives) should be explored in the future.

A strength of this study is the generalizability of the results to the entire three-county area. The GIS-based survey site selection toolkit allowed for random selection of households in the second stage of sampling. This ensured that selected households were independent and represented the totality of the households in the cluster (Lemeshow and Robinson 1985). This modification also prevented the selection bias that may have been introduced by allowing interviewers to select households for subsequent interviews after beginning at a random starting point (e.g., if interviewers avoided homes that appeared to be poorly maintained or had unrestrained pets). Additional strengths of the study include the strong local partnerships with public health and emergency management officials, which contributed to very high response rates, and the use of handheld technology for data collection, which has been demonstrated to improve data quality (Fletcher et al. 2003; Lal et al. 2000).

This study has several limitations. If those who are at highest risk for evacuation failure were also more likely to be missed in this survey, there is potential for response bias. To minimize this problem, interviews were conducted on weekends and weekdays during both day and evening hours. However, only those who were still living in the same location as they were when Hurricane Isabel made landfall were eligible to participate. Renters, those living in poverty, and other underserved groups may be more likely to move to different addresses or stay with friends or family members for a period of time and therefore would have been ineligible to participate. In addition, due to the nature of the questionnaire, only the characteristics and actions of residents were measured. Therefore the role that local governments and other agencies played in evacuation decision making of residents could not be assessed.

Since Hurricane Isabel made landfall nearly 5 years prior to the survey, recall bias could have been a factor in this study. However, a hurricane is a major event in the life of a community, so it seems unlikely that residents would have trouble remembering the effects of the storm or the actions they took in response to it. Some residents may not have wanted to report to the interviewers that they did not evacuate, particularly since accurate forecasted warnings regarding flooding for Hurricane Isabel were widely available prior to landfall. Additionally, since knowledge and beliefs were self-reported by survey respondents, the associations reported between these variables (e.g., the perception that an evacuation order covered your residence) and evacuation failure may have been the result of differential misclassification due to recall bias. Those who chose to evacuate may be more likely to report an evacuation order covered their home as a justification for their decision. Recall bias would not be a concern for variables that were rated by the interviewer or for self-reported demographic variables. Finally, since evacuation status and exposure to social factors were measured in the same interview, there is a potential for dependent errors that could bias results away from the null even if these errors were non-differential.

6. Conclusion

In this study, demographic characteristics including race, gender, martial status, having pets, or having a special medical need were not significantly associated with hurricane evacuation. However, social cohesion, markers of territoriality, civic engagement, and volunteerism were associated with a decrease in hurricane evacuation. When studying a complex action such as hurricane evacuation, a compositional approach that considers only the demographic characteristics of individuals has many limitations. In addition, it is difficult to develop effective interventions based on demographic factors, many of which are non-modifiable by public health scientists or policymakers (e.g., we can't require pet ownership or marriage to encourage evacuation). Using a contextual approach, targeted interventions - such as house to house visits to encourage evacuation among those with markers of territoriality or in neighborhoods where evacuation rates are traditionally low or the development of educational programs on evacuation planning targeted to civic groups, churches and volunteers - could be developed by policy makers and planners to take advantage of neighborhood ties, civic engagement, and peer influence to encourage protective behavior and empower local residents.

7. References

Aguirre, Benigno E. 1994. Planning, Warnings, Evacuations, and Search and Rescue: A Review of the Social Science Research Literature: Proceedings of the U.S-Russia Seminar on Social Research on Mitigation for and Recovery from Disasters and Large Scale System Hazards. Newark, Delaware: The University of Delaware Disaster Research Center.

Arlikatti, Sudha, Michael K. Lindell, Carla S. Prater, andd Yang Zhang. 2006. "Risk Area Accuracy and Hurricane Evacuation Expectations of Coastal Residents." *Environment and Behavior* 38(2): 226-47.

Baker, Earl J. 1979. "Predicting Response to Hurricane Warnings: A Reanalysis of Data from Four Studies." *Mass Emergencies* 4: 9-24.

———. 1991. "Hurricane Evacuation Behavior." *International Journal of Mass Emergencies and Disasters* 9(2): 287-310.

———. 2000. "Hurricane Evacuation in the United States." Pp. 306-319 in *Storms*, edited by R. Pielke, Jr. and R. Piekle, Sr. London: Routledge.

Bateman, Julie M., and Bob Edwards. 2002. "Gender and Evacuation: A Closer Look at Why Women are More Likely to Evacuate for Hurricanes." *Natural Hazards Review* 3: 107-117.

Bland, Susan H., Erin O'Leary, Eduardo Farinaro, Fabrizo Jossa, Vittoria Krogh, John Violanti, Maurizo Trevisan. 1997. "Social Network Disturbances and Psychological Distress Following Earthquake Evacuation." *Journal of Nervous and Mental Disease* 185(3): 188–95.

Buckland, Jerry, and Matiur Rahman. 1999. "Community Based Disaster Management During the 1997 Red River Floods in Canada." *Disasters* 23(2): 174-191.

Coleman, James S. 1990. *Foundations of Social Theory*. Cambridge, MA: Harvard University Press.

Cordasco, Kristina M. 2006. "The Paradox of Social Capital as a Liability in Disaster Management: Understanding the Evacuation Failure of Hurricane Katrina." *Natural Hazards Observer*: 5–6.

Dash, Nicole and Hugh Gladwin. 2007. "Evacuation Decision Making and Behavioral Responses: Individual and Household." *Natural Hazards Review* 8(3): 69-77.

Drabek, Thomas E. 1969. "Social Processes in Disasters: Family Evacuation." *Social Problems* 16: 336–49.

Drabek, Thomas E. and Keith S. Boggs. 1968. "Families in Disaster: Reactions and Relatives." *Journal of Marriage and Family* 30: 443–51.

Drabek, Thomas E, and John J. Stephenson. 1971. "When Disaster Strikes." *Journal of Applied Social Psychology* 1(1): 187–203.

Edwards, Bob, John Maiolo, Marieke Van Willingen, John Whitehead and Ken Wilson. 2001. *Hurricane Evacuation and Willingness to Purchase Evacuation Insurance in Southeastern North Carolina.* Report Prepared for: Fiduciary Consultants.

Fitzpatrick, C. and Dennis Mileti. "Motivating Public Evacuation." *International Journal of Mass Emergencies and Disasters* 9: 647-53.

Fletcher, Linda A., Darin J. Erickson, Traci L. Toomey, and Alexander C. Wagenaar. 2003. "Handheld Computers: A Feasible Alternative to Paper Forms for Field Data Collection." *Evaluation Review* 27(2): 165-178.

Gerth, H.H. and W. Mills. 1946. "Class, Status, Party." In *Max Weber: Essays in Sociology*, edited by H.H. Gerth and W. Mills. New York: Oxford University Press. 180-194.

Gladwin, Hugh. 2005. *Hurricane Ivan Behavioral Analysis.* Report prepared for the Federal Emergency Management Agency and the U.S. Army Corps of Engineers.

Gladwin, C., Hugh Gladwin, and Walter Peacock. 2001. "Modeling Hurricane Evacuation Decisions with Ethnographic Methods." *International Journal of Mass Emergencies and Disasters* 19(3): 117–43.

Gladwin, Hugh and Walter Peacock. 1997. "Warning and Evacuation: A Night for Hard Houses." Pp. 52-74 in *Hurricane Andrew: Ethnicity, Gender and the Sociology of Disasters*, edited by Walter Peacock, Betty Hearn Morrow, and Hugh Gladwin. New York: Routledge.

Granovetter, Mark. "The Strength of Weak Ties: A Network Theory Revisited." *Sociological Theory* 1983 (1): 201–33.

Golec, Judith A. 1980. *Aftermath of Disaster: The Teton Dam Break* [dissertation]. Columbus, OH: Ohio State University.

Haines, Valerie A., Jeanne S. Hurlbert, and John J. Beggs. 1996. "Exploring the Determinants of Support Provision: Provider Characteristics, Personal Networks, Community Contexts, and Support Following Events." *Journal of Health and Social Behavior* 37(3): 252-64.

Horney, J, MacDonald, PDM, Van Willigen, M, Berke, P, Kaufman, J. Individual action or perceived flood risk: Did it predict evacuation from Hurricane Isabel in North Carolina? *Risk Analysis 30* (3), 501-511.

Houts, Peter S., Michael K. Lindell, Teh Weittu, Paul D. Cleary, George Tokuhata, and Cynthia B. Flynn. 1984. "The Protective Action Decision Model Applied to Evacuation During the Three Mile Island Crisis." International Journal of Mass Emergencies and Disasters 2 (1): 27-39.

Kaniasty, Krzysztof and Fran H. Norris. 1995. "In Search of Altruistic Community: Patterns of Social Support Mobilization Following Hurricane Hugo." *American Journal of Community Psychology* 23(4): 447-77.

Kawachi, Ichiro and Lisa Berkman. 2000. "Social Cohesion, Social Capital, and Health." Pp 174-190 in *Social Epidemiology*, edited by Lisa Berkman and Ichiro Kawachi. New York: Oxford University Press.

Killian, Lewis M. 1954. *Evacuation of Panama City Residents Before Hurricane Florence.* Washington, DC: National Academy of Science.

Kleinbaum, David G. and Mitchel Klein. 2005. *Logistic Regression.* 2nd ed. New York, NY: Springer.

Lal, Sophia, FW Smith, JP Davis, HY Castro, DW Smith, DL Chinkes, and RE Barrow. 2000. "Palm Computer Demonstrates Fast and Accurate Means of Burn Data Collection." Journal of Burn Care and Rehabilitation 21 (6): 559-561.

Lemeshow, Stanley, and David Robinson. 1985. "Surveys to Measure Programme Coverage and Impact: A Review of the Methodology Used by the Expanded Programme on Immunization." *World Health Statistics Quarterly* 38: 65-75.

Lin, Nan, Stephen C. Woelfel and Mary W. Light. 1985. The Buffering Effect of Social Support Subsequent to an Important Life Event. *Journal of Health and Social Behavior* 26(3): 247-263.

Lin, Nan. 1999. "Social Networks and Status Attainment." *Annual Review of Sociology* 25: 467-487.

Lindell, Michael K. and Seong Nam Hwang. 2008. "Households' Perceived Personal Risk and Responses in a Multi-Hazard Environment." *Risk Analysis* 28: 539-556.

Lindell, Michael K, Jing-Chein Lu, and Carla S. Prater. 2005. "Household Decision Making and Evacuation in Response to Hurricane Lili." *Natural Hazards Review* 6 (4): 171-179.

Lindell, Michael K. and Ronald W. Perry. 2004. *Communicating Environmental Risk in Multiethnic Communities*. Thousand Oaks, CA: Sage.

Maiolo, John R., John C. Whitehead, Monica McGee, Lauriston King, Jeffery Johnson, Harold Stone, editors. 2001. *Facing Our Future: Hurricane Floyd and Recovery in the Coastal Plain*. Wilmington, NC: Coastal Carolina Press.

Maldonado, George and Sander Greenland. 1993. "Simulation Study of Confounder-Selection Strategies." *American Journal of Epidemiology* 138(11): 923–36.

Mileti, Dennis S., J.H. Sorensen and P.W. O'Brien. 1992. "Towards an Explanation of Mass Care Shelter Use in Evacuations." *International Journal of Mass Emergencies and Disasters* 10 (1): 25-42.

Moore, Harry Estill, Frederick L. Bates, Marvin V. Layman and Vernon J. Parenton. 1963. "Before the Wind: A Study of the Response to Hurricane Carla." (Disaster Study # 19). Washington, DC: National Academy of Science.

Moore, Spencer, Mark Daniel, Laura Linnan, Marci Campbell, Sally Benedict and Andera Meier. 2004. "After Hurricane Floyd Passed: Investigations the Social Determinants of Disaster Preparedness and Recovery." *Journal of Community Health* 27(3): 204–17.

Morrow, Betty Hearn. 1999. "Identifying and Mapping Community Vulnerability." *Disasters* 23(1): 1–18.

Perry, Ronald W. and Michael K. Lindell. 1997. "Aged Citizens in the Warning Phase of Disasters: Re-Examining the Evidence. *International Journal of Aging and Human Development* 44(4): 257-67.

Perry, Ronald W., Michael K. Lindell and Marjorie R. Greene. 1981. *Evacuation Planning in Emergency Management*. Lexington, MA: DC Health.

Portes, Andrea. 1988. "Social Capital: Its Origins and Applications in Modern Sociology." *Annual Review of Sociology* 24: 1–24.

Putnam, Robert D. 2000. Bowling Alone: The Collapse and Revival of American Community. New York: Simon and Schuster.

Quarantelli, E. L. 1980. *Evacuation Behavior and Problems: Findings and Implications from the Research Literature*. Columbus, OH: Disaster Research Center, Ohio State University.

Riad, Jasmin K., and Fran H. Norris. 1998. Hurricane Threat and Evacuation Intentions: An Analysis of Risk Perception, Preparedness, Social Influence, and Resources. Newark, DE: Disaster Research Center, University of Delaware.

Riad, Jasmin K., Fran H. Norris, and R. Barry Ruback. 1999. "Predicting Evacuation in Two Major Disasters: Risk Perception, Social Influence, and Access to Resources." *Journal of Applied Social Psychology* 29 (5): 918-934.

Sampson, Robert J., Stephen W. Raudenbush and Felton Earls. 1997. "Neighborhoods and Violent Crime: A Multilevel Study of Collective Efficacy." *Science* 277: 918-924.

Sattler, David N. 2001. *Repeated Exposure to Hurricanes and Willingness to Evacuate: Implications of the Hurricane Floyd Experience*. Paper presented at: Conference on Social, Physical, and Economic Impacts on Groups and Individuals, Greenville, NC.

Solomon, Susan D. 1987. "Social Involvement as a Mediator of Disaster-Induced Stress." *Journal of Applied Social Psychology* 17(12): 1092–1112.

Strope, Walmer E., John F. Devaney and Jiri Nehnevajsa. 1977. "Importance of Preparatory Measures in Disaster Evacuations." *Mass Emergencies* 2: 1–17.

U.S. Army Corps of Engineers, National Oceanic and Atmospheric Administration, and Federal Emergency Management Agency. 2005. "Hurricane Isabel Assessment: Review of Hurricane Evacuation Study Products and Other Aspects of the National Hurricane Mitigation and Preparedness Program (NHMPP) in the Context of Hurricane Isabel Response." Retrieved February 8, 2008 (http://www.csc.noaa.gov/hes/docs/postStorm/Isabel_PostStorm_Summary.pdf).

Van Heerden, I and Kathryn Streva. *Assessment and Remediation of Public Health Impacts due to Hurricanes and Major Flooding Events: Year 3*. Report prepared for: Center for the Study of Public Health Impacts of Hurricanes; 2004; Baton Rouge, LA. Available at: http://www.publichealth.hurricane.lsu.edu/. Retrieved September 24, 2008.

Van Willigen, Marieke, Bob Edwards, Terri Edwards, and Shaun Hessee. 2002. "Riding Out the Storm: The Experiences of the Physically Disabled During Hurricanes Bonnie, Dennis, and Floyd." *Natural Hazards Review* 3 (3): 98-106

Van Willigen, Marieke, Bob Edwards, Stephanie Lormand, Ken Wilson, and John Maiolo. 2005. "A Comparative Assessment of Impacts and Recovery from Hurricane Floyd among Student and Non- Student Households" *Natural Hazards Review* 6: 180-190.

Whitehead, John, Bob Edwards, Marieke Van Willigen, John Maiolo and Ken Wilson. 2000. "Hurricane Evacuation Behavior: A Preliminary Comparison of Bonnie, Dennis, and Floyd." In *Facing Our Future: Hurricane Floyd and Recovery in the Coastal Plain*, edited by John Maiolo. Wilmington, NC: Coastal Carolina Press. 89-98.

Whitehead, John, John Maiolo, Bob Edwards, Ken Wilson, Marieke Van Willigen, C. Williams and M. Meekins. 2001. *A Socioeconomic Impact Analysis for Hurricanes Bonnie, Dennis and Floyd.* Report prepared for: North Carolina Department of Emergency Management Grant HMGP 1240-0012; Raleigh, NC.

Wilkinson, K.P. and P.J. Ross. 1970. *Citizens' Response to Warnings of Hurricane Camille.* Report prepared for: Mississippi State University Social Science Research Center; State College, MS.

Windham, G.O., P.J. Ross and B. Spencer. 1977. *Reactions to Storm Threat During Hurricane Eloise.* Report prepared for: Mississippi State University Social Science Research Center; 1977; State College, MS.

23

The Role of Non-Recurring Congestion in Massive Hurricane Evacuation Events

Daniel J. Fonseca, Gary P. Moynihan and Huston Fernandes
The University of Alabama
USA

1. Introduction

The response to a potential disaster can require the evacuation of personnel from a specified area. Generally, such efforts are restricted to the orderly mass departure of individuals across pre-planned and well maintained transportation routes. In the U.S., evacuations of up to 1,000 subjects take place every two to three weeks, with more extreme evacuations involving two million or more every one to three years (TRB, 2008).

While evacuation routes are designed to accommodate normal traffic movements, congestion and gridlock can occur as the design capacity of the road system is overwhelmed by the magnitude of vehicles leaving the affected area. The resulting traffic patterns affect the safety and mobility of subjects moving to more secure areas. Adding to this disarray, potential non-recurring incidents congest traffic patterns even more. Estimates indicate that between fifty and sixty-five percent of traffic congestion is caused by non-recurring traffic incidents with an additional ten percent related to construction and weather (Coifman, 2007). A non-recurring traffic incident is any event that both causes a reduction of roadway capacity, or an abnormal increase in demand, and requires first responders to be dispatched. Stalled vehicles, roadway debris, spilled loads, and crashes fall into this category of incidents.

Non-recurring traffic incidents can cause secondary traffic incidents. These incidents further congest the traffic stream and cause delays in clean-up efforts by first-responders. Studies indicate that twenty percent of traffic incidents are secondary incidents, with one out of five resulting in a fatality. In addition to crashes, secondary incidents can include overheated vehicles, out of fuel conditions, and engine stalls.

The delay and traffic gridlock associated with traffic incidents is compounded during the evacuation process due to the large numbers of subjects leaving the affected area. These delays and backups result in:

- Increased response time by first responders
- Lost time resulting in a wider evacuation window
- Increased fuel consumption
- Reduced air quality and other adverse environmental conditions
- Increased potential for more serious secondary incidents resulting from rear end collisions, traffic exiting the route, or exiting to the shoulder of the road
- Increased potential of crashes by incidents involving personnel responding to traffic incidents
- Negative public image of first responders involved in incident management activities.

This paper discusses the dynamics involved in non-recurrent congestion and its potential impact on massive evacuation events, as in the case of a hurricane evacuation. It also presents a comprehensive literature review of the most widely used incident detection algorithms (AID's) developed throughout the years to identify non-recurring congestion conditions.

2. Non-recurring congestion

Roadway incidents refer to non-recurring events resulting in traffic congestion or disruptions. During an incident, the normal capacity of the roadway is restricted, leading to queues and delays. Incidents have far-reaching consequences for safety, congestion, pollution, and the cost of travel.

Previous studies indicated that incidents are one of the major causes of loss of time and increases in avoidable costs in transportation networks in the U.S. For example, in 2003, it was determined that more than 60% of urban freeway congestion was caused by incidents, and that indicator was estimated to be 70% by 2005 (Schrank & Lomax, 2003). Prompt and reliable incident detection is vital in reducing incident congestion, post-incident delay, and the potential for additional incidents.

Incident management is a crucial function in the design and deployment of Advanced Transport Management Systems (ATMS) and Advanced Traveller Information Services (ATIS). It primarily includes incident detection, verification, validation, response, and clearance. Incident detection is a critical step in incident management. It affects consequent actions and determines the reliability and efficiency of the whole system. The procurement of real-time incident detection information is an integral element of and supports the realization of many other functions in traffic management (Presley & Wyrosdick, 1998). Nevertheless, incident detection is one of the weakest links in implementing advanced traffic controls.

Early detection of traffic incidents can both reduce the time to return traffic to normal rates of flow and reduce the potential for secondary incidents (Busch, 1987), thus increasing the number of vehicles leaving the affected area. It would be expected, therefore, that traffic management techniques oriented to reporting and identification of changes in traffic flow patterns would have dramatic effects on the reduction of the impact of traffic incidents in emergency evacuations, such as in the case of a massive evacuation prompted by a hurricane event. Among such techniques incident detection paradigms have been the most widely sought by practitioners in the field.

3. Incident detection

3.1 Traffic data collection

Despite substantial research, the implementation of effective incident detection algorithms has been hampered by limited performance reliability, substantial implementation needs, and strong data requirements (Takaba & Matsuno, 1985).

Traffic information is typically collected from loop detectors and includes occupancy and volume data gathered over a 20- to 60-second interval. Inductive loop detectors (ILD) consist of one or more loops of wire embedded in the pavement and connected to a control box, excited by a signal ranging in frequency from 10 KHz to 200 KHz. When a vehicle passes over or rests on the loop, the inductance of the loop is reduced, showing the presence of a vehicle. The data supplied by inductive loop detectors are vehicle passage, presence, count, and occupancy.

The benefits of inductive loop detectors come from the fact that they are an established technology in the United States and in most parts of the world, they have a well defined zone of detection, and they are generally reliable. ILDs are an established technology in the United States and in most parts of the world. They have a well defined zone of detection, and they are considered reliable. ILDs are generally used for the following reasons (Hualiang & Qi, 2003):

- When properly installed and maintained, ILDs continue to be the best in all weather applications.
- They are the most consistently accurate detectors in terms of vehicle counts.
- ILDs perform well in both high and low volume traffic and in different weather conditions.
- Even with crosstalk problems and a high proportion of lane changes, ILDs have over counts of only around 0.6 percent.
- ILDs meet even the most stringent vehicle flow error specifications required by some ITS applications.

Some reported ILD disadvantages are (Hualiang & Qi, 2003):

- The loop detector system may suffer from poor reliability, primarily from improper connections made in the pull boxes and in the application of sealants over the saw cut. These problems are accentuated when loops are installed in poor pavement or in areas where utilities frequently dig up the roadbed.
- Sources of loop malfunction, such as stuck sensors, can produce erroneous data and may lead to inaccurate detection.
- Another disadvantage of loops is their inability to directly measure speed. If speed is required, then a two-loop speed trap is employed, or an algorithm involving loop length, average vehicle length, time over the detector, and number of vehicles counted is used with a single loop detector.

3.2 Performance measures of incident detection algorithms

Different algorithms may be suitable on different stretches of a road. A gauging factor to be considered is the detection rate and the false alarm rate. The detection rate (DR) is the number of incidents detected as a percentage of the number of incidents occurred. The false alarm rate (FAR) is the number of false alarm signals as a percentage of tests performed by the algorithm. The DR and the FAR are defined as follows:

$$Detection\ rate = \frac{number\ of\ detected\ incidents}{total\ number\ of\ incidents\ detected} \times 100\% \qquad (1)$$

$$False\ Alarm\ Rate\ (FAR) = \frac{Number\ of\ False\ Alarms}{Total\ Number\ of\ Applications\ of\ the\ Algorithm} \times 100 \qquad (2)$$

Most algorithms use thresholds for incident detection. Thresholds for the algorithms are generally calculated by trial and error and from empirical experimentation on historical data and performance curves. These curves are obtained from multiple runs of the respective algorithm on data with incrementally changing thresholds. Several factors affect the performance of all types of incident detection algorithms. The key factors are shown in Table 1 (Busch, 1987).

3.3 Classification of incident detection algorithms

The concept of incident detection algorithms is not new. Algorithms have been developed as early as the 1970's and new algorithms are being developed even now. Depending on how an algorithm analyzes the traffic data in order to detect incidents, it is usually classified into one of five major categories: comparative algorithms, statistical algorithms, time-series and filtering-based algorithms, traffic theory based-algorithms, and advanced algorithms (Dudek, Messer, & Nuckles, 1974).

3.3.1 Comparative algorithms

Comparative algorithms evaluate the tracking variables against certain thresholds or against one another in order to identify anomalies. The tracking variable is usually one of the traffic parameters or a variable derived from the traffic parameters. Occupancy is the most common tracking variable (Chow et al., 1977). The comparative algorithms are also sometimes referred to as the pattern recognition algorithms, as the process is analogous to identification of patterns of behavior of the variables under incident conditions. The California Algorithm is a classic example of this category (Courage & Levin, 1968).

3.3.2 Statistical algorithms

Statistical algorithms use standard statistical techniques to identify sudden changes and other unusual behavior in the variable. Incidents generally result in unusual behavior of the traffic variables. These algorithms are based on the premise that the reverse is also true under most circumstances and that such behavior indicates incidents. The standard traffic variables - flow, average speed, and lane occupancy - as well as variables derived from these primary variables have been used as tracking variables. Examples of the statistical approach include the Standard Normal Deviate and the Bayesian Algorithm (Payne & Tignor, 1978).

Operating conditions of the highway, e.g. at full capacity
Duration of the Incident
Geometric Factors: • Grade • Lane Drops • Ramps
Environmental Factors: • Snow, Ice or Fog • Road Surface - Dry or Wet
Severity of the Incident
Detector Spacing
Location of the Incident Relative to the Detector Station
Heterogeneity of the Vehicle Fleet

Table 1. Factors Affecting Incident Detection

3.3.3 Time series and filtering algorithms

Time series and filtering algorithms treat the tracking variable as a time-series variable. Deviation of the variable from the modeled time-series behavior is used to indicate incidents. The challenge is to differentiate random variations from variations due to incidents. These models include the Auto-Regressive Integrated Moving Average (ARIMA)-based algorithm, the Exponential Smoothing-based Algorithm, and the Kalman Filtering-based Algorithm (Chow et al., 1977).

3.3.4 Traffic theory-based algorithms

The traffic theory-based algorithms depend on the relationship between the traffic variables for their analysis. For example, the McMaster Algorithm, which is based on catastrophe theory, determines the state of traffic based on its position in the flow-density plot. It detects incidents based on the transition from one state to another. The GLR algorithm and the McMaster algorithm are examples of traffic theory-based algorithms (Chow et al.,1977).

3.3.5 Advanced algorithms

In some cases, algorithms with advanced mathematical formulation-based techniques, as well as algorithms that incorporate inexact reasoning and uncertainty into the detection logic, have been developed. These algorithms are based on Artificial Intelligence techniques such as Fuzzy Adaptive Resonance Theory and Probabilistic Neural Networks. For example, in Neural Network-based algorithms, the traffic data is input into a black box of learning layers and a binary decision is generated (Isaksen & Payne, 1973).

3.4 Evolution of incident detection algorithms

Review of the literature has noted considerable work on the development of mathematical models which focus on evacuation route planning. Network representations of evacuation problems are extensions of the classical operations research assignment problem. For these problems, the basic form of the network is that of the more general minimal cost transhipment (or flow) network (Hillier & Lieberman, 2005). In the network, the arcs represent the flow of people, the source nodes represent initial source inventories (points of entrance into the evacuation network), and the sink nodes represent the final inventories (in this case, destinations). Optimization models (e.g. linear programming, goal programming or dynamic programming) are another category of mathematical models. The model is formulated to either maximize (or minimize) the objective function (depending upon the purpose of the model) within the context of available resources and constraints (Hillier & Lieberman, 2005).

The most commonly reported incident detection algorithms in the literature that have been developed to identify non-recurring congestion conditions in transit networks are discussed next. They are presented in the temporal chronology of their associated creations. The emphasis is on providing the outline of algorithms that are significantly methodologically different rather than describing all the available algorithms in meticulous detail.

3.4.1 Standard Normal Deviate algorithm

The Standard Normal Deviate (SND)-based incident detection algorithm (Dudek et al., 1974) was developed at the Texas Transportation Institute (TTI). The SND of a variable is computed as the difference of the given variable from its mean, divided by the standard deviation of the data set. A high value for the SND of a control variable would indicate a major change in the operational conditions in the system. Lane occupancy and energy (kinetic energy) were evaluated as control variables in the belief that tracking the SND of these variables would allow identification of the passage of shockwaves through the detection stations and consequently identification of incidents. Tests were performed with 3- and 5-minute time bases for computation of the mean and standard deviation used in calculating the SND. The TTI researchers tested two strategies - one requiring only one SND value to be critical, and another requiring two successive values to be critical. The performance of the occupancy variable was observed to be superior in the first method. The effect of the time base on performance was not significant. The second method gave a higher detection ratio with occupancy but a lower detection ratio with the energy variable when compared to the first method. The occupancy variable was not sensitive to the time base in the second method either, but the energy variable showed an increase in a detection ratio with a larger time base. Dudek et al. (1974) reported a 92% detection ratio with a 1.3% false alarm rate during peak periods. The time to detect incidents was 1.1 minutes on average. Comparison with the other algorithms developed by Courage and Levin (1968) showed the SND algorithm to be as good as the composite model, which was supposedly the best existing model at the time.

3.4.2 Exponential Smoothing Algorithm

The Exponential Smoothing Algorithm (Cook & Cleveland, 1974) was developed using data from the John C. Lodge Freeway in Detroit. This method uses double exponential smoothing for generating a forecast variable. A tracking signal is generated as the algebraic sum of all the previous estimate errors to the present minute, divided by the current estimate of the standard deviation. When the tracking signal deviates from zero, beyond a pre-specified threshold, detection is indicated. A set of 13 traffic variables, which were derived from the basic traffic variables of volume, occupancy, and speed, were used to test algorithm performance. The thirteen variables are: station volume; station occupancy; station speed (volume/occupancy); station volume-occupancy (root of squared sum of errors for both was used for the error values); station speed-occupancy (analogous to volume-occupancy); station kinetic energy; station discontinuity; subsystem volume; subsystem occupancy; subsystem speed; subsystem kinetic energy; subsystem volume-occupancy discontinuity; and subsystem speed-occupancy discontinuity. Station discontinuity is computed in the same manner as in Courage and Levin (1968). Kinetic energy computations use the surrogate for speed. Station occupancy, volume, and discontinuity were found to give better performance in terms of detection levels at different levels of false alarms.

3.4.3 Low Volume-Based algorithm

An algorithm (Dudek et al., 1974) was specifically developed at the TTI for detecting vehicles under low volume conditions. This algorithm tracks individual vehicle input-output. The time of exit of a vehicle from the control section, the edges of which are defined by detectors, is projected as the summation of the time of entry, with the ratio of distance between detectors to speed of vehicle at the time of entry. The TTI researchers made a preliminary assumption of the constant speed of the vehicle over the section. Two different approaches are defined here: a time-scan operation system and an event-scan operation system. In the time scan operation, fixed sized account intervals are considered and the numbers of vehicles entering and exiting during these intervals are balanced. Projected times of exit are computed for each vehicle entering the control section within an accounting interval, and if the projected time falls within the interval, the vehicle is expected to exit within that interval. If the vehicle fails to do so, an alarm is raised. If nothing had happened to the vehicles and it was just a reduction of speed that delayed the exit, the alarm would then be a false one. Waiting for one more accounting period does not alleviate the problem, because a similar situation may arise in the next interval, and the accounting will still show one less vehicle exiting than expected. This problem is addressed in the event scan approach. It uses a variable time interval for vehicle accounting. For each vehicle, a set of three computations are executed: the shortest possible time the vehicle can take to arrive based on an upper speed limit of 100 miles per hour (mph), the expected arrival time of the vehicle based on the constant speed assumption, and a late expected exit time based on a speed with a 10% factor of safety. If a second vehicle does not arrive at the beginning of the section before the late expected exit time, the accounting interval is closed. If a vehicle does arrive, the process is repeated until such a situation arises in which no vehicle arrives at the upstream detector before the late expected exit time at the downstream detector. If a vehicle is not accounted for at the close of the accounting period, an alarm is raised. Some results pertaining to detector spacing requirements for event scan operations were obtained from simulation runs. Actual data was used to validate the claim that pattern recognition of headway, occupancy, and speed has to supplement volume counts, for the algorithm to work satisfactorily. An average of one false alarm per 10 minutes at 200 vehicles per hour (vph) on a three lane directional freeway was observed during the use of this algorithm (Chow, Willsky, Gershwin, & Houpt, 1977).

3.4.4 Dynamic Model based algorithms (MM and GLR)

Chow et al. (1977) proposed an incident detection approach based on a dynamic model that would make use of all information about the dynamic and stochastic evolution of traffic variables in time and space. Two algorithms resulted from this approach: the Multiple Model (MM) algorithm and the Generalized Likelihood Ratio (GLR) algorithm. The dynamic model uses the Payne equations (Isaksen & Payne, 1973). The MM algorithm models different scenarios, one of them being the occurrence of an incident. Constant-gain Kalman filters are used on the output of the model for the different scenarios to compare against the observations. The residuals from these filters are fed into a probability calculator that is subsequently used in a set of detection rules to isolate incidents. In the GLR algorithms, only one extended Kalman filter is used, corresponding to the normal operations scenario. Using Incident Innovations Signatures (IIS) that are pre-determined from simulations, a correlation is drawn between the residuals of the filter and the corresponding IIS in order to obtain the likelihood of different events. Unlike other algorithms that perform well only in heavy traffic, these algorithms were found to perform well under light and moderate traffic conditions.

3.4.5 The California Algorithms

The California Algorithms (Payne & Tignor, 1978) are a set of 10 algorithms that are based on the same principle. They use a decision tree based on traffic states for incident detection. In this set, Algorithm #8 and Algorithm #7 are the most popular ones. The California Algorithms, developed using data from the Los Angeles roadway system, are one of the first full-scale incident detection algorithms developed. They are normally used as benchmarks for evaluating the performance of other algorithms. At present, several modified forms of the original California algorithms exist and are implemented in several TMCs. The algorithms use 20- and 30-second occupancies and volumes averaged over all lanes at a particular station. Several variables are derived based on the occupancy values at the concerned station and the station downstream at different time points. Some of the most prominent variables are: Downstream Occupancy (DOCC), Spatial Difference in Occupancies (OCCD), Relative Spatial Difference in Occupancies (OCCRD), and Relative Temporal Difference in Downstream Occupancy (DOCCTD). These derived variables are evaluated at every time-step at each station in the concerned section of roadway and compared to thresholds at different points in a decision tree to determine whether an incident has occurred in the system. The thresholds are determined during calibration of the algorithm by minimizing the false alarm rate for a given level of detection rate. The algorithms in this set that used derived variables based on volume and volume-to-occupancy ratios were found to be inferior to algorithms based purely on occupancy based measures. Algorithm #7 uses a persistence requirement and replaces the variable DOCCTD in the last stage of the decision tree with the DOCC variable. This is done in order to account for two observations: 1 - non-incident-related compression waves traveling upstream cause false alarms; and 2 - drops in downstream occupancies are much greater in magnitude in cases of incidents than in normal compression waves generated by recurrent congestion. Algorithm #8, in addition to this, turns off incident detection for 5 minutes after the detection of a compression wave at the downstream detector (Payne & Tignor, 1978). This is supposed to give a better suppression of false alarms.

3.4.6 Bayesian-based algorithm

An incident detection algorithm based on Bayesian probability theory was developed by the Illinois DOT. This approach can be used on top of any algorithm to decrease the false alarm rate of the algorithm. The basic idea uses values of probability of occurrence of an incident for a given tracking signal. The signal can be any traffic variable or a variable derived from a traffic parameter, such as those used by the California algorithms. The requirement of the variable is its stability during the occurrence of the incident. The frequency distribution functions of the variable during incident and non-incident conditions are derived based on historical data during the algorithm calibration. These frequency distributions are used to derive values of Bayesian probability of the occurrence of an incident for strings of signals from the variable. The strings consist of a series of "ones" and "zeros," depending on the presence and absence of the signals, respectively. A signal is generated when the value of the variable crosses a calibrated threshold (Levin & Krause, 1978). There can be several thresholds for operation under different traffic conditions and different geometric conditions. The requirement for the length of the string (string of consecutive "ones") is determined from the probability values associated with the string of signals. It was found that a string length of four was sufficient for the section over which the algorithm was tested. In other words, a string of four consecutive "ones" indicated the occurrence of an incident. Determination of proper thresholds for the signal and the frequency distribution of the variables is critical to the proper functioning of the algorithm. The main drawback of this algorithm is its increased detection time. The logic ensures a reduction of false alarms, and if the base signal variable is stable enough, the detection ratio would not depreciate with the use of this logic (Levin & Krause, 1978). Depending on the length required to obtain a high value of probability of the occurrence of an incident, the time required to detect the incident would increase. The tests conducted during the validation of the algorithm observed an increase of 2 to 2.5 minutes in detection time. This algorithm provides a statistical way of creating a persistence test. The persistence test can reduce the false alarm occurrence, but the detection ratio and the time to detect values depended primarily on the base signal or the base algorithm that feeds this logic (Levin & Krause, 1978).

3.4.7 Committee Decision logic units-based algorithm

Tsai and Case (1979) proposed two techniques designed to operate on top of the basic incident detection algorithm to improve detection performance. The first technique, the Incident Detection Persistence Test, proposed in their study is a methodology for reducing false alarms by distinguishing false alarms from true alarms. The logic is developed on top of a modified California Algorithm. It uses Bayes optimal decision rule to determine a duration threshold that maximizes the likelihood that an alarm with a duration less than the threshold duration turns out to be a false alarm. Alarm duration data for both false alarms and true alarms are used to determine such a threshold. The reduction of false alarms using this technique was observed to have an adverse effect on the detection ratio, which decreased (adversely) with a reduction in false alarms by the introduction of the persistence interval logic. The second technique uses a committee-machine approach to determine the lane of the multilane freeway on which the incident has occurred. The output of several detection algorithms in the form of the incident lane number is used in the first layer, in which the individual decision units are designated as committee decision logic units (CLDU). The second layer of the committee machine structure consists of a vote-taking logic unit (VTLU) that uses the decision outputs from the first layer and determines the lane where the incident has occurred, according to the majority decision principle (Tsai & Case, 1979).

3.4.8 HIOCC and PATREG Algorithms

The High Occupancy (HIOCC) and Pattern Recognition (PATREG) algorithms were developed at the Transport and Road Research Laboratory in Berkshire, UK. The HIOCC algorithm is primarily a congestion detection algorithm. Slow moving or stopped vehicles are detected by using the resulting high occupancy values. Instantaneous occupancy values, at a one tenth of a second sampling rate, are smoothed using an exponential filter before employing the algorithm. The threshold is so chosen that an alarm will be indicated when the passenger-car speed is less than 6 mph or the long-vehicle speed is less than 14 mph. To avoid multiple alarms resulting from fluctuations of the observations, the occupancy values are artificially raised to a 90% level at the beginning of the congestion so that the high is maintained until the occupancy comes back to the level before the congestion. Also, to account for stop-and-go traffic, an 8-second threshold of zero instantaneous occupancy is used to prevent the case of stopped traffic from triggering an end of congestion indication. The PATREG algorithm identifies incidents using patterns of significant speed changes. If the speed lies outside the pre-determined lower and upper thresholds specific to a lane for the duration of the pre-set persistence interval, an alarm is indicated. The PATREG algorithm works efficiently under low to medium volume conditions, whereas the HIOCC algorithm deals with the high volume conditions (Collins, Hopkins, & Martin, 1979; Collins, 1983).

3.4.9 ARIMA

The prediction of freeway traffic variables with an ARIMA (0, 1, and 3) model has been successfully used in the development of an incident detection algorithm (Ahmed & Cook, 1982). The 95% confidence intervals for predictions of occupancy are computed and used to classify traffic state as an incident condition or a non-incident condition, based on the occurrence of the observed value outside or inside the confidence intervals, respectively. Time-to-detect incidents were reported under one minute. One hundred percent detection rates were obtained, with false alarm rates ranging between 1.4 and 2.6 percent (Ahmed & Cook, 1982).

3.4.10 DELOS

The Detection Logic with Smoothing (DELOS) algorithm was developed at the University of Minnesota. In this scheme, values for spatial occupancy differences across two consecutive stations are compared at two time windows. The values from each of the time windows can be obtained by using any one of the three smoothing schemes - moving average, median, and exponential smoothing. Algorithm performance is tested for different smoothing schemes: moving average in both past and current periods, median in both periods, exponential smoothing in both periods, and moving average for the current period with exponential smoothing for the past period. Window sizes range from 5 to 20 terms for the past period and 3 to 10 for the current period in the moving average method, and for exponential smoothing, factors of 0.03 to 0.10 are considered. The size of the windows for smoothing is limited by the excessive delays in algorithm response associated with longer windows. The moving average and exponential smoothing schemes provided better performance than the statistical median. Comparison with the performance of the Double Exponential algorithm and the California algorithms showed better performance results from the DELOS algorithms (Chassiakos & Stephanedes, 1993).

3.4.11 Principal Component Analysis-based algorithm

An incident detection algorithm based on the statistical technique of Novelty detection using Principal Component Analysis was developed at the University of Leeds. The flow, speed, and occupancy at two adjacent detectors form the 6-dimensional input vector. The principal components computed from this input data normally have a much lesser dimension than the input data, masking unwanted noise effects, and, at the same time, preserving the generality of the data. A calibrated threshold value is used to distinguish the novel input vectors from normal data. The novel input vectors, when identified, are indicative of incidents. Encouraging results were reported when using this detection scheme on simulated data sets (Chen, 1997).

3.4.12 Cumulative sum of occupancy-based algorithm

This detection algorithm, developed at the University of California, Berkeley, is based on a comparison of cumulative occupancy data for the two detectors on both sides of a hypothetical incident. Thereby, this is a two-detector algorithm, unlike most of the others, which are single detector algorithms. However, this scheme still relies on the road being more crowded upstream than downstream for an extended period of time - a situation which is also the case in recurrent congestion at bottlenecks. Consequently, this algorithm, like most other algorithms, is prone to generating false alarms under recurrent congestion.

Incidents are detected by tracking the fluctuation of the difference of the cumulative sum of occupancies at the two detectors. If the fluctuation is more than a time-variant threshold, which increases linearly with the passage of time, then an incident is indicated. Effects of variations in occupancy induced by individual driving patterns and faulty detector reporting can be absorbed by carefully choosing an appropriate threshold value (Lin & Daganzo, 1997).

3.4.13 Probabilistic Neural Network algorithm

In the Probabilistic Neural Network (PNN) algorithm (Abdulhai & Ritchie, 1999a; Abdulhai & Ritchie, 1999b; Jin, Cheu, & Srinivasan, 2002), the transfer function of the hidden layer is a radial-basis function, and that for the output layer is a competitive-transfer function. The PNN consists of four layers – the input layer, the pattern layer, the summation layer, and the output layer. The input layer distributes the input vector to the pattern layer. The neurons in the pattern layer are divided into two groups representing incident and incident-free conditions. The summation layer consists of two neurons, one for each class (i.e., incident and non-incident). Each of the summation neurons computes an average output signal for the associated pattern units and the scales it. The output neuron selects the higher value between the two and determines the class (i.e., incident or non-incident). Compared to MLF, PNN has been shown to have lower detection rates (95-100%) and higher false alarm rates (less than 0.33%). However, PNN has a better adaptation potential. Postprocessor feature extractors and postprocessor probabilistic output interpreters have been used successfully (Abdulhai & Ritchie, 1999b) to improve performance. The use of DWTs (Roy & Abdulhai, 2003) has also been explored for training the PNN, with encouraging results.

3.4.14 Fuzzy Wavelet Radial Function Neural Network algorithm

Adeli and Karim (2000) proposed an algorithm using a Discrete Wavelet Transform (DWT) for noise reduction and feature extraction, followed by a fuzzy c-mean clustering to reduce the dimensionality of the input vector, finally using a Radial Basis Function Neural Network (RBFNN) to classify the input pattern as an incident pattern or a non-incident pattern. Sixteen consecutive data-points for occupancy and speed from the immediate past are used to form the input signal. This signal is normalized and the DWT is computed using Daubechies wavelet system of length 8. The wavelet coefficients are filtered using soft-thresholding nonlinearity, followed by an inverse DWT to obtain the de-noised, normalized sequence. A fuzzy c-mean clustering is used to reduce the dimensionality of the pattern. The extracted eight elements (four for occupancy and four for speed) are fed into a trained RBFNN. The output is compared against a preset threshold to indicate either an incident condition or otherwise. This algorithm was compared with the California Algorithm #8 and was found to produce very low false alarms (on the order of 0.07%), as compared to the California algorithm (on the order of 3.82%) under the same detection rate scenarios when tested with simulated data. Limited tests with real data gave 0% false alarms at 95% detection ratio for this algorithm, whereas the California algorithm produced 0.63% false alarms at a 90% detection ratio (Adeli & Karim, 2000; Karim & Adeli, 2002a).

3.4.15 Adaptive conjugate gradient Neural Network-Wavelet algorithm

Another algorithm developed based on the DWT (Adeli & Samant, 2000; Samant & Adeli, 2000; Samant & Adeli, 2001) used an Adaptive Conjugate Gradient Neural Network

(ACGNN) and a Linear Discriminant Analysis (LDA). The DWT and LDA operations were used to filter and preprocess the data, and the ACGNN was used as the state classifier (i.e., incident or non-incident). High incident detection rates of 97.8% and low false alarm rates of around 1% were obtained based on simulated data.

3.4.16 Wavelet Energy with Radial Basis Function Neural Network algorithm

An algorithm using Wavelet Energy representation of traffic patterns was proposed (Karim & Adeli, 2002b, 2003) as an enhancement (in terms of the detection time) over the Fuzzy Wavelet RBFNN Incident Detection model proposed earlier by the same authors (Karim & Adeli, 2002a). The algorithm is based on an advanced energy representation of the time series pattern developed using Wavelet Theory. The desirable features of the traffic pattern are enhanced, and at the same time, a denoising of the traffic pattern is achieved by performing a DWT operation to break the input signal into several time-frequency components. This enables the extraction of features desirable for signal identification and recognition. A RBFNN is then used to classify the pattern as an incident-induced or a non-incident-induced traffic pattern. A sixteen data-point series of the occupancy data and a similar set with flow data are used to provide the input signal. Each signal is normalized to remove the effects of magnitude, followed by padding at both ends to extend the series size to 32 data-points. A two stage low pass filter (Daubechies filter) is applied. Next, the sequence is enhanced with the squared scaling coefficients - a measure of energy in the Wavelet domain which is subsequently extracted. The extracted four element sequences of the occupancy and flow data are then concatenated to form the input vector of the RBFNN with eight input vectors, twelve hidden nodes with Gaussian transfer functions, and one output node with a linear transfer function. If the output is greater than a pre-selected threshold (e.g., 0.2), then an incident is indicated. The algorithm was tested extensively with simulated data and, to a limited extent, with real data. The simulated data gave a 0% false alarm rate, and the real data performed well with false alarms rates going from a minimum of 0.13 % to a maximum of 1.04%.

3.4.17 Discrete Wavelet Transform algorithm

Another incident detection scheme based on the DWT technique involves a different approach to the problem using the same tool. Unlike the previous algorithms (Adeli & Karim, 2000; Adeli & Samant, 2000; Ghosh-Dastidar & Adeli, 2003; Karim & Adeli, 2002a, 2002b, 2003; Samant & Adeli, 2000, 2001), which use the DWT mostly as a tool for de-noising the dataset, this approach implies the direct use of the extracted features in the detection of changes in traffic flow. The difference of occupancies between two stations is used as the input signal. A search for large absolute values in the finest scale level (third stage) of the DWT of the signal comprises the first check. A subsequent check of the direction of change using the scale coefficients of the DWT is used to confirm the incident. This algorithm was compared with the MLF Neural Network algorithm, PNN algorithm, FRBFNN algorithm, Low Pass Filtering algorithm, and the California Algorithms (Teng & Qi, 2003b).

3.4.18 CUSUM-based algorithm

A detection delay-based optimization problem formulation approach to incident detection was proposed, along with a simplified procedure for its implementation. The algorithms developed are based on the Cumulative Sum of Deviations of Subgroups statistic (the CUSUM statistic). Three algorithms, DCUSUM2, CUSUM1, and CUSUM2, involving different assumptions and different treatments of the problem, were developed in the process. The DCUSUM uses differences of occupancy in its analysis. CUSUM1 assumes that the correlation between individual observations (of occupancy) is zero, while CUSUM2 does not make such an assumption. A substantial change in the difference between the cumulative sum of the log-likelihood ratio for the existing time period and the minimum cumulative sum up to the existing time period is used to indicate a change in state of the process, and thereby, to indicate incident conditions. The DCUSUM algorithm was found to perform the best. A comparison with the Low Pass Filtering algorithm, the MLF algorithm, and California Algorithm #7 showed that the DCUSUM algorithm outperformed all of them (Teng & Qi, 2003a).

3.4.19 Support Vector Machine

An algorithm using the Support Vector Machine (SVM) pattern classifier was proposed by Chen, Kwon, Rice, Skabardonis, and Varaiya (2003). The SVM pattern classifier sorts out an input vector into one of two classes with a decision boundary developed based on the concept of structural risk minimization of classification error, using statistical learning theory. Three different SVM models were implemented with distinct embedded Kernel functions. A linear function, a polynomial function, and a radial basis function were the three Kernel functions used for this purpose. Comparative results of these three implementations, along with comparisons with the MLF algorithm and the PNN algorithm, were presented as applied to arterial data and freeway data. SVMs were shown to produce lower misclassification rates, higher detection ratios, lower false alarm rates, and in some cases, shorter detection times compared to the other algorithms (Yuan & Cheu, 2003).

4. Non-recurring congestion in hurricane evacuation preparedness

As it was previously explained, early detection and assessment of instances of non-recurring congestion can both reduce the time to return traffic to normal rates of flow and reduce the potential for secondary incidents (Busch, 1987), thus increasing the number of vehicles leaving an area that is frequently evacuated due to recurring hurricane events. The Alabama's coastline is one of such areas.

Both the Atlantic and Gulf coasts of the United States are periodically subjected to severe tropical storms and hurricanes. Such storms are becoming more violent and destructive. The normal response to these occurrences is to evacuate inland from the coast. Normal traffic flows turn into congestion, frustration and gridlock. Such congestion and gridlock occur as the magnitude of traffic overwhelms the available road capacity, which is further impaired by instances of non-recurring congestion. The resulting stalled traffic reduces the number of vehicles that can leave an area subject to evacuation.

This was the situation in September of 1998 with Hurricane Floyd, when extensive unexpected traffic delays occurred along most inland evacuation routes. Subsequently, the United States Federal Emergency Management Administration conducted regional meetings to identify approaches for better traffic planning, management, and coordination. A variety of simulation tools were applied to project traffic flows during extreme events requiring evacuation. For example, Pal et al. (2005) applied the Oak Ridge Evacuation Modeling System (OREMS) to evaluate traffic conditions resulting from evacuation from Baldwin and Mobile counties in Alabama.

As an initial step in this research, a comprehensive literature search was conducted. The purpose of this literature search was to identify and understand current theory regarding incident detection. The literature reviewed is identified in Section 3 of this paper, and provides a vast array of potential algorithmic models for use as the basis of non-recurrent traffic incident detection. As noted in Table 2, the corresponding detection and false alarm rates vary considerably, as reported in the literature (Fernandes, 2009).

Algorithms	Detection Rate	False Alarm Rate	Average Detection Time
Basic California Algorithm	82 %	1.73 %	0.85 min.
California Algorithm #7	67 %	0.134 %	2.91 min.
California Algorithm # 8	68 %	0.177 %	3.04 min.
Standard Normal Deviate	92 %	1.3 %	1.1 min.
Bayesian Algorithm	100 %	0 %	3.9 min.
Time-Series ARIMA	100 %	1.5 %	0.4 min.
Exponential Smoothening	92 %	1.87 %	0.7 min.
Low-Pass Filter	80 %	0.3 %	4.0 min.
Modified McMaster	68 %	0.0018 %	2.2 min.
Multi-layer Feedback Forward	89 %	0.01 %	0.96 min.
Probabilistic Neural Network	89 %	0.012 %	0.9 min.

Table 2. Comparison of Incident Detection Algorithms

Efforts were conducted to carry out a comprehensive review of the majority of these traffic incident algorithms identified, and determine a subset of the most effective for this problem domain. The authors have shown that the previously listed incident detection algorithms, in particular the California and Exponential Smoothing algorithms, can be successfully applied to assess and characterize instances of non-recurring congestions in highly transited urban areas (Fonseca, *at el.*, 2009).

As the progression of traffic incidents are quantified, and even predicted, the overall roadway capacity of the affected areas can be greatly increased by strategically choosing traffic evacuation measures oriented to the avoidance of travel routes that have been identified, through the detection algorithms, to have a high incidence of non-recurring congestion. Effective traffic management policies, as the ones suggested by the authors to improve the overall evacuation process of the Alabama Gulf region during a hurricane situation (Moynihan, *at el.*, 2008) can then be devised by traffic officials around the negative impact that non-recurring congestion has on massive evacuation events.

5. Conclusion

In this paper, the authors have shown the importance of studying the negative effects of non-recurring congestion on massive evacuation events, as in the case of a hurricane evacuation. Non-recurring congestion can be detected by indentifying the presence of traffic incidents along the roadways of a transit network. Several incident detection algorithms have been successfully developed and implemented, throughout the years, to help traffic officials to assess and predict the occurrence of non-recurring congestion. These detection algorithms represent a formidable tool in the devising of effective planning strategies for massive hurricane-related evacuation events.

6. References

Abdulhai, B., & Ritchie, S. G. (1999a). "Enhancing the universality and transferability of freeway incident detection using a Bayesian-based neural network." Transportation Research, 7(5), 261-280.

Abdulhai, B., & Ritchie, S. G. (1999b). "Preprocessor feature extractor and post processor probabilistic output interpreter for improved freeway incident detection." Transportation Research Record, 1678, 277-286.

Adeli, H., & Karim, A. (2000). "Fuzzy-wavelet rbfnn model for freeway incident detection." Journal of Transportation Engineering, 126(6), 464-471.

Adeli, H., & Samant, A. (2000)."An adaptive conjugate gradient neural network-wavelet model for traffic incident detection." Computer-Aided Civil and Infrastructure Engineering, 15(4), 251-260.

Ahmed, S. A., & Cook, A. R. (1982). "Application of time-series analysis techniques to freeway incident detection." Transportation Research Record, 841, 19-21.

Busch, F. (Eds). (1987). “Incident Detection”. In M. Papageorgiou (Ed). Concise Encyclopedia of Traffic and Transportation Systems. pp. 219-225. Oxford.

Chassiakos, A. P., & Stephanedes, Y. J. (1993). "Smoothing algorithms for incident detection." Transportation Research Record, 1394, 8-16.

Chen, H. (1997). “Motorway incident detection using principal component analysis.” Proceedings from Colloquium on Incident Detection and Management. London, UK: 25-32.

Chen, C., Kwon, J., Rice, J., Skabardonis, A., & Varaiya, P. (2003). "Detecting errors and imputing missing data for single-loop surveillance systems." Transportation Research Record, 1855, 160-167.

Chow, E. Y., Gershwin, S. B., Greene, C. S., Houpt, P. K., Kurkijian, A., & Willsky, A.S. (1977). Dynamic detection and identification of incidents on freeways (Electronic Systems Laboratory ESL-R-764). Cambridge, MA: Massachusetts Institute of Technology.

Chow, E. Y., Willsky, A. S., Gershwin, S. B., & Houpt, P. K. (1977). Dynamic detection and identification of incidents on freeways: The generalized likelihood ratio method (Electronic Systems Laboratory ESL-R-767). Cambridge, MA: Massachusetts Institute of Technology.

Coifman, B.A. (2007). "Distributed Surveillance on Freeways Emphasizing Incident Detection and Verification". Transportation Research. Part A, Policy and Practice. Volume 41, Issue 8, pp. 750-762.

Collins, J. F. (1983). Automatic incident detection - experience with TRRL algorithm HIOCC (TRRL Supplementary Report 775). Crowthorne, Berkshire: Transport and Road Research Laboratory.

Collins, J. F., Hopkins, C. M., & Martin, J. A. (1979). Automatic incident detection -TRRL algorithms HIOCC and PATREG (TRRL Supplementary Report 526). Crowthorne, Berkshire: Transport and Road Research Laboratory.

Cook, A. R,. & Cleveland, D. E. (1974). "Detection of freeway capacity-reducing incidents by traffic-stream measurements." Transportation Research Record, 495, 1-11.

Courage, K. G., & Levin, M. (1968). A freeway corridor surveillance, information and control system (Research Report 488-8). College Station, TX: Texas A&M University, Texas Transportation Institute.

Dudek, C. L., Messer, C. J., & Nuckles, N. B. (1974). "Incident detection on urban freeway." Transportation Research Record, 495, 12-24.

Fernandes, H. (2009). "Development of a platform for traffic incident detection," Masters Thesis, Department of Mechanical Engineering, The University of Alabama.

Fonseca, D.J., Brumback, T., Moynihan, G.P., and Fernandes, H. (2009). "Development of a Multi-Algorithmic Platform for Traffic Incident Detection," Journal of Management and Engineering Integration, Vol.2: no.1, pp. 42-49.

Ghosh-Dastidar, S., & Adeli, H. (2003). "Wavelet-clustering-neural network model for freeway incident detection." Computer-Aided Civil and Infrastructure Engineering, 18(5), 325-338.

Hillier, F. and Lieberman, G. (2005), "Introduction to operations research" (8th Edition), McGraw-Hill, New York.

Hualiang, T., & Qi, Y. (2003). "Application of wavelet technique to freeway incident detection." Transportation Research, 11(3-4): 289-308.

Isaksen, L., & Payne, H. J. (1973). "Freeway traffic surveillance and control." Proceedings from Proceedings of the IEEE: 526-536.

Jin, X., Cheu, R. L., & Srinivasan, D. (2002). "Development and adaptation of constructive probabilistic neural network in freeway incident detection." Transportation Research, 10(2), 121-147.

Karim, A., & Adeli, H. (2002a). "Comparison of fuzzy-wavelet radial basis function neural network freeway incident detection model with California algorithm." Journal of Transportation Engineering, 128(1), 21-30.

Karim, A., & Adeli, H. (2002b). "Incident detection algorithm using wavelet energy representation of traffic patterns." Journal of Transportation Engineering, 128(3), 232-242.

Karim, A., & Adeli, H. (2003). "Fast automatic incident detection on urban and rural freeways using wavelet energy algorithm." Journal of Transportation Engineering, 129(1), 57-68.

Levin, M., & Krause, G. M. (1978). "Incident detection: A Bayesian approach." Transportation Research Record, 682, 52-58.

Lin, W. H., & Daganzo, C. F. (1997). "A simple detection scheme for delay-inducing freeway incidents." Transportation Research, 31(2), 141-155.

Moynihan, G.P., Fonseca, D.J., Cui, Q., Johnston, J., and Jennings, J. (2008). "Development of a Simulation Model for Hurricane Evacuation Planning of the Gulf Coast," Proceedings of the 2008 Industrial Engineering Research Conference, Vancouver, Canada, pp. 64-68.

Pal, A., Triche, M., Graettinger, A., Rao, K., McFadden, J. and Turner, D. (2005). "Enhancements to emergency evacuation procedures", Final research report to the University of Alabama Transportation Center.

Payne, H. J., and Tignor, S. C. (1978). "Freeway incident-detection algorithms based on decision trees with states." Transportation Research Record, 682, 30-37.

Presley, M. W. and Wyrosdick, K. G. (1998). Calculating benefits for NaviGator - Georgia's intelligent transportation system (Internal Report). Atlanta, GA: Georgia Department of Transportation.

Roy, P., & Abdulhai, B. (2003). "Gaid: Genetic adaptive incident detection for freeways." Transportation Research Record, 1856, 96-105.

Samant, A., and Adeli, H. (2000). "Feature extraction for traffic incident detection using wavelet transform and linear discriminant analysis." Computer-Aided Civil and Infrastructure Engineering, 15(4), 241-250.

Samant, A., and Adeli, H. (2001). "Enhancing neural network traffic incident-detection algorithms using wavelets." Computer-Aided Civil and Infrastructure Engineering, 16(4), 239-245.

Schrank, D. and Lomax, T. (2003). The annual urban mobility report. College Station, TX: Texas Transportation Institute.

Takaba, S., and Matsuno, H. (1985). "Traffic incident detection using correlation analysis." Proceedings from Summer Computer Simulation Conference. North-Holland, Amsterdam: 529-534.

Teng, H., and Qi, Y. (2003a). "Application of wavelet technique to freeway incident detection." Transportation Research, 11(3), 289-308.

Teng, H., and Qi, Y. (2003b). "Detection-delay-based freeway incident detection algorithms." Transportation Research, 11(3), 265-287.

Transportation Research Board -TRB-. (2008). The Role of Transit in Emergency Evacuation. Special Report No. 24. Transportation Research Board, Washington, D.C.

Tsai, J., and Case, E. R. (1979). "Development of freeway incident-detection algorithms by using pattern-recognition techniques." Transportation Research Record, 722, 113-116.

Yuan, F., and Cheu, R.L, (2003). "Incident detection using support vector machines." Transportation Research Part C : Emerging Technologies, 11(3), 309-328.

Zifeng, J. (1997). "Macro and micro freeway automatic incident detection (AID) methods based on image processing." Proceedings from IEEE Conference on Intelligent Transportation Systems. Boston, MA: 344-349.

24

Hurricanes Katrina and Rita - Effects on Children's Healthcare

Gensure, Robert Charles, Ponnapakkam and Adharsh
Children's Hospital at Montefiore, Albert Einstein School of Medicine
Bronx, NY,
United States

1. Introduction

Hurricanes Katrina and, to a lesser extent, Rita caused significant damage to a major population center and medical referral center for children's healthcare in the southeastern United States. As such, there were disruptions in children's healthcare not only at the site of maximal damage, but also extending to the greater region served by these referral centers, the effects of which continue to linger even now, after the fifth anniversary of the storms. The damage and disruption of the storms was not uniform, and the challenges to the healthcare system varied both geographically and temporally. In this chapter I will review these challenges, and how the healthcare system responded to each one, from the perspective of a pediatric specialist practicing in the area, using my first-hand experiences and data collected from my practice before and after the hurricane events. Also important to consider is that the healthcare system in the United States is a free-market system, and as such much of this response was not directed by an organized plan, but rather the result of reaction by individual hospitals and practitioners to the circumstances as they presented themselves. These special circumstances go beyond the initial first response - the focus of this chapter will be the short- and long-term challenges faced by the healthcare system after a major natural disaster.

2. The geography

Most are familiar with the damage to downtown New Orleans from Hurricane Katrina. The statistics speak for themselves, over 60,000 homes destroyed, flooding in excess of 10 feet of water, no power, no phone lines, no running water, and no working sewerage. In this region, none of the hospitals were able to continue to function. Patient records were either inaccessible or destroyed. Any local practices were closed, and the physicians and staff evacuated, usually without working contact information. For those who had forwarding contact information, phone numbers (including cell phones) with the local area code could not receive calls due to damage to the phone switchboard. Healthcare consisted of providing first aid followed by evacuation. The hospitals themselves were evacuated and closed. Likewise, nearly the entire population was evacuated.

This portion of the population lost all their usual connections with the health care system. Their medical records were mostly lost. Physicians were themselves evacuated and could not be reached. Prescription renewals could not be obtained. Care for these individuals was conducted according to models of other natural disasters - evacuation to shelters, emergency response teams providing temporary care until the individuals were ultimately relocated.

To the east of New Orleans, the situation was not much different. The eye of the hurricane had passed through this region, again causing massive damage. Importantly, the bridges connecting to New Orleans from the east had all been destroyed, making the city accessible only from the west. This became extremely important in later stages of the recovery phases, as emergency response teams and returning residents would have to approach the city from this direction.

Immediately to the west of New Orleans is Jefferson Parish. Jefferson Parish is an industrial center with a population similar to that of New Orleans (around 400,000). This region had been spared much of the worst of the flooding from the storm surge, but did sustain significant damage from wind and rain flooding. Residents were again mostly evacuated and businesses were closed, but hospitals were still functioning, albeit on emergency power and with reduced staffing. My own hospital, Ochsner Hospital, is located in this region, along with East Jefferson Hospital. While these hospitals were open, they were not in a position to provide much more than continued care for the few patients on the hospital floors. Smaller offices and local practices were all closed.

West of Jefferson Parish is St. Charles Parish. St. Charles Parish is a rural community with a much smaller population (around 60,000), but does have significant heavy industry (oil refineries, chemical plants, and a nuclear reactor). This region sustained disruptions from power outages and significant debris, but little structural damage and minimal flooding. Hospitals may have temporary closed for the storm itself, but were quickly reopened. Business disruptions were greater, with most businesses closed or functioning on emergency plans. Most residents evacuated for the storm itself but quickly returned.

Importantly, the three regions discussed here have independent parish governments and emergency operations centers, which operated independently of each other. Different decisions by these entities through the various phases of response/recovery had almost as much impact on the recovery process and healthcare delivery as did the extent of hurricane damage. It is fair to say that early in the response, neither the state nor the federal agencies had significant influence in the decisions of these independent parish governments.

3. The first stage – disaster response

While Orleans, Jefferson, and St. Charles parish had each ordered mandatory evacuations prior to the arrival of Hurricane Katrina, there was no enforcement of this order. Thus, while most residents had evacuated, there were a significant number of residents present during the storm itself. Emergency response teams thus focused on 'search and rescue' operations in the immediate poststorm period, particularly in the regions with the worst flooding.

3.1 Orleans Parish

Many of the residents who did not evacuate prior to Hurricane Katrina were highly visible, having congregated in the Superdome before the storm, or in the Louisiana Convention Center afterwards. Others were not, having stayed in their homes. Those in the city proper were certainly faced with difficult living situations - no power, no water, no sewerage, no open businesses, and as was well publicized, almost no law and order. The major challenges to the healthcare system were not so much how to provide care, but how to make an orderly withdrawal. Hospitals could not provide care under these circumstances; they were systematically closed and evacuated. What healthcare was provided consisted of immediate first-aid and, if necessary, evacuation to a treatment center. Most common ailments included dehydration and wound infections from floodwaters. This is a well-established model, used in any major natural disaster or man-made 'disaster' (more commonly referred to as combat). Field hospitals were established at convenient gathering points (the New Orleans International Airport) or at the nearest mostly unaffected facilities in Baton Rouge. The only unique challenge was the difficulty of conducting these evacuations through the flooding without the benefit of amphibious vehicles - mostly this was accomplished by helicopter or boat. These well-coordinated efforts have been well documented extensively elsewhere.

3.2 Jefferson Parish

The majority of residents of Jefferson Parish had also evacuated prior to the arrival of Katrina, but again some residents remained behind during the storm. By geography and by chance, most of these were spared major flooding. The now-famous 17th street canal is actually located at the boundary between Jefferson and Orleans Parish; that only the Orleans Parish side breached was a matter of chance. However, Jefferson Parish is also mostly below sea level, and while the levee's held, the pumping stations that would normally pump out the rainwater were not manned during the storm. This curious decision led to more minor but equally widespread flooding throughout the parish. This combined with wind damage and debris from fallen trees effectively shut down normal business operations. The water was not drinkable, and sewerage systems were not operating. Remaining residents were not trapped, but were in a difficult living situation. Again, the most common ailments were dehydration and wound infections from floodwaters.

The major centers in the healthcare system were themselves on life support but remained functioning. Hospitals were operating with skeleton crews on emergency power, and could only provide minimal services for existing patients. It is not clear if first responders in Orleans Parish were aware that these facilities were open and functioning, and even if they were, the facilities were in no condition to serve as triage points. Any practices or facilities not tied to the major hospitals were closed. Overall, the healthcare system in this region was neither an asset nor a burden. First responders passed through on their way into New Orleans, and evacuated residents passed through on their way out of New Orleans.

3.3 St. Charles Parish

It is not clear how many of the residents of St. Charles Parish evacuated prior to Katrina, but regardless, even those that did evacuate did not stay evacuated for very long. The populated areas of this parish are mostly above sea level. The levees did not breach and pumping stations continued to function, so there was minimal flooding. Access was limited initially due to an excess of downed foliage (a bigger problem in a rural parish), but this was quickly cleared. Importantly, generators were utilized at water purification plants and sewerage treatment plants, providing both potable water and working sewerage systems to residents. The local hospital was quickly reopened, and there were even some business open (supermarkets, gas stations). Pharmacies and smaller physician practice sites remained closed. Again, the healthcare system in the region was neither an asset nor a burden, although this was about to change in a most dramatic way.

4. The second stage – an extended evacuation

Hurricane evacuations are a way of life in southern Louisiana. People grab the essentials and a few day's supplies and head out on the road, fully expecting to return in a few days time. This applies to medications as well. Katrina made landfall at the end of August, and most people renew their prescriptions on the first of each month. As such, most people evacuated with only those few pills left from their August prescriptions. When it became clear that they would not be returning to their homes in the next few days, this became a very big problem. rdinarily, one would simply have one's physician call in a prescription renewal to a local pharmacy. However, the physicians offices were closed or damaged, and the physicians themselves were evacuated, most with no way for their patients to contact them. As New Orleans had served as a regional medical referal center, this affected not just residents of the city proper, but of the surrounding regions as well.

4.1 Evacuated residents

The biggest healthcare-related problem for evacuated residents was obtaining medication refills and continued care for chronic medical problems. Ordinarily, a new prescription would be provided by the original prescribing physician, and a referral (complete with past medical records) would be sent to the designated new physician. As neither original prescribing physicians nor medical records were available, a new system developed. Individual patients would arrive with empty pill bottles at pharmacies around the country, pleading their cases for additional medications. Pharmacists technically needed to contact the prescribing physician for refill authorization. They would of course make this attempt, but, knowing the situation, would generally provide the patient with a refill despite the lack of a response. When I arrived back at my office 3 days after the storm, the last thing I expected to see was a stack of faxed medication refill requests, from literally all over the country, but indeed that is what I found. Likewise, physicians offices were generally very accommodating in providing temporary care for these individuals. Even insurance companies suspended all in-network requirements for their plans. As most hurricane evacuees were scattered around the country, staying with friends or relatives, this did not necessarly provide a strain on the local healtcare system.
Included in this group were children with specialty care needs. As the only available pediatric specialist, I found myself attempting to coordinate care for these children, most often by coordinating evacuation to another regional center.

4.2 Orleans Parish – flooded regions

It is fair to consider Orleans parish in two parts, those that flooded and those that did not. The flooded regions were absolutely devastated. Any residents who did not evacuate before the storm were evacuated soon afterwards. The hospitals had all been closed and evacuated by this time. The only healthcare provided was by emergency medical technicians (and a now-famous pathologist) on the scene prior to evacuation.

4.3 Jefferson Parish and non-flooded regions of orleans parish

While Jefferson Parish and the non-flooded regions of Orleans parish did not suffer the devestating damage seen in the heavily flooded areas, there remained widespread minor damage and disruption of critical services, including electricity, water, and sewerage. Most debris had not been cleared from the streets, limiting access. As such, residents were generally not allowed into their homes for a month or more after the storm. This did not, however, stop people from trying, and rather than being safely esconced at a distant evacuation site, most had gathered on the western border of Jefferson Parish waiting to be allowed back into their homes; a few managed to sneak back in. These combined with the residents who had never evacuated made for a very small population to serve. The most common reason for seeking medical care was running out of medications. Other common complaints were dehydration, wound infections from floodwaters, and minor injuries during home repairs.

While the individual health care practices and smaller facilites remained closed, the hospitals were gradually re-establishing normal operations. The Ochsner Clinic reopened to much fanfare but few patients, as the parish residents were held up at checkpoints on the parish borders. Those few who did make it were well cared for – their records were fully accessible, prescription refills could be easily provided and filled at a hospital pharmacy. Personnel at the hospital was an issue, as most of the empolyees had to be housed in hospital beds. There were also lingering questions of safety making hospital workers hesitant to return. Interestingly, security was probably the least issue, with private contractors having been employed by the hospitals throughout the storm period. Underlying it all was a concern if normal operations would ever resume, a concern which was on some levels realized during the next stage of response.

4.4 St. Charles Parish

St. Charles Parish had suffered significant damage from the storm, but was in an excellent position to mount a rapid recovery. The early decisions to remove debris from roads and establish potable water and a working sewerage system proved critical. These allowed residents to return earlier, minimizing ongoing damage to property (roof leaks were repaired or covered, etc), and further hastening the removal of debris (everyone cleared their own lawns). Some businesses were able to reopen on generator power, with their regular employees available for work and their regular customers lining up at the doors. Shortly thereafter, convoys of power trucks arrived to repair the power lines. As Jefferson Parish and Orleans Parish were still inaccessible, they repaired the damaged power lines in St. Charles Parish instead. My own home in St. Charles Parish had electrical power restored exactly one week after the storm (by comparision, I had no power for 3 weeks after the much weaker Hurricane Gustav past over and caused considerably less damage). Hospitals were open and some medical practices re-opened, although most pharmacies remained closed, apparently due to issues with supply chains.

However, St. Charles Parish faced a challenge a new, particularly to its healthcare system, that would have pressed it under the best of circumstances. Those residents who were not allowed into Jefferson Parish or Orleans Parish were stranded in St. Charles Parish. An infrastructure designed to serve a population of ~60,000 was strained by literally hundreds of thousands of people waiting to be allowed back into their homes. The strain on the healthcare system was particularly severe, as these patients developed minor ailments or ran out of their chronic medications. In response, the local hospital, St. Charles Parish Hospital, set up an outpatient triage center, where around 1000 patients each day could quickly be seen and provided prescriptions for their medications. Physician volunteers from around the country, or in my case from down the street, came to help provide care at this center. Of course, the lack of pharmacies also became an issue, as we could provide prescription refills, but there was no place for the patients to fill them. The savior for this problem came in the unlikely form of caravans of dark sedans driven by what we refer to as "drug reps". These sales representatives from the pharmaceutical companies would normally provide medicaion samples to physicians; after Katrina, they had evidently tapped into regional and national stores of these samples to provide free medications to hospitals in the hurriane zone, quite literally one trunkload at a time. To my knowledge, the importance of this effort to the healthcare response teams has not been acknowledged, and I would like to do so now.

In addition to patients out of medications for chronic illnesses, the major illnesses treated at this triage center included minor illnesses and wound infections from cuts while wading through floodwaters, still present in parts of Jefferson Parish and much of Orleans Parish. It must also be mentioned the disappointingly large number of patients on chronic narcotics for back pain, who had been cared for in "pain clinics" in Jefferson Parish.

Healthcare in the United States is usually not free, but no patients were charged for the services at this temporary outpatient center. To do so would have been totally impractical. Happily, this is one good deed which was rewarded - the hospital did keep records of the patients treated (including copies of photo ID's where possible), as was ultimately compensated by the federal government. I had joked about coming back someday to see the new hospital wing, and in fact did just that as part of my grand jury service 2 years later.

5. The third stage – "Where are the patients?"

As people were eventually allowed to return to their homes, new challenges emerged to healthcare delivery. These challenges were more difficult to anticipate, and less outside assistance was provided in meeting them. The challenge to the healthcare system at this stage was primarily economic, from reduced patient volumes seen despite the reduction in the total number of hospitals. The average daily census of hospital beds in the region fell by 50% during the 4 month period after the storm, from 2,500 patients to 1,237 (Louisiana Pubic Health Institute, "NOLA Dashboard", and U.S. Government Accountability Office, 2006). In a free market health care system, reduced patient volume directly translates into reduced revenue, and hospitals began reporting significant financial losses. Bond ratings were lowered, limiting hospitals' ability to borrow money. While the reduced patient volumes and financial losses are easy to document, they were harder to explain. Facing this same challenge in my practice, I prepared a compilation of the patient visits and new referrals before and after the storm, and with this can postulate an explanation which may be relevant to the recovery phase following other natural disasters, hurricanes or otherwise.

5.1 The new demographics

A new term, "Demographic Shift", has been applied to the populations changes of the greater New Orleans area following Hurricane Katrina. Initially this was a euphemism for a reduced number of African Americans living in the city, made famous (infamous?) by Mayor Ray Naggin's comments about New Orleans being a "chocolate city". In fact, the population shifts were far more complex. The reported numbers are staggering, with >75% reduction in the population of Orleans Parish (based on United States Census data). However, there was not a corresponding 75% reduction in traffic, nor was there a 75% reduction in business volume, or in patient volume, for that matter. Clearly, there were fewer residents of public assistance housing, as these were mostly destroyed in the storm and were not reopened. However, this appears to have been offset by an influx of temporary workers attracted by job opportunities in construction. I say "appears" because there are no statistics that can be referenced for this. Many of the workers were undocumented aliens.

Those that were not mostly continued to claim their home address as where they came from, even though they may have spent years in the New Orleans area. Thus the US Census would not show them as residents of New Orleans. The post office could use mail forwarding out of New Orleans to determine who had left, but there was no system for tracking mail forwarding into Orleans Parish. These uncertainties made appropriate resource allocation and long-term planning extremely difficult, particularly with regards to healthcare. Interestingly, they also made crime statistics unreliable, as these are reported as incidents per unit population.

5.2 The response of the healthcare industry

Absent compelling evidence to the contrary, most assumed that the census information was correct, that the population of the greater New Orleans area was significantly lower. This led to assumptions that the reduced patient volumes were permanent, causing hospitals to scramble to consolidate services and reduce operating costs. Hospital acquisitions became quite commonplace - Ochsner acquiring Memorial Hospital, Meadowcrest Hospital, Kenner Regional Hospital, and entering into operational agreements with several other hospitals (Summa Hospital in Baton Rouge, St. Anne's Hospital in Raceland, and St. Charles Parish Hospital). Children's Hospital acquired Touro Medical Center. Salaries were reduced and benefits were cut. Ochsner Hospital's new policies included charging vacation time during hurricane evacuations, decreasing subsidies for workers' health insurance, eliminating the pension plan, and conducting a full audit of eligibility for health insurance of workers' dependents (ironically this audit occurred shortly before new rules in the health care reform act caused most of these dependents' health care coverage to be reinstated). Executives were affected as well, taking a substantial pay cut. Other hospitals (most notably Tulane Hospital and the state hospital, the Medical Center of New Orleans), took even more severe measures, laying off staff or placing them on furlough, including physicians and tenured faculty.

Unfortunately, these cutbacks came at a time when the cost of living in New Orleans had increased dramatically. Destruction of 75% of the housing in Orleans Parish had a predictable effect on the housing market - rents and new home prices skyrocketed. Likewise, house insurance rates increased 3-4 fold, when available at all. Inventory losses and increased demand led retailers to raise prices. Worker attrition rates predictably skyrocketed as well, as high as 20% per year even among the physician staff. Of course, this was by design, as hospitals were trying to reduce overhead by reducing staffing.

6. A closer look at the post-storm changes in my practice

In my own practice, I did not sense that the patient volumes had not fallen in pace with hospital-wide data. There did, however appear to be some unusual trends emerging, which with the assistance of a student working in my laboratory I decided to investigate further. It was round 6 months after the storm when, while attempting to get a routine confirmatory test for early puberty, I was informed that I would have to wait, as the required medication had been in the clinic during power shutdowns with Hurricane Katrina, and was no longer usable. The surprising thing was it had been that long since I had last performed one of these tests - very unusual for a pediatric endocrinologist. Gonadal function is influenced by stress, and I suspected that the stress of the hurricane recovery was affecting the children, and thus reducing the prevalence of stress-associated endocrine disorders. Evidence of increased stress has been reported after Hurricane Katrina, for both parents and children

(Coker et al., 2006; Desalvo et al., 2007). I became curious what other effects the hurricane event had on referral patterns in my practice.

6.1 The study design

I had recently moved my practice to Ochsner Hospital, 4.5 months prior to Hurricane Katrina to be specific. The patient base of my practice was mostly insured patients (a more upscale patient mix), with the majority of patients living in Jefferson Parish or further west of New Orleans. Wait times for new appointments never exceeded 2 weeks in either the pre- or post-storm periods. I had already confirmed that the total number of patients seen during those 4.5 months, compared to the 4.5 months following the start of normal clinical operations after the storm, had not changed (approximately 450 in each time period), as part of my normal tracking of clinic volumes. We therefore looked at new referrals, both the total number and the number of referrals for specific disorders, some expected to be affected by stress (precocious puberty, pubertal delay, short stature, alopecia areata, 'spells'), and some not (thyroid disease, premature adrenarche). The data were compiled and statistical significance determined by risk analysis. The raw numbers are shown in Table 1.

	Pre Storm		Post Storm	
Diagnosis	Number	Percentage	Number	Percentage
Total Referred	177	-	169	-
Precocious Puberty	11	6.2	5	3.0
Premature Adrenarche	10	5.6	10	5.9
Pubertal Delay	24	13.6	25	14.8
Short Stature	45	25.4	30	17.8
Thyroid Disorders	33	18.6	32	18.9
Alopecia	0	0.0	2	1.2

Table 1. Effects of hurricane stress on the pubertal development of the children

6.2 Total referrals

The total number of new patients referred to my practice 4.5 months before and after hurricane Katrina was actually quite similar (Fig. 1). There were no significant differences noted on risk analysis. This was a surprising finding; I had assumed that the total visits were maintained by continued care for existing patients, most of whom were not permanently displaced by the storm. However, there did not appear to be a reduction of new patient referrals, either. Now, it is fair to say I had expanded to a new practice location in Baton Rouge after the hurricane, and this may be the reason the total number of referrals did not decline.

6.3 Effects of stress in children

To look for evidence of stress in the children, I took advantage of the physiology of normal puberty. Puberty refers to the physical and hormonal changes which typically begin in early adolescence and lead to reproductive maturity and completion of growth. In females the physical changes include growth of the breasts, development of pubic hair, and onset of menstrual periods (menarche). In males the physical changes include growth of the penis and testes, pubic hair, increased muscle mass and strength, and increased body and facial hair. The body changes are triggered by rising levels of the sex steroids (androgens and

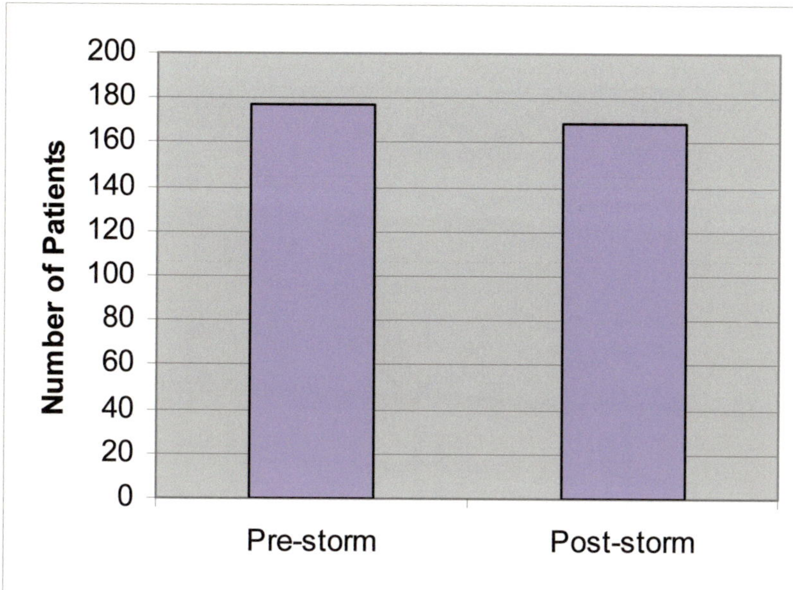

Fig. 1. Total Number of New Referrals. The total number of new patients referred between April 15, 2005-August 28, 2005 (Pre-storm) and those referred between October 1, 2005-February 14, 2006 (Post-storm) are shown.

estrogens). These arise from parallel hormonal processes termed "adrenarche" and "gonadarche." Adrenarche refers to maturation of the adrenal cortex with rising levels of adrenal androgens. These can produce early stages of pubic hair, underarm hair, adult-type body odor, and acne. This process is at least partly independent of gonadarche, which is initiated by the central nervous system via pulsatile release of gonadotropin-releasing hormone from the hypothalamus. This in turn causes the release amplitude of gonadotropins from the pituitary gland, which activates hormone producing cells of the testes or ovaries, resulting in fertility. The two can be distinguished by physical exam and by measuring hormone levels. Importantly, gonadarche is likely to be affected by stress, as stress is well know to induce hypogonadism in adults and may slow pubertal development in children (Charmandari et al., 2003). Interestingly, a recent pediatric study indicated that stress has the opposite effect, accelerating pubertal development (Tremblay & Frigon, 2005), although in this survey-based study the pubertal signs themselves may have been the cause of the associated stress. Adrenarche is not influenced by stress (Zukauskaite S., et. al., 2005), and referrals for this disorder served as an internal control for the study. The details of this study have been published (Ponnapakkam, 2008).

The data shows evidence of health effects from stress in the children - there was a 52% reduction in new patient referrals for gonadarche, while there was only a 4% decrease in new referrals for adrenarche (Fig. 2). Referrals for thyroid disorders, which would also not be expected to be influenced significantly by stress, were also similar in the pre- and post-storm periods. Alopecia areata (a condition of localized hair loss) is another condition which has been shown to be associated with increased stress (Gulec, AT, et. al., 2004). There were 2 referrals for hair loss (alopecia areata) after the hurricane, while there were none in the pre-storm period, further supporting the hypothesis.

6.4 Effects of stress in parents

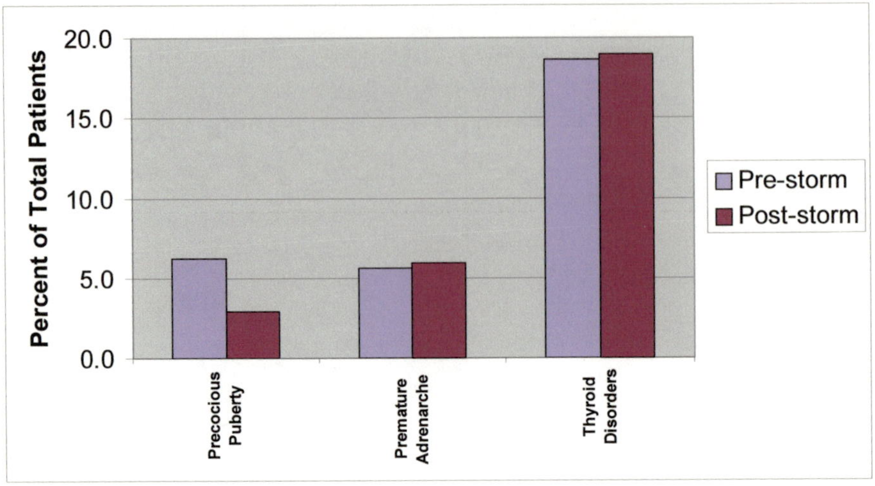

Fig. 2. New Diagnoses - Precocious Puberty. The percent of new patients whose final diagnoses were precocious puberty, premature adrenarche, or thyroid disorders in the pre-storm or post-storm periods are shown.

While these data show evidence of stress in the children, some of the other disease-specific referral rates are more difficult to explain by this finding alone. If stress effects on gonadal function is the reason for the reduced incidence of early gondarche, we should also have seen an increase in the incidence of pubertal delay. However, there was only a very modest (9%) increase in referrals for this disorder after the storm, which did not achieve statistical significance (Fig 3). Furthermore, stress has been previously reported to slow growth in stature (Gilmour, J, et. al., 1999, Charmandari, et. al., 2003), but we actually saw a 23% decrease in referrals for short stature after the hurricane, a result which did achieve statistical significance.

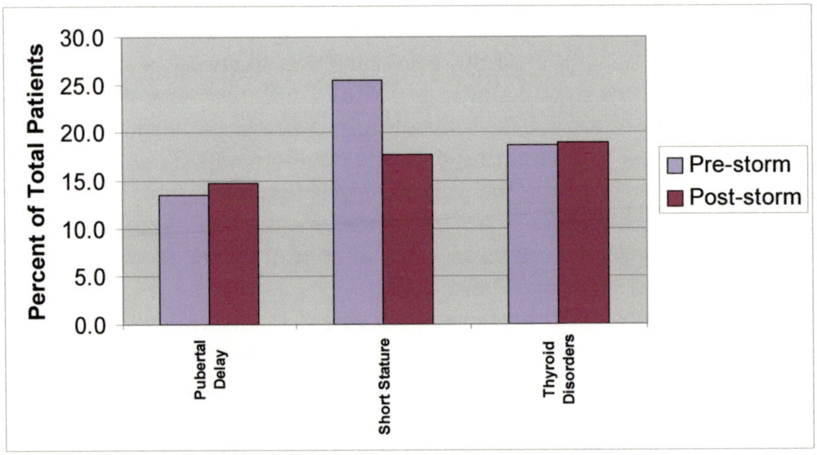

Fig. 3. New Diagnoses - Delayed Puberty. The percent of new patients whose final diagnoses were either pubertal delay, short stature (not associated with pubertal delay), or thyroid disorders in the pre-storm or post-storm periods are shown.

Now, it may be that stress effects on growth might not be as evident in such a relatively short time frame, but it still does not explain why we saw a decrease in referrals. It appears that the explanation is not medical but rather social. Early puberty in children is distressing for parents, so much so that I even saw a case of it while volunteering at the outpatient triage center in St. Charles Parish Hospital. Parents are concerned about early menarche in girls, and even more concerned that the early changes may be caused by a tumor in either sex. Likewise, thyroid disorders cause symptoms which are likely to motivate the patient and parents to seek immediate medical attention (excessive sleeping, constipation, low energy with hypothyroidism; hyperactivity, weight loss with hyperthyroidism). Pubertal delay and short stature do not provoke the same level of anxiety, in either the parent or the physician, and it appears that in the stress of the post-hurricane recovery period these problems were more likely to be ignored. The percent change in incidence for each disorder is summarized in Figure 4.

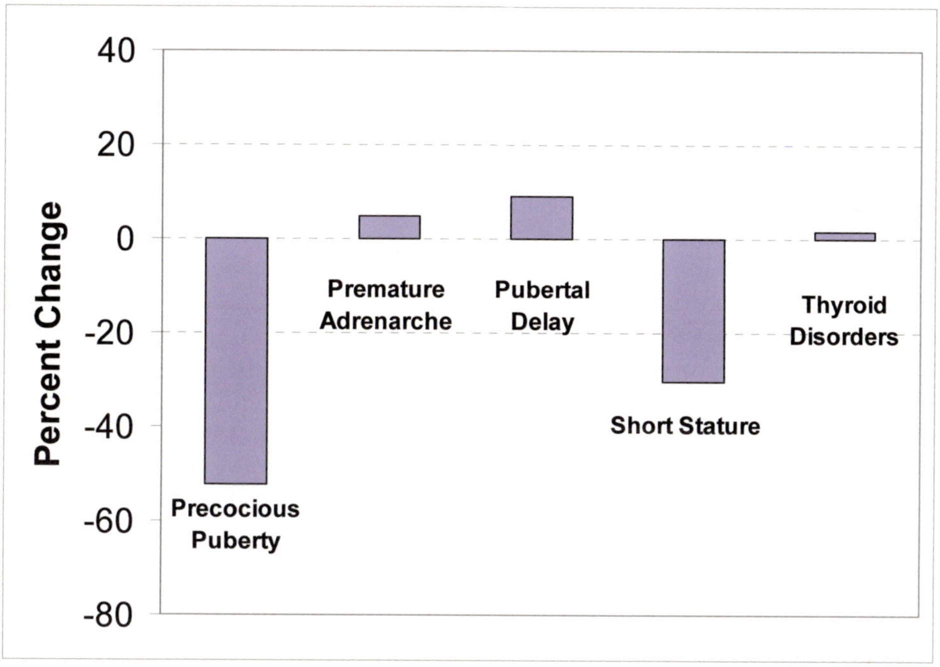

Fig. 4. Percent Change After Hurricane Stress. The percent reduction in the post-storm period of new patients with each of the indicated final diagnoses is shown.

6.5 Study conclusions

The data from the study confirm the hypothesis that chronic stress was influencing the health of children in the New Orleans area after Hurricane Katrina. While I examined illnesses relevant to my specialty, it is likely that other stress-associated illnesses (i.e. inflammatory bowel disease) would have been influenced as well. Of course, effects of chronic stress go well beyond health states, and can have implications in social settings and school performance as well. These have been document in the setting of other hurricane events (La Greca, et. al., 1996 and 1998), and after other natural disasters such as earthquakes (Kitayama , et. al., 2000, Karakaya, et. al., 2004, Uemoto, et. al., 2000).

The findings about parent stress and family situations affecting referral patterns to my practice is perhaps even more important. It suggests that there was an ongoing deferral of aspects of healthcare that might seem non-essential. Of course, this could have contributed to the reduced patient volumes seen during the recovery period, and would have different degrees of effect for different specialties. It also would predict an increase in patient volumes to follow, which a consolidated healthcare system would find difficulty accommodating.

7. The sixth stage – long-term effects

If the above analysis predicted that patient volumes would increase as the population stopped postponing basic healthcare, it is certainly interesting to follow up with what did actually happen. Unfortunately, presenting data on patient volumes is not helpful, as my practice was soon functioning at maximum capacity, and these numbers therefore do not change. What did change was the time interval to the next available appointment. At the time of the analysis, this was 1-2 weeks. By the time of my departure, this was beyond 3 months. To be able to preserve appointment times for existing patients, I was forced to close the practice to new patients on two occasions. Not surprisingly, I was not the only physician in New Orleans experiencing dramatically increased patient volumes. There was a 52% increase in daily hospital census for the region (from 1,237 to 1,877) by 1 year after the hurricane (Louisiana Pubic Health Institute, "NOLA Dashboard", and U.S. Government Accountability Office, 2006). This had a beneficial effect on hospital finances, and Ochsner Hospital was no longer losing money. It had become problematic for patients, however, who found access to care extremely poor.

In addition to having effectively strangled the health care system, this temporary 'boycott' of healthcare services unfortunately had direct consequences to the patients as well. Not all medical problems can be taken care of at one's leisure; some conditions provide a "window of opportunity" for therapy. At one point, I saw 10 patients in 2 weeks for short stature whose growth phase had already completed (I typically might see one or two a year). Had they come to me sooner, I might have been able to help ,but as it was I could not. I also had a patient deceased from a combination thyroid storm and a heart condition (technically not my patient, as I never got the opportunity to see or treat her).

The reductions in healthcare spending during the post-storm period also ultimately had their consequences as well. The uniformly reduced salary and benefits, paired with an increase in cost of living, made the region a fertile recruiting ground for hospitals from around the country that did not suffer these financial setbacks. While all of the physicians in my specialty returned to their practices after the storm (with one physician who had planned to retire postponed those plans during the storm recovery period), by the fifth anniversary of the storm two of the three pediatric endocrine practices (including my own practice) in New Orleans had closed. New Orleans no longer functions as a referral center for pediatric endocrinology; rather, patients are currently being directed out of the city for care.

8. Conclusions – lessons to be learned

The impact of Hurricane Katrina on the New Orleans area healthcare system serves as an important case study for recovery from a natural disaster. Beyond the immediate effects, there will be transient alterations in healthcare utilization, as families prioritize immediate needs over health maintenance. There will be stress in the population, which will have health consequences of their own. The stress of the parents will be felt by the children as well, affecting their health.

Unfortunately, the free market nature of the healthcare system in the United States appears to have amplified these problems rather than dampening them. A central body might have been able to compensate for the temporarily reduced earnings from patient care during the time period of under-utilization. On the other hand, where the central body could have had the most immediate impact, the results were even more dire. The public hospital in New Orleans, Medical Center of New Orleans was closed after the storm, and remains closed to this day. The Veterans Administration hospital (a hospital network providing care for military veterans in the United States, provided by the federal government) is also closed. It is important for healthcare and government officials to realize the strains that will be faced by the healthcare system during the recovery period, and to plan accordingly.

9. References

Charmandari, E., Kino, T., Souvatzoglou, E., & Chrousos, G. P. (2003). Pediatric stress: hormonal mediators and human development. *Horm Res*, 59(4), 161-179.

Coker, A. L., Hanks, J. S., Eggleston, K. S., Risser, J., Tee, P. G., Chronister, K. J., et al. (2006). Social and mental health needs assessment of Katrina evacuees. *Disaster Manag* Response, 4(3), 88-94.

Desalvo, K. B., Hyre, A. D., Ompad, D. C., Menke, A., Tynes, L. L., & Muntner, P. (2007). Symptoms of Posttraumatic Stress Disorder in a New Orleans Workforce Following Hurricane Katrina. *J Urban Health.*, E-publication.

Gilmour, J., & Skuse, D. (1999). A case-comparison study of the characteristics of children with a short stature syndrome induced by stress (Hyperphagic Short Stature) and a consecutive series of unaffected "stressed" children. *J Child Psychol Psychiatry*, 40(6), 969-978.

Gulec, A. T., Tanriverdi, N., Duru, C., Saray, Y., & Akcali, C. (2004). The role of psychological factors in alopecia areata and the impact of the disease on the quality of life. *Int J Dermatol*, 43(5), 352-356.

Karakaya, I., Agaoglu, B., Coskun, A., Sismanlar, S. G., & Yildiz Oc, O. (2004). [The symptoms of PTSD, depression and anxiety in adolescent students three and a half years after the Marmara earthquake]. *Turk Psikiyatri Derg*, 15(4), 257-263.

Kitayama, S., Okada, Y., Takumi, T., Takada, S., Inagaki, Y., & Nakamura, H. (2000). Psychological and physical reactions on children after the Hanshin-Awaji earthquake disaster. *Kobe J Med Sci*, 46(5), 189-200.

La Greca, A., Silverman, W. K., Vernberg, E. M., & Prinstein, M. J. (1996). Symptoms of posttraumatic stress in children after Hurricane Andrew: a prospective study. *J Consult Clin Psychol*, 64(4), 712-723.

La Greca, A. M., Silverman, W. K., & Wasserstein, S. B. (1998). Children's predisaster functioning as a predictor of posttraumatic stress following Hurricane Andrew. J *Consult Clin Psychol*, 66(6), 883-892.

Tremblay, L., & Frigon, J. Y. (2005). Precocious puberty in adolescent girls: a biomarker of later psychosocial adjustment problems. *Child Psychiatry Hum Dev*, 36(1), 73-94.

Ponnapakkam, A, Gensure, R, Effects of Stress After Hurricanes Katrina and Rita on Pubertal Disorders in Children, *The Ochsner Journal* 8:000–000, 2008.

Uemoto, M., Shioyama, A., Koide, K., Honda, M., Takamiya, S., Shirakawa, K., et al. (2000). [The mental health of school children after the Great Hanshin-Awaji Earthquake: I. Epidemiological study and risk factors for mental distress]. *Seishin Shinkeigaku Zasshi*, 102(5), 459-480.

U.S. Government Accountability Office, Status of the Health Care System in New Orleans, Pub. no. GAO-06-576R (Washington: GAO, March 2006)

Zukauskaite, S., Seibokaite, A., Lasas, L., Lasiene, D., Urbonaite, B., & Kiesylyte, J. (2005). Serum hormone levels and anthropometric characteristics in girls with hyperandrogenism. Medicina (Kaunas), 41(4), 305-312.

25

Hurricane Evacuations of the University of Texas Medical Branch at Galveston

Marc Oliver Maybauer MD PhD, Michael Megna FACHE,
Sven Asmussen MD and Dirk Manfred Maybauer MD PhD
The University of Texas Medical Branch at Galveston
Texas, USA

1. Introduction

Hurricanes develop in the trade wind zone over sea, when the water temperature rises above 79.7 °F. A temperature difference between sea level and high heights might lead into a tropical whirlwind. A hurricane develops air speeds of more than 75 mph (rotation speed) according to the wind force of 12 Beaufort. The moving speed of the swirl is only 9.32 - 18.64 mph. The circumference of the storm can be hundreds of square miles. Hurricanes can stay active for weeks and devastate areas of thousands of square miles. The yearly time of origin is limited from May to December (Kim et al., 2009).

The City of Galveston is settled in the greater Houston area, about 50 miles southeast of Houston, Texas on an elongate offshore island in the Gulf of Mexico. This island is the south boundary of the Galveston Bay and the Houston Ship Channel. In September 1900 a hurricane devastated Galveston, destroyed much of the city and claimed the lives of approximately 6,000 to 8,000 people. At that time the average elevation of the island was only slightly above sea level. After the storm Galveston was rebuilt at a raised level and further protected from the sea by a concrete seawall that rises 17 feet above mean low tide (Alperin, 1977).

The University of Texas Medical Branch (UTMB) was founded in the year of 1891. Today UTMB consists of four schools (Medicine, Nursing, Health Professions and Graduate School), seven hospitals on campus, as well as three institutes and a network of more than 100 campus and community based clinics. Although not owned or operated by UTMB, the campus is also home to the Shirner's Burn Hospital, which is staffed by UTMB faculty and researchers. The number of employees is approximately 13,000, plus 2,900 students. Furthermore, the Galveston National Laboratory, which is one of two biodefense laboratories of the US government is located on campus. UTMB is also the site of a regional Level-1 trauma center and the University's annual budget is approximately $1.6 billion (Sexton et al., 2007) .

In September 2005 UTMB was threatened by hurricane Rita. Abandoning its historic practice of sheltering patients and staff in-place or clearing the hospital of all but the sickest patients, UTMB rapidly organized and conducted the first total evacuation in its 114-year history. Threatened for the first time by the possibility of a category five hurricane with a 20-feet storm surge that carried the potential to inundate the island. UTMB's leadership was determined to avoid the disastrous effect of hurricane Katrina seen in New Orleans only weeks earlier. Three days before landfall the incident command center which functioned as the decision-making and communication hub throughout the storm, was activated as specified in the UTMB emergency operations plan. The primary spokesperson and sole person responsible for the evacuation, was the Incident Commander who lead UTMB through this challenging event. However, the evacuation went smoothly and was successful. Hurricane Rita caused only minor damage on Galveston Island and the UTMB campus. Therefore, the clinical and scientific staff could resume work less than one week after the landfall of Rita. Just three years later in September 2008, UTMB was again threatened by hurricane Ike and again had to be evacuated (Maybauer et al., 2009). In the following we will report how this was accomplished, what impact Ike had on Galveston Island and UTMB, and what lessons were learned from both hurricane evacuations.

2. The hurricane Ike evacuation

UTMB emergency management officials and executive leadership began tracking hurricane Ike late in the week of September 1st, and participating in the Texas Division of Emergency Management State Operations Center conference calls with the National Weather Service (NWS) on Sunday September 7th. Predicted paths varied from the lower Rio Grande Valley to the Florida panhandle early on, but it was assumed that a Texas landfall was likely. By Tuesday it was felt that the storm track would make landfall at Corpus Christi, Texas and concern over a Galveston landfall was reduced somewhat. On Wednesday new NWS data indicated a likely upper Texas Coast strike and UTMB leadership were re-engaged in preparations for this storm. It was determined that a full-scale patient evacuation would be necessary and assessments were quickly conducted to position transportation assets to arrive in Galveston on Thursday September 11th. By late Wednesday the UTMB leadership was committed to an evacuation decision early Thursday morning, and this was in fact the case, given the overnight weather information.

UTMB evacuated 469 patients from the inpatient towers, correctional hospital and behavioral health facility. Of these, some 80 were neonates including Infant Special Care/Intensive Care level 2 and level 3 babies. Patients were transported to facilities in Austin, San Antonio and Dallas-Fort Worth metropolitan areas.

According to the special situation of the City of Galveston, a hindering factor for an extra-hospital ground transport evacuation is the 2 mile long causeway leading to Galveston County. Ambulances can only pass this bridge safely if wind-speeds are below 40 mph. Such wind speeds may occur as early as 12 hours before the hurricane makes landfall. As such, planning efforts need to assure that an evacuation should be finished no later than this time point of 12 hours. This specific time was calculated for Friday, September 12th 2008 at 12:00 noon. The evacuation plan of UTMB envisions for such kind of storm tide events, and a countdown of 120 hours (5 days) begins and a "countdown clock" from H-120 hours tracks significant events against the time remaining before tropical storm force winds arrive. The phase of 120 to 73 hours pre arrival of the hurricane estimates a normal hospital duty roster and two daily conferences in the Emergency Operations Center, under the leadership of the Incident Commander. A final decision for an evacuation must be made by 72 hours pre arrival, in order to request evacuation assets in an adequate number and with appropriate staff and equipment, before this time point. Simultaneously, elective admissions and transfers of patients to UTMB hospitals were stopped. Locally arriving patients received only ambulatory or stabilizing treatment in the Emergency Department (ED) and were discharged or transferred to other hospitals. Four hundred, sixty-nine (469) patients were evacuated during Thursday September 11th 2008, during the period of 36 to 24 hours before the estimated arrival of sustained 40 mile per hour winds (H-0). All mobile patients and those without need for monitoring were discharged directly and picked up by family members. Eighty-two patients were transported by coach. One hundred, forty-three (143) ambulances were sent to Galveston from a staging area in San Antonio, Texas, some 225 miles away. Forty-eight patients were transferred by ambulance to the nearest airports. Following these shuttle-transports all available 143 ambulances transported patients to the above mentioned hospitals. An additional 23 patients would be picked up by helicopters, directly from the landing deck of the UTMB Emergency Room. One hundred and two correctional patients were transferred from the hospital operated jointly by UTMB and the Texas Department of Criminal Justice. The patient evacuation was completed within approximately 11 hours with the last patient arriving in Austin just before midnight Thursday night.

UTMB had released students on Wednesday, and non-essential personnel on Thursday, as well as essential personnel who were not required to ride-out the storm as evacuations were completed. The UTMB Emergency Operations Center was fully activated on Thursday morning to support the patient evacuation and finalize planning for the hurricane. Contact was established with the State Operations Center/Department of State Health Services Liaison, to keep them apprised of UTMB's status, and with the Regional Catastrophic Medical Operations Center operating at the City of Houston Emergency Operations Center.

Tropical storm force winds were recorded beginning in the early evening hours of Friday, and sustained winds of 70+ miles per hour by 22:00. Storm surge and wind driven rain resulted in flooding of lower located areas of the campus by 17:00 and flooding was reported in the McCullough Building basement. Power outages and transfers to generators were reported sporadically beginning at 19:00. Water was first noticed in the lobby of John Sealy Tower by 23:00, and the first floor was cordoned off at the stairwells. At 21:45 John Sealy Annex transferred to generator power and at 23:00 the command center was relocated to the 5th floor of John Sealy Tower. At this time the tower had lost primary power and was being supported by emergency generators. By 01:45, shortly before the eye of the hurricane was over land, water was reported to be over two feet deep in the tower lobby. By 09:30 on Saturday morning the winds had diminished, and water had begun to recede. At this time a roster check revealed all personnel were accounted for and safe.

By early Saturday afternoon it was safe to venture outside and damage assessments were begun. An all-staff meeting was called for 15:00 and reports were received from police, facilities, environmental health and command personnel. Early damage assessments indicated that the flooding and interruption of essential services (domestic water, sewer and power) would mean healthcare, research and educational activities would be interrupted for an undetermined period of time. UTMB entered a "preserve and protect" mode and began the process of ordering reserve generators, pumps, chillers and electrical equipment to initiate the recovery phase.

Recovery was expected to continue for several months.

3. Hurricane Ike information

Hurricane Ike was a storm that had significant impact over a wide area of the Caribbean and the US. While categorized as a Category 2 storm when it made landfall in Galveston, its size, timing, duration and the associated coastal surge had the impact of a much greater

storm. Hurricane Ike formed September 1 st, 2008 and dissipated on September 16th, 2008. Highest sustained winds were reported to be 145 mph prior to landfall at Galveston. The eye of Ike passed over UTMB in the early morning hours of Saturday September 13th, at which time the severe flooding across virtually all of the campus was recognized. While wind speeds were diminished to Category 2 force at landfall in Galveston, the size of the storm stretched those winds over a 510 miles area, much larger than the size typically associated with a Category 2 storm. The result was a significant storm surge and a much longer period of wind and surge impact as compared to other Category 2 storms that have hit the Texas Coast. Just before landfall and at landfall, Ike's winds ranged from 92-110 mph based on National Oceanic and Atmospheric Administration/NWS Galveston's report.
The eye of the storm made landfall in Galveston at approximately 02:10, about 2 hours before the morning high tide. The combination of surge and tide resulted in a storm surge typically associated with a Category 4 hurricane. Ongoing studies on storm surge, predict that the maximum surge expected from a hurricane at Galveston is 19 feet at the shoreline. While official reports show a 7 feet surge at the east end of Galveston Island. The back wash wave resulting from release of the surge being pushed up into Galveston Bay as the eye of the storm passed over, resulted in a wave of water from the north pushed by the shifting winds that rose in excess of 14 feet in certain areas of the UTMB campus. The variation in high water elevations experienced, are thought to be due to the characteristics of the path traveled and the strong wind gusts. In some cases the movement of the water into a building wall was accelerated by strong winds pushing the water to elevations higher than the general surge.

4. Emergency preparedness, response and recovery preparation

Preparation for the storm started as the storm approached the Gulf of Mexico. Standard preparations start at the beginning of each hurricane season and the Emergency Standard Operating Procedures served the campus well. Hurricanes Dolly, Edouard and Gustav earlier in the year each served to help prepare for Ike. The situational awareness was at a very high level. There were lessons learned in each previous storm that served to prevent even greater impacts to the campus facilities and helped prepare staff for this event. For Ike, the decision to evacuate patients and non-essential personnel, students and staff came Thursday (36hrs) before the storm. The evacuation went smoothly. Other preparations included ensuring the supply stock was full, preparation of the command center, cleaning up and tying down loose equipment, and hardening certain areas for wind. Reserve supplies of diesel fuel, food, potable water and linens were ordered and delivered in the days before the storm. A mobile kitchen facility was reserved on Wednesday before the storm, and was delivered and installed during the week after Ike. This kitchen served meals for the university staff and recovery workers immediately after the storm, and continued in operation for many months due to the loss of the main kitchen and dining area on the first floor of the hospital complex.

5. Response

The initial storm impact to the campus was the rising water due to the storm surge. Water started accumulating in low located areas Friday morning and by afternoon had completely covered the plaza in front of the Ashbel Smith Building. As afternoon came, the winds picked up and started pushing water up against various buildings. A fire broke out at the Galveston Yacht Basin's dry boat storage facility Friday afternoon. This building was located several hundred yards north of the UTMB campus. Due to the high water in the area, the fire department was unable to access the fire and extinguish it. The smoke from the fire was blown into the Emergency Department causing the ED to relocate their operations to the Waverley Smith Pavilion Hospital on the eastern edge of the campus. The move was a pre-planned action following the campus contingency plan and went smoothly. This location served as the primary entrance point for EMS-transported patients throughout Sunday and into Monday after the storm. Some lessons were learned in using the Waverley Smith Pavilion as the interim ED such as understanding access issues through corridors that were not designed for patient movement, but the operations were maintained without significant impact. As the winds increased and the water surge levels rose, normal power was lost just before 22:00 Friday night. The emergency generators came on to provide power to the areas designed to receive emergency power. Other areas remained dark. The loss of normal power was three hours before the eye of the storm reached the island, evidence of the size of the storm. The overall period of extreme wind and rain lasted over eight hours. The UTMB Incident Command Center operated throughout the storm, despite loss of communications for a period of time and relocation to a previously designated alternate location on the 5th floor of John Sealy Towers. The loss of communications was due to the flooding of the communication lines splicing (in the low lying junction boxes), the loss of cellular antennas and of the short wave radio boosters. The first communications to come back was cellular telephones aided by the arrival on Sunday of an AT&T mobile wireless cell system.

The storm surge was the critical factor in hurricane Ike. Water entered the ground floor, crawl spaces and basements of most buildings. The most substantial water incursion was into the McCullough Building where the linear accelerators were flooded. The wind damage was minimal with some windows lost and some roof damage. Mud and debris covered all areas that had been underwater. In a prior flood assessment it was noted that the highest flood water level recorded was at elevation 14 feet. The impact from the storm resulted in high water ranging from around 8 feet to 14 feet across the campus. The back wash from Galveston Bay as the eye of the storm passed pushed water up in certain areas breaching the entry points to the buildings.

The catastrophic damages to Galveston Island resulted in loss of all municipal and franchise utility services. Normal electrical power was lost during the storm and damages to equipment and campus systems hindered the restoration of power until the buildings were capable of receiving electricity. Normal power started to come back a week after the storm and was not completely restored until October 6th. All but three generators continued to operate in certain buildings while the equipment and the systems were being restored. Natural gas to the island was shut off for fear of fire and explosions after the storm. The municipal water system was lost for almost 2 weeks due to the destruction of pumps and generators at the main water plant, affecting fire protection, use of toilets and municipal drinking water. The municipal wastewater system was lost as well, providing no treatment for sewage until the city's treatment facilities could be restored. Loss of municipal water also meant that air conditioning chilled water systems that relied on city-supplied water for cooling tower make up could not operate. Debris clogged storm drains. The restoration of the systems just to start to serve the campus took over three weeks.

Galveston Island was closed to all traffic except emergency first responder personnel for several days and residents were only allowed to return to inspect property and leave the island before the 06:00 pm curfew that lasted for over two weeks. As a result, essential

UTMB personnel were instructed to remain evacuated, and Disaster Medical Assistance Teams (DMAT) were requested via the State Operations Center. DMAT personnel from more than four states arrived within 48 hours of landfall and provided medical care in the UTMB Emergency Department for over two weeks. At that time UTMB personnel were able to resume emergency treatment, but no patient admissions were possible due to lack of inpatient facilities. All patients were treated and released or stabilized and transported to more appropriate facilities. It was expected that local EMS continued to divert patients to other facilities for life threatening, or complicated conditions for at least 90 days after Ike. UTMB reopened for admissions – limited to Obstetrics and Gynecology patients – 30 days after the hurricane, and admissions were expanded in a phased fashion, as staffing and critical services would allow. By mid-January 2009 over 200 beds were in operation and by the end of May, 370 inpatient beds were staffed on a daily basis.

6. Damages at UTMB due to hurricane Ike

- Loss of function for all offices, support areas, clinics, and mechanical spaces on the first floor of approximately 80 buildings, including (but not limited to):
- 3 Emergency Generators and automated switching equipment
- Hospital Pharmacy Robotics System
- ALL elevators (132)
- All Chilled Air Supply
- All Domestic Water
- All Local Power
- Sterile Processing
- Day Care Center
- Outpatient Clinics
- Materials Management / Central Supply
- Hospital Admitting Office
- Warehouse (surplus equipment and storage)
- Blood Bank (Donor Center and Cross match Lab)
- Food Services
    - Main Kitchen, Retail Food Service, Cafeteria
- Steam due to flooded pits and link seal breach.
    - Critical to Animal Cage washing
    - Sterilization
    - Hot water
    - Condensate, domestic water, sump and other pumps due to flooded conditions
- Medical Gas Vacuum Pumps
- Student Housing
- Main Hospital Supply, Laundry/Linen distribution center
- Linear Accelerators
- Various Research and Clinical Laboratory Analyzers susceptible to high humidity
- Hospital Clinical Equipment Services
- Research Fabrication and Machine Shop
- ALL physical plant / maintenance tools, spare parts, etc.
- 37 Motor Vehicles, including 2 Bloodmobile and 1 Mobile Mammography Coaches

- UTMB Chapel
- UTMB Bookstore, Gift Shop and 2 Starbucks kiosks.

7. Impact and long term recovery

Hurricane Ike's biggest impact was to UTMB's critical core buildings and operations. The buildings on the campus are categorized by the critical nature of each buildings function. Those buildings that must remain operational at all times are addressed differently than those buildings that can be evacuated without significant impact to the operations of the campus.

The impact on these buildings was assessed as follows:

- Over 1 million square feet of building space flooded to depths ranging from 6 inches to 6 feet
- Of the 36 critical Buildings in the Research and Healthcare Complex Core, 32 (89%) buildings were damaged due to flooding.
- Of the critical core Healthcare Buildings only, 10 out of 11 (91%) buildings sustained flood damage
- Of the Academic and Primary Support buildings 20 of a total 25 buildings sustained flood damage

8. Damage estimate

The damage to the UTMB campus was significant. The preliminarily estimated cost to UTMB was $710 million, and is likely to grow to over $1 billion, including but not limited to costs for the following:

- Patient evacuation
- Student relocation
- Building damage
- Campus cleanup,
- Infrastructure and equipment repair
- Business interruption
- Research equipment
- Reconstruction to pre-hurricane conditions
- Mitigation against future flood events (water resistant materials and finishes)

Damage estimates elsewhere in the impacted areas were substantial. By November 2009 over 780,000 damage claims had been received by the Texas Department of Insurance with a gross loss value of $10.64 billion, and net (after insurance payments) of nearly $5 billion. In Galveston County gross damages were estimated at $2.3 billion and over $591 million after insurance payment - NOT including damages at UTMB (Texas House of Representatives Interim Report)

9. Discussion and conclusions

Disaster preparedness plans in hospitals are indispensable for a successful completion of a hospital evacuation. Disasters like the flooding of 2002 in the Oder-Elbe-region in Germany, the South-East Asia tsunami in 2004, or the flooding of New Orleans, Louisiana, USA during hurricane Katrina in 2005, and Galveston, Texas during hurricane Ike in 2008 has let us become aware that flood catastrophes are not only documented far away from us, but they can happen everywhere and could also occur in our neighborhood. Catastrophes do not follow any rules and are barely predictable according to time and location. To become ready for catastrophic cases, it is relevant to develop localized structures, disaster plans and to establish drills. The following functional sections should be established: incident commander, incident planning, logistics, administration, communication (internal, with committees and media), and security (Born et al., 2007a, 2007b; Zane & Prestipino, 2004).

At UTMB in Galveston, valuable lessons have been learned from the sequential evacuation of the university hospital. An adequate disaster preparedness plan and logistics were crucial for the success. The lessons learned from both experiences included especially the following:

1. Appointment of an Incident Commander. This person must be given sole authority for decision making. He or she should also have clinical experience, and occupy an executive leadership position (Sexton et al., 2007).
2. Having an emergency operations center or Incident Command center utilizing National Incident Management System (NIMS) principles (DHS, 2008) to support the incident commander is crucial for success. The center helps to communicate decisions and developments internally and externally, and serves as advisor for the incident commander. It is the responsibility of this leadership to not delay the decision to evacuate or worry about anyone second guessing the decision. The sooner an evacuation may be started, the higher the likelihood that the evacuation will proceed accordingly to the emergency plans. The Incident Command center has to keep track of evacuated patients and on-site personnel, to utilize expert clinical staff (MDs and RNs) to coordinate loading of ambulances and helicopters for patient transfer, or to reassign staff as necessary to care for transferred patients (Zane & Prestipino, 2004).
3. Having reliable in-house communication system not dependent on telephone lines or electricity; During hurricane Rita it was discovered that some of the communication devices such as walkie-talkies or cell-phones were either outdated or did not work in too many „dead" areas of the campus. These issues had been addressed and could be solved during drills before hurricane Ike approached Galveston (Sexton et al., 2007).
4. Having a reliable telephone system for contacting outside facilities is very helpful to keep track on patients, because ambulances may become stuck in traffic on evacuation routes or have to take alternative routes to make their way to the admitting hospital; or even reroute patients to another health care center. The lack of this information may hinder the ability to communicate effectively with family members (Maybauer et al., 2009).
5. Maintaining a paper record of all patients, and patient transfers as well as all other business transactions; The data of electronic medical records may not be accessed in a different hospital because of software or system differences (Sexton et al., 2007).
6. Every healthcare facility - hospitals, urgent care facilities, ambulatory care facilities, long-term care and skilled nursing facilities etc. - should develop evacuation plans beyond the typical fire evacuation plan, that anticipate interruptions in services. Any internal or external cause should be anticipated including power, water or medical gas failures, wind and/or rain damage, loss of critical internal services such as pharmacy, food preparation and/or distribution, non-functional operating rooms, and any other

vulnerability identified for the facility or location. The regular testing of these plans is an essential component of preparedness and ultimately successful execution of the plan. Flooding will occur in coastal areas resulting from a hurricane surge. Emergency power and temporary cooling systems are essential after the evacuation and after the storm moves inland. Emergency evacuation of a large university hospital requires extensive effort from both the hospital staff and the community. It is to remark positively that no patient inside UTMB was hurt directly by the hurricane or indirectly by the evacuation. The UTMB leadership and emergency preparedness officers continue to refine the emergency plan and related procedures on the basis of the experience during the hurricanes Rita and Ike. The authors encourage physicians, as essential members of the health care team, to become prepared to respond to disasters. A significant part of this preparation is formal training in Incident Command System and National Incident Management System concepts, which can by accessed through local emergency management officials, regional training centers or electronically from the Federal Emergency Management Agency.

10. Acknowledgement

The authors thank Donald S. Prough, M.D., F.C.C.P., Professor and Chairman, Department of Anesthesiology, as well Karen H. Sexton, Ph.D., former EVP & CEO Health System Interim and Associate Dean for Clinical Affairs, Department of Health System Administration for their support, and Mr. Andrew W. Hall for outstanding editorial assistance.

11. Correspondence

Marc O. Maybauer, M.D., Ph.D.
Department of Anesthesiology
Division of Critical Care Medicine
University of Texas Medical Branch
301 University Boulevard
Galveston, TX, 77555-0591, USA
Phone: +1 409 772 1221
FAX: +1 409 772 1224
momaybau@utmb.edu

12. References

Kim, H.M.; Webster, P.J.; Curry, J.A. (2009). Impact of shifting patterns of Pacific Ocean warming on North Atlantic tropical cyclones, *Science,* 325, 77-80

Alperin, L.M. (1977). Custodians of the Coast: History of the United States Army Engineers at Galveston, *Galveston, Tex.,* Galveston District

Sexton, K.H.; Alperin, L.M.; Stobo, J.D. (2007). Lessons from Hurricane Rita: the University of Texas Medical Branch Hospital's evacuation, *Acad Med,* 82, 792-796

Born, C.T.; Briggs, S.M.; Ciraulo, D.L.; Frykberg, E.R.; Hammond, J.S.; Hirshberg, A; Lhowe, D.W.; O'Neill, P.A. (2007,a). Disasters and mass casualties: I. General principles of response and management, *J Am Acad Orthop Surg* 15, 388-396

Born, C.T.; Briggs, S.M.; Ciraulo, D.L.; Frykberg, E.R.; Hammond, J.S.; Hirshberg, A; Lhowe, D.W.; O'Neill, P.A.; Mead, J. (2007,b). Disasters and mass casualties: II. Explosive, biologic, chemical, and nuclear agents, *J Am Acad Orthop Surg* 15, 461-473

Zane, R.D.; Prestipino, A.L. (2004). Implementing the Hospital Emergency Incident Command System: an integrated delivery system's experience, *Prehosp Disaster Med,* 19, 311-317

Maybauer, D.M.; Megna, M; Kafka, G; Maybauer, M.O. (2009). [Hurricane Ike and the University of Texas Medical Branch Hospital's evacuation], *Anaesthesist,* 58, 1244-1251

House Select Committee on Hurricane Ike Devastation to the Texas Gulf Coast. (Feb 2009). Texas House of Representatives Interim Report, State Representative Sylvester Turner, Chairman, *Appendix B*, p. 409

United States Depatment of Homeland Security, DHS. (Dec 2008). National Incident Management System

26

Case Studies and Lessons Learned in Chemical Emergencies Related to Hurricanes

Perri Zeitz Ruckart, MPH and Maureen F. Orr, MS
Agency for Toxic Substances and Disease Registry
United States

1. Introduction

The 2005 hurricane season in the United States was one of the most active seasons documented. Hurricane Dennis, the first major hurricane of the season, weakened to Category 3 strength before making landfall over the western Florida Panhandle on July 10, 2005. Hurricane Dennis caused considerable damage across the western Florida Panhandle, including widespread utility and communications outages and storm surge-related damage far to the east of the landfall location (National Oceanic and Atmospheric Administration, 2005a).

Hurricane Katrina entered southern Florida on August 25, 2005 as a Category 1 hurricane before rapidly gaining strength in the Gulf of Mexico. The storm made its second landfall in south eastern Louisiana on August 29, 2005, as a Category 3 hurricane (Knabb et al., 2005a). Due to levy failures in the New Orleans area, Hurricane Katrina was one of the worst natural disasters to ever strike the United States. Katrina resulted in an estimated 1,464 deaths in Louisiana, numerous illnesses and injuries, and substantial infrastructure damage (Centers for Disease Control and Prevention, 2005e; Knabb et al., 2005a; Louisiana Department of Health and Hospitals 2006).

Hurricane Rita was also classified as a Category 3 hurricane when it struck the Louisiana-Texas border on September 24, 2005 (Knabb et al., 2005b). Although the impact from Hurricane Rita was not as severe as that from Hurricane Katrina, the approach of Hurricane Rita generated one of the largest evacuations in United States history; estimates exceed 2 million evacuees in Texas (Knabb et al., 2005b).

Hurricane Wilma, another Category 3 hurricane, struck the southwest coast of Florida on October 24, 2005 (Taylor et al., 2007). Hurricane Wilma caused extensive damage to homes and power lines and also temporarily displaced thousands of residents (Centers for Disease Control and Prevention, 2006b).

Much of the public health response to the hurricanes was focused on assessing health-related needs and surveillance of injuries, infectious diseases, and other illnesses (Centers for Disease Control and Prevention, 2005b; Centers for Disease Control and Prevention, 2005c; Centers for Disease Control and Prevention, 2005e; Knabb et al., 2005a; Knabb et al., 2005b; Tayor et al., 2007; Todd, 2006). However, the hurricanes also resulted in unintended chemical releases in the affected states. When natural disasters, such as hurricanes, trigger technological disasters, the resulting event is called a natural-technologic or "natech" event. Natech events may involve releases of household chemicals or large industrial releases, such as an oil leak from a severed pipeline (Young et al., 2004). Industrial natech releases can involve the release of extremely toxic substances and have the potential to expose large numbers of people. Hurricane Katrina resulted in small- and large-scale natech events. There were unintentional chemical releases with widespread health implications for affected populations (Picou, 2009).

This chapter summarizes chemical incident surveillance data associated with the 2005 hurricanes from three states participating in the Hazardous Substances Emergency Events Surveillance (HSEES) system at the time (Florida, Louisiana, and Texas). Additionally, case studies and lessons learned from hurricanes developed as part of a chemical emergencies workgroup are presented to give more detailed insight into the types and causes of the hurricane natech releases, and their environmental public health impact.

2. Methods

2.1 HSEES data analysis

HSEES was an active multi-state web-based surveillance system maintained by the United States Agency for Toxic Substances and Disease Registry (ATSDR) from 1990-2009. HSEES collected information on uncontrolled or unpermitted acute releases of hazardous substances meeting specific pre-established criteria. The surveillance system enabled identification of factors related to the public health impact of these acute events and activities to lessen the impact of and potential for future events. Events involving petroleum only were excluded because of the Petroleum Exclusion clause of the Comprehensive Environmental Response, Compensation, and Liability Act which prohibits ATSDR from becoming involved with incidents where any form of petroleum was released that had not been refined to the point of becoming a specific chemical product, such as pure xylene (CERCLA/SARA). HSEES collected data on all other hazardous substance incidents, including ones with petroleum where another hazardous substance is present (ATSDR, 2005a). HSEES was a surveillance system for a wide range of hazardous materials incidents and therefore is not specifically designed to collect data on hurricane chemical emergencies.

State health department personnel used a variety of sources (e.g., records and oral reports of state environmental agencies, police and fire departments, and hospitals) to collect information about the acute hazardous substances events. Information collected for each event included the location and industry involved in the event, hazardous substances released, number of victims, evacuations, and contributing causal factors for the event. The HSEES system collected data on the primary contributing factor and the secondary contributing factors related to an event. Information on contributing causal factors was either reported by the notification source or determined by the state HSEES coordinator using various reports. Data were entered into a Web-based application that enabled ATSDR to access the data.

A victim was defined as a person experiencing at least one documented adverse health effect (such as respiratory irritation or chemical burns) that was likely associated with the event and occurred within 24 hours after the release (ATSDR, 2005a). Hazardous substances released were grouped into 13 categories: acids, ammonia, bases, chlorine, hetero-organics, hydrocarbons, mixture across categories, carbon monoxide, pesticides, polymers, volatile organic compounds, other inorganic substances, and other substances. Mixtures across categories consisted of hazardous substances that were mixed before release, including hazardous substances from more than one of the other 12 categories used. The category "other inorganic substances" comprised all inorganic substances—except for acids, bases, ammonia, and chlorine—and includes hazardous substances such as nitrogen oxide and hydrogen sulfide. The "other" category consisted of hazardous substances, such as asbestos, that could not be classified into any of the other 12 categories.

For this analysis, data from the three participating states affected by the hurricanes (Florida, Louisiana, and Texas) were analyzed to describe the characteristics of hazardous substances releases associated with the 2005 hurricanes. Events were identified as hurricane-related based on the date of the incident and state reports. Descriptive statistics are presented including contributing causal factors, hazardous substances and industries/locations involved in the releases, victim information, and event scenarios, with emphasis on those events with actual or potential public health impact.

2.2 Case study derivation

Case studies and lessons learned about chemical emergencies during hurricanes were derived from a review of the literature on three prominent hazardous materials incidents relating to the hurricanes. Only one of the three case studies was captured by HSEES; the other two did not qualify for HSEES because they were either petroleum-related or were a chronic release. The case study of the incident captured by HSEES is based on details from the literature as these were more extensive that the information captured by HSEES.

The three case studies were originally prepared for the Agency for Toxic Substances and Disease Registry Chemical Emergencies workgroup of the National Conversation on Public Health and Environmental Exposures and have been modified for this chapter. These case studies are reviews and interpretations of the current literature concerning these incidents and are presented to give a broader perspective on hurricane-related hazardous substance issues than HSEES data alone can provide.

3. Results

3.1 Data from HSEES

In 2005, 245 hurricane-related chemical events were reported to HSEES from the coastal states of Florida, Louisiana, and Texas. Figure 1 is a timeline of the four hurricanes with information on category, area where they made landfall, and number of HSEES events. Of the 245 events, 133 occurred in Texas, 68 occurred in Florida, and 44 occurred in Louisiana. These events represented 7% of all HSEES events combined in Florida, Louisiana, and Texas in 2005 (ATSDR, 2005a). Fifty-four percent of the events were related to Hurricane Rita, 24% were related to Wilma, 19% were related to Katrina, and 3% were related to Dennis.

The two most frequently reported industries/locations involved in hurricane-related events were manufacturing (159, 65%) and private households (48, 20%) (Table 1). Chemical manufacturing (n=115) and petroleum and coal manufacturing (n=42) accounted for 99% of the manufacturing events.

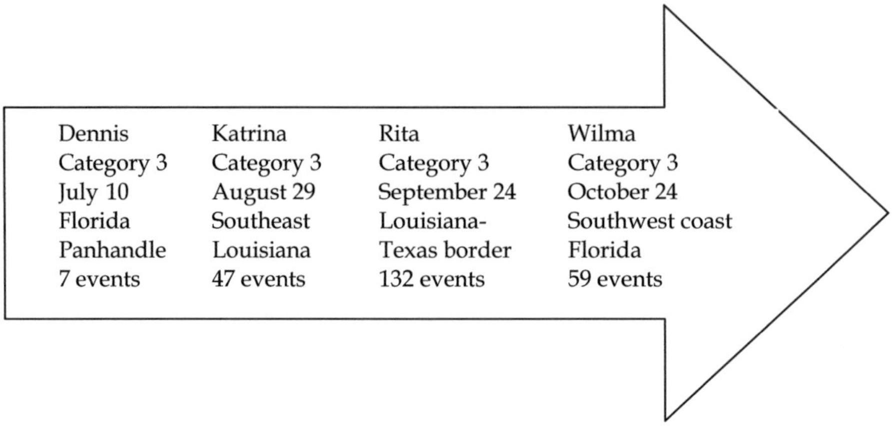

Fig. 1. Timeline of 2005 Atlantic Hurricane and HSEES incidents.

Industry/location	**Number of events**	**Percent**
All manufacturing	159	65
Private households	48	20
Mining	11	4
Unknown	11	4
Utilities	4	2
Wholesale or retail trade	6	2
Other	6	2
Total	245	99

Table 1. Industry/location of hurricane-related events, Hazardous Substances Emergency Events Surveillance (HSEES) System, 2005.

3.2 HSEES events involving evacuation, response, and nearby areas

Activities conducted to protect public health during hurricane-related events included health advisories (n=14) and environmental sampling (n=10). One building was evacuated because of a hurricane-related hazardous substances event; the evacuation lasted 2 hours. No response personnel were involved in 80 of the events; multiple types of responders responded to the remaining 165 events.

Personnel most frequently responding to hurricane-related events included company response teams (n=128) and hospitals or poison control centers (n=22). Other responders included fire departments (n=8), law enforcement officials (n=7), certified HazMat teams (n=5), third-party clean-up contractors (n=5), emergency medical services (n=4), environmental agencies (n=2), and departments of public works (n=1).

Most events (n=220) were contained inside the facility or within 200 feet of the release. The area impacted was missing for 19 events. In six events, the chemical extended beyond the facility and affected greater than 200 feet beyond the release. All of these events were in Louisiana, in predominately industrial areas with little or no residences, schools, daycare facilities, nursing homes, hospitals, or recreational parks within ¼ mile of the release. Descriptions of these 6 incidents are as follows:

Event 1- Approximately 980 pounds of ammonia were released with 490 pounds of nitrogen oxides after a power failure caused by Hurricane Katrina at a nitrogen fertilizer manufacturer. The power failure resulted in a loss of refrigeration to the ammonia storage tanks. This caused an emergency release of ammonia to a flare. The ammonia was only partially combusted by the flare and thus formed the nitrogen oxides.

Event 2- Seven hundred fifty nine pounds of zinc bromide were released from storage tanks that were washed away from an oil and gas support operation during Hurricane Katrina.

Event 3- Ten pounds of nitrogen oxide and 500 pounds of sulfur dioxide were released when a petroleum refinery shut down its plant in preparation for Hurricane Katrina and there was a release to the stack.

Event 4- A chemical product and preparation manufacturer released 270 pounds of ammonia and 75 pounds of nitrogen oxides when the ammonia storage tank routed to the flare after the compressors were shut down in preparation for Hurricane Katrina, thus causing a release of nitrogen oxides as a combustion product.

Event 5- Seven hundred eighty pounds of ammonia were released from a nitrogen fertilizer manufacturer when a power outage caused the loss of key monitoring equipment. The flare on the ammonia tank was blown out by the high winds sustained during Hurricane Rita.

Events 1-5 occurred in industrial areas with no nearby residences, nursing homes, schools, or daycare facilities.

Event 6- One thousand eighty two pounds of chlorine were released from an alkali and chlorine manufacturing plant when a power failure due to Hurricane Rita caused excess pressure in the chlorine tank. The tank had to be manually vented to reduce pressure and protect the tank integrity. Approximately 493 persons lived within ¼ mile, and a licensed daycare center was within ¼ mile of the release; no information was available about whether people were in the homes or daycare center when the release occurred.

3.3 HSEES causal factors

Bad weather conditions (n=165, 67%) were the most frequent primary causal factor contributing to hurricane-related events. Other primary causal factors were human error (n=61, 25%), equipment failure (n=11, 4%), and intentional release (n=8, 3%). Of the 225 (92%) events where a secondary causal factor was also reported, the most frequently reported were system start-up or shut down (n=120, 53%) and power failure (n=70, 31%). Other secondary causal factors included equipment failure (n=19, 8%), system/process upset (n=7, 3%), and fire (n=4, 2%). Of the 120 events where system start-up or shut down was a contributing causal factor, 6 (5%) were shutdowns in preparation for Hurricane Katrina, 59 (49%) were shutdowns in preparation for Hurricane Rita, and 55 (46%) were start-ups after Hurricane Rita.

3.4 HSEES incidents hazardous substances involved

A total of 283 hazardous substances were released in the 245 hurricane-related events. The number of hazardous substances released per event ranged from 1 to 8. Most events were air releases (88%), 9% were spills, 1% were fires, 1% were threatened releases, and <1% involved an air release with a fire.

The categories of hazardous substances most frequently released in these events were mixture across categories (34%), carbon monoxide (21%), volatile organic compounds (16%), and other inorganic substances (15%) (Table 2). The most frequently released individual hazardous substances were carbon monoxide (n=59, 21%), oxides of nitrogen (includes nitrogen oxide and nitrogen dioxide) (n=16, 6%), a mixture of carbon monoxide/nitrogen dioxide/propylene (n=8, 3%), and ammonia (n=8, 3%).

Substance Category	Frequency	Percent
Mixture across categories	97	34
Carbon monoxide	59	21
Volatile organic compounds	45	16
Other inorganic substances	42	15
Other	10	4
Ammonia	8	3
Acids	7	2
Hydrocarbons	4	1
Pesticides	4	1
Bases	2	1
Chlorine	2	1
Polymers	2	1
Hetero-organics	1	<1
Total	283	

Table 2. Substance categories released in hurricane-related events, Hazardous Substances Emergency Events Surveillance (HSEES) System, 2005.

3.5 HSEES victims

There were 160 victims in 62 hurricane-related events (25% of all hurricane-related events). Florida reported 116 of the victims, Louisiana reported 7, and Texas reported 37. Ninety-nine additional people in 49 events were observed at a medical facility but did not require treatment, so are not counted as victims.

Although manufacturing events accounted for 65% of the hurricane-related events, less than 1% of the events in this industry category resulted in victims. Most of the victims were injured in carbon monoxide events in private households (n=103, 64%). All 59 of the carbon monoxide events were in Florida and all had victims. A bus accident and fire involving nursing home residents and their oxygen tanks in Texas resulted in 36 (23%) victims. The industry/location was unknown for 18 (11%) victims, and there was 1 victim each in retail, manufacturing and the postal service. Most (n=139, 87%) victims were members of the general public, followed by employees (n=11, 7%). However there were also 10 (6%) responder victims as follows:

1. Six career firefighters suffered carbon monoxide-related symptoms while helping with post-Katrina recovery efforts. Generator exhaust fumes from a nearby motor home entered the camper in which they were sleeping.
2. One responder was injured when a barrel containing a mixture of hydrocarbon, hydrogen sulfide, and water leaked during post-Katrina orphan drum recovery operations.
3. Three fire rescue responders were among six people injured by carbon monoxide exposure as a result of improper generator use in a power outage after hurricane Wilma. One non-responder in this event died.

The ratio of male to female victims was approximately 50/50. The victims ranged in age from less than 1 year old to 100 years old, but most victims were 20-44 years old (41%) (median age was 34 years). The most frequent injuries/symptoms were dizziness/central nervous system effects (38%), headache (22%), gastrointestinal problems (15%), and thermal burns (11%) (Table 3).

Injury/Illness	Number of victims	Percent
Dizziness/central nervous system symptoms	81	38
Headache	48	22
Gastrointestinal problems	32	15
Thermal burns	24	11
Respiratory irritation	12	6
Trauma	7	3
Eye irritation	4	2
Shortness of breath	4	2
Heart Problems	3	1
Total	215	100

Table 3. Injuries/illnesses sustained in hurricane-related events, Hazardous Substances Emergency Events Surveillance (HSEES) System, 2005.

Twenty-nine persons died on scene or after arrival at a hospital. Treatment information was available for 115 of the remaining 131 victims. These victims were treated at a hospital and released (n=62), admitted to a hospital (n=19), received first aid only (n=4), or were observed at the hospital (n=1). Only one victim was decontaminated.

Two victims wore personal protective equipment. One wore Level C personal protective equipment which requires a full-face or half-mask, air-purifying respirator and chemical resistant hooded clothing, inner and outer gloves, and steel toe boots. The other wore Level D personal protective equipment which is the lowest level of protection and requires coveralls and safety shoes/boots. Personal protective equipment levels are determined by the U.S. Occupational Safety and Health Administration's Standard (29 Code of Federal Regulations 1910.120 (q)).

3.6 HSEES events with fatalities

Twenty-three nursing home residents died from thermal burns in a charter bus fire and medical oxygen tank explosion in Texas during an evacuation in anticipation of Hurricane Rita.

There were six deaths among members of the general public in five events in Florida from carbon monoxide exposure. One event occured in July after Hurricane Dennis, one in August after Hurricane Katrina, and three in October after Hurricane Wilma. Five events were due to generator misuse and one event was due to charcoal grill misuse because of hurricane-related power loss.

4. Case studies

4.1 Case study background

HSEES collected a broad array of incidents and was not specifically tailored to collect data on chemical emergencies during hurricanes or other natural disasters. Therefore, the case study section reviews the literature on three incidents of chemical emergencies during hurricanes for issues that were broader than the typical HSEES data to elucidate lessons learned. These case studies were chosen to bring up issues that HSEES does not focus data collection on, including vulnerable populations, petroleum emergency clean-up, and chemical waste disposal issues. These were three very prominent cases that arose from the 2005 hurricane season.

4.2 Case study 1

On September 23, 2005, at 3 p.m. a nursing home began evacuating prior to Hurricane Rita. A charter bus was secured when the regular bus was unavailable. The nursing home did not have time to carefully select a bus and had to take the first available bus. The bus was carrying 44 nursing home residents and nursing staff from Bellaire, Texas to Dallas, Texas. Although the bus had originally been going to a much closer shelter, the shelter was full and they had to go to a much farther one. Early the next morning, a motorist alerted the bus driver that the right-rear tire hub was glowing red. Upon stopping the bus, the driver and nursing staff observed flames emanating from the right-rear wheel well and they began to evacuate the bus.

The bus was quickly engulfed in flames when the residents' oxygen tanks exploded and further fueled the fire. Twenty-three passengers died, 2 passengers were seriously injured and 19 received minor injuries. The bus driver also received minor injuries (National Transportation Safety Board, 2007). This incident was also contained in the HSEES database. Contributing factors for this event were the charter bus company's failure to conduct proper vehicle maintenance, failure to do pre-trip inspections, and previous violations of several United States safety regulations pertaining to its drivers and vehicles; the lack of fire retardant construction materials on the bus exterior and adjacent to the wheel well; and no guidance for emergency transportation of medical oxygen cylinders on a bus. Release valves on the cylinders were designed to release during a fire only if they were fully pressurized. Therefore, cylinders that were in use and partially pressurized exploded and became dangerous projectiles during the fire. Additionally, emergency responders had difficulty rescuing passengers because of window height and top-hinge design window exits that are not optimally designed for elderly or children; the latches are difficult to open and the drop to the ground was too far (National Transportation Safety Board, 2007).

4.3 Case study 2

Storm surge from Hurricane Katrina placed an oil tank facility and surrounding neighborhoods in St. Bernard Parish, Louisiana under water for several days. When the water receded, it was discovered that the hurricane had dislodged a 250,000 barrel above-ground storage tank containing about 65,000 barrels of mixed crude oil (Agency for Toxic Substances and Disease Registry, 2005b). Approximately 25,110 barrels (slightly over a million gallons) of oil spilled from the ruptured tank.

The initial response was delayed because of high water, debris, barricades by the National Guard or local police, downed telephone lines, and lack of satellite phones. When the area was accessible, the United States Environmental Protection Agency on-scene commander directed the facility to secure the tank, identify the extent of the release, and begin recovery operations. The facility immediately began pumping out the containment canals and recovered approximately 72% of the oil.

In October 2005, long-term remediation was initiated with oversight by the United States Environmental Protection Agency and the Louisiana Department of Environmental Quality, including clean up on land, residential areas, and non-commercial waterways. Approximately 1,800 affected properties in an area of about one square mile were identified through a house-to-house visual survey conducted from the street. The Environmental Protection Agency classified contamination on 114 properties as heavy (more than 50% of the yard, sidewalks, and home were covered with oil), 286 properties as medium (about 50% of the yard and sidewalks were covered in oil), and the balance as light to oil line only (small percent of oil was visible on horizontal surfaces or a "bathtub ring" of visible product band approximately 3 to 6 inches wide was seen on the residence, with no visible oil on the yard, sidewalks, and home (Agency for Toxic Substances and Disease Registry, 2005c). However, some affected properties may have been missed because properties that were not visible from the street or public sidewalk were not surveyed due to legal access requirements. The more heavily affected areas were immediately to the west of the facility.

The 25 month long clean-up of the contaminated properties within the impacted area began with the facility removing oil-stained sediment and soil. After removal, the remaining soil was analyzed to ensure that the Louisiana Department of Environmental Quality risk evaluation/corrective action program residential soil standards for High Public Use Areas were met. If the standards were not met, additional soil was removed and the process was repeated.

Residential clean-up was complex and involved two phases. In Phase 1, property owners requested clean-up from the facility and granted them access to the property, and the facility obtained wipe and sediment samples (10% of the samples were split with the Environmental Protection Agency) and washed home exteriors. In Phase 2, the homeowner was responsible for gutting the house to the studs, and the facility removed the oiled part of the debris and transported it to an industrial landfill. The homeowner then requested an interior cleaning from the facility and granted them a second access to the property which involved the facility power washing the home's interior and exterior and replacing the yard. Reoccupation of the property was determined by the parish based on results of a final air sample (U.S. Environmental Protection Agency Region 6, 2006).

Several factors impeded residential clean-up, including class actions lawsuits filed against the facility that restricted homeowner contact with the facility and therefore barred remediation by the facility; temporary or permanent relocation of many residents after the spill; and lack of funds to complete the clean-up because the facility was only responsible for the oil-damaged part of the cleanup. This resulted in a less efficient clean-up and an "island effect" where oiled homes were next to cleaned homes because crews could not clean up whole contiguous blocks of neighborhoods (U.S. Environmental Protection Agency Region 6, 2006).

The Environmental Protection Agency shared the results of more than 800 sediment/soil samples collected from properties between September 19 and November 8, 2005 with the Agency for Toxic Substances and Disease Registry and requested an assessment of potential health hazards posed by the contamination. In December 2005, the Agency for Toxic Substances and Disease Registry released a health consult for the site which concluded that for the properties sampled, there were no short or long term risks from oil-related chemicals in sediment and soil for most properties. Recommendations were made that properties should be evaluated and remediated if necessary for other potential health hazards, such as indoor mold and structural damage, prior to re-occupancy. The recommendation was also made that properties which exceeded recommended soil standards should be remediated to be protective of public health for re-occupancy. Additionally, the health consult recommended that residents avoid bare skin contact with sediment, soil, and indoor surfaces with visible oil contamination and that homes with visible indoor oil contamination or noticeable petroleum odors be tested to determine if concentrations of chemicals in indoor air were of health concern prior to re-occupancy (Agency for Toxic Substances and Disease Registry, 2005b).

Oil companies were not required to plan to withstand storm surges that resulted from Hurricane Katrina. However, in 2007, a buyout program to create a buffer around the facility was approved to minimize the threat of future spills. In August 2009, the Louisiana Department of Environmental Quality determined that the area's shallow groundwater was unaffected by the spill and concluded that the area impacted by the spill had been remediated to acceptable levels.

Recovery and response related to this spill were complicated by competing priorities (local, state, and federal) and high background levels of contamination in that area of Louisiana which interfered with the sampling. Communication issues were also a major barrier and included delays in the public receiving information about contaminant levels which was often on a website that was not accessible to the affected population; difficulties in interpreting data and comparing it with general drinking water or ambient air quality standards which were not appropriate for an acute exposure event; and adequately conveying re-occupancy policies because federal agencies had a different opinion than the parish who ultimately made the decision (Manuel, 2006; Johnson, et al. 2005).

4.4 Case study 3

As a result of Hurricane Katrina, the city of New Orleans needed to quickly find a disposal site for the approximately 55 million cubic yards of hurricane-related debris and waste contaminated with hazardous substances that were created. In response, the Mayor of New Orleans issued an executive decree which reopened three closed landfills and granted a conditional permit to urgently convert the Chef Menteur site from a light industrial zone to a landfill. The Chef Menteur landfill began accepting waste in April 2006 (Choi et al., 2006).

The approximately 100 acre Chef Menteur site was created about 40 years ago by construction companies for use as a local source of sand for building. In July 1994, an application for a permit to begin construction of a landfill at this site was submitted to the Louisiana Department of Environmental Quality. Because of extensive resistance and pressure from the local community and the proximity of the site to the nation's largest urban wildlife refuge, the city council denied the permit in March 1997 (Choi et al., 2006).

The Chef Menteur Landfill is located in the Versailles neighborhood and is next to the 23,000 acre Bayou Sauvage Wildlife Refuge. It is less than a mile from the center of a mostly Vietnamese- and African-American community of approximately 15,000 residents (Choi et al., 2006). The unlined landfill has a large storage capacity and was approved to accept 6.5 million cubic yards (about 2.6 million tons) of construction and demolition debris (Louisiana Department of Environmental Quality, 2006). Besides trees and sheetrock, the site also accepted asbestos-containing materials and the moldy contents of gutted homes, including household pesticides, electronics, personal-care products, cleaning solutions, paint, and bleach.

Although the landfill lacked the liners required to stop leachate from leaking into the water table and contaminating surface water and groundwater, the state department of environmental quality stated that the landfill had a 10 foot clay bed, which they maintained was enough to stop leachate from reaching the community and wildlife refuge. Opponents of the landfill argued that this was not the industry standard, nor the environmental standard, which calls for multiple layers of both composite and clay (Colby, 2008). Other environmental and public health concerns included that landfill leachate would affect the watering source for the community gardens and locally caught seafood, as well as exposure to air pollution from the over 1,000 trash carrying trucks per day that drove along the main road leading into the community (Colby, 2008; Citizens for a Strong New Orleans East, 2006).

The landfill was supposed to be capped with two feet of clay and six inches of topsoil. This type of cap may not be effective in preventing precipitation from entering the landfill, mixing with wastes, and forming leachate, especially in southern Louisiana, which receives approximately 60-80 inches of rain per year (Choi et al., 2006) Additionally, the landfill is surrounded by permeable soil, and much of the site is a wetlands area where the water table is at or near the surface of the ground. Soil borings showed that groundwater is located between 1.6 and approximately 12 feet below the surface (Choi et al., 2006).

Several air and water samples taken by the state department of environmental quality in May and June 2006 shortly after the landfill opened showed that contaminants were below levels of concern (Choi et al., 2006). However, the community is concerned about potential long-term effects. Amid mounting pressure and lawsuits by the community and environmental groups, the Mayor of New Orleans did not renew the landfill's permit, and the landfill was closed in August 2006.

5. Discussion

5.1 HSEES incident distribution

HSEES recorded 245 hurricane-related events in 2005. All of the Hurricane Dennis and Hurricane Wilma events were reported by Florida because this was the only U.S. state affected by these hurricanes. Texas did not have any Hurricane Katrina-related events because of the path of the hurricane. Texas reported most of the Hurricane Rita-related events because Beaumont and Port Arthur, as well as several other areas in the state, shut down their plants in preparation for Hurricane Rita. Louisiana reported fewer events from Hurricane Rita because of the path of the hurricane and many of their plants were still not operating when Hurricane Rita struck. Florida did not report any events associated with Hurricane Rita, likely because of the mandatory evacuation orders for the Florida keys and because the Florida Panhandle escaped most of the land effects from the hurricane (National Oceanic and Atmospheric Administration, 2005b).

5.2 HSEES limitations

The HSEES system had several limitations. As in any disaster, local and state public health and emergency response infrastructure was severely disrupted by the hurricanes. Because of this some events may not have been reported and some data may not have been captured during follow-up. In Louisiana, agencies were only notified about the most severe releases that resulted from Hurricane Katrina, so some events may not have been included in HSEES. Furthermore, HSEES collected information on acute, not chronic, releases, and releases of only petroleum were excluded. Lastly, since HSEES is a surveillance system for a variety of hazardous material releases, it is not specifically tailored for hurricanes and therefore does not capture detailed information on these incidents.

Carbon monoxide exposure was traditionally underreported in the HSEES system because of the lack of existing reporting mechanisms. A case review of medical charts and emergency medical services records coded as carbon monoxide poisoning found 41 non-fatal cases from 11 incidents and 10 deaths from 4 incidents in Texas that were not reported to HSEES while analysis of hyperbaric oxygen facilities reports detected 16 non-fatal and 5 fatal cases among Louisiana residents that were not reported to HSEES (Centers for Disease Control and Prevention, 2006b; Centers for Disease Control and Prevention, 2005d). Additionally, after data closeout, Florida identified 31 cases of carbon monoxide exposure through hospitals and medical examiners reports.

5.3 Manufacturing industry

Most events involved the manufacturing industry (65%), however, only one injured person was associated with this industry. This is likely because there were few people in the area to be harmed, most facilities had already shut down and were operating with reduced crews and many residents evacuated the areas before the hurricanes hit (Knabb et al, 2005a; Knabb et al, 2005b). The immediate contributing causal factor in over half of the events was system start-up or shut down. Most releases were air releases (88%), and about a third of the events involved the release of mixtures of chemicals.

About a quarter of the events were caused when complex industrial processes were shut down in preparation for the hurricanes. The shutdowns that occurred while preparing for Hurricanes Katrina and Rita were more massive and involved numerous simultaneous activities and rapidly changing process conditions compared with one process or unit during normal shutdowns. Additionally, these large massive shutdowns had not been done before.

There is a need for different shutdown procedures that involve massive shutdowns of entire plants, such as those that occur during hurricanes. One lesson learned from Hurricanes Katrina and Rita is that it is critical for chemical facilities to better coordinate with state and local emergency preparedness agencies, especially for decisions concerning mandatory evacuation orders which can directly impact plant shutdown sequence and timing (Challener, 2006). The U.S. Environmental Protection Agency advises that all industry sectors review past events associated with shutdowns during hazardous weather conditions and make administrative/procedural, operational/process equipment and hardware/software safety improvements as needed (U. S. Environmental Protection Agency, 2010). Chemical facilities should also establish staff responsibilities and procedures to shut down process operations safely (U. S. Department of Energy, 2008).

Almost a quarter of the events were caused when major industrial processes started up after the hurricanes. The start-ups that occurred following the massive shutdowns in preparation for the hurricanes were also large-scale. Many plants used this opportunity to conduct massive maintenance or repairs on the idled plants, as equipment in some facilities in Texas dates back to the 1940s. The maintenance resulted in releases. Additionally, releases are more likely to occur when processes are shut down for more than one day (U.S. Chemical Safety and Hazard Investigation Board, 2005.)

The U.S. Chemical Safety and Hazard Investigation Board issued a safety bulletin for precautions needed during oil and chemical facility start-up following hurricanes. The Chemical Safety and Hazard Investigation Board recommends that as facilities resume operations, established and up-to-date start-up procedures and checklists should be followed and pre-start-up safety reviews should be carefully performed. Specific recommendations include using appropriate management-of-change processes before making any modifications; having adequate staffing and expertise available before starting up; and evacuating nonessential personnel from nearby process units that are starting up.

The Chemical Safety and Hazard Investigation Board also recommends that equipment, tanks, and instrumentation be thoroughly evaluated for damage. Particular attention should be given to examining large bulk storage tanks and pressure vessels for evidence of floating displacement or damage, and examining sewers, drains, furnace systems, electric motors and drives, switchgear, conduits, electrical boxes, electronic and pneumatic instrumentation, emergency warning systems, emergency equipment, and insulation systems for piping, vessels, and tanks for trapped floodwater and debris-impact damage (U.S. Chemical Safety and Hazard Investigation Board, 2005.)

Industrial releases resulting from power failures may benefit from improved backup power generation (Ruckart et al., 2004). Generators and backup lights should be tested in preparation for a hurricane, extra fuel should be on-hand, and generators should be located in areas of the facility that are not likely to be flooded (U. S. Environmental Protection Agency, 2006). Other efforts include filling all storage tanks to prevent floating or falling during hurricane-force winds, adequately securing equipment and piping to withstand high winds, and properly labeling all chemical bulk storage tanks to aid identification if these items are washed or wind-blown away (U. S. Environmental Protection Agency, 2006).

5.4 Carbon monoxide incidents

Carbon monoxide was the most frequently released single hazardous substance. Carbon monoxide poisoning, particularly from misuse of generators in residences, was the largest source of the morbidity and mortality; over two-thirds of the 160 victims were injured in carbon monoxide–related events. In another study of 27 incidents of carbon monoxide poisoning that resulted in 10 fatal and 78 nonfatal cases in Alabama and Texas after Hurricane Katrina and Rita, the majority of incidents were caused by portable generators. Most of the generators were placed outside but close to the home in order to power window air conditioners or to connect to central electric panels.

Interviews of 18 of the 27 incident households showed that only 6 (33%) had a carbon monoxide detector present, but only one alarm went off. The other 4 had dead batteries and one sent a signal to a remote security system that was unable to alert the household by telephone. The Centers for Disease Control and Prevention recommends that because there has shown to be no safe distance for generator placement, there should be functional carbon monoxide detectors in all households (Centers for Disease Control and Prevention, 2006b). Our data show that responders were injured by carbon monoxide while sleeping in temporary housing (camper). Therefore this recommendation should be expanded to temporary housing as well.

In January 2007, the Consumer Product Safety Commission required manufacturers to place a danger label on all new generators and the generator packaging. The commission began to explore various strategies to reduce consumers' exposure to carbon monoxide including generator engines with substantially reduced carbon monoxide emissions, and interlocking or automatic shutoff devices (Consumer Product Safety Commission, 2009). These measures could potentially reduce harm in the future.

5.5 Case studies

Lessons learned in case study 1 are the importance in the haste of preparing for an oncoming hurricane or responding after one occurs of adhering to important health and safety measures. Almost 25% of the 2005 hurricane victims reported to HSEES were injured in a single bus accident when oxygen tanks exploded on a bus carrying senior citizens who were being evacuated prior to Hurricane Rita. They were placed on a bus that had numerous safety violations in the haste of trying to secure a bus when their dedicated bus was not available. Hospital and nursing home administrators face several challenges during hurricanes, including deciding whether to evacuate or shelter inside their facilities until the outside danger is over.

According to a the National Transportation Safety Bord, the charter bus fire incident showed how ill prepared the United States was to evacuate those who are most vulnerable, particularly the elderly, disabled, and those in hospitals and nursing homes. The loss of life was exacerbated by the frailty of the passengers. This incident highlights the need for special plans to be developed for nursing home residents (as well as other institutionalized residents, such as hospitalized persons or prisoners) because they are not able to quickly escape from hazardous substance events and have special considerations.

Motor coaches were used for this evacuation and will be used again should a similar emergency arise. Following this incident, the National Transportation Safety Board issued several recommendations to other agencies to protect the travelling public that should be urgently implemented (National Transportation Safety Board, 2007). The U.S. National Transportation Safety Board also investigated the risk that medical oxygen tanks posed to rescuers during the fire. Three days after the accident, the Department of Transportation issued "Guidance for the Safe Transportation of Medical Oxygen for Personal Use on Buses and Trains" which recommends that medical oxygen be securely stored upright and be limited to one canister per patient in the passenger compartment (U.S. Department of Transportation, 2006).

There are several lessons learned from case study 2, when an oil spill contaminated a large residential area. One lesson is the need for improved and more rapid identification of environmental hazards and their communication to emergency responders and the public. Another lesson learned is the need during a hurricane and aftermath for greater collaboration among Federal, State, and local officials, as well as an enhanced public communication program. All of these measures could have improved the effectiveness of the Federal response (Manuel, 2006; The White House, 2006)

Case Study 3 highlights that in the haste of rebuilding, proper health and safety measures for handling the hurricane hazardous substance contaminated debris were not followed. The "hub and spoke" method of debris handling is supported by the U.S. Environmental Protection Agency and the Army Corps of Engineers. It involves collecting debris on the curbside and taking it to temporary staging sites in more central locations in the day time, followed by night-time hauling to permitted landfills outside the city (Citizens for a Strong New Orleans East, 2006). This would have avoided the use of the unpermitted landfill that was not designed properly to handle hazardous waste.

6. Conclusion

Because preventing hurricanes is not possible, increased attention must be focused on preventing and minimizing acute releases of hazardous substances during hurricanes and preplanning in case releases do occur. Because of the urgency of a hurricane, preplanning for those most vulnerable and increased diligence to health and safety during and after a hurricane are called for.

Many of the incidents occurred due to power interruption. Industries, particularly in hurricane prone areas, can take steps to minimize their risks for chemical releases in future power outages. Additionally, public health campaigns that emphasize placement of generators as far away from the home as possible should continue. Since no safe distance has been determined, the use of battery operated, functional carbon monoxide detectors should be stressed for all sleeping quarters, temporary or permanent.

7. References

Agency for Toxic Substances and Disease Registry. (2005a). *Hazardous Substances Emergency Events Surveillance Annual Report 2005.* U.S. Department of Health and Human Services, Public Health Service, ATSDR, Atlanta. http://www.atsdr.cdc.gov/HS/HSEES/annual2005.html.

Agency for Toxic Substances and Disease Registry. (2005b). *Health Consultation: Murphy Oil Spill.* U.S. Department of Health and Human Services, Public Health Service, ATSDR, Atlanta.

Agency for Toxic Substances and Disease Registry. (2005c). *Health Consult II-Murphy Oil Spill.* U.S. Department of Health and Human Services, Public Health Service, ATSDR, Atlanta.

Centers for Disease Control and Prevention. (2005a). Carbon Monoxide poisonings from hurricane-associated use of portable generators- Florida, 2004. *MMWR* 2005 54(28); 697-700.

Centers for Disease Control and Prevention. (2005b). Vibrio illnesses after Hurricane Katrina --- multiple states, August--September 2005. *MMWR* 2005 54(37);928-931.

Centers for Disease Control and Prevention. (2005c). Infectious disease and dermatologic conditions in evacuees and rescue workers after Hurricane Katrina --- Multiple States, August--September, 2005. *MMW* 2005, 54(38);961-964.

Centers for Disease Control and Prevention. (2005d). Carbon monoxide poisoning after Hurricane Katrina—Alabama, Louisiana, and Mississippi, August—September 2005. *MMWR* 2005 54(39);996-998.

Centers for Disease Control and Prevention. (2005e). Surveillance for illness and injury after Hurricane Katrina --- New Orleans, Louisiana, September 8--25, 2005. *MMWR* 2005 54(40);1018-1021.

Centers for Disease Control and Prevention. (2006a). Public Health Response to Hurricanes Katrina and Rita --- United States, 2005. *MMWR* 2006 55(9);229-231.

Centers for Disease Control and Prevention. (2006b). Carbon Monoxide Poisonings after two major hurricanes—Alabama and Texas, October 2005. *MMWR* 2006 55(9);236-239.

CERCLA/SARA. Comprehensive Environmental Response, Compensation, and Liability Act of 1980, Pub. L. No. 95 510 (Dec. 11, 1980), as amended by the Superfund Amendments and Reauthorization Act of 1986, Pub. L. No. 99 499 (Oct. 17, 1986), 42 U.S.C. 9604(i).

Challener, C. Hurricane Season Rapidly Approaches: Lessons Learned and Tips for Weathering the Next Storm, 2006. http://www.chemalliance.org/featured/?subsec=0&id=7017. Accessed April 16, 2007.

Choi, J.; Bhatt, A. & Chen, F. (2006). In the Aftermath of Hurricane Katrina: The Chef Menteur Landfill & Its Effects on the Vietnamese American Community. August 2006. Asian American Justice Center, Washington, D.C.

Colby (2008). Chef Menteur Lanfill. https://wiki.colby.edu/display/es298b/Chef+Menteur+Landfill). Accessed July 27, 2010.

Consumer Product Safety Commission. Generator Danger Warning 2009. http://www.cpsc.gov/generator.html Accessed August 16,2010.

Citizens for a Strong New Orleans East. (2006). Fact About Landfill in New Orleans East.http://neworleans.indymedia.org/news/2006/05/7620.php.

Johnson, S, Frumkin H, Rovery, C - Bradshaw D. EPA, ATSDR & LDEQ Discuss Recent Environmental Sampling Results from Hurricanes Katrina/Rita Washington, DC 11/10/2005. http://yosemite.epa.gov/opa/admpress.nsf/a883dc3da7094f978 52572a00065d7d8/dc2c7a54ed2867d9852572b5006b4fa8! OpenDocument Accessed July 27, 2010.

Knabb R.; Rhome, J. & Brown, D. Tropical cyclone report: Hurricane Katrina, 23-30 August 2005. Miami, FL: National Hurricane Center; 2005, http://www.nhc.noaa.gov/pdf/TCR-AL122005_Katrina.pdf. Accessed May 5, 2006.

Knabb, R.; Brown, D. &. Rhome, J. Tropical cyclone report: Hurricane Rita, 18-26 September 2005 Miami, FL: National Hurricane Center; 2005,http://www.nhc.noaa.gov/pdf/TCR-AL182005_Rita.pdf Accessed August 16, 2010.

Louisiana Department of Environmental Quality. (2006). Chef Menteur Landfill Ok'd to Accept C&D Waste. http://www.deq.louisiana.gov/portal/portals/0/news/pdf/ChefMenteurLandfillfinal.pdf. Accessed July 27, 2010.Louisiana Department of Health and Hospitals. Hurricane Katrina, Deceased Reports.
http://www.dhh.louisiana.gov/offices/page.asp?ID=192&Detail=5248. Aug. 2, 2006 Accessed July 26, 2010.

Manuel, J. (2006). In Katrina's Wake. *Environ Health Perspect*. 114, 1, (January 2006) A32–A39.

National Oceanic and Atmospheric Administration (2005a), National Weather Service, National Hurricane Center. Hurricane Preparedness, Hurricane History. November 22, 2005 http://www.nhc. noaa.gov/HAW2/english/history.shtml#dennis. Accessed July 27, 2010.

National Oceanic and Atmospheric Administration (2005b), National Climatic Data Center. Hurricane Rita. November 22, 2005 http://www.ncdc.noaa.gov/special-reports/rita.html. Accessed July 27, 2010.National Transportation Safety Board. (2007). Motorcoach Fire on Interstate 45 During Hurricane Rita Evacuation Near Wilmer, Texas, September 23, 2005. Highway Accident Report NTSB/HAR-07/01. Washington, DC.

Picou, J. Katrina as a Natech Disaster. http://www.stevenpicou.com/pdfs/katrina-as-a-natech-disaster.pdf.

Ruckart, P.; Borders, J.; Villanacci, J.; Harris, R. & Samples-Ruiz, M. (2004). The role of adverse weather conditions in acute releases of hazardous substances, Texas, 2000-2001. *J Haz Mat,* 115, (2004) 27-31.

Taylor, M.; Stokes, W.; Bajuscak, R,; Serdula, M.; Siegel, K. ; Griffin, B.; Keiser, J.; Agate, L.; Kite-Powell, A.; Roach, D.; Humbert, N.; Brusuelas, K. & Shekar, S. (2007). Mobilizing Mobile Medical Units for Hurricane Relief: The United States Public Health Service and Broward County Health Department Response to Hurricane Wilma, Broward County, Florida. *Journal of Public Health Management & Practice,* 13, 5, (September-October 2007) 447-452.

The White House. February 2006. The Federal Response to Hurricane Katrina-Lessons Learned. http://library.stmarytx.edu/acadlib/edocs/katrinawh.pdf. Accessed July 27, 2010.

Todd, B. (2006). Infection Control and Hurricane Katrina: What nurses can learn in the aftermath of the disaster. *Am J Nurs,* 106, 3, (2006) 29-31.

U.S. Chemical Safety and Hazard Investigation Board, Safety bulletin after Katrina: precautions needed during oil and chemical facility startup, No. 2005-01-S, September 2005, http://www.csb.gov/safety_publications/docs/CSBKatrina SafetyBulletin.pdf. Accessed May 8, 2006.

U.S. Department of Energy, Energy Efficiency and Renewable Energy Industrial Technologies Program. (2008). Hurricane Ready, http://www1.eere.energy.gov/industry/hurricaneready/preparation_steps.html. Accessed April 16, 2007.

U.S. Department of Transportation Pipeline and Hazardous Materials Safety Administration. DOT Guidance for the Safe Transportation of Medical Oxygen for Personal Use on Buses and Trains. (2006). http://www.phmsa.dot.gov/news/gstomo.pdf. Accessed July 27, 2010.

U.S. Environmental Protection Agency Region 6 Response and Prevention Branch Oil Team. (2006). Murphy Oil USA Refinery Spill Chalmette and Meraux, LA. http://www.epa.gov/oem/docs/oil/fss/fss06/franklin_2.pdf. Accessed July 27, 2010.

U.S. Environmental Protection Agency, Ground water & drinking water. (2006). Suggested pre-hurricane activities for water and wastewater facilities, http://www.epa.gov/ safewater/hurricane/pre-hurricane.html. Accessed May 8, 2006.

U.S. Environmental Protection Agency. (2010). As Hurricane Season Begins: A Reminder to Minimize Process Shutdown Related Releases and to Report Releases in a Timely Manner, http://www.epa.gov/region4/r4_hurricanereleases.html. Accessed April 16, 2007.

Young, S.; Balluz, L.; Malilay J. (2004). Natural and technologic hazardous material releases during and after natural disasters: a review. *Sci. Total Environ.* 322, (2004) 3–20, ISSN: 0048-9697.

Part 6

Ecological Impacts and Modeling

27

The Complexity of Catastrophic Wind Impacts on Temperate Forests

Weimin Xi[1,2] and Robert K. Peet[1]
[1]*University of North Carolina at Chapel Hill*
[2]*University of Wisconsin-Madison*
USA

1. Introduction

Catastrophic wind disturbance events have profound impacts on forests in many parts of the world. As an ecological factor, catastrophic wind events not only cause extensive damage to trees, but also affect many aspects of the disturbed forests including community structure, individual tree growth, tree regeneration, species diversity, and ecosystem function (Coutts & Grace, 1993; Ennos, 1997; Martin & Ogden, 2006; Bellingham, 2008; Hoeppner et al., 2008; Zeng et al., 2009). Although catastrophic windstorms are easily seen to have major impacts on forest structure, the longer-term effects on less conspicuous ecosystem attributes such as species composition and diversity are more complex, and at smaller scales of observation are relatively unpredictable (DeCoster, 1996; McMaster, 2005; Xi et al., 2008a; Oswalt & Oswalt, 2008). Many factors, meteorologic, topographic and biologic, simultaneously interact to influence the complexity of patterns of damage and dynamics of recovery. A deep understanding of wind disturbance effects is essential for effective forest management and biodiversity conservation. This information is particularly important as ongoing climate change is likely to sustain the recent increased incidence of major windstorms for the foreseeable decades (Goldenberg et al., 2001; Emanuel, 2005; Xi, 2005; Xi & Peet, 2008a; Stanturf et al., 2007).

The effects of wind damage have long been recognized and observed by foresters and ecologists (e.g., Baker, 1915; Bromely, 1939; Curtis, 1943; Spurr, 1956; Webb, 1958) and extensive research has been conducted on the ecological impacts of catastrophic windstorms (Canham & Loucks, 1984; Foster, 1988; Webb, 1988, 1989; Boucher et al., 1990; Brokaw & Grear, 1991; Walker, 1991; Peterson & Pickett, 1991; Merrens & Peart, 1992; Bellingham et al., 1992, 1994, 1995; Boose et al., 1994; Vandermeer et al., 1995; Imbert et al., 1996; Turner et al., 1997; Herbert et al., 1999; Sinton et al., 2000; Burslem et al., 2000; Boose et al., 2001; Platt et al., 2002; Woods, 2000; Peterson, 2004; Uriarte et al., 2004; Zhao et al., 2006; Uriarte & Papaik, 2007; Xi et al., 2008b; Prengaman et al., 2008; Zeng et al., 2009). This work has greatly increased our understanding of the importance of wind disturbance for community composition and ecosystem function, and has led to the wide acceptance among researchers of a nonequilibrium perspective (Reice, 1994, 2001). As a consequence of this and related work, the traditional view of wind as a simple damage force has evolved into the contemporary view of wind as a spatially heterogeneous, multi-scale disturbance agent that affects forest structure, diversity, dynamics, and ecosystem processes (Xi, 2005).

Several reviews of windstorm impacts have collectively provided a general framework for viewing how various windstorm disturbances might influence forest patterns and processes, and several generalizations have emerged from those reviews (Brokaw & Walker, 1991; Tanner et al., 1991; Foster & Boose, 1995; Everham & Brokaw, 1996; Whigham et al., 1999; Webb, 1999; Peterson, 2000). In particular, important reviews by Webb (1999) and Peterson (2000) have shown highly variable forest responses to windstorm disturbances in temperate forests, but there has been a continuing increase in knowledge about the complexity of the impacts (Table 1). In this review we focus on the complex effects of large, infrequent windstorm disturbances in temperate forests and provide information for improving forest management that helps to minimize the timber loss under the increasing risk of catastrophic damage in temperate forest regions.

The purpose of this review is to present a synthesis of the complex array of forest responses to catastrophic windstorm disturbances and a framework for its interpretation and future study. We particularly focus on large, infrequent hurricane disturbances in temperate forests. The extensive literature cited in this review documents complex patterns of forest response to the highly variable windstorm disturbance regimes in temperate forests. We attempt to combine in one common conceptual framework several important concepts and theories pertaining to wind disturbance effects that have emerged in recent years. This synthesis is structured around four questions: 1) Are there consistent patterns in the damage exhibited by forest communities? 2) What factors influence damage patterns and predict damage risk? 3) How do forests respond to and recover from the catastrophic wind damage? 4) What are the long-term effects of wind disturbances on species diversity and succession?

2. Understanding catastrophic wind disturbance

2.1 Concepts

Despite extensive previous work on catastrophic wind disturbance and subsequent ecological effects, there has been no specific definition of wind disturbance. Defining catastrophic wind disturbance is difficult because wind varies within/between events in intensity, size and frequency. In this review, we refer to catastrophic wind disturbance as including high wind events, mainly hurricanes, tornados, downbursts, gales and severe windstorms that may result in substantial tree damage or mortality. In most cases, the catastrophic wind disturbances we focus on in this synthesis represent a form of large, infrequent disturbance (LID) such as described by Turner and others (1998) as natural, catastrophic events that are 'large in spatial extent and infrequent in occurrence'.

Catastrophic wind disturbances can be identified from their high wind intensity and extreme maximum gusts (Foster & Boose, 1995; Everham & Brokaw, 1996; Peterson, 2000; Lugo, 2008). The strongest winds (maximum wind speed about 125 m/s and average speed about 100 m/s) characterize tornadoes. A hurricane is a tropical storm when its wind speed is higher than 35 m/s and a typical hurricane has an average wind speed of 70 m/s. Gales (average wind speed about 50 m/s) and severe windstorms (average wind speed about 30-50 m/s) more often produce winds of only moderate intensity, but in some cases, they can generate winds as destructive as tornadoes. A downburst is a straight-direction catastrophic surface wind in excess of 17 m/s caused by a small-scale, strong downdraft from the base of convective thundershowers and thunderstorms (Fujita, 1985), and can exceed 50 m/s (or even 75m/s) and cause tornado-like damage.

Location	Forest type	Windstorm type	Reference
Asia			
Taiwan	hardwoods	typhoon	Lin et al., 2003
China	northern forests	wind and snow	Zhu et al., 2005
Japan	Subtropical forests	typhoon	Xu et al., 2003b, 2004
Australia-New Zealand			
Australia	Australian rain forests	cyclones	Bellingham, 2008
New Zealand	planted forests	windstorms	Moore & Quine, 2003
Europe			
United Kingdom	spruce forests	windstorm	Quine, 2003
Finland	boreal forests	wind and snow damage	Pellikka &Järvenpää, 2003
North America			
Canada	boreal forests	windstorm	Mitchell et al., 2001
United States			
Alaska	coastal temperate rainforests	severe windstorm	Kramer et al., 2001
Colorado	*Picea-Abies-Pinus* subalpine forests	severe windstorm	Veblen et al., 2001; Lidemann & Baker, 2002
Massachusetts	hardwoods	hurricanes	Wilson et al., 2005
Michigan	hardwoods	windstorms	Woods, 2000, 2001
Minnesota	hardwoods	severe windstorm	Peterson, 2004
New England	hardwoods	hurricanes	Boose et al., 2001
Wisconsin	hemlock-north hardwoods	severe windstorm	Schulte & Mladenoff, 2005
New York	mixed hardwood forests	severe windstorm	McMaster, 2005
Pennsylvania	hemlock-hardwood forests	tornados	Peterson, 2000
North Carolina	Piedmont pine-hardwoods, Appalachian hardwoods	hurricanes & downburst	Elliott et al., 2002; Xi et al., 2008a, 2008b
Florida	slash pine savannas; mixed-hardwoods	hurricanes	Platt et al., 2000, 2002; Batista & Platt, 2003
Texas	southern mixed hardwood forests	hurricane & severe storm	Harcombe et al., 2002

Table 1. Example case studies of catastrophic windstorms in temperate forests since 1998 by geographic locality, forest type, and windstorm type

Occurrences of catastrophic wind events vary greatly in frequency and return times among windstorm types and localities. Hurricanes are tropical, high-wind events and can be common in near-coast tropical regions, but are less frequent in the inland tropics. Catastrophic hurricanes (defined as Saffir-Simpson as category 4 or 5) reoccur for a particular area of the coastal tropics on average every 20-60 years (Brokaw, 1991). The frequency of hurricanes decreases from tropical coasts to inland temperate regions. Major hurricanes only occasionally achieve landfall in temperate areas, and rarely reach the inland temperate areas (Webb, 1999).

The reoccurrence intervals of major hurricanes in temperate forests vary greatly from less than 20 years in the Southeastern US coastal regions (Gresham et al., 1991; Doyle, 1997; Platt et al., 2000), to about 50 years in the temperate Piedmont of the Southeastern US (Xi, 2005), to about 70-100 years in the Northeastern US (Foster & Boose, 1986, 1995). Storms like the 1938 hurricane (Category 5) that caused disastrous forest damage in the Northeastern US typically occur in the region only once per century. The reoccurrence rates of major hurricanes on a geological time scale for a specific location in a temperate region might be even longer. A sediment core study used to quantify hurricane activity in the Lake Shelby region of coastal Alabama showed a recurrence interval of about 300 years for catastrophic hurricanes during the last 5,000 years, and about 600 years during the last 10,000 years (Liu & Fearn, 1993).

Compared to hurricanes, the frequencies of other types of catastrophic wind events (e.g., tornado, gales, downburst and severe storms) are highly variable in the temperate zone. Tornadoes have been reported widely in temperate North America, especially in the central Great Plains of the United States, where they can be particularly violent. In the Tornado Alley region (Oklahoma-Kansas, USA), the number of tornadoes can reach 38 per 100 km^2 per year (Fujita, 1985). Downbursts are more frequent than tornadoes, but due to their isolated and sudden nature (lasting several minutes to half an hour), their recurrence rates are rarely reported in the literature. To date, few studies have reported the occurrences of gales, although Gallagher (1974) and Fraser (1971) reported 34 years and 75 years for return times in forest regions of Ireland and Scotland respectively. For windthrow, Zhang and others (1999) reported the average rotation period over the Upper Peninsula of Michigan to be 541 years. In the northern temperate forests of Wisconsin, severe windstorm return periods vary greatly with estimates ranging from 450 to 1200 years (Canham & Loucks, 1984; Schulte & Mladeoff, 2005).

The spatial extent or magnitude of catastrophic wind disturbances, which can be expressed as mean affected area per disturbance event, varies significantly among windstorm types. Sizes of hurricanes are generally large. The eye of a hurricane is normally 30-60 km in diameter and its influence often extends over an area 300-500 km in diameter along its path (Baldwin, 1995). One example that illustrates the potential large size of a hurricane is the 1989 Hurricane Hugo, one of the most catastrophic windstorms in United States history and which significantly damaged more than 18,210 km^2 of timberland in South Carolina (Sheffield & Thompson, 1992). Along its path in North Carolina, Hugo damaged more than 10,926 km^2 of forests, with almost complete destruction of 275 km^2 (Barnes, 2001). Another example was the 1938 hurricane, which blew down more than 2,400 km^2 of forestland in central New England (Spurr, 1956). In contrast, a tornado usually causes substantial damage only along its long and narrow path. A typical tornado path is normally several dozen to several hundred meters wide, and 15-20 km long (Ruffner & Bair, 1984). The actual surface damaged by a tornado may be much less than its path owing to the way tornadoes skip across the landscape (Peterson, 2000).

2.2 Scales

Catastrophic wind damage and subsequent forest recovery are scale-dependent phenomena and both spatial and temporal scales are important in understanding effects of catastrophic winds. As Levin (1992) pointed out, "no single mechanism explains pattern on all scales." Consequently, it is essential to clarify both the spatial and temporal scale over which wind damage and recovery patterns occur and are examined.

Windstorms are often distributed over a broad range of spatial scales, and certain damage effects and recovery patterns can only be observed at a specific spatial scale in the context of specific processes (Foster & Boose, 1992, 1994, 1995, 2000). For example, the geographic and meteorological factors that control the formation and movement of hurricanes can be only be understood on a continental scale (~5000 km), whereas wind velocity, local topography (variation in site exposure), and individual stand attributes are the controlling factors of hurricane damage at the landscape scale (~10 km). At small scales biotic factors become more significant. For example, Peterson (2004) found that within-stand variation in damage can be largely explained in the context of tree size and species. Our study in the Piedmont forests of the southeastern United States also showed that at the stand scale, tree size (i.e., its vertical stratum) and resistance to wind are the most important indicators of mortality probability and damage type during a major hurricane (Xi, 2005; Xi et al., 2008a).

It is important to clarify the temporal scale across which the research is conducted and ecological patterns are compared. Recovery time from catastrophic windstorms varies tremendously between forests from a few years to a predicted period of several hundred years, depending on wind intensity and the regeneration capability of the damaged forest. Ecologists often divide windstorm impacts and post-disturbance forest responses into three temporal categories: immediate (a few months to one year, e.g., Walker et al., 1992), short-term (few months to several years, e.g., Vandermeer et al., 2000; Pascarella et al., 2004) and long-term (few decades to centuries, e.g., Hibbs, 1983; Foster, 1988; Burslem et al., 2000). Moreover, forest recovery processes also vary with time. For example, during and immediately after a hurricane, mortality processes dominate, whereas the recruitment process becomes important in the years immediately after the wind damage. Consequently, the timing of surveys of wind-disturbed forests is critical for understanding the damage, mortality and recovery.

The predictability of forest damage from catastrophic winds and the subsequent recovery pattern generally is scale-dependant. Although wind conditions are highly variable in all aspects during a windstorm, wind gusts are more random at smaller scales. The predictability of forest damage at the stand scale (~1 km) is, therefore, relatively low due to the random effects of wind gusts and the complex interactions among their neighbour individuals. The larger-scale forest damage patterns and recovery processes (e.g., at landscape and regional scale) can be predicted reasonably well (Fig. 1). For example, forest damage patterns across post-hurricane landscapes are predictable based on wind speeds, topography (site exposure), stand structure, disturbance, and land-use history (e.g., Foster, 1998; Foster & Boose, 1992; DeCoster, 1996; Xi, 2005; Xi et al., 2008a).

2.3 A framework for understanding large, infrequent catastrophic winds

Before reviewing past work, we briefly provide a conceptual framework for understanding the impact of large, infrequent catastrophic winds on temperate forests. Our proposed framework for understanding large, infrequent catastrophic winds includes the following key elements:

1. The catastrophic-wind-induced changes (including changes in community structure, tree mortality, and species diversity) are complex and highly variable. In temperate forests, the most conspicuous changes caused by catastrophic winds are structural changes. The effects on tree composition and species diversity vary greatly. The

predictability of damage varies among wind events across temporal and spatial scales of observation (Fig. 1).

2. The roles of factors influencing forest damage and tree mortality risks are scale-dependent, including wind characteristics (wind types, timing of events), topography, species and forest characteristics (tree size, resistance to wind, stand density etc.). The importance of variation in risk factors with wind intensity and other factors needs to be examined at ecologically relevant scales (Fig. 2).
3. Species differences in vulnerability to damage from wind are less important in higher wind events, but species traits are important in the recovery process.
4. Past catastrophic wind events have had significant, but highly variable long-term impacts on community composition and species diversity.

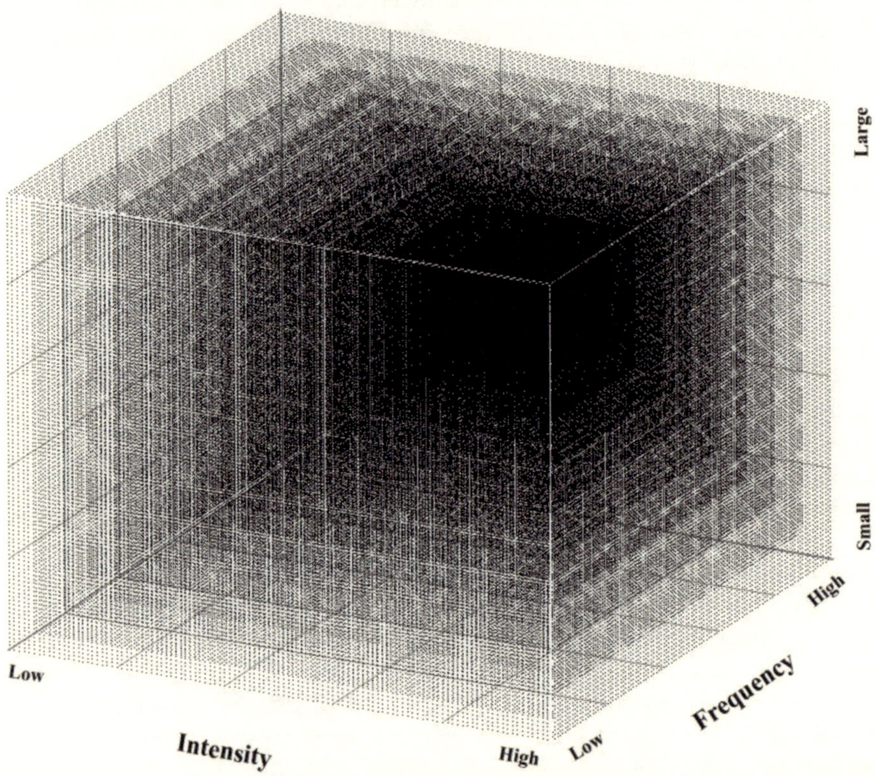

Fig. 1. A conceptual model of temperate forests in response to varied wind regime (wind intensity, frequency and size). The predictability of wind damage on forests varies with event and scale. The darker areas show lower predictability of forest response. Post-damage responses of forest structure, species composition, and diversity are more predicable (lighter in this graph) when wind frequency is high but wind intensity is low, and become less predictable (darker in this graph) when wind intensity increases and frequency decreases.

As we described in the previous section, wind intensities of any catastrophic wind event can be complex and highly variable in space and time due the interactions between the unstable turbulence and the complex ground surface features over which the air moves (Barnes et al.,

1998). To understand the patterns of wind disturbance in forests, the risk factors need to be examined at relevant spatial and temporal scales and in the context of specific site conditions and stand history. Both abiotic (e.g., winds, topography, soil) and biotic factors (e.g., individual tree characteristics, tree species, stand attributes) interact to generate complex damage and mortality patterns. The features of a storm, forest location relative to the windstorm, pre-disturbance community attributes, disturbance history, and species susceptibility to wind all play a role in generating the complex and subtle patterns of damage.

The importance of species traits and pre-disturbance community attributes decreases as wind intensity increases (Fig. 2). Moreover, the occurrence of windstorms may also interact with other disturbance forces such as subsequent wildfires, insect outbreaks, and fungal infections in complex ways to increase the degree of the complexity and unpredictability (Pickett & White, 1985; Webb, 1999; Platt et al., 2003; Peterson, 2007).

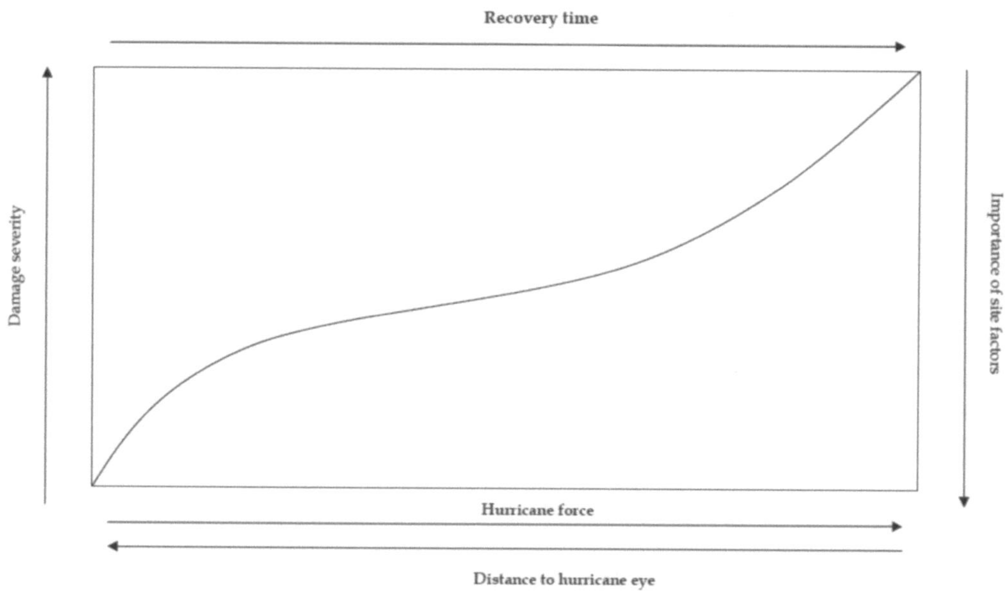

Fig. 2. Hypothesized relationships among hurricane force, forest damage severity, recovery time, and importance of site factors. Both forest damage and recovery time increase with hurricane force. The importance of site factors decreases when hurricane force increases (After Ackerman et al., 1991).

Catastrophic windstorms have various long-term effects on forest composition, dynamics and successional development. Those effects vary greatly from setting back forest succession by allowing establishment of early successional species to speeding up succession by releasing later-succession species already established in the understory. Change in species diversity following catastrophic wind disturbance ranges from increasing to decreasing to no change, depending on many factors such as damage intensity as well as the scale of the investigation. As a consequence, the long-term effects of catastrophic windstorms on forest composition, diversity, and succession are not well known and appear not particularly predictable. Additional long-term monitoring and predictive model development will be necessary to improve predictability.

3. Complexity of forest damage resulting from catastrophic wind disturbances

3.1 Impacts on community structure

The most conspicuous forest changes caused by catastrophic winds are structural changes, which are often measured in terms of the changes in tree size or age distributions, basal area or biomass, stem density, or canopy heterogeneity. Three relatively consistent patterns in structural change that have been reported in both wind-damaged tropical and temperate forests are: 1) immediate increase in canopy heterogeneity, 2) short-term decrease in biomass, and 3) immediate decrease in density of all tree sizes followed by a dramatic increase in understory density a few years after wind damage. In temperate forests, the degree of structural change varies greatly depending on many abiotic and biotic factors including wind intensities, rainfall associated with the storm, community attributes, site conditions, and susceptibility to windstorm damage (DeCoster, 1996; Peterson, 2007; Xi et al., 2008a).

Studies of forest damage have reported loss of stand biomass following catastrophic wind disturbances to be highly variable and to depend on wind intensity, forest type, site exposure to wind, pre-disturbance species composition, and interactions of these major factors with subsequent risk factors such as fires and insect inflections. Reported losses of stand biomass vary greatly from 2% to 94% among forests and wind events. In several reported extreme cases, temperate forests have experienced high biomass loss due to the extreme intensities of windstorms and the high vulnerability of temperate forests to windstorm disturbances. Localized windstorms (e.g., downbursts) cause intensive damage across forest landscapes. For example, the 1977 Independence Day Storm, along a path 266 km long and up to 27 km wide, virtually leveled 240,000 ha mesic temperate forests in eastern Minnesota and northern central Wisconsin (Canham & Loucks, 1984). Hurricanes, on the other hands, often cause forest damage in much larger areas. The 1938 hurricane, for example, resulted in about 94% basal area loss in a 2000-ha survey area in New England (Spurr, 1956; Foster, 1988). Another example is the 1998 Hurricane Hugo, which reduced stocking by an average 66% in moderately to heavily damaged hardwood and oak-pine stands, and reduced the inventory of timber growing stock by about 20% in South Carolina as a whole (Sheffield & Thompson, 1992).

Catastrophic windstorms can substantially alter forest structure by simultaneously decreasing overall canopy height, increasing canopy patchiness, and increasing understory light heterogeneity. For example, hurricanes often result in a substantially increased gap size and a dramatic rise in understory light. Among studies of forest structural changes, canopy damage varies greatly from slight defoliation to about 90% increases in understory light (Turton, 1992; Veblen et al., 1989; Bellingham et al., 1996). In addition, catastrophic winds can increase within-stand spatial heterogeneity through clumped distribution of hurricane-induced tree mortality and aggregation patterns of surviving trees within the wind-damaged forest stands as a result of the uneven uprooting and stem snapping among different species and tree size classes (McDonald et al., 2003; Xi, 2005; Xi et al., 2008b).

Catastrophic winds have profound impacts on the size distribution of trees and can induce substantial increases in the relative abundance of small size-class trees in the damaged forests during the subsequent years. Although catastrophic windstorms usually cause immediate reduction in tree densities of all sizes, especially for large canopy trees, they often result in a dramatic increase in the density of understory seedlings and saplings several years after the windstorms due to subsequent release of suppressed understory stems and widespread sprouting. Sprouting is undoubtedly an important mechanism of tree recovery following windstorms in temperate forests. Studies have shown sprouting rates in the 20-80% percent range to be typical for temperate forests (Harcombe & Mark, 1983; Peterson & Pickett, 1991; DeCoster, 1996).

Our study on the effects of the 1996 Hurricane Fran on the Duke Forest in North Carolina has shown that hurricanes significantly diversify the live-tree size distribution in damaged forest stands. Overall, the predominant tree species of the upper canopy layer in both pine and hardwood forests decreased substantially due to the higher mortality of large-size trees. In the damaged pine stands, the mean size of the most dominant tree species (*Pinus taeda*) was increased and the density of pines decreased in all size classes. The hurricane also greatly affected pine stands by decreasing the relative abundance of small sized oaks (*Quercus* spp.) and hickories (*Carya* spp.). Several light-demanding and shade-intolerant hardwood species, such as tuliptree (*Liriodendron tulipifera*) and sweetgum (*Liquidambar styraciflua*), increased dramatically in density in the smallest size class (1-3 cm) during the 5 years following the hurricane, whereas dogwood (*Cornus florida*), the most damaged tree in the pine stands, decreased in stem density in all tree sizes (Xi, 2005; Xi & Peet, 2008a).

3.2 Complex patterns of tree mortality

The most obvious effect of catastrophic wind is tree mortality. Tree morality in general is positively related to wind intensity and inversely related to frequency, whereas no clear relationship has been identified between tree species and tree mortality (Everham & Brokaw, 1996; Xi et al., 2008b; Keeland & Gorham, 2009). Wind-induced mortality can be subtle, complex, and delayed, depending on several contributing factors such as the wind intensity, species of interest, individual size, and life form. In the literature, wind-induced tree mortality rates in temperate forests vary greatly among forest types and wind events, ranging up to around 80%.

In the tropics, tree mortality rates after a severe hurricane tend to be low. Walker (1991), for example, only recorded 7% mortality one year followed Hurricane Hugo (a category 3 hurricane) in Puerto Rica. Bellingham (1991) found 8% tree mortality 23 months after Hurricane Gilbert in Jamaica. Whigham and others reported 11.2 % in a Mexican forest 17 months after Hurricane Gilbert. These forests experience high hurricane return rates and the tree species that occupy them appear well adapted to these frequent disturbances.

Wind-induced tree mortality in temperate forests varies from low to extremely high. For example, Batista and Platt (2003) reported 7% mortality for the overstory trees after the relatively modest 1985 Hurricane Kate in an old-growth forest. However, high tree mortality by catastrophic winds have been reported for a number of temperate forests. Foster (1988) reported about 30% tree mortality for the 1938 hurricane in central New England, USA. Similarly, Hook and others (1991) found that Hurricane Hugo caused over 80% tree mortality in the Santee Experimental forest, South Carolina. In Piedmont forests, we found tree mortality of large-size trees to double in the period that spanned the hurricane event, in comparison to the pre-hurricane, although this increased mortality was not uniformly distributed across species. In addition, there was widespread delayed mortality of hardwood tree species following the hurricane (Xi et al., 2008b; Keeland & Gorham, 2009). These significant structural and dynamic changes appear likely to have a great and continuing influence on stand regeneration and forest development.

Tree mortality may vary among species. Several studies have assessed species-specific mortality caused by hurricanes in temperate forests (Foster, 1988, 1992; Bellingham et al., 1995, 1996; Batista & Platt, 2003; Chapman et al., 2008). In a comprehensive study of response of trees to the 1938 hurricane in central New England, Foster (1988, 1992) found large differences among tree species in their susceptibility to windstorm damage. However, species-specific mortality may not always be clearly distinguished since other mortality risk factors may interact to contribute to the complex patters of tree mortality. For example, in a study of the impact of a typhoon on Japanese warm temperate forests, Bellingham and others (1996) found that there was no consistent mortality pattern for most common species, but they found a few species, such as *Symplocos prunifolia*, sustained a high level of basal area loss, while others, such as *Podocarpus nagi*, had low mortality.

Understory morality patterns are less documented than those of the overstory, both in tropical and temperate forests. In some cases understory mortality may be low due to the shielding effects from high canopy trees (Imbert, 1996), but these effects vary among forests. Other factors such as leaf liter, woody debris, and light may also contribute to the mortality patterns of seedlings and saplings. In temperate Piedmont forests, the most rapid changes following catastrophic winds were seen in the understory seedling layer (Xi, 2005). Seedling density and species richness experienced an immediate drop. This was followed by a rapid rebound in seedling density and more gradual recovery and enhancement in richness and diversity. Seedling recruitment did not increase continuously over time and overall seedling density was relatively low compared to the pre-hurricane level. These disturbance-induced changes in the understory must be viewed in the context of variation in pre-disturbance tree species composition resulting from differences in habitat and stand history.

Observations of tree mortality are needed not just between forests but also across a time interval of several years following the event. One reason is the need to correct for variable background mortality rates among tree species, forest types, and successional phases. Another is that mortality following large catastrophic windstorms is often delayed (Walker, 1991, 1995; Sharitz et al., 1992; Xi, 2005; Xi & Peet, 2008a; Xi et al., 2008b). Temperate forest researchers have noticed that most damaged deciduous hardwood trees can remain alive for many years while still suffering enhanced mortality, plus a certain portion of the damaged trees might grow back through sprouting (e.g. Peterson & Pickett, 1991; DeCoster, 1996; Paciorek et al., 2000). Consequently, tree mortality must be examined over a long time period and in the context of background mortality of the specific species and successional phases. An immediately survey after a catastrophic wind event could significantly underestimate wind-induced tree death rates. We concur with the suggestion of Everham and Brokaw (1996) that "Mortality should be tracked for several years after catastrophic wind events to determine the extent of elevated mortality." We further suggest that the 5-10 years of observation of the damaged plots is critical for a better understanding of long-term recovery process, particularly the underlying mechanisms of forest recovery from large disturbances.

3.3 Change in species composition and diversity

Changes in species composition and diversity following wind damage in temperate forests are often gradual and complex. Such subtle compositional changes can only be understood through longer-term observation, and in the context of baseline data at specific spatial and temporal scales. To a large extent, these changes are difficult to detect without baseline data, which are rarely available.

A variety of patterns of change in species composition and diversity following large wind events has been reported in the literature. Relatively large changes in species composition and diversity are often, though not always, reported in temperate forests following catastrophic winds. Species diversity enrichment may occur during long periods of recovery in places where a canopy species has been heavily damaged, thereby releasing species present in the understory and perhaps allowing establish of new species in the less competitive environment (Spurr, 1956; Abrams & Scott, 1989). Severe wind intensities are needed to create large patches and to reconfigure the limiting resources such as light and soil nutrients. In these cases species diversity is enriched at the scales of the multiple-patch mosaic, and succession is set back (Webb, 1999).

Changes in species composition in temperate forests following wind disturbance can be modest if the same species that regenerate in disturbed patches are most heavily damaged. For example, after examining changes in two low canopy diversity Minnesota forests during 14 years following a catastrophic windthrow, Palmer and others (2000) concluded that the windstorm only affected understory species composition and that the forests increased in understory species richness, although the magnitude of the changes was modest. This is also the case for positive neighborhood effects suggested by Frelich and Reich (1995). Where the positive neighborhood effect is strong, little compositional change will occur because wind-thrown trees are often replaced by the same species (Webb, 1999).

The third possible outcome of wind disturbance commonly seen in temperate forests is loss of species diversity following large wind disturbance. This outcome results when shade-intolerant species sustain heavy mortality owing to concentration in the canopy and are unable to colonize disturbed patches because of a pre-established understory of shade-tolerant species. Sharitz and others (1992), for example, found that Hurricane Hugo reduced the tree diversity in the slough forest communities in a South Carolina riparian area by having disproportionably larger negative effects on shade-intolerant and transition species of the canopy than on the shade-tolerant species that dominated the subcanopy.

In Piedmont temperate forests, changes in sapling diversity following the 1996 Hurricane Fran were varied. Mostly, sapling diversity increased slightly following the hurricane. However, a decrease of sapling diversity was also observed where canopy damage was extremely high, though this may ultimately be compensated for by increased establishment of new seedlings of shade-intolerant species. The density of saplings initially decreased in most damaged plots, but sapling recruitment subsequently increased due to release of previously established seedlings. This observation is consistent not only with the hypothesized relaxation of competition, but also the hypothesis that windthrow can contribute greatly to tree diversity in temperate forests (Xi, 2005; Xi et al, 2008b).

4. Factors influencing mortality and their interactions

Severity of tree damage and mortality is related to both abiotic factors (e.g., winds, topography, and soil) and biotic factors (e.g., individual tree characteristics, tree species, stand attributes). Although wind speeds are the primary determinant of tree damage and mortality, topographic exposure, soil moisture and community attributes are the most important factors influencing differential damage across landscapes. Exposure to winds, saturated soil, and high stand density are all associated with high tree damage and mortality risks. Tree species mixtures are also important for predicting landscape and stand-level damage severity, but evidence of species-specific damage and mortality is often less clear as species effects often interact with tree size.

4.1 Abiotic factors

Wind speed: Various studies have examined the relationship between wind speed and tree damage. In a broad sense, tree damage severity can be considered to be a function of wind speed. Fraser (1962) found that tree damage increases linearly with wind speed. Powell and others (1991) reported that little damage occurred below wind speeds of 17.5 m/s, and that trunk snapping and uprooting generally occurred at wind speeds above 33 m/s. Peltola (1996) found that the wind speed required to uproot a tree was much less than that required to cause the stem to break, and wind speeds of 12-14 m/s can to be strong enough to uproot Scots pines (slender individuals) located along a stand edge. Since even in flat terrain wind speed can vary substantially at scales of less than a kilometer, the local variation in wind speeds must be taken into account in examining landscape- and region-level wind damage (Foster & Boose, 1992; DeCoster, 1996; Peterson, 2000).

Topography: Topographic exposure has been shown to have major effects on wind damage at the landscape scale. In a Jamaican forest Bellingham (1991) found higher damage on southern slopes and ridge crests that were exposed to the hurricane-face winds, while minor damage occurred on protected northern slopes. Boose and others (1992) found a similar pattern of hurricane damage in New England, USA; higher damage occurred on southwestern slopes exposed to the hurricane winds, whereas minor damage occurred in a protected deep valley. They concluded that topographic exposure, combined with wind intensity and forest stand attributes, could largely explain damage patterns at landscape scale.

Soil features: Pre-hurricane soil moisture has been found to be a major factor in controlling whether uprooting or stem breakage is the dominant damage type (DeCoster, 1996). Where the soil is dry, uprooting is more difficult, and trees more common experience stem breakage. When the soil is wet, uprooting is more common (Xi, 2005). In the cold temperate forest zone such as in Finland, soil frost can reduce uprooting, and a decrease in the period or depth of frost can make trees more vulnerable to windthrow (Peltola, 1996).

4.2 Biotic factors

Individual tree architecture: Although not always true, the largest canopy trees often experience the most severe damage. Damage severity tends to increase approximately linearly with increasing tree height (e. g., Putz, 1983; Walker et al., 1992). Peltola (2006) found that the wind speed needed to cause uprooting or stem breakage of trees will decrease as the tree height or the tree height to dbh (diameter at breast height) ratio increases or the stand density decreases. For example, trees with tall, slim stems are usually extremely vulnerable. In addition to tree height, the shape and size of the tree crown and the shape of bole are important. Open-grown trees with large crowns could be extra vulnerable to high winds (Barry et al., 1993).

Species susceptibility: Tree species vary in their ability to withstand wind damage, their resistance depending on the interaction of several factors such as strength of wood, shape and size of the crown, extent and depth of root systems, shape of the bole (Barry et al., 1993), canopy characteristics, leaf features, and characters of root systems. Species with weaker wood (Webb, 1989), lower leaf flexibility (Vogel, 1996, 2009), and shallower root systems (Lorimer, 1977; Whitney, 1986; Gresahm et al., 1991; Putz & Sharitz, 1991) generally suffer greater damage and mortality, although it is difficult to distinguish the effects of species from effects of tree size (Falinski, 1978; Johnsen et al., 2009). In the Duke Forest on the North Carolina Piedmont Hurricane Fran caused a higher incidence of damage in canopy hardwoods than pines. This was because hardwoods usually have broad spreading canopies and flat leaves that can catch the force of the wind much more readily than the smaller canopies and the needle leaves of pine trees. Moreover, hardwoods often have shallow, spreading root systems that increase their susceptibility to uprooting during hurricanes (Xi, 2005).

Tree species can be classified into different groups based on their susceptibility to wind disturbance (Xi & Peet, 2008a). Bellingham and Tanner (1995) studied tree damage and responsiveness in a Jamaican montane forest following Hurricane Gilbert. Based on indices of hurricane-caused damage (including short-term change in morality and percent of stem that lost crown) and species response following the hurricane (including change in recruitment rate, change in growth rate, and frequency of sprouting), they classified 20 tree species into four groups: resistant (low damage, low response), susceptible (high damage, low response), resilient (high damage, high response), and usurpers (low damage, high response). They further predicted that species classified as usurpers would increase their relative abundance in the forest in the next decades, whereas the susceptible tree species would decrease in relative abundance of adults. Similarly, in an old-growth forest damaged by hurricanes in southeastern USA, Batista and Platt (2003) classified 10 tree species into four similar syndromes of response to disturbance according to observed mortality, recruitment, and growth patterns: resilient, usurper, resistant and susceptible. Barry et al. (1993) have provided a rank of resistance of tree species to hurricane-related damage for the major tree species in the southern United States. Although a more complete classification is needed, these classifications provide helpful information for forest managers.

Community attributes: Community attributes such as stand height and age, stand density, and stand edge inevitably influence tree damage risk. Taller forests are generally subject to greater damage and mortality risk than shorter ones. This increase is thought to be primarily a result of greater exposure to wind in the canopy and the increased leverage achieved with canopy movement. Because wind speeds are much higher at and above the crown level than within the stand, taller trees are subject to higher damage risk than shorter ones (Fraser, 1964; Somerville, 1980). Another reason for increasing damage with increasing stand height is that smaller, younger trees are generally more flexible to wind flows (Vogel, 1996). Foster (1988, 1992) found where severe windthrow of more than 75% of the trees was reached, it mostly occurred in stands of > 25 m height. Similarly, DeCoster (1996) reported a positive relationship between stand height and tree damage for 1989 Hurricane Hugo in South Carolina and for a separate severe tornado event on the Carolina Piedmont.

Literature reports on the effect of stand density on tree damage risk have been variable. Most studies have shown a trend of increasing damage with decreasing stand density (Prior, 1959; Busby, 1965; Thomson, 1983; Jane, 1986; Foster, 1988b; Hook et al., 1991; Peterson & Leach, 2008a, 2008b), but there are contrasting results, in part because denser stands often consist of younger and more flexible trees. For example, Fraser (1965) found a dense stand would decrease the lateral spread of roots and thereby increase tree damage. Overall, the complex effect of stand density on tree damage is unclear, perhaps because the confounding effects of stand density, tree size, tree species, and tree architectural characteristics have generally not been adequately separated. These relationships need to be examined through more comprehensive field experiments (e.g., Peltola, 1995; Vogel, 1996, 2009).

Interactions of factors: Much of the complexity of tree damage and mortality is caused by meteorological, topographical, and biological factors simultaneously interacting to create patterns of damage. Consequently, the interactions among factors must to be taken into account to better understand wind-damage relationships. Wind-induced effects and their interactions (e.g., insect breakouts, subsequent fires) need to be considered in evaluating indirect damage. For example, smaller trees sustain wounds caused by the falling tops of adjacent uprooted trees and the major branch breakages during the windstorm are often attacked by insects or affected by diseases (Barry et al., 1993). Similarly, trees with damaged root systems are often invaded by root rot organisms and subjected to higher risk to subsequent windstorms (Pickett & White, 1985).

In temperate forests, large wild fires often interact with hurricanes to cause greater forest damage (Platt et al., 2002). Myers and Lear (1998) in a literature review found that in temperate forests, conditions after exceptionally strong hurricanes promote the occurrence of fires of higher than normal intensity. Paleotempestological records also support this hurricane-fire interaction in the Holocene maritime pine-oak forests of the Gulf coast region (Liu, 2003). Conversely, Kulakowski and Veblen (2002), working in montane forests of Colorado, found fire history and topography can influence the severity of wind blowdown and the susceptibility of forest stands to wind damage.

Ackerman and others (1991) developed a graphic model depicting expected variation in forest damage and recovery following hurricanes (Fig. 2). The force exerted by a hurricane increases as a function of wind velocity and storm duration, and decreases with distance from the eye of the hurricane. Forest damage severity increases with intensity of a hurricane (i.e. wind speed), but the amplitude of the relationship depends on the physical and biotic factors of a given site, such as topography, geomorphology, soil moisture, species composition, vegetation structure, state of recovery since last disturbance, plant architecture, size, age, and anatomy. The influence of site factors on the extent of forest damage decreases as the magnitude of the hurricane increases.

Multiple factors simultaneously interact to contribute the observed damage complexity. Canham and others (2001), for example, examined the specific variation in susceptibility to windthrow as a function of tree size and storm severity for northern temperate tree species. In future studies, research should address the interplay of multiple factors, pre- and post wind disturbance event, through experiments, modeling, and cross-site comparison to separate the confound effects.

5. Forest responses

The distinct feature of wind-damaged forests, as compared with forests that have experienced other large, infrequent disturbances such as wild fires and volcano eruption, is that wind-damaged forests often have relatively rapid recovery through multiple recovery pathways. Foster and others (1991) identify two major regeneration pathways: 1) from surviving vegetation through advanced regeneration (advanced growth) and vegetative reproduction (sprouting), and 2) from seedling dispersal, recruitment and establishment (Fig. 3). The rapid recovery of wind-damaged forests largely results from stem sprouting and the advanced growth of the surviving trees in the new environment of increased light, soil moisture, and nutrient resources. In addition, windthrow creates more diverse soil substrates and allows active seedling and sapling regeneration. Here we review studies of surviving trees and the understory response to canopy tree gaps and disturbed soil.

5.1 Regrowth of surviving trees by sprouting

Regrowth plays an important role in tree recovery from catastrophic wind disturbances, especially in temperate hardwood forests. After damage by intensive winds, a high portion of hardwood trees can regrow from sprouts. Although several researchers have reported differences among species in sprouting ability in both tropical (Walker et al., 1992; Zimmerman et al., 1994; Bellingham et al., 1994) and temperate forests (Perterson & Pickett, 1991; DeCoster, 1996; Busby et al., 2009), this capability appears common. In Piedmont forests of North Carolina, resprouting of damaged individuals and vegetative production of additional shoots were common for most hardwoods (Xi, 2005).

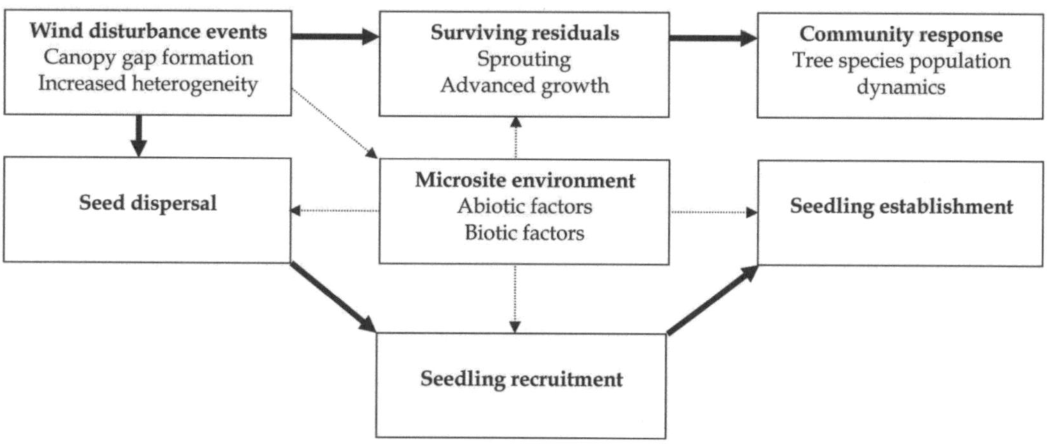

Fig. 3. Conceptual model of temperate forest regeneration following hurricane disturbance. Two major recovery pathways are represented by large arrows. The microsite environment influences each stage of the pathway of regeneration from seed but exerts less influence on the pathway of regeneration from surviving vegetation (Modified from Foster et al., 1991).

5.2 Understory response to canopy gaps

The understory of damaged forests plays a major part in forest response to windstorms in temperate forests (Webb, 1999). Three mechanisms have been often reported in the wind disturbance literature: release of understory plants, recruitment, and repression. "Release" refers to the rapid growth of suppressed understory plants following catastrophic disturbances. Reduced competition often allows an increased growth of established seedlings and saplings of primarily shade-tolerant species that were present in the understory at the time of disturbance. Most work on plant "release" after catastrophic winds has been done on samplings and small trees, though the release of established seedlings should also be expected (Fajvan et al., 2006). Piedmont forests have remarkable resilience to hurricane damage because of widespread advanced regeneration. In Piedmont North Carolina, most tree seedlings and saplings approximately doubled their relative growth rates after the 1996 Hurricane Fran, although not uniformly across tree species (Xi, 2005; Xi & Peet, 2008a).

Recruitment is the addition of new individuals into a community (Ribbens et al., 1995). Previous post-disturbance observations on seedling establishment have shown an increase in seedling density following hurricanes, due probably to increased light and soil nutrient availability (Guzman-Grajales & Walker, 1991). In Puerto Rican forests, recruitment from seeds was promoted by the large increase in area of gaps and the increased understory light following Hurricane Hugo (Everham et al., 1998).

5.3 Ground features: mounds and pits, leaf litter, and woody debris

In addition to increasing light, windstorms generate a highly diverse substrate with treefall mounds and pits, stumps, leaf litter, and rotting logs. With increased light, the microsites play important roles, influencing understory composition, species diversity, growth, and dynamics (Peterson et al., 1990; Webb, 1999; Busing et al., 2009). These newly formed microsites often differ from intact forests in their greater soil moisture and nutrient availability, thereby allowing rapid establishment of species that require not only increased light, but also more abundant soil water and nutrients than typically found in an intact stand.

Although several studies have examined the roles of mounds and pits following windstorm disturbance, the results have varied greatly between forests. Walker and others (2000) examined seedling and saplings dynamics in treefall pits in a Puerto Rican rain forest and found that treefall pits significantly alter recruitment and mortality of many understory species, but not species richness. In some cases, mounds support more species than pits or un-damaged forests (e.g., Collin & Pickett, 1982). However, Peterson and others (1990), working in a temperate forest, found lower species richness on mounts than in pits.

Increased leaf litter can be an important factor influencing seed germination and seedling establishment after windstorm disturbance. In addition, woody debris can provide important sites for germination and establishment (Webb, 1999). Guzman-Grajales and Walker (1991) examined the effects of three litter treatments on seedling emergence, growth, density, and mortality during the year following Hurricane Hugo in a Puerto Rican forest. Their conclusion was that leaf litter is a major constraint to seedling recruitment. The role of leaf litter in temperate forests is still less known.

6. Long-term effects of catastrophic wind disturbance

Despite the fact that much has been learned about immediate damage patterns and short-term impacts of catastrophic winds, less is known regarding long-term effects on forest composition, diversity, and succession. Study of long-term effects of historical wind events is difficult because rarely have ecologists been able to combine long-term pre-event and long-term post-event data. Moreover, the few long-term datasets that are available for this purpose were generally not designed or initiated with disturbance events in mind (e.g., Xi & Peet, 2008b). Nonetheless, sufficient information is available to indicate that hurricanes can have long lasting effects on tree growth, species composition, diversity, and succession, and that these effects can vary greatly with wind intensities, pre-disturbance community attributes and the timing of the winds (Fig. 4).

6.1 Long-term effects on species composition and diversity

A widely accepted view among forest ecologists is that severe hurricanes have relatively minor long-term effects on species composition and diversity in tropical forest regions and coastal temperate regions where hurricanes are common. Many case studies in the tropics, including studies in Puerto Rico, Nicaragua, Jamaica, and Kolombangara, support this general conclusion (but see Vandermeer, 2000). For example, Burslem and others (2000) found that historical hurricanes had only limited effects on species composition after 60 years of forest recovery.

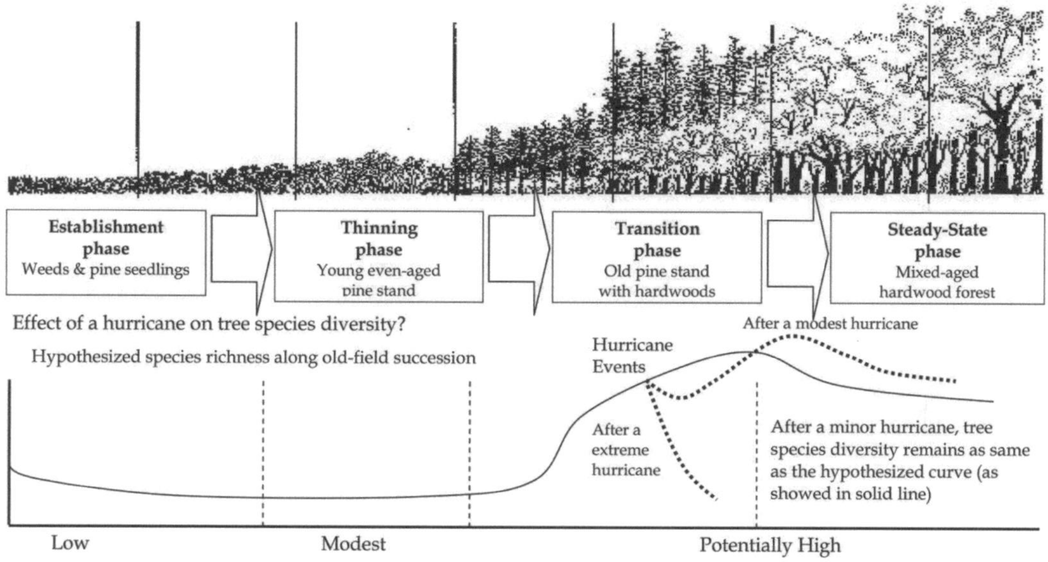

Fig. 4. Old-field succession on Piedmont and four-stage forest succession model and hypothesized tree species diversity curve (as showed in solid line) over time. The effect of a hurricane on tree species diversity is relatively low during the establishment and thinning phases due to the low species diversity at these two stages, whereas impacts are potentially higher at the transition and steady-state phases due to the increased species diversity. Hypothesized post-hurricane changes in species richness are showed as dash-lines. In the extremely damaged areas, local species diversity likely immediately drops to early succession levels due to direct elimination of species by high winds and gradually recovery over time; When wind intensity is low, tree richness changes little due to the resilience of the forests. Modest wind intensity may cause temporary reduction in species diversity following a hurricane but increase tree diversity over time due to more fragmented habitats for new species. These hypotheses need to be further examined by long-term field studies.

In contrast with results from most coastal tropical studies, significant but highly variable results regarding long-term change in community composition and species diversity have been reported for temperate forests (Table 2). Large, infrequent wind disturbance events have played an important role in shaping regional vegetation and influencing dynamics in many temperate forests (Foster & Boose, 1995; Webb, 1999). Change in species diversity following catastrophic wind disturbance varies from increasing to decreasing, depending on many factors such as damage intensity and scale of the investigation. In the extremely damaged areas, local species diversity likely immediately drops to early succession levels due to direct elimination of species by high winds and gradually recovers over time (Fig. 4). Large temperate-zone hurricanes generally have had a stronger impact on species richness in heavily damaged stands (Peet & Christensen, 1980; Foster et al., 1998; Boose et al., 2001). For example, Peet and Christensen (1980) reported increased species richness in a comparison study of two hardwood plots in the Duke Forest, North Carolina Piedmont, 23 years after the 1954 Hurricane Hazel. The permanent plots that were severely damaged had twice as many as tree species saplings as compared with the number before the 1954 Hurricane Hazel. This post-disturbance increase in regeneration of multiple species following an intense windstorm is consistent with a general pattern of dynamic, patch-driven regeneration and diversity maintenance in temperate forests.

	Damage patterns and forest responses	Reference
Temperate forests	Although geographically variable, generally a low frequency of hurricane damage, but less intense. Windstorms are frequent. Trees are more susceptible to windthrows. In some cases damage severity can be extremely high. Release of advanced regeneration is common. Greater portion of uprooting than in other types.	Forster, 1992; Peterson, 2002; Xi et al., 2008b
Coastal tropical forests	Geographically variable, but in general more frequent catastrophic hurricanes. Trees are more wind-resistant. Less composition and diversity change; high and relatively stable tree species diversity. Regrowth and sprouting are common.	Walker, 2002; Whigham, 2003
Early succession forests	The young trees of secondary forests are typically more resistant to wind throw than the larger and more brittle trees of old growth. Increased pioneer species in the damaged forests.	Webb, 2000; Xi, 2005; Xi et al., 2008a; Xi & Peet, 2008a
Late succession forests	Forests are susceptible to windthrows. Diversified forest structure and dynamics; maintenance, increased or decreased tree species.	Mitchell et al., 2004; Xi et al., 2002, 2008b

Table 2. Comparison of temperate forests and tropical forests, early succession forests and late succession forests in their responses to catastrophic wind disturbance events.

Species dominance may shift substantially after wind disturbance because early successional species thrive in the hurricane-created gaps, although the long-term term effects are less evident. Nonetheless, the addition of early succession species in those successional patches in some cases leads to short-term increases in landscape diversity. Moreover, the results may be scale dependent. For example, following the 1989 Hurricane Hugo, Everham (1996) found that the number of plant species increased in some sites when observed at an intermediate spatial scale (i.e. hectares), but was essentially constant at both larger and smaller scales. Over the several decades following a hurricane, the short life span of the early successional species, coupled with the self-thinning process, may again result in reduced dominance and landscape diversity. Thus, overall, catastrophic wind disturbance may have a limited small-scale effect on species diversity over time, while enhancing diversity at intermediate scales.

6.2 The lasting effects of windstorms on forest succession

Extreme windstorms tend to differentially remove the oldest and largest trees in a stand. As a consequence, large, catastrophic wind events have been concluded to significantly change forest structure and alter the rates of various processes in the temperate forests, even though their long-term effects on forest succession are uncertain (Waring & Schlesinger, 1985; Foster & Boose, 1995). Studies of long-term wind effects on temperate forest succession to date have shown that windstorms can have all possible effects from setting succession back to advancing successional stages, to initiating multiple-stages of succession, depending on wind intensity, frequency, forest types and pre-disturbance successional stage.

The traditional idea that wind disturbance sets back succession to some earlier seral stage may apply in temperate forests where extreme high winds create large forest openings and initiate secondary succession. The mechanism for this change is that severe windstorms substantially damage the late-successional, canopy-dominant tree species and lead to establishment of early successional species. Therefore, 'setting back of succession' often occurs in the later successional hardwood forests exposed to extreme wind intensity. The New England hurricane of 1938, for example, leveled many thousands of acres of mature and semi-mature hardwood forests and initiated new forest associations over a large area with the long-lasting effects (Wilson et al., 2005).

Wind disturbance can also accelerate succession when early successional canopy tree species are heavily disturbed (White & Jentsch, 2004; Xi, 2005). In temperate forests where early successional tree species such as various pines and oaks are dominants, instantaneous death of the even-aged canopy by intensive winds tends to advance forest succession and differentially favor the shade-tolerant understory species. Abrams and Scott (1989) in particular showed that windstorms, among other disturbances, can accelerate forest succession in some North American forest communities. The 1938 hurricane that caused in excess of 30% tree mortality and large areas of windthrow in New England heavily damaged successional Pinus *strobus* forests, accelerating successional turnover to hardwood species that were in some cases already present in the understory (Foster & Boose, 1992). Arevola and others (2000) examined the changes in both pine and hardwood stands 14 years following a catastrophic windstorm in Minnesota and concluded that the wind disturbance acted to accelerate the successional process in both forest types by increasing the rate of compositional change from early successional pines and hardwoods to late-successional hardwoods. Although this pattern may be somewhat simplistic, the patterns they found appear common in temperate forests.

When the dominants in temperate forests are damaged by windstorms but are replaced by same type of species, succession can be held at the same stage. In this regard, biotic factors such as propagule supply may strongly influence long-term forest recovery and succession following a large disturbance. In the case of intensive wind, the interactions of survivors and the pre-disturbance understory species (small trees and saplings) may determine the initial state in which the forest develops and the recovery pathways from the catastrophic wind event. Turner and others (1998) have argued that the abundance and spatial arrangement of the survivors and the arrival pattern of propagules may be the pivotal factors determining how succession differs between catastrophic disturbances of large and small extent. However, few studies actually examine this effect and the role of propagule availability in influencing forest regeneration and succession largely remains a matter of conjecture (Webb, 1999).

7. The role of the predictive models for evaluating wind impacts

Ecologists and foresters have increasingly used modeling approaches to evaluate damage-risk factors and predict forest responses to large disturbances such as windstorms. A major focus of such modeling work has been integration of remote sensing, aerial photo, and ground field data with GIS software to assess damage risk factors at various spatial scales. For instance, Foster and Boose (1992, 1994) took an integrative approach through analysis of remotely sensed, historical and field data to assess actual forest damage in both tropical and temperate forests. They also developed meteorological and topographic exposure models to reconstruct wind conditions and site exposure to windstorms. Pleshikov and others (1998) developed a computer system for evaluating and predicting pine stand resistance to hurricane-force winds in central Siberia. They attempted to analyze risk factor at landscape, stand, and single-tree scales. Lindemann and Baker (2002) used GIS with CART (Classification and Regression Tree) and logistic regression to analyze a severe forest blowndown in the Southern Rocky Mountains and found that the blowdown was most influenced by the factors pertaining to the physical setting. However, McMaster (2005) suggested that detailed site-specific factors such as average stem diameter, species, canopy height, and stand age are critical for improved accuracy of forest blowdown prediction.

Several studies have focused on modeling forest dynamics after large hurricanes. Doyle (1997) developed the HURISIM model for modeling hurricane effects on mangrove forests. He used historical simulations that included actual hurricane tracks and tree conditions and found hurricanes account for much of the structural composition of modern-day mangrove forests across south Florida. He suggested that the occurrence of major storms with a contemporary recurrence interval of 30 years may be the most important factor controlling mangrove ecosystem dynamics in south Florida. Canham and others (2001) developed maximum-likelihood models for simultaneously estimating both local storm severity and the parameters of functions that define species-specific variation in susceptibility to windthrow.

Development of spatially-explicit and landscape-scale models is becoming an active research arena of forest disturbance dynamics. These models have proven especially useful for examination of windstorm impacts. Kramer and others (2001) built such a spatially-explicit model to examine abiotic controls on windthrow and forest dynamics in southeast Alaska. More recently, Schumacher and others (2004) developed a modified LANDIS landscape model to examine the interaction among species-specific responses, intra- and inter-specific competition, and exogenous disturbance regimes including winds. Landscape models have an important role as tools for synthesizing existing information and making projections of possible future vegetation dynamics at large spatial scales.

In summary, developing and applying predictive models provides a promising opportunity for evaluating and projecting windstorm-induced forest damage. The predictive models can project the loss/alteration of habitat and the resulting impact on species diversity, and thus can be effective evaluation tools that when used properly and in conjunction with other assessment techniques could be a valuable aid in understanding forest damage patterns and controlling factors at various temporal and spatial scales. These models can also be an effective tool for post-damage forest management decision-making (Kupfer et al., 2008).

8. Synthesis and future directions

A general framework is needed for understanding the complexity of windstorm effects on temperate forests and subsequent forest response. In this paper, we combine illustrative examples to present a conceptual framework and then link them to several important themes that have emerged in recent years. Two relatively separated lines of investigation are apparent in the literature review, one focused on the complexity of forest damage patterns and their risk factors, and the other focused on the high degree of variation among forests in their structural and compositional responses to windstorm disturbances.

The variation among wind regimes and forest responses makes generalization a challenge. The literature here reviewed shows the complexity of pattern in forest damage and tree mortality following catastrophic wind, as well as the significant variation among forests in structural and composition responses. Many factors interact to influence the patterns of damage and dynamics of recovery. Therefore, evaluating the relative importance of multiple-factors and various recovery patterns across the full spectrum of disturbance severity levels will help elucidate these factors and their interactions. Nonetheless, there remains a clear need for additional studies that quantify wind disturbance severity and complexity of impact in high-wind damaged forests.

Windstorm-induced dynamics may vary at the different spatial and temporal scales. The ecological consequences of catastrophic winds are complex, subtle, and at smaller scales relatively unpredictable. Consequently, wind-induced changes must be viewed in the context of interaction and variations among multiple factors, especially species composition resulting from differences in habitat and stand history. Remarkably few studies have actually examined multiple factors and multiple-scale wind damage and forest recovery. Windstorm-induced effects should be examined across a gradient of spatial and temporal scales if we are to understand and explore these complicated and scale-dependent processes and patterns.

Long-term studies of forest response to different combinations of the wind disturbance severity are needed. The variable effects of windstorms on temperate forests largely depend on the wind intensity, size, specificity, frequency of individual windstorms in a given location, pre-disturbance species composition, and successional stage. The complex impacts of winds and variable forest recovery are more readily discerned when detailed, long-term pre-disturbance and long-term post-disturbance data are available. Certainly, more extensive long-term studies on permanent research sites will be important for understanding the long-term impacts.

Finally, better and more generally applicable models are needed for predicting the impacts of future catastrophic windstorm events on forests. Both population-based gap models and spatially explicit landscape models provide powerful tools for predicting forest disturbance and dynamics. Recent progress has been made in constructing such models applicable to temperate forests (Doyle, 1997; Schumacher et al., 2004), but parameterization of these models for species-rich systems presents considerable challenges. Direct estimates of colonization and mortality rates from long-term studies in temperate forests could be highly valuable for improving these models. Predictive models will ultimately provide the knowledge essential for understanding the role of windstorm disturbances in forest communities, in guiding conservation efforts, and in informing forest management decisions.

9. Acknowledgements

This work was supported by a grant from the National Science Foundation (DEB-97-07551). We thank Peter White, Seth Reice, Dean Urban, and Aaron Moody for providing valuable comments on earlier versions of this manuscript. We thank Jie Zhao for her considerable help in refining the figures.

10. References

Abrams, M. D. & Scott, M. L. (1989). Disturbance-mediated accelerated succession in 2 Michigan forest types. *Forest Science*, Vol. 35, No. 1, 42-49, ISSN:0015-749X

Abrams, M. D.; Sprugel, D. G. & Dickmann, D. I. (1985). Multiple successional pathways on recently disturbed jack pine sites in Michigan. *Forest Ecology and Management*, Vol. 10, No. 1, 31-48, ISSN: 0378-1127

Ackerman, J. D.; Walker, L. R.; Scatena, F. N. & Wunderle, J. (1991). Ecological effects of hurricanes. *Bulletin of the Ecological Society of America*, Vol. 72,178-180, ISSN: 0012-9623

Akachuku, A. E. (1993). Recovery and morphology of *Pinus resinosa* trees 50 years after they were displaced by a hurricane. *Forest Ecology and Management*, Vol. 56, No. 1-4, 113-129, ISSN: 0378-1127

Allen, B. P.; Pauley, E. F. & Sharitz, R. R. (1997). Hurricane impacts on liana populations in an old-growth southeastern bottomland forest. *Journal of the Torrey Botanical Society*, Vol. 124, No. 1, 34-42, ISSN: 1095-5674

Arevalo, J. R.; DeCoster, J. K.; Mcalister, S. D. & Palmer, M. W. (2000). Changes in two Minnesota forests during 14 years following catastrophic windthrow. *Journal of Vegetation Science*, Vol. 11,833-840, ISSN: 1100-9233

Baker, G. T. (1915). A windfall problem. *Forestry Quarterly*, Vol. 13,317-324, ISSN: 0022-1201

Baker, W. L.; Flaherty, P. H.; Lindemann, J. D.; Veblen, T. T.; Eisenhart, K. S. & Kulakowski, D. W. (2002). Effect of vegetation on the impact of a severe blowdown in the southern Rocky Mountains, USA. *Forest Ecology and Management*, Vol. 168,63-75, ISSN: 0378-1127

Barnes, J. (2001). *North Carolina's hurricane history*. The University of North Carolina Press, ISBN 978-0-8078-4969-9, Chapel Hill and London. USA

Barry, P.; Doggett, C.; Anderson, R. & Swain, K. (1993). How to evaluate and manage storm-damaged forest areas. United States Forest Service, Southern Region Management Bulletin R8-MB 63. 11 p

Batista, W. B. & Platt, W. J. (2003). Tree population responses to hurricane disturbance: syndromes in a south-eastern USA old-growth forest. *Journal of Ecology*, Vol. 91,197-212, ISSN: 0022-0477

Batista, W. B.; Platt, W. J. & Macchiavelli, R. E. (1998). Demography of a shade-tolerant tree (*Fagus grandifolia*) in a hurricane-disturbed forest. *Ecology*, Vol. 79,38-53, ISSN: 0012-9658

Battaglia, L. L.; Sharitz, R. R. & Minchin, P. R. (1999). Patterns of seedling and overstory composition along a gradient of hurricane disturbance in an old-growth bottomland hardwood community. *Canadian Journal of Forest Research*, Vol. 29,144-156, ISSN:0045-5067

Bazzaz, F. A. (1983). Characteristics of populations in relation to disturbance in natural and man-modified ecosystems, In: *Disturbance and Ecosystems - Components of response*, Mooney, H. A. & Godron, M. (Eds.), page 259-275, Springer-Verlag, ISBN-10: 3540124543, Berlin, Germany

Beatty, S. W. & Sholes, O. D. V. (1988). Leaf litter effect on plant-species composition of deciduous forest treefall pits. *Canadian Journal of Forest Research*, Vol. 18,553-559, ISSN:0045-5067

Beckage, B. & Clark, J. S. (2003). Seedling survival and growth of three forest tree species: the role of spatial heterogeneity. *Ecology*, Vol. 84,1849-1861, ISSN: 0012-9658

Beckage, B.; Clark, J. S.; Clinton, B. D. & Haines, B. L. (2000). A long-term study of tree seedling recruitment in southern Appalachian forests: the effects of canopy gaps and shrub understories. *Canadian Journal of Forest Research*, Vol. 30,1617-1631, ISSN: 0045-5067

Bellingham, P. J. (2000). Resprouting as a life history strategy in woody plant communities. *Oikos*, Vol. 89,409-416, ISSN: 0030-1299

Bellingham, P. J. (2008). Cyclone effects on Australian rain forests: An overview. *Austral Ecology*, Vol. 33,580-584, ISSN: 0012-9658

Bellingham, P. J.; Kapos, V.; Varty, N.; Healey, J. R.; Tanner, E. V. J.; Kelly, D. L.; Dalling, J. W.; Burns, L. S.; Lee, D. & Sidrak, G. (1992). Hurricanes need not cause high mortality - the effects of Hurricane Gilbert on forests in Jamaica. *Journal of Tropical Ecology*, Vol. 8,217-223, ISSN: 0266-4674

Bellingham, P. J.; Kohyama, T. & Aiba, S. (1996). The effects of a typhoon on Japanese warm temperate rainforests. *Ecological Research*, Vol. 11,229-247, ISSN: 0912-3814

Bellingham, P. J.; Tanner, E. V. J. & Healey, J. R. (1994). Sprouting of trees in Jamaican montane forests, after a hurricane. *Journal of Ecology*, Vol. 82,747-758, ISSN: 0022-0477

Bellingham, P. J.; Tanner, E. V. J. & Healey, J. R. (1995). Damage and responsiveness of Jamaican montane tree species after disturbance by a hurricane. *Ecology*, Vol. 76,2562-2580, ISSN: 0012-9658

Boose, E. R.; Chamberlin, K. E. & Foster, D. R. (2001). Landscape and regional impacts of hurricanes in New England. *Ecological Monographs*, Vol. 71,27-48, ISSN: 0012-9615

Boose, E. R.; Foster, D. R. & Fluet, M. (1994). Hurricane impacts to tropical and temperate forest landscapes. *Ecological Monographs*, Vol. 64,369-400, ISSN: 0012-9615

Boose, E. R.; Serrano, M. I. & Foster, D. R. (2004). Landscape and regional impacts of hurricanes in Puerto Rico. *Ecological Monographs*, Vol. 74,335-352, ISSN: 0012-9615

Bormann, F. H. & Likens, G. E. (1979). *Pattern and process in a forested ecosystem*. Springer-Verlag, ISBN:0387903216, New York, New York, USA

Bormann, F. H. & Likens, G. E. (1979). Catastrophic disturbance and the steady state in northern hardwood forests. *American Scientist*, Vol. 67,660-669, ISSN: 0003-0996

Boucher, D. H.; Vandermeer, J. H.; De La Cerda, I. G.; Mallona, M. A.; Perfecto, I. & Zamora, N. (2001). Post-agriculture versus post-hurricane succession in southeastern Nicaraguan rain forest. *Plant Ecology*, Vol. 156,131-137, ISSN: 1385-0237

Boucher, D. H.; Vandermeer, J. H.; Yih, K. & Zamora, N. (1990). Contrasting hurricane damage in tropical rain-forest and pine forest. *Ecology*, Vol. 71,2022-2024, ISSN: 0012-9658

Brewer, R. & Merritt, P. G. (1978). Windthrow and tree replacement in a climax beech-maple forest. *Oikos*, Vol. 30,149-152, ISSN: 0030-1299

Brokaw, N. V. L. & Walker, L. R. (1991). Summary of the effects of Caribbean hurricanes on vegetation. *Biotropica*, Vol. 23,442-447, ISSN: 0006-3606

Bromley, S. W. (1939). Factors influencing tree destruction during the New England hurricane. *Science*, Vol. 90, 15-16, ISSN: 0036-8075

Burslem, D. & Whitmore, T. C. (1999). Species diversity, susceptibility to disturbance and tree population dynamics in tropical rain forest. *Journal of Vegetation Science*, Vol. 10,767-776, ISSN: 1100-9233

Burslem, D.; Whitmore, T. C. & Brown, G. C. (2000). Short-term effects of cyclone impact and long-term recovery of tropical rain forest on Kolombangara, Solomon Islands. *Journal of Ecology*, Vol. 88,1063-1078, ISSN: 0022-0477

Busby, P.E.; Motzkin, G. & Boose, E.R. (2008). Landscape-level variation in forest response to hurricane disturbance across a storm track. *Canadian Journal of Forest Research*, Vol. 38,2942-2950, ISSN: 0045-5067

Busby, P.E.; Canham, C. D.; Motzkin, G. & Foster, D.R. (2009). Forest response to chronic hurricane disturbance in coastal New England. *Journal of Vegetation Science*, Vol. 20,487-497, ISSN: 1100-9233

Busing, R.T.; White, R.D.; Harmon, M.E.; White, P.S. (2009). Hurricane disturbance in a temperate deciduous forest: patch dynamics, tree mortality, and coarse woody detritus. *Plant Ecology*, Vol. 201,351-363, ISSN: 1385-0237

Canham, C. D. & Loucks, O. L. (1984). Catastrophic windthrow in the presettlement forests of Wisconsin. *Ecology*, Vol. 65,803-809, ISSN: 0012-9658

Canham, C. D.; Papaik, M. J. & Latty, E. F. (2001). Interspecific variation in susceptibility to windthrow as a function of tree size and storm severity for northern temperate tree species. *Canadian Journal of Forest Research*, Vol. 31,1-10, ISSN: 0045-5067

Carlton, G. C. & Bazzaz, F. A. (1998). Regeneration of three sympatric birch species on experimental hurricane blowdown microsites. *Ecological Monographs*, Vol. 68,99-120, ISSN: 0012-9615

Carlton, G. C. & Bazzaz, F. A. (1998). Resource congruence and forest regeneration following an experimental hurricane blowdown. *Ecology*, Vol. 79,1305-1319, ISSN: 0012-9658

Carpino, E. (1998). *Ecological determinants of hurricane damage in a southeastern piedmont forest*. Master thesis. Duke University, Durham, North Carolina, USA

Chapman, E.L.; Chambers, J.Q.; Ribbeck, K.F.; Baker, D.B.; Tobler, M.A.; Zeng, H.C. & White, D.A. (2008). Hurricane Katrina impacts on forest trees of Louisiana's Pearl River basin. *Forest Ecology and Management*, Vol. 256,883-889, ISSN: 0378-1127

Christensen, N. L. (1988). Succession and natural disturbance: paradigms, problems, and preservation of natural ecosystems, In: Agee, J. K. & Johnson, D. R. (Eds.), pp. 62-81, *Ecosystem Management for Parks and Wilderness*. University of Washington Press, ISBN-10:0295968176, Seattle, Washington, USA

Cooper-Ellis, S.; Foster, D. R.; Carlton, G. & Lezberg, A. (1999). Forest response to catastrophic wind: results from an experimental hurricane. *Ecology*, Vol. 80,2683-2696, ISSN: 0012-9658

Coutts, M. P. & Grace, J. (1995). *Wind and Trees*. Cambridge University Press, ISBN-10: 0521460379, Cambridge, UK

Curtis, J. D. (1943). Some observations of wind damage. *Journal of Forestry*, Vol. 41,877-882, ISSN: 0022-1201

DeCoster, J. K. (1996). *Impacts of tornados and hurricanes on the community structure and dynamics of north and South Carolina forests*. Ph.D. Dissertation. University of North Carolina at Chapel Hill, Chapel Hill, North Carolina, USA

Denslow, J. S. (1980). Patterns of plant-species diversity during succession under different disturbance regimes. *Oecologia*, Vol. 46,18-21, ISSN: 0029-8549

Desteven, D.; Kline, J. & Matthiae, P. E. (1991). Long-term changes in a Wisconsin *Fagus-Acer* forest in relation to glaze storm disturbance. *Journal of Vegetation Science*, Vol. 2,201-208, ISSN: 1100-9233

Doyle, T. W. (1997). Modeling hurricane effects on mangrove ecosystems. USGS Biological Resources Division USGS FS-095-97. 2 pp

Dunn, C. P.; Guntenspergen, G. R. & Dorney, J. R. (1983). Catastrophic wind disturbance in an old-growth hemlock hardwood forest, Wisconsin. *Canadian Journal of Botany*, Vol. 61,211-217, ISSN: 0008-4026

Elliott, K. J.; Hitchcock, S. L. & Krueger, L. (2002). Vegetation response to large scale disturbance in a southern Appalachian forest: hurricane opal and salvage logging. *Journal of the Torrey Botanical Society*, Vol. 129,48-59, ISSN: 1095-5674

Emanuel, K. A. (2005). Increasing destructiveness of tropical cyclones over the past 30 years. *Nature*, Vol. 486,686-688, ISSN: 0028-0836

Ennos, A. R. (1997). Wind as an ecological factor. *Trends in Ecology & Evolution*, Vol. 12,108-111, ISSN: 0169-5347

Everham, E. M. & Brokaw, N. V. L. (1996). Forest damage and recovery from catastrophic wind. *Botanical Review*, Vol. 62,113-185, ISSN: 0006–8101

Fajvan, M.A.; Plotkin, A. B. & Foster, D. R. (2006). Modeling tree regeneration height growth after an experimental hurricane. *Canadian Journal of Forest Research*, Vol. 36,2003-2014, ISSN: 0045-5067

Foster, D. R. (1988). Disturbance history, community organization and vegetation dynamics of the old-growth Pisgah forest, southwestern New Hampshire, USA. *Journal of Ecology*, Vol. 76,105-134, ISSN: 0022-0477

Foster, D. R. (1988). Species and stand response to catastrophic wind in central New England, USA. *Journal of Ecology*, Vol. 76,135-151, ISSN: 0022-0477

Foster, D. R. & Boose, E. R. (1992). Patterns of forest damage resulting from catastrophic wind in central New England, USA. *Journal of Ecology*, Vol. 80, 79-98, ISSN: 0022-0477

Fraser, A. I. (1964), Wind tunnel and other related studies on coniferous trees and tree crops, *Scottish Forestry*, Vol. 18,84-92, ISSN: 0036-9217

Fujita, T. T. (1985). *The Downburst, Microburst and Macroburst*. SMRP Research Paper No. 210 (NTIS No. PB85-148880), University of Chicago, Chicago, Illinois, USA

George, L. O. & Bazzaz, F. A. (1999). The fern understory as an ecological filter: growth and survival of canopy tree seedlings. *Ecology*, Vol. 80,846-856, ISSN: 0012-9658

Glitzenstein, J. S. & Harcombe, P. A. (1988). Effects of the December 1983 tornado on forest vegetation of the Big Thicket, southeast Texas, USA. *Forest Ecology and Management*, Vol. 25,269-290, ISSN: 0378-1127

Glitzenstein, J. S.; Harcombe, P. A. & Streng, D. R. (1986). Disturbance, succession, and maintenance of species-diversity in an east Texas forest. *Ecological Monographs*, Vol. 56,243-258, ISSN: 0012-9615

Goldenberg, S. B.; Landsea, C. W.; Mestas-Nunez, A. M. & Gray, W. M. (2001). The recent increase in Atlantic hurricane activity: causes and implications. *Science*, Vol. 293,474-479, ISSN: 0036-8075

Greenberg, C. H. & McNab, W. H. (1998). Forest disturbance in hurricane-related downbursts in the Appalachian mountains of North Carolina. *Forest Ecology and Management*, Vol. 104,179-191, ISSN: 0378-1127

Gresham, C. A.; Williams, T. M. & Lipscomb, D. J. (1991). Hurricane Hugo wind damage to southeastern United-States coastal forest tree species. *Biotropica*, Vol. 23,420-426, ISSN: 0006-3606

Guzman-Grajales, S. M. & Walker, L. R. (1991). Differential seedling responses to litter after Hurricane Hugo in the Luquillo Experimental Forest, Puerto Rico. *Biotropica*, Vol. 23,407-413, ISSN: 0006-3606

Harcombe, P. A.; Bill, C. J.; Fulton, M.; Glitzenstein, J. S.; Marks, P. L. Elsik, & I. S. (2002). Stand dynamics over 18 years in a southern mixed hardwood forest, Texas, USA. *Journal of Ecology*, Vol. 90,947-957, ISSN: 0022-0477

Harcombe, P. A.; Leipzig, L. E. M. & Elsik, I. S. (2009). Effects of hurricane Rita on three long-term forest study plots in east Texas, USA. *Wetlands*, Vol. 29,88-100, ISSN: 0277-5212

Hibbs, D. E. (1983). 40 years of forest succession in central New England. *Ecology*, Vol. 64,1394-1401, ISSN: 0012-9658

Hoeppner, S. S.; Shaffer, G. P. & Perkins, T. E. (2008). Through droughts and hurricanes: Tree mortality, forest structure, and biomass production in a coastal swamp targeted for restoration in the Mississippi River Deltaic Plain. *Forest Ecology and Management*, Vol. 56,937-948, ISSN: 0378-1127

Hook, D. D., Buford, M. A. & Williams, T. M. (1991), Impact of Hurricane Hugo on the South Carolina coastal plain forest: *Journal of Coastal Research* (Special Issue), Vol. 8, 291-300, ISSN: 0749-0208

Hubbell, S. P.; Foster, R. B.; O'brien S. T.; Harms, K. E.; Condit, R.; Wechsler, B., S.; Wright, J. & De Lao S. L. (1999). Light-gap disturbances, recruitment limitation, and tree diversity in a Neotropical forest. *Science*, Vol. 283,554-557, ISSN: 0036-8075

Imbert, D.; Rousteau, A. & Labbe, P. (1998). Hurricanes and biological diversity in tropical forests - the case of Guadeloupe. *Acta Oecologica*, Vol. 19,251-262, ISSN: 1146-609X

Jarrell, J. D.; Mayfield, M. & Rappaport, E. N. (2001). The deadliest, costliest, and most intense United States hurricanes from 1900 to 2000. NOAA Technical Memorandum NWS TPC-1

Johnsen, K.H.; Butnor, J. R.; Kush, J.S.; Schmidtling, R.C. & Nelson, C.D. (2009). Hurricane Katrina winds damaged longleaf pine less than Loblolly Pine. *Southern Journal of Applied Forestry*, Vol. 33,178-181, ISSN: 0148-4419

Kapustka, L. A. & Koch, R. G. (1979). The acceleration of succession of a birch-maple forest due to high wind. *The Michigan Botanist*, Vol. 18,83-87, ISSN: 0026-203X

Keeland, B.D. & Gorham, L. E. (2009). Delayed tree mortality in the Atchafalaya basin of southern Louisiana following Hurricane Andrew. *Wetlands*, Vol. 29,101-111, ISSN: 0277-5212

Kramer, M. G.; Hansen, A. J.; Taper, M. L. & Kissinger, E. J. (2001). Abiotic controls on long-term windthrow disturbance and temperate rain forest dynamics in southeast Alaska. *Ecology*, Vol. 82,2749-2768, ISSN: 0012-9658

Kulakowski, D. & Veblen, T. T. (2002). Influences of fire history and topography on the pattern of a severe wind blowdown in a Colorado subalpine forest. *Journal of Ecology*, Vol. 90,806-819, ISSN: 0022-0477

Kulakowski, D. & Veblen, T. T. (2003). Subalpine forest development following a blowdown in the Mount Zirkel Wilderness, Colorado. *Journal of Vegetation Science*, Vol. 14,653-660, ISSN: 1100-9233

Kupfer, J. A.; Myers, A. T.; McLane, S. E. & Melton, G. N. (2008). Patterns of forest damage in a southern Mississippi landscape caused by Hurricane Katrina. *Ecosystems*, Vol. 11,45-60, ISSN: 1432-9840

Kwit, C.; Platt, W. J. & Slater, H. H. (2000). Post-hurricane regeneration of pioneer plant species in south Florida subtropical hardwood hammocks. *Biotropica*, Vol. 32,244-251, ISSN: 0006-3606

Levin, S. A. (1992). The problem of pattern and scale in ecology. *Ecology*, Vol. 73,1943-1967, ISSN: 0012-9658

Lindemann, J. D. & Baker, W. L. (2002). Using GIS to analyze a severe forest blowdown in the southern Rocky Mountains. *International Journal of Geographical Information Science*, Vol. 16,377-399, ISSN: 1365-8816

Liu, C. X.; Glitzenstein, J. S.; Harcombe, P. A. & Knox, R. G. (1997). Tornado and fire effects on tree species composition in a savanna in the Big Thicket National Preserve, southeast Texas, USA. *Forest Ecology and Management*, Vol. 91,279-289, ISSN: 0378-1127

Liu, K. B. & Fearn, M. L. (2000). Reconstruction of prehistoric landfall frequencies of catastrophic hurricanes in northwestern Florida from lake sediment records. *Quaternary Research*, Vol. 54,238-245, ISSN: 0033-5894

Lodge, D. J. & Mcdowell, W. H. (1991). Summary of ecosystem-level effects of Caribbean hurricanes. *Biotropica*, Vol. 23,373-378, ISSN: 0006-3606

Loope, L.; Duever, M.; Herndon, A.; Snyder, J. & Jansen, D. (1994). Hurricane impact on uplands and fresh-water swamp forest. *BioScience*, Vol. 44,238-246, ISSN:0006-3568

Lorimer, C. G. (1989). Relative effects of small and large disturbances on temperate hardwood forest structure. *Ecology*, Vol. 70,565-567, ISSN: 0012-9658

Lorimer, C. G. & Frelich, L. E. (1994). Natural disturbance regimes in old-growth northern hardwoods - implications for restoration efforts. *Journal of Forestry*, Vol. 92,33-38, ISSN: 0022-1201

Lugo, A. E. (2000). Effects and outcomes of Caribbean hurricanes in a climate change scenario. *Science of the Total Environment*, Vol. 262,243-251, ISSN: 0048-9697

Lugo, A. E. (2008). Visible and invisible effects of hurricanes on forest ecosystems: an international review. *Austral Ecology*, Vol. 33,368-398, ISSN: 1442-9985

Lugo, A. E. & Waide, R. B. (1993). Catastrophic and background disturbance of tropical ecosystems at the Luquillo Experimental Forest. *Journal of the Biosciences*, Vol. 18,475-481, ISSN: 0250-5991

Martin, T. J. & Ogden, J. (2006). Wind damage and response in New Zealand forests: A review. *New Zealand Journal of Ecology*, Vol. 30,295-310, ISSN: 01106465

McMaster, K. J. (2005). Forest blowdown prediction: A correlation of remotely sensed contributing factors. *Northern Journal Applied Forestry*, Vol. 22,48-53, ISSN: 0742-6348

McDonald, R. I., Peet, R. K. & Urban, D. L. 2003. Spatial pattern of *Quercus* regeneration limitation and *Acer rubrum* invasion in a piedmont forest. *Journal of Vegetation Science*, Vol. 14,441-450, ISSN: 1100-9233

Merrens, E. J. & Peart, D. R. (1992). Effects of hurricane damage on individual growth and stand structure in a hardwood forest in New Hampshire, USA. *Journal of Ecology*, Vol. 80,787-795, ISSN: 0022-0477

Mitchell, S. J. (1995). The windthrow triangle - a relative windthrow hazard assessment procedure for forest managers. *Forestry Chronicle*, Vol. 71,446-450, ISSN: 0015-7546

Myers, R. K. & Van Lear, D. H. (1998). Hurricane-fire interactions in coastal forests of the south: a review and hypothesis. *Forest Ecology and Management*, Vol. 103,265-276, ISSN: 0378-1127

Nobel, P. S. (1981). Wind as an ecological factor, In: Lange, O. L., Nobel, P. S.; Osmond, C. B. & Ziegler, H (Eds.), *Physiological plant ecology I: Responses to the physical environment*, Encyclopedia of plant physiology new series vol. 12A, pp. 475–500, Springer-Verlag, ISBN: 0387107630, Berlin, Germany

Oswalt, S. N. & Oswalt, C. M. (2008). Relationships between common forest metrics and realized impacts of Hurricane Katrina on forest resources in Mississippi. *Forest Ecology and Management*, Vol. 255,1692-1700, ISSN: 0378-1127

Palmer, M. W.; Mcalister, S. D.; Arevalo, J. R. & DeCoster, J. K. (2000). Changes in the understory during 14 years following catastrophic windthrow in two Minnesota forests. *Journal of Vegetation Science*, Vol. 11,841-854, ISSN: 1100-9233

Palmer, M. W.; Peet, R. K.; Reed, R. A.; Xi, W. & White, P. S. (2007). A multiscale study of vascular plants in a North Carolina Piedmont forest. *Ecology*, Vol. 88, 2674, ISSN: 0012-9658

Pascarella, J. B., Aide, T. M. & Zimmerman, J. K. (2004). Short-term response of secondary forests to hurricane disturbance in Puerto Rico, USA. *Forest Ecology and Management*, Vol. 199,379-393, ISSN: 0378-1127

Peart, D. R.; Cogbill, C. V. & Palmiotto, P. A. (1992). Effects of logging history and hurricane damage on canopy structure in a northern hardwoods forest. *Bulletin of the Torrey Botanical Club*, Vol. 119,29-38, ISSN: 0040-9618

Peet, R. K. & Christensen, N. L. (1980). Succession - a population process. *Vegetatio*, Vol. 43,131-140, ISSN: 1385-0237

Peet, R. K. & Christensen, N. L. (1987). Competition and tree death. *BioScience*, Vol. 37,586-595, ISSN: 0006-3568

Peltola, H. (1996). Model computations on the wind flow and turning moment for Scots pine along the margins of clearcut areas. *Forest Ecology and Management*, Vol. 83, 203-215, ISSN: 0378-1127

Peltola, H. (2006). Mechanical stability of trees under static loads. *American Journal of Botany*, Vol. 93:1501-1511, ISSN: 0002-9122

Peterson, C. J. (2000). Catastrophic wind damage to North American forests and the potential impact of climate change. *Science of the Total Environment*, Vol. 262,287-311, ISSN: 0048-9697

Peterson, C. J. (2000). Damage and recovery of tree species after two different tornadoes in the same old growth forest: a comparison of infrequent wind disturbances. *Forest Ecology and Management*, Vol. 35,237-252, ISSN: 0378-1127

Peterson, C. J. (2004). Within-stand variation in windthrow in southern boreal forests of Minnesota: is it predictable? *Canadian Journal of Forest Research*, Vol. 34,365-375, ISSN: 0045-5067

Peterson, C. J. (2007). Consistent influence of tree diameter and species on damage in nine Eastern North America tornado blowdowns. *Forest Ecology and Management*, Vol. 250, 96-108, ISSN: 0378-1127

Peterson, C. J. & Rebertus, A. J. (1997). Tornado damage and initial recovery in three adjacent, lowland temperate forests in Missouri. *Journal of Vegetation Science*, Vol. 8,559-564, ISSN: 1100-9233

Peterson, C. J. & Pickett, S. T. A. (1990). Microsite and elevational influences on early forest regeneration after catastrophic windthrow. *Journal of Vegetation Science*, Vol. 1,657-662, ISSN: 1100-9233

Peterson, C. J. & Pickett, S. T. A. (1991). Treefall and resprouting following catastrophic windthrow in an old-growth hemlock hardwoods forest. *Forest Ecology and Management*, Vol. 42,205-217, ISSN: 0378-1127

Peterson, C. J. & Pickett, S. T. A. (2000). Patch type influences on regeneration in a western Pennsylvania, USA, catastrophic windthrow. *Oikos*, Vol. 90,489-500, ISSN: 0030-1299

Peterson, C. J. & Leach, A.D. (2008a). Limited salvage logging effects on forest regeneration after moderate-severity windthrow. *Ecological Applications*, Vol. 18, No. 2, 407-420, ISSN: 1051-0761

Peterson, C. J. & Leach, A.D. (2008b). Salvage logging after windthrow alters microsite abundance and environment, but not vegetation. *Forestry*, Vol. 81, No. 3, 361-376, ISSN: 0015-752X

Pimm, S. L.; Davis, G. E.; Loope, L.; Roman, C. T.; Smith, T. J. & Tilmant, J. T. (1994). Hurricane Andrew. *BioScience*, Vol. 44,224-229, ISSN:0006-3568

Platt, W. J.; Beckage, B., Doren, R. F. & Slater, H. H. (2002). Interactions of large-scale disturbances: prior fire regimes and hurricane mortality of savanna pines. *Ecology*, Vol. 83,1566-1572, ISSN: 0012-9658

Platt, W. J.; Doren, R. F. & Armentano, T. V. (2000). Effects of hurricane Andrew on stands of slash pine (*Pinus elliottii var. densa*) in the everglades region of south Florida (USA). *Plant Ecology*, Vol. 146,43-60, ISSN: 1385-0237

Pleshikov, F. I.; Ryzkova, V. A.; Kaplunov, V. Y. & Usoltseva, J. V. (1998). A computer system for evaluating and predicting hurricane impact on forest. *Safety Science*, Vol. 30,3-8, ISSN: 0925-7535

Prengaman, K.A.; Kribel, J. R. G. & Ware, S. (2008). Effects of Hurricane Isabel on a maturing hardwood forest in the Virginia Coastal Plain. *Journal of the Torrey Botanical Society*, Vol. 135,360-366, ISSN: 1095-5674

Putz, F. E., Coley, P. D.; Lu, K.; Montalvo, A. & Aiello, A. (1983). Uprooting and snapping of trees - structural determinants and ecological consequences. *Canadian Journal of Forest Research*, Vol. 13, 1011-1020, ISSN: 0045-5067

Putz, F. E. & Sharitz, R. R. (1991). Hurricane damage to old-growth forest in Congaree Swamp National Monument, South Carolina, USA. *Canadian Journal of Forest Research*, Vol. 21,1765-1770, ISSN: 0045-5067

Reice, S. R. (1994). Nonequilibrium determinants of biological community structure. *American Scientist*, Vol. 82, 424-435, ISSN: 0003-0996

Reice, S. R. (2001). *The silver lining: the benefits of natural disasters*. Princeton University Press, Princeton, ISBN13: 978-0-691-11368-5, New Jersey, USA

Ruel, J. C. (1995). Understanding windthrow - silvicultural implications. *Forestry Chronicle*, Vol. 71,434-445, ISSN: 1499-9315

Ruel, J. C. (2000). Factors influencing windthrow in balsam fir forests: from landscape studies to individual tree studies. *Forest Ecology and Management*, Vol. 135,169-178, ISSN: 0378-1127

Ruel, J. C.; Mitchell, S. J. & Dornier, M. (2002). A GIS based approach to map wind exposure for windthrow hazard rating. *Northern Journal of Applied Forestry*, Vol. 19,183-187, ISSN: 0742-6348

Ruel, J. C. & Pineau, M. (2002). Windthrow as an important process for white spruce regeneration. *Forestry Chronicle*, Vol. 78,732-738, ISSN: 1499-9315

Runkle, J. R. (1982). Patterns of disturbance in some old-growth mesic forests of eastern North America. *Ecology*, Vol. 63,1533-1546, ISSN: 0012-9658

Runkle, J. R. (1985). Disturbance regimes in temperate forests, In Pickett, S. T. A. & White, P. S. (Eds.). *The Ecology of Natural Disturbance and Patch Dynamics*, pp. 17-33, Accademic Press, ISBN-10: 0125545215, Orlando, Florida. USA

Schaetzl, R. J.; Burns, S. F.; Johnson, D. L. & Small, T. W. (1989). Tree uprooting - review of impacts on forest ecology. *Vegetatio*, Vol. 79,165-176, ISSN: 1385-0237

Schaetzl, R. J.; Burns, S. F.; Small, T. W. & Johnson, D. L. (1990). Tree uprooting - review of types and patterns of soil disturbance. *Physical Geography*, Vol. 11,277-291, ISSN: 0272-3646

Schulte, L. A. & Mladenoff, D. J. (2005). Severe wind and fire regimes in northern forests: historical variability at the regional scale. *Ecology*, Vol. 86,431-445, ISSN: 0012-9658

Sharitz, R.R.; Vaitkus, M. R. & Cook, A. E. (1992). Hurricane damage to an old-growth floodplain forest in the southeast, *Proceddings of the Seventh Biennial Southern Silvicultural Research Conference*, pp. 203-210, Mobile, Alabama, USA

Schumacher, S.; Bugmann, H. & Mladenoff, D.J. (2004). Improving the formulation of tree growth and succession in a spatially explicit landscape model. *Ecological Modelling*, Vol. 180,175-194, ISSN: 0304-3800

Seischab, F. K. & Orwig, D. (1991). Catastrophic disturbances in the presettlement forests of western New York. *Bulletin of the Torrey Botanical Club*, Vol. 118,117-122, ISSN: 0040-9618

Sheffield, R. M. & Thompson, M. T. (1992). Hurricane Hugo: effects on South Carolina's forest resource. Res. Pap. SE284. Asheville, NC: U.S. Department of Agriculture, Forest Service, Southeastern Forest Experiment Station

Sinton, D. S. & Jones, J. A. (2002). Extreme winds and windthrow in the Western Columbia River Gorge. *Northwest Science*, Vol. 76,173-182, ISSN: 0029-344X

Sinton, D. S.; Jones, J. A.; Ohmann, J. L. & Swanson, F. J. (2000). Windthrow disturbance, forest composition, and structure in the Bull Run basin, Oregon. *Ecology*, Vol. 81,2539-2556, ISSN: 0012-9658

Slater, H.; Platt W. J.; Baker, D. B. & Johnson, H. A. (1995). Effects of Hurricane Andrew on damage and mortality of trees in subtropical hardwood hammocks on long pine key, everglades national park, Florida, USA. *Journal of Coastal Research*, Vol. 21,197-207, ISSN: 0749-0208

Smith, G. F.; Nicholas, N. S. & Zedaker, S. M. (1997). Succession dynamics in a maritime forest following Hurricane Hugo and fuel reduction burns. *Forest Ecology and Management*, Vol. 95,275-283, ISSN: 0378-1127

Sousa, W. P. (1984). The role of disturbance in natural communities. *Annual Review of Ecology, Evolution, and Systematics*, Vol. 15,353-391, ISSN: 0066-4162

Spurr, S. H. (1956). Natural restocking of forests following the 1938 hurricane in central New England. *Ecology*, Vol. 37,443- 451, ISSN: 0012-9658

Stanturf, J.A.; Goodrick, S. L. & Outcalt, K.W. (2007). Disturbance and coastal forests: A strategic approach to forest management in hurricane impact zones. *Forest Ecology and Management*, Vol. 250,119-135, ISSN: 0378-1127

Turner, M. G.; Baker, W. L.; Peterson, C. J. & Peet, R. K. (1998). Factors influencing succession: lessons from large, infrequent natural disturbances. *Ecosystems*, Vol. 1,511-523, ISSN: 1432-9840

Turner, M. G.; Dale, V. H. & Everham, E. H. (1997). Fires, hurricanes, and volcanoes: comparing large disturbances. *BioScience*, Vol. 47,758-768, ISSN:0006-3568

Turton, S. M. (1992). Understorey light environments in a north-east Australian rain-Forest before and after a tropical cyclone. *Journal of Tropical Ecology*, Vol. 8,241-252, ISSN: 0266-4674

Vandermeer, J., De La Cerda, I. G.; Boucher, D.; Perfecto, I. & Ruiz, J. (2000). Hurricane disturbance and tropical tree species diversity. *Science*, Vol. 290,788-791, ISSN: 0036-8075

Vandermeer, J.; Delacerda, I. G. & Boucher, D. (1997). Contrasting growth rate patterns in eighteen tree species from a post-hurricane forest in Nicaragua. *Biotropica*, Vol. 29,151-161, ISSN: 0006-3606

Vandermeer, J.; Mallona, M. A.; Boucher, D.; Yih, K. & Perfecto, I. (1995). 3 years of ingrowth following catastrophic hurricane damage on the Caribbean coast of Nicaragua - evidence in support of the direct regeneration hypothesis. *Journal of Tropical Ecology*, Vol. 11,465-471, ISSN: 0266-4674

Veblen, T. T.; Kulakowski, D.; Eisenhart, K. S. & Baker, W. L. (2001). Subalpine forest damage from a severe windstorm in northern Colorado. *Canadian Journal of Forest Research*, Vol. 31, 2089-2097, ISSN: 0045-5067

Veblen, T. T.; Hadley, K. S.; Reid, M. S. & Rebertus, A. J. (1989). Blowdown and stand development in a Colorado subalpine forest. *Canadian Journal of Forest Research*, Vol. 19, 1218-1225, ISSN: 0045-5067

Vogel S. (1996). Blowing in the wind: storm-resisting features of the design of trees. *Journal of Arboriculture*, Vol. 22, 92–98, ISSN: 02785226

Vogel, S. (2009). Leaves in the lowest and highest winds: temperature, force and shape. *New Phytologist*, Vol. 183, No. 1, 13-26, ISSN: 0028-646X

Uriarte, M.; Canham, C. D.; Thompson, J. & Zimmerman, J. K. (2004). A neighborhood analysis of tree growth and survival in a hurricane-driven tropical forest. *Ecological Monographs*, Vol. 74, No. 4, 591-614, ISSN: 0012-9615

Uriarte, M. & Papaik, M. (2007). Hurricane impacts on dynamics, structure and carbon sequestration potential of forest ecosystems in Southern New England, USA. *Tellus Series A-Dynamic Meteorology and Oceanography*, Vol. 59,519-528, ISSN: 0280-6495

Xi, W.; Peet, R. K. & Urban, D. L. (2002). Immediate impacts of Hurricane Fran on the structure of a North Carolina Piedmont forest. *Southeastern Biology*, Vol. 49, 162, ISSN: 1533-8436

Xi, W. (2005). *Forest response to natural disturbance: Changes in structure and diversity on a North Carolina piedmont forest in response to catastrophic wind events*. Ph.D. Dissertation. The University of North Carolina at Chapel Hill, Chapel Hill, North Carolina, USA

Xi, W.; Peet, R. K.; DeCoster, J. K. & Urban, D. L. (2008a). Tree damage risk factors associated with large, infrequent wind disturbances of Carolina forests. *Forestry*, Vol. 81, No. 3,317-334, ISSN: 0015-752X

Xi, W.; Peet, R. K. & Urban, D. L. (2008b). Changes in forest structure, species diversity, and spatial pattern following hurricane disturbance in a Piedmont North Carolina forest, USA. *Journal of Plant Ecology*, Vol. 1, No. 1, 43-57, ISSN: 1752-9921

Xi, W. & Peet, R. K. (2008a). Hurricane effects on the Piedmont forests: Patterns and implications. *Ecological Restoration*, Vol. 26, No. 4, 295-298, ISSN: 1543-4060

Xi, W. & Peet, R. K. (2008b). Long-Term Forest Dynamics and Ecological Monitoring Study in the Duke Forest of the Eastern United States: A Synthesis. *Chinese Journal of Plant Ecololgy*, Vol. 32, No. 2,299-318, ISSN: 1005-264XCN 11-3397/Q

Walker, L. R. (1991). Tree damage and recovery from Hurricane Hugo in Luquillo Experimental Forest, Puerto Rico. *Biotropica*, Vol. 23,379-385, ISSN: 0006-3606

Walker, L. R. (1995). Timing of post-hurricane tree mortality in Puerto Rico. *Journal of Tropical Ecology*, Vol. 11,315-320, ISSN: 0266-4674

Walker, L. R. (2000). Seedling and sapling dynamics in treefall pits in Puerto Rico. *Biotropica*, Vol. 32,262-275, ISSN: 0006-3606

Walker, L. R.; Voltzow, J.; Ackerman, J. D.; Fernandez, D. S. & Fetcher, N. (1992). Immediate impact of Hurricane Hugo on a Puerto-Rican rain-forest. *Ecology*, Vol. 73,691-694, ISSN: 0012-9658

Waring, R. H. & Schlesinger, W. H. (1985). *Forest ecosystems: concepts and management*. Academic Press, ISBN-10: 0127354409, Orlando, Florida, USA

Webb, S. L. (1988). Windstorm damage and microsite colonization in 2 Minnesota forests. *Canadian Journal of Forest Research*, Vol. 18, 1186-1195, ISSN: 0045-5067

Webb, S. L. (1989). Contrasting windstorm consequences in 2 forests, Itasca State Park, Minnesota. *Ecology*, Vol. 70,1167-1180, ISSN: 0012-9658

Webb, S. L. (1999). Wind disturbances in temperate forests, In: Walker, L. R. (Ed.), *Ecosystems of disturbed ground,* pp. 187-222. Elsevier, ISBN-10: 0444824200, Amsterdam, The Nethelands

Webb, S. L. & Scanga, S. E. (2001). Windstorm disturbance without patch dynamics: twelve years of change in a Minnesota forest. *Ecology,* Vol. 82,893-897, ISSN: 0012-9658

Whigham, D. F.; Dickinson, M. B. & Brokaw, N. V. L. (1999). Background canopy gap and catastrophic wind distubance in tropical forests, In: Walker, L. R. (Ed.), *Ecosystems of Disturbed Ground,* pp. 223-252. Elsevier, ISBN-10: 0444824200, Amsterdam, The Nethelands

White, P. S. (1979). Pattern, process, and natural disturbance in vegetation. *Botanical Review,* Vol. 45,229-299, ISSN: 0006-8101

White, P. S. & Jentsch, A. (2001). The search for generality in studies of disturbance and ecosystem dynamics. *Progress in Botany,* Vol. 62,399-450, ISSN: 0340-4773

White, P. S. & Jentsch, A. (2004). Disturbance, succession, and community assembly in terrestrial plant communities. In: *Assembly Rules and Restoration Ecology,* Temperton, V.; Hobbs, R. & Halle, S. (Eds.), pp. 342-366, Island Press, ISBN-10: 1559633751, California. USA

White, R. D. (1999). *The impacts of hurricane Fran on a North Carolina piedmont woodland.* Master thesis. University of North Carolina at Chapel Hill, Chapel Hill, North Carolina, USA

Wilkinson, R. C.; Britt, R.W., Spence, E.A. & Seiber, S.M. (1978). Hurricane-tornado damage, mortality, and insect infestations of slash pine. *Southern Journal of Applied Forestry,* Vol. 2,132-134, ISSN: 0148-4419

Zeng, H.C.; Chambers J.Q.; Negron-Juarez, R.I.; Hurtt, G.C.; Baker, D.B.; & Powell, M.D. (2009). Impacts of tropical cyclones on U.S. forest tree mortality and carbon flux from 1851 to 2000. *Proceedings of the National Academy of Sciences,* Vol. 106, ISSN: 7888-7892

Zhao, D.H.; Allen, B.; Sharitz, R. R. (2006). Twelve year response of old-growth southeastern bottomland hardwood forests to disturbance from Hurricane Hugo. *Canadian Journal of Forest Research,* Vol. 36,3136-3147, ISSN: 0045-5067

Zimmerman, J. K.; Aide, T. M.; Rosario, M.; Serrano, M. & Herrera, L. (1995). Effects of land management and a recent hurricane on forest structure and composition in the Luquillo Experimental Forest, Puerto Rico. *Forest Ecology and Management,* Vol. 77, 65-76, ISSN: 0378-1127

Zhang, Q.; Pregitzer,K.S. & Reed, D. D. (1999). Catastrophic disturbance in the presettlement forests of the Upper Peninsula of Michigan. *Canadian Journal of Forest Research,* Vol. 29,106-114, ISSN: 0045-5067

28

Hurricane Georges Accelerated Litterfall Fluxes of a 26 yr-old Novel Secondary Forest in Puerto Rico

Ariel E. Lugo[1], Carlos M. Domínguez Cristóbal[1], and Noemí Méndez[2]
[1]*International Institute of Tropical Forestry, USDA Forest Service Ceiba 1201, Jardín Botánico Sur, Río Piedras, PR 00926-1115*
[2]*Distrito Escolar de Utuado, Escuela Superior Vocacional Luis Muñoz Rivera Utuado, Puerto Rico*

1. Introduction

Defoliation and fallen debris are common visible effects of hurricanes and are accompanied by invisible effects such as altered element fluxes, changes in element concentration of plant parts, and variability of processes (Lugo 2008). While ecologists are increasingly documenting visible effects of hurricanes and other disturbances (Everham and Brokaw 1996, Lugo 2008, Turton 2008) less work has been conducted on the invisible effects of hurricanes, which center on the functional attributes of forests. One such functional attribute with both visible and invisible components is litterfall and associated mass and element fluxes. Examples of studies of tropical forests where litterfall mass and element fluxes were monitored before and after a hurricane include work in Puerto Rico (Lodge et al. 1991, Frangi and Lugo 1991, Scatena et al. 1996, Beard et al. 2005, Van Bloem et al. 2005), Hawaii (Herbert et al. 1999), and Mexico (Whigham et al. 1991). These studies cover a range of climates (dry, moist, and wet) and focused on mature native vegetation. To avoid repetitive citations to these papers, we will refer to them by country when the information cited is supported by all of them.

The expansion of human populations and changes in land cover due to deforestation, use, and land abandonment, cause increases in the cover of tropical secondary forests (Brown and Lugo 1990) and there is a need to learn about their functioning and resilience (Lugo and Helmer 2004, Chazdon 2008, Lugo 2009). We had an opportunity to contribute to this research need when Hurricane Georges passed over a study area (Fig. 1) where we were monitoring litterfall and element fluxes of a secondary forest with a novel composition of introduced and native tree species (below). The hurricane moved at 24 km/h over Puerto Rico between 21 and 22 September 1998 as a Category 3 hurricane in the Saffir Simpson scale with maximum sustained winds of 184 km/h and occasional gusts of 240 km/h.

We addressed two questions: How resilient to hurricane winds are the fluxes of litterfall mass and chemical elements in a young secondary forest and how do resilience and fluxes in this forest type compare with those of mature forest vegetation in dry, moist, and wet climates? The comparison with mature forests sheds light on the effects of age in forest response to hurricane winds. Moreover, since our study forest had a mixed composition of introduced and native species and the mature ones were dominated by native species, the comparison adds value to the question of the effects of introduced species on ecosystem functioning.

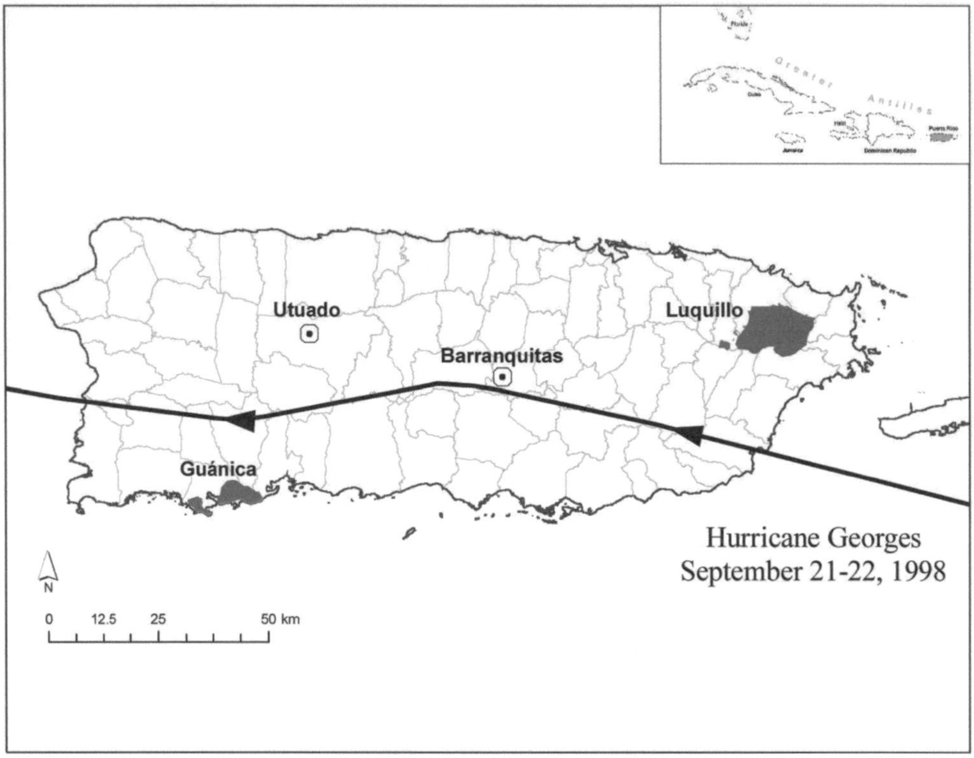

Fig. 1. Location of the study site in relation to the path of Hurricane Georges.

2. Methods

Study area - The study area is located at 416 m elevation in the Jácanas sector, Caguana ward, Utuado, Puerto Rico (lat 18°17’07.995” N, long 66°43’57.446”W NAD 83 (CONUS) CORS96). The site is in the subtropical warm wet moist forest life zone (*sensu* Holdridge 1967) with an annual rainfall of 2331 mm (NOAA 1998). The soil was classified as an Inceptisol and is now classified as a vertic Eutopeps, specifically Múcara clay (Lugo López 1977). The Múcara soil group is a fine, montmorillonitric soil. The Múcara clay sub order is moderately deep, well drained, and occurs on side slopes of humid volcanic uplands. The parent material at the site is non-carbonate sedimentary rock, adjacent to intrusive rocks.

The site was used for coffee production between 1910 and January 1972. Thus, the forest was 26 yr-old when the hurricane passed, although some of the canopy trees were older as they provided shade for coffee trees during the period of active production. Before the hurricane, the stand had tall coffee shade canopy trees (mean of the tallest three trees was 26 m) dominated by native (*Miconia prasina, Schefflera morototoni*) and naturalized (*Syzygium jambos*) tree species. Average tree height was 9.4 $\pm$ 0.4 m, n = 168 (Popper *et al.* 1999). *Syzygium jambos, Miconia prasina,* and *Coffea arabica* (also a naturalized species) dominated the understory. There were 15 to 20 (three years before and four years after the hurricane, respectively) canopy tree species, 18 to 23 understory tree species, and 26 seedling and sapling species in 0.1 ha (21 m^2 for the seedling and sapling species count). Tree basal area ranged from 20 to 25 m^2/ha (Lugo et al. 2005). Because of the high Importance Value of introduced and naturalized species (23 percent for canopy trees, 67 percent for understory trees, and 34 percent for seedlings and saplings), the forest is an example of new or novel forests on abandoned agricultural lands (*sensu* Lugo and Helmer 2004, Hobbs et al. 2006). These forests are highly fragmented over the landscape, with most fragments being < 1 ha (Lugo 2002).

A grid of 10 m x 10 m plots was overlaid over a 50 m x 20 m forest area and on February 25, 1998, we randomly placed two 50 cm x 50 cm wire baskets with plastic lining in each plot to collect litterfall. Our measurements do not estimate coarse wood fall (macro litter), which requires larger plots (Newbould 1967). The 20 baskets were set above the soil and emptied every two weeks over a period of two years (February 25, 1998 when baskets were placed in the forest to February 15, 2000, the last time baskets were emptied). The sampling consisted of 47 biweekly collections, 20 baskets per collection. For each collection we averaged the weights of each of the 20 baskets. However, for the statistical analysis we used all baskets or 940 observations.

The collection interval was affected by the hurricane, as we could not gain access to the site for 61 days, *i.e.*, between the last pre-hurricane collection of September 10, 1998 and the hurricane collection of November 10, 1998. Fortunately, all baskets were intact and collected at that time. We estimated the instantaneous hurricane-peak fluxes of mass and elements by subtracting from the hurricane collection the litter that fell previous to September 21 and after September 22, and assumed the site was exposed for 24 hr to hurricane winds. We based this assumption on the direct observations of Van Bloem *et al.* (2005) at Guánica Forest just south of our site. For the material that fell before September 21, we extrapolated the rate of the last pre-hurricane collection by 11 days. Similarly, we estimated the post-hurricane fall by extrapolating the first post-hurricane rate between November 10 and December 1, 1998 by 49 days (the time between September 23 and November 10, 1998).

Samples were sorted in the laboratory into leaves, wood ≤ 2 cm diameter, wood > 2 cm diameter, flowers and fruits, and miscellaneous. Sorted litter was oven-dried to constant weight at 65 °C and ground through an 18-mesh (1 mm) sieve with a Wiley Mill. All ground litter samples were analyzed for P, K, Ca, Mg, Al, Mn, and Fe with a Beckman plasma emission spectrometer (Spectra Span V). Samples were digested with concentrated HNO_3 and 30 percent H_2O_2 using the digestion method recommended by Luh Huang and Schulte (1985). Ash was determined by igniting 1 g samples in a muffle furnace at 490 °C for at least 10 hr (Wilde *et al.* 1979). Total N, total C, and total S were analyzed using the dry combustion method in LECO Corp. (1995) by means of a LECO CNS-2000 analyzer. In this method, a sample of known weight is combusted by heating it to a high temperature (1300 °C) inside a resistance furnace and in a stream of purified oxygen. Precision for all analyses was assured by running samples of known chemical composition every forty determinations. These control samples (citrus leaves [NBS-1572], peach leaves [NIST-1547] and pine needles [NIST-1575]) were obtained from the National Institute of Standards and Technology, USA. The calibration standards (tobacco leaves, orchard leaves, and alfalfa) used in the total C, N, and S analysis were obtained from LECO Corporation in St. Joseph, MI.

Our chemical analysis of 91 litterfall samples were distributed as follows: 22 leaf fall, 20 wood fall with diameter ≤ 2 cm, 11 wood fall with diameter > 2 cm, 21 flowers and fruits fall, and 17 miscellaneous fall. The number of samples depended on the availability of sufficient material for all chemical determinations. For the purpose of estimating element return, we combined samples with low amounts for analysis such that we analyzed litterfall samples representing all the months between the first collection in March 10, 1998 and September 21, 1999. This pooling of samples corresponded to 38 of the 47 biweekly collection of litterfall. We did not have chemical analyses for litterfall samples collected after September 21, 1999. For the estimate of element return in these remaining 9 biweekly collections of litterfall (those collected after September 21, 1999), we used the mean element concentration of all analyses for the particular element and litterfall component.

We converted element concentration data into units of element mass per ground area by multiplying element concentration by the corresponding mass of litterfall. The annual weighted element concentration was estimated by dividing the annual element mass return in a litterfall compartment by the annual mass fall of that compartment. Within-stand nutrient use-efficiency is the inverse of the weighted nutrient concentration of litterfall (Vitousek 1982). To estimate the C/N and N/P element molar ratios, we divided the concentrations of C, N, and P by 12, 14, and 30.97, respectively, to express them in molar units.

Litterfall mass and element mass fluxes, and element concentration data were tested for normality and homogeneity of variance prior to further statistical analysis. We used the Shapiro-Wilk W test for goodness of fit of the normal distribution and four tests (O'Brien [.5], Brown-Forsythe, Levene, and Bartlett) for equal variances. When the assumptions of normality and equal variances were met, we used Oneway ANOVA to determine significant difference at $p < .05$.

A repeated measures analysis of variance (ANOVA) with a generalized linear model (GLM) procedure was used for the analysis of mean differences of litterfall rates (total, wood, leaf) among collection periods (using all 940 individual basket observations coded for date of collection). When significance differences were found, the post hoc Ryan-Einot-Gabriel-Welsch multiple range (REGWQ) test was used to compare means among collection periods. These statistics were done with SAS Version 9.1, SAS Institute 2003 at alpha <.05. We also used version 7.0 of the SAS Institute Inc. (2007) JMP program for Macintosh for calculating descriptive statistics.

3. Results

Litterfall Mass—Total litterfall rates spanned four orders of magnitude, and two large pulses dominated the pattern of biweekly litterfall (Fig. 2a). Peak rates of litterfall occurred in the hurricane collection and on the December 15, 1998 collection (post-hurricane fall). These two rates were the only significantly different rates among the 47 bi-weekly collections. Two weeks after the hurricane collection and one month after the second pulse, total litterfall rate reached a low value. After the second low value, litterfall rates increased until the end of the study. Leaf fall was 57 percent of total litterfall until September 10, 1998 when the proportion changed and wood fall became the highest fraction of litterfall (77 percent). Leaf fall followed a similar temporal pattern to that of total litterfall with the exception that leaf fall did not exhibit a post-hurricane pulse (Fig. 2a). After the hurricane collection, leaf fall rate decreased to low values and then increased steadily for a year. Both the hurricane pulse and the low value after the pulse were statistically different from other leaf fall rates in the record. The proportion of leaf fall increased rapidly after the hurricane from a low of 16 percent reaching 86 percent of total litterfall at the end of the study.

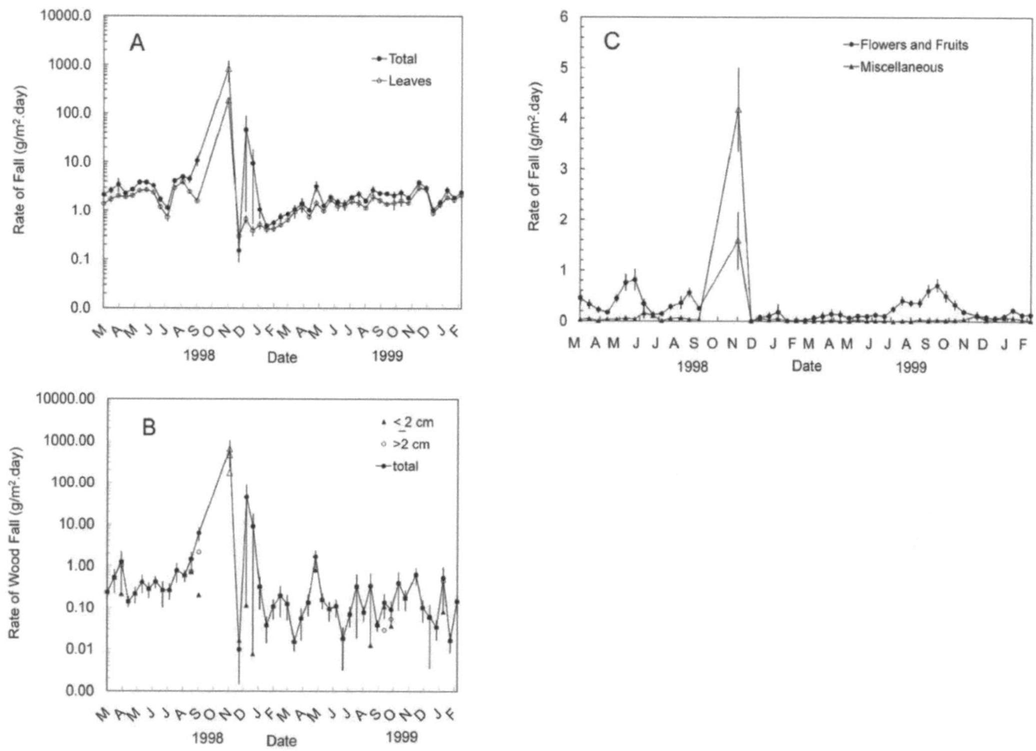

Fig. 2. Leaf and total litterfall (A), wood fall (B) and flowers and fruit fall and miscellaneous fall (C) in a secondary forest at Utuado, Puerto Rico. Vertical lines are 1 standard error of the mean, n = 20. Note the logarithmic scale for A and B. The datum reflecting peak hurricane fall is shown with a large open triangle. The line in 2B connects total wood fall.

Wood fall ranged five orders of magnitude with significantly high pulses in the hurricane collection and the December 15, 1998 collection (Fig. 2b). One month after the post-hurricane pulse, wood fall rates were not different from pre-hurricane rates. Most of the peak wood fall fluxes were due to wood > 2 cm. Flowers and fruits fall had four significant pulses: two pre-hurricane, the hurricane flux, and a post-hurricane pulse a year later (Fig. 2c). Flowers and fruits fall rates decreased after the hurricane and were different from a peak flowers and fruits fall eleven months after the hurricane. Miscellaneous fall was low throughout the study period but had a significant pulse in the hurricane collection and quickly returned to pre-hurricane levels (Fig. 2c).

The annual quantity of total, leaf, and wood fall combined was four times higher in the first year of measurement (February 25, 1998 to February 23, 1999), which included the hurricane, than in the second year (February 23, 1999 to February 15, 2000) (Table 1). The first year, 69 percent of the litterfall was wood fall, while leaf fall was 76 percent of the total litterfall in the second year. For the second year of study, flowers and fruits comprised 11 percent of the total annual litterfall.

		Mass	N	P	K	Ca	Mg	Mn	Al	S	Fe	Ash
Leaves												
	98-99	8.7	181	4	33	55	19	6	24	27	2	444
	99-00	5.5	105	3	29	38	13	4	11	15	1	287
Wood												
	98-99	22.0	259	8	62	229	35	12	152	33	2	1045
	99-00	0.8	11	0.2	2	5	1	1	2	1	0.1	25
Flowers and fruits												
	98-99	0.9	20	0.8	5	3	1	0.3	2	2	0.2	28
	99-00	0.8	17	0.8	4	3	1	0.4	1	2	0.2	26
Miscellaneous												
	98-99	0.2	0.2	0.0	0.1	0.1	0.0	0.0	0.0	0.0	0.0	1
	99-00	0.1	0.1	0.0	0.0	0.1	0.0	0.0	0.0	0.0	0.0	0.5
Total												
	98-99	31.9	459	13	100	287	55	18	178	62	3	1518
	99-00	7.2	133	4	35	46	15	5	13	18	1	338

Table 1. Annual litterfall (mass in Mg/ha) and element return (kg/ha) in litterfall for a secondary forest in Utuado, Puerto Rico. The time interval is from March to February for both years (1998 to 1999 and 1999 to 2000). Hurricane Georges struck in the 1998 to 1999 interval. Element return columns might not add to the total because of rounding to the nearest kg or 0.1 kg.

Element Concentrations and Molar Ratios—Leaf fall had the highest concentrations of Ca, Mg, S, and ash among litterfall components (Table 2). Flowers and fruits fall had the highest concentrations of N, P, K, Fe, and C among litterfall components. Wood fall had the highest C/N, and with leaf fall, shared the highest N/P and Al and Mn concentrations. Wood fall had the lowest concentrations of N, P, K, Mg, S, and ash among litterfall components (Table 2).

Element or Combination	Leaf	Wood < 2cm	Flowers and Fruits
Nitrogen	19.5 (0.5)	13.2 (0.6)	22.4 (0.8)
Phosphorus	0.48 (0.04)	0.32 (0.02)	1.04 (0.06)
Potassium	5.0 (0.3)	2.6 (0.1)	6.1 (0.4)
Calcium	6.5 (0.2)	5.6 (0.5)	4.2 (0.4)
Magnesium	2.3 (0.1)	1.6 (0.1)	1.9 (0.1)
Sulphur	2.9 (0.1)	1.7 (0.1)	2.4 (0.1)
Aluminum	2.5 (0.4)	2.2 (0.5)	1.4 (0.3)
Manganese	0.70 (0.04)	0.70 (0.01)	0.46 (0.05)
Iron	0.13 (0.01)	0.09 (0.01)	0.33 (0.05)
Carbon	530 (2)	527 (3)	542 (4)
Ash	52 (1)	31 (2)	38 (3)
C/N	32 (1)	49 (2)	29 (1)
N/P	96 (5)	101 (8)	49 (2)

Table 2. Mean element and ash concentrations (mg/g), and mean molar ratios of litterfall components of a novel secondary forest in Utuado, Puerto Rico. Standard error is in parenthesis and the n values were 19, 19, and 21 for leaf, wood, and flowers and fruits, respectively. Results are rounded.

Except for two instances, wood fall and flowers and fruits fall did not exhibit differences in the temporal pattern of element concentration, but for leaf fall, there were several significant patterns of temporal change in element concentration (Fig. 3). The two exceptions were the pattern for changes in P concentration in flowers and fruits fall, which was similar to the one in leaf fall (Fig. 3b) and the pattern of N/P in wood fall, which was similar to that of leaf fall (Fig. 3l).

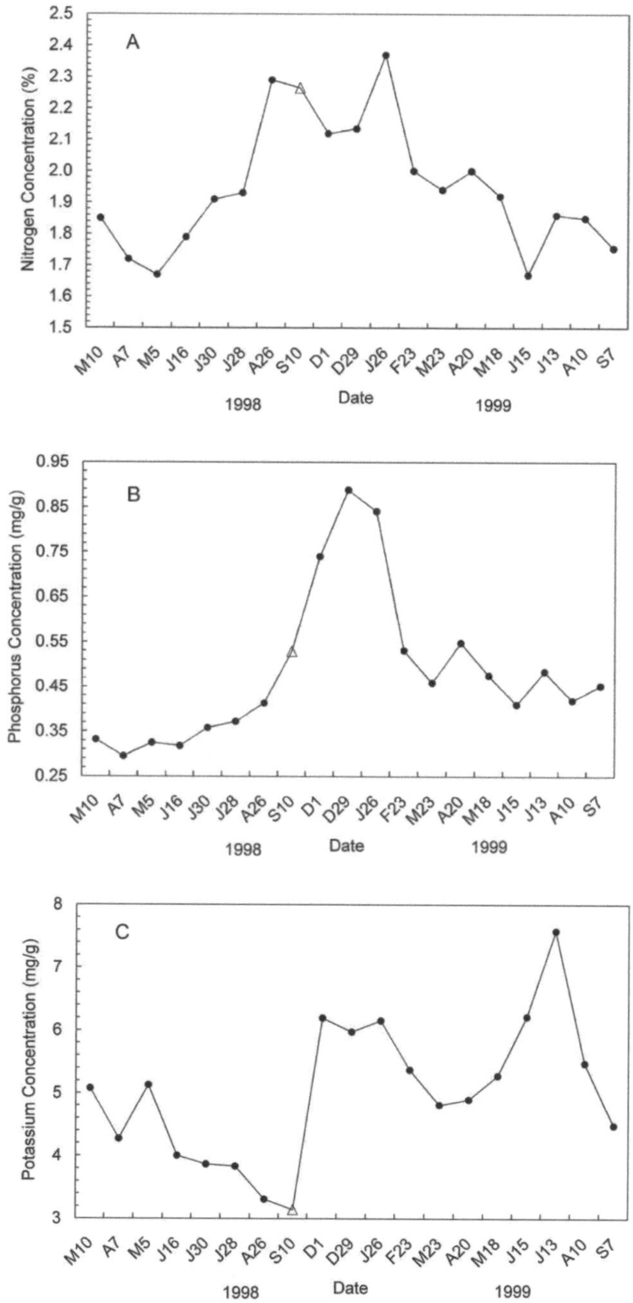

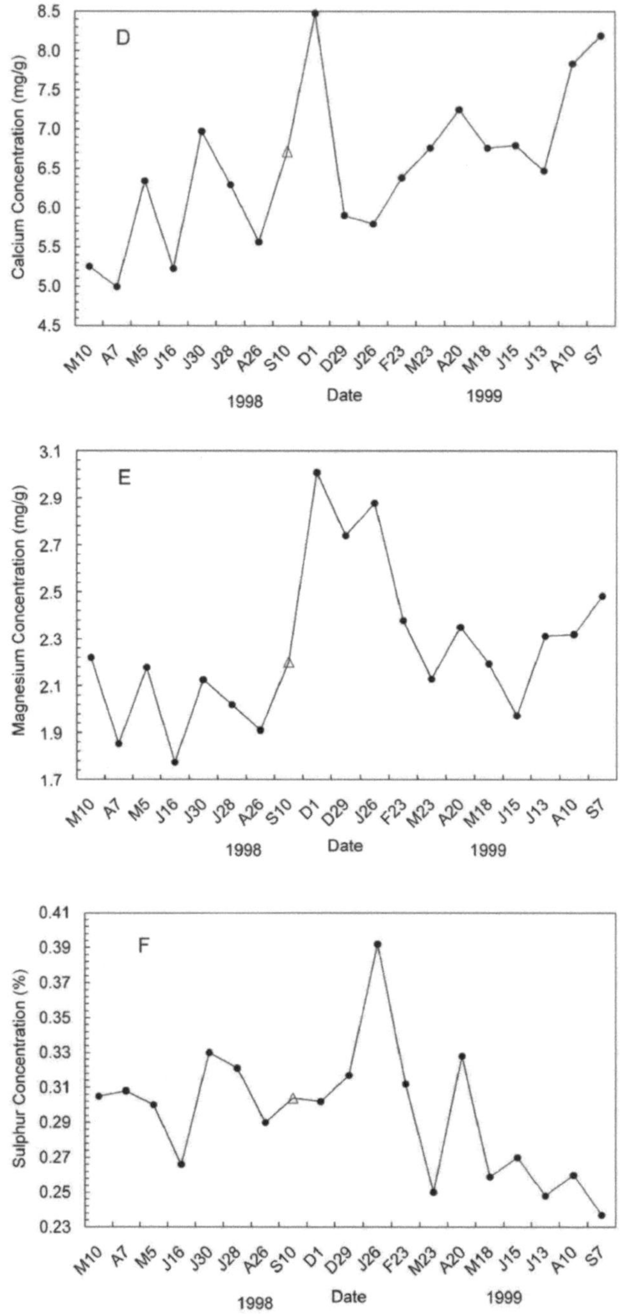
D
Calcium Concentration (mg/g)
8.5
8.0
7.5
7.0
6.5
6.0
5.5
5.0
4.5
M10 A7 M5 J16 J30 J28 A26 S10 D1 D29 J26 F23 M23 A20 M18 J15 J13 A10 S7
1998
Date
1999
E
Magnesium Concentration (mg/g)
3.1
2.9
2.7
2.5
2.3
2.1
1.9
1.7
M10 A7 M5 J16 J30 J28 A26 S10 D1 D29 J26 F23 M23 A20 M18 J15 J13 A10 S7
1998
Date
1999
F
Sulphur Concentration (%)
0.41
0.39
0.37
0.35
0.33
0.31
0.29
0.27
0.25
0.23
M10 A7 M5 J16 J30 J28 A26 S10 D1 D29 J26 F23 M23 A20 M18 J15 J13 A10 S7
1998
Date
1999

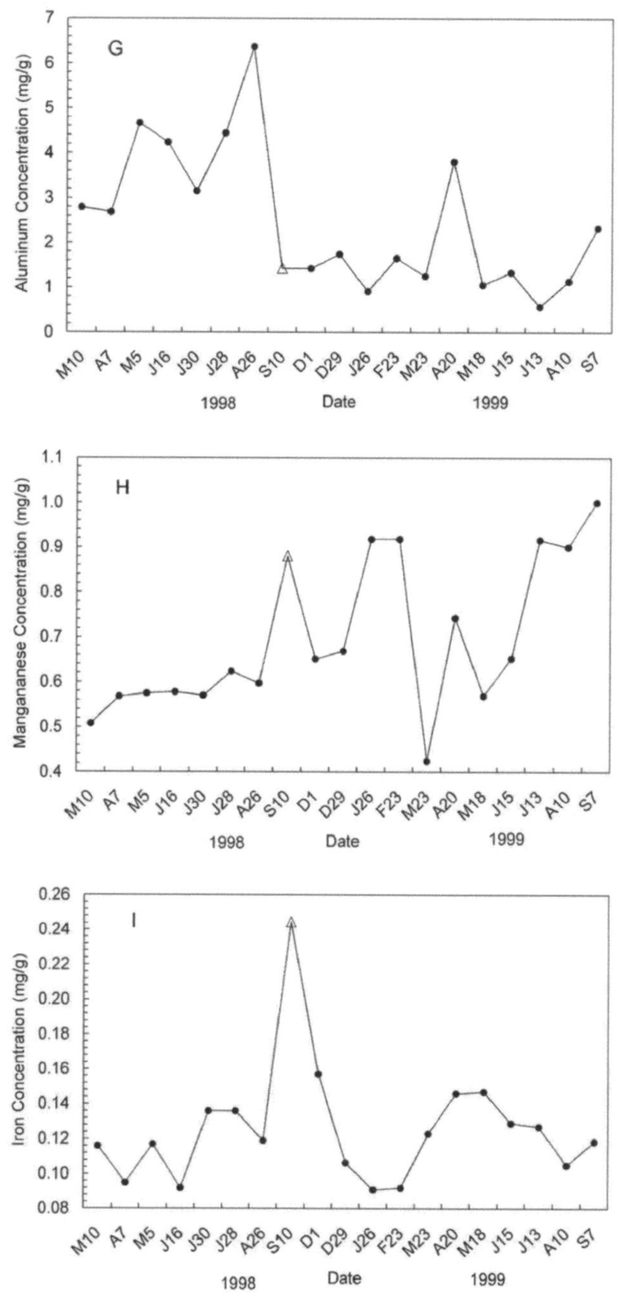
G
Aluminum Concentration (mg/g)
0
1
2
3
4
5
6
7
M10 A7 M5 J16 J30 J28 A26 S10 D1 D29 J26 F23 M23 A20 M18 J15 J13 A10 S7
1998
Date
1999
H
Mangananese Concentration (mg/g)
0.4
0.5
0.6
0.7
0.8
0.9
1.0
1.1
M10 A7 M5 J16 J30 J28 A26 S10 D1 D29 J26 F23 M23 A20 M18 J15 J13 A10 S7
1998
Date
1999
I
Iron Concentration (mg/g)
0.08
0.10
0.12
0.14
0.16
0.18
0.20
0.22
0.24
0.26
M10 A7 M5 J16 J30 J28 A26 S10 D1 D29 J26 F23 M23 A20 M18 J15 J13 A10 S7
1998
Date
1999

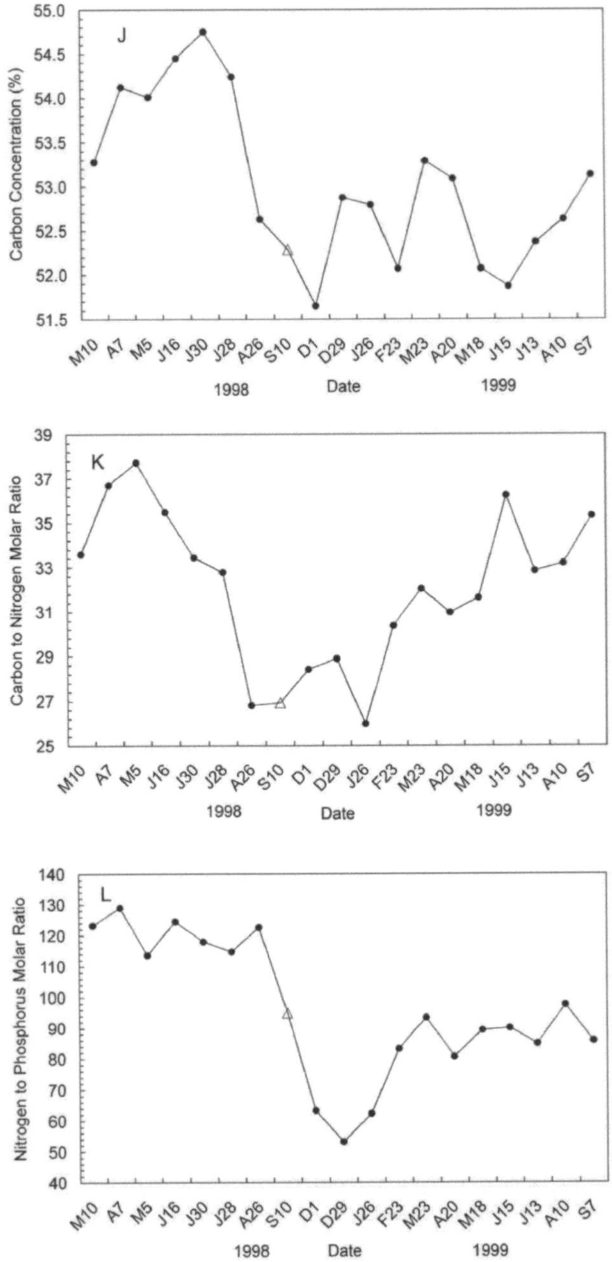

Fig. 3. Element concentrations and C/N and N/P molar ratios in leaf fall of a secondary forest at Utuado, Puerto Rico. The elements are: nitrogen (A), phosphorus (B), potassium (C), calcium (D), magnesium (E), sulphur (F), aluminum (G), manganese (H), iron (I), and carbon (J). Molar ratios are: carbon to nitrogen (K) and nitrogen to phosphorus (L). Data for the first collection after the passage of Hurricane Georges are shown with an open triangle in each graph. The date shows the abbreviation for the month followed by the day of collection.

Phosphorus (Fig. 3b), K (Fig. 3c), and Mg (Fig. 3e) concentrations in leaf fall had the same pattern of temporal change, which involved a rapid increase in concentration after the hurricane followed by a decreasing trend to pre-hurricane concentrations. We found a rising post-hurricane trend in leaf fall Ca (Fig. 3d) and Mn (Fig. 3h) concentrations. In contrast, S (Fig. 3f) and Al (Fig. 3g) concentrations exhibited a declining post-hurricane trend, while Fe (Fig. 3i) had a high concentration in the hurricane collection and quickly returned to pre-hurricane concentrations. Leaf fall concentrations of C (Fig. 3j) and N (Fig. 3a) contrasted in their temporal patterns. Nitrogen concentrations were rising before the hurricane and declined afterwards, while C concentrations were declining before the hurricane and roused after the event. The C/N molar ratio (Fig. 3k) behaved similar to the C concentration, while the N/P molar ratio (Fig. 3l) declined sharply after the hurricane and then increased somewhat in the following months. All elements with the exception of Al had achieved (N, P, K, Mg, S, Fe, C, C/N, ash) or exceeded (Ca, Mn) pre-hurricane concentrations or molar ratios by November 23, 1999 (Table 2). Only Al and the N/P molar ratio remained below pre-hurricane values.

Element Return Flux—Annual element return flux by miscellaneous fall was insignificant, and we will not address it any further (Table 1). Flowers and fruits fall returned a low but steady amount of elements to the forest floor. Element return was higher in the hurricane year than in the year after. Element return by wood fall was higher than those of leaf fall during the hurricane year, but lower the year after the hurricane (Table 1). The biweekly pattern of total element return to the forest floor is not shown but was similar to the biweekly pattern of total litterfall (Fig. 2).

4. Discussion

Our study documented both the instantaneous and short-term effects (months to years) of Hurricane Georges on litterfall mass and element fluxes of a young novel secondary forest. We first discuss six such effects of this hurricane followed by the reasons why the forest behaved as it did, and the implications of our results to the functioning of tropical forests in the hurricane belt.

Effects on the Pattern of Litterfall—The temporal pattern of litterfall is modified by hurricanes. The modified pattern usually involves a rapid decline in litterfall after the hurricane pulse and a trajectory to pre-hurricane rates in subsequent months (Fig. 2). We also measured a second litterfall pulse (mostly wood fall), which reflects delayed mortality or late return of debris retained in the canopy and caught in branches and the lower canopy where it remains suspended until wind and rain cause it to fall. Lodge et al. (1991) found 9.3 Mg/ha of suspended litter after Hurricane Hugo in the Luquillo Experimental Forest (LEF).

The most obvious effect of a hurricane on litterfall is the instantaneous pulse associated with hurricane winds and crown defoliation (Fig. 2) as measured in Puerto Rico, Hawaii, and Mexico. Pulses of litterfall are also common after heavy rains and wind events associated with low-pressure atmospheric systems (Lugo 1992, Lugo et al. 2007). An example of a rain-induced litterfall event in our data is the peak wood flux in the collection of April 7, 1998, before Hurricane Georges passed over the forest (Fig. 2). That pulse was statistically different from rates before and after the event. However, rainfall events do not reach the high rates that accompany hurricanes because hurricanes have the highest velocity winds and litterfall rates are linearly related to wind speed (Beard et al. 2005).

Effects on the Proportion of Litterfall Components—The hurricane changed the proportion of litterfall components from a predominance of leaves to wood and then back to leaves (Table 1, Fig. 2a, b). This is another common effect of hurricanes (Frangi and Lugo 1991, Whigham et al. 1991, Scatena et al. 1996, Herbert et al. 1999). The norm is the predominance of leaf fall as observed in a similar forest type in nearby Barranquitas (Fig. 1), which had 64 to 82 percent leaf fall in a replicated three-year study (Lugo et al. 1999). The change in the relative fractions of wood and leaves in litterfall has implications to litter chemistry, the quality of material available to soil organisms, and post-hurricane nutrient availability (below). This is reinforced by the predominance of >2 cm wood fall (Fig. 2b). Large wood pieces require longer processing by soil organisms than leaves, flowers and fruits, or miscellaneous materials (Swift et al. 1979). Moreover, one study at the LEF predicted nutrient immobilization, reduced decomposition, and reduced tree growth as a result of the large fraction of wood (including coarse woody debris) returned to the forest floor by Hurricane Hugo (Zimmerman et al. 1995, but see Lugo 2008).

Effects on Litterfall Mass and Element Fluxes—The instantaneous pulse of leaf and total litter fall in the hurricane collection was 176 and 449 times the respective pre-hurricane rates. On an annual basis, total litterfall the year of the Hurricane was 4.4 times higher than that of the year after the hurricane (Table 1). The hurricane transferred a large mass of material to the forest floor, thus temporarily increasing the accumulation of litter. Similar pulses of litter input to the forest floor have been documented in Hawaii, Mexico, and Puerto Rico. In all examples, the instantaneous hurricane pulse exceeds or approaches the normal annual flux of mass.

Pulses of element return accompany the hurricane pulse of litterfall mass. The annual quantities of N, P, K, Ca and Mg returned to the forest floor by total litterfall the year of Hurricane Georges were 3.5, 3.2, 2.9, 6.2 and 3.7 times the quantities returned the year after the hurricane (Table 1). These levels of element return, with the exception of Ca, were close but not proportional to the increase in litterfall mass. The differences are caused by differences in concentration (Fig. 3, Table 2) and the relative proportions of leaf and wood fall in total litterfall mass (Table 1). The higher proportion of Ca return, for example, is attributed to the large wood fall with high Ca concentration.

The ratios of hurricane pulse to pre-hurricane fluxes of mass and element return in litterfall were similar for mass but lower for nutrients in our study (with the exception of Ca) than those measured elsewhere in Puerto Rico after Hurricanes Hugo and Georges (Table 3). The similarity in the mass pulse ratio (except for Guánica and elfin forests) and the differences in the nutrient pulse ratios, suggest ecophysiological differences among sites rather than differences in forest resistance to hurricane winds. The reason is that the mass pulse is the direct effect of winds, while the nutrient content of litterfall reflects in addition the concentration in plant parts, which is unrelated to the wind storm. We will address this issue below.

Effects on the Chemistry of Litterfall and Nutrient Use-Efficiency—The element concentration of the material that falls instantaneously during a hurricane is high because it is mostly composed of live plant parts. Element concentration is higher in live materials than in dead material because of retranslocation of nutrients associated with senescence particularly in leaves (Noodén and Leopold 1988). The higher nutrient concentration in downed hurricane material has already been documented in Puerto Rico, Mexico, and Hawaii. A hurricane or windstorm interrupts senescence and retranslocation processes, which in turn affects element concentrations of falling parts and processing of debris by forest floor organisms (Berg and McClaugherty 2003).

Forest Type	Mass	N	P	K	Ca	Mg	Mn	Al	S	Fe
Palm floodplain[1]	36									
Palm floodplain[2]	447		1726							
Tabonuco forest, El Verde[3]	436	448	800	466	336	319				
Tabonuco forest, Bisley[3]	455	475	610	1021	465	510				
Elfin forest[3]	682	430	219	997	417	635				
Guánica dry forest[4]	767	842	1460	520						
This study	449	278	487	316	935	478	442	992	246	522

1. Subtropical wet forest at 750 m elevation after Hurricane Allen (Frangi and Lugo 1985).
2. Same as above after Hurricane Hugo (Frangi and Lugo 1991).
3. Subtropical wet forest after Hurricane Hugo (Lodge *et al.* 1991).
4. Subtropical dry forest after Hurricane Georges (Van Bloem *et al.* 2005).

Table 3. Ratio of instantaneous mass or element return to the forest floor due to a hurricane to pre-hurricane daily average mass or element return to the forest floor. Values used to calculate the ratio are for total litterfall expressed on a daily basis. Empty cells mean there are no data. All sites are from Puerto Rico and all but the study site are mature forests.

Element concentration of leaf fall in our study changed for about a year after the passage of the hurricane probably due to the development and turnover of new leaves in fast-growing trees. For example, both in our study (Fig. 3e) and the LEF (Scatena *et al.* 1996), post-hurricane falling leaves had high Mg concentration. Magnesium is required for chlorophyll production, and it is expected that high quantities of Mg are required for the re-growth of the canopy. The leaves that fell on our litterfall baskets after Hurricane Georges had higher elemental concentrations than did senesced leaves falling before the hurricane. In some instances, leaf fall months after the hurricane had higher nutrient concentrations than live leaves that fell during the hurricane pulse. At the end of the study, leaf litter fall had higher Ca and Mn concentrations than pre-hurricane leaf fall. Ostertag et al. (2003) showed that after Hurricane Georges these high concentrations in leaf fall flux are reflected in the concentration of standing litter where they remain high for several months while soil organisms process the material. They found that N, P, Ca, and Mg increased in concentration in the standing litter of a moist forest but in other forest types this increase was not observed. In our study P concentration increased after the hurricane while N concentration increased just before the hurricane and remained high for three collection intervals before returning to pre-hurricane values. Our concentration data is consistent with Ostertag et al. (2003) observations on ground litter but we have no explanation for the pre-hurricane N concentration increase.

The within-stand nutrient use-efficiency reflects a low nutrient use-efficiency in our study site when compared to other forests and tree plantations in Puerto Rico (Table 4). Brown and Lugo (1990) found that tropical secondary forests in general had lower within-stand nutrient use-efficiency than mature forests and this general pattern applies to the comparison of secondary forests with mature forests in Puerto Rico (Lugo 1992). However, as first demonstrated by Scatena et al. (1996), the within-stand nutrient use-efficiency of forests decreases even more with hurricanes. This means that the weighted nutrient concentrations (the inverse of the within stand nutrient use efficiency in Table 4) in post-hurricane leaf fall in our forest were high, thus reflecting high nutritional quality. Wood fall had lower weighted nutrient concentrations than leaf fall, but it also had a post-hurricane increase in nutritional quality (Table 4). Nevertheless, the predominance of wood in litterfall during and immediately after the hurricane (Fig 2b) dominated nutrient inputs to the forest floor and because of the lower nutritional quality of this material it probably immobilized nutrients relative to the effect of the higher quality material in leaf fall.

Site	N		P		K		Source
	Leaf	Total	Leaf	Total	Leaf	Total	
			Secondary Forests				
Guzmán	132		2652	2747	584		Lugo 1992
Cubuy	117	123	2366	2382	447	477	Lugo 1992
Sabana	66	69	1329	1310	434	490	Lugo 1992
El Verde	91	92	3222	2999	527	565	Lugo 1992
Guánica	97	99	6056	6076	123	134	Lugo and Murphy 1986
Barranquitas	56	57	1553	1544	164	159	Lugo *et al.* 1999
Utuado (H)	48	70	2175	2454	263	319	This study
Utuado (PH)	52	54	1833	1800	190	206	This study
				Plantations			
Pine	198	134	4997	4153	974	992	Lugo 1992
Mahogany	82	92	140	163	537	611	Lugo 1992
Eucalyptus	165		5100				Cuevas and Lugo 1998

Table 4. Nutrient use efficiency in leaf and total litterfall of various secondary forests and plantations in Puerto Rico. Utuado data are for the hurricane (H) year (March 1998 to February 1999), and the post-hurricane (PH) year (March 1999 to February 2000). Empty cells = data not available. Data are in kg organic matter falling as litter/kg element in litter fall.

The lowering of the N/P molar ratio of leaf fall after the hurricane implies an improved availability of P in relation to N to soil organisms (Table 2). The N/P ratio of leaf fall was lowest during the first three months after the hurricane, and remained below pre-hurricane values through the end of the study (Fig. 3l). This, coupled to a decrease in the C/N ratio (Fig. 3k), and lowered within-stand nutrient use-efficiency (Table 4) suggests that conditions for the decomposition of falling leaf litter, and thus subsequent plant uptake, are favorable post-hurricane, particularly when leaves become the dominant component of litterfall. Nutrients such as P, K, Ca, and Mg (Figs. 3b-e) had increased concentrations in post-hurricane leaf fall. Part of the high resilience exhibited by the processing of leaf litter by forest floor organisms in a moist forest was attributed to the high nutrient concentration of incoming leaf fall (Ostertag et al. 2003). Our results support this suggestion.

The Influence of Species Composition on Litterfall Element Fluxes—The young forest that we studied was composed of pioneer and secondary vegetation, including a significant component of introduced species. Although we did not analyze leaf fall chemistry by species, we believe that the species composition influenced the high litterfall element fluxes that we measured. We base our inference on studies of leaf fall chemistry that show the influence of species composition on element concentration and flux data (Cuevas and Lugo 1998, Scatena et al. 1996, Lugo 2004, Lugo et al. 2004, available unpublished data).

It is thus possible that the dominance of pioneer species may be responsible for observed nutrient use-efficiency differences between secondary and mature forests (Brown and Lugo 1990). Fast growing introduced species, which are common in our forest, may accelerate nutrient fluxes even further. Beyond the secondary forest/mature forest comparison; hurricanes influence the nutrient dynamics and use-efficiency of forests in two ways. They favor the fast growth of pioneer species, and they enable the growth of nutrient-enriched leaves. At the LEF, Scatena et al. (1996) measured the highest net primary productivity so far measured at that site and the lowest nutrient use-efficiency after Hurricane Hugo. In our study, the result was that the low nutrient use-efficiency typical of secondary successional vegetation was lowered even more by the hurricane (Table 4). Nutrient use-efficiency was lower than that measured by Scatena et al. (1996), which we attribute to species composition (mostly pioneer and introduced species) and young age (26 yr-old).

Effects on Post-Hurricane Litterfall Rates—The post-hurricane production of flowers and fruits is of significance to the survival of wildlife. Several studies in diverse locations show that nectarivorous and frugivorous wildlife populations experience significant reductions after hurricanes as populations migrate to locations where nectar and fruit remain available (Waide 1991a and b, Askins and Ewert 1991, Lynch 1991, Will 1991, Wunderle *et al.* 1992). Months can pass before fruit production returns to some normalcy and allow recovery of wildlife. At the LEF, flowers and fruits fall remained below average for about 2.5 years after Hurricane Hugo (F.N. Scatena, University of Pennsylvania, Personal Communication), while at our site the post-hurricane reduction lasted 10 months (Fig. 2c). Other litterfall fluxes had different levels of resilience (Fig. 4). Leaf fall did so in about eight months while wood fall and total litterfall returned to pre-hurricane rates after the pulses of fall during and within a month of the hurricane. Phosphorus, K, Ca, and Mn in leaf fall reached pre-hurricane fluxes in about 5 months or faster than leaf fall mass and faster than N, Mg, S, and ash fluxes, which required about 8 months to reach pre-hurricane rates.

Leaf expansion after the hurricane is rapid and new leaves have higher element concentrations (Table 2), thus accelerating the recovery of these fluxes. The required time to reach pre-hurricane fluxes in our forest, with one exception, oscillates between 5 and 8 months for leaf fluxes. This level of resilience is shorter than the 20 months in the mature forest studied by Scatena et al. (1996), a forest that had a higher reduction in litterfall fluxes after Hurricane Hugo (Table 3). The exception is Al, the element most affected by the passage of the hurricane at our site, which had a positive slope at the end of our measurements but had not reached pre-hurricane rates after 17 months. Thus, while the hurricane caused a wide oscillation in the pattern of element return, it appears that our forest might reach pre-hurricane litterfall fluxes in one or two years after the event.

Two factors might explain the high litterfall resilience of the study forest to hurricane winds. One is the low level of hurricane effects on the structure of the forest. For example, most of the effects to understory trees were caused by fallen debris, which in most cases bent and injured but did not kill trees (Lugo et al. 2005). As a result, loss of canopy was low in this forest and most trees exhibited rapid increases of growth rate after the hurricane. The other explanation for our results is the nature of the affected species. The tree species that dominated the canopy of our forest were secondary forest species characterized by rapid growth and frequent and profuse flowering (Little and Wadsworth 1964) and the chemical characteristics discussed above. Therefore, in spite of the passage of the hurricane, the forest had the capacity to respond quickly. When fast-growing species are dominant in a forest, they circulate more chemical elements per unit mass (Table 4) and enrich litter and soil (Brown and Lugo 1990, Cuevas and Lugo 1998, Lugo *et al.* 2004). These accelerated fluxes contribute to resilience after hurricanes and other disturbances.

Comparison of Secondary Forests With Mature Forests—The immediate effect of hurricane-force winds on litterfall is the ensuing pulse of mass and element return to the forest floor (Fig. 2). The relative magnitude of this pulse is its ratio to pre-hurricane rates (Table 3). Our results, when compared to mature natural forests elsewhere in Puerto Rico, reflect lower ratios for nutrients and similar ratios for mass flux, except for the Guánica dry forest and the elfin forest. Part of the difference may be due to the timing in the collection of litterfall samples. Both Lodge et al. (1991) and Van Bloem et al. (2005) collected hurricane litterfall quicker after the event than we did. Our hurricane pulse is underestimated due to the decomposition of litter during the 61 days between its fall and collection. However, this underestimate was small in comparison with the hurricane pulse (5 g/m^2 based on the decomposition rate of leaves after Hurricane Hugo in Ostertag *et al.* 2003). This correction does not explain the differences between the forests.

We believe two circumstances explain our results and the differences between forests in Table 3. One is the high ratio of wood fall to leaf fall at our site (3:1) compared to 2:1 for Guánica Forest (Van Bloem et al. 2005) and 1.5:1 for Bisley (Lodge et al. 1991). The larger fraction of wood in our samples lowers the nutrient return of the pulse due to the lower element concentration of wood (Table 2). Calcium concentration in wood is the exception and for that element our forest had the highest pulse ratio of any forest in Table 3. Thus, the relative proportion of leaves and wood in the hurricane pulse, influences the ratio of nutrients in the hurricane pulse to pre-hurricane nutrient fluxes.

The second circumstance is the difference in nutrient concentration in pre-hurricane litter compared to hurricane litter in the various forests. Both Guánica dry forest and the elfin forest produce nutrient-poor litterfall, which is reflected in the within-stand nutrient use-efficiency. For the Guánica dry forest, the N and P nutrient use-efficiency under normal conditions is three and two times, respectively, higher than for our site (Table 4). A lower nutrient return in the absence of hurricanes would accentuate the flux difference between pre- and post-hurricane element return. That is, the nutrient turnover in our study site is normally high and the increase due to the hurricane flux is not as dramatic as in native mature forests with pre-hurricane conservative nutrient fluxes.

Some Implications to Tropical Forests on the Hurricane Belt—A common denominator to our results is the recurrence of pulses of mass fall and element return to the forest floor. These pulses occur at different intensities and times, depending on atmospheric conditions. The highest rates occur after hurricanes. These pulses increase the amount of nutrients entering the litter/soil compartments of the forest and change the quality of decomposing litter through altered C/N and N/P ratios and high leaf fall nutrient concentrations. Sayer et al. (2006) found that increasing litter inputs to the forest floor in a tropical moist forest in Barro Colorado accelerated the decay of wood but not the meso arthropod population abundance in litter. Thus, pulses of litter fall and nutrient return have proven effects on decomposition and recycling processes of tropical forests. They also found greater accumulation of P and N in litter as did Ostertag et al. (2003) after Hurricane Georges in Puerto Rico. However, our study shows that the nutritional quality of the pulse depends on the plant part that dominates the flux. Wood fall pulses have a different nutritional quality than a leaf fall pulse.

The implications to soil processes of rapid changes in litterfall quantity and chemistry, coupled to the disproportional behavior of nutrients and organic mass is poorly known, but is receiving increased attention by ecologists. For example, Lodge et al. (1994) suggest that responses to pulsed inputs to the soil/litter compartment involve complex non-lineal phenomena. Our results are in agreement with previous observations of the nutrient-rich litter environment in tropical secondary forests (Brown and Lugo 1990). However, this high nutrient richness, low C/N ratios, and low N/P ratios apply to leaf fall and not to wood fall, which contains lower quality material and can immobilize nutrients. Falling leaves after a hurricane constitute a favorable environment for supporting high rates of litter decomposition and nutrient cycling (Ostertag et al. 2003). In turn, these ecosystem processes have a practical application for supporting the rehabilitation of soil fertility in degraded lands colonized by new forest ecosystems (Lugo and Helmer 2004, Lugo et al. 2004).

5. Acknowledgements

This study was part of a cooperative agreement between the USDA International Institute of Tropical Forestry and the Luis Muñoz Rivera Vocational High School in Utuado. The work is done in cooperation with the University of Puerto Rico and is part of the contribution of the Forest Service to the Long-Term Ecological Research Program and the School Yard Program of the National Science Foundation (Grant No. BSR-8811902 to the University of Puerto Rico and the International Institute of Tropical Forestry). We thank students in the school's research program for their assistance in the collection of field samples and Mary Jean Sánchez, Edwin López, and other technicians and professionals at the Institute for their assistance in the collection and analysis of samples. Skip Van Bloem and Olga Ramos collaborated with the development of Fig. 1, and Eileen Helmer classified the site according to her new life zone map for Puerto Rico. The following colleagues and two anonymous reviewers reviewed the manuscript: M. Alayón, J. Hernández, E. Medina, F.N. Scatena, F.H. Wadsworth, Skip Van Bloem, and D.J. Lodge. Stephanie Whitmire, Michael Willig, Tamara Heartsill, and Skip Van Bloem helped us with the statistical analysis of data.

6. References

Askins, R. A. & Ewert, D.N. (1991). Impact of Hurricane Hugo on bird populations on St. John, U.S. Virgin Islands. *Biotropica* 23: 481-487.

Beard, K. H., Vogt, K.A., Vogt, D.G., Scatena, F.N., Covich, A.P., Sigurdardottir, R., Siccama, T.G. & Crowl, T. A. (2005). Structural and functional responses of a subtropical forest to 10 years of hurricanes and droughts. *Ecological Monographs* 75: 345-361.

Berg, B. & McClaugherty, c. (2003). *Plant litter: decomposition, humus formation, carbon sequestration*. Springer-Verlag, Berlin, Germany.

Brown, S. & Lugo, A.E. (1990). Tropical secondary forests. *Journal of Tropical Ecology* 6: 1-32.

Chazdon, R. L. (2008). Beyond deforestation: restoring forests and ecosystem services on degraded lands. *Science* 320: 1458-1460.

Cuevas, E. & Lugo, A.E. (1998). Dynamics of organic matter and nutrient return from litterfall in stands of ten tropical tree plantation species. *Forest Ecology and Management* 112: 263-279.

Everham-III, E. M. & Brokaw, N.V.L. (1996). Forest damage and recovry from catastrophic wind. *The Botanical Review* 62: 113-185.

Frangi, J. L. & Lugo, A.E. (1985). Ecosystem dynamics of a subtropical floodplain forest. *Ecological Monographs* 55: 351-369.

Frangi, J. L. & Lugo, A.E. (1991). Hurricane damage to a flood plain forest in the Luquillo Mountains of Puerto Rico. *Biotropica* 23: 324-335.

Herbert, D. A., Fownes, J.H & Vitousek, P. M. (1999). Hurricane damage to a Hawaiian forest: nutrient supply rate affects resistance and resilience. *Ecology* 80: 908-920.

Hobbs, R. J., Arico, S., Aronson, J., Baron, J.S., Bridgewater, P., Cramer, V.A., Epstein, P.R., Ewel, J. J., Klink, C. A.,

Lugo, A. E., Norton, D., Ojima, D, Richardson, D.M., Sanderson, E. W., Valladares, F., Vilà, M. , Zamora, R. & Zobel. M. (2006). Novel ecosystems: theoretical and management aspects of the new ecological world order. *Global Ecology and Biogeography* 15: 1-7.

Holdridge, L. R. (1967). *Life zone ecology*. Tropical Science Center, San José, Costa Rica.

LECO Corp. (1995). Carbon, nitrogen, and sulphur testing in soil/plant tissue. Organic Application Note CNS-2000, No. 203-821-002, St. Joseph, MI.

Little, E. L. & Wadsworth, F. H. (1964). Common trees of Puerto Rico and the Virgin Islands. Agriculture Handbook 249. USDA Forest Service, Washington DC, U.S.A.

Lodge, D. J., Scatena, F.N., Asbury, C. E. & Sánchez, M. J. (1991). Fine litterfall and related nutrient inputs resulting from Hurricane Hugo in subtropical wet and lower montane rain forests of Puerto Rico. *Biotropica* 23: 336-342.

Lodge, D. J., McDowell, W. H. & McSwiney, C. P. (1994) The importance of nutrient pulses in tropical forests. *TRENDS in Ecology & Evolution* 9: 384-387.

Lugo, A. E. (1992). Comparison of tropical tree plantations with secondary forests of similar age. *Ecological Monographs* 62: 1-41.

Lugo, A. E. (2002). Can we manage tropical landscapes? An answer from the Caribbean perspective. *Landscape Ecology* 17: 601-615.

Lugo, A. E. (2004). The outcome of alien tree invasions in Puerto Rico. *Frontiers in Ecology and the Environment* 2: 265-273.

Lugo, A. E. (2008). Visible and invisible effects of hurricanes on forest ecosystems: an international review. *Austral Ecology* 33: 368-398.

Lugo, A. E. (2009). The emerging era of novel tropical forests. *Biotropica* 41:589-591.

Lugo, A. E., Domínguez Cristóbal, C., Santos, A. & Torres Morales, E. (1999). Nutrient return and accumulation in litter of a secondary forest in the coffee region of Puerto Rico. *Acta Científica* 13: 43-74.

Lugo, A. E., Domínguez Cristóbal, C.M. & Méndez Irizarry, N. (2005). Efectos del huracán Georges sobre la composición de especies y estructura de un bosque secundario en el interior de Puerto Rico. *Acta Científica* 19 (1-3) 41-61.

Lugo, A. E. & Helmer, E. (2004). Emerging forests on abandoned land: Puerto Rico's new forests. *Forest Ecology and Management* 190: 145-161.

Lugo, A. E., Medina, E., Cuevas, E., Cintrón, G., Laboy-Nieves E. N. & Schäeffer Novelli, Y. (2007). Ecophysiology of a mangrove forest in Jobos Bay, Puerto Rico. *Caribbean Journal of Science* 43: 200-219.

Lugo, A. E. & Murphy, P.G. (1986). Nutrient dynamics of a Puerto Rican subtropical dry forest. *Journal of Tropical Ecology* 2: 55-76.

Lugo, A. E., Silver, W. L. & Molina Colón, S. (2004). Biomass and nutrient dynamics of restored neotropical forests. *Water, Air and Soil Pollution*: Focus 4: 731-746.

Lugo López, M. A. (1977). Updated taxonomic classification of the soils of Puerto Rico. University of Puerto Rico Agriculture Experiment Station, Mayagüez, PR.

Luh Huang, C. Y. & Schulte E.E. (1985). Digestion of plant tissue for analysis by ICP emission spectroscopy. *Communications in Soil Science and Plant Analysis* 16: 943-958.

Lynch, J. F. (1991). Effects of Hurricane Gilbert on birds in a dry tropical forest in the Yucatan Peninsula. *Biotropica* 23: 488-496.

Newbould, P.J. (1967). *Methods for estimating primary production of forests.* IBP Handbook No. 2 Blackwell Scientific Publications, Oxford.

NOAA. (1998). Puerto Rico and the Virgin Islands. *Climatological data annual summary* 44: 1-13.

Noodén, L. D & Leopold, A. C., Eds. (1988). *Senescence and aging in plants.* Academic Press Inc., New York, NY.

Ostertag, R., Scatena, F.N. & W. L. Silver, W.L. (2003). Forest floor decomposition following hurricane litter inputs in several Puerto Rican forests. *Ecosystems* 6: 261-273.

Popper, N., Domínguez Cristóbal, C., Santos, A., Méndez Irizarry, N., Torres Morales, E., Lugo, A. E., Rivera Lugo, Z. Z., Soto Toledo, B., Santiago Irizarry, M., Rivera, I. L., Zayas, L. A. & Colón, C. (1999). A comparison of two secondary forests in the coffee zone of central Puerto Rico. *Acta Científica* 13: 27-41.

SAS-Institute-Inc. (2007). JMP TM Release 7. SAS Institute Inc., Cary, NC.

Sayer, E. J., Tanner, E. V. J. & Lacey, A. L. (2006). Effects of litter manipulation on early-stage decomposition and meso-arthropod abundance in a tropical moist forest. *Forest Ecology and Management* 229: 285-293.

Scatena, F. N., Moya, S., Estrada, C. & Chinea, J.D. (1996). The first five years in the reorganization of aboveground biomass and nutrient use following Hurricane Hugo in the Bisley Experimental Watersheds, Luquillo Experimental Forest, Puerto Rico. *Biotropica* 28: 424-440.

Swift, M. J., Heal, O. W. & Anderson, J. M. (1979). *Decomposition in terrestrial ecosystems.* University of California Press, Berkeley, CA.

Turton, S. M., editor. (2008). Ecological impacts of tropical cyclones on Australian terrestrial ecosystems: insights from Cyclones Larry and Monica. *Austral Ecology* 33: 1-584.

Van Bloem, S. J., Murphy, P. G., Lugo, A. E. , Ostertag, R., Rivera Costa, M., Ruiz Bernard, I., Molina Colón, S. & Canals Mora, M. (2005). The influence of hurricane winds on Caribbean dry forest structure and nutrient pools. *Biotropica* 37: 571-583.

Vitousek, P. M. (1982). Nutrient cycling and nutrient use efficiency. *American Naturalist* 119: 553-572.

Waide, R. B. (1991a). The effect of Hurricane Hugo on bird populations in the Luquillo Experimental Forest, Puerto Rico. *Biotropica* 24: 475-480.

Waide, R. B. (1991b). Summary of the response of animal populations to hurricanes in the Caribbean. *Biotropica* 23:508-512.

Whigham, D. F., Olmsted, I., Cano, E. C. & Harmon, M. E. (1991). The impact of Hurricane Gilbert on trees, litterfall and woody debris in a dry tropical forest in the northeastern Yucatan peninsula. *Biotropica* 23: 434-441.

Wilde, S. A., Corey, R. B., Iyer, J. G. & Voigt, G. K. (1979). *Soil and plant analysis for tree culture.* Oxford & IBH Publishing Co., Oxford, England.

Will, T. (1991). Birds of a severely hurricane-damaged Atlantic coast rain forest in Nicaragua. *Biotropica* 23: 497-507.

Wunderle, J. M., Lodge, D. J. & Waide, R. B. (1992). Short-term effects of Hurricane Gilbert on terrestrial bird populations on Jamaica. *Auk* 109: 148-166.

Zimmerman, J. K., Pulliam, W. M., Lodge, D. J., Quiñones-Orfila, V., Fetcher, N., Guzmán-Grajales, S., Parrotta, J. A., Asbury, C. E., Walker, L. & Waide, R. B. (1995). Nitrogen immobilization by decomposing woody debris and the recovery of tropical wet forest from hurricane damage. *Oikos* 72: 314-322.

29

Hurricane-Induced Phytoplankton Blooms: Satellite Observations and Numerical Model Simulations

Menghua Wang[1], Xiaoming Liu[1,2], and Wei Shi[1,3]
[1]*NOAA National Environmental Satellite, Data, and Information Service Center for Satellite Applications and Research, Camp Springs, MD,*
[2]*SP Systems Inc., Greenbelt, MD,*
[3]*CIRA at Colorado State University, Fort Collins, CO,*
USA

1. Introduction

Hurricane-force winds have significant effects on the upper ocean. Hurricane-induced vertical mixing, entrainment, and upwelling bring up cold and nutrient-rich sub-surface water to the surface. As a result, sea surface temperature (SST) cools several degrees Celsius as the mixed layer deepens by tens of meters, and conversely, downward mixing of heat also warms the upper seasonal thermocline waters. The nutrient-rich water in the upper euphotic zone can also stimulate biological production and phytoplankton bloom. The potential impact of hurricanes on upper ocean physics and biogeochemistry is quite complex, and their observation and study have benefited greatly from ocean satellite remote sensing. The Advanced Very High Resolution Radiometer (AVHRR) has been used for measuring SST for decades, and hurricane-induced SST drops have been reported (Price, 1981 and references therein). In comparison, the development of ocean color instruments onboard satellites observing ocean biological properties is relatively recent. With the advancement of satellite ocean color remote sensing from the Sea-viewing Wide Field-of-view Sensor (SeaWiFS) (Hooker *et al.*, 1992) and the Moderate Resolution Imaging Spectroradiometer (MODIS) (Salomonson *et al.*, 1989) in recent years, hurricane-induced phytoplankton blooms have been observed and reported (Babin *et al.*, 2004; Lin *et al.*, 2003; Miller *et al.*, 2006; Walker *et al.*, 2005; Shi and Wang, 2007).

The SeaWiFS, which has operated since August 1997, and MODIS on the Terra platform (December 1999–present) and Aqua platform (May 2002–present) have provided us a view of chlorophyll patterns and biospheres on global scales by using the advanced atmospheric correction algorithm for the data processing (Gordon, 1997; Gordon and Wang, 1994; IOCCG, 2010). For the global open oceans, both SeaWiFS and MODIS have been producing high-quality ocean color products (Bailey and Werdell, 2006; McClain *et al.*, 2004), and these data have been used by researchers and scientists worldwide to study and understand the ocean's physical, optical, and biological changes and their effects on climate processes, as discussed in Behrenfeld *et al.* (2001), Chavez *et al.* (1999), and others.

However, in NASA's standard ocean color data processing system for MODIS, turbid waters caused by sediment mix-up and re-suspension may be misidentified as an enhanced chlorophyll-a (Chl-a) concentration and phytoplankton bloom due to very significant ocean contributions in near-infrared (NIR) wavelengths (Siegel *et al.*, 2000; Lavender *et al.*, 2005; Wang and Shi, 2005). To distinguish a phytoplankton bloom from hurricane-induced turbid waters, the shortwave infrared (SWIR) atmospheric correction algorithm (Wang and Shi, 2005; Wang, 2007; Wang *et al.*, 2009) is used to derive the normalized water-leaving radiance spectra from the blue to the NIR bands, as well as the Chl-a concentration following Hurricane Katrina (Shi and Wang, 2007). Compared with the NIR bands, waters are much strongly absorbing at the SWIR wavelengths (Hale and Querry, 1973). Thus, the black pixel assumption (i.e., zero ocean radiance contribution) at the SWIR wavelengths can be used for turbid and productive waters in deriving satellite ocean color products (Shi and Wang, 2009).

Satellite observations of SST and sea surface height (SSH) anomalies are also very useful for the investigation of hurricane impacts on upper ocean physical processes. Hurricanes are known to lower SST significantly during their passage, and this upper-ocean process depends on many factors such as mixed-layer depth, thermocline depth, air-sea heat fluxes, and the storm's intensity and transition speed. Despite its advantage of high spatial resolution, AVHRR SST measurement is available only in clear-sky conditions. On the other hand, Advanced Microwave Scanning Radiometer EOS (AMSR-E) on Aqua provides microwave-derived SST under both cloudy and clear-sky conditions, and is therefore very useful for continuous observations for periods before, during, and after hurricane passage. *Shi and Wang* (2007) used AMSR-E data to study the time series of SST in the Hurricane Katrina-induced phytoplankton bloom event in the Gulf of Mexico (GOM). SSH anomaly data derived from satellite altimetry measurements can provide mesoscale features like warm- or cold-core eddies in the ocean, which are sometimes not available in the SST measurement. Pre-existing mesoscale eddies on the hurricane track in the ocean have significant impacts on the upper ocean response to hurricanes. Since ocean eddies modulate the upper ocean's vertical and horizontal structures and heat balance, they are also very important in the oceanic response process to hurricanes. When a hurricane enters the warm (cold) core eddy, SST cooling may be greatly suppressed (intensified) due primarily to deep (shallow) mixed-layer and thermocline depth.

The GOM is greatly influenced by its unique ocean circulation feature, i.e., the Loop Current (LC). Warm water from the Caribbean Sea enters the Gulf of Mexico through the Yucatan Straits and then forms the LC. The LC penetrates northward into the Gulf of Mexico and then turns anti-cyclonically to flow southward along the steep west Florida shelf before turning eastward into the southern Straits of Florida (SSF). The Straits of Florida forms a conduit for the Florida Current (FC), which connects the LC with the Gulf Stream in the North Atlantic. The instability processes that are associated with LC shed warm core eddies (WCEs), which propagate westward and eventually dissipate along the shelf break. Cyclonic, cold-core eddies (CCEs), called Loop Current Frontal Eddies (LCFEs), are also formed along the shoreward boundary of the LC (Paluszkiewicz *et al.*, 1983; Vukovich, 1988; Fratantoni *et al.*, 1998; Romanou *et al.*, 2004). LCFEs travel rapidly southward downstream with the LC at speeds of 20–50 cm/s. Since the pre-existing oceanic features such as LC, WCE, and CCE can modulate the upper ocean vertical and horizontal structures and heat balance, they are also very important in the oceanic response process to hurricanes. When a hurricane enters the warm (cold) core eddy, SST cooling may be greatly suppressed (intensified) due primarily to deep (shallow) mixed-layer and thermocline depth. Using the sea surface height (SSH) anomaly data derived from satellite altimetry measurements, *Liu et al.* (2009) reported that the Hurricane Katrina-induced phytoplankton bloom was coincident with a cyclonic eddy near the Dry Tortugas, and the pre-existing eddy played an important role in the phytoplankton bloom event.

Numerical simulation is an important approach to studying the upper ocean response to hurricanes. With a three-dimensional numerical model, the variations of ocean surface properties, such as SST and SSH, can be projected to water columns below the surface. *Price* (1981 and 1994) used a three-dimensional, primitive-equation, hydrostatic model that represents ocean upper mixed-layer and vertical structure on a fixed grid to simulate the physical processes of the hurricane-induced upwelling and mixing. *Liu et al.* (2009) used the Hybrid Coordinate Ocean Model (HYCOM) (Chassignet *et al.*, 2003; 2007) to study the effect of the pre-existing eddy in the Hurricane Katrina-induced phytoplankton bloom in the GOM in 2005. HYCOM is a primitive-equation, ocean general circulation model that was evolved from the Miami Isopycnic-Coordinate Ocean Model (MICOM) (Bleck *et al.*, 1992). In the study, the upper layer heat budget of the physical process was analyzed, and the nutrient enhancement in the upper layer was also simulated.

2. Upper-ocean physical response to hurricanes

Early observations of upper-ocean response to hurricanes have been reported and studied by *Leipper* (1967), *Brooks* (1983), *Sanford et al.* (1987), *Shay and Elsberry* (1987), *Shay et al.* (1989; 1998), *Brink* (1989), *Dickey et al.* (1998), and *Jacob et al.* (2000). *Price* (1981) summarized early hydrographic surveys and found two prominent features of the SST response to hurricanes: (1) hurricane-induced SST cooling increases as a hurricane's moving speed decreases and intensity increases; and (2) the SST response is markedly asymmetrical about the hurricane track. Hurricane-force winds (speeds greater than 33 m s^{-1}) have dramatic effects on the upper ocean. Frequently, near-surface waters cool several degrees Celsius as the mixed layer deepens by tens of meters with the most intense changes occurring under the more intense winds on the right-hand side (left-hand side in the Southern Hemisphere) of the storm track (Hazelworth, 1968; Dickey and Simpson, 1983; Stramma *et al.*, 1986; Sanford *et al.*, 1987). Conversely, downward mixing of heat causes the upper seasonal thermocline waters to warm. High-amplitude, near-inertial internal gravity waves (with vertical displacements of isotherms of a few tens of meters) and currents (around 1 m s^{-1}) are induced, which persist for several days, and geostrophic flows may also be produced (Shay and Elsberry, 1987; Shay *et al.*, 1989, 1998; Dickey *et al.*, 1998b; Zedler *et al.*, 2002). The SST decrease is a result of both enhanced vertical mixing and upwelling induced by a near-inertial response of the oceanic mixed-layer to the asymmetric surface wind stress (Price, 1981; Shay and Elsberry, 1987), as well as flux-induced cooling.

The upper ocean response to a particular hurricane depends on several parameters involving atmospheric and oceanic variables (e.g., Price, 1981; Dickey *et al.*, 1998b). Some of the important atmospheric variables include hurricane size (e.g., radius of tropical storm force winds, radius of hurricane-force winds), strength (wind speed), and transit speed. Intense, slowly moving hurricanes (~4 m s^{-1}) cause the most significant upwelling and the largest SST response (Price, 1981). Local hydrodynamic conditions, i.e., pre-existing stratification and near-inertial currents, also play an important role in the oceanic response to a hurricane, while the vertical distributions of nutrients and phytoplankton are primary factors in defining the resulting biogeochemical response.

Brink (1989) collected current data at depths from 159 to 1059 m at a location on the path of Hurricane Gloria in 1985 in the western North Atlantic and characterized the response of the deep thermocline and the downward propagation of near-inertial energy. *Dickey et al.* (1998) studied the Bermuda testbed mooring data during the passage of Hurricane Felix in August 1995. It was found that as Hurricane Felix passed the mooring, large inertial currents were generated within the upper layer by the onset of intense localized wind stress and lasted a period of approximately 22.8 hours. The current shear was the greatest at the base of the mixed layer, where deep cooler waters were entrained into the mixed layer resulting in cooling of the surface layer. The mixed-layer depth was about 15 m prior to the arrival of Felix, and deepened to about 45 m within three days after Felix's passage, according to the mooring. The temperature at 25 m decreased by 3.5–4.0°C and the temperature at 45 m increased by about 2.0°C through the mixing process. Temperatures at 71 m and greater depths decreased slightly.

Three-dimensional ocean numerical models were also used to study the physical process (Price, 1981; Price *et al.*, 1994; Prasad and Hogan; 2007). *Price et al.* (1994) used a three-dimensional model to simulate ocean response to a moving hurricane and found that the ocean's response to hurricanes can be divided into two stages: forced and relaxation. In the forced stage, hurricane-force winds drive the mixed-layer currents, SST cooling by vertical mixing (entrainment), and air-sea heat exchanges (mainly due to loss of latent heat flux). The barotropic response consists of a geostrophic current and an associated trough in sea surface height. The relaxation stage response following a hurricane's passage is primarily due to inertial-gravity oscillations excited by the storm. The mixed-layer velocity oscillates with a near-inertial period, as do the divergence and associated upwelling and downwelling.

Entrainment has been emphasized as the dominant term in lowering the SST beneath a moving hurricane. Based on observations, *Jacob et al.* (2000) suggested that entrainment at the mixed-layer base generally accounts for ~75% to 90% of the cooling, while the *Price* (1981) model indicated that ~85% of heat flux into the mixed layer was through entrainment. Only about 10% to 15% of the cooling in the upper ocean is due to surface heat fluxes, which would range between 2000 and 3000 W m^{-2}. Estimations from *Jacob and Shay* (2003) ranged from ~10% to ~30% in the directly forced region. Horizontal advection is also found to be important in the mixed-layer heat balance during and subsequent to the passage of hurricanes (Price, 1981; Jacob *et al.*, 2000). This contribution is particularly significant in the eddy region, where maximum cooling due to geostrophic advection (-0.69 °C d^{-1}) was as large as the surface heat flux term in the overall heat budget (Jacob *et al.*, 2000).

3. Satellite observations of hurricane-induced phytoplankton blooms

With the advancement of satellite ocean color remote sensing from SeaWiFS and MODIS, hurricane-induced phytoplankton blooms were also observed and reported in the past decade (Babin *et al.*, 2004; Lin *et al.*, 2003; Miller *et al.*, 2006; Walker *et al.*, 2005). Hurricane-induced mixing, entrainment, and upwelling result in an increase in near-surface phytoplankton because phytoplankton is brought closer to the surface so that their photosynthetic system receives greater solar irradiance (Babin *et al.*, 2004) and also because requisite nutrients are transported into the euphotic zone. On the other hand, the change of the phytoplankton in the ocean may lead to a time-lagged response of the colored dissolved organic matter (CDOM) (Hu *et al.*, 2006). Enhanced CDOM induced by the storm in the upper ocean layer has also been reported (Hoge and Lyon, 2002). In coastal regions, however, entrainment and upwelling can also result in sediment re-suspension and transport from nearby locations (Acker *et al.*, 2002; Wren and Leonard, 2005). *Hu and Muller-Karger* (2007) also reported the hurricane-induced sediment re-suspension in the eastern Gulf of Mexico. To distinguish the phytoplankton bloom from turbid waters resulting from hurricane-induced sediment re-suspension and transportation, *Shi and Wang* (2007) used the SWIR atmospheric correction algorithm to derive the Chl-a concentration as well as the spectral normalized water-leaving radiance from the blue to the NIR bands. The normalized water-leaving radiance spectra were also used to discriminate the phytoplankton bloom from the storm-induced CDOM.

Babin et al. (2004) examined the passages of 13 hurricanes through the Sargasso Sea region of the North Atlantic during the years of 1998 through 2001, and the remotely sensed chlorophyll-a data from SeaWiFS were used to analyze the hurricane-induced phytoplankton blooms. The SeaWiFS data shows increased concentrations of surface chlorophyll within the cool wakes of the hurricanes, apparently in response to the injection of nutrients and/or biogenic pigments into the oligotrophic surface waters. The observed increases in Chl-a in the wakes of these hurricanes ranged from a minimum of ~5% to a maximum of ~91%. This maximum is comparable to the spring bloom near Bermuda, which shows an increase of over 100% but with absolute magnitudes more pronounced than our observations of hurricane wakes. This increase in post-storm surface chlorophyll-a concentration usually lasted about two to three weeks before it returned to its nominal pre-hurricane level.

Lin et al. (2003) presented evidence of the Tropical Cyclone Kai-Tak-induced phytoplankton bloom in the South China Sea (SCS) in July 2000. During its short three-day stay, Kai-Tak triggered an average 30-fold increase in surface chlorophyll-a concentration. The estimated carbon fixation resulting from this event alone is 0.8 Mt, or ~2–4% of SCS's annual new production. Given an average of ~14 cyclones passing over the SCS annually, it was suggested that the long-neglected contribution of tropical cyclones to SCS's annual new production may be as much as ~20–30%. In addition, *Walker et al.* (2005) analyzed the remotely sensed SST and Chl-a data subsequent to Hurricane Ivan's passage across the Gulf of Mexico in September 2004, and revealed sea surface cooling of 3–7°C and elevated Chl-a concentrations over two large areas along Ivan's track. Within the northern area of cooling, concentrations increased from 0.36 mg/m^3 (before Ivan) to 0.81 mg/m^3 (after Ivan) over a region of ~27,000 km^2. In the southern region, mean Chl-a concentrations increased from 0.24 mg/m^3 to 0.99 mg/m^3 over a region of ~13,000 km^2. Peak concentrations lagged Ivan's passage by 3 and 4 days in the northern and southern features, respectively. *Walker et al.* (2005) also reported that the increases of hurricane-forced upwelling and chlorophyll-a concentrations could be enhanced within cold-core cyclones in the Gulf of Mexico.

4. Hurricane Katrina-Induced phytoplankton bloom

Shi and Wang (2007) reported a Hurricane Katrina-induced SST cooling and a phytoplankton bloom in the Gulf of Mexico in August 2005. In their study, combined data sets of sea surface wind from the National Center for Environmental Prediction (NCEP), SST from AMSR-E, and ocean color products from MODIS-Aqua are used to analyze physical, optical, and biological processes after Hurricane Katrina. MODIS-Aqua data were processed using the SWIR atmospheric correction algorithm (Wang, 2007; Wang and Shi, 2005). It has been demonstrated that, compared to using the NASA standard algorithm (Gordon and Wang, 1994), use of the SWIR algorithm ocean color products can be improved over productive ocean waters (Wang *et al.*, 2007; 2009). Therefore, the phytoplankton bloom can be distinguished from sediment re-suspension and transportation, which can also be induced by a hurricane in coastal waters. The AMSR-E measurements produce the microwave SST under both cloudy and clear sky conditions (Wentz *et al.*, 2000). Therefore, AMSR-E data are quite useful for continuous observations of the SST for the periods before, during, and after hurricane passage.

Hurricane Katrina formed as a tropical depression over the southeastern Bahamas on August 23, 2005, and was then upgraded to tropical storm status on August 24. The tropical storm continued to move towards Florida, and became a hurricane near 2100 UTC on August 25, 2005, only two hours before it made landfall on the southeastern coast of Florida. The storm weakened over land, but it regained hurricane status about one hour after entering the Gulf of Mexico. The storm rapidly intensified after entering the Gulf because of the storm's movement over the LC warm waters. On August 27, the storm reached Category 3 intensity, becoming the third major hurricane of the season. Katrina again rapidly intensified, attaining Category 5 status on August 28. The pressure measurement made Katrina the fourth most intense Atlantic hurricane on record, and it was also the strongest hurricane ever recorded in the Gulf of Mexico at the time. Katrina made its second landfall on August 29 as a Category 3 hurricane with sustained winds of 125 mph (205 km/h) near Buras-Triumph, Louisiana.

Shi and Wang (2007) analyzed Hurricane Katrina wind fields from NCEP and estimated the upwelling speed induced by the wind stress curl. The track of Hurricane Katrina is outlined in red as shown in Figure 1 (Shi and Wang, 2007), and the hurricane eye location is marked by a red square at 6-hour intervals. Also shown in Fig. 1 is a snapshot of the Katrina wind field at 0600 UTC, August 27th from NCEP, when the hurricane eye was located at a latitude and longitude of (24.4°N, 84.0°W), and the maximum sustained wind speed reached ~30 m/s. At this location, Hurricane Katrina was also found moving slowly with a speed of ~3 m/s, about a half of the moving speed before landing in New Orleans. The wind stress curl-induced ocean upwelling was calculated by *Shi and Wang* (2007) as shown in Fig. 2. The highest upwelling velocity is estimated to be ~$2.0x10^{-3}$ m/s, and the strongest upwelling is located almost in the hurricane center.

A notable phytoplankton bloom centered at a location of (24°N, 84°W) was observed four days after Katrina's passage (Shi and Wang, 2007). Before Katrina, Chl-a concentration was low at ~0.1-0.2 mg/m^3 for the whole Gulf region. Two days after the passage of Hurricane Katrina, enhanced Chl-a concentration can be observed in the large regions east of the Gulf of Mexico. A notable patch of high Chl-a concentration with value ~1.5 mg/m3 was observed. The elevated Chl-a region coincides with the location of the hurricane eye and the area with a strong upwelling. On September 7, while the patch of phytoplankton bloom can still be observed, Chl-a magnitude decreased significantly with values at ~0.2 to 0.3 mg/m^3. Basin-wide, the Chl-a concentration level dropped back to the pre-Katrina level.

Figure 3 provides, quantitatively, the time series of the wind speed, SST, and Chl-a concentration at the center of the bloom located at (24.19°N, 84.06°W). Due to frequent cloud coverage, only nine Chl-a data were obtained. Obviously, the ocean physical response (SST) and the biological response (Chl-a) lag the hurricane winds. Before Katrina, SST at this location was ~31°C and Chl-a concentration ~0.1 mg/m^3. The wind speed started to pick up on August 25, and reached the maximum on August 27. However, there was no substantial

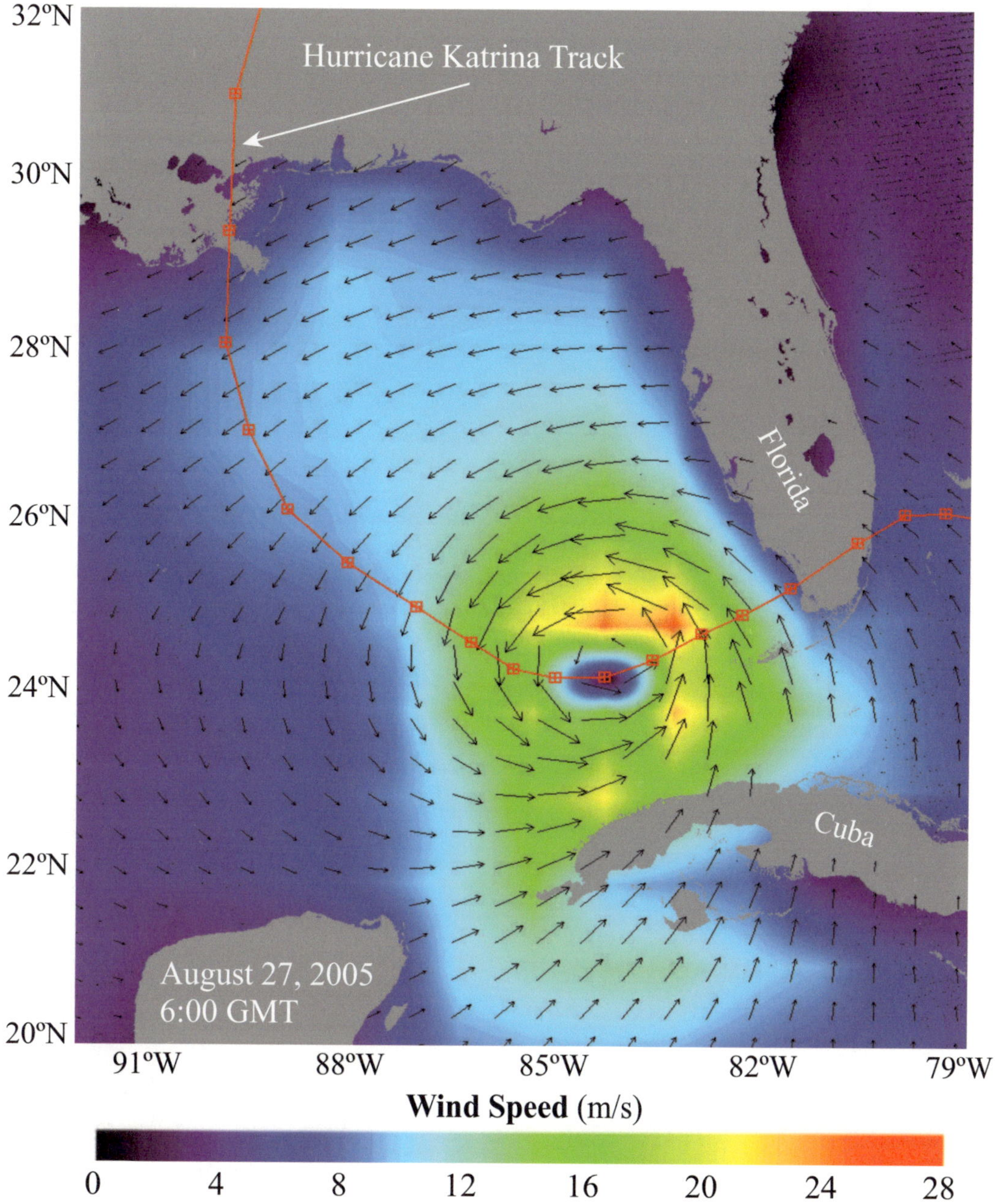

Fig. 1. Snapshot of Hurricane Katrina wind fields at 0600 UTC, August 27, 2005, with the track of the hurricane outlined in red (reproduced from *Shi and Wang* (2007)). The hurricane eye location is marked by red squares at intervals of every 6 hours.

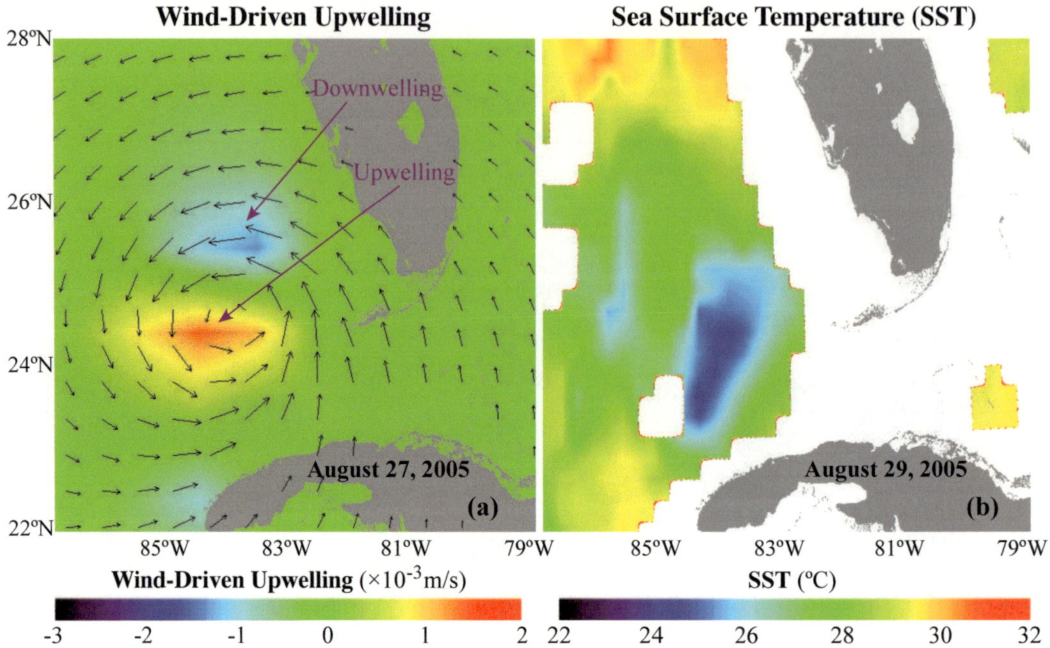

Fig. 2. Hurricane Katrina-induced ocean physical responses as presented in images of (a) the wind-driven upwelling on August 27, 2005 and (b) the microwave SST on August 29, 2005 (reproduced from *Shi and Wang* (2007)).

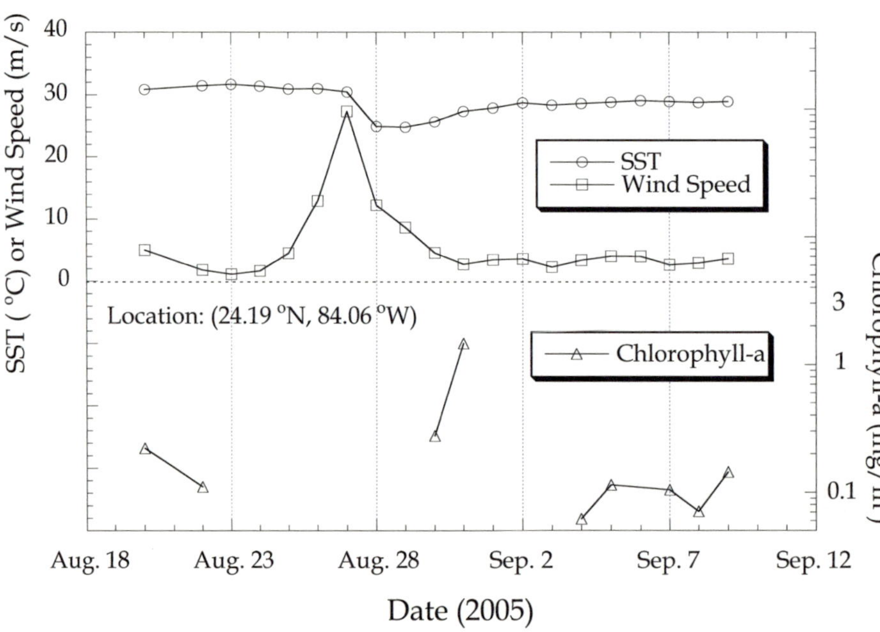

Fig. 3. The time series of the wind speed, SST, and Chl-a concentration at the center of the bloom located at (24.19°N, 84.06°W) from August 20 to September 9, 2005 (reproduced from *Shi and Wang* (2007)).

SST drop during this period. The SST at this location reached the lowest level on August 28, one day after the hurricane passed. Unfortunately, there was no ocean color data available during the hurricane period. The phytoplankton bloom can be seen picking up slightly on August 30 with Chl-a concentration of ~0.3 mg/m^3 from the pre-Katrina value of ~0.1 mg/m^3. It reached ~1.5 mg/m^3 on the next day, August 31. One week later on September 7, SST gradually increased to ~29°C, but still below the pre-Katrina SST. The phytoplankton bloom also decayed quickly, and Chl-a values returned to pre-Katrina levels as early as September 4.

Nutrients such as nitrate and phosphate are critical to phytoplankton growth in the ocean. *Shi and Wang* (2007) analyzed the water temperature and nutrient fields from the climatology data set (Levitus *et al.*, 1994), and found from the after-Katrina SST value that the waters were between a depth of ~80 and 100 m at the phytoplankton bloom location. At that depth, the nitrate and phosphate concentrations are nearly four to five times higher than that at the surface values, according to the climatology data (Levitus *et al.*, 1994). Following strong upwelling and vertical entrainment mixing, however, nutrient concentration could be elevated on the surface and provide the required nutrient supply for the phytoplankton bloom. It was concluded that the phytoplankton bloom was triggered by abnormally high nutrient concentrations due to a strong entrainment mixing and upwelling following the passage of Hurricane Katrina.

5. Numerical simulation of hurricane Katrina-induced phytoplankton bloom

Liu et al. (2009) presented some further evidence from SSH anomaly data derived from satellite altimetry measurements obtained from the Colorado Center for Astrodynamics Research (CCAR) at the University of Colorado, which shows that the Hurricane Katrina-induced phytoplankton bloom near (24°N, 84°W) was coincident with a cyclonic eddy near the Dry Tortugas (Fig. 4). These observations indicate that the cyclonic eddy might have played a critical role in the phytoplankton bloom event. Near the Dry Tortugas, the Tortugas gyre often occurs when the southward flowing LC overshoots entry into the Straits of Florida and approaches the coast of Cuba before abruptly turning to the east (Lee *et al.*, 1992; 1995). This offshore position of the Florida Current, combined with the strong cyclonic curvature of the flow field, can cause a counterclockwise recirculation off the Tortugas, i.e., the "Tortugas gyre" (Lee *et al.*, 1995). In a cyclonic eddy, the cold nutrient-rich waters from the deep layer stimulate primary production.

To further investigate the mechanism of the Katrina-induced phytoplankton bloom and the role of the pre-existing cyclonic eddy, *Liu et al.* (2009) simulated the process of the upper ocean response to Hurricane Katrina using the HYCOM model (Bleck, 2002). Particularly, two experiments with and without the presence of a pre-existing cyclone eddy near the Dry Tortugas were simulated and the results were compared. In addition to the standard physical parameters, such as temperature and salinity in HYCOM, three nutrient parameters—nitrate, phosphate, and silicate—were also included in the simulation to study the process of the Katrina-induced nutrient enhancement in the upper ocean.

To evaluate the importance of the cold-core eddy on the physical, chemical, and biological responses of the ocean to hurricane forcing, and to identify the mechanism of the hurricane-driven bloom, the north Atlantic HYCOM model configuration developed by *Xie et al.* (2007) was used to spin up the HYCOM model. Two cases with and without cold-core front eddy were then identified. The ocean conditions for these two cases were set up as the initial

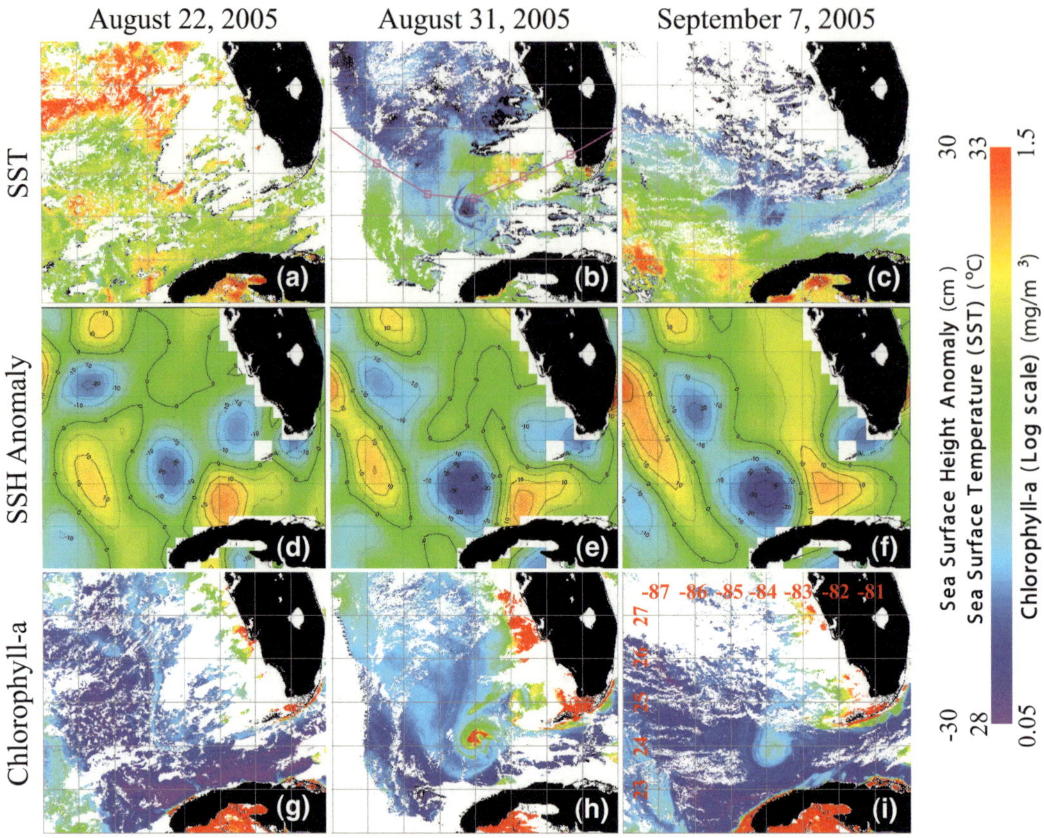

Fig. 4. Satellite-derived (a–c) MODIS SST, (d–f) SSH anomaly, and (g–i) MODIS chlorophyll-a concentration on August 22, August 31, and September 7, 2005, respectively. SSH anomaly data is from Center for Astrodynamics Research (CCAR). Plot (b) also shows the Hurricane Katrina eye locations (marked by a pink square) at 6-hour intervals. Latitude and longitude for the region are indicated in plot (i). Figure 4 was reproduced from *Liu et al.* (2009).

condition, and then wind forcing associated with Hurricane Katrina was applied to these two cases. The model domain covers the north and equatorial Atlantic (98°W–14°E, 17°S–52°N). The horizontal resolution is 0.18° and the vertical resolution is 22 layers that move vertically as a function of total depth. The model was spun up for one year with an initial condition constructed from the Levitus Data Set (Levitus, 1982) and forced by the Comprehensive Ocean Atmosphere Data Set's (COADS) monthly climatology atmospheric forcing including wind, air-temperature, relative humidity, and short-wave and long-wave radiations. The results were then used as the initial conditions of the hurricane response for simulations.

In the study, two cases of LC states near the Dry Tortugas on two different model days were chosen: Experiment I, with the presence of a cyclone eddy near (24°N, 84°W); and Experiment II, without the presence of a cyclone eddy near the Dry Tortugas. Since the differences between the mean monthly climatology of July and August in the vertical structure are small when compared with the differences between the two cases with and without the presence of the eddy, it was assumed that the differences in the simulations between the two cases were mainly caused by the eddy. The annual mean fields of nitrate, phosphate, and silicate were extracted from the *World Ocean Atlas 2005* and set as the initial conditions of the nutrients at the beginning of both experiments. The two cases of hurricane simulations were carried out from August 25-31, 2005, and driven by the NCEP atmospheric forcing fields of wind, air-temperature, and relative humidity at the ocean surface.

Figure 5 shows the SST map near the LC before and after Hurricane Katrina's passage in Experiment I (Figs. 5a and 5b) and Experiment II (Figs. 5c and 5d). In both cases, the area was dominated by warm SSTs of ~30°C on August 27 before Hurricane Katrina (Figs. 5a and 5c). However, after Hurricane Katrina, the SST cooling shows quite large differences between the two cases: in Experiment I (Fig. 5b), the maximum SST drop (~4°C) is at the location of the cold-core eddy, and the SST cooling inside of the LC frontal eddy is much more significant than other areas; in Experiment II (Fig. 5d), SST cooling shows no significant difference in the observed phytoplankton bloom location than other areas on the hurricane track.

The time series of vertical temperature profile and mixed-layer depth near the location of the observed phytoplankton bloom (24°N, 84°W) in the two experiments are shown in Figure 6. Before Hurricane Katrina, the surface waters down to ~30 m depth were dominated by warm temperature >29°C in both cases. However, the vertical structure below

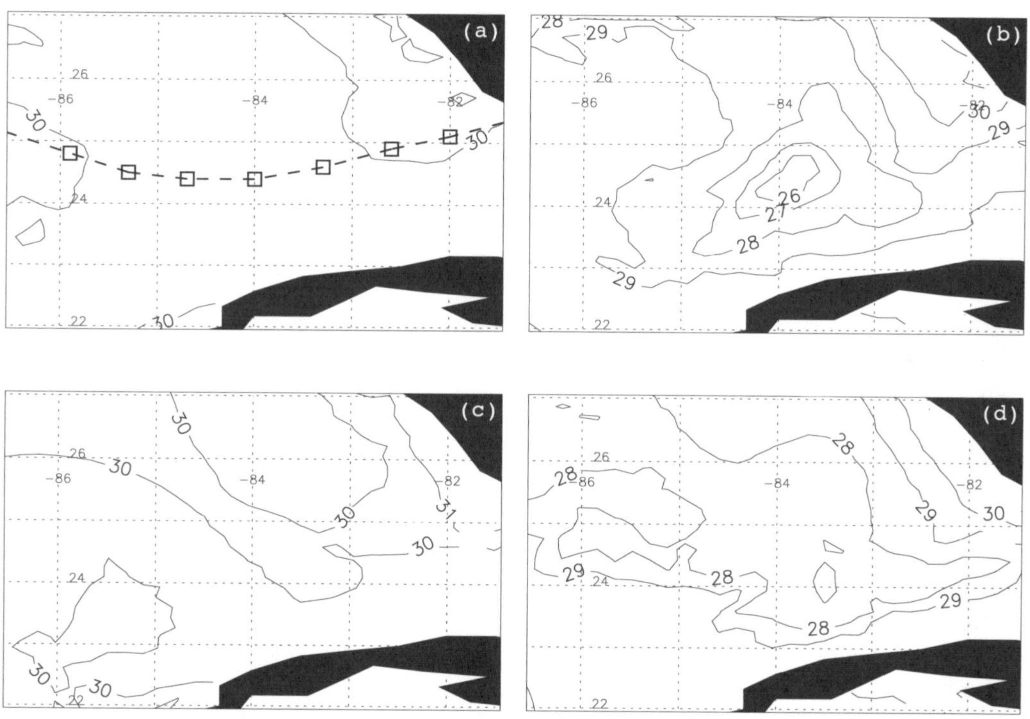

Fig. 5. Model simulated SST in (a–b) Experiment I and (c–d) Experiment II (units: °C). Plots (a) and (c) are on August 27, and plots (b) and (d) are on August 29, respectively. Katrina's track is overlaid in plot (a) as a dashed line with square symbols. Figure 5 was reproduced from *Liu et al.* (2009).

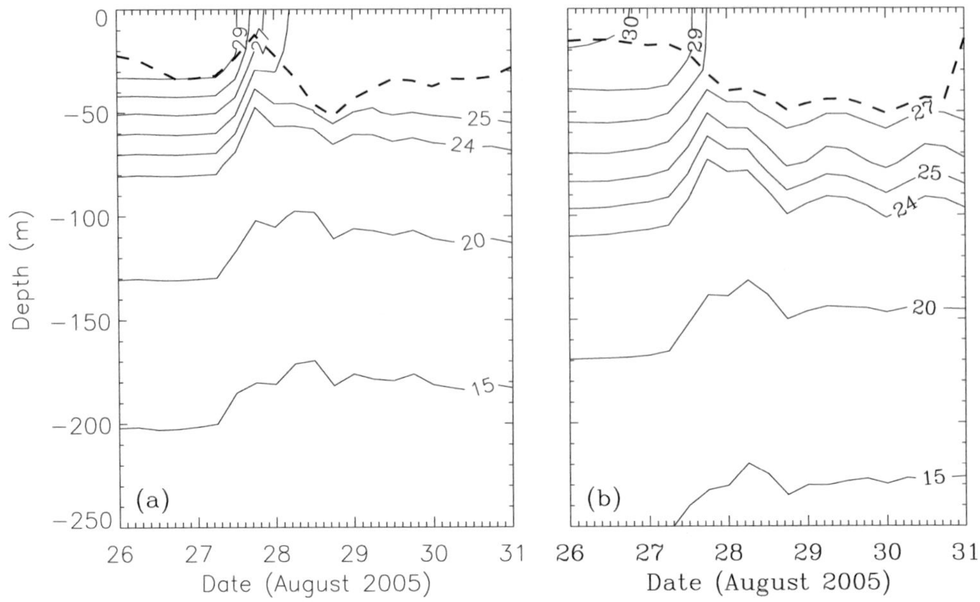

Fig. 6. Time series of vertical temperature profile at (24°N, 84°W) in (a) Experiment I and (b) Experiment II (units: °C). Mixed-layer depth is indicated as the dashed line (reproduced from *Liu et al.* (2009)).

the surface shows a large difference between the two cases due to different LC states. Without the presence of the cyclone eddy in Experiment II (Fig. 6b), the location of the observed phytoplankton bloom is on the path of the LC, and, thus, is dominated by warm LC waters. In this case, the isotherms are deep, e.g., the 26°C isotherm is at ~85 m depth. With the presence of the cyclone eddy in Experiment I (Fig. 6a), the isotherms are elevated, and the 26°C isotherm is now at ~60 m, which is 25 m shallower than in Experiment II. In Experiment I, strong hurricane-induced upwelling lifted up the isotherms by ~30 m on August 27, and the mixed-layer depth was also decreased. Due to strong vertical diffusion, the mixed-layer depth deepened by ~40 m on August 28, and then it started to decrease gradually. The maximum mixed-layer depth was about 50 m, and coldest temperature below the mixed-layer depth was 25°C, which was originally located at ~70 m depth before Katrina. Thus, the deepest layer that contributed to the surface cooling was at ~70 m depth. The significant mixed-layer temperature cooling occurred between August 27 and 28, which was similar to satellite SST observations. In Experiment II, because the location was under the influence of LC advection, changes of mixed-layer depth and vertical profiles were different from those in Experiment I. The upwelling and vertical diffusion processes were not as strong as in Experiment I, and the isotherms were deeper. The 27°C isotherm was well below the bottom of the mixed layer during and after Katrina's passage, never reaching the surface.

The initial surface nitrate, phosphate, and silicate concentrations were about 0.8μM, 0.1μM and 1.6μM, respectively, in both cases before Hurricane Katrina. The basic mechanism of nutrient response was the same as temperature: deep nutrient-rich waters were brought up to the surface by hurricane-induced strong upwelling and vertical diffusion processes during the Katrina passage. Figure 7 shows the post-Katrina/prior-Katrina ratio of surface

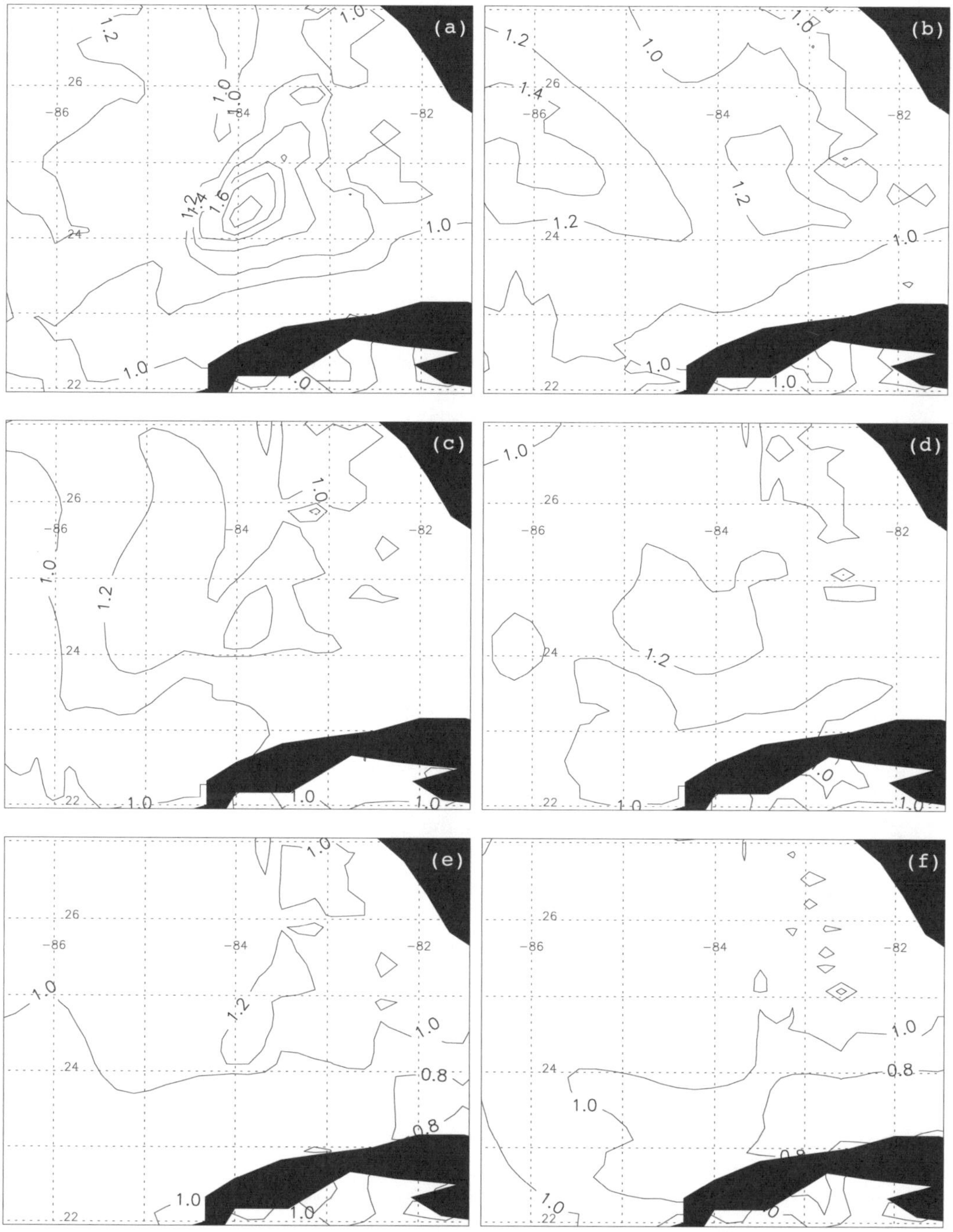

Fig. 7. Post-Katrina/prior Katrina ratio of (a–b) nitrate, (c–d) phosphate, and (e–f) silicate. Plots (a), (c) and (e) are from Experiment I, and plots (b), (d), and (f) are from Experiment II. Data from August 27 and August 29 are used as the prior-Katrina and post-Katrina data, respectively. Figure 7 was reproduced from *Liu et al.* (2009).

nitrate, phosphate, and silicate concentrations. In Experiment I, all three types of nutrient concentrations within the cyclone eddy increased more significantly than in other areas on the hurricane track. The maximum nutrient concentration increase occurred at the location of the observed phytoplankton bloom (24°N, 84°W). The nitrate concentration increased most significantly by ~100% at the location of the observed phytoplankton bloom, while phosphate and silicate concentration increased ~40% and ~20%, respectively. In Experiment II, nitrate, phosphate, and silicate concentrations at (24°N, 84°W) did not show more significant increases than in other areas on the hurricane track. In general, the enhancement of all three nutrient concentrations was less than 20%.

Figure 8 shows the time series of nitrate, phosphate, and silicate vertical profiles at (24°N, 84°W). Before Katrina, the subsurface nitrate concentrations in Experiment I were higher than in Experiment II at the same depth (Figs. 8a and 8b). For example, at a 70 m depth, the nitrate concentration was ~2.0 μM in Experiment I, while it was only ~1.2 μM in Experiment II. This was primarily due to cyclone eddy-induced upwelling, which elevated the nitrate levels in Experiment I. The differences in the initial vertical structure between the two cases consequently led to different results in the nitrate response to Katrina. In Experiment I, waters at a 70 m depth with a nitrate concentration of ~2.0 μM were lifted up by ~30 m and further entrained into the mixed layer, which significantly increased the nitrate concentrations at the surface. The surface nitrate concentration increased dramatically between August 27 and 28, and the increase was in phase with SST cooling as shown in Fig. 6a. In Experiment II, however, the 2.0 μM contour line is well below the bottom of the mixed-layer, and does not contribute to the surface nitrate enhancement. The maximum nitrate concentration at the mixed-layer base is about 1.2 μM, and thus the change of the nitrate concentration at the surface is less significant compared with Experiment I.

The phosphate vertical profile (Figs. 8c and 8d) shows a similar pattern to that of nitrate in the two experiments. In Experiment I, the surface phosphate concentration increased from ~0.1 μM to ~0.14 μM after Katrina passed near the location of the observed phytoplankton bloom. Before Katrina, the 0.14 μM contour line was originally at ~70 m depth, and it was uplifted and further entrained into the mixed layer during the Katrina passage. The 0.16 μM contour line, which was originally at ~80 m depth, never reached the surface. Therefore, the impact of the hurricane-induced upwelling and vertical diffusion can reach ~70 m depth, which was consistent with the results of nitrate concentration. In Experiment II, the contour lines are much deeper than in Experiment I. In this case, waters with 0.14 μM phosphate concentration were originally at 90 m depth and well below the bottom of the mixed layer during the Katrina passage. Before Katrina, the phosphate concentration at a 70 m depth was ~0.12 μM. Although it was brought up to the surface by the upwelling and vertical diffusion, it did not enhance the surface phosphate concentration due to little vertical gradient in the top 70 meters.

The surface silicate concentrations show much less significant changes after Katrina's passage compared with nitrate and phosphate (Figs. 8e and 8f). Before Katrina, although the silicate concentration was higher in Experiment I than at the same depth in Experiment II, it can be seen that in both cases, the vertical gradient of silicate concentration was minimal in the top 70 m depth. Consequently, the upwelling and vertical diffusion process does not significantly increase the surface silicate levels in either of the experiments.

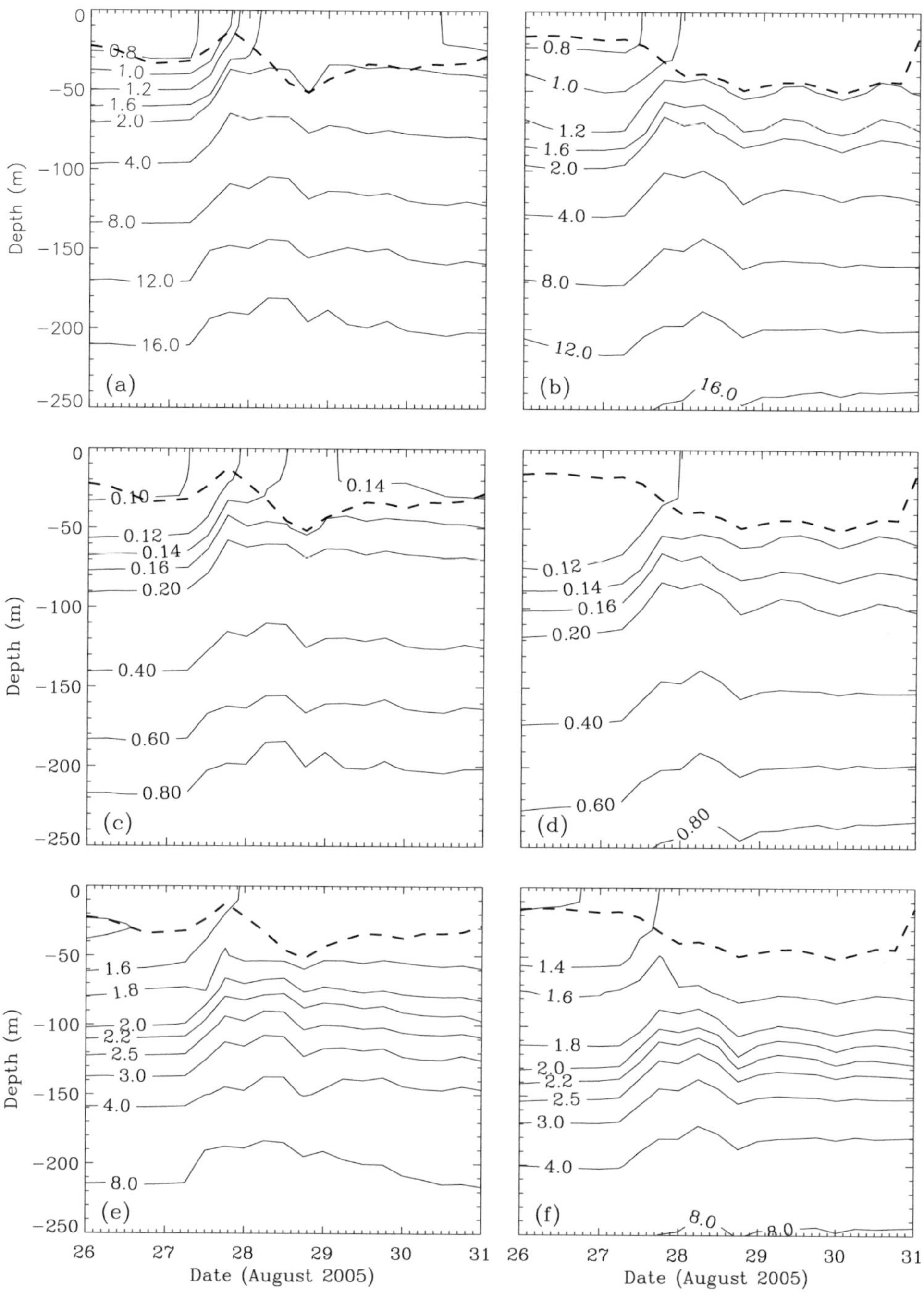

Fig. 8. Time series of vertical profile of (a–b) nitrate, (c–d) phosphate, and (e–f) silicate at (24°N, 84°W). Plots (a), (c), and (e) are from Experiment I, and plots (b), (d), and (f) are from Experiment II (units: µM). Mixed-layer depth is indicated as the dashed line. Figure 8 was reproduced from *Liu et al.* (2009).

6. Discussions and summary

In this chapter, we reviewed the status of research on the upper-ocean physical response to hurricanes and the hurricane-induced phytoplankton blooms observed by the satellite remote sensing. Using the phytoplankton bloom triggered by Hurricane Katrina as an example, a notable phytoplankton bloom triggered by hurricane wind-driven upwelling and vertical mixing is analyzed and quantified with combined data sets of the NCEP winds, AMSR-E microwave SST, and MODIS ocean color products. MODIS-Aqua ocean color observation shows that the hurricane-driven phytoplankton bloom lasted less than one week. By looking at both the blue and the NIR ocean contributions, we conclude that the phytoplankton bloom is the only source for the ocean surface optical property change after Hurricane Katrina. In addition, AMSR-E SST and MODIS-Aqua ocean color observations show that physical and biological responses to the hurricane are not synchronized with the hurricane winds. Indeed, it is found that the ocean physical response as represented by SST lags the hurricane peak wind speed by about one day, while the maximum biological response could be observed nearly four days later after the passing of the hurricane.

Indeed, satellite observations (e.g., AMSR-E and MODIS-Aqua) provide us with effective tools to monitor the physical, optical, geochemical, and biological changes of the ocean environments following an extreme weather event such as a hurricane. They also extend further evidence that the tropical cyclone is an important mechanism to pump nutrients into the upper euphotic zone and result in significant phytoplankton blooms, thereby leading to an increase of the ocean's primary production.

Even though satellite observations provide insights on the phytoplankton and its relationship with the physical response and nutrient supply changes following a hurricane, the location of the phytoplankton bloom at (24°N, 84°W) cannot be explained only by upwelling and vertical diffusion associated with Hurricane Katrina. In fact, it has been well documented that the maximum SST drop usually occurs to the right of the hurricane track because of the asymmetry in turning direction of the wind-stress vector that drives a very strong asymmetry in the mixed-layer velocity (Price, 1981). On the contrary, the observed Katrina-induced maximum SST drop and phytoplankton bloom at (24°N, 84°W) occurred on the left of the hurricane track. According to a National Hurricane Center report (www.nhc.noaa.gov/pdf/TCR-AL122005_katrina.pdf), the center of Hurricane Katrina was at (24.5°N, 84°W) while moving towards the west on August 27 at 0600 UTC. Thus, the location of the observed maximum SST drop and the center of the phytoplankton bloom is 20–30 km left on the hurricane track. The co-location of the cyclone eddy and the phytoplankton bloom explains this contradiction to the classic rightward-bias response theories: the pre-existing cold-core eddy at (24°N, 84°W) played an important role in the bloom event.

The numerical simulations in this study have further supported this hypothesis. The simulation results show that the SST cooling is more significant in the presence of the cyclone eddy at the location of the observed phytoplankton bloom. Although the response of the surface nitrate, phosphate, and silicate concentrations is due to the same mechanism of upwelling and vertical diffusion that induces SST cooling, the enhancement of the three types of nutrient concentrations at the surface are quite different. The surface nitrate and phosphate concentrations increase by ~100% and 40%, respectively, while the surface silicate concentration increase only ~20%. Given the same strength of upwelling and vertical diffusion, the enhancement of the surface nutrient levels depends on the initial vertical gradient of the subsurface nutrient concentrations within the maximum depth that the upwelling and vertical diffusion processes can reach. Generally, the vertical gradient of the nitrate and phosphate concentrations within the top 70 m depth are larger than that of silicate, and thus lead to much more significant surface enhancement. In addition, the pre-existing cyclonic eddy greatly enhances the vertical gradient of nitrate and phosphate in the top 70 m depth, and consequently leads to more significant surface nitrate and phosphate concentrations in Experiment I. However, because of deep nutricline in the silicate vertical profile, the eddy does little enhancement on its vertical gradient of the top 70 m depth. Thus, there is little difference in the surface silicate enhancement between the two simulated cases.

Satellite observations revealed that Hurricane Katrina led to a significant SST cooling and a phytoplankton bloom near (24°N, 84°W) in the Gulf of Mexico. The SSH anomaly derived from satellite altimetry also showed that a southward-propagating Loop Current frontal eddy co-located with the phytoplankton bloom, which suggested that the eddy was an important factor in the phytoplankton bloom event. In this study, HYCOM is used to study the process of upper ocean responses to Hurricane Katrina. Two different cases with and without the presence of a cyclone eddy at the location of the observed phytoplankton bloom are simulated. The model run demonstrates that the Katrina-induced phytoplankton bloom is attributed to the pre-existing eddy, as well as to a slow-moving hurricane. The cyclonic eddy plays a critical role in the development of the Katrina-induced bloom, and it significantly strengthens the upper ocean dynamics and nutrient responses. In the simulations, the Katrina-induced upwelling and vertical diffusion process can reach ~70 m depth. The cyclonic eddy in the phytoplankton bloom area uplifted the isotherms and increased vertical temperature, nitrate, and phosphate gradients in the upper 70 m depth, thus leading to more significant SST cooling and increases in nitrate and phosphate concentrations than in the non-eddy case. Our model results also suggest that the nitrate concentration plays a dominant role in the development of the phytoplankton bloom with over 100% increase in the surface concentration, while the phosphate increases ~40%. The silicate has a minimal effect on the bloom with the smallest increase for its surface concentration.

7. Acknowledgments

The views, opinions, and findings contained in this article are those of the authors and should not be construed as an official NOAA or U.S. Government position, policy, or decision.

8. References

Acker, J. G., C. W. Brown, A. C. Hine, E. Armstrong, and N. Kuring (2002), Satellite remote sensing observations and serial photography of storm-induced neritic carbonate transport from shallow carbonate platforms, *Int. J. Remote Sens.*, 23, 2853- 2868.

Babin, S. M., J. A. Carton, T. D. Dickey, and J. D. Wiggert (2004), Satellite evidence of hurricane-induced phytoplankton blooms in an oceanic desert, *J. Geophys. Res.*, 109, C03043, doi:03010.01029/02003JC001938.

Bailey, S. W. and P. J. Werdell (2006), A multi-sensor approach for the on-orbit validation of ocean color satellite data products. *Remote Sensing of Environment*, 102, 12–23.

Behrenfeld, M. J., J. T. Randerson, C. R McClain, G. C. Feldman, S. O. Los, C. J. Tucker, et al. (2001), Biospheric primary production during an ENSO transition, *Science*, 291, 2594–2597.

Bleck, R., (2002), An oceanic general circulation model framed in hybrid isopycnic-Cartesian coordinate. *Ocean Modell.*, 4, 55–88.

Bleck, R., C. Rooth, D. Hu, and L. T. Smith (1992), Salinity-driven thermocline transients in a wind and thermohaline-forced isopycnic coordinate model of the North Atlantic, J. *Phys. Oceanogr.*, 22, 1486–1505.

Brink, K. H. (1989), Observations of the response of thermocline currents to a hurricane, *J. Phys. Oceanogr.*, 19, 1017–1022.

Brooks, D. A. (1983), The wake of Hurricane Allen in the western Gulf of Mexico, *J. Phys. Oceanogr.*, 13, 117–129.

Chassignet, E. P., H. E. Hurlburt, O. M. Smedstad, G. R. Halliwell, P. J. Hogan, A. J. Wallcraft, R. Baraille, and R. Bleck (2007), The HYCOM (Hybrid Coordinate Ocean Model) data assimilative system, *J. Mar. Syst.*, 65, 60– 83.

Chassignet, E., L. Smith, G. Halliwell, and R. Bleck (2003) North Atlantic simulations with the Hybrid Coordinate Ocean Model (HYCOM): Impact of coordinate choice, reference pressure, and thermobaricity, *J. Phys. Oceanogr.*, 33, 2504–2526.

Chavez, F. P., P. G. Strutton, G. E. Friederich, R. A. Feely, G. C. Feldman and D. G. Foley, et al. (1999), Biological and chemical response of the Equatorial Pacific Ocean to the 1997 - 98 El Niňo, *Science*, 286, 2128–2131.

Dickey, T. and J. J. Simpson (1983), The sensitivity of upper ocean structure to time varying wind direction, *Geophys. Res. Lett.*, 10, 133–136.

Dickey, T., et al. (1998), Upper-ocean temperature response to Hurricane Felix as measured by the Bermuda Testbed Mooring, *Mon. Weather Rev.*, 126, 1195–1200.

Fratantoni, P. S., T. N. Lee, G. P. Podesta and F. Muller-Karger (1998), The influence of Loop Current perturbations on the formation and evolution of Tortugas eddies in the southern Straits of Florida, *J. Geophys. Res.*, 103(C11): 24759–24799.

Gordon, H. R. (1997), Atmospheric correction of ocean color imagery in the Earth Observing System era, *J. Geophys. Res.*, 102, 17,081–017,106.

Gordon, H. R., and M. Wang (1994), Retrieval of water-leaving radiance and aerosol optical thickness over the oceans with SeaWiFS: A preliminary algorithm, *Appl. Opt.*, 33, 443–452.

Hale, G. M., and M. R. Querry (1973), Optical constants of water in the 200nm to 200μm wavelength region, Appl. Opt., 12, 555-563.

Hazelworth, J. B. (1968), Water temperature variations resulting from hurricanes, *J. Geophys. Res.*, 73, 5105–5123.

Hoge, F. E. and P. E. Lyon (2002), Satellite observation of Chromophoric Dissovled Organic Matter (CDOM) variability in the wake of hurricanes and typhoons, *Geophys. Res. Lett.*, 29(19), 1908, doi:10.1029/2002GL015114.

Hooker, S. B., W. E. Esaias, G. C. Feldman, W. W. Gregg and C. R. McClain (1992), An overview of SeaWiFS and Ocean Color, Vol. 1 of SeaWiFS Technical Report Series, In S. B. Hooker & E. R. Firestone (Eds), NASA Tech. Memo, 104566 Greenbelt, Maryland: NASA Goddard Space Flight Center.

Hu, C., and F. E. Muller-Karger (2007), Response of sea surface properties to Hurricane Dennis in the eastern Gulf of Mexico, *Geophys. Res. Lett.*, 34, L07606, doi:10.1029/2006GL028935.

Hu, C., et al. (2006), Ocean color reveals phase shift between marine plants and yellow substance, *IEEE Geosci. Remote Sens. Lett.*, 3, 262–266.

IOCCG (2010), Atmospheric correction for remotely-sensed ocean colour products, Wang, M. (ed.), *Reports of the International Ocean-Colour Coordinating Group*, No. 10, IOCCG, Dartmouth, Canada.

Jacob, S. D. and L. K. Shay (2003), The role of mesoscale features on the tropical cyclone-induced mixed layer response: A case study. J. Phys. Oceanogr., 33:649–676.

Jacob, S. D., L. K. Shay, and A. J. Mariano (2000), The 3D oceanic mixed layer response to Hurricane Gilbert, *J. Phys. Oceanogr.*, 30, 1407–1429.

Lavender, S. J., M. H. Pinkerton, G. F. Moore, J. Aiken and D. Blondeau-Patissier (2005), Modification to the atmospheric correction of SeaWiFS ocean color images over turbid waters. Continental Shelf Research, 25, 539 - 555.

Lee, T. N., C. Rooth, E. Williams, M. McGowan, A. F. Szmant and M. E. Clarke (1992), Influence of Florida Current, gyres and wind-driven circulation on transport of larvae and recruitment in the Florida Keys coral reefs, *Cont. Shelf. Res.*, 12: 971–1002.

Lee, T.N., K. Leaman, E. Williams, T. Berger and L. Atkinson (1995), Florida Current meanders and gyre formation in the southern Straits of Florida, *J. Geophys. Res.*, 100(C5): 8607–8620.

Leipper, D. F. (1967), Observed ocean conditions and Hurricane Hilda, 1964, *J. Atmos. Sci.*, 24, 182–196.

Levitus, S. (1982), Climatological atlas of the world ocean, NOAA Prof. Pap. 13, U.S. Govt. Print. Off., Washington, D. C.

Levitus, S., and T. Boyer (1994), World Ocean Atlas 1994 Volume 4: Temperature, U.S. Department of Commerce, Washington, D.C.

Lin, I., W. T. Liu, C-C. Wu, G. T. F. Wong, C. Hu, Z. Chen, W-D. Liang, Y. Yang, and K-K. Liu (2003), New evidence for enhanced ocean primary production triggered by tropical cyclone, *Geophy. Res. Lett.*, 30, 1718, doi:10.1029/2003GL017141.

Liu, X, M. Wang, and W. Shi (2009), A study of a Hurricane Katrina-induced phytoplankton bloom using satellite observations and model simulations, *J. Geophy. Res.*, 114, C03023, doi:10.1029/2008JC004934.

McClain, C. R., G. C. Feldman and S. B. Hooker (2004), An overview of the SeaWiFS project and strategies for producing a climate research quality global ocean bio-optical time series. *Deep Sea Research, Part 2, Topical studies in oceanography*, 51, 5–42.

Miller, W. D., L. W. H. Jr., and J. E. Adolf (2006), Hurricane Isabel generated an unusual fall bloom in Chesapeake Bay, *Geophy. Res. Lett.*, 33, L06612, doi:06610.01029/02005GL025658.

Paluszkiewicz, T., L. P. Atkinson, E. S. Posmentier and C. R. McClain (1983), Observations of a Loop Current Frontal Eddy intrusion onto the West Florida Shelf. *J. Geophys. Res.*, 88: 9639–9652.

Prasad, T. G., and P. J. Hogan (2007), Upper-ocean response to Hurricane Ivan in a 1/25° nested Gulf of Mexico HYCOM, J. Geophys. Res., 112, C04013, doi:10.1029/2006JC003695.

Price, J. F. (1981), Upper ocean response to a hurricane, *J. Phys. Oceanogr.*, 11, 153–175.

Price, J. F., T. B. Sanford, and G. Z. Forristall (1994), Forced stage response to a moving hurricane, *J. Phys. Oceanogr.*, 24, 233–260.

Romanou, A., E. P. Chassignet, and W. Sturges (2004), Gulf of Mexico circulation within a high-resolution numerical simulation of the North Atlantic Ocean, *J. Geophys. Res.*, 109, C01003, doi:10.1029/2003JC001770.

Salomonson, V. V., W. L. Barnes, P. W. Maymon, H. E. Montgomery and H. Ostrow (1989), MODIS: Advanced facility instrument for studies of the Earth as a system, *IEEE Transactions on Geoscience and Remote Sensing*, 27, 145–152.

Sanford, T. B., P. G. Black, J. R. Haustein, J. W. Feeney, G. Z. Forristall, and J. F. Price (1987), Ocean response to a hurricane, part I: Observations, *J. Phys. Oceanogr.*, 17, 2065–2083.

Shay, L. K., A. J. Mariano, S. D. Jacob, and E. H. Ryan (1998), Mean and near-inertial ocean current response to Hurricane Gilbert, *J. Phys. Oceanogr.*, 28, 858–889.

Shay, L. K., and R. L. Elsberry (1987), Near-inertial ocean current response to Hurricane Frederic, *J. Phys. Oceanogr.*, 17, 1249–1269.

Shay, L. K., R. L. Elsberry, and P. G. Black (1989), Vertical structure of the ocean current response to a hurricane, *J. Phys. Oceanogr.*, 19, 649–669.

Siegel, D. A., M. Wang, S. Maritorena, and W. Robinson (2000), Atmospheric correction of satellite ocean color imagery: the black pixel assumption, *Appl. Opt.*, 39, 3582–3591.

Shi, W., and M. Wang (2007), Observations of a Hurricane Katrina induced phytoplankton bloom in the Gulf of Mexico, *Geophys. Res. Lett.*, 34, L11607, doi:10.1029/2007GL029724.

Shi, W., and M. Wang (2009), An assessment of the black ocean pixel assumption for MODIS SWIR bands, *Remote Sens. Environ.*, 113, 1587–1597.

Stramma, L., P. Cornillon, and J. F. Price (1986), Satellite observations ofsea surface cooling by hurricanes, *J. Geophys. Res.*, 91, 5031–5035.

Vukovich, F. M. (1988), On the formation of elongated cold perturbations off the Dry Tortugas, *J. Phys. Oceanogr.*, 18: 1051–1059.

Walker, N. D., R. R. Leben, and S. Balasubramanian (2005), Hurricane-forced upwelling and chlorophyll a enhancement within cold-core cyclones in the Gulf of Mexico, *Geophy. Res.* Lett., 32, L18610, doi:18610.11029/12005GL023716.

Wang, M., and W. Shi (2005), Estimation of ocean contribution at the MODIS near-infrared wavelengths along the east coast of the U.S.: Two case studies, *Geophy. Res. Lett.*, 32, L13606, doi:13610.11029/12005GL022917.

Wang, M. (2007), Remote sensing of the ocean contributions from ultraviolet to near-infrared using the shortwave infrared bands: simulations, *Appl. Opt.*, 46, 1535–1547.

Wang, M., J. Tang, and W. Shi (2007), MODIS-derived ocean color products along the China east coastal region, *Geophys. Res. Lett.*, 34, L06611, doi:10.1029/2006GL028599.

Wang, M., S. Son, and W. Shi (2009), Evaluation of MODIS SWIR and NIR-SWIR atmospheric correction algorithms using SeaBASS data, *Remote Sens. Environ.*, 113, 635–644.

Wentz, F. J., C. Gentemann, D. Smith, and D. Chelton (2000), Satellite measurements of sea surface temperature through clouds, Science, 288, 847-850.

Wren, P. A., and L. A. Leonard (2005), Sediment transport on the midcontinental shelf in Onslow Bay, North Carolina during Hurricane Isabel, *Estuarine Coastal Shelf Sci.*, 63, 43–56.

Xie, L., X. Liu, and L.J. Pietrafesa (2007), Effect of Bathymetric Curvature on Gulf Stream Instability in the Vicinity of the Charleston Bump. *J. Phys. Oceanogr.*, 37, 452–475.

Zedler, S. E., T. D. Dickey, S. C. Doney, J. F. Price, X. Yu, and G. L. Mellor (2002), Analyses and simulations of the upper ocean's response to Hurricane Felix at the Bermuda Testbed Mooring site: August 13–23, 1995, *J. Geophys. Res.*, 107, 3232, doi:10.1029/2001JC000969.

30

Using Regional-scale Pre- and Post Hurricane Katrina Lidar for Monitoring and Modeling

Jason Stoker, D. Phil Turnipseed and K. Van Wilson
United States Geological Survey
USA

1. Introduction

Hurricane Katrina caused the largest natural disaster in U.S. history, with the eye of the storm hitting about 55 km east of New Orleans (Turnipseed et al., 2007). Although the storm initially brought more destruction to areas along the Mississippi and Louisiana coast, several levees protecting New Orleans failed the following day, and the city, about 80% of which is below sea level, was flooded. The flooding in New Orleans and the hurricane storm tide that impacted the Gulf of Mexico coast in Mississippi and southeast Louisiana killed hundreds, made tens of thousands more homeless, and triggered a massive relief effort that is still ongoing (Travis, 2005). It has been estimated that Hurricane Katrina caused the loss of more than 1,800 human lives and about $81 billion in damage (Turnipseed et al., 2007). The insured losses moved Katrina into the number one spot globally of all disasters since 1970. Over 200 miles of the 350 miles of levees and flood walls were damaged and had to be repaired prior to the next hurricane season that started in June 2006. In addition to the many rescue and recovery efforts that were initiated immediately after the storm, there was a strong need to quickly ascertain the areas inundated by storm tide water for rescuers and assess wind and flood damage (Raber and Tullis, 2007). First responders, emergency managers, city planners, insurance companies, and other groups have critical needs for geospatial information relating to the extent and effect of Hurricane Katrina's devastation to make rapid and efficient emergency response decisions.
In the late evening of August 25, 2005, less than two hours before Tropical Storm Katrina made first landfall on the southeastern Atlantic coast of Florida, the storm was upgraded to a Category 1 (Saffir-Simpson Hurricane Scale) hurricane after forming as a tropical depression over the Bahamas on August 19 (Knabb et al., 2005). After spending only six hours over land in southern Florida, Tropical Storm Katrina reentered open water in the southeastern Gulf of Mexico in the early morning of August 26, just north of Cape Sable (Knabb et al., 2005). During the next three days, Katrina rapidly intensified from a tropical storm to a Category 5 hurricane by late in the afternoon of August 28 with a maximum peak wind speed intensity of greater than 170 miles per hour (mph). This maximum intensity occurred about 170 nautical miles southeast of the mouth of the Mississippi River and helps explain the extreme magnitude of the storm tide height that occurred when Hurricane Katrina made landfall (Knabb et al., 2005). During Katrina's maximum intensity, tropical storm and hurricane force winds extended 200 and 90 nautical miles from the eye, respectively (Knabb et al., 2005). These conditions defined Hurricane Katrina as one of the

most intense and largest storms to ever form in the northern region of the Gulf of Mexico (Knabb et al., 2005). Katrina created the highest measured surge and equaled the highest wave height recorded on a NOAA buoy in North America (Link, 2010). After some erosion of the eye wall late on August 28, Hurricane Katrina turned northward to make landfall near Buras, Louisiana, with sustained winds of about 125 mph, making the storm a strong Category 3 hurricane. Hurricane Katrina made initial landfall on the northern Gulf of Mexico coast early on August 29, 2005, first slamming into the Mississippi River delta near Buras, Louisiana, and later, the Pearl River delta at the Louisiana-Mississippi border with an estimated intensity of 120 mph sustained winds. Knabb et al. (2005) explained that although Hurricane Katrina had weakened from a Category 5 to a Category 3 hurricane in the last eighteen hours before landfall, the radial extent of tropical storm and hurricane force winds remained about the same, which further explains the extreme storm tide in southeastern Louisiana and the Mississippi Gulf coastal region.

Katrina weakened rapidly after its final landfall near the Louisiana-Mississippi border, becoming a Category 1 storm by 1800 UTC on August 29 in central Mississippi. The storm was downgraded to tropical storm status early on August 30, after five days as a hurricane in the Gulf of Mexico (Knabb et al., 2005).

1.2 Katrina Storm surge

Hurricane force winds change the water level, a process called storm surge generation. Storm surge is defined here as the abnormally high still-water levels attributable to the presence of the storm itself. That is, storm surge for the purposes of this chapter does not include the discussion of the destructive wave action that occurs on top of storm surge and usually is evident in these events. The water level during a storm is influenced by the storm surge, the timing of the storm with the astronomical tidal variations that normally occur without the presence of a storm, and the areal extent of the storm. The word "still" to describe the water level is intended here to differentiate between the slower rise and fall of the water surface due to the storm surge/astronomical tide level that occurs over time scales of hours, and changes in water surface that occur at much higher frequencies associated with water surface motion due to wave action that occur over time scales from seconds to tens of seconds.

Hurricane winds exert stress at the water surface that pushes the water. Wind effect on storm surge has a higher contribution in shallow water. The shallower the water at the land-water interface, the more effect wind can have in developing storm surge. Like waves, the storm surge generation potential for a storm is also related to the surface shear stress of the water. Broad, shallow continental shelf regions are the most effective areas for generating storm surge, and therefore most vulnerable to damage from storm surge. As winds push water, it moves freely until it encounters a coastal land mass or other obstruction where it then begins to amass. Indentations, irregularities, and pockets along the U.S. coast are particularly prone to catching water pushed toward and in to these geographic features by the wind. The Mississippi River delta is a coastal land feature that acts to catch water being pushed toward it along the Mississippi and Alabama continental shelves. Since winds in hurricanes rotate in counterclockwise direction, hurricanes in the northern Gulf of Mexico tend to create winds that blow from the east in the northern gulf.

In the case of Hurricane Katrina, these winds from the east acted to push water toward Southeast Louisiana, and toward the Mississippi River delta. As storm surge grew and inundated the wetlands of Southeast Louisiana, the Mississippi River levees in the lower basin became a topographic control. Along the Southeast Louisiana coast, for Katrina, the storm surge contribution due to wind and geographic/topographic controls exceeded 15 ft in places, and along the Mississippi coast it exceeded 20 ft in places (IPET, 2008). This contribution of the water level can vary greatly with location.

Precipitation can increase the storm-induced water level, either by falling directly on the local water bodies, or by falling on the adjacent watershed and then running off into the water bodies. The effect of direct precipitation can be significant to the peak storm surge. Typically, some of the direct precipitation will be felt after the storm surge peak is experienced; and often the effect of runoff from the adjacent watershed will be experienced after the peak storm water level has been experienced.

One of the primary efforts in determining the physical performance of flood control structures in the area in and around New Orleans after the Hurricane Katrina disaster was the Interagency Performance Evaluation Task Force (IPET). A principal objective of the IPET was to determine the effects of Hurricane Katrina on flow control structures and to provide a framework for the subsequent repair and rebuilding of hurricane protection in New Orleans. The IPET was made up of experts from 25 universities, 25 private companies and 10 government agencies (Link, 2010). The IPET study used a combination of measured data and model-simulated data to characterize water level conditions over time. Measured data fell into two categories: High Water Mark (HWM) measurements which capture peak water levels, and hydrographs, which capture the water level as a function of time (IPET, 2008). An extensive post-storm effort was undertaken to identify and survey HWMs following passage of the storm. While HWMs can capture the peak water levels, they do not record the temporal variability in the water level and generally do not include wave action. HWMs also have their own inherent quality issues, such as whether they reflect a peak condition, and whether or not water surface motion due to wind-caused waves is reflected in a HWM. Measured hydrographs are a reliable source of data for capturing both the temporal variation and the maximum water level. Water level fluctuations were measured with instrumentation during the build-up stage of the storm at a number of sites throughout the study region; however, few instruments operated throughout the storm. Most of them failed prior to the peak. Consequently, there are little continuously measured data that capture peak conditions. In a few cases, photographs and other visual observations were utilized to provide information about the temporal variation of water level to supplement the recorded hydrographs. These constructed hydrographs proved to be extremely valuable for characterizing conditions along the south shore of Lake Pontchartrain.

1.3 Lidar

After Hurricane Katrina, the USGS quickly began the task of obtaining and processing detailed existing pre-Katrina Light Detection and Ranging (lidar) data throughout the affected region. Farris et al. (2007) documented many technologies used in the USGS response to the devastation caused by Katrina, including lidar mapping of New Orleans. Lidar mapping is an effective and accurate technology for producing high-resolution elevation data for bare earth, vegetation, and structures (Lefsky et al. 2002; Stoker et al., 2006). High level spatial detail and vertical accuracy of elevation measurements make lidar remote sensing an excellent mapping technology for use in low-relief hurricane-prone coastal areas (Farris et al., 2007; Gesch, 2009). By converting lidar data into usable information, extremely accurate, high-resolution models can be used to visualize and quantitatively represent scenes in three dimensions.

In addition to high-resolution bare earth digital elevation models (DEMs) used in this effort, other lidar-derived products possible include quantitative estimates of vegetative features, such as canopy height, canopy closure, and biomass (Lefsky et al., 2002), and models of urban areas such as building footprints and three-dimensional city models (Maas, 2001). However, due to the fact that the topographic lidar laser is usually in the near-infrared wavelength (around 1064 nm), bare ground topography accuracy can be poor if flooding of stream channels and floodplains occurs during lidar acquisition.

Due to the sheer size of Hurricane Katrina, an unprecedented regional analysis at very high resolution and accuracy was needed to properly quantify and understand the effects of the hurricane and the entire storm tide. Many disparate sources of lidar data were acquired and processed for varying environmental reasons by pre- and post- Katrina projects (Stockdon et al., 2009). Lidar data from disparate sources were available for Baldwin and Mobile Counties in Alabama, for Jackson, Hancock, and Harrison Counties in Mississippi, and for the southeastern parishes affected by the storm in Louisiana. These separate datasets encompassed the most significantly affected areas from Hurricane Katrina. These datasets were in several formats and projections and originally obtained by the USGS in varying phases of completion. As a result the task of producing a seamless digital elevation dataset required a high level of coordination, research, and revision. To create a seamless digital elevation dataset, many technical issues had to be resolved before producing the desired 1/9-arc-second (3 meter) grid needed as the map base for projecting the Katrina peak storm tide in the affected coastal region. As a result of these complex datasets, a methodology was developed to construct seamless DEMs from multipurpose, multiuse, and disparate lidar datasets that serves as a template for use by other agencies. The output of this work included an easily accessible Web application for viewing the maximum storm tide caused by Hurricane Katrina in southeastern Louisiana, Mississippi, and Alabama (Stoker et al., 2009).

2. Methods

2.1 High water Mark analyses

The passage of hurricanes often results in short-period wind waves on top of the much longer-period storm surge that couple to create various types of debris lines including vegetation, seeds, dirt, man-made trash, and dislodged building material. These debris lines can be deposited on or adhere to some surfaces after the peak water level has been reached and it begins to fall. The deposited debris typically leaves a linear feature that is referred to as a High Water Mark (HWM) and this mark with experience and interpretation can be used to approximate the local peak water level. The highest quality marks for estimating storm still-water levels are those that have little or no wave effect. Some HWMs are collected where significant wave or backwater effects are present, but those effects should be noted in the field surveying notes and documentation. In this analysis, the focus was on use of HWMs as indicators of storm water level, without the effects of water level fluctuations due to wave crests wave effects, or riverine flooding.

The U.S. Geological Survey (USGS) began flagging and surveying HWMs along the Mississippi coast on September 1, 2005 to compare the peak surge to the peak surge of Hurricane Camille in 1969. An extensive post-storm effort was undertaken to identify and survey HWMs following passage of the storm in concert with the IPET. While certain

HWMs captured the peak water levels well, they did not contain information about the temporal variation of water level. It is critical in the analysis of HWMs to determine if the HWM reflects the maximum crest of the storm. An experienced field hydrologist /hydrographer can usually determine if the HWM has been affected by waves, riverine flooding or other factors.

Acquisition of HWMs following Katrina was principally performed by three federal agencies: USGS, the U.S. Army Corps of Engineers (USACE), and the Federal Emergency Management Agency (FEMA) (or a FEMA contractor). The State of Louisiana through the Louisiana State University (LSU) also helped recover HWM data. All entities participating in the IPET shared the data. Marks identified by USACE, FEMA, and LSU were also recovered by that respective agency. Most of the marks identified by USGS were recovered by FEMA (or a FEMA contractor). A subset of approximately 50 marks was also recovered by USGS field crews to confirm elevations provided by FEMA contractors. All HWMs were reviewed and assigned a reliability rating. The reliability of each HWM was assessed as "Excellent," "Good," "Fair/Poor," or "Unknown" if there was no information provided regarding the type of mark or setting in which it was acquired. Currently, there is not a standard method for determining HWM reliability. Moreover, assignment of reliability values to HWMs is not a totally objective process, but by its nature involves both objective and subjective elements. HWM recovery and determination requires experience and engineering judgment. Discussion by the IPET assigning the reliability values led to a consensus that the mark should reflect, as closely as possible, the stable ("still") or mean, storm water level. That is, the physical setting where the mark was located should approximate a tide gauge stilling-well if possible. The basis for this consensus is that storm surge models do not explicitly include wave crest or other wave effects, and one of the important uses of the HWM data is validation and verification of surge model simulations.

2.2 Lidar processing

The USGS and its partners began acquiring, organizing, and preprocessing lidar datasets immediately following the landfall of Hurricane Katrina. Lidar data were obtained from multiple sources, including private, local and state agencies, and Federal bureaus, such as the National Oceanic and Atmospheric Administration (NOAA), the USGS, and the USACE. Many disparate sources of lidar data were acquired and processed for individual projects to accomplish various tasks pre- and post-Katrina. The original datasets were in several different formats and projections and were processed to varying phases of completion before they were acquired for this project (Table 1). Some of these datasets already existed in a bare earth, ready-to-use format, and were easily downloadable from the USGS National Elevation Dataset (NED) (Gesch et al., 2002). Some of the raw point cloud data were already available and being processed for the USGS Center for Lidar Information Coordination and

Area	Format	Projection	Elevation Units	Metadata?
Eastern Louisiana Parishes	5-m DEM (raster)	UTM	Feet	Yes
Hancock County, MS	LAS binary (points): Reprocessed with breaklines added	MS State Plane (meters)	Feet	Yes
Harrison County, MS	ASCII XYZ (points) & EBN binary: Reprocessed with breaklines added	MS State Plane (feet)	Meters	No
Jackson County, MS	LAS binary (points): Reprocessed with breaklines added	MS State Plane (meters)	Feet	Yes
Mobile County, AL	LAS binary (points) w/ CAD breaklines	AL State Plane (feet)	Feet	Partial
Baldwin County, AL	ESRI Shapefiles: Mass points and breaklines	Al State Plane (feet)	Meters	Partial

Table 1. Dataset differences (from Stoker et al., 2009)

Knowledge (CLICK) database (Stoker et al., 2006). The lidar data were in various file formats and projections and represented different levels of processing. Differences in datum, projection, units, and file types created a need to standardize the inputs in order to create a regional product. The end product was a raster in a geographic coordinate system, which is the standard deliverable format for the NED.

Southeastern Louisiana Datasets

Lidar data that were already bare-earth processed and gridded into an acceptable Digital Elevation Model (DEM) format were available from southeastern Louisiana parishes. The Louisiana data were projected in Universal Transverse Mercator (UTM) Zone 15-North-Meters, North American Datum of 1983 (NAD 83). The original source DEM resolution was 5 m x 5 m, and the data were resampled to 1/9-arc-second grids to be consistent in resolution (3 meters) with the other datasets. Although the datasets were resampled to a 3-m resolution, the data were originally acquired and processed at a 5-m resolution.

Mississippi Gulf Coast Datasets

A seamless dataset for the three counties that border the Gulf of Mexico in Mississippi (i.e., Hancock, Harrison, and Jackson Counties) was created by processing lidar data of each county separately and then merging them.. The Hancock and Jackson County lidar data were flown on February 25th and March 1, 10, 11 and 12th, 2005. The sensor pulse rate was set at 29 kHz with a field of view of 45 degrees. The aircraft was flown at an altitude of 12,000 feet, and the point spacing was approximately 5 meters. The data were delivered as binary laser file format (LAS) files in State Plane Meters Mississippi East NAD 83, with elevation data (z-values) in feet. Initially, these LAS files were preprocessed and point data were classified as bare or non-bare. Because of the large size of the LAS files, the data were tiled to more efficiently manage the processing. Each tile was interpolated to a raster format to convert the data from points to 1/9-arc-second grids.

The Harrison County lidar data were flown on March 8-9th, 2004. They were obtained at an altitude of 11,100 feet above mean terrain. The sensor pulse rate was set at 20 kHz with a field of view of 45 degrees. Average swath width of the raw data was 9,112 feet, and the point spacing was approximately 5 meters. The final data were delivered as bare earth ASCII comma-delimited text files, as well as proprietary files. These data were projected in State Plane Feet Mississippi East feet NAD 83, although the z-values were in meters and not feet. These data were converted to feet for compliance with vertical units in the Jackson and Hancock Counties' datasets before being preprocessed, tiled, and interpolated into 1/9-arc-second grids. These data were processed using software to process the data to surfaces, and then were converted into 1/9-arcsecond grids.

Alabama Datasets

Both coastal Alabama counties (Baldwin and Mobile) were significantly affected by the Hurricane Katrina storm tide. The Mobile County, Alabama lidar data obtained by the USGS were LAS binary files with breaklines as computer-aided drafting files. A breakline is a hard-edge digitized line that is delineated to have a constant defined elevation. The data were in State Plane Feet Alabama West NAD83 with z-values in feet. The files were processed to create surfaces with breaklines, and then output into grid formats for conversion into ArcGIS compliant grids. The Baldwin County, Alabama lidar data were acquired as mass points and breaklines using ArcGIS shapefile format. These data were projected in State Plane Feet Alabama West NAD83 with z-values in feet. The data were converted to Triangulated Irregular Networks (TINs), and breaklines were added before they could be converted to grids.

After the Mississippi data were completely processed once, the lidar points were completely reprocessed to add in breakline information. Each binary dataset was converted to an ArcGIS shapefile, and then converted to TINs. The breakline data were then incorporated into the TINs, and the files were converted to 1/9-arc-second raster grids. In all, 182 gigabytes (GB) of lidar data were processed for five counties in Mississippi and Alabama. The datasets for Mississippi and Louisiana were seamlessly integrated using ASSEMBLE, a custom program used for processing the NED (Gesch et al., 2002). All datasets were projected into a geographic coordinate system for use in the ASSEMBLE program. The ASSEMBLE program mitigated all noticeable dataset seams and other edge boundary anomalies (Stoker et al., 2009). The final output was a seamless digital elevation model of the region at a resolution of 3 meters available for viewing on the web.

2.3 Storm surge analyses

Mapped storm surge data are critical for emergency managers and decision makers of affected coastal communities and other areas, but also provide future design criteria for engineers and planners in the rebuilding phase after a coastal zone is devastated by a storm. Historically these data have been relatively scarce. Following the Hurricane Katrina disaster, high-resolution digital elevation data, HWMs, and mapping interpolation methods were used to estimate storm tide throughout the Katrina affected region in Louisiana, Mississippi and Alabama. Turnipseed et al. (2007) used a spline interpolation method along with interpretative analysis to complete a thorough storm surge analysis. The maximum Hurricane Katrina storm tide was digitally mapped, using water-level data collected by the USGS, FEMA, USACE, and others.

Maps are inherently in error, however, and the error varies with the scale and resolution of the map created. High-resolution digital maps generated for the Katrina storm tide also have error, and in places, ignore the effects of topography, roads, and other hydraulic features that resist inland flowing storm tide. This analysis is also subject to erroneous extrapolations beyond the data locations. Simply interpolating a surface from the data may not result in reliable maps of storm tide, but clearly the data provide engineers, planners, and emergency officials with a tool to prepare, warn, and respond to coastal disasters such as Hurricane Katrina.

Many published interpolation methods permit depictions of storm tide inundation, but often ignore the effects of topographic and hydraulic features that influence the relative depths and extent of storm tide inundation. These storm tide analyses were used to create an improved digital map coverage of the maximum storm tide produced by Hurricane Katrina and to compare the improved map to a flood-inundation map prepared by FEMA.

Storm surge analysis and the creation of storm-tide maps involve iterative steps of interpretation, refinement, extrapolation, and often further interpretation and refinement of the algorithm-derived coverages. Automated feature extraction and even generalized interpolation methods without the experienced and trained eye of the scientist cannot accurately evaluate mapping features, especially in regions of low topographic relief such as found in the north Gulf of Mexico coast. Figure 1 depicts the maximum storm-tide elevation of the Katrina storm and is a digital coverage created by using hundreds of HWMs, interpretation, a high-resolution digital elevation model created from LiDAR data and a spline interpolation of topographic and hydraulic influences from the storm tide. Berenbrock et al. (2009) used a barrier mapping technique to effectively map the Hurricane Rita storm tide in Louisiana and Texas from over 30 continuously recording storm tide pressure transducers.

Identification and insertion of barriers to simulate natural features such as islands, elevated highways and railroads, dunes and bluffs, natural levees, canal berms and other features significantly affect how inland flow from storm tides inundates the landscape. A general description of a typical step-wise analysis to insert flow barriers into a high-resolution minimally included:

- Assembly of a storm-surge HWMs or other method to simulate the maximum extent of flooding; For the Hurricane Katrina storm surge coverage, over 1500 HWMs were analyzed, of which over 800 HWMs were used to prepare the coverage.
- Creation of a first-cut digital storm-tide elevation surface using the spline method or other method (Berenborck et al., 2009; Turnipseed et al., 2007).
- Interpretation and analysis of the storm-tide elevation surface to create an estimation of the extent of flooding from the storm-tide inundated area (boundary of flooding) from the interpolated data. This should also include an identification of zero elevation contour;
- Identification and digitization of natural hydraulic barriers from topographical maps or other data that serve as natural impedance of storm tide inundation on the landscape. These might include islands, large bridge structures, elevated highways and railroads, dunes and bluffs, natural levees, canal berms, and any other feature that could significantly affect the flow of storm-tide inland;
- Creation of a maximum storm-surge map utilizing all HWM data or other data used to represent storm tide elevation across the landscape

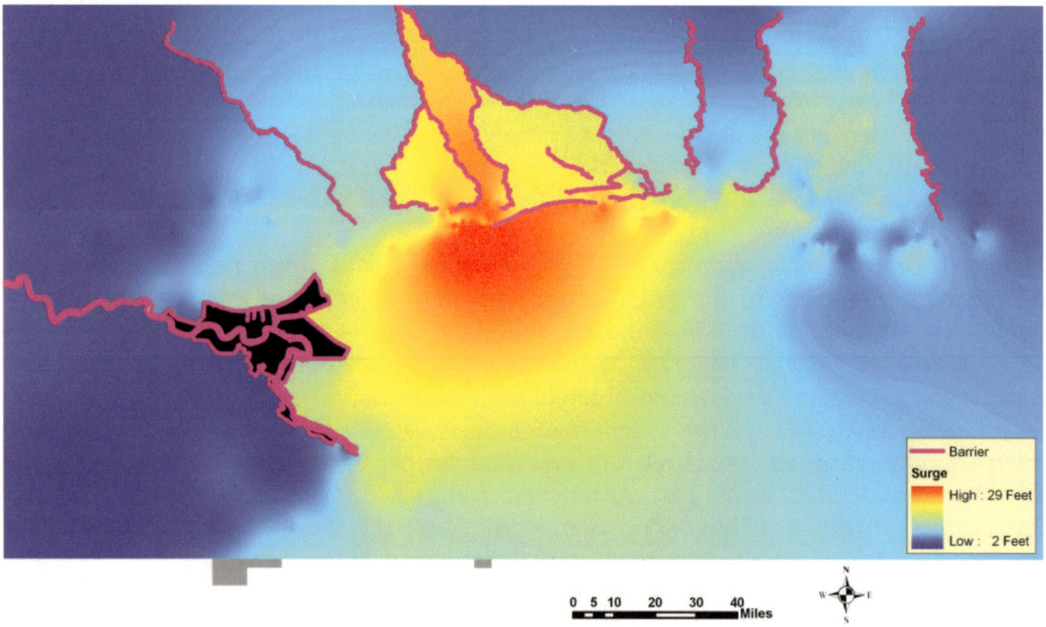

Fig. 1. Improved maximum storm-tide elevation with barriers (from Stoker et al., 2009)

2.4 Multi-temporal analyses

The assessment of a hurricane-impacted forest region and forest damage severity at regional scales has been traditionally based on a ground survey, aerial photography (Gardner et al. 1992), satellite imagery (Kovacs et al. 2001), topographic exposure models (Boose et al. 1994), or simulated by ecological models or storm models. Hurricanes frequently damage and fragment the forested landscape (Boose et al. 1994; Foster and Boose 1992), and Hurricane Katrina was no exception. Large areas of the region were severely damaged, both by the storm surge and by the wind. The spatial patterns of the hurricane's disturbance are essential for managers and scientists to perform efficient management practices such as rebuilding structures, insurance estimations, forest salvage harvesting, habitat protection, as well as to make assessments on long-term environmental impacts and forest ecosystem recovery. Due to the sheer size of areas affected by Katrina, a full-scale ground-based assessment is technically difficult and not very economical (Wang and Xu, 2009). As a result, remote sensing methods are the best option for quantifying disturbed areas over large regions, such as Katrina affected areas. Remote sensing techniques using change detection methods in concert with vegetation indices have been shown to be an effective means for high resolution forest disturbance estimations caused by insect defoliation, clearcuts, and forest fires (Ayala-Silva and Twumasi 2004; Coppin and Bauer 1994; Franklin et al. 2000; Hegarat-Mascle et al. 2006; Kwarteng and Chavez 1998; Lyon et al. 1998; Nackaerts et al. 2005; Ramsey et al. 1997). Change detection with remote sensing typically involves the use of two or more multi-temporal aerial or satellite images to estimate two-dimensional spectral changes between time periods. Three-dimensional change detection based on lidar has been used to help quantify structural changes of damaged forests (Dwyer et al., 1999; Wiesmann et al., 2001; Fransson et al., 2002; Boutet and Weishampel, 2003).

Hurricane damage to forests has been mapped via satellite-based remote sensing in a number of studies. Landsat Thematic Mapper (TM) data at 30-m resolution were used by Kovacs (2001) to map disturbance in mangrove forests and by Clark et al. (2006) to map Hurricane Katrina's impacts on pines and hardwoods in the DeSoto National Forest in southern Mississippi. Ramsey et al. (1998) used 1-km Advanced Very High Resolution Radiometer (AVHRR) imagery to map Katrina's impacts to bottomland hardwood and bald-cypress/tupelo (*Taxodium distichum/Nyssa aquatica*) stands in coastal Louisiana. AVHRR imagery was also used by Ayala-Silva and Twumasi (2004) to investigate spatial patterns of damage related to distance away from the path of Hurricane Georges in Puerto Rico. All of these studies found change in the Normalized Difference Vegetation Index (NDVI) to be a useful indicator of forest damage, but none of the lower-resolution studies systematically validated their estimates against reference data.

Wind speed is the variable probably most influential to forest damage. Doyle et al. (1995) found that the severity and spatial extent of hurricane damage declined at greater distances from the storm track, although (Kelly 1993) found damaging winds may extend farther from the center on the eastern side of the track. Ramsey et al. (2001) compared forest damage in Louisiana with modeled wind speeds and found evidence that the severity of damage depended on the duration for which forests are subjected to wind speeds exceeding a certain critical threshold. Jacobs (2006) used relationships between maximum sustained wind speed and forest damage to create rapid assessment maps of forest damage resulting from Hurricanes Katrina and Rita.

The topographic slope and aspect associated with a forest stand determine its degree of exposure to peak winds (e.g., McNab et al. 2004). Topography also influences the soil conditions available for tree rooting. Damage has been found to be greater in valleys than on ridges and slopes, probably because of poor soil drainage, shallow rooting depths, and lack of root anchorage (Basnet et al. 1992).

Lidar was collected by NOAA after Katrina for the southern portions of Hancock, Harrison and Jackson Counties in Mississippi. This data was flown along the coastline in late September / early October 2005. The availability of lidar point cloud data from before and after Katrina allow for an assessment of forest, structural and topographic change in three dimensions, rather than spectral changes in passive optical imagery. By understanding the vertical errors associated with the pre- and post Katrina lidar data, we can evaluate three-dimensional changes outside of the error of the instruments to determine if these differences are indeed true changes, such as destroyed trees, houses and roads or are simply artifacts of the data.

To evaluate three-dimensional changes due to Hurricane Katrina, we created raster images or grids interpolated from the lidar point clouds of these three counties for the bare earth points and the first reflective surface points. Differencing these raster images provided us with an image we have coined the height above ground (HAG) raster, which includes the heights of both vegetation and structures above the bare ground. Employing these techniques for both the pre- and post-Katrina lidar at 3 meters, we were then able to difference the post-Katrina lidar from the pre-Katrina lidar for first reflective surfaces, bare earth surfaces, and the height above ground surfaces. Any negative values indicate a loss in elevation after Katrina (such as from destruction), and any positive elevations indicate increases in elevation (such as from new construction or tree growth). Because of the vertical errors stated in the analysis of the instrumentation used in the projects (18.5 - 40 cm RMSE) any vertical changes within +/- 56.6 cm (Root Sum of Squares +/- 40 cm) cannot be statistically proven as true change; however, any vertical changes outside of these error bounds need to be looked at more closely as possible bare earth, first reflective surface or HAG changes.

3. Results

3.1 Lidar elevation model results

After the data were seamlessly integrated, a shaded relief image was created and used for quality assurance and quality control of the processing methods. Initial quality assurance checks revealed that a few tiles needed to be reprocessed, and some differences in the Geoid models used by the disparate projects were detected and corrected. In the final datasets, all elevations are in feet above NAVD88 (Fig. 2).

3.2 Storm surge results

The final output of the storm surge model incorporating HWMs and barriers to flow are depicted in Figure 3. The created map is only as accurate and precise as the data used to create it, and for the storm surge analysis used to create the maximum storm surge digital elevation coverage for Hurricane Katrina, the accuracy and precision is no better than the HWMs used to make that coverage. Datasets constructed from field data collected by the USGS, FEMA, USACE, and other scientists and engineers post-Katrina accomplished a significant hindcast of the Katrina devastation that will be used extensively in the future in the rebuilding and redevelopment phases of the north Gulf of Mexico coastal region.

As importantly, the development and use of the storm surge analysis techniques and mapping methods in defining the extent of the Hurricane Katrina storm tide inundation

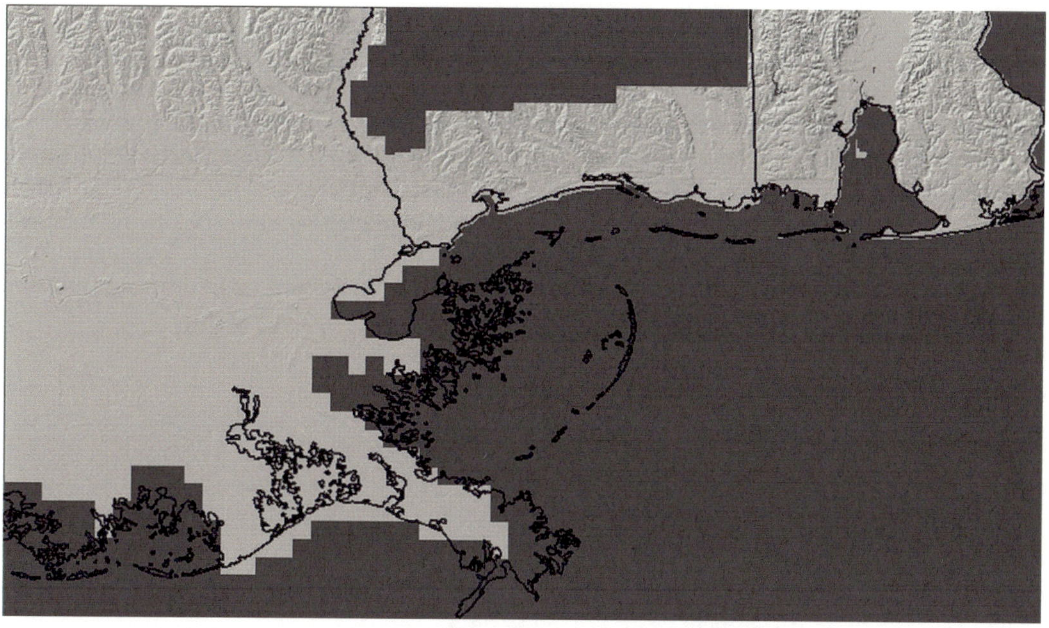

Fig. 2. Final seamless elevation surface from disparate lidar datasets

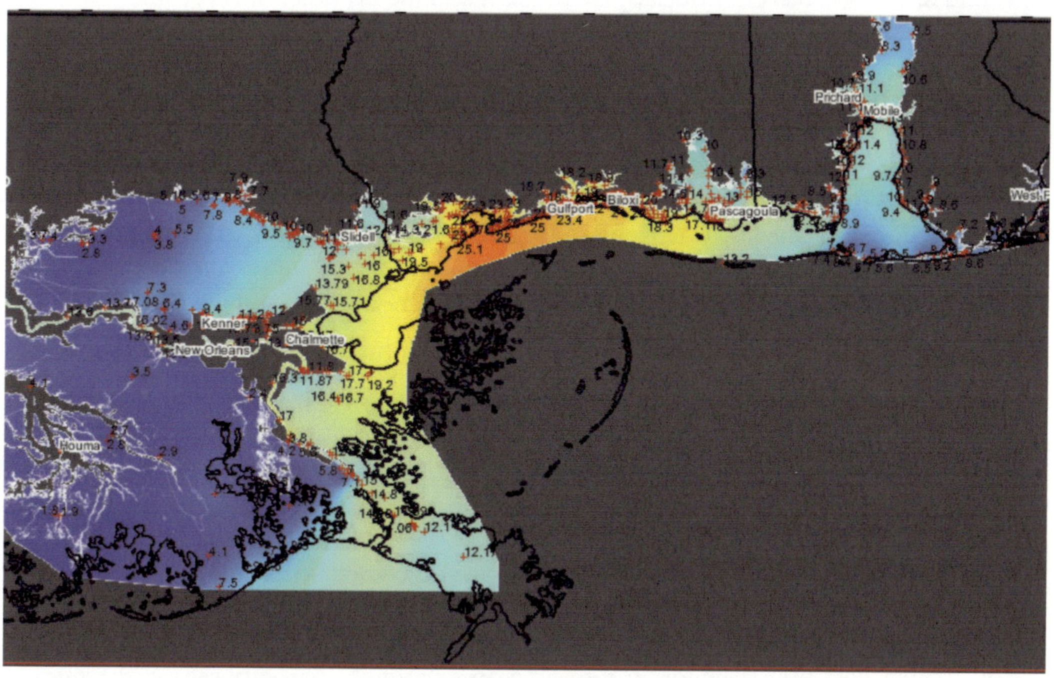

Fig. 3. Final storm surge model incorporating barriers and HWMs

links this critical water-level dataset to the natural topographic and hydraulic features found in southeastern Louisiana, and the Mississippi and Alabama coastlines, and improves the overall depiction of the storm tide for the future of this region. This analysis also provides a realistic, consistent, and robust digital map for use by managers and decision makers charged with the recovery of infrastructure in the region. This analysis does not provide the detail or accuracy that could be produced from an appropriately calibrated 2- or 3-dimensional hydrodynamic storm surge model, but instead documents a much simpler, but conservative depiction of the extent of flooding represented by the Hurricane Katrina storm tide in Alabama, Mississippi, and southeast Louisiana. Given the need for these critical data before, during, and after such devastating events as Hurricane Katrina, studies such as this may represent the most practical and efficient solution for emergency managers and decision makers that must prepare, warn, and respond to the coastal and inland devastation that intense tropical storms can cause.

3.3 Multi-temporal results

After creating the post-Katrina minus pre-Katrina lidar layers, the next task was filtering out the ranges of data that could have been attributed to errors in the two instruments used to collect the lidar. Since the maximum error specified by the vendors for all data was around +/- 40 cm, we assumed that the maximum instrument error between the two temporal datasets was around 2 x +/- 40 cm, or +/- 80 cm. Any vertical differences outside of this 80 cm range were looked at to determine if these differences were caused by Katrina processes, or were simply other differences or errors in the data.

In parts of coastal Hancock and Harrison Counties, once sensor errors were filtered out, a thin band of changes became quite apparent. To determine the possible cause of this artifact, ancillary data were brought in for comparison. The High Water Marks, post-Katrina imagery, and building footprints digitized in 2007 provided excellent context to the possible nature of this band (Fig 4). The Mississippi Coordinating Council for Remote Sensing and Geographic Information Systems (MCCRSGIS) is part of the Gulf Region Base Mapping Ownership Data Development Project, funded through the Community Development Block Grant (CDBG) program. These vector data were developed for five Coastal Area counties – Pearl River, Stone, Hancock, Harrison, and Jackson. The spring 2007 natural color digital orthophotos at 6 inch and 1 foot resolution were used in creation of the data for the project area. Urban areas and the coastline were typically 6 inch resolution with the remaining areas at 1 foot.

Putting this bare earth change layer in context with the high water marks, post-Katrina imagery and 2007 digitized building footprints strongly suggest that the differences in bare earth change along this band relates to debris that was pushed inland due to storm surge. Further analyses of this data needs to be performed before a definitive statement can be made; however, qualitative visual inspection of this artefact strongly suggests this conclusion.

In addition to the bare earth change evaluation, we also compared the height above ground models from before Katrina to after Katrina. After a quick assessment of results, several issues became apparent. Comparing results from the changes in the vertical to various optical change detection methods in the same area highlight the importance of temporal consistency in using lidar for detecting and monitoring non-bare earth vertical change over time (Fig. 5). One of the primary challenges to detect large-scale abrupt vegetation modification is to eliminate or reduce errors caused by vegetation phenology. Due to the fact that this change detection assessment was opportunistic in nature, differences in phenology for data collected in early spring/late winter of 2004 for Harrison County, the early spring of 2005 for Hancock and Jackson Counties, and the fall of 2005 after Katrina are apparent. It appears that for the Hancock/Jackson Counties data, there were decreases in the Height Above Ground models, while for northern Harrison County data, there were actually increases in heights. The primary cause of this is believed to be the leaf on/leaf off status of the data due to the differences in dates for each of the collections, and the differences in heights due to high scan angles.

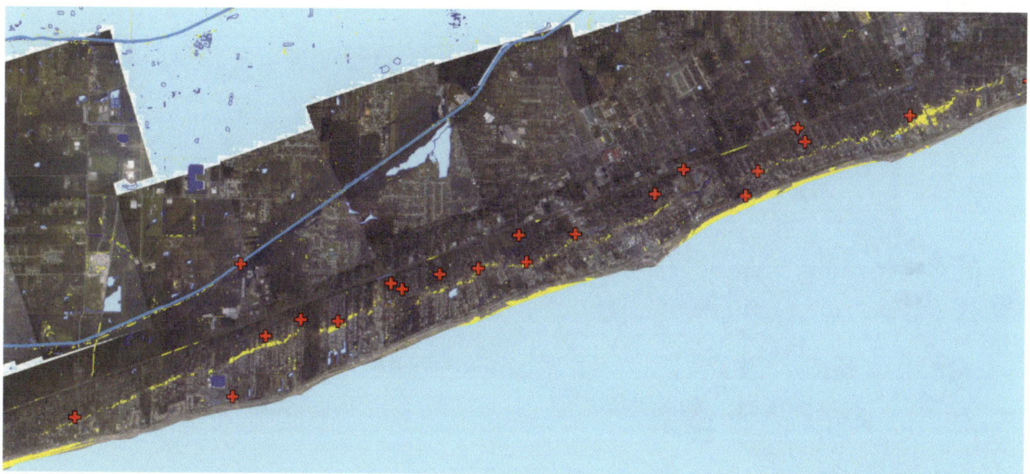

(a)

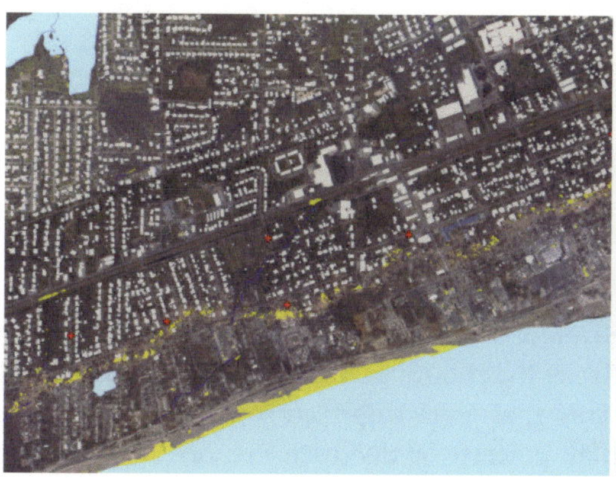

(b) (c)

Fig. 4a, 4b, 4c. Outside-sensor-error vertical changes (yellow = increase, blue = decrease) for bare earth versus HWMs (red crosses) and 2006 MCCRSGIS digitized building footprints

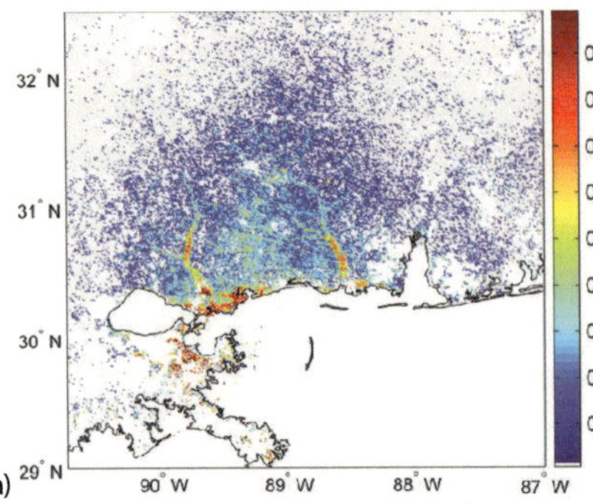

a)

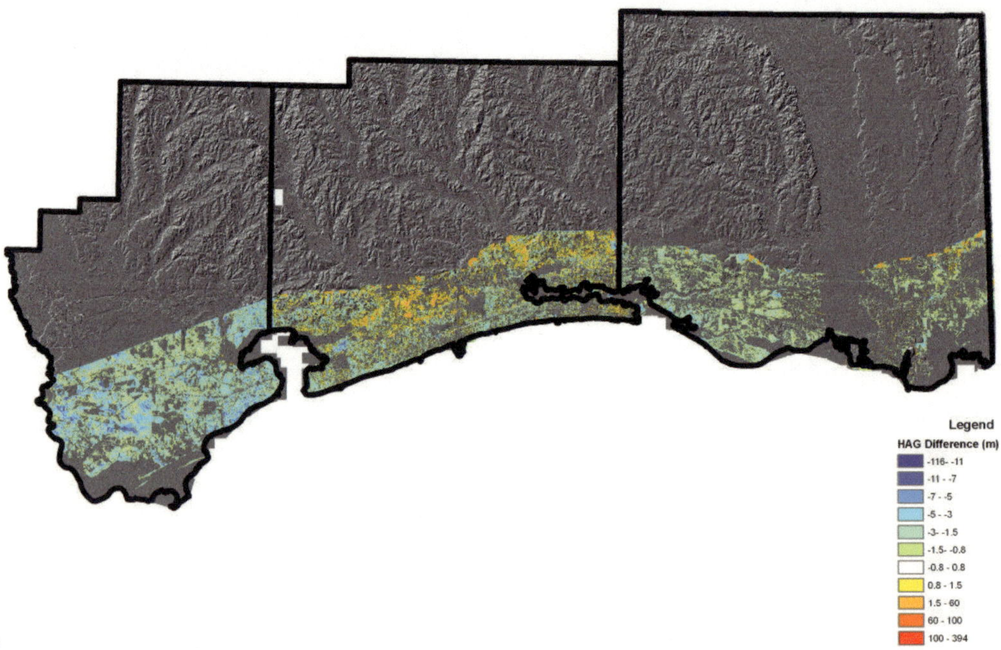

b)

Fig. 5. a) Maps of forest damage levels mapped by Wang et al., 2010, and b) A HAG difference grid from lidar (blue = decrease, red = increase)

For buildings, qualitative comparison demonstrated that the differences in heights related well to the 2007 building footprints (Fig. 6). Areas that show no changes in vertical height correlate well with outlines of building footprints that were digitized in 2007. Areas along the nearshore coastline that show a rectangular pattern of decrease are not digitized as buildings in the 2007 footprint layer. These areas are assumed to be buildings that have been destroyed between the two datasets. There are also areas where digitized polygons are associated with rectangular increases in elevation inside that polygon. These areas are assumed to be buildings that were built between the times of the pre- and post-Katrina datasets.

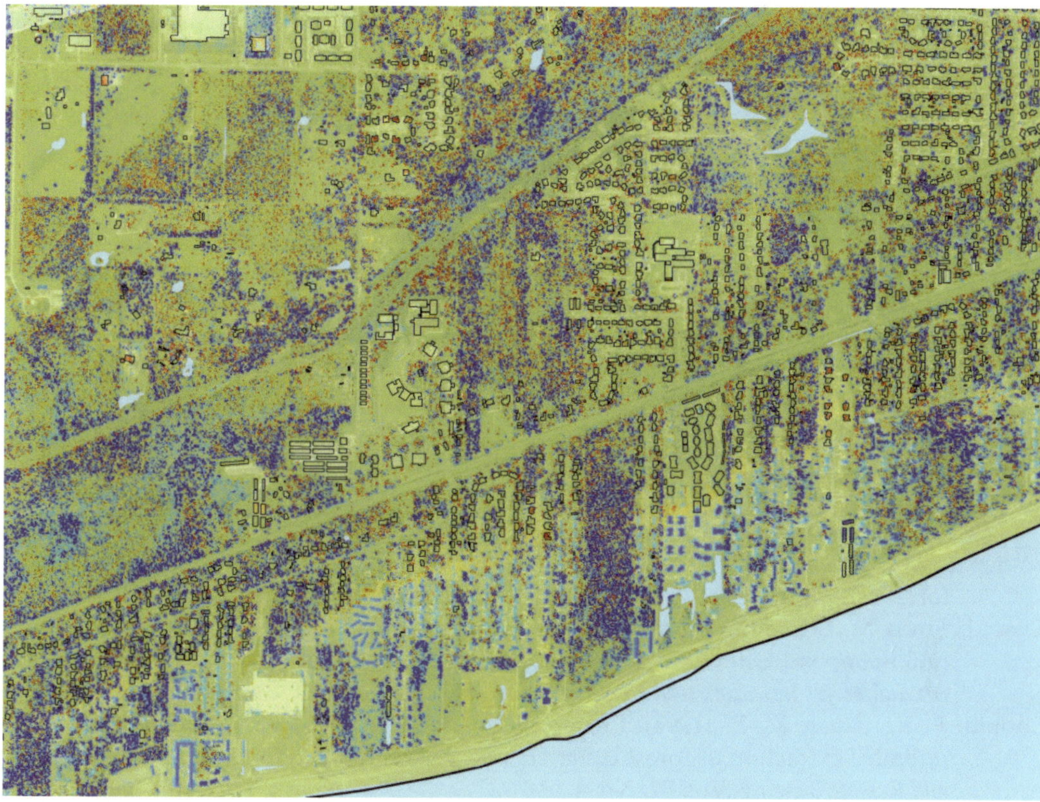

Fig. 6. HAG difference overlaid with 2007 building footprints (red = increase, green = no change, blue = decrease)

Due to the lack of temporal consistency in the pre-and post Katrina datasets, quantifying the amount of vegetation that was affected by Katrina is very difficult. Also, the high scan angles may have contributed to the differences in heights in forested areas. Further work needs to be done with multi-temporal lidar datasets to be confident in our assesments of detecting vertical changes over time. For future monitoring efforts, we would want to actively define consistent specifications in order to minimize the contributions of variability in the pre- and post-event data.

4. Conclusion

Hurricane Katrina was one of the largest natural disasters in U.S. history. Due to the sheer size of the affected areas, an unprecedented regional analysis at very high resolution and accuracy was needed to properly quantify and understand the effects of the hurricane and the storm tide. Many disparate sources of lidar data were acquired and processed for varying environmental reasons by pre- and post-Katrina projects. The datasets were in several formats and projections and were processed to varying phases of completion, and as a result the task of producing a seamless digital elevation dataset required a high level of coordination, research, and revision. This completed integration allowed for regional-scale storm surge modeling based on very high-resolution elevation information. Taking advantage of lidar flown after Katrina allowed for an assessment of the capability of lidar for change detection in the vertical dimension. Many artifacts that were outside of the error bounds attributable to the sensors themselves actually suggested vertical changes related to storm-induced debris, building creation and destruction, and vegetative change. However, further analysis and better controlled pre- and post-event data needs to be done to fully assess the capability of lidar to detect these three-dimensional changes.

5. References

Ayala-Silva, T., Twumasi, Y.A. (2004). Hurricane Georges and vegetation change in Puerto Rico using AVHRR satellite data. *International Journal of Remote Sensing* Vol. 25, No. 9, 1629–1640.

Basnet, K. Likens, G.E. Scatena, F.N. Lugo, A.E. (1992). Hurricane Hugo: damage to a tropical rainforest in Puerto Rico. *Journal of Tropical Ecology*. Vol. 8:, 47-55.

Berenbrock, C., Mason, R.R. & Blanchard, S.F. (2009). Mapping Hurricane Rita inland storm tide, *Journal of Flood Risk Management*, Vol. 2, No. 1, 76-82.

Boose, E.R., Foster, D.R., Fluet, M. (1994). Hurricane impacts to tropical and temperate forest landscapes. Ecological Monographs 64, 369–400.

Boutet Jr., J.C., Weishampel, J.F. (2003). Spatial pattern analysis of pre- and posthurricane forest canopy structure in North Carolina, USA. *Landscape Ecology*, 18, 553–559.

Clark, J., Finco, M., Schwind, B., Megown, K. (2006). Rapid assessment of forest damage using multi-temporal Landsat TM imagery and high resolution aerial photography. *In: 8th Annual Forest Inventory and Analysis Symposium*, Monterey, California, October 16–19.

Coppin, P. R., Bauer, M. E. (1994). Processing of multitemporal Landsat TM imagery to optimize extraction of forest cover change features. *IEEE Transactions on Geoscience and Remote Sensing*, Vol. 32, No. 4, 918-927.

Doyle, T.W. Keeland, B.D. Gorham, L.E. Johnson, D.J. (1995). Structural impact of Hurricane Andrew on the forested wetlands of the Atchafalaya Basin in South Louisiana. *Journal of Coastal Research*. Vol. 21, 354-364.

Dwyer, E., Pasquali, P., Holecz, F., Arino, O. (1999). Mapping forest damage caused by the 1999 Lothar Storm in Jura (France), using SAR Interferometry. *ESA Earth Observation Quarterly*. Vol. 65, 28–29.

Farris, G.S. Smith, G.J. Crane, M.P. Demas, C.R. Robbins, L.L., and Lavoie, D.L., eds. (2007). Science and the storms—the USGS response to the hurricanes of 2005: *U.S. Geological Survey Circular 1306*, 283p.

Foster DR, Boose ER. (1992). Patterns of forest damage resulting from catastrophic wind in central New England, USA. *Journal of Ecology*. Vol. 80, 79–98.

Franklin, S. E., Moskal, L. M., Lavigne, M. B., & Pugh, K. (2000). Interpretation and classification of partially harvested forest stands in the Fundy model forest using multitemporal Landsat TM digital data. *Canadian Journal of Remote Sensing*. Vol. 26, 318–333.

Fransson, J.E., Walter, F., Blennow, K., Gustavsson, A., Ulander, L.M.H. (2002). Detection of storm-damaged forested areas using airborne CARABAS-II VHF SAR image data. *IEEE Transactions on Geoscience and Remote Sensing*. Vol. 40, No.10, 2170–2175.

Gardner, L.R., Michener, W.K., Williams, T.M., Blood, E.R., Kjerfve, B., Smock, D.J., Lipscomb, L.A., Gresham, C. (1992). Disturbance effects of Hurricane Hugo on a pristine coastal landscape: North Inlet, South Carolina, USA. *Netherlands Journal of Sea Research*. Vol. 30, 249-263.

Gesch, D. (2009). Analysis of lidar elevation data for improved identification and delineation of lands vulnerable to sea-level rise. *Journal of Coastal Research*, SI(53), 49-58.

Gesch, D. Oimoen, M., Greenlee, S., Nelson, C., Steuck, M., and Tyler, D. (2002). The National Elevation Dataset. *Photogrammetric Engineering and Remote Sensing.* Vol. 68, No. 1, 5-11.

Hegarat-Mascle, S. L., Seltz, R., Hubert-Moy, L., Corgne, S., & Stach, N. (2006). Performance of change detection using remotely sensed data and evidential fusion: Comparison of three cases of application. *International Journal of Remote Sensing.* Vol 27, 3515–3532.

Interagency Performance Evaluation Taskforce (IPET), (2008). *Performance evaluation of the New Orleans and southeast Louisiana hurricane protection system: Final report of the Interagency Performance Evaluation Task Force.* Vicksburg, Mississippi: U.S. Army Corps of Engineers, 9v. URL: https://ipet.wes.army.mil/

Jacobs, D.M. (2006). Rapid assessment model for determining extent of hurricane damage. Poster presentation at the *Forest Health Monitoring Program work group workshop;* 2006 January-February; Charleston, SC.

Kelly, J.F. (1993). Hurricane Andrew forest damage assessment. *World Resource Review.* Vol. 5, 401-408.

Knabb, R.D. Rhome, J.R., and Brown, D.P. (2005). *Tropical cyclone report: Hurricane Katrina 23-30 August 2005.* Miami, Florida: National Oceanic and Atmospheric Administration, National Hurricane Center, 43 p.
URL: http://www.nhc.noaa.gov/pdf/TCR-AL122005_Katrina.pdf

Kovacs, J.M. (2001). Mapping disturbances in a mangrove forest using multi-date Landsat TM imagery. *Environmental Management.* Vol. 27, 763-776.

Kovacs, J.M., Blanco-Correa, M., Flores-Verdugo, F. (2001). A logistic regression modelof hurricane impacts in a mangrove forest of the Mexican Pacific. *Journal of Coastal Research.* Vol. 17, No. 1, 30–37.

Kwarteng, A. Y., & Chavez, P. S. (1998). Change detection study of Kuwait City and environs using multi-temporal Landsat Thematic Mapper data. *International Journal of Remote Sensing.* Vol. 19, 1651–1662.

Lefsky, M.A.,Cohen, W.B., Parker, G.G., and Harding, D.J. (2002). Lidar remote sensing for ecosystem studies. *Bioscience.* Vol. 52, No. 1, 19-30.

Link, L. E. (2010). The anatomy of a disaster, an overview of hurricane katrina and new orleans. *Ocean Engineering.* Vol. 37, No. 1, 4-12.

Lyon, J. G., Yuan, D., Lunetta, R. S., & Elvidge, C. D. (1998). A change detection experiment using vegetation indices. *Photogrammetric Engineering and Remote Sensing.* Vol. 64, No. 2, 143–150.

Maas, H.G. (2001). The suitability of airborne laser scanner data for automatic 3D object reconstruction. *In*: Baltsavias, E.P., Gruen, A. and Van Gool, L. (eds.), *Automatic Extraction of Man-Made Objects from Aerial and Space Images (III).* Ascona, Switzerland: A.A. Balkema Publishers, pp. 291-296.

McNab, W.H. Greenberg, C.H. Berg, E.C. (2004). Landscape distribution and characteristics of large hurricane-related canopy gaps in a southern Appalachian watershed. *Forest Ecology and Management.* Vol. 196, 435-447.

Nackaerts, K., Vaesen, K., Muys, B., & Coppin, P. (2005). Comparative performance of a modified change vector analysis in forest change detection. *International Journal of Remote Sensing. Vol. 26,* 839–852.

National Oceanic and Atmospheric Administration (NOAA). (2008). *Center for Operational Oceanographic Products and Services: What is the difference between storm surge and storm tide?* Silver Spring, Maryland: National Oceanic and Atmospheric Administration, National Ocean Service.
URL: http://co-ops.nos.noaa.gov/quicklook_faqs.shtml

Raber, G.T. and Tullis, J.A. (2007). Rapid assessment of storm-surge inundation after Hurricane Katrina utilizing a modified distance interpolation approach. *GIScience and Remote Sensing*. Vol. 44, No. 3, 220-236.

Ramsey, E.W., Chappell, D.K., Baldwin, D.G. (1997). AVHRR imagery used to identify hurricane damage in a forested wetland of Louisiana. *Photogrammetric Engineering and Remote Sensing*. Vol. 63, No. 3, 293–297.

Ramsey, E.W. Chappell, D.K. Jacobs, D.M. Sapkota, S.K. Baldwin, D.G. (1998). Resource management of forested wetlands: hurricane impact and recovery mapped by combining Landsat TM and NOAA AVHRR data. *Photogrammetric Engineering and Remote Sensing*. Vol. 64, No. 7, 733-738.

Ramsey E.W., Hodgson M.E., Sapkota S.K., Laine S.C., Nelson, G.A., Chappell D.K. (2001). Forest impact estimated with NOAA AVHRR and Landsat TM data related to an empirical hurricane wind-field distribution. *Remote Sensing of Environment*. Vol. 77, No. 3, 279–92.

Stockdon, H. Doran, K.S., and Sallenger, A.H., Jr. (2009). Extraction of lidar-based dune-crest elevations for use in examining the vulnerability of beaches to inundation during hurricanes. *Journal of Coastal Research*, SI(53), 59-65.

Stoker, J.M., Greenlee, S.K., Gesch, D.B., and Menig, J.C. (2006). CLICK: The new USGS center for lidar information coordination and knowledge. *Photogrammetric Engineering and Remote Sensing*, Vol. 72, No. 6, 613-616.

Stoker, J.M. Tyler, D.J. Turnipseed, D.P. Van Wilson, K., Jr., and Oimoen, M.J., 2009. Integrating disparate lidar datasets for a regional storm tide inundation analysis of Hurricane Katrina. *Journal of Coastal Research*, SI(53), 66–72.

Travis, J. (2005). Hurricane Katrina: Scientists' fears come true as hurricane floods New Orleans. *Science*, Vol. 309, 1656-1659.

Turnipseed, D.P., Wilson, K. V., Stoker, J., and Tyler, D. (2007). Mapping the Hurricane Katrina storm surge crest in Louisiana, Mississippi, and Alabama, *Proceedings of the 37th Annual Mississippi Water Resources Conference (April 24-25, 2007)*, Jackson, MS, 10 p.

Wang, F., Xu, Y. (2009). Hurricane Katrina-induced forest damage in relation to ecological factors at landscape scale. *Environmental Monitoring and Assessment*. Vol. 156, No. 1-4, 491–507.

Wang, W., Qu, J.J., Hao, X., Liu, Y., Stanturf, J.A. (2010). Post-hurricane forest damage assessment using satellite remote sensing *Agricultural and Forest Meteorology*, Vol. 150, No. 1, 122-132.

Wiesmann, A., Wegmuller, U., Honikel, M., Strozzi, T., Werner, C.L. (2001). Potential and methodology of satellite based SAR for hazard mapping. In: *Proceedings of Geoscience and Remote Sensing Symposium, IGARSS 2001*, Sydney, Australia, July 9–13.

31

Segmentation Methods for Synthetic Aperture Radar

Shuang Wang
Xidian University, Xi'an, China

1. Introduction

Synthetic aperture radar (SAR) image segmentation is a fundamental problem in SAR image interpretation. SAR images often contain non-texture object and texture object. Level set method, known as deformable model, is a powerful image segmentation technique. It can get accurate contours of non-texture object, but has poor performance in getting contours of texture object.

In recent years, parametric active contour model (snake), which was proposed by Kass in 1987 [1], has become one of the most studied techniques for image segmentation. The snake approach is based on deforming an initial contour or surface towards the boundary of the object to be detected. The deformation is obtained by minimizing a global energy designed such that its minimum is obtained at the boundary of the object. The energy is basically composed by a term which controls the smoothness of the deforming curve and another one which attracts it to the boundary. However, the classical active contour model presents several limitations. In particular, it is sensitive to initial contour placement, and most importantly, it can't handle topological changes of the curves during their evolution [2-4].

Geometric active contour model was subsequently proposed by Osher and Sethian in 1988 [5]. This model is based on the theory of curve evolution and geometric flows, which implemented using level set. Level set is designed to handle problems in which the evolving interfaces can develop sharp corners, change topology and become very complex. Level set approach has been widely applied to image processing [6-9].

SAR image segmentation is an important, challenging problem and a necessary first step in image analysis and interpretation. However, segmentation of distinct areas, such as city and river, is a challenging task due to their complex topologies. So we use level set approach to solve the topology problem. One class of image segmentation is object detection, where certain objects in the image are to be singled out. In this case, the image is basically divided into two sets: objects and background. Some non-texture objects in SAR image, such as river, ravine and railway, can be detected easily by level set method because of their distinctness with background and lack of texture. But in particular, objects in SAR image often contain texture, such as the city zone, which may cause considerable difficulties when applying level set approach. It makes the detection result too minute, and loses the consistency of objects. We would like to use level set method to get accurate contour of objects, whereas the consistency of objects is also needed. As a result, a new technique which could make level set method adaptable to detection of texture object is required.

Here, a new modified model of level set based on clonal selection algorithm [10] is proposed. We use clonal selection algorithm to choose some pixels near the contour, and then perform a neighborhood modification on these points of the level set function during its evolution. Their region texture information, supervising the modification process, is incorporated into the level set framework. This new method is particularly well adapted to segmentation of texture object of interesting. We illustrated the performance of the new method on SAR images. Furthermore, we compared our method with level set method and the modified model of level set based on standard genetic algorithm (SGA) in texture object detection result and image segmentation result. The experimental results show that incorporating region texture information into the level set framework, consistent texture objects are obtained, and accurate and robust segmentations can be achieved.

The level set method was devised by Osher and Sethian in 1988. The main idea is to describe a closed curve $\Gamma(t)$ in the image plane as the zero level set of a higher dimensional function $\Phi(x,t)$ in R^3.

$$\begin{cases} \Phi(x,t) > 0, \text{ if } x \text{ is inside } \Gamma(t) \\ \Phi(x,t) = 0, \text{ if } x \text{ is at } \Gamma(t) \\ \Phi(x,t) < 0, \text{ if } x \text{ is outside } \Gamma(t) \end{cases} \tag{1}$$

In numerical implementations, often regularity is imposed on Φ to prevent the level set function to be too steep or flat near the interface. This is normally done by requiring Φ being the signed distance function (SDF) to the interface.

$$\begin{cases} \Phi(x,t) = d(\Gamma(t),x), & \text{if } x \text{ is inside } \Gamma(t) \\ \Phi(x,t) = 0, & \text{if } x \text{ is at } \Gamma(t) \\ \Phi(x,t) = -d(\Gamma(t),x), & \text{if } x \text{ is outside } \Gamma(t) \end{cases} \tag{2}$$

Where $d(\Gamma(t),x)$ denotes Euclidian distance between x and $\Gamma(t)$. We emphasize that requiring (2) is a technical processing to prevent instabilities in numerical implementations. Having defined the level set function, there is a one-to-one correspondence between the curve Γ and the function Φ.

The evolution of $\Phi(x,t)$ can be modeled as

$$\frac{\partial \Phi}{\partial t} + V\left|\nabla \Phi\right| = 0 \tag{3}$$

with a given $\Phi(x,t=0)$. At any instant, the position of Γ shall be given as the zero level set of evolving function Φ. The speed function V depends on factors as the image gradient and local curvature.

The above motion equation (3) is a partial differential equation in a higher dimension than the original problem. Given the initial value, it can be solved by means of the following discretization and linearization.

$$\Phi(n+1) = \Phi(n) + \Delta\Phi(n) \tag{4}$$

$$\Delta\Phi(n) = \Delta t \cdot \delta_\varepsilon(\Phi(n))\left[\mu \,\mathrm{div}\left(\frac{\nabla\Phi(n)}{|\nabla\Phi(n)|}\right) - \nu + \lambda_1(u_0 - c_1(n))^2 - \lambda_2(u_0 - c_2(n))^2\right] \quad (5)$$

In formula (4) n is the iterative time, $\Phi(n)$ is the level set function value at time n, and $\Delta\Phi(n)$ is the update quantity at time n. And Δt is the time step, $\delta_\varepsilon(\Phi(n))$ is the Dirac measure, u_0 is the image to be segmented, $c_1(n)$ and $c_2(n)$ are the averages of u_0 outside And inside the curve Γ respectively, the constant μ is length parameters, ν is a correction term, and λ_1, λ_2 are fixed parameters in formula (5). This implementation allows the function Φ to automatically follow topological changes and corners during evolution [6, 7, 11, 12].

We all know that level set method can solve the complex topology problem and get accurate contour of objects. But it is lack of region information in its implementation, which makes its performance in getting contours of texture object not as well as we expect. So we must introduce region information into our work to supervise level set function during its evolution.

We perform modification repeatedly on the level set function in every iteration when the function is evolving. Clonal selection algorithm is used to choose pixels near the contour. According to the region texture information of each chosen point, we decide how to modify (positive or negative) and how much to modify (the parameter value). To effectively shift the contour and change topology, we do neighborhood modification instead of point modification. The principle of the modified model is shown in figure 1.

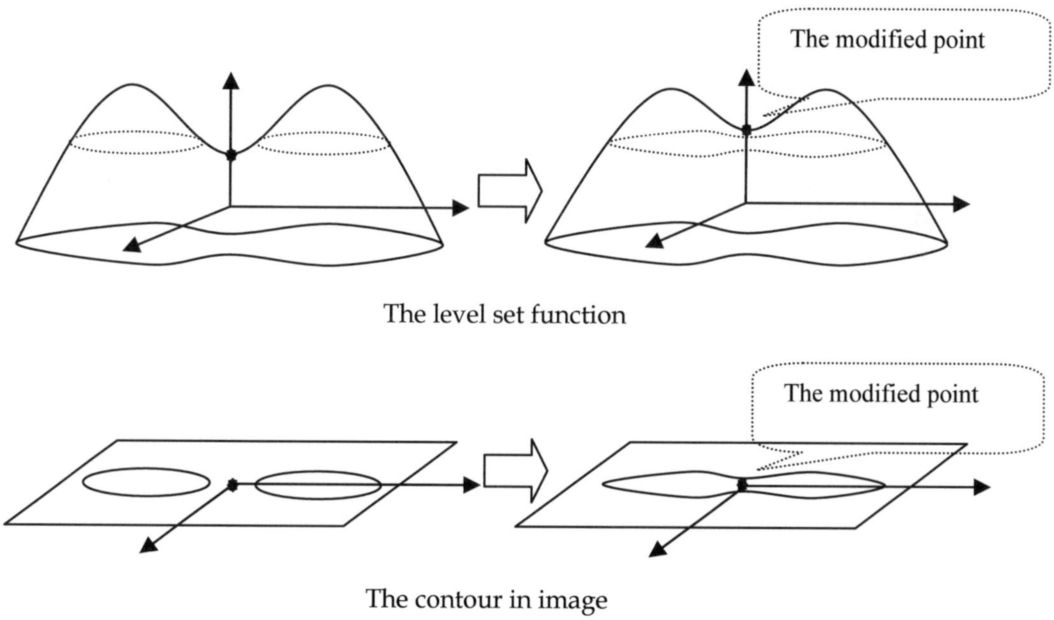

Fig. 1. The principle of the modified model, the point who has a region texture feature approximate to the object feature is given a positive modification. As a result, the nearby contours are connected and the object becomes more consistent.

To interpret the modification strategy, we note the area $(i-\delta,i+\delta;j-\delta,j+\delta)$ as the δ neighborhood of the pixel (i,j), and note $\Phi_{i,j}(\delta)$ as the δ neighborhood of $\Phi_{i,j}$.
The modification of $\Phi_{i,j}$ can be modeled as

$$\Phi_{i,j}(n+1,\delta)=\Phi_{i,j}(n,\delta)+\Delta\Phi_{i,j}(\delta) \tag{6}$$

where n is the iterative time, $\Phi_{i,j}(n,\delta)$ is the level set function value of δ neighborhood of pixel (i,j) at time n, and $\Delta\Phi_{i,j}(\delta)$ is the modification value matrix. This implementation allows the function Φ to follow region feature information automatically during modification. Here, $\Delta\Phi_{i,j}(\delta)$ is defined as:

$$\Delta\Phi_{i,j}(\delta)=\sigma c_{i,j}T(\delta) \tag{7}$$

where σ is the global modification coefficient, $c_{i,j}$ is the local modification coefficient of pixel (i,j), $T(\delta)$ is the neighborhood template, which is expressed by a matrix.
We define d_n as the Euclidian distance between background n and the certain object in the feature space, $d_{i,j}$ as the Euclidian distance between pixel (i,j) to be modified and the certain object in the feature space. Then, the parameter in formula (7) can be denoted as

$$c_{i,j}=-\frac{\left|d_{i,j}-d\right|}{d}\bullet(d_{i,j}-d) \tag{8}$$

$$T(\delta)=T(1)=\begin{bmatrix}0.5 & 1 & 0.5\\ 1 & 2 & 1\\ 0.5 & 1 & 0.5\end{bmatrix} \tag{9}$$

$$d=\min(d_1,d_{2,}\cdots,d_n)/2 \tag{10}$$

In order to make the modification steady, we introduce an inhibiting factor during the evolution of level set function and replace formula (4) by the following one.

$$\Phi_{i,j}(n+1)=\Phi_{i,j}(n)+h_{i,j}\Delta\Phi_{i,j}(n) \tag{11}$$

$$h_{i,j}=\lambda(1-\frac{\left|d_{i,j}-d\right|}{d}) \tag{12}$$

where λ is the inhibiting coefficient.
In conclusion, the new method can be summarized as follows:

Step 1. Initialization.

1.1 Extract the texture feature. Sample the object and each background to calculate the mean feature value.

1.2 Initialize the level set function using SDF.

1.3 Initialize the parameters of clonal selection algorithm. The affinity is the level set function.

1.4 Initialize the generation $n=0$. Initialize the maximal evolution generation.

Step 2. Evolve the level set function by formula (4) and (5).
Step 3. Modification.
3.1 Implement the clone and mutation operators, evaluate the affinity. And then select the points whose level set value are near zero level set.
3.2 Perform the modification by formula (6).
Step 4. Evolve the level set function by formula (11) for ten times.
Step 5. Let $n = n + 1$, return Step 3 until the maximal evolution generation is achieved.

Experimental results

Three experiments have been carried out to test the efficiency of the proposed method. We regard the image segmentation problem as an object detection problem. That is to say, we detect texture and non-texture objects from the background separately, and then combine the detection results to obtain the final segmentation results.
In the following experiments, three different SAR images with size 256×256 are used which contain both texture and non-texture objects. We combine the non-texture object detected by

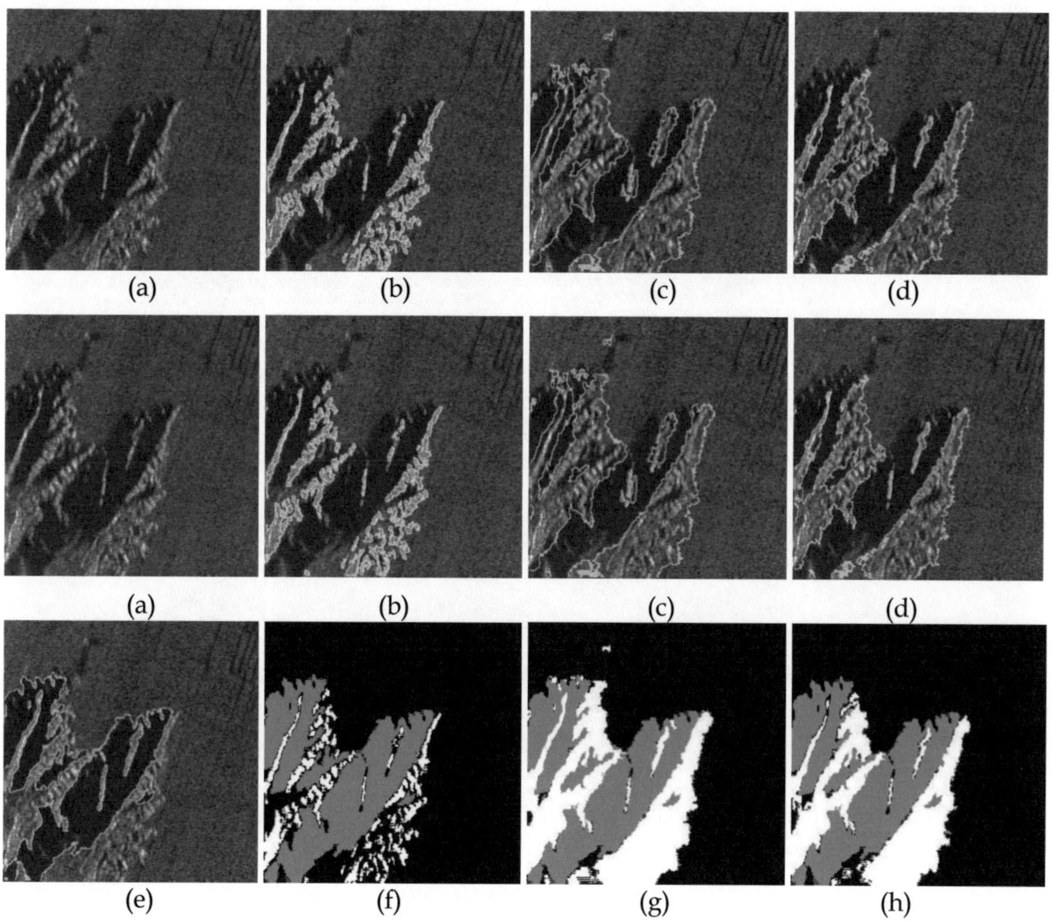

Fig. 2. SAR image experimental result 1, three methods were used and corresponding segmentation results were given.

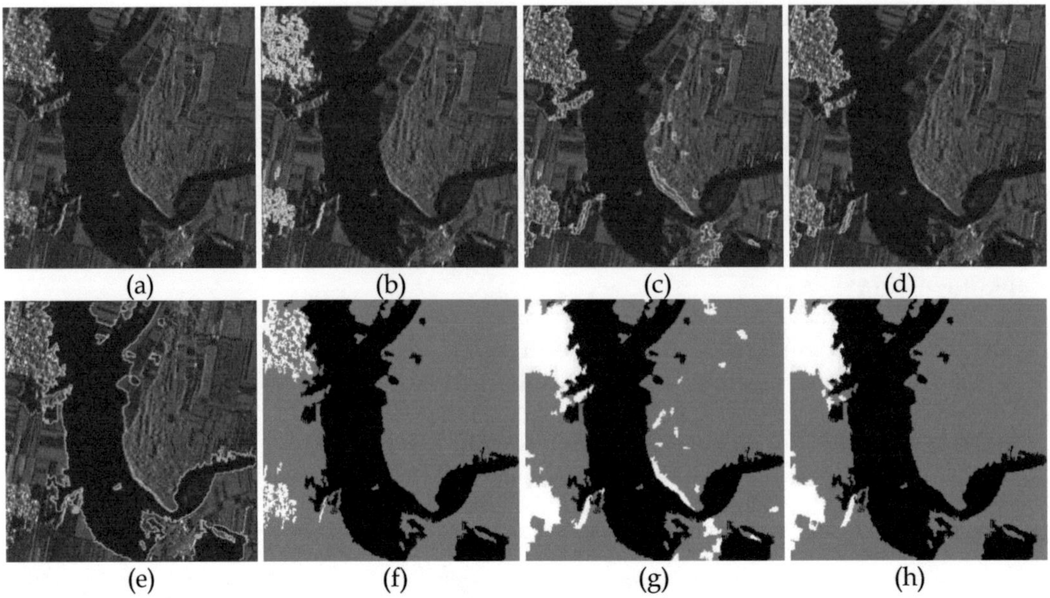

Fig. 3. SAR image experimental result 2, three methods were used and corresponding segmentation results were given.

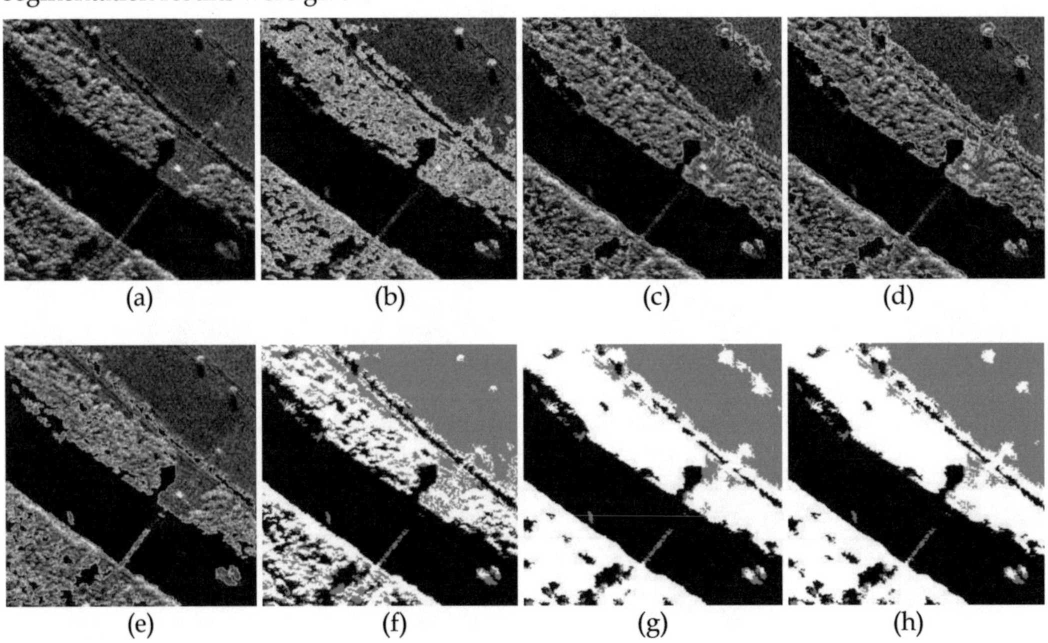

Fig. 4. SAR image experimental result 3, three methods were used and corresponding segmentation results were given.

level set method and the texture object detected by three different methods to obtain the segmentation results separately. First we compared our method with level set method without modification and the modified model of level set based on standard genetic algorithm (SGA) in texture object detection results. Then a further comparison of the three methods was made in the final segmentation results. Feature vector with dimension 10 is extracted using the wavelet analysis. In clonal selection algorithm, a binary encoding scheme is used to represent the coordinate (i, j). The initial antibody population size is 300 and the antibody length is 16. The clonal size is 5. The maximal evolution generation is 500. In SGA, the chromosome length, population size and the maximal evolution generation are as same as clonal selection algorithm. The experimental results are shown in figure 2 to 4.

In the experiments above, (a) is the original images, (b) is the texture object detection results by level set method, (c) is the texture object detection results by the modified model of level set based on SGA, (d) is our texture object detection results by the new modified model of level set method based on clonal selection algorithm, (e) is the non-texture object detection results by level set method, (f) is the segmentation results based on level set method, which is obtained by combine (b) and (e), (g) is the segmentation results by the modified model of level set based on SGA, which is obtained by combine (c) and (e), (h) is our finally segmentations by the new modified model of level set method based on clonal selection algorithm, which is obtained by combine (d) and (e).

From (b) to (d), we can see that (b) has the most accurate contour position, but it is too minute to be considered as a whole object, (c) is not accurate enough in contour and has some unfavorable modification, (d) is our result, which is more accurate than (c) and more consistent than (b). It is clear that the proposed method is efficient on the texture object detection. From (f) to (h), we give the corresponding segmentation results. From the object detection results acquired by three different methods referred above, it is clear that the proposed method is efficient on the texture object contour detection. And comparing the corresponding segmentation results, our method shows better performance. As a result, we can draw the conclusion that the modified model of level set based on clonal selection algorithm is more effective than the modified model of level set based on SGA and traditional level set methods in detecting texture objects, and therefore leads to better segmentation results.

We analyzed the modified model based on SGA and the modified model based on clonal selection algorithm especially. Two methods take the approximately same time, but show quite different performance. The modified model based on SGA is tend to be trapped in local optimization and induce excess or wrong modification, while the modified model based on clonal selection algorithm can jump out of local optimization and find proper points to be modified. As a result, we can draw the conclusion that the modified model of level set based on clonal selection algorithm is more effective than the modified model of level set based on SGA and traditional level set methods in detecting texture objects, and therefore leads to better segmentation results.

A modified model of level set based on clonal selection algorithm is formed by incorporating region texture information into the level set framework. The experiments show that this new method is of great efficiency in detecting consistent texture objects and proves to be particularly well adapted to accurate and robust segmentation of texture object of interest. If we choose our method to detect texture object and level set method to non-texture object, then accurate and robust segmentations can be achieved.

2. An improved watershed-based SAR image segmentation algorithm

In this section, we give another method about image segmentation which is based on the watershed algorithm because sometimes we need to keep the information such as boundaries. The watershed algorithm is a well established morphological segmentation tool, it commonly segments an image into a set of non overlapping regions. The watershed algorithm has the advantage of region growing algorithm, the regions are spatially consistent, with boundaries forming a closed, connected set, it also makes use of edge information, as captured by the gradient surface. This kind of mathematical morphology segmentation method respectively considers light and dark image areas as the hills and valleys of an image landscape. To segment a given image the "landscape" is flooded, whereby water flows from high altitude areas (areas with high gray scale values) along lines of steepest descent until it reaches some regional minimum (low grayscale regions). The watersheds or catchment basins of the image are the draining areas of its regional minima. These areas are separated by lines called watershed lines.

Unfortunately, the segmentation produced by a naive application of the watershed algorithm is oftentimes inadequate: the image is usually over-segmented into a large number of minuscule regions. As a result, several extensions have been proposed in order to produce more natural image segmentation (e.g., hierarchical watersheds or region split/merge [14]). The most common remedy is to use markers for identifying relevant region minima (e.g., [15], [16] and[17]). By setting marker locations as the only local minima within the watershed image, the number of regions can be automatically controlled. Also, particular approach to finding and utilizing markers can be found in [18], [19]and[20], where researchers used a naive Bayes classifier to identify (i.e., classify) pixel groups as internal markers. Of course,these particular approach have good performance in controlling over-segmentation, but they are complex, low speed. So we show an improved watershed algorithm by referring to that in paper [17]. Its marker extraction method is one of the focal points of this paper.

Watershed segmentation is not effective for Synthetic Aperture Radar (SAR) images which are generally corrupted by coherent speckle noise, the image is usually over-segmented into a large number of minuscule regions due to speckle noise which tends to be amplified by the gradient operator. Gaussian filter is involved to smooth the image in many steps in our algorithm. Otsu algorithm is used to produce inner mark and external mark . It's unsupervised, low complexity and high speed. Then we use the mark to modify the gradient image of the SAR image. Additionally, rather than flooding the gradient image, we use the gradient image modified by the mark as input to the watershed algorithm.

First, we provide a brief overview of Gaussian filter. Next, we present the details of our algorithm. The following section presents experimental results carried out on a set of SAR images and discusses the significance of the results. The last section presents future research directions and concludes the paper.

Gaussian filter [21][22] has been intensively studied in image processing and computer vision. It's a kind of transform in frequency domain .Since frequency is directly related to rate of change, it is not difficult intuitively to associate frequencies in the Fourier transform with patterns of intensity variations in an image. It's that the slowest varying frequency component (u=v=0) corresponds to the average gray level of an image. As we move away from the origin of the transform, the low frequencies correspond to the slowly varying components of an image while high frequencies correspond to detail, such as noise. A filter that attenuates high frequencies while "passing" low frequencies is called lowpass filter. A filter that has the opposite characteristic is appropriately called a highpass filter. We would expect a lowpass-filtered image to have less sharp detail than original because the high frequencies have been attenuated. Similarly, a highpass-filtered image would have less gray level variations in smooth areas and emphasized transitional gray-level detail. Such an image will appear sharper.

In this paper, Gaussian lowpass filter is used to suppress noise before extracting gradient image. So there will be less regional minima in the gradient image. It's a good help to control over-segmentation. It consists of the following steps[23]: (1) Multiply the input image by $(-1)^{x+y}$ to center the transform; (2) compute $F(u,v)$,the DFT of the image;(3) Multiply $F(u,v)$ by filter function $H(u,v)$ (4) compute the inverse DFT of the result in (3); (5) Obtain the real part of the result in(4) (6)Multiply the result in (5) by $(-1)^{x+y}$
The Fourier transform of the output image after Gaussian lowpass filtering can be expressed as:

$$G(u,v) = H(u,v)F(u,v)\,,\, H(u,v) = \exp(-(u^2+v^2)/2\sigma^2)$$

Otsu algorithm can be said to be adaptive to calculate the single threshold (for converting images to gray scale image) simple Efficient methods. Algorithm for the importation of gray image histogram analysis of histogram is divided into two parts, the method achieved a optimal threshold in many cases. Using the optimal segmentation threshold, we segment the SAR image into the target and background classes. In this paper, we extract mark from the gradient of the segmented image for watershed algorithm.
The Otsu algorithm is proposed on the principle of least squares. Firstly, its basic principle is to compute the inter-classes variance of each gray value, then segment the image according to the gray value with the maximum inter-classes variance value. Suppose the total pixel number of the image is M, the range of the gray values is [1, *L*], the number of the pixel gray value i is m_i, the corresponding probability is $P_i = m_i / M$ .Let C_0 and C_1 as the target and background classes respectively, which are divided by the value *k*.The gray value range of C_0 is [1, *k*], the gray value range of C_1 is [*k*+1, *L*]. The gray value probability C_0 is $\omega_0 = \sum_{i=1}^{k} p_i$,the gray value probability of C_1 is $\omega_1 = \sum_{i=k+1}^{L} p_i$. In order to evaluate optimal segmentation threshold, we shall introduce the following discriminant criterion measures (or measures of class separability) used in the discriminant.
Analysis. Let σ_w^2 , σ_B^2 and σ_T^2 as the within-class variance, between-classes variance and total variance of levels respectively.

$$\lambda = \frac{\sigma_B^2}{\sigma_w^2},\quad \kappa = \frac{\sigma_T^2}{\sigma_w^2},\quad \eta = \frac{\sigma_B^2}{\sigma_T^2}$$

there $\sigma_w^2 = \omega_0\sigma_0^2 + \omega_1\sigma_1^2$, $\sigma_B^2 = \omega_0(\mu_0 - \mu_T)^2 + \omega_1(\mu_1 - \mu_T)^2 = \omega_0\omega_1(\mu_0 - \mu_1)^2$

$$\sigma_T^2 = \sigma_w^2 + \sigma_B^2,\quad \mu_T = \sum_{i=0}^{L-1} iP_i\,,\quad \mu_t = \sum_{i=0}^{t} iP_i,\quad \mu_0 = \frac{\mu_t}{\omega_0},\quad \mu_1 = \frac{\mu_T - \mu_t}{1-\omega_0}$$

the optimal segmentation threshold that maximizes η, or equivalently maximizes σ_B^2 is $t^* = \arg\max_{t\in G} \sigma_B^2$.

The watershed algorithm was originally developed by Lantuéjoul [25] and is extensively described together with its many applications by Beucher and Meyer [26]. Since its original development with grey-scale images [25] it has been extended to a computationally efficient form (using FIFO queues) [27] and applied to color images [28]. The main advantages of the watershed method over other previously developed segmentation methods are [26]:

- The resulting boundaries form closed and connected regions.Traditional edge based techniques most often form disconnected boundaries that need post-processing to produce closed regions.
- The boundaries of the resulting regions always correspond to contours which appear in the image as obvious contours of objects. This is in contrast to split and merge methods where the first splitting is often a simple regular sectioning of the image leading sometimes to unstable results.
- The union of all the regions form the entire image region.

Direct application of the watershed segmentation algorithm leads to overs-egmentation due to noise and other local irregularities of the gradient. A practical solution to this problem is to limit the number of allowable regions by incorporating a preprocessing stage designed to bring additional knowledge into segmentation procedure. To be effective these methods require object markers.

Instead of using the image directly, watershed transform uses a gradient image extracted from the original image. The initial stage of any watershed segmentation method is therefore to produce a gradient image from the actual image. So we need to extract gradient image of original image by "priwitt" operator. Soille's algorithm use markers to modifies it using morphological reconstruction so it only has regional minima when internal image and external image are nonzero. A marker is a connected component belonging to an image. In this paper, there are two kinds of markers, an internal markers and external markers, marker based watershed was transposed to marker selection problem. In this paper, internal markers is defined in this case as (1) a region that is surrounded by points of higher 'altitude';(2)such that the points in the region form a connected component; and(3)in which all the points in the region form a connected component have the same gray level value.

However, Soille's algorithm is useless for SAR image segmentation, coherent speckle noise and texture information also cause so much over-segmentation. So we propose our improved algorithm. To obtain fewer internal markers , Firstly, we segment the SAR image using Otsu method. Smoothing the initial segmentation by Gaussian filter is also necessary, because it can reduce small regions which will produce redundant markers. Secondly, we obtain the gradient of the segmentation by gradient operator and we smooth it again using low pass filter to emphasize the significant gradient within the image and reduce the gradient caused by coherent speckle noise or other minor structures, such as texture in SAR images. Thirdly, we select internal marker from the smoothed gradient image. Selecting internal marker is to find regional minima in the gradient image, regional minima are connected components of pixels with a constant intensity value, and whose external boundary pixels all have a higher value. The value is called fall threshold. The higher fall threshold, the fewer the number of region. One marker per region is necessary since there will be a one-to-one correspondence between the marker and segments of the final partition. The external markers are the watershed of internal marker image which effectively partition the image into regions, with each region containing a single internal marker. At last, we obtain the gradient of original SAR image by gradient operator, the gradient image is modified by internal marker and external marker. The watershed algorithm applied to the modified gradient. The markers controlled the number of regions, and the original gradient image insures that the boundaries of segmentation result are accurate.

The program in detail is shown in Fig 1.

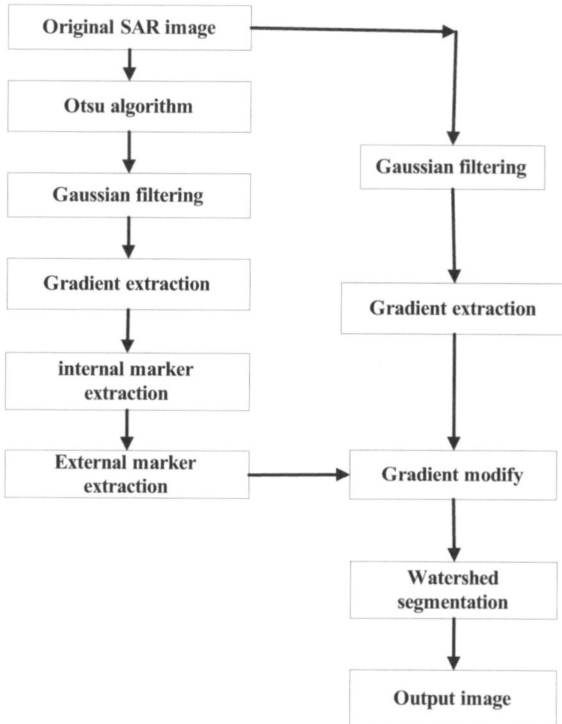

Fig. 1

Experimental results

The proposed algorithm has been executed on a set of SAR images. The test images used in the experiment are shown in Fig.1a to Fig.4a. The segmentation results of our proposed algorithm, watershed algorithm and marker-controlled watershed alogorithm with different threshold are shown in (d)- (i) Fig.1. to Fig.4. The average number of block of the segmentation results of different methods are shown in Table.1. In addition, we use different fall thresholds in marker-controlled watershed and our algorithm. In comparison with the initial segmentation results of watershed algorithm , the edges obtained by the proposed method are quite close to the real edges . From Table.1, we can see that the block number of our algorithm are lesser than that of watershed method and the algorithm of [17].

Original image		Watershed algorithm	Algorithm of [5]			Our algorithm		
			1	10	20	30	40	50
SAR image I	Block number	3187	3187	1166	611	230	84	30
SAR image II	Block number	4932	4932	3096	2475	22	6	4
SAR image III	Block number	2121	2121	1313	1107	57	33	19
SAR image IV	Block number	1434	1434	565	358	110	71	47

Table1. experiment result

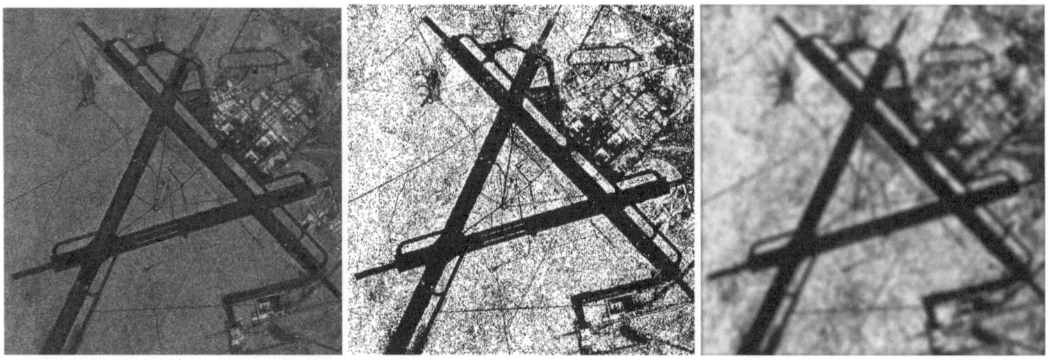

(a) original SAR image (b) watershed segmentation (c) final result

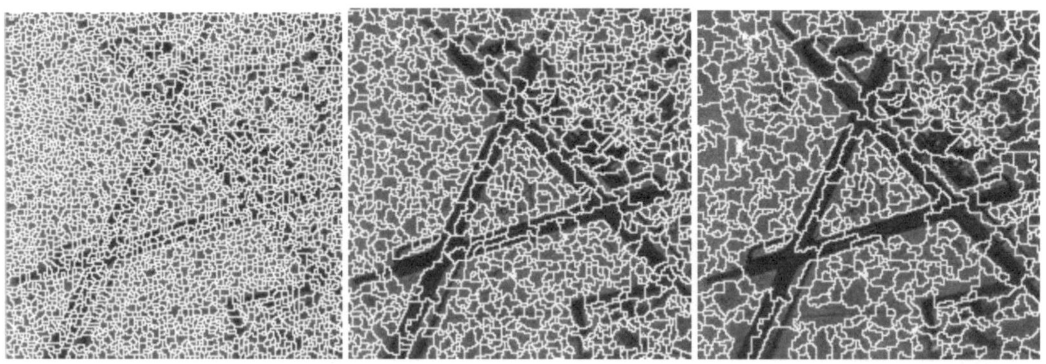

(d) watershed (e) marker-controlled watershed (threshold 10) (f) marker-controlled watershed (threshold 20)

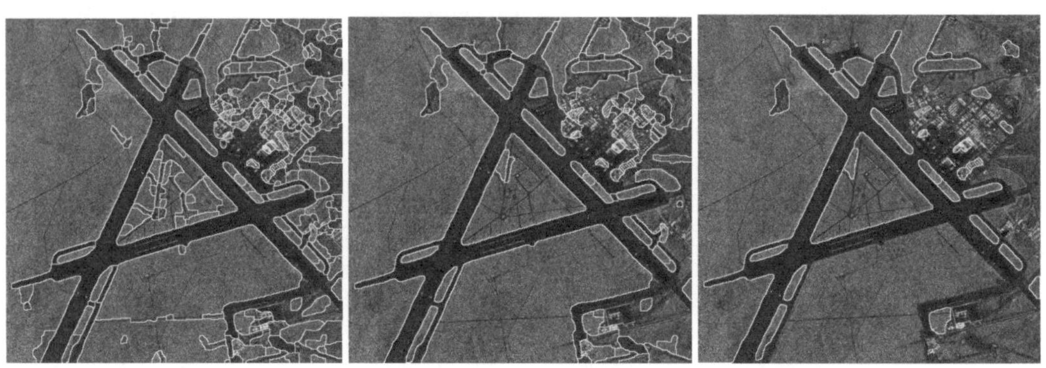

(g) our algorithm (threshold 30) (h) our algorithm (threshold 40) (i) our algorithm (threshold 50)

Fig. 2.

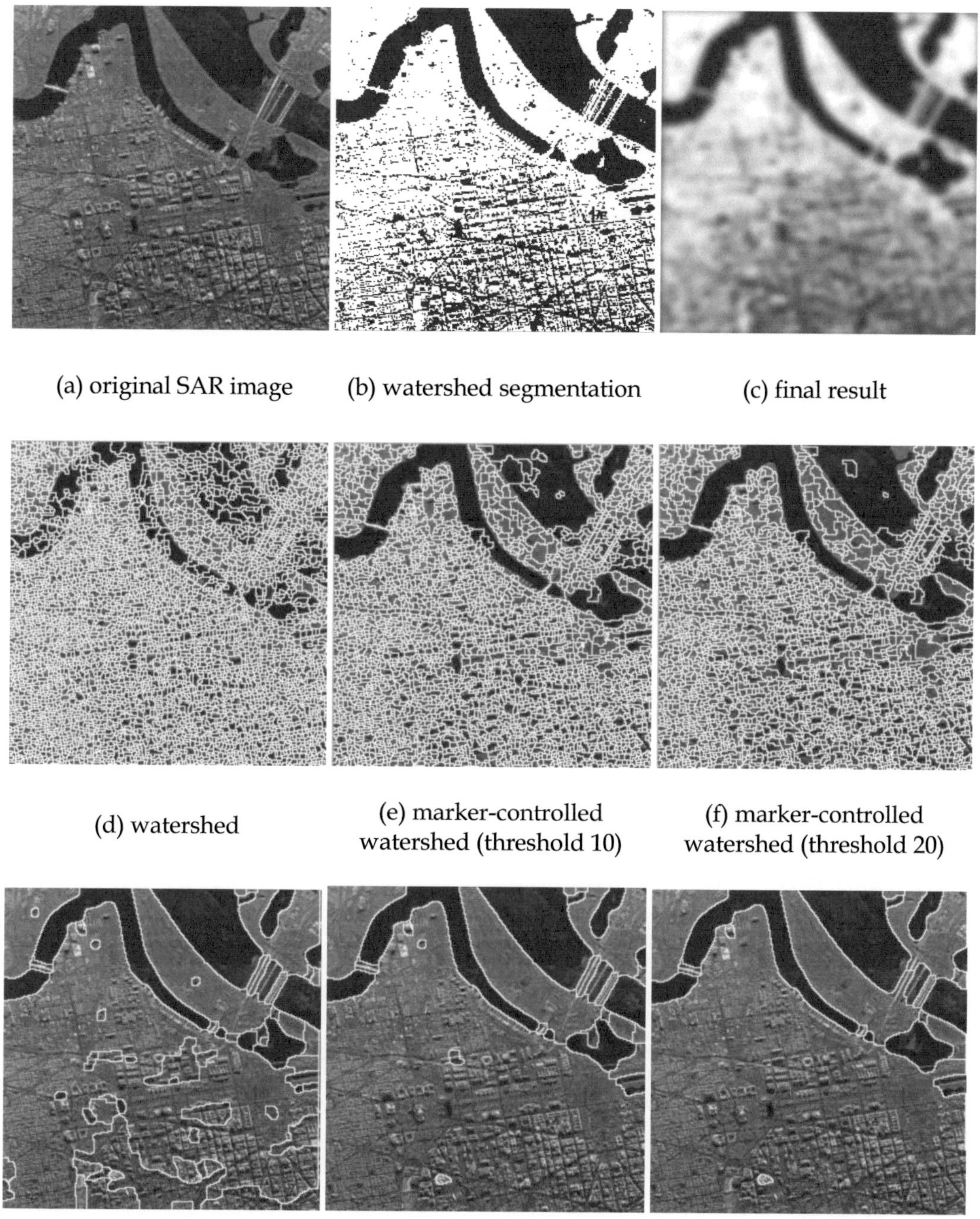

(a) original SAR image
(b) watershed segmentation
(c) final result
(d) watershed
(e) marker-controlled watershed (threshold 10)
(f) marker-controlled watershed (threshold 20)
(g) our algorithm (threshold 30)
(h) our algorithm (threshold 40)
(i) our algorithm (threshold 50)

Fig. 3

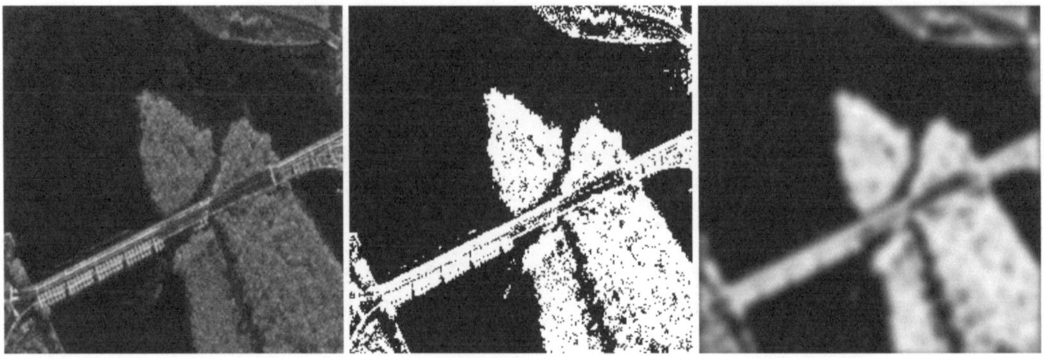

(a) original SAR image (b) watershed segmentation (c) final result

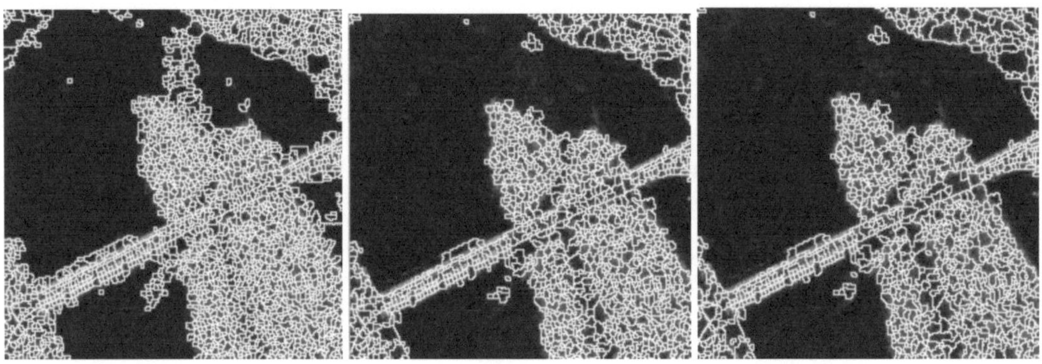

(d) watershed (e) marker-controlled watershed (threshold 10) (f) marker-controlled watershed (threshold 20)

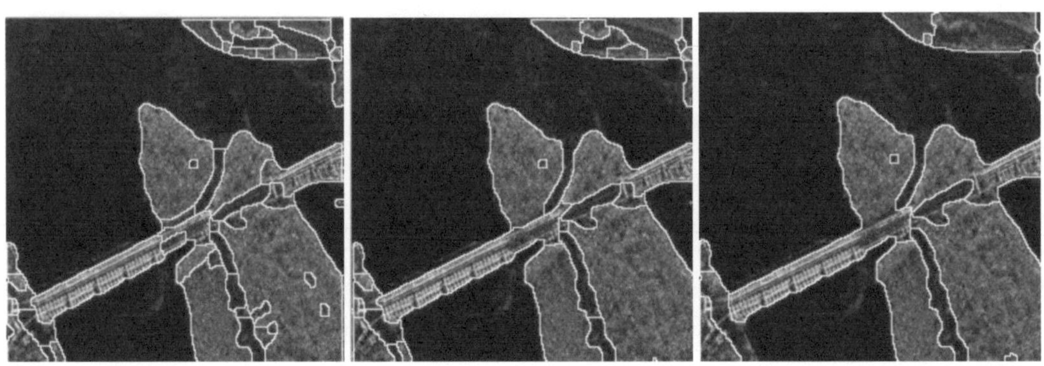

(g) our algorithm (threshold 30) (h) our algorithm (threshold 40) (i) our algorithm (threshold 50)

Fig. 4

(a) original SAR image (b) watershed segmentation (c) final result

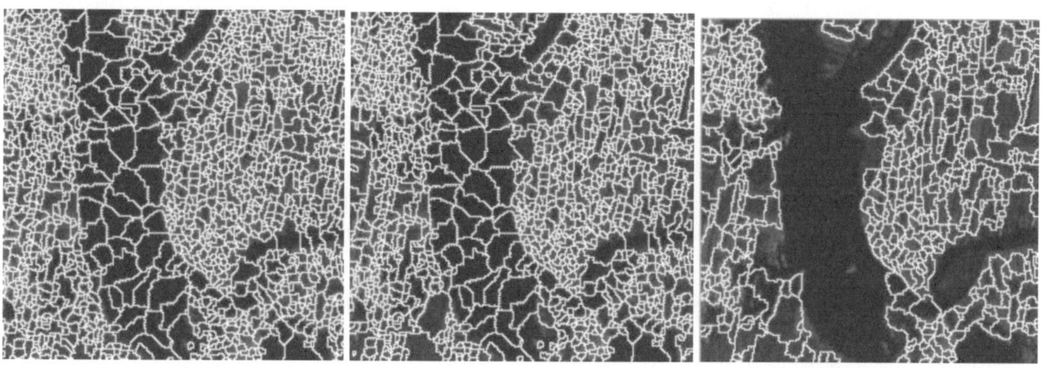

(d) watershed (e) marker-controlled watershed (threshold 10) (f) marker-controlled watershed (threshold 20)

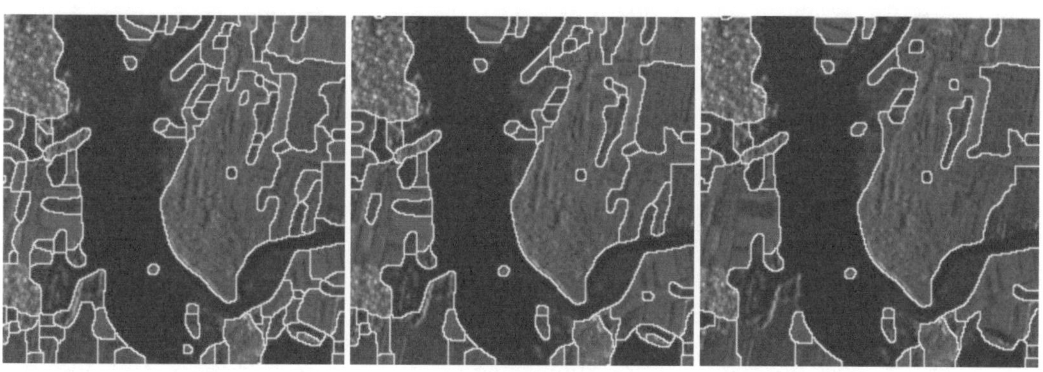

(g) our algorithm (threshold 30) (h) our algorithm (threshold 40) (i) our algorithm (threshold 50)

Fig. 5

An algorithm for improving the quality of the SAR images segmentation based on watershed was proposed. Finding markers is the focal points of this paper. We use markers to solve the over-segmentation of watershed. Otsu method has been used during marker extraction. By using Gradient image modified by the markers as the input of watershed , the number of regions can be automatically controlled, more natural image segmentation can be produced. Our results show that the proposed technique is robust to SAR images segmentation.

However, further research is necessary to concentrate on finding markers, such as some texture feature method can be used. These methods may be more effective to SAR image segmentation since they contain more texture information.

3. SAR images segmentation based on undecimated wavelet transform and mean shift

Synthetic Aperture Radar (SAR) has been a useful tool in domains such as disaster detection, cartography and crop monitoring. The images usually have inferior illumination quality mainly due to different type of environmental distributions. So the segmentation of different land covered regions is an important but difficult step in SAR image processing. It is also significant to understand how information changes over different scales of SAR image. Undecimated wavelet transform is good in describing a scene in terms of the scale; hence we choose this transformation's sub-band as features.

Based on feature space analysis, there are many image segmentation algorithms, which map the pixels into a feature space and cluster, with each cluster delineating a homogeneous region in the image. Although the popular clustering algorithms such as FCM and K-means are simple and fast, they have a major drawback: they seek to minimize the trace of the within-group dispersion matrix and consequently can only generate spherical boundaries. In contrast, mean shift can avoid this drawback. It does not assume any prior shape on data clusters and can handle arbitrary feature spaces.

Besides, texture is an important characteristic used to identify objects or regions of interest in a SAR image. Those textural properties of a SAR image are likely to provide valuable information for analysis, where different object regions are treated as different texture classed. There are numerous texture analysis methods over the past decades. Tuceryan and Jain[30]identifies five major categories of features for texture identification: statistical, geometrical, structural, model-based, and signal processing features. And many people use those methods to obtain supervised segmentation. In other words, many classical segmentation methods can not segment SAR images successfully without prior knowledge of the number of clusters. Fortunately, mean shift can meet our needs, which can determine the number of cluster automatically.

This SAR image segmentation algorithm include four parts: extracting features using undecimated wavelet decomposition; selecting suitable feature subset; filtering feature subset using a Kuwahara filter; at last, applying mean shift algorithm to gain final results. This algorithm does not require prior knowledge of the number of clusters, and does not constrain their shapes.

Discrete wavelet transform (DWT) is a popular tool for feature extraction. Its ability to repeatedly decompose an image in the low-frequency channels makes it ideal for image analysis since the lower frequencies tend to dominate real images. Decimation of the wavelet coefficients is an intrinsic property of the discrete wavelet transform

(DWT).Nevertheless, the decimation is causing shift variance of the wavelet transform. The shift variance means that the DWTs of a signal and its shifted version are not the same. This is because there are downsampling in pyramid discrete wavelet transform. Consequently, we can hardly obtain accurate and robust segmentation using wavelet coefficients as features. Fortunately, undecimated wavelet transform (UWT) can solve this problem.

Undecimated wavelet transform (UWT) is known redundant, shift-invariant and stationary. The number of the wavelet coefficients does not shrink between the transform levels. This additional information can be very useful for the better analysis and understanding of the image properties. Undecimated wavelet transform (UWT) W using the filter bank (h, g) of a 1D signal c_0 leads to a set $W=\{w_1,...,w_J,c_J\}$ where w_j are the wavelet coefficients at scale j and c_J are the coefficients at the coarsest resolution. The passage from one resolution to the next one is obtained using the "`a trous" algorithm:

$$c_{j+1}[l]=(\bar{h}^{(j)}*c_j)[l]=\sum_k h[k]c_j[l+2^j k] \tag{1}$$

$$w_{j+1}[l]=(\bar{g}^{(j)}*c_j)[l]=\sum_k g[k]c_j[l+2^j k] \tag{2}$$

Where $h^{(j)}[l]=h[l]$ if $l/2^j$ is an integer and 0 otherwise. For example, we have

$$h^{(1)}=(...,h[-2],0,h[-1],0,h[0],0,h[1],0,h[2],...) \tag{3}$$

The reconstruction is obtained by

$$c_j[l]=\frac{1}{2}[(\tilde{h}^{(j)}*c_{j+1})[l]+(\tilde{g}^{(j)}*\omega_{j+1})[l]] \tag{4}$$

The filter bank $(h,g,\tilde{h},\tilde{g})$ needs only to verify the exact reconstruction condition:

$$H(z^{-1})\tilde{H}(z)+G(z^{-1})\tilde{G}(z)=1 \tag{5}$$

The `a trous algorithm can be extended to 2D, by:

$$\begin{aligned}
c_{j+1}[k,l]&=(\bar{h}^{(j)}\bar{h}^{(j)}*c_j)[k,l]\\
\omega^1_{j+1}[k,l]&=(\bar{g}^{(j)}\bar{h}^{(j)}*c_j)[k,l]\\
\omega^2_{j+1}[k,l]&=(\bar{h}^{(j)}\bar{g}^{(j)}*c_j)[k,l]\\
\omega^3_{j+1}[k,l]&=(\bar{g}^{(j)}\bar{g}^{(j)}*c_j)[k,l]
\end{aligned} \tag{6}$$

where $hg*c$ is the convolution of c by the separable filter hg (i.e. convolution first along the columns by h and then convolution along the rows by g). At each scale, we have three wavelet images, $\omega^1,\omega^2,\omega^3$, and each has the same size as the original image. The redundancy factor is therefore 3(J − 1) + 1

This procedure is implemented by the similar filter bank as wavelet transform just without down sampling performance. In Fig.1, a typical filter bank of undecimated wavelet transform is shown.

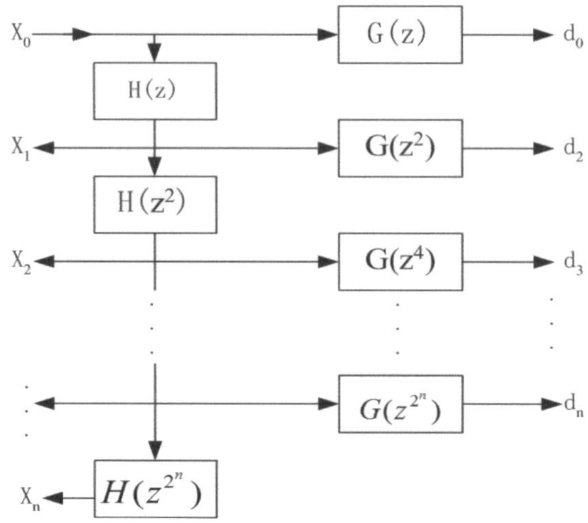

Fig. 1. Filter bank of undecimated wavelet transform

Accordingly, the undecimated wavelet decomposition provides robust texture features at the expense of redundancy. We use undecimated coefficients as features for SAR image analysis.

The simple combination of these features with each other is usually not suitable in SAR image segmentation due to the resulting redundancy and the additive computation complexity. Accordingly, it is necessary to find the most suitable feature subset for SAR image segmentation. This procedure is always considered as an optimization problem, and sequential forward selection (SFS) algorithm can be used.

Sequential forward selection is a traditional feature selection algorithm. It starts with an empty feature subset. On each iteration, exactly one feature is added to the feature subset. To determine which feature to add, the algorithm tentatively adds to the candidate feature subset one feature that is not already selected and tests the criterion function built on the tentative feature subset. The feature that results in the highest value of criterion function is definitely added to the feature subset. In experiments, the termination condition is the maximal iterations. In this paper, we choose Euclidean distance as criterion function and use the coarsest resolution's coefficient as first selected feature.

To avoid noise in SAR, we filter those selected features through a Kuwahara filter. Owing to the mask which contain edges having bigger variance, Kuwahara filter [7] is designed as Fig.2. Calculate variances in those four masks respectively, and then change the center feature with features' mean in which mask has the smallest variance. The realization step is as follows

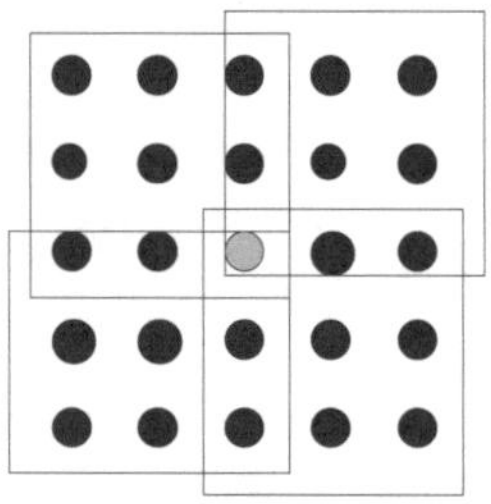

Fig. 2. Kuwahara filter

- Calculate variances in four masks respectively.
- Choose the mask which has the smallest variance, and calculate features' mean in this mask.
- Change center feature with this mean.
- Carry on this procedure to every feature.

Mean shift was proposed in 1975 by Fukunaga and Hostetler and largely forgotten until Dorin Comaniciu's paper[34]rekindled interest in it. Dorin Comaniciu proved that the mean shift procedure is an extremely versatile tool for feature space analysis and can provide reliable solutions for many vision tasks. It tries to obtain the modes of the probability density function of the feature space, using a nonparametric estimate of the density function. And the number of clusters is obtained automatically by finding the centers of the densest regions in the feature space.

To estimate the density function, we use kernel density estimation (known as the Parzen window technique in pattern recognition literature[33]). Assuming the reduced feature dimension is d and a set of point $X=\{X_1,...,X_N\}\subset\Re^d$ drawn from a probability density $f(x)$. Then, the multivariate kernel density estimate obtained with kernel function $K(x)$ and window radius h, computed in the point x is defined as:

$$\hat{f}(x)=\frac{1}{nh^d}\sum_{i=1}^{n}K(\frac{x-x_i}{h}) \tag{7}$$

Where the kernel function k(x) can be different types such as Gaussian kernel, unit kernel, Epanechikov kernel and so on. Using Epanechikov kernel, the density estimator (7) can be rewritten as

$$\hat{f}_{h,k}(x)=\frac{C_{k,d}}{nh^d}\sum_{i=1}^{n}k(||\frac{x-x_i}{h}||^2) \tag{8}$$

Where $C_{k,d}$ is the volume of unit d-dimensional sphere.

The density gradient estimator is obtained as gradient of the density estimator by exploiting the linearity

$$\begin{aligned}\hat{\nabla}f_{h,k}(x)\equiv\hat{\nabla}f_{h,k}(x)&=\frac{2C_{k,d}}{nh^{d+2}}\sum_{i=1}^{n}(x-x_i)k'(||\frac{x-x_i}{h}||^2)\\&=\frac{2C_{k,d}}{nh^{d+2}}[\sum_{i=1}^{n}g(||\frac{x-x_i}{h}||^2)][\frac{\sum_{i=1}^{n}x_ig(||\frac{x-x_i}{h}||^2)}{\sum_{i=1}^{n}g(||\frac{x-x_i}{h}||^2)}-x]\end{aligned} \tag{9}$$

Where $g(x)=-k'(x)$. And the second term is the mean shift

$$m_{h,G}(x)=\frac{\sum_{i=1}^{n}x_ig(||\frac{x-x_i}{h}||^2)}{\sum_{i=1}^{n}g(||\frac{x-x_i}{h}||^2)}-x \tag{10}$$

I.e., the difference between the weighted mean, using the kernel G for weight, and x, the center of the kernel (window). This vector thus always points toward the direction of maximum increase in the density. Since the mean shift vector is aligned with the local gradient estimate, it can define a path leading to a stationary point of the estimated density. Denote by $\{y_j\}_{j=1,2...}$ the sequence of successive locations of the kernel G, where, from(10),

$$y_{j+1} = \frac{\sum_{i=1}^{n} x_i g(\| \frac{x - x_i}{h} \|^2)}{\sum_{i=1}^{n} g(\| \frac{x - x_i}{h} \|^2)} \quad j = 1,2,... \tag{11}$$

Is the weighted mean at y_j computed with kernel G and y_1 is the center of the initial position of the kernel. The corresponding sequence of density estimates computed with K, $\{\hat{f}_{h,K}(j)\}_{j=1,2...}$, is given by

$$\hat{f}_{h,K}(j) = \hat{f}_{h,K}(y_j) \quad j = 1,2... \tag{12}$$

Dorin Comaniciu[34] proved that if the kernel K has a convex and monotonically decreasing profile, the sequences $\{y_j\}_{j=1,2...}$ and $\{\hat{f}_{h,K}(j)\}_{j=1,2...}$ converge and $\{\hat{f}_{h,K}(j)\}_{j=1,2...}$ is monotonically increasing.

An image is typically represented as a two-dimensional lattice of p-dimensional vectors (features). The space of the lattice is known as the spatial domain, while the features are represented in the range domain. In image analysis problem, mean shift incorporate this spatial coordinates of a feature in to its feature space representation. Thus, the multivariate kernel is defined as the product of two radically symmetric kernels and the Euclidean metric allows a single bandwidth parameter for each domain

$$K_{h_s^2, h_r^p}(x) = \frac{C}{h_s^2 h_r^p} k(\| \frac{x^s}{h_s} \|^2) k(\| \frac{x^r}{h_r} \|^2) \tag{13}$$

Where x^s is the spatial part, x^r is the range part of a feature vector, k(x) the common profile used in both two domains, h_s and h_r the employed kernel bandwidths, and C the corresponding normalization constant.

Using gray level of the image as features, this method performs well in natural image segmentation. But we proved that it could not segment SAR images directly, as there were too many texture regions which could not be described effectively just using gray level features. To solve this problem, mean shift algorithm is applied to undecimated wavelet coefficients features. In this feature space, the mean shift vector is aligned with the local gradient estimate, it can define a path leading to the local maxima of the density, i.e. the detected modes. This procedure is applied recursively to every point in the feature space, and the number of clusters present in the feature space is automatically determined by the number of significant modes. This algorithm need to be setted only the bandwidth parameter $h = (h_s, h_r)$, which, by controlling the size of the kernel, determines the resolution of the mode detection. In this paper, we set it just through experiments.

Let x_i and z_i , $i = 1,...,n$, be the d-dimensional input and filtered image features. For each pixel,

1. Initialize $j = 1$ and $y_{i,1} = x_i$.
2. Compute $y_{i,j+1}$ according to (11)until convergence, $y = y_{i,c}$.
3. Store all the information about the d-dimensional convergence point in z_i , i.e., $z_i = y_{i,c}$.
4. Delineate in the joint domain the clusters $\{C_p\}_{p=1...m}$ by grouping together all z_i which are closer than h_s in the spatial domain and h_r in the range domain, i.e., concatenate the basins of attraction of the corresponding convergence points.
5. For each $i = 1,...,n$, assign $L_i = \{p \mid z_i \in C_p\}$.
6. Eliminate spatial regions containing less than M pixels.

The algorithm has several steps as follows:

1. Extract features using undecimated wavelet transform.
2. Select suitable feature subset using SFS algorithm.
3. For all reserve features, filter them through a Kuwahara filter.
4. Use mean shift in feature space to gain the segmentation result.

Experimental results

The proposed method has been applied to several SAR images as Fig.3.(a) and Fig.4.(a) with $(h_s, h_r, M) = (5, 0.2, 20)$. The segmentation of these image presented in Fig.3.(d) and Fig.4.(d) are satisfactory.

We compare our algorithm with Salari E's [36]. Salari E developed an algorithm that segments image using hierarchical wavelet decomposition. In his method, an original image is decomposed into three detail coefficients and one approximate coefficients. The decomposition can be recursively applied to the approximate image to generate a lower resolution of the pyramid. The segmentation starts at the lowest resolution using the K-means clustering scheme and the result is propagated through the pyramid to a higher one with continuously improving the segmentation. Results based on this method using 5*5 and 7*7 feature windows is showed in Fig.3.(b) and Fig.4.(b). Compared with Salari E's method, the proposed algorithm obtains the number of clusters automatically by finding the centers of the densest regions in the feature space. It succeeds in segmenting SAR images without any prior knowledge. On the other hand, using undecimated wavelet coefficients as features, Salari E's method can catch images' texture information, but K-means cluster can only generate spherical boundaries. Our method can avoid this drawback and delineate arbitrarily shaped of clusters.

We also using traditional mean shift approach obtain another team of results, Fig.3.(c). and Fig.4.(c). The traditional mean shift algorithm can delineate arbitrarily shaped clusters but can not performance well just using gray-level as features. Consequently, we can see that some parts of boundaries in our method's results are much more distinct than the other two's. Our method succeeds in overcoming the inherent limitations of methods based on undecimated wavelet features and feature clustering which always oversegment SAR image.

Here, we presente a novel image segmentation algorithm, including three parts: extracting features using undecimated wavelet decomposition; selecting suitable feature subset by SFS algorithm; filtering feature subset using a Kuwahara filter; at last, applying mean shift algorithm to gain final results. In our method, prior knowledge of the number of clusters is not necessary as those classical algorithm., and clusters' shapes are not constrained.

Experiments demonstrated that the proposed method leads to a successful unsupervised segmentation. Compared with traditional meanshift segmentation and FCM based segmentation using undecimated wavelet decomposition, our algorithm had higher performance.

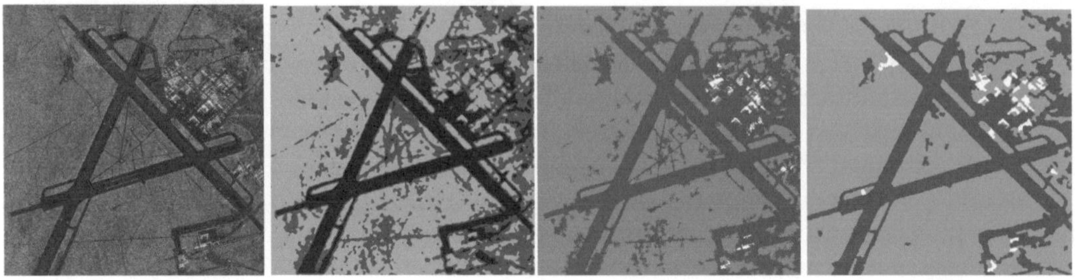

(a) (b) (c) (d)

Fig. 3. 512*512 SAR image segmented by different algorithms. (a) Original image, (b) Segmentation result using Salari E's method, (c) Segmentation result using mean shift, (d) Segmentation result using proposed algorithm.

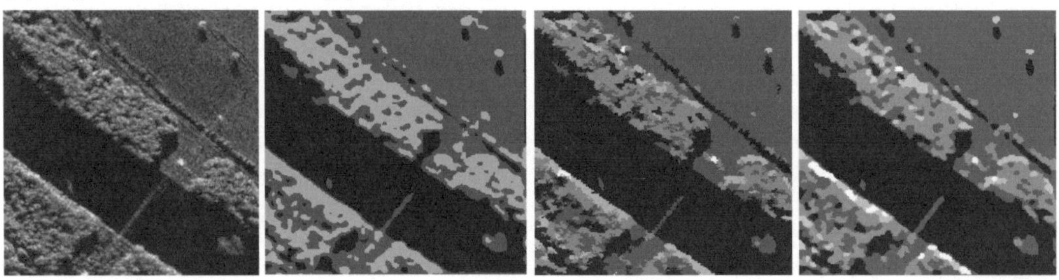

(a) (b) (c) (d)

Fig. 4. 256*256 SAR image segmented by different algorithms.(a) Original image, (b) Segmentation result using Salari E's method, (c)Segmentation result using mean shift, (d)Segmentation result using proposed algorithm.

4. References

[1] M. Kass and A. Witkin, "Snakes: Active Contour Models", *International Journal of Computer Vision*, 1(4), 321-331(1988).

[2] C. Xu and J.L. Prince, "Snakes, Shapes, and Gradient Vector Flow", *IEEE Transactions on Image Processing*, 7(3), 359-369(1998).

[3] Z. Yu and C. Bajaj, "Image Segmentation Using Gradient Vector Diffusion and Region Merging", *Proceedings of 16th International Conference on Pattern Recognition (ICPR'02)*, 941-944(2002).

[4] C. Xu and J.L. Prince, "Gradient Vector Flow: A New External Force for Snakes", *Proceedings of IEEE Conference on Comuter Vision and Pattern Recognition (CVPR'97)*, 66-71(1997).

[5] S. Osher and J. A. Sethian, "Fronts Propagating with Curvature Dependent Speed: Algorithms Based on Hamilton- Jacobi Formulations", *Journal of Computational Physic*, 79(1), 12-49(1988).
[6] R. Malladi and J.A. Sethian, "Shape Modeling with Front Propagation: A Level Set Approach", *IEEE Transactions on Pattern Analysis and Machine Intelligence*, 17(2), 158-175(1995).
[7] P. Lin and C.X. Zhen, "Model-based Medical Image Segmentation: A Level Set Approach", *Proceedings of the 5th World Congress on Intelligent Control and Automation*, 5541-5544(2004).
[8] I.B. Ayed and C. Vazquez, "SAR Image Segmentation with Active Contours and Level Set", *International Conference on Image Processing (ICIP)*, 2717-2720(2004).
[9] K. Wang and C. A. Taylor, "Level Set Methods and MR Image Segmentation for Geometric Modeling in Computational Hemodynamics", *Proceedings of the 20th Annual International Conference of the IEEE Engineering in Medicine and Biology Society*, 20(6), 3079-3082(1998).
[10] H.F. Du and L.C. Jiao "Clonal Operator and Antibody Clone Algorithms", *Proceeding of the First International Conference on Machine Learning and Cybernetics*, 506-510(2002).
[11] J. Lie and M. Lysaker "A Binary Level Set Model and Some Applications to Mumford-Shah Image Segmentation", *IEEE Transactions on Image Processing*, 15(5), 1171-1181(2006).
[12] T. F. Chan and L. A. Vese, "Active Contours Without Edges", *IEEE Transactions on Image Processing*, 10(2), 266-277(2001).
[13] S. Beucher and F. Meyer, "The morphological approach to segmentation: The watershed transformation," in Mathematical Morphology in Image Processing, E. Dougherty, Ed. New York: Marcel Dekker,1992.
[14] A. Bleau and L. J. Leon, "Watershed-based segmentation and region merging," CVIU, vol. 77, no. 3, pp. 317–370, Mar. 2000.
[15] R. Adams and L. Bischof, "Seeded region growing," Pattern Anal. Mach. Intell., vol. 16, no. 6, pp. 641–647, Jun. 1994.
[16] J. Fan, G. Zeng, M. Body, and M. S. Hacid, "Seeded region growing: An extensive and comparative study," PRL, vol. 26, no. 8, pp. 1139–1156, 2005.
[17] Pierre Soille. Morphological image analysis （Second Edition） [M].springer Verlag, 2002. 277－281
[18] O. Lezoray and H. Cardot, "Bayesian marker extraction for color watershed in segmenting microscopic images," in Proc. 16th Int. Conf. Pattern Recognition, 2002, pp. 739–742.
[19] O. Lezoray and H. Cardot, "Cooperation of color pixel classification schemes and color watershed: A study for microscopic images," IEEE Trans. Image Process., vol. 11, no. 7, pp. 783–789, Jul. 2002.
[20] Ilya Levner and Hong Zhang "Classification-Driven Watershed Segmentation" IEEE TRANSACTIONS ON IMAGE PROCESSING, VOL. 16, NO. 5, MAY 2007 1437-1445
[21] F. Bergholm, "Edge focusing," IEEE Trans. Putt Anal. Mach. Intellig. vol. PAMI-9, pp. 726-741, Nov. 1987.
[22] D. Marr and E. Hildreth, " Theory of edge detection," Proc. of the Royal Society London B, vol. 207, pp. 187-217, 1980.

[23] Gonzalez C R, Woods R E. Digital Image Processing(Second Edition)[M]. Publishing House of Electronics Industry, 2002.

[24] Otsu N. A threshold selection method from gray-level histogram [J]. IEEE Trans SMC, 1979, 9(1):62-66.

[25] C. Lantuéjoul, "La Squelettisatoin et son Application aux Mesures Topologiques des Mosaiques Polycristalines," Ph.D. dissertation, School of Mines, Paris, France, 1978.

[26] S. Beucher and F. Meyer, "The morphological approach to segmentation: the watershed transformation," in Mathematical Morphology and its Applications to Image Processing, E.R. Dougherty, Ed. New York: Marcel Dekker, 1993, ch. 12, pp. 433–481.

[27] L. Vincent and P. Soille, "Watersheds in digital spaces: an efficient algorithm based on immersion simulations," IEEE Trans. Pattern Anal. Machine Intell., vol. 13, no. 6, pp. 583–598, 1991.

[28] M. Petrou, L. Shafarenko, and J. Kittler, "Automatic watershed segmentation of randomly textured color images," IEEE Trans. Image Processing, vol. 6, pp. 1530–1544, Nov. 1997.

[29] "Unsupervisede Multiscale Color Imgae Segmentation Based on MDL Principle", IEEE, 2006

[30] Dorin Comaniciu, Peter Meer, "Mean Shift: A Robust Approach Toward Feature Space Analysis", IEEE transactions on pattern analysis and machine intelligence, vol.24, NO.5, May 2002, 603~619.

[31] Christopher, "Synergism in low level vision"

[32] Richard O. Duda, Peter E. Hart, David G. Stork. Pattern Classification[M]. Li H D, Yao T X, Beijing: China Machine Press, 2003:132-164

[33] Dorin Comaniciu, Peter Meer, "Mean Shift Analysis and Application"

[34] J.-L. Starck, J. Fadili, "The Undicimated Wavelet Decomposition and its Reconstruction", 2006

[35] Salari E, Ling Z. Texture segmentation using hierarchical wavelet decomposition [J]. Pattern Recognition, 1995, 28(12): 1819-1824.